Lecture Notes in Artificial Intelligence 8189

Subseries of Lecture Notes in Computer Science

LNAI Series Editors

Randy Goebel
University of Alberta, Edmonton, Canada
Yuzuru Tanaka
Hokkaido University, Sapporo, Japan
Wolfgang Wahlster
DFKI and Saarland University, Saarbrücken, Germany

LNAI Founding Series Editor

Joerg Siekmann
DFKI and Saarland University, Saarbrücken, Germany

Hendrik Blockeel   Kristian Kersting
Siegfried Nijssen   Filip Železný (Eds.)

Machine Learning and Knowledge Discovery in Databases

European Conference, ECML PKDD 2013
Prague, Czech Republic, September 23-27, 2013
Proceedings, Part II

 Springer

Volume Editors

Hendrik Blockeel
KU Leuven, Department of Computer Science
Celestijnenlaan 200A, 3001 Leuven, Belgium
E-mail: hendrik.blockeel@cs.kuleuven.be

Kristian Kersting
University of Bonn, Fraunhofer IAIS, Department of Knowledge Discovery
Schloss Birlinghoven, 53754 Sankt Augustin, Germany
E-mail: kristian.kersting@iais.fraunhofer.de

Siegfried Nijssen
Universiteit Leiden, LIACS, Niels Bohrweg 1, 2333 CA Leiden, The Netherlands
and KU Leuven, Department of Computer Science, 3001 Leuven, Belgium
E-mail: snijssen@liacs.nl

Filip Železný
Czech Technical University, Department of Computer Science and Engineering
Technicka 2, 16627 Prague 6, Czech Republic
E-mail: zelezny@fel.cvut.cz

Cover image: © eleephotography

ISSN 0302-9743
e-ISSN 1611-3349
ISBN 978-3-642-40990-5
e-ISBN 978-3-642-40991-2
DOI 10.1007/978-3-642-40991-2
Springer Heidelberg New York Dordrecht London

Library of Congress Control Number: 2013948101

CR Subject Classification (1998): I.2.6, H.2.8, I.5.2, G.2.2, G.3, I.2.4, I.2.7, H.3.4-5, I.2.9, F.2

LNCS Sublibrary: SL 7 – Artificial Intelligence

Typesetting: Camera-ready by author, data conversion by Scientific Publishing Services, Chennai, India

Printed on acid-free paper

Springer is part of Springer Science+Business Media (www.springer.com)

Preface

These are the proceedings of the 2013 edition of the European Conference on Machine Learning and Principles and Practice of Knowledge Discovery in Databases, or ECML PKDD for short. This conference series has grown out of the former ECML and PKDD conferences, which were Europe's premier conferences on, respectively, Machine Learning and Knowledge Discovery in Databases. Organized jointly for the first time in 2001, these conferences have become increasingly integrated, and became one in 2008. Today, ECML PKDD is a world–leading conference in these areas, well–known in particular for having a highly diverse program that aims at exploiting the synergies between these two different, yet related, scientific fields.

ECML PKDD 2013 was held in Prague, Czech Republic, during September 23–27. Continuing the series' tradition, the conference combined an extensive technical program with a variety of workshops and tutorials, a demo track for system demonstrations, an industrial track, a nectar track focusing on particularly interesting results from neighboring areas, a discovery challenge, two poster sessions, and a rich social program.

The main technical program included five plenary talks by invited speakers (Rayid Ghani, Thorsten Joachims, Ulrike von Luxburg, Christopher Re and John Shawe-Taylor) and a record–breaking 138 technical talks, for which further discussion opportunities were provided during two poster sessions. The industrial track had four invited speakers: Andreas Antrup (Zalando), Ralf Herbrich (Amazon Berlin), Jean-Paul Schmetz (Hubert Burda Media), and Hugo Zaragoza (Websays). The demo track featured 11 software demonstrations, and the nectar track 5 talks. The discovery challenge, this year, focused on the task of recommending given names for children to soon–to–be–parents. Twelve workshops were held: Scalable Decision Making; Music and Machine Learning; Reinforcement Learning with Generalized Feedback; Languages for Data Mining and Machine Learning; Data Mining on Linked Data; Mining Ubiquitous and Social Environments; Tensor Methods in Machine Learning; Solving Complex Machine Learning Problems with Ensemble Methods; Sports Analytics; New Frontiers in Mining Complex Pattern; Data Analytics for Renewable Energy Integration; and Real–World Challenges for Data Stream Mining. Eight tutorials completed the program: Multi–Agent Reinforcement Learning; Second Order Learning; Algorithmic Techniques for Modeling and Mining Large Graphs; Web Scale Information Extraction; Mining and Learning with Network–Structured Data; Performance Evaluation of Machine Learning Algorithms; Discovering Roles and Anomalies in Graphs: Theory and Applications; and Statistically Sound Pattern Discovery.

The conference offered awards for distinguished papers, for the paper from ECML / PKDD 2003 with the highest impact after a decade, and for the best demonstration. In addition, there was the novel Open Science Award. This award was installed in order to promote reusability of software, data, and experimental setups, with the aim of improving reproducibility of research and facilitating research that builds on other authors' work.

For the first time, the conference used a mixed submission model: work could be submitted as a journal article to Machine Learning or Data Mining and Knowledge Discovery, or it could be submitted for publication in the conference proceedings. A total of 182 original manuscripts were submitted to the journal track, and 447 to the proceedings track. Of the journal submissions, 14 have been published in the journal, as part of a special issue on ECML PKDD 2013, and 14 have been redirected to the proceedings track. Among the latter, 13 were accepted for publication in the proceedings. Finally, of the 447 submissions to the proceedings track, 111 have been accepted. Overall, this gives a record number of 629 submissions, of which 138 have been scheduled for presentation at the conference, making the overall acceptance rate 21.9%.

The mixed submission model was introduced in an attempt to improve the efficiency and reliability of the reviewing process. Reacting to criticism on the conference–based publication model that is so typical for computer science, several conferences have started experimenting with multiple reviewing rounds, continuous submission, and publishing contributions in a journal instead of the conference proceedings. The ECML PKDD model has been designed to maximally exploit the already existing infrastructure for journal reviewing. For an overview of the motivation and expected benefits of this new model, we refer to *A Revised Publication Model for ECML PKDD*, available at `arXiv:1207.6324`.

These proceedings of the 2013 European Conference on Machine Learning and Principles and Practice of Knowledge Discovery in Databases contain full papers of work presented at the main technical track, abstracts of the journal articles and invited talks presented there, and short papers describing the demonstrations and nectar papers. We thank the chairs of the demo track (Andreas Hotho and Joaquin Vanschoren), the nectar track (Rosa Meo and Michèle Sebag), and the industrial track (Ulf Brefeld), as well as the proceedings chairs Yamuna Krishnamurthy and Nico Piatkowski, for their help with putting these proceedings together. Most importantly, of course, we thank the authors for their contributions, and the area chairs and reviewers for their substantial efforts to guarantee and sometimes even improve the quality of these proceedings. We wish the reader an enjoyable experience exploring the many exciting research results presented here.

July 2013

Hendrik Blockeel
Kristian Kersting
Siegfried Nijssen
Filip Železný

Organization

ECML PKDD 2013 was organized by the Intelligent Data Analysis Research Lab, Department of Computer Science and Engineering, of the Czech Technical University in Prague.

Conference Chair

Filip Železný — Czech Technical University in Prague, Czech Republic

Program Chairs

Hendrik Blockeel — KU Leuven, Belgium & Leiden University, The Netherlands

Kristian Kersting — University of Bonn & Fraunhofer IAIS, Germany

Siegfried Nijssen — Leiden University, The Netherlands & KU Leuven, Belgium

Filip Železný — Czech Technical University in Prague, Czech Republic

Local Chair

Jiří Kléma — Czech Technical University in Prague, Czech Republic

Publicity Chair

Élisa Fromont — Université Jean Monnet, France

Proceedings Chairs

Yamuna Krishnamurthy — TU Dortmund, Germany
Nico Piatkowski — TU Dortmund, Germany

Workshop Chairs

Niels Landwehr — University of Potsdam, Germany
Andrea Passerini — University of Trento, Italy

Tutorial Chairs

Kurt Driessens   Maastricht University, The Netherlands
Sofus A. Macskassy   University of Southern California, USA

Demo Track Chairs

Andreas Hotho   University of Würzburg, Germany
Joaquin Vanschoren   Leiden University, The Netherlands

Nectar Track Chairs

Rosa Meo   University of Turin, Italy
Michèle Sebag   Université Paris-Sud, France

Industrial Track Chair

Ulf Brefeld   TU Darmstadt, Germany

Discovery Challenge Chair

Taneli Mielikäinen   Nokia, USA

Discovery Challenge Organizers

Stephan Doerfel   University of Kassel, Germany
Andreas Hotho   University of Würzburg, Germany
Robert Jäschke   University of Kassel, Germany
Folke Mitzlaff   University of Kassel, Germany
Jürgen Müller   L3S Research Center, Germany

Sponsorship Chairs

Peter van der Putten   Leiden University & Pegasystems,
                          The Netherlands
Albert Bifet   University of Waikato & Yahoo, New Zealand

Awards Chairs

Bart Goethals   University of Antwerp, Belgium
Peter Flach   University of Bristol, UK
Geoff Webb   Monash University, Australia

Open-Science Award Committee

Tias Guns   KU Leuven, Belgium
Christian Borgelt   European Center for Soft Computing, Spain
Geoff Holmes   University of Waikato, New Zealand
Luis Torgo   University of Porto, Portugal

Web Team

Matěj Holec, Webmaster   Czech Technical University in Prague,
                           Czech Republic
Radomír Černoch   Czech Technical University in Prague,
                           Czech Republic
Filip Blažek   Designiq, Czech Republic

Software Development

Radomír Černoch   Czech Technical University in Prague,
                           Czech Republic
Fabian Hadiji   University of Bonn, Germany
Thanh Le Van   KU Leuven, Belgium

ECML PKDD Steering Committee

Fosca Gianotti, Chair   Universitá di Pisa, Italy
Jose Balcazar   Universitat Polytécnica de Catalunya, Spain
Francesco Bonchi   Yahoo! Research Barcelona, Spain
Nello Cristianini   University of Bristol, UK
Tijl De Bie   University of Bristol, UK
Peter Flach   University of Bristol, UK
Dimitrios Gunopulos   University of Athens, Greece
Donato Malerba   Universitá degli Studi di Bari, Italy
Michèle Sebag   Université Paris-Sud, France
Michalis Vazirgiannis   Athens University of Economics and Business,
                           Greece

Area Chairs

Henrik Boström   Stockholm University, Sweden
Jean-François Boulicaut University of Lyon, France
Carla Brodley   Tuft University, USA
Ian Davidson   University of California, Davis, USA
Jesse Davis   KU Leuven, Belgium
Tijl De Bie   University of Bristol, UK
Janez Demšar   University of Ljubljana, Slovenia
Luc De Raedt   KU Leuven, Belgium

Pierre Dupont — UC Louvain, Belgium
Charles Elkan — University of California, San Diego, USA
Alan Fern — Oregon State University, USA
Johannes Fürnkranz — TU Darmstadt, Germany
Joao Gama — University of Porto, Portugal
Thomas Gärtner — University of Bonn and Fraunhofer IAIS, Germany
Aristides Gionis — Aalto University, Finland
Bart Goethals — University of Antwerp, Belgium
Geoff Holmes — Waikato University, New Zealand
Andreas Hotho — University of Würzburg, Germany
Eyke Hüllermeier — Philipps-Universität Marburg, Germany
Manfred Jaeger — Aalborg University, Denmark
Thorsten Joachims — Cornell University, USA
George Karypis — University of Minnesota, USA
Stefan Kramer — University of Mainz, Germany
Donato Malerba — University of Bari, Italy
Dunja Mladenic — Jožef Stefan Institute, Slovenia
Marie-Francine Moens — KU Leuven, Belgium
Bernhard Pfahringer — University of Waikato, New Zealand
Myra Spiliopoulou — Magdeburg University, Germany
Hannu Toivonen — University of Helsinki, Finland
Marco Wiering — University of Groningen, The Netherlands
Stefan Wrobel — University of Bonn and Fraunhofer IAIS, Germany

Program Committee

Leman Akoglu
Mohammad Al Hasan
Aris Anagnostopoulos
Gennady Andrienko
Annalisa Appice
Cedric Archambeau
Marta Arias
Hiroki Arimura
Ira Assent
Martin Atzmüller
Chloe-Agathe Azencott
Antonio Bahamonde
James Bailey
Jose Balcazar
Christian Bauckhage
Roberto Bayardo
Aurelien Bellet

Andras Benczur
Bettina Berendt
Michael Berthold
Indrajit Bhattacharya
Albert Bifet
Mario Boley
Francesco Bonchi
Gianluca Bontempi
Christian Borgelt
Zoran Bosnic
Abdeslam Boularias
Kendrick Boyd
Pavel Brazdil
Ulf Brefeld
Björn Bringmann
Wray Buntine
Robert Busa-Fekete

Toon Calders
Andre Carvalho
Francisco Casacuberta
Michelangelo Ceci
Loic Cerf
Duen Horng Chau
Sanjay Chawla
Weiwei Cheng
Fabrizio Costa
Sheldon Cooper
Vitor Costa
Bruno Cremilleux
Tom Croonenborghs
Boris Cule
Tomaz Curk
James Cussens
Martine De Cock
Colin de la Higuera
Juan del Coz
Francois Denis
Jana Diesner
Wei Ding
Janardhan Doppa
Devdatt Dubhashi
Ines Dutra
Sašo Džeroski
Tina Eliassi-Rad
Tapio Elomaa
Seyda Ertekin
Floriana Esposito
Ines Faerber
Fazel Famili
Hadi Fanaee Tork
Elaine Faria
Ad Feelders
Stefano Ferilli
Carlos Ferreira
Jordi Fonollosa
Antonino Freno
Elisa Fromont
Fabio Fumarola
Patrick Gallinari
Roman Garnett
Eric Gaussier
Ricard Gavalda

Pierre Geurts
Rayid Ghani
Fosca Giannotti
David Gleich
Vibhav Gogate
Michael Granitzer
Dimitrios Gunopulos
Tias Guns
Jiawei Han
Daniel Hernandez Lobato
Frank Hoeppner
Thomas Hofmann
Jaako Hollmen
Arjen Hommersom
Vasant Honavar
Tamás Horváth
Dino Ienco
Elena Ikonomovska
Robert Jäschke
Frederik Janssen
Szymon Jaroszewicz
Ulf Johansson
Alipio Jorge
Kshitij Judah
Hachem Kadri
Alexandros Kalousis
U Kang
Panagiotis Karras
Andreas Karwath
Hisashi Kashima
Samuel Kaski
Latifur Khan
Angelika Kimmig
Arno Knobbe
Levente Kocsis
Yun Sing Koh
Alek Kolcz
Andrey Kolobov
Igor Kononenko
Kleanthis-Nikolaos Kontonasios
Nitish Korula
Petr Kosina
Walter Kosters
Georg Krempl
Sergei Kuznetsov

Helge Langseth
Pedro Larranaga
Silvio Lattanzi
Niklas Lavesson
Nada Lavrač
Gregor Leban
Chris Leckie
Sangkyun Lee
Ping Li
Juanzi Li
Edo Liberty
Jefrey Lijffijt
Jessica Lin
Francesca Lisi
Corrado Loglisci
Eneldo Loza Mencia
Peter Lucas
Francis Maes
Michael Mampaey
Giuseppe Manco
Stan Matwin
Michael May
Mike Mayo
Wannes Meert
Ernestina Menasalvas
Rosa Meo
Pauli Miettinen
Bamshad Mobasher
Joao Moreira
Emmanuel Müller
Mohamed Nadif
Alex Nanopoulos
Balakrishnan Narayanaswamy
Sriraam Natarajan
Aniruddh Nath
Thomas Nielsen
Mathias Niepert
Xia Ning
Niklas Noren
Eirini Ntoutsi
Andreas Nürnberger
Gerhard Paass
David Page
Rasmus Pagh
Spiros Papadimitriou

Panagiotis Papapetrou
Andrea Passerini
Mykola Pechenizkiy
Dino Pedreschi
Jian Pei
Nikos Pelekis
Ruggero Pensa
Marc Plantevit
Pascal Poncelet
Aditya Prakash
Kai Puolamaki
Buyue Qian
Chedy Raïssi
Liva Ralaivola
Karthik Raman
Jan Ramon
Huzefa Rangwala
Umaa Rebbapragada
Jean-Michel Renders
Steffen Rendle
Achim Rettinger
Fabrizio Riguzzi
Celine Robardet
Marko Robnik Sikonja
Pedro Rodrigues
Juan Rodriguez
Irene Rodriguez-Lujan
Ulrich Rückert
Stefan Rüping
Jan Rupnik
Yvan Saeys
Alan Said
Lorenza Saitta
Antonio Salmeron
Scott Sanner
Raul Santos-Rodriguez
Sam Sarjant
Claudio Sartori
Taisuke Sato
Lars Schmidt-Thieme
Christoph Schommer
Michèle Sebag
Marc Sebban
Thomas Seidl
Giovanni Semeraro

Demo Track Program Committee

Nectar Track Program Committee

Maria Florina Balcan
Christian Böhm
Toon Calders
Luc De Raedt
George Karypis

Hugo Larochelle
Donato Malerba
Myra Spiliopoulou
Vicenc Torra

Additional Reviewers

Rohit Babbar
Aubrey Barnard
Christian Beecks
Alejandro Bellogin
Daniel Bengs
Souhaib Ben Taieb
Mansurul Bhuiyan
Sam Blasiak
Patrice Boizumault
Teresa Bracamonte
Janez Brank
George Brova
David C. Anastasiu
Cécile Capponi
Annalina Caputo
Jeffrey Chan
Anveshi Charuvaka
Claudia d'Amato
Xuan-Hong Dang
Ninh Dang Pham
Lucas Drumond
Wouter Duivesteijn
François-Xavier Dupé
Ritabrata Dutta
Pavel Efros
Dora Erdos
Pasqua Fabiana Lanotte
Antonio Fernandez
Georg Fette
Manoel França
Sergej Fries
Atsushi Fujii
Patrick Gabrielsson
Esther Galbrun
Michael Geilke

Christos Giatsidis
Robby Goetschalckx
Boqing Gong
Michele Gorgoglione
Tatiana Gossen
Maarten Grachten
Xin Guan
Massimo Guarascio
Huan Gui
Amaury Habrard
Ahsanul Haque
Marwan Hassani
Kohei Hayashi
Elad Hazan
Andreas Henelius
Shohei Hido
Patricia Iglesias Sanchez
Roberto Interdonato
Baptiste Jeudy
Hiroshi Kajino
Yoshitaka Kameya
Margarita Karkali
Mehdi Kaytoue
Fabian Keller
Mikaela Keller
Eamonn Keogh
Umer Khan
Tushar Khot
Benjamin Kille
Dragi Kocev
Jussi Korpela
Domen Kosir
Hardy Kremer
Tanay K. Saha
Gautam Kunapuli

Guest Editorial Board (Journal Track)

Luc De Raedt	KU Leuven, Belgium
Luis Torgo	University of Porto, Portugal
Marie-Francine Moens	KU Leuven, Belgium
Matthijs van Leeuwen	KU Leuven, Belgium
Michael May	Fraunhofer IAIS, Germany
Michael R. Berthold	Universität Konstanz, Germany
Nada Lavrač	Jožef Stefan Institute, Slovenia
Nikolaj Tatti	University of Antwerp, Belgium
Pascal Poupart	University of Waterloo, Canada
Pierre Dupont	UC Louvain, Belgium
Prasad Tadepalli	Oregon State University, USA
Roberto Bayardo	Google Research, USA
Soumya Ray	Case Western Reserve University, USA
Stefan Wrobel	University of Bonn and Fraunhofer IAIS, Germany
Stefan Kramer	University of Mainz, Germany
Takashi Washio	Osaka University, Japan
Tamás Horváth	Fraunhofer IAIS, Germany
Tapio Elomaa	Tampere University of Technology, Finland
Thomas Gärtner	University of Bonn and Fraunhofer IAIS, Germany
Tijl De Bie	University of Bristol, UK
Toon Calders	Eindhoven University of Technology, The Netherlands
Willem Waegeman	Ghent University, Belgium
Wray Buntine	NICTA, Australia

Additional Reviewers (Journal Track)

Babak Ahmadi	Emma Brunskill
Amr Ahmed	Michelangelo Ceci
Leman Akoglu	Sanjay Chawla
Mohammad Al Hasan	Weiwei Cheng
Massih-Reza Amini	KyungHyun Cho
Bart Baesens	Tom Croonenborghs
Andrew Bagnell	Florence d'Alché-Buc
Arindam Banerjee	Bhavana Dalvi
Christian Bauckhage	Kurt De Grave
Yoshua Bengio	Bolin Ding
Albert Bifet	Chris Ding
Andrew Bolstad	Jennifer Dy
Byron Boots	Sašo Džeroski
Karsten Borgwardt	Alan Fern
Kendrick Boyd	Luis Ferre
Ulf Brefeld	Daan Fierens

Arik Friedman
Mark Gales
Joao Gama
Roman Garnett
Rainer Gemulla
Marek Grzes
Tias Guns
Gregor Heinrich
Katherine Heller
Francisco Herrera
McElory Hoffmann
Andreas Hotho
Christian Igel
Mariya Ishteva
Manfred Jaeger
Kshitij Judah
Ata Kaban
Lars Kaderali
Alexandros Kalousis
Panagiotis Karras
Hendrik Blockeel
Kristian Kersting
Siegfried Nijssen
Filip Železný
Marius Kloft
Hans-Peter Kriegel
Gautam Kunapuli
Niels Landwehr
Ni Lao
Agnieszka Lawrynowicz
Honglak Lee
Nan Li
Yu-Feng Li
James Robert Lloyd
Manuel Lopes
Haibing Lu
Michael Mampaey
Dragos Margineantu
Stan Matwin
Francisco Melo
Pauli Miettinen
Alessandro Moschitti
Uwe Nagel
Mirco Nanni
Sriraam Natarajan

Mathias Niepert
John Paisley
Ankur Parikh
Srinivasan Parthasarathy
Alessandro Perina
Jan Peters
Massimiliano Pontil
Foster Provost
Ricardo Prudencio
Tao Qin
Novi Quadrianto
Ariadna Quattoni
Chedy Raïssi
Alexander Rakhlin
Balaraman Ravindran
Jesse Read
Chandan Reddy
Achim Rettinger
Peter Richtarik
Volker Roth
Céline Rouveirol
Cynthia Rudin
Stefan Rüping
Scott Sanner
Lars Schmidt-Thieme
Jeff Schneider
Stephen Scott
Michèle Sebag
Mathieu Serrurier
Sohan Seth
Jude Shavlik
Le Song
Peter Stone
Peter Sunehag
Charles Sutton
Johan Suykens
Nima Taghipour
Emmanuel Tapia
Nikolaj Tatti
Graham Taylor
Manolis Terrovitis
Grigorios Tsoumakas
Tinne Tuytelaars
Antti Ukkonen
Laurens van der Maaten

Marcel van Gerven
Martijn van Otterlo
Lieven Vandenberghe
Joaquin Vanschoren
Michalis Vazirgiannis
Aki Vehtari
Byron Wallace
Thomas J. Walsh
Chao Wang
Pu Wang
Shaojun Wang
Randy Wilson

Han-Ming Wu
Huan Xu
Zhao Xu
Jieping Ye
Yi-Ren Yeh
Shipeng Yu
Dengyong Zhou
Shenghuo Zhu
Arthur Zimek
Albrecht Zimmermann
Indre Zliobaite

Sponsors

Gold Sponsor
Winton Capital

http://wintoncapital.com

Silver Sponsors
Cisco Systems, Inc.
Deloitte Analytics
KNIME
Yahoo! Labs

http://www.cisco.com
http://www.deloitte.com
http://www.knime.com
http://www.yahoo.com

Bronze Sponsors
CSKI
Definity Systems
DIKW Academy
Google
Xerox Research Centre Europe
Zalando

http://www.cski.cz
http://www.definity.cz
http://dikw-academy.nl
http://research.google.com
http://www.xrce.xerox.com
http://www.zalando.de

Prize Sponsors
Data Mining and Knowledge Discovery

Deloitte Analytics
Google
Machine Learning

Yahoo! Labs

http://link.springer.com/
journal/10618
http://www.deloitte.com
http://research.google.com
http://link.springer.com/
journal/10994
http://www.knime.com

Abstracts of Invited Talks

Using Machine Learning Powers for Good

Rayid Ghani

The past few years have seen increasing demand for machine learning and data mining—both for tools as well as experts. This has been mostly motivated by a variety of factors including better and cheaper data collection, realization that using data is a good thing, and the ability for a lot of organizations to take action based on data analysis. Despite this flood of demand, most applications we hear about in machine learning involve search, advertising, and financial areas. This talk will talk about examples on how the same approaches can be used to help governments and non-prpofits make social impact. I'll talk about a summer fellowship program we ran at University of Chicago on social good and show examples from projects in areas such as education, healthcare, energy, transportation and public safety done in conjunction with governments and non-profits.

Biography

Rayid Ghani was the Chief Scientist at the Obama for America 2012 campaign focusing on analytics, technology, and data. His work focused on improving different functions of the campaign including fundraising, volunteer, and voter mobilization using analytics, social media, and machine learning; his innovative use of machine learning and data mining in Obama's reelection campaign received broad attention in the media such as the New York Times, CNN, and others. Before joining the campaign, Rayid was a Senior Research Scientist and Director of Analytics research at Accenture Labs where he led a technology research team focused on applied R&D in analytics, machine learning, and data mining for large-scale & emerging business problems in various industries including healthcare, retail & CPG, manufacturing, intelligence, and financial services. In addition, Rayid serves as an adviser to several start-ups in Analytics, is an active organizer of and participant in academic and industry analytics conferences, and publishes regularly in machine learning and data mining conferences and journals.

Learning with Humans in the Loop

Thorsten Joachims

Machine Learning is increasingly becoming a technology that directly interacts with human users. Search engines, recommender systems, and electronic commerce already heavily rely on adapting the user experience through machine learning, and other applications are likely to follow in the near future (e.g., autonomous robotics, smart homes, gaming). In this talk, I argue that learning with humans in the loop requires learning algorithms that explicitly account for human behavior, their motivations, and their judgment of performance. Towards this goal, the talk explores how integrating microeconomic models of human behavior into the learning process leads to new learning models that no longer reduce the user to a "labeling subroutine". This motivates an interesting area for theoretical, algorithmic, and applied machine learning research with connections to rational choice theory, econometrics, and behavioral economics.

Biography

Thorsten Joachims is a Professor of Computer Science at Cornell University. His research interests center on a synthesis of theory and system building in machine learning, with applications in language technology, information retrieval, and recommendation. His past research focused on support vector machines, text classification, structured output prediction, convex optimization, learning to rank, learning with preferences, and learning from implicit feedback. In 2001, he finished his dissertation advised by Prof. Katharina Morik at the University of Dortmund. From there he also received his Diplom in Computer Science in 1997. Between 2000 and 2001 he worked as a PostDoc at the GMD Institute for Autonomous Intelligent Systems. From 1994 to 1996 he was a visiting scholar with Prof. Tom Mitchell at Carnegie Mellon University.

Unsupervised Learning with Graphs:
A Theoretical Perspective

Ulrike von Luxburg

Applying a graph–based learning algorithm usually requires a large amount of data preprocessing. As always, such preprocessing can be harmful or helpful. In my talk I am going to discuss statistical and theoretical properties of various preprocessing steps. We consider questions such as: Given data that does not have the form of a graph yet, what do we loose when transforming it to a graph? Given a graph, what might be a meaningful distance function? We will also see that graph–based techniques can lead to surprising solutions to preprocessing problems that a priori don't involve graphs at all.

Biography

Ulrike von Luxburg is a professor for computer science/machine learning at the University of Hamburg. Her research focus is the theoretical analysis of machine learning algorithms, in particular for unsupervised learning and graph algorithms. She is (co)–winner of several best student paper awards (NIPS 2004 and 2008, COLT 2003, 2005 and 2006, ALT 2007). She did her PhD in the Max Planck Institute for Biological Cybernetics in 2004, then moved to Fraunhofer IPSI in Darmstadt, before returning to the Max Planck Institute in 2007 as a research group leader for learning theory. Since 2012 she is a professor for computer science at the University of Hamburg.

Making Systems That Use Statistical Reasoning Easier to Build and Maintain over Time

Christopher Re

The question driving my work is, how should one deploy statistical data–analysis tools to enhance data–driven systems? Even partial answers to this question may have a large impact on science, government, and industry—each of whom are increasingly turning to statistical techniques to get value from their data.

To understand this question, my group has built or contributed to a diverse set of data–processing systems: a system, called GeoDeepDive, that reads and helps answer questions about the geology literature; a muon filter that is used in the IceCube neutrino telescope to process over 250 million events each day in the hunt for the origins of the universe; and enterprise applications with Oracle and Pivotal. This talk will give an overview of the lessons that we learned in these systems, will argue that data systems research may play a larger role in the next generation of these systems, and will speculate on the future challenges that such systems may face.

Biography

Christopher Re is an assistant professor in the department of Computer Sciences at the University of Wisconsin-Madison. The goal of his work is to enable users and developers to build applications that more deeply understand and exploit data. Chris received his PhD from the University of Washington, Seattle under the supervision of Dan Suciu. For his PhD work in the area of probabilistic data management, Chris received the SIGMOD 2010 Jim Gray Dissertation Award. Chris's papers have received four best papers or best–of–conference citations (best paper in PODS 2012 and best–of–conference in PODS 2010, twice, and one in ICDE 2009). Chris received an NSF CAREER Award in 2011.

Deep–er Kernels

Kernels can be viewed as shallow in that learning is only applied in a single (output) layer. Recent successes with deep learning highlight the need to consider learning richer function classes. The talk will review and discuss methods that have been developed to enable richer kernel classes to be learned. While some of these methods rely on greedy procedures many are supported by statistical learning analyses and/or convergence bounds. The talk will highlight the trade–offs involved and the potential for further research on this topic.

Biography

John Shawe-Taylor obtained a PhD in Mathematics at Royal Holloway, University of London in 1986 and joined the Department of Computer Science in the same year. He was promoted to Professor of Computing Science in 1996. He moved to the University of Southampton in 2003 to lead the ISIS research group. He was Director of the Centre for Computational Statistics and Machine Learning at University College, London between July 2006 and September 2010. He has coordinated a number of European wide projects investigating the theory and practice of Machine Learning, including the PASCAL projects. He has published over 300 research papers with more than 25000 citations. He has co-authored with Nello Cristianini two books on kernel approaches to machine learning: "An Introduction to Support Vector Machines" and "Kernel Methods for Pattern Analysis".

Abstracts of Industrial
Track Invited Talks

ML and Business: A Love–Hate Relationship

Andreas Antrup

Based on real world examples. the talk explores common gaps in the mutual understanding of the business and the analytical side; particular focus shall be on misconceptions of the needs and expectations of business people and the resulting problems. It also touches on some approaches to bridge these gaps and build trust. At the end we shall discuss possibly under–researched areas that may open the doors to a yet wider usage of ML principles and thus unlock more of its value and beauty.

Bayesian Learning in Online Service: Statistics Meets Systems

Ralf Herbrich

Over the past few years, we have entered the world of big and structured data— a trend largely driven by the exponential growth of Internet–based online services such as Search, eCommerce and Social Networking as well as the ubiquity of smart devices with sensors in everyday life. This poses new challenges for statistical inference and decision–making as some of the basic assumptions are shifting:

- The ability to optimize both the likelihood and loss functions
- The ability to store the parameters of (data) models
- The level of granularity and 'building blocks' in the data modeling phase
- The interplay of computation, storage, communication and inference and decision–making techniques

In this talk, I will discuss the implications of big and structured data for Statistics and the convergence of statistical model and distributed systems. I will present one of the most versatile modeling techniques that combines systems and statistical properties—factor graphs—and review a series of approximate inference techniques such as distributed message passing. The talk will be concluded with an overview of real–world problems at Amazon.

Machine Learning in a Large diversified Internet Group

Jean-Paul Schmetz

I will present a wide survey of the use of machine learning techniques across a large number of subsidiaries (40+) of an Internet group (Burda Digital) with special attention to issues regarding (1) personnel training in state of the art techniques, (2) management buy–in of complex non interpretable results and (3) practical and measurable bottom line results/solutions.

Some of the Problems and Applications of Opinion Analysis

Hugo Zaragoza

Websays strives to provide the best possible analysis of online conversation to marketing and social media analysts. One of the obsessions of Websays is to provide "near–man–made" data quality at marginal costs. I will discuss how we approach this problem using innovative machine learning and UI approaches.

Abstracts of Journal Track Articles

The full articles have been published in *Machine Learning* or *Data Mining and Knowledge Discovery*.

Fast sequence segmentation using log–linear models

Nikolaj Tatti
Data Mining and Knowledge Discovery
DOI 10.1007/s10618-012-0301-y

Sequence segmentation is a well–studied problem, where given a sequence of elements, an integer K, and some measure of homogeneity, the task is to split the sequence into K contiguous segments that are maximally homogeneous. A classic approach to find the optimal solution is by using a dynamic program. Unfortunately, the execution time of this program is quadratic with respect to the length of the input sequence. This makes the algorithm slow for a sequence of non–trivial length. In this paper we study segmentations whose measure of goodness is based on log–linear models, a rich family that contains many of the standard distributions. We present a theoretical result allowing us to prune many suboptimal segmentations. Using this result, we modify the standard dynamic program for 1D log–linear models, and by doing so reduce the computational time. We demonstrate empirically, that this approach can significantly reduce the computational burden of finding the optimal segmentation.

ROC curves in cost space

Cesar Ferri, Jose Hernandez-Orallo and Peter Flach
Machine Learning
DOI 10.1007/s10994-013-5328-9

ROC curves and cost curves are two popular ways of visualising classifier performance, finding appropriate thresholds according to the operating condition, and deriving useful aggregated measures such as the area under the ROC curve (*AUC*) or the area under the optimal cost curve. In this paper we present new findings and connections between ROC space and cost space. In particular, we show that ROC curves can be transferred to cost space by means of a very natural threshold choice method, which sets the decision threshold such that the proportion of positive predictions equals the operating condition. We call these new curves rate–driven curves, and we demonstrate that the expected loss as measured by the area under these curves is linearly related to AUC. We show that the rate–driven curves are the genuine equivalent of ROC curves in cost space, establishing a point–point rather than a point–line correspondence. Furthermore, a decomposition of the rate–driven curves is introduced which separates the loss due to the threshold choice method from the ranking loss (Kendall

τ distance). We also derive the corresponding curve to the ROC convex hull in cost space: this curve is different from the lower envelope of the cost lines, as the latter assumes only optimal thresholds are chosen.

A framework for semi–supervised and unsupervised optimal extraction of clusters from hierarchies

Ricardo J.G.B. Campello, Davoud Moulavi, Arthur Zimek and Jörg Sander
Data Mining and Knowledge Discovery
DOI 10.1007/s10618-013-0311-4

We introduce a framework for the optimal extraction of flat clusterings from local cuts through cluster hierarchies. The extraction of a flat clustering from a cluster tree is formulated as an optimization problem and a linear complexity algorithm is presented that provides the globally optimal solution to this problem in semi–supervised as well as in unsupervised scenarios. A collection of experiments is presented involving clustering hierarchies of different natures, a variety of real data sets, and comparisons with specialized methods from the literature.

Pairwise meta–rules for better meta–learning–based algorithm ranking

Quan Sun and Bernhard Pfahringer
Machine Learning
DOI 10.1007/s10994-013-5387-y

In this paper, we present a novel meta–feature generation method in the context of meta–learning, which is based on rules that compare the performance of individual base learners in a one–against–one manner. In addition to these new meta–features, we also introduce a new meta–learner called Approximate Ranking Tree Forests (ART Forests) that performs very competitively when compared with several state–of–the–art meta–learners. Our experimental results are based on a large collection of datasets and show that the proposed new techniques can improve the overall performance of meta–learning for algorithm ranking significantly. A key point in our approach is that each performance figure of any base learner for any specific dataset is generated by optimising the parameters of the base learner separately for each dataset.

Block coordinate descent algorithms for large–scale sparse multiclass classification

Mathieu Blondel, Kazuhiro Seki and Kuniaki Uehara
Machine Learning
DOI 10.1007/s10994-013-5367-2

Over the past decade, ℓ_1 regularization has emerged as a powerful way to learn classifiers with implicit feature selection. More recently, mixed–norm (e.g., ℓ_1/ℓ_2) regularization has been utilized as a way to select entire groups of features. In this paper, we propose a novel direct multiclass formulation specifically designed for large–scale and high–dimensional problems such as document classification. Based on a multiclass extension of the squared hinge loss, our formulation employs ℓ_1/ℓ_2 regularization so as to force weights corresponding to the same features to be zero across all classes, resulting in compact and fast–to–evaluate multiclass models. For optimization, we employ two globally–convergent variants of block coordinate descent, one with line search (Tseng and Yun in Math. Program. 117:387423, 2009) and the other without (Richtrik and Tak in Math. Program. 138, 2012a, Tech. Rep. arXiv:1212.0873, 2012b). We present the two variants in a unified manner and develop the core components needed to efficiently solve our formulation. The end result is a couple of block coordinate descent algorithms specifically tailored to our multiclass formulation. Experimentally, we show that block coordinate descent performs favorably compared to other solvers such as FOBOS, FISTA and SpaRSA. Furthermore, we show that our formulation obtains very compact multiclass models and outperforms ℓ_1/ℓ_2–regularized multiclass logistic regression in terms of training speed, while achieving comparable test accuracy.

A comparative evaluation of stochastic–based inference methods for Gaussian process models

Maurizio Filippone, Mingjun Zhong and Mark Girolami
Machine Learning
DOI 10.1007/s10994-013-5388-x

Gaussian process (GP) models are extensively used in data analysis given their flexible modeling capabilities and interpretability. The fully Bayesian treatment of GP models is analytically intractable, and therefore it is necessary to resort to either deterministic or stochastic approximations. This paper focuses on stochastic–based inference techniques. After discussing the challenges associated with the fully Bayesian treatment of GP models, a number of inference strategies based on Markov chain Monte Carlo methods are presented and rigorously assessed. In particular, strategies based on efficient parameterizations and efficient proposal mechanisms are extensively compared on simulated and real data on the basis of convergence speed, sampling efficiency, and computational cost.

Probabilistic topic models for sequence data

Nicola Barbieri, Antonio Bevacqua, Marco Carnuccio, Giuseppe Manco and Ettore Ritacco
Machine Learning
DOI 10.1007/s10994-013-5391-2

Probabilistic topic models are widely used in different contexts to uncover the hidden structure in large text corpora. One of the main (and perhaps strong) assumptions of these models is that the generative process follows a bag–of–words assumption, i.e. each token is independent from the previous one. We extend the popular Latent Dirichlet Allocation model by exploiting three different conditional Markovian assumptions: (i) the token generation depends on the current topic and on the previous token; (ii) the topic associated with each observation depends on topic associated with the previous one; (iii) the token generation depends on the current and previous topic. For each of these modeling assumptions we present a Gibbs Sampling procedure for parameter estimation. Experimental evaluation over real–word data shows the performance advantages, in terms of recall and precision, of the sequence–modeling approaches.

The flip–the–state transition operator for restricted Boltzmann machines

Kai Brügge, Asja Fischer and Christian Igel
Machine Learning
DOI 10.1007/s10994-013-5390-3

Most learning and sampling algorithms for restricted Boltzmann machines (RBMs) rely on Markov chain Monte Carlo (MCMC) methods using Gibbs sampling. The most prominent examples are Contrastive Divergence learning (CD) and its variants as well as Parallel Tempering (PT). The performance of these methods strongly depends on the mixing properties of the Gibbs chain. We propose a Metropolis–type MCMC algorithm relying on a transition operator maximizing the probability of state changes. It is shown that the operator induces an irreducible, aperiodic, and hence properly converging Markov chain, also for the typically used periodic update schemes. The transition operator can replace Gibbs sampling in RBM learning algorithms without producing computational overhead. It is shown empirically that this leads to faster mixing and in turn to more accurate learning.

Differential privacy based on importance weighting

Zhanglong Ji and Charles Elkan
Machine Learning
DOI 10.1007/s10994-013-5396-x

This paper analyzes a novel method for publishing data while still protecting privacy. The method is based on computing weights that make an existing dataset, for which there are no confidentiality issues, analogous to the dataset that must be kept private. The existing dataset may be genuine but public already, or it may be synthetic. The weights are importance sampling weights, but to protect privacy, they are regularized and have noise added. The weights allow statistical queries to be answered approximately while provably guaranteeing differential privacy. We derive an expression for the asymptotic variance of the approximate answers. Experiments show that the new mechanism performs well even when the privacy budget is small, and when the public and private datasets are drawn from different populations.

Activity preserving graph simplification

Francesco Bonchi, Gianmarco De Francisci Morales, Aristides Gionis and Antti Ukkonen
Data Mining and Knowledge Discovery
DOI 10.1007/s10618-013-0328-8

We study the problem of simplifying a given directed graph by keeping a small subset of its arcs. Our goal is to maintain the connectivity required to explain a set of observed traces of information propagation across the graph. Unlike previous work, we do not make any assumption about an underlying model of information propagation. Instead, we approach the task as a combinatorial problem.
We prove that the resulting optimization problem is **NP**–hard. We show that a standard greedy algorithm performs very well in practice, even though it does not have theoretical guarantees. Additionally, if the activity traces have a tree structure, we show that the objective function is supermodular, and experimentally verify that the approach for size–constrained submodular minimization recently proposed by Nagano et al (2011) produces very good results. Moreover, when applied to the task of reconstructing an unobserved graph, our methods perform comparably to a state–of–the–art algorithm devised specifically for this task.

ABACUS: frequent pattern mining based community discovery in multidimensional networks

Michele Berlingerio, Fabio Pinelli and Francesco Calabrese
Data Mining and Knowledge Discovery
DOI 10.1007/s10618-013-0331-0

Community Discovery in complex networks is the problem of detecting, for each node of the network, its membership to one of more groups of nodes, the communities, that are densely connected, or highly interactive, or, more in general, similar, according to a similarity function. So far, the problem has been widely studied in monodimensional networks, i.e. networks where only one connection between two entities may exist. However, real networks are often multidimensional, i.e., multiple connections between any two nodes may exist, either reflecting different kinds of relationships, or representing different values of the same type of tie. In this context, the problem of Community Discovery has to be redefined, taking into account multidimensional structure of the graph. We define a new concept of community that groups together nodes sharing memberships to the same monodimensional communities in the different single dimensions. As we show, such communities are meaningful and able to group nodes even if they might not be connected in any of the monodimensional networks. We devise ABACUS (frequent pAttern mining–BAsed Community discoverer in mUltidimensional networkS), an algorithm that is able to extract multidimensional communities based on the extraction of frequent closed itemsets from monodimensional community memberships. Experiments on two different real multidimensional networks confirm the meaningfulness of the introduced concepts, and open the way for a new class of algorithms for community discovery that do not rely on the dense connections among nodes.

Growing a list
Benjamin Letham, Cynthia Rudin and Katherine A. Heller
Data Mining and Knowledge Discovery
DOI 10.1007/s10618-013-0329-7

It is easy to find expert knowledge on the Internet on almost any topic, but obtaining a complete overview of a given topic is not always easy: Information can be scattered across many sources and must be aggregated to be useful. We introduce a method for intelligently growing a list of relevant items, starting from a small seed of examples. Our algorithm takes advantage of the wisdom of the crowd, in the sense that there are many experts who post lists of things on the Internet. We use a collection of simple machine learning components to find these experts and aggregate their lists to produce a single complete and meaningful list. We use experiments with gold standards and open–ended experiments without gold standards to show that our method significantly outperforms the state of the art. Our method uses the ranking algorithm Bayesian Sets even when its underlying independence assumption is violated, and we provide a theoretical generalization bound to motivate its use.

What distinguish one from its peers in social networks?
Yi-Chen Lo, Jhao-Yin Li, Mi-Yen Yeh, Shou-De Lin and Jian Pei
Data Mining and Knowledge Discovery
DOI 10.1007/s10618-013-0330-1

Being able to discover the uniqueness of an individual is a meaningful task in social network analysis. This paper proposes two novel problems in social network analysis: how to identify the uniqueness of a given query vertex, and how to identify a group of vertices that can mutually identify each other. We further propose intuitive yet effective methods to identify the uniqueness identification sets and the mutual identification groups of different properties. We further conduct an extensive experiment on both real and synthetic datasets to demonstrate the effectiveness of our model.

Spatio–temporal random fields: compressible representation and distributed estimation

Nico Piatkowski, Sangkyun Lee and Katharina Morik
Machine Learning
DOI 10.1007/s10994-013-5399-7

Modern sensing technology allows us enhanced monitoring of dynamic activities in business, traffic, and home, just to name a few. The increasing amount of sensor measurements, however, brings us the challenge for efficient data analysis. This is especially true when sensing targets can interoperate—in such cases we need learning models that can capture the relations of sensors, possibly without collecting or exchanging all data. Generative graphical models namely the Markov random fields (MRF) fit this purpose, which can represent complex spatial and temporal relations among sensors, producing interpretable answers in terms of probability. The only drawback will be the cost for inference, storing and optimizing a very large number of parameters—not uncommon when we apply them for real–world applications.

In this paper, we investigate how we can make discrete probabilistic graphical models practical for predicting sensor states in a spatio–temporal setting. A set of new ideas allows keeping the advantages of such models while achieving scalability. We first introduce a novel alternative to represent model parameters, which enables us to compress the parameter storage by removing uninformative parameters in a systematic way. For finding the best parameters via maximum likelihood estimation, we provide a separable optimization algorithm that can be performed independently in parallel in each graph node. We illustrate that the prediction quality of our suggested method is comparable to those of the standard MRF and a spatio–temporal k–nearest neighbor method, while using much less computational resources.

Table of Contents – Part II

Social Network Analysis

Natural Language Processing and Information Extraction

Ranking and Recommender Systems

Matrix and Tensor Analysis

Structured Output Prediction, Multi-label and Multi-task Learning

Transfer Learning

Bayesian Learning

Graphical Models

Nearest-Neighbor Methods

Incremental Local Evolutionary Outlier Detection for Dynamic Social Networks

Tengfei Ji, Dongqing Yang, and Jun Gao

School of Electronics Engineering and Computer Science,
Peking University, Beijing, 100871 China
tfji@pku.edu.cn

Abstract. Numerous applications in dynamic social networks, ranging from telecommunications to financial transactions, create evolving datasets. Detecting outliers in such dynamic networks is inherently challenging, because the arbitrary linkage structure with massive information is changing over time. Little research has been done on detecting outliers for dynamic social networks, even then, they represent networks as un-weighted graphs and identify outliers from a relatively global perspective. Thus, existing approaches fail to identify the objects with abnormal *evolutionary behavior only* with respect to their *local neighborhood*. We define such objects as *local evolutionary outliers*, LEOutliers. This paper proposes a novel *incremental* algorithm IcLEOD to detect LEOutliers in *weighted* graphs. By focusing only on the time-varying components (*e.g.*, node, edge and edge weight), IcLEOD algorithm is highly efficient in large and gradually evolving networks. Experimental results on both real and synthetic datasets illustrate that our approach of finding local evolutionary outliers can be practical.

Keywords: Outlier detection, Dynamic Social Networks, Weighted evolving graphs, Local information.

1 Introduction

Outlier detection is a task to uncover and report observations which appear to be inconsistent with the remainder of that set of data [1]. Since outliers are usually represented truly unexpected knowledge with underlying value, research has been widely studied in this area, often applicable to static traditional strings or attribute-value datasets [2].

Little work, however, has focused on outlier detection in dynamic graph-based data. With the unprecedented development of social networks, various kinds of records like credit, personnel, financial, medical, etc. all exist in a graph form, where vertices represent objects, edges represent relationships among objects and edge weights represent link strength [3]. Graph-based outlier detection problem is specially challenging for three major reasons as follows:

Dynamic Changes: Vertices, the relationships among them as well as the weight of the relationships are all continuously evolving. For example, users join friendship networks (*e.g.* Facebook), friendships are established, and communication

H. Blockeel et al. (Eds.): ECML PKDD 2013, Part II, LNAI 8189, pp. 1–15, 2013.

becomes increasingly frequent. To capture outliers in evolving networks, detecting approaches should obtain temporal information from a collection of snapshots instead of a particular instant. For example, snapshots of the Facebook graph should be taken periodically, forming a sequence of snapshot graphs [4].

Massive Information: Compared with average data sets, social networks are significantly larger in size. The volume is even larger when the network is dynamic, massive information involved in a series of snapshots with millions of nodes and billions of edges[5]. In this case, it is difficult for algorithms to obtain full knowledge of the entire networks.

Deeply Hidden Outliers: Recent studies suggest that social networks usually exhibit hierarchical organization, in which vertices are divided into groups that can be further subdivided into groups of groups, and so forth over multiple scales [21]. Therefore, outliers are more difficult to distinguish from normal ones if they are hidden deeply among their neighbored but not globally.

However, outlier detection in social networks has not yet received as much attention as some other topics, e.g. community discovery [9,10]. Only a few studies have been conducted on graph-based outlier detection (e.g. [3], [6], [7], [8]). While a more detailed discussion on these approaches will be provided in section 2, it suffices to point out here that most of these approaches identify outliers in *unweighted* graphs from a more *global* perspective. For example, community-based algorithms [3,6] identify objects whose evolving trends are different with that of entire community. All such global outlier detection algorithms require the entire structure of the graph be fully known, which is impractical when dealing with large evolving networks. Furthermore, the local abnormality may be highly covered by global evolution trend. Thus, existing global methods fail to identify the objects with abnormal evolutionary behavior *only* relative to their *local neighborhood*. We define such objects as local evolutionary outliers, LEOutliers. The following example is adopted to illustrate directly the feature of LEOutliers.

Example: Who Should be Liable for Examination Leakage

Figure 1 shows a communication network with two communities, teacher community C_1 and student community C_2. Different colors are used to distinguish between members of two communities. Because of space constraints, links between nodes have been omitted. It is worthwhile to note that we use the **overlapping area** of two communities to denote the interactions between teachers and students. The more they are connected, the larger the overlapping area becomes.

Figure 1(a) contains two snapshots at time T-1 and T and we suppose that the Entrance Examination time is near at T. It is obvious that, from T-1 to T, the evolution trend of entire teacher community is communicating more frequently with student community, which is reasonable since more guidance is needed before examination. According to the global-view algorithms, objects that follow the entire community evolution trend are regarded as normal ones. Interestingly, once local neighborhood is taken into account, as illustrated in Figure 1(b), the black node v is an example of local evolutionary outlier. We suppose v

and its neighbors at time T-1 (blue triangles) are a special kind of teachers, paper setters. The blue triangles avoid communicating with students as the examination approaches for confidential reasons. On the contrary, node v is behaving abnormally as he frequently interacts with students at T, which is a violation of principle. Therefore, although node v evolving consistently with entire community, he is the most likely suspect in examination leakage.

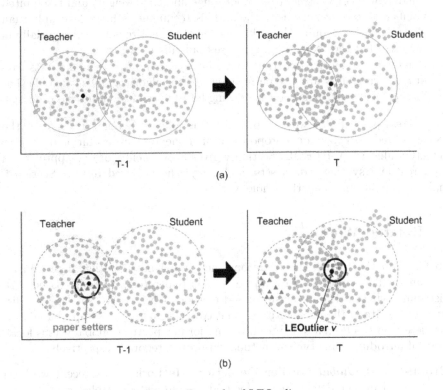

Fig. 1. Example of LEOutlier

The above example shows that the global-view algorithm is adequate under certain conditions, but not satisfactory for the case when evolutionary outliers are hidden deeply among their neighborhood. In this paper, we propose a novel method named IcLEOD to effectively detect LEOutlier in weighted graphs from a local perspective. The technical contributions of our work can be summarized as follows:

- Besides descriptive concept, we put forward a novel measurable definition of *local evolutionary outlier*. To the best of our knowledge, this is the first straightforward concept of a local evolutionary outlier which quantifies how outlying an object's evolving behavior is from a local perspective.
- We propose an incremental local evolutionary outlier detection algorithm (IcLEOD), which fully considers the varying temporal information and the

complex topology structure of social networks. Our algorithm consists of two stages: In stage I, a local substructure named $Corenet(v)$ is constructed for every object v according to structure information and edge weights; In stage II, we detect local evolutionary outliers by carefully analyzing and comparing the $Corenet(v)$ at different snapshots.

- Our algorithm greatly increases the efficiency by incrementally analyzing the dynamic components (e.g., node, edge and edge weight) and the limited number of nodes affected by them. This technique is more favorable than algorithms that require global knowledge of the entire network, especially in the case that the snapshot graphs are gradually evolving.
- Finally, the extensive experiments on both real and synthetic datasets confirm the capability and the performance of our algorithm. We conclude that finding local evolutionary outliers using IcLEOD is meaningful and practical.

The rest part of this work is organized as follows: Section 2 discusses the recent related work; Section 3 proposes our incremental local evolutionary outlier detection algorithm, IcLEOD; Section 4 gives experiments for our approach on both real and synthetic data sets, and shows the achieved results. Section 5 makes a conclusion about the whole work.

2   Related Work

To focus on the theme, the traditional non-graph based outlier detection algorithms will no more be introduced in this paper (e.g., distance-based [17], distribution-based [1] and density-based methods [15,16]). We are eager to discuss some state-of-the-art algorithms that conduct on graphs. Graph-based outlier detection has been studied from two major perspectives: global versus local. We will introduce some typical methods in both categories respectively.

Graph-Based *Global* Outlier Detection Methods: Most recent work on graph-based outlier detection has focused on unweigted graphs from a more global perspective (i.e. entire graph, community). For example, a stream-based outlier detection algorithm [14] takes a global view of entire graph to identify graph objects which contain unusual bridging edges. Community-based outlier detection methods [7,13] detect outliers within the context of communities such that the identified outliers deviate significantly from the rest of the community members. Some methods [3,6] capture the dynamic anomalous objects whose evolution behaviors are quite different from that of their respective communities. All global outlier detection algorithms require that the entire graph should be obtained, which may be impractical if networks are too large or too dynamic.

Graph-Based *Local* Outlier Detection Methods: Saligrama [11] proposes a statistical method based on local K-nearest neighbor distances to identify anomalies localized to a small spatial region, which is used mainly to deal with spatial data and cannot be easily generalized to non-spatial networks. OddBall algorithm [12] takes a egocentric view to search weighted graphs based upon a set of power laws, and determines four types of anomalous subgraphs centered on individual nodes: near-cliques, near-stars, heavy vicinities and dominant heavy links. Los Alamos National Laboratory [20] explores local areas and paths in the network which are least likely to occur under normal conditions by combining anomaly scores from edges in a neighborhood. Most methods in this category utilize only single snapshot data to find unexpected nodes/edges/sub-structures and hence they cannot detect temporal changes.

In summary, most of existing methods represent social networks (static and dynamic) as unweighted graphs, and find outliers from a global point of view. Thus the outliers detected by previous algorithms are not local evolutionary outliers as proposed in this paper.

3 IcLEOD Algorithm

Consider a dynamic social network as a sequence of snapshots G_1, G_2, ...,G_T, each snapshot is represented by weighted graphs G = (V, E), where V is the set of objects (nodes) and E is the set of weighted edges. The weight of an edge denotes the link strength (connecting times). In this paper, we focus on the problem of detecting local evolutionary outliers from any of the two snapshots G_i and G_j. Local evolutionary outliers across multiple snapshots can be obtained by simple post-processing. More specifically, input for our problem thus consists two snapshots of a weighted evolving network, and meaningful LEOutliers are output.

Our LEOD algorithm involves two major phases. In the first phase, *Corenet* for individual object is formed according to local topology structure and edge weights information. In the second phase, local evolutionary outliers are identified by comparing individual's *Corenets* of different snapshots. We will present two phases in Subsection 3.1 and 3.2 respectively.

3.1 Phase I: Discovering Corenet for Individual Object

As noted above, the evolutionary behavior of a LEOutlier is extremely different from that of its "closest" neighbors. Thus, the primary goal in phase I is to reasonably measure the closeness between objects, so as to determine which nodes could be regarded as the closest ones. There are two basic concepts usually used to group local nodes in un-weighted graph [19]. We will briefly introduce them before providing the notion of *Corenet*.

Definition 1 (Egonet). Given a node(ego) $v_i \in V$, the egonet of v_i is defined as egonet(v_i)={v_i} $\cup$ {v_j | $v_j \in V, e_{ij} \in E$}

Where e_{ij} is the edge between v_i and v_j.

Definition 2 (Super-egonet). Given a node(ego) $v_i \in V$, the super-egonet of v_i is defined as super-egonet(v_i)={$ego(v_i)$} $\cup$ {$ego(v_j)$ | $v_j \in V, e_{ij} \in E$}

Obviously, these two concepts are very simple in obtaining the local substructure: they just regard 1-hop neighbors(egonet) or neighbors within 2-hop(super-egonet) as the ego's closest neighbors. However, they will encounter problems when dealing with weighted graphs. As in the case of a friendship network with edge-weights representing interactions between friends, one is likely to be closer to his intimate friend's intimate friend instead of his nodding acquaintances. Consider the situation in Figure 2, where node X is the ego, Y_1, Y_2,Y_3 are 1-hop neighbors of X, Z_1 is its 2-hop neighbor. By following the definition of egonet, as Figure 2(b) shows, Y_1, Y_2 and Y_3 are the 3-closest neighbors of X. The concept of egonet focuses only on structural connection but ignores the power of closeness transmission. Therefore, it requires a forceful measurement considering both *connectivity* and *closeness*.

First, we propose the following two notions to assess the *closeness* between ego and its neighbors. We call the node of interest *core* to differentiate it from egonet.

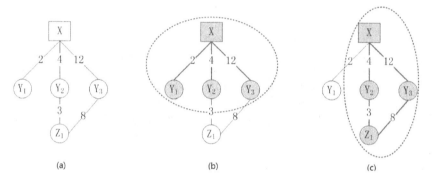

(a)   (b)   (c)

Fig. 2. Comparison of Egonet and Corenet

Definition 3 (Closeness related to the core). Let node v_0 be core, $v_0 \in V$. For $\forall v_l \in V$, we assume that there are d paths connecting v_0 and v_l. The jth path (l in length) passes through nodes $\{v_0, v_1, v_2, ..., v_l\}$ in sequence, where $1 \leqslant j \leqslant d$. Then the closeness between v_0 and v_l is defined as:

$$Closeness(v_0, v_l) = \max_{1 \leq j \leq d} \prod_{i=0}^{l-1} \frac{w_{v_i v_{i+1}}}{w_{v_i}} \tag{1}$$

Where $w_{v_i v_{i+1}}$ is the weight of the edge between v_i and v_{i+1}, and w_{v_i} is the sum of the weights of the edges connected to node v_i. Obviously, $\forall v_l \in V$, $Closeness(v_0, v_l) \in [0, 1]$. The higher the value, the more intimate the relation is. It is possible that a node directly connected with the core owns a smaller closeness. For example, in Figure 2, $Closeness(X, Y_1) = \frac{2}{2+4+12} = \frac{1}{9}$ and $Closeness(X, Z_1) = \frac{12}{2+4+12} \times \frac{8}{12+8} = \frac{4}{15}$.

In the case that two (or more) identical values of closeness are obtained from two (or more) different paths, to avoid closeness drift, we prefer the path that includes the edge directly connecting the core with maximum weight.

Definition 4 (k-closeness of the core). Let node v_0 be core, $v_0 \in V$. For $\forall$ $k>0$, the k-closeness of the core, denoted as k-closeness(v_0), is defined as :

(i) For at least k nodes $v_p \in V \backslash \{v_0\}$, it holds that Closeness(v_0,v_p)$\geq$ k-closeness(v_0), and

(ii) For at most k-1 nodes $v_p \in V \backslash \{v_0\}$, it holds that Closeness(v_0,v_p)$>$ k-closeness(v_0).

Different with the concepts of Egonet and Super-egonet, the definition 4 considers the top-k "closest" neighbors of the core only based on closeness transmission, instead of linking relationships. In this definition, the "closest" neighbors are those nodes with larger value of closeness, rather than directly connecting with the core.

Definition 5 (k-closeness neighborhood of the core). Given the k-closeness of core v_0, the k-closeness neighborhood of v_0 contains every node whose closeness related to v_0 is not smaller than the k-closeness(v_0). Formally, $N_k(v_0) = \{v_p \in V \backslash \{v_0\} \mid$ Closeness(v_0,v_p) $\geq$ k-closeness(v_0)$\}$.

As mentioned above, egonet concerns only the nodes directly connected with the node of interest, while the closeness measurement (Def. 3-5) mainly consider closeness transmission. The former completely ignores the edge-weight information, similarly, the latter ignores the risk that the reliability may reduce after successive transmissions. Thus, for the purpose of discovering the local context for the core, we propose a notion named *Corenet* that balances the topology structure and the closeness transmission.

Definition 6 (Corenet). Given the k-closeness of core, k-closeness(v_0), the Corenet of v_0 contains nodes that satisfy the conditions: (i) the closeness related to v_0 is not smaller than the k-closeness(v_0), and (ii) they are in the super-egonet of v_0. Formally, $v_p \in$ super-egonet(v_0)$\backslash \{v_0\}$, Corenet(v_0) is defined as:

$$\text{Corenet}(v_0) = \begin{cases} \text{super-egonet}(v_0), & \textit{if } \min_{v_p} \text{Closeness}(v_0, v_p) \geq k-\text{closeness}(v_0) \\ N_k(v_0), & \text{others} \end{cases}$$

So far, we have defined *corenet* as the local context of the core, which fully takes closeness transmission into account and avoids meaningless excessive transmissions by imposing a structural restriction. It is obvious that only the nodes in super-egonet(v_0) need to be calculated closeness related to the core and the maximum size of corenet is the number of the core's neighbors within 2-hop.

3.2   Phase II: Measuring Outlying Score

In this subsection, we will discuss how to detect LEOutliers by comparing *Corenets* at different snapshots. Since most real social networks are gradually evolving, which means successive snapshots are likely similar to each other (sharing more than 99% of their edges [4]). We utilize this property to exploit redundancies among similar snapshots and focus only on the components changing over time. The time-varying components and their notations are listed in Table 1. The changes of these components will affect their neighbors in a certain range. For example, if Z_1 is deleted from Figure 2(c), it will affect the *Corenet* of X. Thus, the *Corenet* of X need to be redetermined and X has to be examined for any anomalous evolving behavior. The following definition describes the influence of the time-vary components.

Table 1. Time-varying Components and their Notations

Time-varying Component	Event at time $t+1$	Symbol
Node	insertion of a new object	v^+
	deletion of an old object	v^-
Edge	generation of a new edge	e^+ with endpoints v_{e+}
	deletion of an old edge	e^- with endpoints v_{e-}
Edge-weight	increase weight of an edge	w^+ with endpoints v_{w+}
	decrease weight of an edge	w^- with endpoints v_{w-}

Definition 7 (Incremental nodes collection: *IC*). Given two snapshots G_{T-1} and G_T, the differences between them are time-vary components, as illustrated in Table 1. The range of nodes that could be affected by time-varying components is defined as:

$$IC = superegonet^T(v^+) \cup superegonet^{T-1}(v^-)$$
$$\cup\ egonet^T(v_{e+}) \cup egonet^{T-1}(v_{e-})$$
$$\cup\ egonet^T(v_{w+}) \cup egonet^{T-1}(v_{w-})$$

Where $superegonet^T(v^+)$ is the super-egonet of time-varying node v^+ in graph G_T, and other five are similar.

From definition 7, the time-vary components influence only limited number of their neighbors, namely nodes in *IC*. Thus our algorithm only need to examine the nodes in *IC* instead of the total number of nodes in the social network.

Before we present the particular measuring function, we first analyze the signs that a node is evolving abnormally. Consider we have two snapshots G_{T-1} and G_T, and the node of interest is v, there are two major signs to show that v is likely to be a LEOutlier:

(1) The members of Corenet(v) in G_{T-1} no longer belong to Corenet(v) or their closeness related to v is getting weaker from G_{T-1} to G_T;

(2) The new members added to Corenet(v) at time T have clear distinction with the former members, moreover, their closeness related to v can be unexpected high.

These two anomalous indication can be measured by Score 1 and Score 2 respectively, and the outlying score is the sum.

Definition 8 (Outlying Score). Given two snapshots G_{T-1} and G_T, $\text{Corenet}^{T-1}(v)$ and $\text{Corenet}^{T}(v)$ represent the Corenets of node v in G_{T-1} and G_T respectively. We denote the intersection of $\text{Corenet}^{T-1}(v)$ and $\text{Corenet}^{T}(v)$ except v as C_{old}, which is the set of old neighbors of node v. The elements of $\text{Corenet}^{T-1}(v) \backslash C_{old}$ are the neighbors removed from Corenet(v) at time T, denoted as $C_{removed}$. The elements of $\text{Corenet}^{T}(v) \backslash C_{old}$ are new neighbors of v, denoted as C_{new}. The outlying score of node v is defined as:

$$
\begin{aligned}
OutlyingScore(v) = & \sum_{v_i \in C_{old}} [closeness^{T-1}(v_i, v) - closeness^{T}(v_i, v)] \\
& + \sum_{v_r \in C_{removed}} closeness^{T-1}(v_r, v) \\
& + \sum_{\substack{v_j \in C_{new} \\ v_i \in C_{old}}} [(1 - \frac{w_{v_i v_j}}{w_{v_j}}) \times closeness^{T}(v_j, v)]
\end{aligned}
\tag{2}
$$

Where $w_{v_i v_j}$ is the weight of edge between v_i and v_j, w_{v_j} is the sum of the weights of the edges connected to v_j.

The sum of former summation terms is Score 1, which measures outlying degree caused by situation (1). Similarly, the third summation term represents Score 2, which measures outlying degree caused by new neighbors in situation (2).

Algorithm. IcLEOD Algorithm (High level definition)

Input: Snapshots G_{T-1} and G_T, the number of closet neighbors related to the core k, the number of LEOutliers n;
Output: n LEOutliers

Step 1: Identify the time-varying components by comparing G_{T-1} and G_T;
Step 2: Determine incremental nodes collection IC based on time-varying components;
Step 3: For each node v in IC, compute $\text{Corenet}^{T-1}(v)$ and $\text{Corenet}^{T}(v)$;
Step 4: Compute outlying score for each object according to Eq.2;
Step 5: Select and output the objects with the first n-largest Outlying Score;

---

4   Experiments

In this section, we illustrate the general behavior of the proposed IcLEOD algorithm. Since there is no ground truth for outlier detection, we test the accuracy of our approach on multiple synthetic datasets with injected outliers. We also compare scalability performance of our approach with several baseline methods on synthetic datasets, and we present some meaningful cases obtained by our approach on real data set DBLP.

4.1 Baselines

We compare the proposed algorithm with the following three baseline methods:

- *CEOD*: This baseline is a community-based outlier detection method [3,6], which takes three necessary procedures to detect outliers evolving differently with their communities, including community discovery, community matching and outlier detection.
- *EGO*: In this approach, we regard the nodes in egonet are the closest neighbors of ego (node of interest), and we detect outliers by comparing the egonets at different timestamps.
- *SuperEGO*: This method is similar to *EGO* except that it considers neighbors within 2-hop as the ego's closest neighbors.

4.2 Data Description and Evaluation Measure

Synthetic Data Sets: We generate a variety of synthetic datasets, each of which consists of two snapshots.

First, we use the Butterfly generator [18] in order to generate datasets with normal nodes. The synthetic weighted graph follows WPL(weight power law) and SPL(snapshot power law), i.e., $W(t) = E(t)^\alpha$ and $W_n \propto d_n^\beta$. E(t), W(t) are the number of edges and total weight of a graph respectively at time t, W_n is the total weight of the edges attached to each node and d_n is the degree of the node. We set α and β to be 1.3 and 1.1 respectively.

Next, for each dataset, we inject outliers. We first set the percentage of outliers η, and inject $|V|_{snapshot1} \times \eta$ outliers into datasets. $|V|_{snapshot1}$ is the number of vertices in Snapshot$_1$. Then we choose a random couple of objects e.g. v_1 and v_2, which exist in both Snapshot$_1$ and Snapshot$_2$. If v_1 and v_2 are far apart with common acquaintances few enough, we swap v_1 and v_2 in Snapshot$_2$. Thus, we inject two outliers in the dataset. More detail information about synthetic datasets is shown in Table 2. Change ratio is the percentage of time-varying components.

Table 2. Summary of Synthetic Datesets

Dataset	$|V|_{snapshot1}$	$|E|_{snapshot1}$	$|V|_{snapshot2}$	$|E|_{snapshot2}$	Change Ratio
SYN1	1,000	6,520	1,054	7,093	9.42%
SYN2	5,000	19,762	5,109	20,251	3.56%
SYN2	10,000	29,415	10,184	30,019	2.01%

DBLP: We adopt DBLP as the real dataset (dblp.uni-trier.de/), which contains computer science scientific publications. In our representation, we consider a undirected co-authorship network. The weighted graph W is constructed by extracting author-paper information: each author is denoted as a node in W; journal and conference papers are represented as links that connect the authors together; the edge weight is the number of joint publications by these two authors. We first removed the nodes with too low degree, than we extracted two co-authorship snapshots corresponding to the years 2001-2004 (13,511 authors) and 2005-2008 (14,270 authors).

We measured the performance of different algorithms using well-known metric F1 measure, which is defined as follows.

$$F1 = \frac{2 \times Recall \times Precision}{Recall + Precision}$$

Where recall is ratio of the number of relevant records retrieved to the total number of relevant records in the dataset; precision is ratio of the number of relevant records retrieved to the total number of irrelevant and relevant records retrieved.

4.3 The Accuracy of IcLEOD Algorithm

We evaluate the accuracy of the proposed algorithms on the simulated datasets. The accuracy of the algorithms is measured by detecting the injected outliers as that of the groundtruths. We set the number of closet neighbors k to 30, 15, 10 for SYN1, SYN2 and SYN3, respectively. We vary the percentage of injected outliers η as 1%, 2% and 5%. In fairness to all algorithms, we perform 50 experiments for each parameter setting and report the average F1 of all algorithms. Table 3 illustrates the comparison results.

Table 3. The Accuracy Comparison on the Synthetic Datasets

Dataset	Outlier η	CEOD	EGO	SuperEGO	IcLEOD
SYN1	1%	0.1554	0.2012	0.2965	**0.8940**
	2%	0.2018	0.1912	0.2244	**0.7836**
	5%	0.1614	0.2845	0.3122	**0.9065**
SYN2	1%	0.0867	0.2150	0.4016	**0.7926**
	2%	0.1945	0.2631	0.4936	**0.8012**
	5%	0.2124	0.1983	0.6288	**0.9174**
SYN3	1%	0.2182	0.2064	0.5462	**0.7329**
	2%	0.3796	0.2042	0.4986	**0.7074**
	5%	0.1862	0.3216	0.6032	**0.8922**

As it can be observed from Table 3, the proposed algorithm (IcLEOD) outperforms the others in indicating outliers precisely for all the settings. It is clear that CEOD and EGO fail to find local evolutionary outliers. This is because the former identifies outliers form the view of entire community instead of the local neighborhood substructure, and the latter only consider the neighbors with direct connectivity. The overall performance of SuperEGO is better than other baselines, but it significantly underperforms when the individual object's edge-weight distribution is clearly not uniform, like SYN1. This is due to SuperEGO ignores the edge-weight information. In contrast, the proposed algorithm detects outliers by considering both the local topology structure and the closeness transmission.

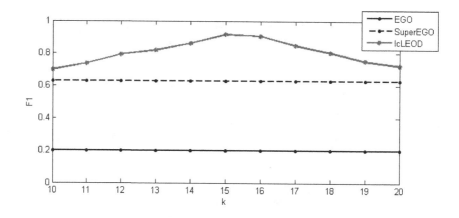

Fig. 3. Sensitivity

Figure 3 shows the sensitivity of the IcLEOD algorithm on parameter k. Two black lines represent the performance of baseline methods EGO and SuperEGO, respectively. We vary k from 10 to 20 for IcLEOD algorithm, as illustrated using the red line. The three algorithms are applied on the same data set, SYN2 and 5% outliers. Obviously, the proposed method is superior to two baseline methods, in spite of some changes caused by parameter variation.

4.4   The Scalability of IcLEOD Algorithm

To evaluate the scalability of IcLEOD, we conduct experiments on generated datasets as they vary the number of nodes. In Figure 4, the X-axis represents the number of nodes, whereas the Y-axis illustrates the computation time. We noticed that the processing time of the proposed approach is obviously lower than CEOD method. This is because the proposed approach only needs to calculate Corenets for nodes in IC (Def. 7), whereas CEOD method has to discover communities for entire network at each snapshot, even when there is no apparent change between two snapshots. Despite the EGO and SuperEGO approaches need less computation time, they have no specific procedure to determine the closeness neighborhood, which is likely to cause unfavorable results. The experiments demonstrate that there is a linear dependency of IcLEOD's processing

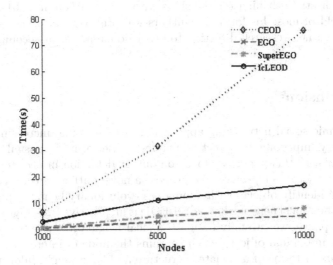

Fig. 4. Scalability Test of algorithms

time on the number of time-varying components in networks. Moreover, we can see that for the 10^4 network, the computation time is less than 20 seconds. This property means that the algorithm is practical in real applications.

4.5   Case Studies for Real Data Set

We will discuss an interesting outlier discovered by our algorithm on DBLP data set, which provides an intuitive perception about the effectiveness of our approach.

LEOutlier Case: [DBLP] Alexander Tuzhilin
We notice that Alexander Tuzhilin is a LEOutlier corresponding to DBLP 2001-2004 and DBLP 2005-2008. In DBLP 2001-2004 he was interested in Association Rules Analysis, and he shifted the focus of his research to Recommendation System in DBLP 2005-2008. We further noticed that his coauthors and the number of joint publications with these coauthors in two snapshots are very different. The principal members of his Corenets in two snapshots are listed as follows:

- Snapshot DBLP 2001-2004, $Corenet_1$(Alexander Tuzhilin): Tianyi Jiang, Hong Zhang, Balaji Padmanabhan, Gediminas Adomavicius etc.
- Snapshot DBLP 2005-2008, $Corenet_2$(Alexander Tuzhilin): Ada wai chee Fu, Cosimo Palmisano, Michele Gorgoglione, David jensen, Tianyi Jiang, Christos Faloutsos, Gueorgi Kossinets etc.

As the number of his publications increased, he established partnership with new researchers in recommendation system domain in the years 2005-2008 instead of keeping or strengthening relationships with his coauthors in 2001-2004. The research field of most his former coauthors was still association rules analysis, still others turned research direction to other domains except recommendation system.

5 Conclusions

Since dynamic social networking applications are becoming increasingly popular, it is very important to detect anomalies in the form of unusual evolutionary behaviors. In this paper, we focus on outlier detection in *evolving weighted* graphs from a *local* perspective. We propose a novel outlier detection algorithm IcLEOD, to identify objects with anomalous evolutionary behavior particularly relative to their local neighborhoods. IcLEOD is an effective two-stage algorithm. In the first phase, we carefully design the local neighborhood subgraph named Corenet for individual object, which contains the node of interest and its closest neighbors in terms of associated structure and edge-weight information. To quantify how outlying an object is, we put forward a measurement in the second phase by analyzing and comparing the Corenets at different snapshots. IcLEOD algorithm is significant efficient for LEOutlier detection in gradually evolving networks, because it could avoid repeated calculations by incrementally analyzing the dynamic components. The experimental results on both real datasets and synthetic datasets clearly ascertain that the proposed algorithm is capable of identifying local evolutionary outliers accurately and effectively.

Future work could will concentrate on further refinement of IcLEOD algorithm for dealing with general evolving datasets with multiple snapshots efficiently.

Acknowledgment. This work was supported by the National High Technology Research and Development Program of China (Grant No. 2012AA011002), National Science and Technology Major Program (Grant No. 2010ZX01042-002-002-02, 2010ZX01042-001-003-05), National Science & Technology Pillar Program (Grant No. 2009BA H44B03), Natural Science Foundation of China 61073018, the Cultivation Fund of the Key Scientific and Technical Innovation Project, Ministry of Education of China (Grant No. 708001) and the Shenzhen-Hong Kong Innovation Cooperation Project (No. JSE201007160004A). We would like to thank anonymous reviewers for their helpful comments.

References

1. Barnett, V., Lewis, T.: Outliers in Statistical Data, 3rd edn. Wiley, New York (1994)
2. Chandola, V., Banerjee, A., Kumar, V.: Anomaly detection: A survey. Technical Report, University of Minnesota (2007)
3. Gupta, M., Gao, J., Sun, Y., Han, J.: Integrating Community Matching and Outlier Detection for Mining Evolutionary Community Outliers. In: KDD (2012)

4. Ren, C., Lo, E., Kao, B., Zhu, X., Cheng, R.: On Querying Historical Evolving Graph Sequences. In: VLDB (2011)
5. Parthasarathy, S., Ruan, Y., Satuluri, V.: Community Discovery in Social Networks: Applications, Methods and Emerging Trends. In: Social Network Data Analytics. Springer, US (2011)
6. Gupta, M., Gao, J., Sun, Y., Han, J.: Community Trend Outlier Detection using Soft Temporal Pattern Mining. In: Flach, P.A., De Bie, T., Cristianini, N. (eds.) ECML PKDD 2012, Part II. LNCS, vol. 7524, pp. 692–708. Springer, Heidelberg (2012)
7. Gao, J., Liang, F., Fan, W., Wang, C., Sun, Y., Han, J.: On Community Outliers and their Efficient Detection in Information Networks. In: KDD (2010)
8. Aggarwal, C.C., Zhao, Y., Yu, P.S.: Outlier Detection in Graph Streams. In: ICDE (2011)
9. Flake, G., Lawrence, S., Giles, C.L.: In: SIGKDD (2000)
10. Bagrow, J.P., Bollt, E.M.: Phys. Rev. E (2005)
11. Saligrama, V., Zhao, M.: Local anomaly detection. In: AISTATS (2012)
12. Akoglu, L., McGlohon, M., Faloutsos, C.: Oddball: Spotting Anomalies in Weighted Graphs. In: Zaki, M.J., Yu, J.X., Ravindran, B., Pudi, V. (eds.) PAKDD 2010, Part II. LNCS (LNAI), vol. 6119, pp. 410–421. Springer, Heidelberg (2010)
13. Ji, T., Gao, J., Yang, D.: A Scalable Algorithm for Detecting Community Outliers in Social Networks. In: Gao, H., Lim, L., Wang, W., Li, C., Chen, L. (eds.) WAIM 2012. LNCS, vol. 7418, pp. 434–445. Springer, Heidelberg (2012)
14. Aggarwal, C.C., Zhao, Y., Yu, P.S.: Outlier Detection in Graph Streams. In: ICDE (2011)
15. Breunig, M.M., Kriegel, H.-P., Ng, R.T., Sander, J.: LOF: Identifying Density-Based Local Outliers. In: SIGMOD (2000)
16. Aggarwal, C.C., Yu, P.S.: Outlier Detection with Uncertain Data. In: SDM (2008)
17. Knorr, E.M., Ng, R.T., Tucakov, V.: Distance-Based Outliers: Algorithms and Applications. The VLDB Journal 8, 237–253 (2000)
18. Mcglohon, M., Akoglu, L., Faloutsos, C.: Weighted graphs and disconnected components: patterns and a generator. In: KDD (2008)
19. Wasserman, S., Faust, K.: Social Network Analysis. Cambridge University Press (1994)
20. Neil, J.C., Fisk, M., Storlie, C., Brugh, A.: Graph-Based Network Anomaly Detection. In: JSM (2010)
21. Clauset, A., Moore, C., Newman, M.E.J.: Hierarchical structure and the prediction of missing links in networks. Nature 453, 98–101 (2008)

How Long Will She Call Me? Distribution, Social Theory and Duration Prediction*

Yuxiao Dong[1,5], Jie Tang[2], Tiancheng Lou[3],
Bin Wu[4], and Nitesh V. Chawla[1,5]

[1] Department of Computer Science and Engineering, University of Notre Dame
[2] Department of Computer Science and Technology, Tsinghua University
[3] Google Inc., USA
[4] Beijing University of Posts and Telecommunications
[5] Interdisciplinary Center for Network Science and Applications,
University of Notre Dame
{ydong1,nchawla}@nd.edu, jietang@tsinghua.edu.cn, acrush@google.com,
wubin@bupt.edu.cn

Abstract. Call duration analysis is a key issue for understanding underlying patterns of (mobile) phone users. In this paper, we study to which extent the duration of a call between users can be predicted in a dynamic mobile network. We have collected a mobile phone call data from a mobile operating company, which results in a network of 272,345 users and 3.9 million call records during two months. We first examine the dynamic distribution properties of the mobile network including periodicity and demographics. Then we study several important social theories in the call network including strong/weak ties, link homophily, opinion leader and social balance. The study reveals several interesting phenomena such as people with strong ties tend to make shorter calls and young females tend to make long calls, in particular in the evening. Finally, we present a time-dependent factor graph model to model and infer the call duration between users, by incorporating our observations in the distribution analysis and the social theory analysis. Experiments show that the presented model can achieve much better predictive performance compared to several baseline methods. Our study offers evidences for social theories and also unveils several different patterns in the call network from online social networks.

1 Introduction

Analysis of mobility-based usage patterns can not only help understand users' requirements but also reveal underlying patterns behind user behaviors. The discovered patterns can be used to evaluate traffic demand and forecast call volumes, and also as a tool for infrastructure monitoring (such as switches and

* This work was done when the first author was vising Tsinghua University. Jie Tang is supported by the Natural Science Foundation of China (No.61222212, 61073073). Yuxiao Dong and Nitesh V. Chawla are supported by the Army Research Laboratory (W911NF-09-2-0053), and the U.S. Air Force Office of Scientific Research and the Defense Advanced Research Projects Agency (FA9550-12-1-0405).

H. Blockeel et al. (Eds.): ECML PKDD 2013, Part II, LNAI 8189, pp. 16–31, 2013.

cables). There is a lot of work on mobile call network analysis, e.g., scaling properties analysis [22,5], distribution analysis [21,19,1], behavior prediction [27,28], social ties analysis [3,2,25], and link prediction [14,20].

Vaz de Melo et al. [19] studied the call duration distributions of individual users in large mobile networks. They found that the call duration distribution of each user follows the log-logistic distribution, a power-law-like distribution and further designed a model for modeling the behavior of users based on their call duration distributions. The work has mainly focused on studying the call duration distributions of individual users. In [21], Seshadri et al. examined the distributions of the number of phone calls per customer; the total talk minutes per customer; and the distinct number of calling partners per customer. They found that the distributions significantly deviate from the expected power-law or lognormal distributions. However, both papers do not answer questions like what is the call duration distribution between two different users? How the distributions depend on the status (e.g., position, age, and gender) of the communicating users? And how the call duration reflects different properties of social ties between (or among) mobile users?

We focus on the call duration analysis. We understand and model the intricacies of social theory with the predictability of call duration between given two nodes in a network. What are the fundamental patterns underlying the call duration between people? What is the difference of call duration patterns between different groups of people? To which extent can we predict a call's duration between two users?

Contribution. We conduct systematic investigation of the call duration behaviors in mobile phone networks. Specifically, the paper makes the following contributions:

1. We first present a study on the call duration distributions. In particular, we focus on the dynamic properties of the duration distributions.
2. Second, we study the call duration network from a sociological point of view. We investigate a series of social theories in this network including social balance [4], homophily [13], two-step information flow [10], and strong/weak ties hypothesis [6,11]. Interestingly, we have found several unique patterns from the call duration network. For example, different from the online instant messaging networks, where people with more interactions would stay longer in each communication, while in the mobile call network, it seems that people who are familiar with each other tend to make shorter calls.
3. Based on the discovered patterns, we develop a time-dependent factor graph model, which incorporates those patterns into a semi-supervised machine learning framework. The discovered patterns of social theories are defined as social correlation factors and the dynamic properties of call duration are defined as temporal correlation factors. The model integrates all the factors together and learns a predictive function for call duration forecast.

Experimental results show that the presented model incorporating the discovered social patterns and the dynamic distributions significantly improves the

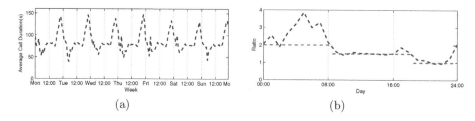

Fig. 1. Duration Periodicity. (a). X-axis: Time in one week. Y-axis: Average call duration. (b). X-axis: Time in one day. Y-axis: The ratio between call times (<60s) and call times (>60s).

prediction performance (5-18%) by comparing with several baseline methods using Support Vector Machine, Logistic Regression, Bayesian Network, and Conditional Random Fields.

2   Mobile Data and Characteristics

The data set used in this paper is made of a large collection of non-America call records provided by a large mobile communication company[1]. The data set contains 3.9 million call records during two months (December 2007 and January 2008). Each call record contains the phone numbers of the caller and the callee, and the start and end time of the call. Based on this, we construct a social network by viewing each phone number as a user and creating a relationship between user A and B if A made a call to B. The weight of the relationship is quantified by the number of calls between the two related users. In this way, the resultant network contains around 272,345 nodes and 521,925 edges.

We first study the distribution of calls between users. Clearly, the distribution fits a "power-law" distribution in the network. We also found that some users had intensive communications (more than 10 calls in 8 weeks) with each other. About 20% of the pairs of users produce 80% of the call records, which satisfied the Pareto Principle (also known as 80-20 rule) [18,16]. Thus in this work we mainly focus on the call duration between these pairs of users. For each user, we also extract her/his profile information such as age and gender information. A further statistic shows that there are about 60% calls which are less than 60 seconds (1 minute) and remaining 40% calls (>60s).

2.1   Periodicity

There exist periodic patterns for call duration between human beings. We reach this conclusion by tracking daily calls of mobile phone users. Figure 1(a) shows the average call duration curve on both weekdays and weekends. It clearly shows

---

[1] Data and codes are publicly available at
http://arnetminer.org/mobile-duration

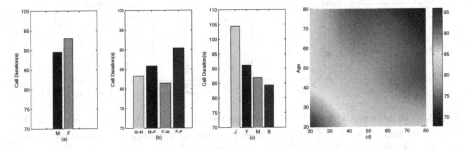

Fig. 2. Call duration of users by gender and age. (a). duration of different genders. (b). duration of different pairwise genders. (c). duration of different age(J: junior, Y: youth, M: middle-age, S: senior). (d). heat-map by plotting age *vs.* age and the color represents the duration of calls.

that there exists obvious week-period and day-period laws for the duration. From Monday to Sunday, we can see that the daily duration curves are very similar: 1) In work hours from 8:00 a.m. to 7:00 p.m., people call each other with stable average duration (75s); 2) After getting off work, the average duration between each other increases to 150 seconds gradually until mid-night; 3) From mid-night to early morning, the duration becomes shorter gradually and reaches to its lowest value (about 50s); 4) It ascends to 75 seconds at 8:00 a.m. Moreover, we perform a temporal analysis by tracking the hourly call duration in one day (see Figure 1). We observe that the curve of ratio between the number of calls (<60s) and calls (>60s) varies unevenly over hours: 1) From mid-night to 8:00 a.m., probability that people call each other with duration (<60s) is at least twice than duration which is greater than 60 seconds; 2) In work hours, the ratio is stable to 1.5. 3) From 18:00 p.m. to mid-night, the number of calls with a duration less than 60 seconds is almost the same as the number of calls with a duration larger than 60 seconds.

2.2 Demographics

How does the call duration distribution depend on the gender and age of callers? In this section, we examine the interplay of communication and user demographic attributes. First, we seek to understand how long males and females call. Figure 2 (a) and (b) represent the duration difference by different genders or between different gender-gender pairs. Figure 2 (a) shows that females tend to make longer calls than males. In Figure 2 (b), it shows that, in male-male calls, 84 seconds are taken per call which is lower than 91 seconds for female-female, whereas male-female calls, per call takes 86 seconds, whereas 81 seconds for female-male. Second, we report the analysis on different duration distribution based on age of users. Figure 2 (c) shows that, the average durations for juniors (0, 25], youths (25, 40], middle-aged people (40, 55], seniors (55, +) are 105, 91, 86, 84 seconds respectively, and they decrease as people get older. Figure 2 (d)

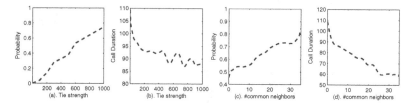

Fig. 3. Tie strength and Link homophily. X-axis: (a)(b). Tie strength as the increase of call times; (c)(d). The number of common neighbors between two callers. Y-axis: (a)(c). Probability that the duration is less than 60s, conditioned on tie strength or #common neighbors; (b)(d). Average call duration.

uses a heat-map visualization to call duration for different age-age pairs. The rows and columns represent the age of both caller and receiver and the color at each age-age cell captures the duration of this pair. The color spectrum extends from blue (short duration) through green, yellow, and onto red (long duration). In this Figure, it is evident that older people tend to have shorter conversations than young users. This trend of obviously descending-order duration in pairwise age fits individual case, as age increases.

3   Social Theory

Besides the dynamic properties of duration distribution, we investigate the interplay of human call behaviors and social theory, and try to answer the question: how social theories, i.e., social tie, homophily, social balance theory etc. are satisfied in the mobile social network? More specifically, we connect the call duration to four classical social psychological theories and focus our analysis on the network based correlation via the following statistics:

1. Strong/weak Ties [6,11]. How long do people with a strong or weak tie call?
2. Link homophily [13]. Do similar users tend to call each other with long or short duration?
3. Opinion leader [10]. How different (or how similar) are the calling behavior patterns between opinion leaders and ordinary users?
4. Social balance [4]. How does the duration-based network satisfy the social balance theory? To which extent?

Social Tie. Interpersonal ties, generally, come in two varieties: strong and weak. It is argued that weak ties are responsible for the majority of the structure of social networks and the transmission of information through the networks [6], but strong ties make people move to the same circles [11]. The strength of tie represents the extent of closeness of social contacts [4]. In mobile network, we define strong ties, representing frequent calls between two users, and weak ties, representing more casual social contacts with less calls between two users. Such a definition suggests a way of thinking about and answering the following question: How long do people call each other with a strong or weak tie? Figure 3 (a) illustrates our interesting finding: weak ties have a lower probability that their

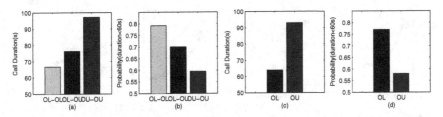

Fig. 4. Opinion leader. OL-Opinion leader; OU-Ordinary user. X-axis: (a)(b). calls between two users; (c)(d). calls made by OL or OU. Y-axis: (b)(d). average call duration; (b)(d). probability that the duration is less than 60s.

duration is less than 60 seconds. The stronger the tie between two users is, the larger the probability that their duration is less than 60 seconds is. When their tie strength reaches to 1,000 (calls before), there is a high probability (approximately 80%) that the duration of their future call is less than 60 seconds. In Figure 3 (b), we can see that the average call duration between strong ties is shorter than the calls between weak ties. This finding from both figures seems to be different from the situation in online instant messaging networks, where people with more interactions would stay longer in each communication.

Link Homophily. The principle of homophily [13] suggests that users with similar characteristics tend to associate with each other. Particularly, we study link homophily and test whether two users who share common links (caller or receiver) will have a tendency to call each other with longer or shorter duration. In Figure 3 (c), we can see clearly that the probability that the duration is less than 60 seconds when people have more common neighbors becomes higher gradually. Figure 3 (d) shows that the average duration between pairwise users becomes shorter and shorter when they have more and more common neighbors. Intuitively, in mobile communication, more homophily (more common neighbors) and stronger ties (more call times) between two people means that they are familiar with each other. In the point of human behaviors, thus, we can say that the call duration between acquaintances has larger probability to make a short call.

Opinion Leader. Opinion leadership is a concept that arises out of the theory of two-step flow of communication propounded by Lazarsfeld [12] and Katz [9,10], which suggests that innovation (idea) usually flows first to *opinion leaders*, and spreads to more people from them. There are several considerable algorithms to detect opinion leaders in social networks. We apply PageRank [17] to sort all users in our mobile phone data, then top 1% users are labeled as opinion leader according to their PageRank score and the others as ordinary users [8]. Figure 4 clearly shows that the calls between two opinion leaders have 30% shorter duration than the calls between two ordinary users in Figure 4 (a), and the average duration made by opinion leaders is also approximately 30% lower than ordinary users in Figure 4 (c). Figure 4 (b) shows that there is 80% possibility that the duration is less than 60 seconds, when an opinion leader calls another opinion leader, and the possibility is 60% when an ordinary user calls an ordinary user. As to individuals, there are the similar patterns in Figure 4 (d).

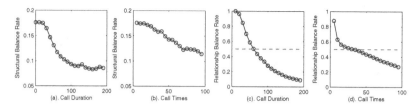

Fig. 5. Social Balance. X-axis: Whether a link is a non-friend(negative) one based on call duration(a)(c) or call times(b)(d). Y-axis: (a)(b) structural balance rate. (c)(d) relationship balance rate.

Social Balance. Now, we connect our work to the social balance theory [4]. For each triad (a group of three users), structural balance property implies that either all three of these users are friends or only one pair of them are friends. We assume two users are friends if they call each other at least once. In Figure 5 (a) and (b), it clearly shows that the mobile call network does not satisfy the structural balance theory and the balance rate decreases when the average duration or call times increases. As to relationship balance, the balance rate is the percentage of triangles with even number of negative ties. To adapt the theory to our problem, we assume whether a tie is a negative one based on either average call duration or call times, where the premise is that there exists at least one call between any two users in the triangle. Figure 5 (c) and (d) show that it is much more likely (more than 50%) for users to be connected with a balanced relationship when their duration is less than 60 seconds or they call each other less than 40 times. It represents that mobile network satisfies relationship balance in lower call times or shorter duration.

4   Duration Prediction

4.1   Problem Definition

Now, we study how to design a machine learning model to infer the call duration in the mobile call network based on the discovered patterns from the analysis of data distribution and social theory. We first give necessary definitions and then present a formal definition of the duration prediction problem. We assume that each user is associated with a number of attributes and thus have the following definition.

Definition 1. *Attributes Matrix: Let X be an $N \times d$ attribute matrix of people in which every row X_i corresponds to a user, each column an attribute, and an element x_{ij} denotes the jth attribute value of user v_i.*

The attributes matrix describes user-specific characteristics and can be defined in different ways. In the call network, an attribute can be defined as night call ratio and the value of an attribute can be defined as the frequency of calls occurring at night. Then, we define a dynamic call network with node attributes and call duration logs, as the input of our problem.

Definition 2. *Dynamic Call Network: A network at time t can be denoted as $G^t = (V, E, \boldsymbol{X}, Y^t)$, where V is the set of users and E is the set of call links between users, and $\boldsymbol{X}$ represents the attribute matrix of all users in the network, and Y^t is a set of the call duration score between two users at time t. Then we can define the dynamic call network $\boldsymbol{G} = \{V, E, \boldsymbol{X}, \boldsymbol{Y}\}$ and $\boldsymbol{Y} = Y^1 \cup Y^2 \cup \cdots \cup Y^T$.*

Based on the above concepts, we can define the problem of call duration prediction.

Problem 1. **Call Duration Prediction.** Given a dynamic call network $\boldsymbol{G} = \{V, E, \boldsymbol{X}, \boldsymbol{Y}\}$, the goal of the prediction is to learn a mapping function $f : (\boldsymbol{G}, \boldsymbol{Y}) \rightarrow Y^{T+1}$to predict the call duration in the next time stamp.

Future call duration can be defined as two cases, first, we use the past call duration to predict the duration of next call; and the second case is to predict the average duration of several calls in a future period (one user can call the other user more than one time) based on the historic call records. Furthermore, there might be more than one call between two users in the next time stamp. We consider two kinds of test cases. The first one is to predict the first call duration in next time stamp, which is called next call duration prediction; The other case is to predict the average call duration in the next time stamp, which is called average call duration prediction. We consider two different scenarios: a binary classification task by setting a threshold $T_{threshold}$ in call duration.

4.2   Prediction Model

Social Time-dependent Factor Graph Model. Tang et. al. [25] first proposed a partially labeled factor graph model to infer social tie. Hopcroft et al. [8] also proposed a triad based factor graph model for reciprocal relationship prediction in the Twitter network. Tan et al. [23] proposed a noise tolerant time-varying factor graph model for predicting users' behavior in social networks. In this work, we come up with a dynamic factor graph model based on previous partially labeled and triad factor graph model. The dynamic factor graph model incorporates both the correlations among latent variables in different timestamps and other social or attribute features for modeling and prediction. We take next call duration prediction as an example to formalize it in a dynamic factor graph model referred as Social Time-dependent Factor Graph Model (STFG) and propose an approach to learn the model for predicting call duration of pairwise callers. The name is derived from the idea that we incorporate social theory into the factor graph model.

Figure 6 illustrates the graphical illustration of STFG model. The top figure shows the dynamic call network of five users with duration and the bottom figure shows the proposed STFG model. The arrows indicate calls between two users and weight indicates the duration. In bottom figure, the model incorporates three different types of information including social theory (social correlation), user attributes and user's historic duration records (temporal correlation).

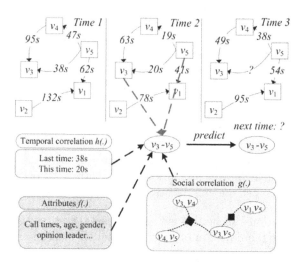

Fig. 6. Model illustration of duration prediction in a dynamic mobile call network

Now, we explain the proposed STFG model in details. Given a dynamic call network $\mathbf{G} = \{V, E, \mathbf{X}, \mathbf{Y}\}$, we can define the joint distribution over the durations Y^{T+1} given $\mathbf{G}$ as

$$p(Y^{T+1}|\mathbf{G}) = \prod f(x_i, y_i)g(X_c, Y_c)h(\mathbf{Y}, y_i^{T+1}) \tag{1}$$

The joint probability has three kinds of factor function, corresponding to the illustration in Figure 6. Specifically,

1. **Attribute factor:** $f(x_i, y_i)$. It represents the influence of an attribute of user v_i.
2. **Social correlation factor:** $g(X_c, Y_c)$. It denotes the influence of social relation Y_c.
3. **Temporal correlation factor:** $h(\mathbf{Y}, y_i^{T+1})$. It represents the dependency of one's duration at time $T + 1$ on its durations at time t ($t \in \{1, \cdots, T\}$), which denotes the difference between our dynamic model with others [25,8].

In principle, the three factors can be instantiated in different ways. In this work, we model them by the Hammersley-Clifford theorem [7] in a Markov random field. For the attribute factor, we accumulate all the attributes and obtain a local entropy for all users:

$$\frac{1}{Z_\alpha} \exp\{\sum_{i=1}^{|E|} \sum_{j=1}^{d} \alpha_j f_j(x_{ij}, y_i)\} \tag{2}$$

where α is the weight of function f_j and Z_α is a normalization factor. It can be defined as either a binary function or a real-value function. For example, for the user's social tie feature, we simply define it as a binary feature, that is if the link between user v_i and v_j is a strong tie and v_i calls v_j with duration (>60s), then a feature $f_j(x_{ij} = 1, y_i = 1)$ is defined and its values is 1, otherwise 0. For social correlation factor, we define a set of correlation feature functions $g_k(X_c, Y_c)$ over each triad Y_c in the network. Then we define a social correlation factor function as follows:

$$\frac{1}{Z_\beta} \exp\{\sum_c \sum_k \beta_k g_k(X_c, Y_c)\} \tag{3}$$

where β_k is the weight of the function, representing the influence degree of kth factor function on Y. We take opinion leader feature as the example to explain social correlation factor. It is defined as a binary function, that is, if a triad contains an opinion leader, then the value of a corresponding triad factor function is 1, otherwise 0.

For temporal correlation factor, we try to use it to model dynamic properties of duration distribution define it as:

$$\frac{1}{Z_\gamma} \exp\{\sum_{i=1}^{|E|} \sum_{t=1}^{T} \sum_m \gamma_m h_m(\mathbf{Y}, y_i^{T+1})\} \tag{4}$$

where $\mathbf{Y}$ is the past durations of the ith pair callers; γ_m represents how strongly the periodicity of the mth pair is. In reality, some users may call each other with similar durations in approximately same time every day or every week. For example, if user v_i and v_j call each other more than ten minutes in every everything, we can define a temporal function with value 1, otherwise 0.

Finally, a factor graph model is constructed by combining Eqs. 2-4 together into Eq. 1, $i.e.,$

$$p(Y^{T+1}|\mathbf{G}) = \frac{1}{Z} \exp\{\sum_{i=1}^{|E|} \sum_{j=1}^{d} \alpha_j f_j(x_{ij}, y_i)$$
$$+ \sum_c \sum_k \beta_k g_k(X_c, Y_c) + \sum_{i=1}^{|E|} \sum_{t=1}^{T} \sum_m \gamma_m h_m(\mathbf{Y}, y_i^{T+1})\} \tag{5}$$

where $Z = Z_\alpha Z_\beta Z_\gamma$ is a normalization factor. Based on Eq. 5, we define the following log-likelihood objective function $\mathcal{O}(\theta) = \log p(Y^{T+1}|\mathbf{G})$:

$$\mathcal{O}(\theta) = \sum_{i=1}^{|E|} \sum_{j=1}^{d} \alpha_j f_j(x_{ij}, y_i) + \sum_c \sum_k \beta_k g_k(X_c, Y_c)$$
$$+ \sum_{i=1}^{|E|} \sum_{t=1}^{T} \sum_m \gamma_m h_m(\mathbf{Y}, y_i^{T+1}) - \log Z \tag{6}$$

where $\theta = (\{\alpha\}, \{\beta\}, \{\gamma\})$ indicates a parameter configuration.

Model Learning. Learning STFG is to estimate the remaining free parameters θ, which maximizes the log-likelihood objective function $\mathcal{O}(\theta)$, $i.e.,$ $\theta^* = \arg\max \mathcal{O}(\theta)$

We use a gradient decent method (or a Newton-Raphson method) to optimize the objective function. We adopt α as the example to explain how we learn the parameters. Specifically, we first write the gradient of each α_j with regard to the objective function:

$$\frac{\partial \mathcal{O}(\theta)}{\partial \alpha_j} = \mathbb{E}[f_j(y_i, x_{ij})] - \mathbb{E}_{P_{\alpha_j}(y_i|x_{ij},G)}[f_j(y_i, x_{ij})] \tag{7}$$

where $\mathbb{E}[f_j(y_i, x_{ij})]$ is the expectation of feature function $f_j(y_i, x_{ij})$ given the data distribution and $\mathbb{E}_{P_{\alpha_j}(y_i|x_{ij},G)}[f_j(y_i, x_{ij})]$ is the expectation of feature function $f_j(y_i, x_{ij})$ under the distribution $P_{\alpha_j}(y_i|x_{ij},G)$ given by the estimated model. Similar gradients can be derived for parameter β_k and γ_m.

Here, there is a challenge that the graphical structure in STFG model can be arbitrary and may contain circles, which makes it intractable to directly calculate the marginal distribution $P_{\alpha_j}(y_i|x_{ij},G)$. Several approximate algorithms have been proposed, such as Loopy Belief Propagation (LBP) [15] and Mean field [26]. Due to the ease of implementation and effectiveness of LBP, in this work, we use LBP to approximate the marginal distribution $P_{\theta_k}(y_i|x_{ij},G)$. We are then able to obtain the gradient by summing over all the factor graph nodes with the marginal probabilities. It is worth noting that we need to perform the LBP twice in each iteration, one time for estimating the marginal distribution of unknown variables $y_i =?$ and the other time for marginal distribution over all features. Finally, we update each parameter with a learning rate η with the gradient. Related algorithms can be found in [25,24].

Prediction. With the estimated parameter θ, we can predict the future call durations. Specifically, the prediction problem can be cast as assigning the value of unknown call durations Y^{T+1} which maximizes the objective function given the learned parameters and network data.

$$Y^* = \arg\max \; \mathcal{O}(Y^{T+1}|\mathbf{G}, \mathbf{X}, \mathbf{Y}, \theta) \tag{8}$$

Obtaining an exact solution is again intractable. The LBP is utilized to calculate the marginal probability for each node in the factor graph. Finally, labels that produce the maximal probability will be assigned to each factor graph node.

5 Experiments

Our goal here is to predict the next call duration and the average duration of calls in next time stamp based on historic call detail records. We use the first 7 week call detail records as historic data, the first call in the 8th week as next call duration and the average duration of calls in the 8th week as average call duration. For binary duration prediction, we present the results with $T_{threshold} = 60s$.

Baseline Methods. We compare our proposed model with four methods.
SVM: it uses the same attributes associated with each edge or node as features to train a classification model and then apply it to predict the call duration label in the test data. For SVM, we use SVM-light[2].

---

[2] http://svmlight.joachims.org/

Table 1. Binary duration prediction performance of different methods. Case 1: Next Call Duration Prediction; Case 2: Average Call Duration Prediction.

	Method	Precision	Recall	F1-Measure
	SVM	0.5057	0.5021	0.5042
	LRC	0.6184	0.5548	0.5173
Case 1.	BNet	0.5812	0.5705	0.5692
	CRF	0.5865	0.5886	0.5871
	STFG	**0.6501**	**0.6375**	**0.6393**
	SVM	0.4869	0.4875	0.4847
	LRC	0.6143	0.6044	0.5996
Case 2.	BNet	0.5943	0.5902	0.5873
	CRF	0.6085	0.6054	0.6049
	STFG	**0.6695**	**0.6707**	**0.6692**

LRC: it uses the same attributes in SVM as features to train a logistic regression classification model and them apply it to predict the label in the test data.

BNet: the method uses the same features as that in SVM. The only difference is that it uses the Naive Bayes classifier.

CRF: it trains a Conditional Random Field model with attributes associated each edge. The difference of this method from our model is that it does not consider structural balance factors.

STFG: our proposed model, which trains a factor graph model with unlabeled data.

5.1   Prediction Performance

We quantitatively evaluate the performance of inferring call duration in terms of *Precision, Recall and F1-Measure*.

Table 1 shows the results for binary duration prediction next call duration prediction and average call duration prediction, set under the threshold. From Table 1, we see that our method clearly outperforms the baseline methods on both cases. For next call duration prediction, the STFG achieves a 5-13% improvement compared with SVM, LRC, BNet, CRF methods in terms of *F1-Measure*. Now, we further validate the effectiveness of our STFG model in the following three aspects: (1) contributions of distribution and social factors in the model; (2) convergence of the learning algorithm; (3) effect of different settings for the duration threshold.

Distribution Factor Contribution Analysis. Now, we analyze how different distribution factors: gender (G), age (A), week periodicity (W), day periodicity (D), can help infer future call duration. We first remove the gender factor (denoted as STFG-G), followed by further removing the age factor (STFG-GA),

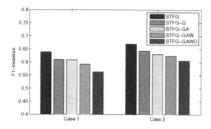 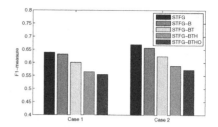

(a) Distribution factor contribution analysis   (b) Social factor contribution analysis

Fig. 7.   Factor contribution analysis

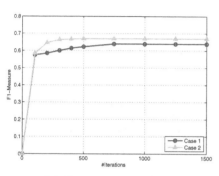 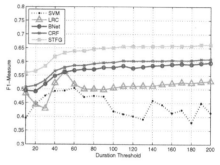

(a) Convergence analysis.   (b) Parameter $T_{threshold}$ analysis.

Fig. 8.   Convergence and parameter analysis

week periodicity (STFG-GAW), and finally removing day periodicity (STFG-GAWD). Figure 7 (a) shows the *F1-Measure* of the different STFG models. Obviously, we can observe clear drops on the performance when removing each of the factors. The result indicates that our model works well by combining the dynamic properties of data distribution, and each factor in our model contributes improvement in the performance.

Social Factor Contribution Analysis.   In STFG, we also consider five different social factors: social balance theory (B), social tie (T), link homophily (H) and opinion leaders (O). Here, we take the analysis to evaluate the contribution of different social factors for the prediction performance. With the same removing operations, we also can see clear drops in *F1-Measure* score in Figure 7 (b). For both two cases, we can find that there is a quick drop when ignoring social tie or link homophily factors. Figure 7 (b) also shows that the other social factors contribute to the prediction of call duration in two cases.

Convergence Analysis.   We conduct an experiment to analyze the convergence property of the loopy belief propagation based learning algorithm. Figure 8(a) illustrates the convergence analysis results of the learning algorithm. For

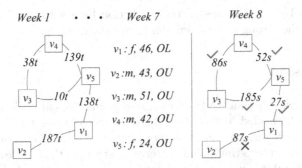

Fig. 9. Case study. Portion of the dynamic call network. The numbers associated with each link in left figure are the number of calls in first 7 weeks. $v_1 : f, 46, OL$ means v_1 is a 46-year female opinion leader. The right figure shows the real average call duration in the 8th week.

case 1, the LBP-based learning algorithm converges in about 300 iterations. For case 2, the learning algorithm reach to convergence in about 750 iterations. This suggests that the learning algorithm is able to reach convergence and its efficiency is acceptable.

Threshold Analysis. Finally, we analyze how different settings for the parameter $T_{threshold}$ influence the performance of call duration prediction. Figure 8 (b) lists the average prediction performance of all methods in case 1 with $T_{threshold}$ varied. There are similar patterns in case 2. In general, most methods have similar patterns with different parameter settings, except SVM which is unstable as $T_{threshold}$ varies. It shows that when setting $T_{threshold}$ less 60 seconds, the prediction performance of all models is not very acceptable, while when setting it more than 60 seconds, the performance varies very little and has a slightly increasing trend. However, when $T_{threshold}$ comes to 180s, the number of calls (<180s) is about 9 times to the number of calls (>180s). It means that only one of ten calls in our daily life tends to be greater than 3 minutes.

5.2   Qualitative Case Study

We present a case study to demonstrate the effectiveness of STFG model, see Figure 9. The left figure shows a portion of the dynamic call network from 1st to 7th week. Green colored sign indicates that our model predicts correctly whether the label of duration (<60s or >60s) between respective users or not. Red colored sign means that our model did not infer the real duration label. In left figure, there exists stronger ties in (v_1, v_2), (v_1, v_5) and (v_4, v_5) than (v_3, v_4) and (v_3, v_5). STFG model predicted (v_1, v_2), (v_1, v_5), (v_4, v_5) as short calls (<60s) and (v_3, v_4), (v_3, v_5) as long calls (>60s) based on social tie theory. Our model predicted four of five labels successfully. User v_5 as a young female tends to make short calls with others, so STFG predicted (v_1, v_5), (v_4, v_5) correctly. As to (v_3, v_5), our model made a compromise between gender factor and social tie factor, and finally predicted it as a long call (>60s) because of the weak tie between v_3 and v_5. STFG missed the prediction between v_1 and v_2, as it was misguided by the strong tie and opinion leader status of v_1.

6   Conclusion

In this paper, we systematically investigated a large mobile call duration network. We first identified and studied the dynamic properties of mobile calling patterns and characteristics, and how they relate to the social network attributes. We discover some interesting social patterns — stronger the ties, lower the probability of call duration; average duration between pairwise users becomes shorter and shorter when they have more and more common neighbors; opinion leaders tend to have shorter call durations; and social balance tends to emerge with shorter call durations. Inspired by these observations, we combined them in to a feature vector to learn a time-dependent factor graph model. Experimental results show that the presented model incorporating the discovered social patterns and the dynamic distributions significantly improves the predictive performance (5-18%) by comparing it with several baseline methods.

Our work has a significant impact in studying the usage patterns of cell phone communication, which can then impact the capacity planning of the communication networks, as well as informing social attitudes and behaviors. Our study can inform cascading effect of information and behavior through a cell phone network, and how duration of phone calls and social topology are closely intertwined.

References

1. Cattuto, C., Van den Broeck, W., Barrat, A., Colizza, V., Pinton, J., Vespignani, A.: Dynamics of person-to-person interactions from distributed rfid sensor networks. Plos One 5 (July 2010)
2. Dasgupta, K., Singh, R., Viswanathan, B., Chakraborty, D., Mukherjea, S., Nanavati, A.A., Joshi, A.: Social ties and their relevance to churn in mobile telecom networks. In: EDBT 2008, pp. 668–677. ACM (2008)
3. Eagle, N., Pentland, A.S., Lazer, D.: Inferring friendship network structure by using mobile phone data. PNAS 106 (2009)
4. Easley, D., Kleinberg, J.: Networks, Crowds, and Markets: Reasoning about a Highly Connected World. Cambridge University Press (2010)
5. Gonzalez, M.C., Hidalgo, C.A., Barabasi, A.-L.: Understanding individual human mobility patterns. Nature (2008)
6. Granovetter, M.: The strength of weak ties. American Journal of Sociology 78(6), 1360–1380 (1973)
7. Hammersley, J.M., Clifford, P.: Markov field on finite graphs and lattices (1971) (unpublished manuscript)
8. Hopcroft, J.E., Lou, T., Tang, J.: Who will follow you back? reciprocal relationship prediction. In: CIKM 2011, pp. 1137–1146 (2011)
9. Katz, E.: The two-step flow of communication: an up-to-date report of an hypothesis. In: Enis, Cox (eds.) Marketing Classics, pp. 175–193 (1973)

10. Katz, E., Lazarsfeld, P.F.: Personal Influence. The Free Press (1955)
11. Krackhardt, D.: The Strength of Strong ties: the importance of philos in networks and organization, Cambridge. Harvard Business School Press, Hershey (1992)
12. Lazarsfeld, P.F., Berelson, B., Gaudet, H.: The people's choice: How the voter makes up his mind in a presidential campaign. Columbia University Press, New York (1944)
13. Lazarsfeld, P.F., Merton, R.K.: Friendship as a social process: A substantive and methodological analysis. In: Freedom and Control in Modern Society, pp. 8–66 (1954)
14. Lichtenwalter, R.N., Lussier, J.T., Chawla, N.V.: New perspectives and methods in link prediction. In: KDD 2010, pp. 243–252. ACM (2010)
15. Murphy, K.P., Weiss, Y., Jordan, M.I.: Loopy belief propagation for approximate inference: An empirical study. In: UAI 1999, pp. 467–475 (1999)
16. Newman, M.E.J.: Power laws, pareto distributions and zipf's law. Contemporary Physics (2005)
17. Page, L., Brin, S., Motwani, R., Winograd, T.: The pagerank citation ranking: Bringing order to the web. Technical Report SIDL-WP-1999-0120, Stanford University (1999)
18. Pareto, V.: Manuale di economia politica. Societa Editrice (1906)
19. Vaz de Melo, P.O.S., Akoglu, L., Faloutsos, C., Loureiro, A.A.F.: Surprising Patterns for the Call Duration Distribution of Mobile Phone Users. In: Balcázar, J.L., Bonchi, F., Gionis, A., Sebag, M. (eds.) ECML PKDD 2010, Part III. LNCS (LNAI), vol. 6323, pp. 354–369. Springer, Heidelberg (2010)
20. Scellato, A., Noulas, S., Mascolo, C.: Exploiting place features in link prediction on location-based social networks. In: KDD 2011, pp. 1046–1054. ACM (2011)
21. Seshadri, M., Machiraju, S., Sridharan, A., Bolot, J., Faloutsos, C., Leskove, J.: Mobile call graphs: beyond power-law and lognormal distributions. In: KDD 2008, pp. 596–604. ACM (2008)
22. Song, C., Qu, Z., Blumm, N., Barabási, A.-L.: Limits of Predictability in Human Mobility. Science (2010)
23. Tan, C., Tang, J., Sun, J., Lin, Q., Wang, F.: Social action tracking via noise tolerant time-varying factor graphs. In: KDD 2010, pp. 1049–1058 (2010)
24. Tang, J., Lou, T., Kleinberg, J.: Inferring social ties across heterogenous networks. In: WSDM 2012, pp. 743–752 (2012)
25. Tang, W., Zhuang, H., Tang, J.: Learning to infer social ties in large networks. In: Gunopulos, D., Hofmann, T., Malerba, D., Vazirgiannis, M. (eds.) ECML PKDD 2011, Part III. LNCS (LNAI), vol. 6913, pp. 381–397. Springer, Heidelberg (2011)
26. Xing, E.P., Jordan, M.I., Russell, S.: A generalized mean field algorithm for variational inference in exponential families. In: UAI 2003, pp. 583–591 (2003)
27. Zhang, Y., Tang, J., Sun, J., Chen, Y., Rao, J.: Moodcast: Emotion prediction via dynamic continuous factor graph model. In: ICDM 2010, pp. 1193–1198 (2010)
28. Zhuang, H., Chin, A., Wu, S., Wang, W., Wang, X., Tang, J.: Inferring geographic coincidence in ephemeral social networks. In: Flach, P.A., De Bie, T., Cristianini, N. (eds.) ECML PKDD 2012, Part II. LNCS, vol. 7524, pp. 613–628. Springer, Heidelberg (2012)

Discovering Nested Communities

Nikolaj Tatti and Aristides Gionis

Helsinki Institute for Information Technology
Department of Information and Computer Science
Aalto University
{nikolaj.tatti,aristides.gionis}@aalto.fi

Abstract. Finding communities in graphs is one of the most well-studied problems in data mining and social-network analysis. In many real applications, the underlying graph does not have a clear community structure. In those cases, selecting a single community turns out to be a fairly ill-posed problem, as the optimization criterion has to make a difficult choice between selecting a tight but small community or a more inclusive but sparser community.

In order to avoid the problem of selecting only a single community we propose discovering a sequence of nested communities. More formally, given a graph and a starting set, our goal is to discover a sequence of communities all containing the starting set, and each community forming a denser subgraph than the next. Discovering an optimal sequence of communities is a complex optimization problem, and hence we divide it into two subproblems: 1) discover the optimal sequence for a fixed order of graph vertices, a subproblem that we can solve efficiently, and 2) find a good order. We employ a simple heuristic for discovering an order and we provide empirical and theoretical evidence that our order is good.

Keywords: community discovery, monotonic segmentation, graph mining, nested communities.

1   Introduction

Discovering communities, tightly connected subgraphs, is one of the most well-studied problems in the field of graph mining. Given some optimization criterion, discovering a community is a computationally challending task, typically **NP**-hard. Additionally, as pointed out by Leskovec et al. [17], in many real applications the underlying graph does not have a clear community structure. Such cases make the community-finding problem inherently ill-posed, as the optimization criterion has to make a difficult, and eventually arbitrary, choice between selecting a tight but small community or a more inclusive but more sparse community. Moreover, the existence of a universal criterion for making such a choice is unlikely as the balance between the size and the density of the desired community will depend on the underlying application.

In order to avoid the problem of selecting only a single community, we propose a problem of discovering a *sequence of nested communities*. More formally, given

H. Blockeel et al. (Eds.): ECML PKDD 2013, Part II, LNAI 8189, pp. 32–47, 2013.

a graph G and a set of source vertices S, our goal is to discover a sequence of k communities around S, such that each community is a subset of the next one. The first community will consist only of S while the last community will contain the whole graph. Inner communities should be tighter than the outer communities. We express this requirement by computing the density of each community and require that the next community should have a lower density than the current community. In addition, we require that each community should be as uniform as possible. We measure uniformity by computing the variance of weights of the edges and requiring it to be small.

Discovering a sequence of communities by optimizing the uniformity criterion is a challenging problem. We will show that several optimization problems related to the optimal solution are **NP**-hard. Hence, we split the problem into two subproblems. We can view a community sequence as a bucket order on the vertices, each bucket consisting of vertices contained in the community and not contained in the previous community. Our first subproblem is to discover a total order on the vertices respecting the optimal bucket order. The second subproblem is to discover the optimal sequence of communities, given an order on the graph vertices. Fortunately, this subproblem can be formulated as a standard sequence-segmentation problem, and thus, it can be solved in polynomial time. In particular, we can solve this problem optimally in quadratic time or we can find an approximate solution in nearly-linear time. Discovering the order is more difficult as this is a complex combinatorial problem. We propose a simple ordering technique used for discovering dense subgraphs: pick iteratively a vertex with the lowest degree, and remove it from the graph. We provide theoretical evidence implying that this is a good order and we also show experimentally that this order outperforms several baselines.

The rest of the paper is organized as follows. We introduce preliminary notation in Section 2 and formalize our optimization problem in Section 3. In section 4 we develop our discovery algorithm and point out theoretical properties of our approach. Section 5 is devoted to related work and Section 6 is devoted to experimental evaluation. We conclude our paper with a short conclusion in Section 7.

2   Preliminaries

We consider a weighted undirected graph $G = (V, E, w)$ over a set of vertices V and edges $E \subseteq \binom{V}{2}$. We use the notation $\binom{V}{2}$ to denote the set of unordered pairs of distinct vertices from V. The function $w : E \to \mathbb{R}$ assigns a weight $w(e)$ to each edge $e \in E$. Also, given a subset of vertices $V' \subseteq V$ we denote by $E(V')$ the set of edges in the *induced* subgraph of G defined by V'.

The definitions and algorithms in this paper rely on a notion of *edge density*, which is defined not only over subsets of vertices, but also over arbitrary *pairs* of subsets of vertices. Even though it is conceptually simple, our edge-density definition requires slightly complex notation for determining the set of potential edges to be used as a denominator in the density ratio. To simplify our presentation we use the notation described below.

Given the graph $G = (V, E, w)$, we consider its *completed* representation $G_0 = (V, E_0, w_0)$, where $E_0 = \binom{V}{2}$, and where w_0 is an extension of w, so that $w_0(e) = w(e)$ if $e \in E$, and $w_0(e) = 0$ if $e \notin E$. In other words, G_0 can be seen as a complete graph, where all non-edges of G become zero-weight edges in G_0. We note again that we use the completed graph representation only to simplify our notation; in our implementation there is no need to store the zero-weight edges.

Now consider the completed representation $G_0 = (V, E_0, w_0)$ of a graph G, and let $F \subseteq E_0$ be a non-empty subset of edges. We define the *weight* and *density* of F as

$$w(F) = \sum_{e \in F} w(e) \quad \text{and} \quad d(F) = \frac{w(F)}{|F|}.$$

Consider now two subsets of vertices $S, T \subseteq V$. We define the set of *cross edges* from S to T as $c(S, T) = \{(x, y) \in E \mid x \in S, y \in T\}$. It is important to note that we do not impose any constraint on the sets S and T; they may overlap in an arbitrary way. For instance, if the sets S and T are disjoint the edges in $c(S, T)$ are the *cut* edges from S to T, while if $S \subseteq T$ the edge set $c(S, T)$ contains, among others, all the edges within S.

Finally, we write $w(S, T)$ as a shorthand of $w(c(S, T))$ and we write $d(S, T)$ as a shorthand of $d(c(S, T))$.

3   Nested Communities

As we discussed in the introduction, our goal is to find the optimal sequence of nested communities, with respect to a set of source vertices of the input graph. We denote this set of source vertices by S. For conceptual simplicity, one may think of S as a singleton set, that is, identifying the sequence of nested communities for a single vertex. However, all our problem definitions, algorithms, and proofs, hold for the general case of S being any subset of V.

Our objective is to find k nested communities, where the parameter k is part of the problem input. Given a set of source vertices S, we represent a sequence of nested communities with respect to S, by the sequence of vertex sets $S = V_0 \subseteq V_1 \subseteq \cdots \subseteq V_k = V$.

Intuitively, the inner sets of the nested-community sequence are expected to be more strongly related to the source set S. This type of relatedness is expressed by the notion of density. So, V_1 is the densest community that contains S, V_2 is the second densest community, and in general, we require that the density of V_i should decrease as i increases.

Considering the requirement of monotonically decreasing density in isolation is not sufficient to determine in a well-defined manner a desirable sequence of nested communities. Indeed, given a graph G, a set of source vertices S, and integer k, there is a potentially exponential number of ways to partition the set of vertices of the graph into a sequence of nested communities $V_0, \ldots, V_k$.

The main question we are facing is to decide where exactly to draw the boundary between each pair of communities V_i and V_{i+1}. To answer this question, we follow an approach inspired by *segmentation problems*. In particular, our approach is as follows: consider the set of vertices $D_{i+1} = V_{i+1} \setminus V_i$ that need to be added to the community V_i in order to form community V_{i+1}. Consider also the set of edges $E_{i+1} = E(V_{i+1}) \setminus E(V_i)$, defined as the additional edges brought in by extending the community V_i to the community V_{i+1}. We can then define the density of the set of edges E_{i+1}. To capture the intuition that the set D_{i+1} should form a coherent extension to V_i we require that the density of E_{i+1} is as *uniform* as possible.

The notion of uniformity for a set of edges, among many ways, can be expressed as a sum of square of difference of the weight of each edge from the average weight of the set. We thus have the following definition.

Definition 1. *Given a set of edges $F \subseteq E$, we define the* density-uniformity score *as*

$$q(F) = \sum_{e \in F} (w(e) - d(F))^2 \, .$$

Our goal is then to find a sequence of nested communities so that the successive segments of added edges are as uniform as possible with respect to their density. Formulating this objective as an optimization problem not only gives meaningful semantics to the nested community detection problem, but it also makes the problem well-defined. Motivated by the discussion above, our main problem definition is given below.

Problem 1. Given a weighted input graph $G = (V, E, w)$, a set of source vertices $S \subset V$, and an integer k, find the sequence of nested communities $\mathcal{V} = \{S = V_0 \subseteq V_1 \subseteq \cdots \subseteq V_k = V\}$ that minimizes the density-uniformity score

$$q(\mathcal{V}) = \sum_{i=1}^{k} q(E(V_i) \setminus E(V_{i-1})) \, ,$$

subject to the constraint $d(V_i) < d(V_{i-1})$ for $i = 2, \ldots, k$.

4   An Algorithm for Discovering Nested Communities

In this section we present our algorithm for discovering nested communities. We begin by demonstrating a necessary condition for the optimal solution based on dense subgraphs. Discovering such subgraphs turns out to be computationally intractable. We then split the original problem into two subproblems: discovering community sequence for a fixed order of vertices, a problem which we can solve efficiently, and discovering such an order. We provide a simple heuristic for discovering an order, and provide theoretical evidence that this order is good.

4.1   Nested Communities and Dense Subgraphs

We start our discussion by demonstrating a connection of the problem of finding the optimal sequence of nested communities, i.e., solving Problem 1, with problems related to finding dense subgraphs of a given graph.

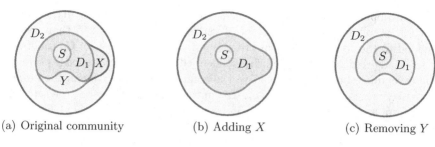

(a) Original community   (b) Adding X (c) Removing Y

Fig. 1. Communities related to Proposition 1. If $d(X, X \cup D_1) > d(Y, D_1)$, then either adding X to D_1 or removing Y from D_1 will yield a better score.

To establish this connection, consider a triple of communities $V_{i-1} \subseteq V_i \subseteq V_{i+1}$ in an *optimal solution* to Problem 1. Consider the two corresponding segments $D_{i+1} = V_{i+1} \setminus V_i$ and $D_i = V_i \setminus V_{i-1}$. Consider also any two subsets of those segments, $X \subseteq D_{i+1}$ and $Y \subseteq D_i$, that is, X is a subset of the outer segment, while Y is a subset of the inner segment, see Figure 1(a) for a visualization. As we will show shortly, adding the outer subset X in the community V_i leads to a situation where the density of the subset X with respect to the overall community V_i is no better than the density of the subset Y with respect to the community V_i. Otherwise, either adding X to V_i (see Figure 1(b)) or removing Y from V_i (see Figure 1(c)) lead to a better solution. This follows from the fact that we require that the densities of the nested communities in any feasible solution of Problem 1 decrease monotonically.

Before proceeding to discussing the implications of this observation, we first give a formal statement and its proof.

Proposition 1. *Consider a graph $G = (V, E, w)$, a set of source vertices $S \subseteq V$, and an integer k. Let $\mathcal{V} = (S = V_0 \subseteq V_1 \subseteq \cdots \subseteq V_k = V)$ be the optimal sequence of nested communities, that is, a solution to Problem 1. Fix i such that $1 \leq i \leq k - 1$ and let $X \subseteq V_{i+1} \setminus V_i$ and $Y \subseteq V_i \setminus V_{i-1}$. Then*

$$d(X, X \cup V_i) \leq d(Y, V_i).$$

For the proof of the proposition we require the following lemma, which states that the mean square error of a set of numbers from a single point, increases with the distance of that point from the mean of the numbers. The lemma can be derived by simple algebraic manipulations, and its proof is omitted.

Lemma 1. *Let $w_1, \ldots, w_N$ and $x_1, \ldots, x_N$ be two sets of real numbers. Let $W = \sum_{i=1}^{N} w_i$ and $\mu = \frac{1}{W} \sum_{i=1}^{N} w_i x_i$. For any real number d it is*

$$\sum_{i=1}^{N} w_i(x_i - d)^2 = \sum_{i=1}^{N} w_i(x_i - \mu)^2 + W(d - \mu)^2.$$

We are now ready to prove the proposition.

Proof (Proposition 1). Let $C_1 = E(V_{i+1}) \setminus E(V_i)$ and $C_2 = E(V_i) \setminus E(V_{i-1})$. Let us break C_1 into two parts, $D_{11} = c(X, X \cup V_i)$ and $D_{12} = C_1 \setminus D_{11}$. Similarly, let us break C_2 into two parts, $D_{21} = c(Y, V_i)$ and $D_{22} = C_2 \setminus D_{21}$. Define the centroids $\mu_{ij} = d(D_{ij})$ and $\lambda_i = d(C_i)$. Lemma 1 now implies that

$$s = q(C_1) + q(C_2) = \text{const} + |D_{11}|(\mu_{11} - \lambda_1)^2 + |D_{21}|(\mu_{21} - \lambda_2)^2,$$
$$s_1 = q(C_1 \cup D_{21}) + q(D_{22}) = \text{const} + |D_{11}|(\mu_{11} - \lambda_1)^2 + |D_{21}|(\mu_{21} - \lambda_1)^2,$$
$$s_2 = q(D_{12}) + q(C_1 \cup D_{11}) = \text{const} + |D_{11}|(\mu_{11} - \lambda_2)^2 + |D_{21}|(\mu_{21} - \lambda_2)^2,$$

where const is equal to

$$\sum_{i=1}^{2} q(D_{i1}) + q(D_{i2}) + |D_{i2}|(\mu_{i2} - \lambda_i)^2 .$$

Since $\mathcal{V}$ is optimal we must have $s \leq s_1$ and $s \leq s_2$. Otherwise, we can obtain a better segmentation by attaching X to V_i or deleting Y from V_i. This implies that $|\mu_{21} - \lambda_2| \leq |\mu_{21} - \lambda_1|$ and $|\mu_{11} - \lambda_1| \leq |\mu_{11} - \lambda_2|$. Since $\lambda_2 \geq \lambda_1$, this implies that $\mu_{21} \geq (\lambda_1 + \lambda_2)/2$ and $\mu_{11} \leq (\lambda_1 + \lambda_2)/2$, which implies $\mu_{11} \leq \mu_{21}$. This completes the proof. □

Proposition 1 implies that in an optimal solution the graph vertices can be *ordered* in such a way so that subgraph density, as specified by the proposition, decreases along this order. This observation motivates the following *greedy* algorithm for solving the problem of discovering nested communities:

Algorithm Outline: Greedy–add–densest–subgraph

1. Start with S, the set of source vertices.
2. Given the current set S, find a subset of vertices T that maximize $d(T, S \cup T)$.
3. Set $S \leftarrow S \cup T$, and repeat the previous step until the set S includes all the vertices of the graph.
4. Consider the vertices in the order discovered by the previous process. Find the optimal sequence of k nested communities that respects this order.

One potential problem with the above greedy approach is that the subroutine that is called iteratively in step 2, is an **NP**-hard problem. This is formalized below as problem DENSESUPERSET.

Problem 2 (DENSESUPERSET). Given a weighted graph $G = (V, E, w)$ and a subset of vertices $S \subseteq V$, find a subset of vertices T maximizing $d(T, S \cup T)$.

Proposition 2. *The DENSESUPERSET problem is **NP**-hard.*

Proof (Sketch). Due to space constraints we will only sketch the proof. The complete proof is available in Appendix.[1] We will reduce CLIQUE to DENSESUPERSET. Given a graph G, we add a vertex s and connect it to each vertex with a weight of $\alpha = 1 - \frac{1}{2|V|^2}$. Let $k < n < m$. It follows that

---

[1] For the appendix, see http://users.ics.aalto.fi/~ntatti/

$$\frac{\binom{n}{2} + \alpha n}{\binom{n}{2} + n} > \frac{\binom{k}{2} + \alpha k}{\binom{k}{2} + k} \quad \text{and} \quad \frac{\binom{n}{2} + \alpha n}{\binom{n}{2} + n} > \frac{\binom{m}{2} + \alpha m - 1}{\binom{k}{2} + m} .$$

The left-hand side term in the first equation is the density of n-clique while the the right-hand side term bounds the density of a graph with k vertices. The right-hand side term in the second equation upper bounds the density of a non-clique with m vertices. Consequently, the largest clique, say X, in G will also have the largest density $d(X, X \cup s)$, which is a sufficient to prove the result. □

Similarly, one can think of solving the problem by working on the opposite direction, that is, start with the whole vertex set V and "peel off" the set V by removing the sparsest subgraph, until left with the set of source vertices S. The corresponding algorithm will be the following.

Algorithm Outline: Greedy–remove–sparsest–subgraph

1. Start with V, the vertex set of G.
2. Given a current set V, find a subset of vertices T that does not include the source vertex set S and minimizes the density $d(T, V)$.
3. Set $V \leftarrow V \setminus T$, and repeat the previous step until left only with the set of source vertices S.
4. Consider the vertices in the order removed by the previous process. Find the optimal sequence of k nested communities that respects this order.

Not surprisingly, the problem of finding the sparsest subgraph, which corresponds to step 2 of the above process is **NP**-hard.

Problem 3 (SPARSENBHD). Given a weighted graph $G = (V, E, w)$ find a set of vertices T minimizing $d(T, V)$.

Proposition 3. *The* SPARSENBHD *problem is* **NP**-*complete.*

Proof (Sketch). Due to space constraints we will only sketch the proof. The complete proof is available in Appendix. We will reduce CLIQUE to SPARSENBHD. Assume that we are given a graph G with l nodes. We extend the graph by adding two vertices s and t with an edge of such high weight that neither s or t will appear in the optimal solution. We then add an edge from s to each vertex v in G with a weight of $p - \deg(v)$, where $p = (l + 1) - \frac{1}{2}(l + 1)^{-2}$. This will make the weighted degree of all vertices in G equal so a dense subgraph X will have a low density $d(X, X \cup \{s, t\})$. Let $k < n < m$. Then a straightforward calculation reveals that

$$\frac{pn - \binom{n}{2}}{(l+1)n - \binom{n}{2}} < \frac{pk - \binom{k}{2}}{(l+1)k - \binom{k}{2}} \quad \text{and} \quad \frac{pn - \binom{n}{2}}{(l+1)n - \binom{n}{2}} < \frac{pm - \binom{m}{2} + 1}{(l+1)m - \binom{m}{2}} .$$

The left-hand side term in the first equation is the density of n-clique while the right-hand side term bounds the density of a graph with k vertices. The right-hand side term in the second equation lower bounds the density of a non-clique with m vertices. Consequently, the largest clique, say X, in G will also have the lowest density $d(X, X \cup \{s, t\})$, which is a sufficient to prove the result. □

4.2 Algorithm for Discovering Nested Communities

Armed with intuition from the previous section, we now proceed to discuss the proposed algorithm. The underlying principle of both of the greedy algorithms described above is to consider the vertices of the graph in a specific order and then find a sequence of nested communities that respects this order. In one case, the order of graph vertices is obtained by starting from S and iteratively adding the densest subgraph, while in the other case, the order is obtained by starting from the full vertex set V and iteratively removing the sparsest subgraph.

Our algorithm is an instantiation of this general principle. We specify in detail (i) how to obtain an order of the graph vertices, and (ii) how to find a sequence of nested communities that respects a given order.

We start our discussion from the second task, i.e., finding the sequence of nested communities given an order. As it turns out, this problem is an instance of sequence segmentation problems. We define this problem below, which is a refinement of Problem 1.

Problem 4 (Sequence of nested communities from a given order). Given a graph $G = (V, E, w)$ *with ordered vertices,* a set of source vertices $S = \{v_1, \ldots, v_s\} \subset V$, and an integer k, find a monotonically increasing sequence of $k + 1$ integers $b = (b_0 = s, \ldots, b_k = |V|)$ such that

$$\mathcal{V} = (S = V_0 \subseteq V_1 \subseteq \cdots \subseteq V_k = V), \quad \text{where} \quad V_k = \{v_1, \ldots, v_{b_i}\},$$

minimizes the density-uniformity score $q(\mathcal{V})$ and satisfies the monotonicity constraint $d(V_i) < d(V_{i-1})$ for $i = 1, \ldots, k$.

It is quite easy to see that Problem 4 can be cast as a segmentation problem. Typical segmentation problems can be solved optimally using dynamic programming, as shown by Bellman [3]. The most interesting aspect of Problem 4, seen as segmentation problem, is the monotonicity constraint $d(V_i) < d(V_{i-1})$, for $i = 1, \ldots, k$. That is, not only we ask to segment the ordered sequence of vertices so that we minimize the density variance on the segments, but we also require that the density scores of each segment decrease monotonically. The situation can be abstracted to the monotonic segmentation problem stated below.

Problem 5 (Monotonic segmentation). Let $a_1, \ldots, a_n$ and $x_1, \ldots, x_n$ be two sequences of real numbers. Given an integer k, find $k + 1$ indices $b_0 = 1, \ldots, b_k = n + 1$ minimizing

$$\sum_{j=1}^{n} \sum_{i=b_{j-1}}^{b_j - 1} a_i(x_i - \mu_j)^2,$$

where μ_j is the weighted centroid of j-th segment such that $\mu_j < \mu_{j-1}$.

In order to express Problem 4 with Problem 5, consider a group of edges, $P_i = c(v_i, \{v_1, \ldots, v_{i-1}\})$ for each vertex $v_i \in V \setminus S$. If we set $a_i = |P_{i+|S|}|$ and $x_i = d(P_{i+|S|})$, we can apply Lemma 1 and show that the score of community sequence is equal to the variance minimized by Problem 5, plus a constant. In fact, this constant is the sum of the variances within each P_i.

Similarly to the unconstrained segmentation problem, the monotonic segmentation problem can be solved *optimally*. The idea is to use as preprocessing step the classic "pool of adjacent violators" algorithm (PAV) [2], which merges points until there are no monotonicity violations, and then apply the classic dynamic-programming algorithm on the resulting sequence of merged points. This algorithm runs in $O(|V|)$ time. By definition the merged points do not contain any monotonicity violations, and thus, the resulting segmentation respects the monotonicity constraint, as well. As shown by Haiminen et al. [14], this two-phase algorithm gives the optimal k segmentation under the monotonicity constraints. As a result of the optimality of the monotonic segmentation problem, Problem 4 can be solved optimally.

We next proceed to discuss the first component of the algorithm, namely, how to obtain an order of the graph vertices. Recall that, according to the principles discussed in the previous section, we can either start from S and iteratively add dense subgraphs, or start from V and remove sparse subgraphs. We follow the latter approach. In order to overcome the **NP**-hard problem of finding the sparsest subgraph and in order to obtain a total order, we use the heuristic of iteratively removing the sparsest subgraph of size one, namely, a single vertex. The sparsest one-vertex subgraph is simply the vertex with the smallest weighted degree. Thus, overall, we obtain the simple algorithm SORTVERTICES, whose pseudocode is given as Algorithm 1.

As an interesting side remark, we note that the algorithm SORTVERTICES is encountered in the context of finding subgraphs with the highest average degree. In particular, it is known that the densest subgraph obtained by the algorithm during the process of iteratively removing the smallest-degree vertex is a factor-2 approximation to the optimally densest subgraph in the graph [4].

The natural question to ask is how good is the order produced by algorithm SORTVERTICES? As we will demonstrate shortly, it turns out that the order is quite good. First, we note that the optimal solution obtained for Problem 4, satisfies an analogous structural property, with respect to subgraph densities, as the optimal solution for Problem 1, We omit the proof of the following proposition as it is similar to the one of Proposition 1.

Proposition 4. *Consider a graph $G = (V, E, w)$ with ordered vertices, a set of source vertices $S \subset V$, and an integer k. Let $\mathcal{V} = (S = V_0 \subseteq V_1 \subseteq \cdots \subseteq V_k = V)$ be the optimal sequence of nested communities with respect to the order, that is, a solution to Problem 1. Fix i such that $1 \leq i \leq k - 1$ and let $b = |V_i|$. Let $X \subseteq V_{i+1} \setminus V_i$ and $Y \subseteq V_i \setminus V_{i-1}$ such that $X = \{v_{b+1}, \ldots, v_{b+|X|}\}$ and $Y = \{v_{b-|Y|+1}, \ldots, v_b\}$. Then $d(X, X \cup V_i) \leq d(Y, V_i)$.*

The only difference between Proposition 1 and Proposition 4 is that in Proposition 4 we require additionally that V_{i+1} starts with X and V_i ends with Y with respect to the order. We want this condition to be redundant, otherwise the given order is suboptimal. For example, consider the adjacency matrix of G given in

Figure 2(a). The given segmentation is optimal with respect to the given order. However if we rearrange the vertices in D_1 and D_2, given in Figure 2(b), then the same segmentation is no longer optimal as X and Y violate Proposition 4. The additional condition in Proposition 4 becomes redundant if V_i ends with the sparsest subset while V_{i+1} starts with densest subset. We will show that the algorithm SORTVERTICES produces an order that satisfies this property *approximately*. The exact formulation of our claim is given as Propositions 5 and 6.

Algorithm 1. SORTVERTICES. Sort vertices of a weighted graph by iteratively removing a vertex with the least weight of adjacent edges.

---

    input   : weighted graph $G = (V, E, w)$, a set S
    output : order on V
1 $W \leftarrow V \setminus S$;
2 $o \leftarrow$ empty sequence;
3 **while** $|W| > 0$ **do**
4 | $x \leftarrow \arg\min_{x \in W} d(x, W \cup S)$;
5 | delete x from W and add x to the beginning of o;
6 add S in an arbitrary order to the beginning of o;
7 **return** o;

Proposition 5. *Consider a weighted graph $G = (V, E, w)$, whose vertices are ordered by algorithm SORTVERTICES. Let $1 \le b < c \le |V|$. Let $U = \{v_b, \dots, v_c\}$ and $W = \{v_1, \dots, v_c\}$. Let $f = d(v_c, W)$. Then $2f \le d(X, W)$ for any $X \subseteq U$.*

Proof. Note that $s = \sum_{x \in X} w(x, W) = 2w(X) + w(X, W \setminus X) \le 2w(X, W)$. Write $m_x = |c(x, W)|$. Since v_c has the smallest $d(v_c, W)$, we have

$$s = \sum_{x \in X} m_x d(x, W) \ge d(v_c, W) \sum_{x \in X} m_x \ge d(v_c, W)\, |c(X, W)| \quad.$$

Combining the inequalities and dividing by $|c(X, W)|$ proves the result.   □

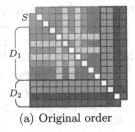 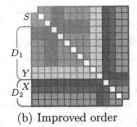

        (a) Original order   (b) Improved order

Fig. 2. Consequences of Proposition 4. If we reorder the vertices in D_1 and D_2, then an optimal solution with respect to the order may become suboptimal with respect to the improved order.

Proposition 6. *Consider a weighted graph $G = (V, E, w)$, whose vertices are ordered by algorithm* SortVertices. *Let $1 \leq b < c \leq |V|$. Let $U = \{v_b, \ldots, v_c\}$ and $W = \{v_1, \ldots, v_{b-1}\}$. Assume that there is $\alpha \geq 0$ such that for all $v \in U$ it is $\alpha w(v, W) \geq w(v, U)$. Let $f = d(v_b, W)$. Then $(1 + \alpha)^2 f \geq d(X, X \cup W)$ for any $X \subseteq U$.*

Proof. Let $A = c(X, W)$ and $B = c(X, X)$. The density of X is bounded by

$$d(X, X \cup W) = \frac{w(A) + w(B)}{|A| + |B|} \leq \frac{w(A) + \alpha w(A)}{|A| + |B|} \leq \frac{(1 + \alpha)w(A)}{|A|} = (1 + \alpha)d(A).$$

Select $x \in X$ with the highest $d(x, W)$. Then $d(A) \leq d(x, W)$. Let us prove that $d(x, W) \leq (1 + \alpha)f$. If $v_b = x$, then we are done. Assume that $v_b \neq x$. Since G is fully-connected, SortVertices always picks the vertex with the lowest weight. Let $Z = \{v_1, \ldots, x\}$. Then $w(x, W) \leq w(x, Z) \leq w(v_b, Z) = w(v_b, W) + w(v_b, U) \leq (1 + \alpha)w(v_b, W)$. Since, G is fully-connected $w(y, W) = |W|d(y, W)$ for any $y \in U$. Hence, dividing the inequality gives us $d(x, W) \leq (1 + \alpha)f$, which proves the proposition.   □

5   Related Work

Finding communities in graphs and social networks is one of the most well-studied topics in graph mining. The amount of literature on the subject is very extensive. This section cannot aspire to cover all the different approaches and aspects of the problem, we only provide a brief overview of the area.

Community Detection. A large part of the related work deals with the problem of partitioning a graph in disjoint clusters or communities. A number of different methodologies have been applied, such as hierarchical approaches [11], methods based on modularity maximization [1, 6, 11, 26], graph-theoretic approaches [8, 9], random-walk methods [21, 24, 28], label-propagation approaches [24], and spectral graph partition [5, 15, 18, 25]. A thorough review on community-detection methods can be found on the survey by Fortunato [10]. We note that this line of work is different than the present paper, since we do not aim at partitioning a graph in disjoint communities.

Overlapping Communities. Researchers in community detection have realized that, in many real situations and real applications, it is meaningful to consider that graph vertices do not belong only to one community. Thus, one asks to partition a graph into overlapping communities. Typical methods here rely on clique percolation [19], extensions to the modularity-based approaches [12, 20], analysis of ego-networks [7], or fuzzy clustering [27]. Again the problem we address in this paper is quite different. First, we find communities centered around a given set of source vertices, and not for the whole graph. Second, the communities output by our algorithm do not have arbitrary overlaps, but they have a specific nested structure.

Centerpiece Subgraphs and Community Search. Perhaps closer to our approach is work related to the centerpiece subgraphs and the community-search problem [16,22,23]. In this class of problems, a set of source vertices S is given and the goal is to find a subgraph so that S belongs in the subgraph and the subgraph forms a tight community. The quality of the subgraph is measured with various objective functions, such as degree [22], conductance [16], or random-walk-based measures [23]. The difference of these methods with the one presented here is that these methods return only one community, while in this paper we deal with the problem of finding a sequence of nested communities.

In summary, despite the numerous research on the topic of community detection in graphs and social networks, to the best of our knowledge, this is the first paper to address the topic of nested communities with respect to a set of source vertices. Furthermore, our approach offers novel technical ideas, such as providing a solid theoretical analysis that allows to decompose the problem of finding nested communities into two sub-problems: (i) ordering the set of vertices, and (ii) segmenting the graph vertices according to that given order.

6   Experimental Evaluation

We will now provide experimental evidence that our method efficiently discovers meaningful segmentations and that our ordering algorithm outperforms several natural baselines.

Datasets and Experimental Setup. In our experiments we used six datasets, five obtained from Mark Newman's webpage,[2] and a bibliographic dataset obtained from DBLP. The datasets are as follows: *Adjnoun*: adjacency graph of common adjectives and nouns in the novel David Copperfield, by Charles Dickens. *Dolphins*: an undirected social graph of frequent associations between 62 dolphins in a community living off Doubtful Sound, New Zealand. *Karate*: social graph of friendships between 34 members of a karate club at a US university in the 1970s. *Lesmis*: coappearance graph of characters in the novel Les Miserables. *Polblogs*: a directed graph of hyperlinks between weblogs on US politics, recorded in 2005. *DBLP*: coauthorship graph between researchers in computer science. The statistics of these datasets are given in Table 1.

For each dataset and a given source set S, we considered three different weighting schemes: First we run personalized PageRank using the source node with a restart of 0.1. Let $p(v)$ be the PageRank weight of each vertex. Given an edge $e = (v, w)$, we set three different weighting schemes,

$$w_n(e) = \frac{p(v)}{\deg(v)} + \frac{p(w)}{\deg(w)}, \quad w_s(e) = p(v) + p(w), \quad w_m(e) = \min(p(v), p(w)).$$

These weights are selected so that the vertices that are hard to reach with a random walk will have edges with small weights, and hence will be placed in outer communities. For *DBLP*, we weighted the edges during PageRank computation with the number of joint papers, each paper normalized by the number of authors. We use the vertex with the highest degree as a starting set.

---

[2] http://www-personal.umich.edu/~mejn/netdata/

Table 1. Basic statistics of graphs (first two columns) and performance over hops baseline. The third column represents a typical running time while the fourth column represents a typical number of entries during the segmentation. The last three columns represent the normalized score compared to the baseline score $q(\mathcal{H})$.

Name	$|V(G)|$	$|E(G)|$	Time	N	performance $q(\mathcal{V})/q(\mathcal{H})$		
					w_n	w_s	w_m
Adjnoun	112	425	2ms	84	0.90/0.95	0.88/0.95	0.77/0.94
Dolphins	62	159	1ms	41	0.67/0.80	0.61/0.78	0.57/0.80
Karate	34	78	1ms	21	0.78/0.91	0.76/0.91	0.60/0.93
Lesmis	77	254	2ms	37	0.77/0.93	0.84/0.94	0.62/0.94
Polblogs	1 222	16 714	84ms	872	0.87/0.96	0.95/0.99	0.57/0.96
DBLP	703 193	2 341 362	23s	1 797	0.87/0.99	0.98/1.00	0.45/0.99

Time Complexity. Our first step is to study the running time of our algorithm. We ran our experiments on a laptop equipped with a 1.8 GHz dual-core Intel Core i7 with 4 MB shared L3 cache, and typical running times for each dataset are given in 3rd column of Table 1.[3] Our algorithm is fast: for the largest dataset with 2 million edges, the computation took only 20 seconds. The algorithm consists of 4 steps, computing PageRank, ordering the vertices, grouping the vertices into blocks such that monotonicity condition is guaranteed, and segmenting the groups. The only computationally strenuous step is segmentation which requires quadratic time in the number of blocks. The number of vertices in *DBLP* is over 700 000, however, grouping according to the PAV algorithm leaves only 2 000 blocks, which can be easily segmented. It is possible to select weights in such a way that there will no reduction when grouping vertices, so that finding the optimal segmentation becomes infeasible. However, in such a case, we can always resort to a near-linear approximation optimization algorithm [13].

Comparison to Baseline. A key part in our approach is discovering a good order. Our next step is to compare the order induced by SORTVERTICES against several natural baselines. For the first baseline we group the vertices based on the length of a minimal path from the source. We then compared these communities, say $\mathcal{H}$, to the (same number of) communities obtained with our method. The scores, given in Table 1, show that our approach beats this baseline in every case, which is expected since this naïve baseline does not take into account density. For our next two baselines we order vertices based on vertex degree and PageRank. We then compute community sequences with 2–10 communities from these orders. Typical scores are given in Figure 3. Out of $6 \times 3 \times 9 = 162$ comparisons, SORTVERTICES wins both orders 158 times, ties once (*Karate*, w_m, 3 communities) and loses 3 times to the degree order (*DBLP*, w_n, 3–5 communities).

---

[3] For the code, see http://users.ics.aalto.fi/~ntatti/

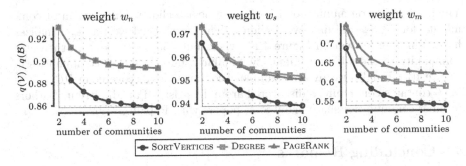

Fig. 3. Quality scores of community sequences based on different orders as a function of number of communities for *Polblogs*. The scores are normalized by the score of a community sequence $\mathcal{B}$ with a single community.

Table 2. Top-3 communities from a sequence of 5 communities for Christos Papadimitriou from *DBLP* set and using w_s

1. segment	D. Johnson	E. Dahlhaus	V. Vianu	G. Gottlob	A. Itai
M. Yannakakis	M. Garey	P. Crescenzi	P. Kanellakis	M. Sideri	A. Schäffer
F. Afrati	R. Karp	P. Seymour	S. Abiteboul	E. Koutsoupias	A. Aho
2. segment	R. Fagin	O. Vornberger	A. Piccolboni	C. Daskalakis	P. Serafini
J. Ullman	**3. segment**	M. Blum	D. Goldman	X. Deng	P. Raghavan
Y. Sagiv	G. Papageorgiou	K. Ross	E. Arkin	P. Goldberg	P. Bernstein
S. Cosmadakis	V. Vazirani	P. Kolaitis	I. Diakonikolas	T. Hadzilacos	

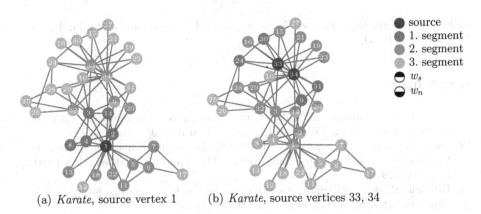

(a) *Karate*, source vertex 1   (b) *Karate*, source vertices 33, 34

Fig. 4. 4 community sequences with 3 communities of *Karate*. Segmentations in Figure 4(a) use 1 as a source and community sequences in Figure 4(b) use 33, 34 as sources. Communities are decoded as colors, the top-half represents w_s, the bottom-half represents w_n.

Examples of Communities. Our final step is to provide examples of discovered communities. In Figure 4 we provide 4 different community sequences with 3 communities using weights w_s and w_n and sources $S = \{1\}$ and $S = \{33, 34\}$.

The inner-most community for 1 contains a near 5-clique. The inner-most community for $33, 34$ contains two 4-cliques. The normalized weight w_n penalizes hubs. This can be seen in Figure 4(a), where hubs 33, 34 move from the outer community to the middle community. Similarly, hub 1 changes communities in Figure 4(b). Finally, we give an example of communities discovered in *DBLP*. Table 6 contains communities discovered around Christos Papadimitriou. Authors in inner communities share many joint papers with Papadimitriou.

7   Concluding Remarks

We considered a problem of discovering nested communities, a sequence of subgraphs such that each community is a more connected subgraph of the next community. We approach the problem by dividing it into two subproblems: discovering the community sequence for a fixed order of vertices, a problem which we can solve efficiently, and discovering an order. We provided a simple heuristic for discovering an order, and provided theoretical and empirical evidence that this order is good.

Discovering nested communities seems to have a lot of potential as it is possible to modify or extend the problem in many ways. We can generalize the problem by not only considering sequences but, for example, trees of communities, where a parent node needs to be a denser subgraph than the child node. Another possible extension is to consider multiple source sets instead of just one.

Acknowledgements. This work was supported by Academy of Finland grant 118653 (ALGODAN).

References

1. Agarwal, G., Kempe, D.: Modularity-maximizing network communities via mathematical programming. European Physics Journal B 66(3) (2008)
2. Ayer, M., Brunk, H., Ewing, G., Reid, W.: An empirical distribution function for sampling with incomplete information. The Annals of Mathematical Statistics 26(4) (1955)
3. Bellman, R.: On the approximation of curves by line segments using dynamic programming. Communications of the ACM 4(6) (1961)
4. Charikar, M.: Greedy approximation algorithms for finding dense components in a graph. In: Jansen, K., Khuller, S. (eds.) APPROX 2000. LNCS, vol. 1913, pp. 84–95. Springer, Heidelberg (2000)
5. Chung, F.R.K.: Spectral Graph Theory. American Mathematical Society (1997)
6. Clauset, A., Newman, M.E.J., Moore, C.: Finding community structure in very large networks. Physical Review E (2004)
7. Coscia, M., Rossetti, G., Giannotti, F., Pedreschi, D.: DEMON: a local-first discovery method for overlapping communities. In: KDD (2012)
8. Flake, G.W., Lawrence, S., Giles, C.L.: Efficient identification of web communities. In: KDD (2000)

9. Flake, G.W., Lawrence, S., Giles, C.L., Coetzee, F.M.: Self-organization and identification of web communities. Computer 35(3) (2002)

10. Fortunato, S.: Community detection in graphs. Physics Reports, 486 (2010)

11. Girvan, M., Newman, M.E.J.: Community structure in social and biological networks. PNAS 99 (2002)

12. Gregory, S.: An algorithm to find overlapping community structure in networks. In: Kok, J.N., Koronacki, J., Lopez de Mantaras, R., Matwin, S., Mladenič, D., Skowron, A. (eds.) PKDD 2007. LNCS (LNAI), vol. 4702, pp. 91–102. Springer, Heidelberg (2007)

13. Guha, S., Koudas, N., Shim, K.: Approximation and streaming algorithms for histogram construction problems. ACM TODS 31 (2006)

14. Haiminen, N., Gionis, A.: Unimodal segmentation of sequences. In: ICDM (2004)

15. Karypis, G., Kumar, V.: Multilevel algorithms for multi-constraint graph partitioning. In: CDROM (1998)

16. Koren, Y., North, S.C., Volinsky, C.: Measuring and extracting proximity graphs in networks. TKDD 1(3) (2007)

17. Leskovec, J., Lang, K.J., Dasgupta, A., Mahoney, M.W.: Statistical properties of community structure in large social and information networks. In: WWW (2008)

18. Ng, A.Y., Jordan, M.I., Weiss, Y.: On spectral clustering: Analysis and an algorithm. In: NIPS (2001)

19. Palla, G., Derényi, I., Farkas, I., Vicsek, T.: Uncovering the overlapping community structure of complex networks in nature and society. Nature 435 (2005)

20. Pinney, J., Westhead, D.: Betweenness-based decomposition methods for social and biological networks. In: Interdisciplinary Statistics and Bioinformatics (2006)

21. Pons, P., Latapy, M.: Computing communities in large networks using random walks. Journal of Graph Algorithms Applications 10(2) (2006)

22. Sozio, M., Gionis, A.: The community-search problem and how to plan a successful cocktail party. In: KDD (2010)

23. Tong, H., Faloutsos, C.: Center-piece subgraphs: problem definition and fast solutions. In: KDD (2006)

24. van Dongen, S.: Graph Clustering by Flow Simulation. PhD thesis, University of Utrecht (2000)

25. von Luxburg, U.: A tutorial on spectral clustering. Statistics and Computing 17(4) (2007)

26. White, S., Smyth, P.: A spectral clustering approach to finding communities in graph. In: SDM (2005)

27. Zhang, S., Wang, R.-S., Zhang, X.-S.: Identification of overlapping community structure in complex networks using fuzzy c-means clustering. Physica A (2007)

28. Zhou, H., Lipowsky, R.: Network brownian motion: A new method to measure vertex-vertex proximity and to identify communities and subcommunities. In: Bubak, M., van Albada, G.D., Sloot, P.M.A., Dongarra, J. (eds.) ICCS 2004. LNCS, vol. 3038, pp. 1062–1069. Springer, Heidelberg (2004)

CSI: Community-Level Social Influence Analysis

Yasir Mehmood[1], Nicola Barbieri[2], Francesco Bonchi[2], and Antti Ukkonen[3]

[1] Pompeu Fabra University, Spain
yasir.mehmood01@estudiant.upf.edu
[2] Yahoo! Research Barcelona, Spain
{barbieri,bonchi}@yahoo-inc.com
[3] Helsinki Institute for Information Technology HIIT, Aalto University, Finland
antti.ukkonen@aalto.fi

Abstract. Modeling how information propagates in social networks driven by peer influence, is a fundamental research question towards understanding the structure and dynamics of these complex networks, as well as developing viral marketing applications. Existing literature studies influence at the level of individuals, mostly ignoring the existence of a community structure in which multiple nodes may exhibit a common influence pattern.

In this paper we introduce CSI, a model for analyzing information propagation and social influence at the granularity of communities. CSI builds over a novel propagation model that generalizes the classic Independent Cascade model to deal with groups of nodes (instead of single nodes) influence. Given a social network and a database of past information propagation, we propose a hierarchical approach to detect a set of communities and their reciprocal influence strength. CSI provides a higher level and more intuitive description of the influence dynamics, thus representing a powerful tool to summarize and investigate patterns of influence in large social networks. The evaluation on various datasets suggests the effectiveness of the proposed approach in modeling information propagation at the level of communities. It further enables to detect interesting patterns of influence, such as the communities that play a key role in the overall diffusion process, or that are likely to start information cascades.

1 Introduction

Understanding the dynamics of influence in online social networks is becoming an interesting point of convergence for different subjects, including social science, statistical analysis and computational marketing. Social influence analysis is receiving a growing attention by both academic and industrial communities, mainly due to the wide range of applications, e.g personalized recommendations, viral marketing, feed ranking, and scenarios in which influence plays an important role in predicting users' behavior. Most of the networks of interests for this analysis are very large, with millions of edges. Therefore, *graph summarization* techniques are needed in order to help the analysis by highlighting the main properties of the influence dynamics and recurring patterns. Most of the research in graph summarization has focused on finding abstraction of a graph (e.g., by aggregating nodes in meta-nodes) that preserves the structural properties of the original graph, or properties defined as aggregates over the node attributes. In this

H. Blockeel et al. (Eds.): ECML PKDD 2013, Part II, LNAI 8189, pp. 48–63, 2013.

paper instead, our goal is to devise a graph summarization paradigm for the analysis of the phenomena of information propagation and social influence. More in concrete, we aim to find an abstraction which, although being coarser than the original data, it still describes well a database of past propagation traces. Our technique provides the data analyst with a compact, and yet meaningful, view of the patterns of influence and information diffusion over the considered network, where members of the same community tend to play the same role in the information propagation process.

Towards this goal, we extend the well known Independent Cascade model [10], to study influence at the level of communities. Briefly, the community structure detected by our approach reflects macro influence propagation patterns. A community is identified by a set of *connected nodes* that share a *similar influence tendency* over nodes belonging to other communities. The strength of influence relationships between communities can be used to understand the importance of their connection. Moreover, by directly modeling community-level influence relationships, we can provide an high level picture of the global diffusion process over the network, and a summary of the main influence patterns that shape the underlying process of information propagation.

The main contributes of this paper are the following:

- We introduce the CSI (Community-level Social Influence) model, which extends the peer-influence relationships that define the Independent Cascade model at the granularity of communities.
- We devise a greedy algorithm which explores a given hierarchical partitioning of the network and provides as output the community structure that achieves a good balance between the accuracy in describing observed propagation data, and a compact representation of the influence relationships.
- Given a set of disjoint communities, we devise an Expectation-Maximization algorithm to effectively learn the strength of their pairwise influence relationships.
- Through an experimental evaluation on three real-world datasets, we show the effectiveness of our approach, which is able to provide a meaningful and compact summary of the influence patterns on the considered networks.

The rest of the paper is organized as follows. We briefly review related prior art in Sec. 2. In Sec. 3, we formally define the problem tackled in this paper while our algorithm is presented in Sec. 4. The experimental evaluation is in Sec. 5. Finally, Sec. 6 concludes the paper with a summary of our major findings and future direction of research.

2   Related Work

Social influence and the phenomenon of influence-driven propagations in social networks have received considerable attention in the last years. One of the key problems in this area is the identification of a set of influential users in a given social network. Domingos and Richardson [4] approach the problem with Markov random fields, while Kempe et al. [10] frame influence maximization as a discrete optimization problem. Another vein of study has focused on the problem of learning the influence probabilities on every edge of a social network given an observed log of propagations over this network [16,18,7,20]. In this paper we use the method by Saito et al. [16].

Although our goal is that of summarizing the social graph, our work could also be collocated in the wide *community detection* literature: for a thorough survey of the topic we refer the reader to [5]. While the bulk of this literature only focuses on the structure of the social graph, a recent paper [1] is the first to define a community-detection mechanism that exploits information propagation traces to find better communities. Our contribution in this paper is different as we aim at modeling community-to-community influence, while the goal of [1] is to find good communities w.r.t. the graph structure and information propagation.

Finally, many tasks in machine learning and data mining involve finding simple and interpretable models that nonetheless provide a good fit to observed data. In graph summarization the objective is to provide a coarse representation of a graph for further analysis. Tian et al. [19] as well as Zhang et al. [21] consider algorithms to build graph summaries based on node attributes, while Navlakha et al. [13] use MDL to find good structural summaries of graphs. In [14] this method is applied to study protein interaction networks. Our work is also related to research that uses a taxonomy to impose the right level of granularity to the model being learned. Garriga et al. [6] consider the problem of feature selection for regression models given a taxonomy over the independent variables, while Bonchi et al. [2] use a hierarchical decomposition a state space to simplify Markov models. Lavrač et al. [11] construct interpretable rules by selecting attributes with the help of a hierarchical ontology.

3   Community-Level Social Influence Model

We first (Sec. 3.1) recall the independent cascade propagation model [10], that is at the basis of our proposal. Then we introduce CSI (Sec. 3.2), our model that generalizes peer-influence to the community level: we devise the procedure for learning the parameters of the model (Sec. 3.3), and we discuss model selection (Sec. 3.4).

3.1   Preliminaries: The Independent Cascade (IC) Model

Let $G = (V, E)$ denote a directed network, where V is the set of vertices and $E \subseteq V \times V$ denotes a set of directed arcs. Each arc $(u, v) \in E$ represents an influence relationship, i.e u is a potential influencer for v, and it is associated with a probability $p(u, v)$ representing the strength of such influence relationship. Let $\mathbb{D} = \{\alpha_1, \cdots, \alpha_r\}$ denote a log of observed propagation traces over the G. For each trace α the log contains the activation time of each node $t_\alpha(v)$, where $t_\alpha(v) = \infty$ if v does not become active in trace α.

We assume that each propagation trace in $\mathbb{D}$ is initiated by a special node $\Omega \notin V$, which models a source of influence that is external to the network. More specifically, we have $t_\alpha(\Omega) < t_\alpha(v)$ for each $\alpha \in \mathbb{D}$ and $v \in V$. Time unfolds in discrete steps. At time $t = 0$ all vertices in V are inactive, and Ω makes an attempt to activate every vertex $v \in V$ and succeeds with probability $p(\Omega, v)$. At subsequent time steps, when a node u becomes active, it makes one attempt at influencing each inactive neighbor v with probability $p(u, v)$. Multiple nodes may try to activate, independently, the same node at the same time. If at least one attempt to activate node v at time t succeeds, then

v becomes active at $t + 1$. Therefore, at a given time step, each node is either active or inactive, and active nodes never become inactive again.

Note that we have not specified the function p in detail. The independent cascade model can be instantiated with an arbitrary choice of p, but these come with different trade-offs. Kempe et al. [10] use a uniform probability q in their experiments, that is, $p(u, v) = q$ for all $(u, v) \in E$. On the other hand, Saito et al. [16] estimate a separate probability $p(u, v)$ for every $(u, v) \in E$ from a set of observed traces. These two approaches can be viewed as opposite ends of a complexity scale. Using a single parameter leads to a simple but potentially inaccurate model, while estimating a different probability for each arc might provide a good fit but at the price of risking to overfit, due to the very large number of parameters [12,8].

Next we introduce our CSI model, that shifts the modeling of influence strength from node-to-node, to community-to-community. In our community-based variant of the IC model, all vertices that belong to the same cluster are assumed to have identical influence probabilities towards other clusters.

3.2 The CSI Model

We start by introducing the likelihood of a single trace α when expressed as a function of single edge probability: this is needed to define the problem that we tackle in this paper.

When it comes to fit real data, some of the assumptions of the the theoretical IC propagation model are hardly met. For instance time is not coarsely discrete, and we cannot assume that in real data, if u activates at time t and it succeeds in influencing some of its peers, then this will happen at time $t+1$. Following [12], we circumvent this problem by adopting a *delay threshold* Δ to distinguish between potential influencers that may have triggered an activation, and those who have certainly failed. Let $F^+_{\alpha,u}$ be the set of u's neighbors that potentially influenced u's activation in the trace α:

$$F^+_{\alpha,u} = \{v \mid (v, u) \in E, 0 \le t_\alpha(u) - t_\alpha(v) \le \Delta\}.$$

Similarly, we define as $F^-_{\alpha,u}$ the set of u's neighbors who definitely failed in influencing u on α:

$$F^-_{\alpha,u} = \{v \mid (v, u) \in E, t_\alpha(u) - t_\alpha(v) > \Delta\}.$$

Let $p : V \times V \to [0, 1]$ denote a function that maps every pair of nodes to a probability. The log likelihood of the traces in $\mathbb{D}$ given p can be expressed as

$$\log L(\mathbb{D} \mid p) = \sum_{\alpha \in \mathbb{D}} \log L_\alpha(p), \quad (1)$$

because the traces are assumed to be i.i.d. The likelihood of a single trace α is

$$L_\alpha(p) = \prod_{v \in V} \left[1 - \prod_{u \in F^+_{\alpha,v}} (1 - p(u, v)) \right] \cdot \left[\prod_{u \in F^-_{\alpha,v}} (1 - p(u, v)) \right]. \quad (2)$$

As we already anticipated, in the CSI model we shift the influence strength estimation from node-to-node, to community-to-community. To this end, we use a hierarchical decomposition $\mathcal{H}$ of the underlying network G. In particular, $\mathcal{H}$ is a *tree* rooted at r, with the nodes in V as leaves, and an arbitrary number of internal nodes. A *cut* h of $\mathcal{H}$ is a set of edges of $\mathcal{H}$, so that for every $v \in V$, one and only one edge $e \in h$ belongs to the path from the root r to v. Therefore, the removal of the edges in h from $\mathcal{H}$ disconnects every $v \in V$ from r.

Let $C(\mathcal{H})$ denote the set of all possible cuts of $\mathcal{H}$. Each $h \in C(\mathcal{H})$ induces thus a partition P_h of the network G, so that all vertices in V that are below the same edge $e \in h$ in $\mathcal{H}$ belong to the same cluster $c_e \subseteq V$. Let $a(v)$ denote the cluster to which the node $v \in V$ belongs to in the partition P_h.

In the CSI model, all vertices that belong to the same cluster are assumed to have identical influence probabilities towards other clusters. Given a function $\tilde{p}_h : P_h \times P_h \to [0, 1]$ that assigns a probability between any two clusters of the partition P_h, we define

$$p_h(u, v) = \tilde{p}_h (a(u), a(v)) .$$

Below, in Section 3.3, we will show that given G, $\mathcal{H}$, the cut h, and $\mathbb{D}$, we can find $\tilde{p}_h$ using an expectation maximization (EM) algorithm. For the moment we can assume that $\tilde{p}_h$ is induced by h in a deterministic way, because our aim is to define our problem in terms of finding an optimal cut $h^* \in C(\mathcal{H})$.

A straightforward observation is that the likelihood defined in equations 1 and 2 is maximized by the cut at the leaf level of $\mathcal{H}$. Reducing the number of pairwise influence probabilities used by the model can only result in a lower likelihood. Therefore, we use a *model selection function* f that takes into account both likelihood as well as the complexity of the model. We discuss choices for f later in Section 3.4. Note that since the network G and the hierarchy $\mathcal{H}$ remain fixed, model complexity is only affected by the cut $h \in C(\mathcal{H})$.

Example 1. Figures 1 and 2 illustrate a possible input for our problem and a possible output, i.e., a CSI model, respectively. In particular in the example, the cut h_1 corresponds to the leaf level model where each single node of the social graph constitutes a state of the CSI model: this is the maximum likelihood cut. However, this would correspond to the standard IC model and is not our goal. In the picture two other cuts are shown, where h_2 corresponds to the clustering $\{\{A, F\}, \{D, E\}, \{BC\}, \{H\}, \{I\},$ $\{J, K\}, \{L, M\}\}$, and the cut h_3 results in the CSI model in Figure 2 , which in our example is the *"best"* model according to the model selection function f.

Now we have all necessary ingredients to formally state the problem addressed in this paper.

Problem 1 (Learning a CSI model). Given a network $G = (V, E)$, a log of propagation traces $\mathbb{D}$ across this network, a hierarchical partitioning $\mathcal{H}$ of G, and a model selection function f, find the optimal cut of $\mathcal{H}$ defined as

$$h^* = \underset{h \in C(\mathcal{H})}{\arg \min} f \left(L(\mathbb{D}, p_h), h \right) .$$

We do not formally address the complexity of the problem in this paper. An exhaustive enumeration of all possible cuts is infeasible, since the size of $C(\mathcal{H})$ can be

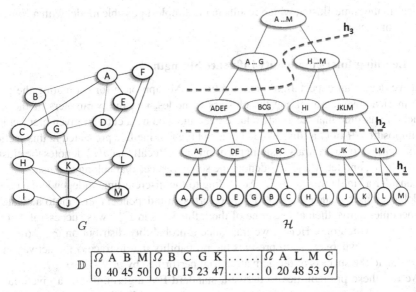

Ω A B M	Ω B C G K	......	Ω A L M C
0 40 45 50	0 10 15 23 47	......	0 20 48 53 97

Fig. 1. An example of input for our problem: a social graph G (here represented as undirected, but we can always consider each undirected edge as the corresponding two directed arcs), a hierarchy $\mathcal{H}$ over G, and a log $\mathbb{D}$ of observed propagation traces over the social graph G

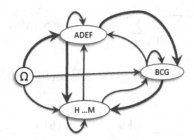

Fig. 2. A possible CSI model resulting from the input of Figure 1 and corresponding to the cut h_3. The arcs tickness represent the strength of the influence along that arc, i.e., the ticker the arc, the larger the associated probability.

exponential in the size of V. Moreover, using the structure of $\mathcal{H}$ is complicated by the possibly complex interplay between the likelihood and the model selection function f. Designing efficient algorithms might be possible, at least for some choices of f, but we leave those as future work.

Finally, it is worth noting that the two extreme cases outlined above, i.e., all links have the same probability, or all links have a different probability can be modeled in our framework. The cut h_1 in Figure 1 places all vertices of G in separate clusters, which corresponds to the most complex model with a separate influence probability on every edge. The cuts h_2 and h_3 induce models with a lower granularity. Finally, a cut right above the root of $\mathcal{H}$ (assuming that there is a "super-root" above r) places all vertices in the same cluster, which results in the simplest possible model with a constant $p(u, v)$ for each arc (u, v).

3.3 Learning Inter-community Influence Strength

Next we devise an expectation-maximization (EM) approach for estimating the pairwise influence strength among the clusters of nodes, i.e., the parameters of the CSI model. We assume that the clusters have been induced by a cut of a given hierarchical decomposition $\mathcal{H}$ of G as discussed above, but the EM method presented in this section can be applied to an arbitrary disjoint partition of V. Recall that $a(v)$ denotes the cluster where $v \in V$, and let $A(x) \subseteq V$ denote the set of vertices that belong to cluster $x \in P$.

Consider a single trace $\alpha \in \mathbb{D}$. According to the discrete-time independent cascade model, each user u such that $u \in F_{\alpha,v}^+$ performs an independent attempt to activate v. If v becomes active then at least one of the influencers in $F_{\alpha,v}^+$ was successful, but we don't know which one. Hence, we introduce a probability distribution $\varphi_{\alpha,v}$ over the nodes in $F_{\alpha,v}^+$, where $\varphi_{\alpha,v,u}$ represents the probability that in trace α the activation of v was due to the success of the activation trial performed by u.

We use these probabilities to derive a standard EM algorithm. For a given cut h, each $u \in F_{\alpha,v}^+$ succeeds in activating v on the considered trace with probability $p_h(a(u), a(v))$ and fails with probability $(1 - p_h(a(u), a(v)))$. By exploiting users' responsabilities $\varphi_{\alpha,v,u}$, we can define the complete expectation (log)likelihood of the observed propagation as follows:

$$
Q(\tilde{p}_h; \tilde{p}_h^{old}) = \sum_{\alpha \in \mathbb{D}} \sum_v \left\{ \sum_{u \in F_{\alpha,v}^+} \left(\varphi_{\alpha,v,u} \log \tilde{p}_h(a(u), a(v)) + \right. \right.
$$

$$
\left. (1 - \varphi_{\alpha,v,u}) \log (1 - \tilde{p}_h(a(u), a(v))) \right) + \tag{3}
$$

$$
\left. \sum_{u \in F_{\alpha,v}^-} \log (1 - \tilde{p}_h(a(u), a(v))) \right\}.
$$

Given an estimate of every $\varphi_{\alpha,v,u}$, we can determine the $\tilde{p}_h$ which maximizes Eq.3 by solving $\frac{\partial Q(\tilde{p}_h; \tilde{p}_h^{old})}{\partial \tilde{p}_h(x,y)} = 0$ for all pair of clusters $x, y \in P_h$. This gives the following estimate of $\tilde{p}_h(x, y)$:

$$
\tilde{p}_h(x, y) = \frac{1}{|S_{x,y}^+| + |S_{x,y}^-|} \sum_{\alpha \in \mathbb{D}} \sum_{v \in A(y)} \sum_{\substack{u \in F_{\alpha,v}^+ \\ u \in A(x)}} \varphi_{\alpha,v,u}, \tag{4}
$$

where

$$
S_{x,y}^+ = \sum_{\alpha} \sum_{v \in A(y)} \sum_{u \in A(x)} \mathbb{I}(u \in F_{\alpha,v}^+) \quad \text{and} \quad S_{x,y}^- = \sum_{\alpha} \sum_{v \in A(y)} \sum_{u \in A(x)} \mathbb{I}(u \in F_{\alpha,v}^-).
$$

We still must provide an estimate for every $\varphi_{\alpha,v,u}$. We do this on the basis of the assumption that the probability distributions $\varphi_{\alpha,v}$ are independent of the partition P. That is, if u is believed to be the activator for v in the trace α, this belief should not change for different ways of clustering the two nodes. Therefore, we estimate the $\varphi_{\alpha,v,u}$s from the model where every $v \in V$ belongs to its own cluster, because this will lead to estimates that only depend on the network structure. Denote this model by $\tilde{p}_l$. We obtain:

$$\varphi_{\alpha,v,u} = \frac{\tilde{p}_l(v,u)}{1 - \prod_{w \in F_{\alpha,u}^+}(1 - \tilde{p}_l(w,u))}. \tag{5}$$

Thus we design the following procedure:

- Run the EM algorithm without imposing a clustering structure (which is equivalent to the estimation proposed in [16]) to obtain $\tilde{p}_l(u,v) \forall (u,v) \in E$.
- Compute each $\varphi_{\alpha,v,u}$ using Equation 5.
- For different partitions P, keep the $\varphi_{\alpha,v,u}$ fixed, and update $\tilde{p}(x,y)$ according to Eq.4.

3.4 Model Selection

Recall that the likelihood $\log L(\mathbb{D} \mid p_h)$ is maximized for the cut h that places every node in its own cluster. We need thus a way to address the trade-off between fit and model complexity. Two principled approaches to this are *Bayesian Information Criterion* (BIC) and *Minimum Description Length* (MDL), which we both use in this paper.

We instantiate BIC [17] as follows: BIC $= -2 \log L(\mathbb{D} \mid p_h) + |h| \log(|\mathbb{D}|)$.

In the basic two-part MDL [15], we first use the model, in our case the cut h, to *encode* the observed data (the traces in $\mathbb{D}$), and then encode the model itself. We denote the encoding length of $\mathbb{D}$ given h by $L(\mathbb{D} \mid h)$, and the encoding length of the cut by $L(h)$.

To apply MDL in our context, we must specify both $L(\mathbb{D} \mid h)$ as well as $L(h)$. A standard result is that we can simply use the log-likelihood of $\mathbb{D}$ given h as $L(\mathbb{D} \mid h)$ (see e.g. [3]). That is, we let $L(\mathbb{D} \mid h) = \log L(\mathbb{D} \mid p_h)$. The encoding length $L(h)$ of the cut h is defined as the number of bits needed to communicate h to a receiving party. We assume that the recipient already has the hierarchy $\mathcal{H}$ as well as the network G. We must send every edge in h using $|h| \cdot \log(|\mathcal{H}|)$ bits, as well as the influence probabilities between all pairs of communities where this probability is nonzero. If there are $X \leq |h|^2$ such pairs, we use $X (2 \log(|h|) + C)$ bits (probabilities are encoded with C bits each) for the probabilities. The total encoding length of h is thus:

$$L(h) = |h| \cdot \log(|\mathcal{H}|) + X (2 \log(|h|) + C).$$

MDL favors the model that minimizes the combined encoding: $L(\mathbb{D} \mid h) + L(h)$.

4   Algorithm

In the previous section, we introduced the CSI model and discussed how to evaluate different cuts of $h \in \mathcal{C}(\mathcal{H})$ of the hierarchical decomposition of the network. However, as mentioned in Sec. 3, the search space $\mathcal{C}(\mathcal{H})$ can in general be exponential in the size of $\mathcal{V}$, making exhaustive search infeasible. Next we present a heuristic algorithm that performs a *bottom-up* greedy visit of $C(\mathcal{H})$, and provides the best solution found as output.

In our implementation, $\mathcal{H}$ is always a binary tree, but the approach applies to general trees as well. The procedure starts from the cut corresponding to the leaf level; at each iteration we compute all the possible cuts which can be obtained from the current one by merging communities that share a same parent in the hierarchy. Since $\mathcal{H}$ is a binary tree, each merge will involve exactly two communities. More formally, given a cut $h = \{e_1, \cdots, e_{|h|}\} \in C(\mathcal{H})$, let $\mathcal{M}(h)$ denote the set of candidate merges available from the cut h:

$$\mathcal{M}(h) = \{\langle y, y\prime \rangle : (x, y) \in h \land (x, y\prime) \in h, x \neq r\}$$

where r is the root of $\mathcal{H}$. A simple greedy heuristic would pick the merge in $\mathcal{M}(h)$ that results in the best value of the objective function. However, evaluating our objective function is computationally intensive, because it involves re-estimating model parameters, and computing the likelihood of $\mathbb{D}$ given those parameters. This is too slow to be useful in practice.

To speed-up the algorithm, we make use of the following observation: in an "ideal merge" (with respect to the outgoing influence patterns) the two communities *exhibit exactly the same influence probabilities with other communities*. That is, if for some y and $y\prime$ we have $p(y, z) = p(y\prime, z)$ and $p(z, y) = p(z, y\prime)$ for every z, merging the communities y and $y\prime$ does not affect the likelihood of $\mathbb{D}$ at all. In practice these influence probabilities are never exactly identical, but we can still find a merge where they are *as similar as possible*. Rather than computing the entire objective function for every possible merge in $\mathcal{M}(h)$, we find the merge that is the best in terms of the above condition. To this end we use a similarity function defined as

$$\text{similarity}(y, y\prime, p) = \sum_z \left(p(y, z)p(y\prime, z) + p(z, y)p(z, y\prime) \right), \tag{6}$$

which can be thought of as the dot product between the influence probability vectors associated with communities y and $y\prime$.

Our whole procedure, summarized in Algorithm 1, finds in each iteration the best merge using Eq. 6 and updates the model given this. The resulting cut as well as the corresponding parameters are stored in a set, denoted L. Once the algorithm reaches the root of $\mathcal{H}$, it evaluates the objective function for every cut in the set L and returns the one having the best value. The function updateModel runs the estimation procedure described in Section 3.3.

5   Experimental Evaluation

CSI provides a compact description of influence patterns in the underlying network; in the following we will describe how this approach can be exploited for several purposes, including data understanding and characterization of information propagation flow.

Datasets. The evaluation focuses on three datasets where each dataset comprises of a network G and the propagation log $\mathbb{D}$. The first dataset has been extracted from Yahoo!

Algorithm 1. CSI model learning

Input : A propagation log $\mathbb{D}$, a network $G = (V, E)$ and hierarchical decomposition $\mathcal{H}$.
Output: The cut $h \in C(\mathcal{H})$ which achieves the best value of the objective function.
$h \leftarrow \text{leafLevel}(\mathcal{H})$
$p \leftarrow \text{updateModel}(h, G, \mathbb{D})$
$L \leftarrow \emptyset$
while $C(h) \neq \emptyset$ **do**
    $\langle x^*, y^* \rangle \leftarrow \arg\max_{\langle x, y \rangle \in C(h)} \{\text{similarity}(x, y, p)\}$
    $h \leftarrow \text{merge}(h, \langle x^*, y^* \rangle)$
    $p \leftarrow \text{updateModel}(h, G, \mathbb{D})$
    $L \leftarrow L \cup \{\langle h, p \rangle\}$
end
$\langle h^*, p^* \rangle \leftarrow \arg\min_{\langle h, p \rangle \in L} \{\text{objFunc}(h, p, \mathbb{D})\}$
return h^*

Table 1. Datasets statistics

	MEME	FLIXSTER	TWITTER
Number of Nodes	$9,523$	$6,354$	$23,537$
Number Links	$759,369$	$97,314$	$1,299,652$
Traces	$9,578$	$7,158$	$6,139$
Activations	$552,732$	$1,439,875$	$383,866$
Avg. number of activations per node	84	221	16
Avg. number of nodes per trace	58	200	62

Meme, a microblogging service[1], in which users can share different kinds of information called "memes". Memes are shared on the main user's stream and a re-post button allows to display an item from another user's stream on the personal one. If the user v posts a meme which is later re-posted by the user u, we say that the meme propagates from v to u, and thus v is a potential influencer of u. The second dataset has been crawled from Flixster[2], one of the main social movie website. It allows users to share ratings on movies and to meet other users with similar tastes. In Flixster, the propagation log records the time at which a user rated a particular movie. In this context, an item or movie is considered to propagate from v to u, if u rates the item shortly after the rating by v. The last dataset was obtained by crawling the public timeline of Twitter[3]. We track the propagation of URLs across the network where an activation corresponds to the instance when a user uses a certain URL for sharing with other friends. In Table 1 we report the main characteristics of the datasets.

Experiment Settings. The optimization algorithm proposed in Sec. 4 requires as input a hierarchical decomposition of the network. We obtain this hierarchy by recursively partitioning the underlying network using *METIS* [9], which reportedly provides high

---

[1] Discontinued in May 25, 2012.
[2] http://www.cs.sfu.ca/~sja25/personal/datasets/
[3] https://dev.twitter.com/docs/api/1/get/
   statuses/public_timeline

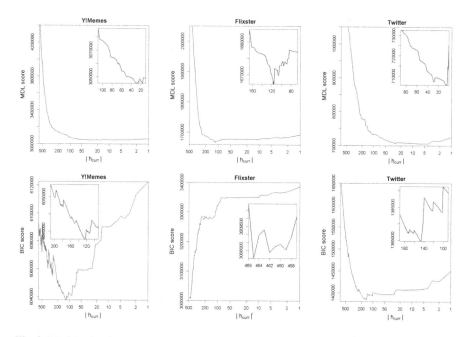

Fig. 3. Model selection: MDL (first row), and BIC (second row) - the subplots show a focused view on the region of the minimum

quality partitions. At each stage in the recursive procedure, we split the network to two components that are roughly equal in size. The resulting hierarchical partitioning is thus a binary tree. In order to reduce the computational overhead of the algorithm, we initialize the hill climbing clustering procedure with a cut slightly above the leaf level. This is obtained by making 5 passes over the leafs, and during each pass we merge leafs that have a common parent. While the choice of the similarity function is somewhat arbitrary, we found that the cosine similarity of Eq. 6 works well in practice. Finally, the delay threshold Δ is set to ∞: i.e., for a given node we consider potential influencer any neighbor active before the node in a given trace. We ran our experiments on a Intel Xeon 2.4 GHz processor and 8 GB memory. The learning time ranges from few hours (Flixster and Y!Meme) to several days for Twitter, where the number of links (approx. 1.3 million) impact the learning time as it increases the computational effort in Eq. 4 where a greater number of potential activators of v needs to be considered. It is worth noting that parallelizing the EM computation of Section 3.3 is possible and it is planned in our future work.

Model Selection. In Figure 3, we compare the BIC and MDL scores that we obtain for each cut found by our algorithm from the lowest (many communities) to the highest (few communities). The two model selection criteria do not agree on the identification of the optimal model. MDL tends to favor less complex models than BIC in our case, which is most likely caused by the quadratic dependence on $|h|$ of $L(h)$. For instance, MDL provides us 115 communities for Flixster dataset whereas BIC provides 454 communities (just after a couple of iterations of the main algorithm). In the rest of

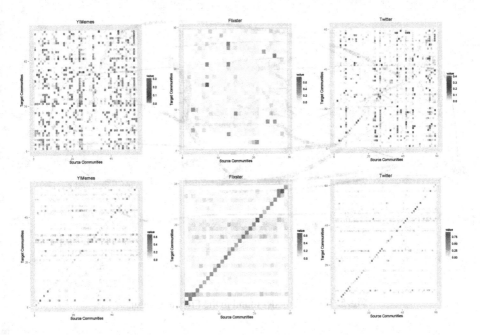

Fig. 4. Community to community influence probabilities (first row) and social links (second row)

this section we will characterize one model found for each dataset. More specifically, for Flixster we select the model provided by MDL since it provides a more compact view of the influence pattern, with (29 communities, after removing singleton node communities). For the other two datasets, we select the model provided by BIC. The number of communities for Twitter and Y!Meme, after removing singletons, are 60 and 53, respectively.

Community-Level Social Influence Analysis. The output of the CSI model can be easily graphically represented according to two different, and complementary, perspectives. The first way to analyze the strength of the influence and social link relationships is by plotting the corresponding heat-maps, as shown in Figure 4. In these figures, we plot the intensity of the influence probability between two communities, and the probability of observing a link between them, respectively. On the whole, we register almost no correlation between influence and link probabilities. From the heat maps corresponding to link probabilities, we can see that the clustering procedure use to find the hierarchy $\mathcal{H}$ (METIS) has correctly identified communities of highly connected nodes. Influence relationships, however, do not in general exhibit any clear structure, although we have a slight diagonal in the Twitter dataset. Interestingly, even if a community is dense, it does not necessarily exhibit strong internal influence.

An alternate and perhaps more effective way of summarizing influence relationships in the network is to consider the community-level influence propagation network. In Figure 5 we show the CSI propagation network for the Flixster dataset, where node

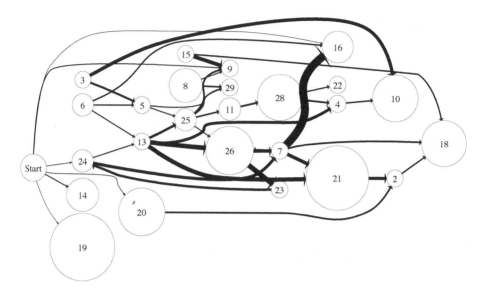

Fig. 5. The CSI model found in the Flixster dataset

size is proportional to community size, "Start" represents the Ω node, and edge width is proportional to influence probability. We preprocessed the graph by pruning all the edges between communities having influence probability less than 0.1, while we use 0.03 as threshold for the links connecting Ω with the rest of the network. Interestingly, the network is almost acyclic, and this suggests a clear directionality pattern in the flow of information. This finding is further confirmed by the analysis of the other two datasets, for which, due to the limited space, we omit the CSI propagation networks.

Both the heat-map representation and the compact propagation network provide an useful tool to understand influence relationships between communities, and at the level of the whole network. We can evaluate the capabilities of this approach in providing a compact and yet accurate description of the real influence process on the underlying network, by setting up two simple tests.

In the first test, the goal is to verify if our approach detects correctly the communities that play a key role in the information propagation process. To quantify the importance of each community, we run the greedy influence maximization algorithm [10] (with a budget of 100 nodes) on the considered networks, where the influence probabilities are the ones estimated at the leaf level and we employ 5000 Monte Carlo simulation to estimate the spread. This procedure provides a ranked list $\mathcal{S}$ of nodes that should be targeted to maximize the expected spread on the network. For each $s \in \mathcal{S}$ we record Δ_s as the gain, in terms of spread, that can be achieved by adding s to the set $\mathcal{S}$. We can measure its importance in the overall diffusion process as the percentage of the overall spread achieved by targeting the seed nodes selected in the considered community. More formally:

$$\text{score}(c) = \frac{\sum_{s' \in \mathcal{S} \wedge s' \in A(c)} \Delta_{s'}}{\sum_{s' \in \mathcal{S}} \Delta_{s'}}.$$

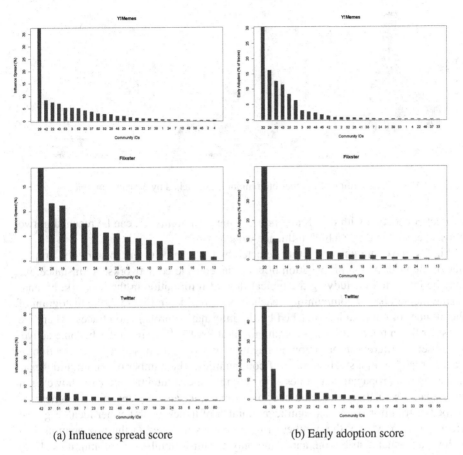

(a) Influence spread score   (b) Early adoption score

Fig. 6. Influence spread and early adoption scores of the communities produced by CSI

As clearly visible in Figure 6(a), a high fraction of the overall spread can be explained by only a few communities. The high score of the most important communities is due to the fact that the multiple nodes having a large spread gain belong to those communities. As instance, the greedy procedure picks 13 nodes in the community 29 for Meme, 17 belonging to the community 21 for Flixster, 13 in the community 42 for Twitter. Interestingly, the community structure provided by our approach gathers "high influential" nodes in the same community.

The second validation test to provide a qualitative evaluation of the CSI model is focused on the identification of early adopter communities. For each trace we consider the communities that have highest number of active nodes during the first quarter of the trace's overall propagation time, and rank communities accordingly. Again, as shown in Figure 6(b), a significant number of traces involves the initial activation of nodes belonging to a small set of communities. The identity of those communities can be easily tracked by considering the length of the path from the "Start" node in the CSI propagation graph. As instance, on Flixster (Figure 5), the communities 14 and 19 exhibit

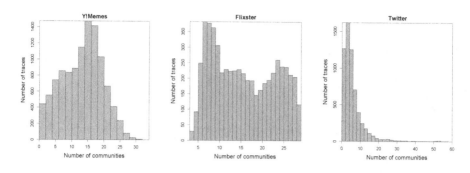

Fig. 7. Distribution of number of communities touched by propagation traces

a direct connection with the "Start" node, while community 21, can be reached in two hops, characterized by an high influence weight.

Finally, the community structure detected by CSI can be useful to study the properties of the information propagation flow on the considered networks. As instance, we may be interested in studying the typical flow of information in the network, by analyzing the number of communities reached. We provide, in Figure 7, the histogram of the number of communities reached by the information propagation traces, where we consider that a trace enters in a community if at least 10% of its nodes become active. We observe different information propagation patterns: in Flixster a larger fraction of traces propagate in a small number of communities. The number of communities participating in a propagation declines as the number of communities increase, however, it rises again for relatively large number of communities and, thereby, making a near U-shaped distribution. Hence, a significant number of traces exhibit either local or global diffusion. In Y!Meme, the number of communities involved in the propagation flow follows a normal distribution, and again only a limited number of communities (15 in average) are typical involved in the propagation. This "local propagation" behavior is emphasized on Twitter, where each trace generally involves less than 5 communities.

6   Conclusions

In this paper we introduce a hierarchical approach to summarize patterns of influence in a network, by detecting communities and their reciprocal influence strength. Our model, dubbed CSI, generalizes the Independent Cascade propagation model, by modeling influence between communities of connected nodes rather than a pairwise node influence. This enables more compact representation of the network of influence, which can be easily plotted and exploited to understand and detect interesting properties in the information propagation flow over the network. Our empirical analysis over real-world networks highlights two interesting observations: (i) the propagation networks found by CSI are almost acyclic, (ii) information propagates in the network mainly "locally", reaching few communities. While the first observation offers interesting insights, since it shows the existence of a clear direction in the propagation of information, the latter confirms a strong relationships between information propagation and the community structure, that might be exploited for community detection [1].

Acknowledgments. This research was partially supported by the Torres Quevedo Program of the Spanish Ministry of Science and Innovation, and partially funded by the European Union 7th Framework Programme (FP7/2007-2013) under grant n. 270239 (ARCOMEM).

References

1. Barbieri, N., Bonchi, F., Manco, G.: Cascade-based community detection. In: WSDM 2013 (2013)
2. Bonchi, F., Castillo, C., Donato, D., Gionis, A.: Taxonomy-driven lumping for sequence mining. Data Mining and Knowledge Discovery 19(2), 227–244 (2009)
3. Cover, T.M., Thomas, J.A.: Elements of information theory. Wiley-interscience (2012)
4. Domingos, P., Richardson, M.: Mining the network value of customers. In: KDD 2001 (2001)
5. Fortunato, S.: Community detection in graphs. Physics Reports 486(3), 75–174 (2010)
6. Garriga, G.C., Ukkonen, A., Mannila, H.: Feature selection in taxonomies with applications to paleontology. In: Boulicaut, J.-F., Berthold, M.R., Horváth, T. (eds.) DS 2008. LNCS (LNAI), vol. 5255, pp. 112–123. Springer, Heidelberg (2008)
7. Goyal, A., Bonchi, F., Lakshmanan, L.V.S.: Learning influence probabilities in social networks. In: WSDM 2010 (2010)
8. Goyal, A., Bonchi, F., Lakshmanan, L.V.S.: A data-based approach to social influence maximization. PVLDB 5(1), 73–84 (2011)
9. Karypis, G., Kumar, V.: A fast and high quality multilevel scheme for partitioning irregular graphs. SIAM J. Sci. Comput. 20(1), 359–392 (1998)
10. Kempe, D., Kleinberg, J.M., Tardos, É.: Maximizing the spread of influence through a social network. In: KDD 2003 (2003)
11. Lavrač, N., Vavpetič, A., Soldatova, L., Trajkovski, I., Novak, P.K.: Using ontologies in semantic data mining with segs and g-segs. In: Elomaa, T., Hollmén, J., Mannila, H. (eds.) DS 2011. LNCS, vol. 6926, pp. 165–178. Springer, Heidelberg (2011)
12. Mathioudakis, M., Bonchi, F., Castillo, C., Gionis, A., Ukkonen, A.: Sparsification of influence networks. In: KDD 2011 (2011)
13. Navlakha, S., Rastogi, R., Shrivastava, N.: Graph summarization with bounded error. In: SIGMOD 2008 (2008)
14. Navlakha, S., Schatz, M.C., Kingsford, C.: Revealing biological modules via graph summarization. Journal of Computational Biology 16(2), 253–264 (2009)
15. Rissanen, J.: A universal prior for integers and estimation by minimum description length. The Annals of Statistics, 416–431 (1983)
16. Saito, K., Nakano, R., Kimura, M.: Prediction of information diffusion probabilities for independent cascade model. In: Lovrek, I., Howlett, R.J., Jain, L.C. (eds.) KES 2008, Part III. LNCS (LNAI), vol. 5179, pp. 67–75. Springer, Heidelberg (2008)
17. Schwarz, G.: Estimating the dimension of a model. The Annals of Statistics 6(2), 461–464 (1978)
18. Tang, J., Sun, J., Wang, C., Yang, Z.: Social influence analysis in large-scale networks. In: KDD 2009 (2009)
19. Tian, Y., Hankins, R.A., Patel, J.M.: Efficient aggregation for graph summarization. In: SIGMOD 2008 (2008)
20. Xiang, R., Neville, J., Rogati, M.: Modeling relationship strength in online social networks. In: WWW 2010 (2010)
21. Zhang, N., Tian, Y., Patel, J.M.: Discovery-driven graph summarization. In: ICDE 2010 (2010)

Supervised Learning of Syntactic Contexts for Uncovering Definitions and Extracting Hypernym Relations in Text Databases

Guido Boella and Luigi Di Caro

Department of Computer Science
University of Turin
{boella,dicaro}@di.unito.it

Abstract. In this paper we address the problem of automatically constructing structured knowledge from plain texts. In particular, we present a supervised learning technique to first identify definitions in text data, while then finding hypernym relations within them making use of extracted syntactic structures. Instead of using pattern matching methods that rely on lexico-syntactic patterns, we propose a method which only uses syntactic dependencies between terms extracted with a syntactic parser. Our assumption is that syntax is more robust than patterns when coping with the length and the complexity of the texts. Then, we transform the syntactic contexts of each noun in a coarse-grained textual representation, that is later fed into hyponym/hypernym-centered Support Vector Machine classifiers. The results on an annotated dataset of definitional sentences demonstrate the validity of our approach overtaking the current state of the art.

1 Introduction

Nowadays, there is a huge amount of textual data coming from different sources of information. Wikipedia[1], for example, is a free encyclopedia that currently contains 4,168,348 English articles[2]. Even Social Networks play a role in the construction of data that can be useful for Information Extraction tasks like Sentiment Analysis, Question Answering, and so forth. From another point of view, there is the need of having more structured data in the forms of ontologies, in order to allow semantics-based retrieval and reasoning. Ontology Learning is a task that permits to automatically (or semi-automatically) extract structured knowledge from plain text. Manual construction of ontologies usually requires strong efforts from domain experts, and it thus needs an automatization in such sense. In this paper, we focus on the extraction of hypernym relations. The first step of such task relies on the identification of what [21] called *definitional sentences*, i.e., sentences that contain at least one hypernym relation. This subtask is important by itself for many tasks like Question Answering [8], construction

---

[1] http://www.wikipedia.org/
[2] February 19, 2013.

H. Blockeel et al. (Eds.): ECML PKDD 2013, Part II, LNAI 8189, pp. 64–79, 2013.

of glossaries [18], extraction of taxonomic and non-taxonomic relations [19,25], enrichment of concepts [13,6], and so forth.

The main contribution of this work is to relax the problem of inducing semantic relations by separating it into two easier subtasks and facing them independently with the most appropriate techniques. Indeed, hypernym relation extraction involves two aspects: linguistic knowlege, and model learning. Patterns uses to collapse both of them, preventing to tackle them separately with the most suitable techniques. On the one hand, patterns have limited expressivity, while linguistic knowledge inside patterns is learned from small corpora so it is likely to have low coverage. On the other hand, classification strictly depends on the learned patterns, so performance decreases and the available classification techniques are restricted to those compatible with the pattern approach. Instead, we use a syntactic parser for the first aspect (with all its native and domain-independent knowledge on language expressivity), and a state-of-the-art approach to learn models with the use of Support Vector Machine classifiers.

2 Motivating Examples

Most of the existing work in this field use manual rather than automatic generation of sequential patterns inducing hypernym relations. Although this approach achieves good results, as demonstrated in well-founded papers like [8] and [21], it is limited in the sense that it exclusively relies on the sequentiality of the expressions. Natural language offers potentially infinite ways of expressing concepts, without any limits on the length and complexity of the sentences.

Definitions can present a great variety of linguistic constructions in natural language. For instance, it is possible to have definitions that make use of punctuation, as in the following sentence:

"IP: a protocol for sending data across a network."

Some work concentrated on the English copular verb *to be*, as in [3], while even other verbs can indicate the presence of a hypernym relation, as in the sentence below:

"The term ontology stands for a formal representation of objects in a specific domain."

Still, in the sentence:

"Browsers, tools for navigating the Web, can also reproduce sound."

the identification of the syntactic apposition is necessary to determine the hypernym relation. Modifiers are other linguistic constructions that can lengthen the sentences, making them more complex to match with trained patterns, as in this example:

"The Aardvark is a medium-sized, burrowing, nocturnal mammal native to Africa"

Notice that, however, a POS-aware pattern mathing technique like the one by [21] can handle this type of complexity. Finally, linguistic coordinations need to be identified to be able to extract all possible hypernyms, as in this last example:

"*Agathon was an Athenian tragic poet and friend of Euripides and Plato*"

where "*Agathon*" is a "*poet*", a "*friend of Euripides*", and a "*friend of Plato*".

3 Related Work

In this section we present the current state of the art concerning the automatic extraction of definitions and hypernym relations from plain text.

3.1 Definition Classification

Considering the initial formal representation proposed by [26], a *definitional sentence* is composed by different information fields:

- a *definiendum* (DF), i.e., the word being defined with its modifiers,
- a *definitor* (VF), i.e., the verb phrase to introduce the definition,
- a *definiens* (GF), i.e., the genus phrase that usually contains the hypernym,
- and the *rest* of the sentence (REST), that can contain additional clauses.

An example of annotated definition is represented by the following sentence:

[In computer science, a $[pixel]_{DF}$ $[is]_{VF}$ [a **dot**]$_{GF}$ [that is part of a computer image]$_{REST}$.

In this paper, we will use the term *definitional sentence* referring to the more general meaning given by [21]: *A sentence that provides a formal explanation for the term of interest*, and more specifically as a sentence containing at least one hypernym relation.

So far, most of the proposed techniques rely on lexico-syntactic patterns, either manually or semi-automatically produced [17,31,28]. Such patterns are sequences of words like "*is a*" or "*refers to*", rather than more complex sequences including part-of-speech tags.

In the work of [28], after a manual identification of types of definitions and related patterns contained in a corpus, the author successively applied Machine Learning techniques on syntactic and location features to improve the results.

A fully-automatic approach has been proposed by [3], where the authors applied genetic algorithms to the extraction of English definitions containing the keyword "*is*". In detail, they assign weights to a set of features for the classification of definitional sentences, reaching a precision of 62% and a recall of 52%.

Then, [8] proposed an approach based on *soft patterns*, i.e., probabilistic lexico-semantic patterns that are able to generalize over rigid patterns enabling partial matching by calculating a generative degree-of-match probability between a test instance and the set of training instances.

[10] used three different Machine Learning algorithms to distinguish actual definitions from other sentences, relying on syntactic features and reaching high accuracy levels.

The work of [18] relies on a rule-based system that makes use of "cue phrases" and structural indicators that frequently introduce definitions, reaching 87% of precision and 75% of recall on a small and domain-specific corpus.

Finally, [21] proposed a system based on Word-Class Lattices (WCL), i.e., graph structures that try to generalize over the POS-tagged definition patterns found in the training set. Nevertheless, these mechanisms are not properly able to handle linguistic exceptions and linguistic ambiguity.

3.2 Hypernym Extraction

According to [2] and [4], the problem of extracting ontologies from text can be faced at different levels of granuarity. According to the former, our approach belongs to the extraction of *terminological ontologies* based on IS-A relations, while for the latter we refer to the *concept hierarchies* of their *Ontology Learning layer cake*.

As for the task of definition extraction, most of the existing approaches use symbolic methods that are based on lexico-syntactic patterns, which are manually crafted or deduced automatically. The seminal work of [16] represents the main approach based on fixed patterns like "NP_x *is a/an* NP_y" and "NP_x *such as* NP_y", that usually imply $< x$ IS-A $y >$. The main drawback of such technique is that it does not face the high variability of how a relation can be expressed in natural language. Still, it generally extracts single-word terms rather than well-formed and compound concepts. The work of [21], as already mentioned in the previous section, is based on graph structures that generalize over the POS-tagged patterns between x and y. [1] proposed similar lexico-syntactic patterns to extract *part-whole* relationships.

[9] proposed a rule-based approach to the extraction of hypernyms that, however, leads to very low accuracy values in terms of Precision.

[23] proposed a technique to extract hypernym relations from Wikipedia by means of methods based on the connectivity of the network and classical lexico-syntactic patterns. [29] extended their work by combining extracted Wikipedia entries with new terms contained in additional web documents, using a distributional similarity-based approach.

Finally, pure statistical approaches present techniques for the extraction of hierarchies of terms based on words frequency as well as co-occurrence values, relying on clustering procedures [5,12,30]. The central hypothesis is that similar words tend to occur together in similar contexts [15]. Despite this, they are defined by [2] as *prototype-based ontologies* rather than formal terminological ontologies, and they usually suffer from the problem of data sparsity in case of small corpora.

4   Approach

In this section we present our approach to identify hypernym relations within plain text. Our methodology consists in relaxing the problem in two different subtasks. Given a semantic relation between two terms $rel(x, y)$ within a sentence, the task becomes to find 1) a possible x, and 2) a possible y. In case of more than one possible x or y, a further step is needed to associate the correct x to the right y.

By seeing the problem as two different classification problems, there is no need to create abstract patterns between the target terms x and y. In addition to this, the general problem of identifying definitional sentences can be seen as to find at least one x and one y in a single sentence.

4.1   Local Syntactic Information

Dependency parsing is a procedure that extracts syntactic dependencies among the terms contained in a sentence. The idea is that, given a hypernym relation, hyponyms and hypernyms may be characterized by specific sets of syntactic contexts. According to this assumption, the task can be seen as a classification problem where each noun in a sentence has to be classified as hyponym, hypernym, or neither of the two.

More in detail, for each noun "a" the system creates one instance composed by textual items describing its syntactic context. Each item can be seen as a classic word, and it represents a single syntactic relation taking a as one of its arguments. To build these items, extracted dependencies are transformed into abstract textual representation in the form of triples. In particular, for each syntactic dependency $dep(a, b)$ (or $dep(b, a)$) of a target noun "a", we create an abstract term $dep\text{-}target\text{-}\hat{b}$ (or $dep\text{-}\hat{b}\text{-}target$), where "$a$" becomes "$target$" and where "$\hat{b}$" is transformed into the generic string "$noun$" in case it is a noun; otherwise it is equal to "b". This way, the nouns are transformed into coarse-grained context abstractions, creating a level of generalization of the feature set that collapses the variability of the nouns involved in the syntactic dependencies. The string "$target$" is useful to determine the exact position of the noun in a syntactic dependency (as a left argument, or as a right argument).

Running Example. In order to describe the process of tranforming the input data to fit with a standard classification problem, we present here a step-by-step concrete example. Let us consider the sentence below:

> The Albedo of an object is the extent to which it diffusely reflects light from the sun.

The result of the Part-Of-Speech tagging procedure will be the following:

> The/DT Albedo/NNP of/IN an/DT object/NN is/VBZ the/DT extent/NN to/TO which/WDT it/PRP diffusely /RB reflects/VBZ light/NN from/IN the/DT sun/NN.

where *DT* stands for *determiner*, *NNP* is a proper name, and so on[3]. Then, the syntactic parsing will produce the following dependencies (the numbers are unique identifiers)[4]:

$$det(\text{Albedo-2, The-1})$$
$$nsubj(\text{extent-8, Albedo-2})$$
$$det(\text{object-5, an-4})$$
$$prepof(\text{Albedo-2, object-5})$$
$$cop(\text{extent-8, be-6})$$
$$det(\text{extent-8, the-7})$$
$$rel(\text{reflects-13, which-10})$$
$$nsubj(\text{reflects-13, it-11})$$
$$advmod(\text{reflects-13, diffusely-12})$$
$$rcmod(\text{extent-8, reflect-13})$$
$$dobj(\text{reflect-13, light-14})$$
$$det(\text{sun-17, the-16})$$
$$prepfrom(\text{reflect-13, sun-17})$$

where the dependency *nsubj* represents a noun phrase which is the syntactic subject of a clause, *dobj* identifies a noun phrase which is the (accusative) object of the verb, and so on[5]. The related parse tree is shown in Figure 1.

At this point, the system creates one instance for each term labeled as "noun" by the POS-tagger. For example, for the noun "*Albedo*", the instance will be represented by three abstract terms, as shown in Table 1.

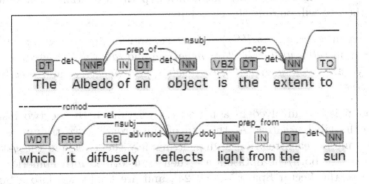

Fig. 1. The resulting parse tree of the example

Once the instance for the noun "*Albedo*" is created, it is passed to the classification process that will decide if "*Albedo*" can be considered as part of a

---

[3] A complete overview of the parts-of-speech can be found at
http://nlp.stanford.edu/software/tagger.shtml

[4] We used the Stanford Syntactic Parser available at
http://nlp.stanford.edu/software/lex-parser.shtml

[5] A complete overview of the Stanford dependencies is available at
http://nlp.stanford.edu/software/dependencies_manual.pdf

Table 1. The instance created for the noun "*Albedo*" is composed by three items (one for each syntactic dependency related to "*Albedo*"). Note that the considered noun "*Albedo*" is replaced by the generic term "**target**", while the other nouns are replaced with "*noun*".

Dependence	Instance Item
det(Albedo, The)	det-**target**-the
nsubj(extent, Albedo)	nsubj-*noun*-**target**
prepof(Albedo, object)	prepof-**target**-*noun*

hypernym relation, as explained in the next section. This is done for each noun in a sentence.

4.2 Learning Phase

Our model assumes a transformation of the local syntactic information into labelled numeric vectors. More in detail, given a sentence S annotated with some terms linked by one hypernym relation, the system produces as many input instances as the number of nouns contained in S. Only the nouns that are involved in the annotated hypernym relation (as x or y in $rel(x, y)$) will be positive instances.

More specifically, for each noun n in S, the method produces two instances S_x^n and S_y^n (i.e., one for each argument of a hypernym relation). The difference between the two will be only the class label:

1. If $n = x$ in $rel(x, y)$, $label(S_x^n) = positive$, and $label(S_y^n) = negative$
2. If $n = y$ in $rel(x, y)$, $label(S_x^n) = negative$, and $label(S_y^n) = positive$
3. If $n \neq x \wedge n \neq y$ in $rel(x, y)$, $label(S_x^n) = negative$, and $label(S_y^n) = negative$

If a noun is not involved in a hypernym relation, both the two instances will have the label *negative*. At the end of this process, two training sets are built, i.e., one for each relation argument, namely the x-set and the y-set. All the instances of both the datasets are then transformed into numeric vectors according to the Vector Space Model [24], and are finally fed into a Support Vector Machine classifier[6] [7]. We refer to the two resulting models as the x-model and the y-model. These models are binary classifiers that, given the local syntactic information of a noun, estimate if it can be respectively an x or a y in a hypernym relation.

Once the x-model and the y-model are built, we can both classify definitional sentences and extract hypernym relations. In the next section we deepen our proposed strategy in that sense.

---

[6] We used the Sequential Minimal Optimization implementation of the Weka framework [14].

Running Example. We present here a complete example of the learning phase. In detail, all the nouns contained in all the sentences of the dataset are transformed into textual instances, as shown in Table 1. The result of the sentence illustrated in the previous section is shown in Table 2.

Table 2. The instances created for the sentence of the example (one for each noun). Note that the nouns "*Albedo*" and "*extent*" are labeled as x and y respectively, as in the annotated dataset. The nouns "*object*", "*light*" and "*sun*" are negative examples for both the x-set and the y-set that will be used by the classifier for learning the models.

noun	Instance	x	y
Albedo	det-**target**-the nsubj-*noun*-**target** prepof-**target**-*noun*	+	-
extent	nsubj-**target**-*noun* cop-**target**-be det-**target**-the rcmod-**target**-reflect	-	+
object	det-**target**-a prepof-*noun*-**target**	-	-
light	dobj-reflect-**target**	-	-
sun	det-**target**-the prepfrom-reflect-**target**	-	-

The whole set of instances of all the sentences are fed into two Support Vector Machine classifiers, one for each target label (i.e., x and y).At this point, it is possible to classify each term as possible x or y by querying the respective classifiers with its local syntactic information.

4.3 Classification of Definitional Sentences

As already mentioned in previous sections, we label as *definitional* all the sentences that contain at least one noun n classified as x, and one noun m classified as y (where $n \neq m$). In this phase, it is not further treated the case of having more than one x or y in one single sentence. Thus, given an input sentence:

1. we extract all the nouns (POS-tagging),
2. we extract all the syntactic dependencies of the nouns (dependency parsing),
3. we classify each noun (i.e., its instance) with the x-model and to the y model,
4. we check if there exist at least one noun classified as x and one noun classified as y: in this case, we classify the sentences as *definitional*.

4.4 Extraction of Hypernym Relations

Our method for extracting hypernym relations makes use of both the x-model and the y-model as for the the task of classifying definitional sentences. If exactly one x and one y are identified in the same sentence, they are directly connected and the relation is extracted. The only constraint is that x and y must be connected within the same parse tree.

Now, considering our target relation $hyp(x, y)$, in case the sentence contains more than one noun that is classified as x (or y), there are two possible scenarios:

1. there are actually more than one x (or y), or
2. the classifiers returned some false positive.

Up to now, we decided to keep all the possible combinations, without further filtering operations[7]. Finally, in case of multiple classification with both x and y (i.e., if there are multiple x and multiple y at the same time, the problem becomes to select which x is linked to which y. To do this, we simply calculate the distance between these terms in the parse tree (the closer the terms, the better the connection between the two). Nevertheless, in the used corpus, only around 1.4% of the sentences are classified with multiple x and y.

5   Evaluation

In this section we present the evaluation of our approach, that we carried out on an annotated dataset of definitional sentences [22]. The corpus contains 4,619 sentences extracted from Wikipedia, where 1,908 are annotated as *definitional*. On a first instance, we test the classifiers on the extraction of hyponyms (x) and hypernyms (y) from the definitional sentences, independently. Then, we evaluate the classification of definitional sentences. Finally, we evaluate the ability of our technique when extracting whole hypernym relations. With the used dataset, the constructed training sets for the two classifiers (x-set and y-set) resulted to have approximately 1.5k features.

5.1   Dataset Problems

In this section we present some problems we encountered in the dataset provided by [22]. In these cases, however, we decided not to remove such sentences from the data, in order to be fully compliant with the results obtained by [21].

Incorrect Relationships. We found relationships that were different from the target one, i.e., IS-A, or that were incorrectly annotated in general. For example, considering the sentence:

→ A *hull* is the **body of a ship** or **boat**.

the annotation indicates two hypernyms for the term "*hull*", namely "**body of a ship**" and "**boat**". First, this can be more correctly seen as a *part-whole* relationship. Then, the second relation <*hull* IS-A **boat**> is incorrect[8].

Incorrect Hypernyms. Some sentences present incorrectly annotated hypernyms. For instance, let us consider the following sentence:

---

[7] We only used the constraint that x has to be different from y.
[8] This is due to the untreated linguistic coordination between "ship" and "boat".

→ An *actor* is defined both as the person who originates or gives existence to anything and as **one** who sets forth written statements in the Oxford English Dictionary.

where *italic* and **bold** represent the hyponym and the hypernym respectively. In this case, the IS-A relationships should have linked the term *actor* with the hypernym **person**, while the chosen hypernym **one** seems quite forced.

Missing Hypernyms. Some sentences provide partial annotations, like:

→ In Greco-Roman mythology, *Aeneas* was a Trojan **hero**, the son of prince Anchises and the goddess Aphrodite.

where only <*Aeneas* IS-A **hero**> has been annotated, while also <*Aeneas* is-a **son of Anchises**> and <*Aeneas* is-a **son of Aphrodite**> can be part of the annotation.

Fixed Hyponym. In each sentence, only one hyponym has been annotated, while the data actually contain sentences with more than one possible hyponym. For this reason, we could not correctly evaluate and compare our Precision and Recall values if we did not fix the hyponym during the automatc construction of the relation, in case of multiple choices.

Misaligned Modifiers and Matching Strategy. The evaluation has been carried out only looking for substring matching between the manual annotation and the estimation given by our system (as also done by [21]), since it seems that no guideline has been given for the annotators during the annotation phase. In fact, identical cases were annotated differently in terms of inclusion/exclusion of noun modifiers. For instance, the following two similar instances present different annotations:

→ *Argon* is a chemical **element** designated by the symbol Ar.

→ An *acid* (often represented by the generic formula HA) is traditionally considered as any **chemical compound** that, when dissolved in water, (...)

where *italic* and **bold** represent the hyponym and the hypernym respectively. In the first case only the noun has been marked as hypernym, while in the second case even the modifier has been included. Notice that, in such case, both the two modifiers present the same degree of information, so they should have been identically annotated.

Finally, since our method is able to extract single nouns that can be involved in a hypernym relation, we included modifiers preceded by preposition "*of*", while the other modifiers are removed, as shown by the extracted hypernym relation of the following sentence:

→ An *Archipelago* is a **chain of islands**.

where the whole chunk **chain of islands** has been extracted, from the single triggered noun **chain**.

5.2 Results

In this section we present the evaluation of our technique on both the tasks of classifying definitional sentences and extracting hypernym relations. Notice that our approach is susceptible from the errors given by the POS-tagger and the syntactic parser. In spite of this, our approach demonstrates how syntax can be more robust for identifying semantic relations. Our approach does not make use of the full parse tree, thus we are not dependent on a complete and correct result of the parser.

The goal of our evaluation is threefold: first, we evaluate the ability of the proposed approach to classify single hypernyms or hyponyms by means of their (Bag-Of-Words tranformed) local syntactic information; then, we evaluate the ability of classifying definitional sentences; finally, we measure the accuracy of the hypernym relation extraction.

Table 3. Accuracy levels for the classification of single hyponyms (x) and hypernyms (y) using their local syntactic context, in terms of Precision (P), Recall (R), and F-Measure (F), using 10-folds cross validation

Target	P	R	F
x	93.85%	79.04%	85.81%
y	82.26%	76.77%	79.42%

In the first phase, no x-to-y linking procedure is evaluated. Table 3 shows the results, in terms of Precision, Recall, and F-Measure. As can be noticed, the approach is able to identify correct x and y with high accuracy. Interestingly, hyponyms seem to have more stable syntactic contexts rather than hypernyms. Moreover, while Recall seems to be quite similar between the two, Precision is much higher (+11.6%) for the extraction of hyponyms.

While these results demonstrate the potential of the approach, it is interesting to analyze which syntactic information frequently reveal hyponyms and hypernyms. Table 4 shows the top 10 most important features for both the x and the y in a $hyp(x, y)$ relation, computing the value of the chi-squared statistics with respect to the class (x and y, respectively). A part from dataset-specific features like *amod-target-geologic* (marked in *italic*), many interesting considerations can be done by looking at Table 4. For example, the syntactic dependency *nsubj* results to be important for the identification of both hyponyms and hypernyms. The formers, in fact, are often syntactic subjects of a clause, and vice versa for the latters. Interestingly, *nsubj-noun-target* (marked in **bold** in Table 4) is important to both identify a correct hyponym and to reveal that a noun *is not a*

Table 4. The top 10 most relevant features for the classification of single hyponyms and hypernyms from a sentence, computing the value of the chi-squared statistic with respect to the class (x and y, respectively). The feature "nsubj-noun-target" (marked in **bold**) is important to identify a correct hyponym and to estimate that a noun *is not* a hypernym, while this seems not true for "nsubj-target-noun". Clear dataset-specific features are marked in *italic*.

Top Features for x	Top Features for y
nsubj-noun-target	cop-target-be
det-target-a	nsubj-target-noun
nsubj-refer-target	det-target-a
cop-target-be	prepin-target-noun
nsubj-target-noun	**nsubj-noun-target**
prepof-noun-target	partmod-target-use
prepof-target-noun	prepto-refer-target
nn-noun-target	prepof-target-noun
det-noun-a	det-target-any
nsubjpass-define-target	*amod-target-geologic*

Table 5. Evaluation results for the classification of definitional sentences, in terms of Precision (P), Recall (R), F-Measure (F), and Accuracy (Acc), using 10-folds cross validation

Algorithm	P	R	F	Acc
WCL-1 [21]	**99.88%**	42.09%	59.22 %	76.06 %
WCL-3 [21]	98.81%	60.74%	75.23 %	83.48 %
Star Patterns [21]	86.74%	66.14%	75.05 %	81.84 %
Bigrams [8]	66.70%	**82.70%**	73.84 %	75.80 %
Our approach	88.09%	76.01%	**81.61%**	**89.67%**

hypernym (*nsubj-noun-target* is present in both the two columns x and y), while this seems not true for *nsubj-target-noun* (it is only important to say if a noun can be a hypernym, and not to say if such noun *is not* a hyponym).

A definitional sentences is extracted only if at least one x and one y are found in the same sentence. Table 5 shows the accuracy of the approach for this task. As can be seen, our proposed approach has a high Precision, with a high Recall. Although Precision is lower than the pattern matching approach proposed by [21], our Recall is higher, leading to an higher F-Measure.

Table 6 shows the results of the extraction of the whole hypernym relations. We also added the performance of a system named "Baseline", which implements our strategy but only using the POS tags of the nouns' neighbor words instead of their syntactic dependencies. Its low effectiveness demonstrates the importance of the syntactic information, independently from the learning phase. Finally, note that our approach reached high levels of accuracy. In particular, our system outperforms the pattern matching algorithm proposed by [21] in terms of both Precision and Recall.

Table 6. Evaluation results for the hypernym relation extraction, in terms of Precision (P), Recall (R), and F-Measure (F). These results are obtained using 10-folds cross validation (* Recall has been inherited from the definition classification task).

Algorithm	P	R	F
WCL-1 [21]	77.00%	42.09% *	54.42%
WCL-3 [21]	78.58%	60.74% *	68.56%
Baseline	57.66%	21.09%	30.76%
Our approach	**83.05%**	**68.64%**	**75.16%**

5.3   Further Considerations

The data provided by [22] also contain a dataset of over 300,000 sentences retrieved from the UkWac Corpus [11]. Unfortunately, Precision was only manually validated, therefore we could not be able to make any fair comparison. Nevertheless, they made available a subset of 99 definitional sentences. On such data, our technique obtained a Recall of 59.6% (59 out of 99), while their approaches reached 39.4%, 56.6%, and 63.6% respectively for WCL-1, WCL-3, and Star Patterns.

In the dataset, the syntactic parser found hundreds of cases of coordinated hyponyms, while the annotation provides only one hyponym for each sentence. For this reason, we were not able to evaluate our method on the extraction of all possible relations with all coordinated hyponyms.

The *really-desired* result of the task of extracting hypernym relations from text (as for any semantic relationships in general) depends on the domain and the specific later application. Thus, we think that a precise evaluation and comparison of any systems strictly depends on these factors. For instance, given a sentence like:

> → In mathematics, computing, linguistics and related disciplines, an *algorithm* is a sequence of instructions.

one could want to extract only "**instructions**" as hypernym (as done in the annotation), rather than the entire chunk "**sequence of instructions**" (as extracted by our technique). Both results can be valid, and a further discrimination can only be done if a specific application or use of this knowledge is taken into consideration.

In this work, we only suggest how syntax can be more robust for identifying semantic relations, avoiding general discussions on the growth of web data and the complexity / noise of the contents deriving from personal blogs and social network communities. Nevertheless, we are not dependent on a complete and correct result of the parser. For example, we could apply our methodology to the result of simple chunk parsers. Still, to the best of our knowledge, no other work considers noisy data on this specific task, and we based our idea thinking on encyclopedic and formal texts, where syntax is less subjected to language inflections and the need to support semantic resources construction is even more tangible.

6 Conclusion and Future Work

We presented an approach to reveal definitions and extract underlying hypernym relations from plain text, making use of local syntactic information fed into Support Vector Machine classifiers. The aim of this work was to revisit these tasks as classical supervised learning problems that usually carry to high accuracy levels with high performance when faced with standard Machine Learning techniques. Our approach demonstrates that relaxing the problem into two different subtasks can actually improve the identification of hypernym relations. Nevertheless, this could not be true for any possible semantic relations, since semantics is independent from syntax to a certain extent. The results of the presented approach highlight its validity by significantly improving current state-of-the-art techniques in the classification of definitional sentences as well as in the extraction of hypernym relations from text. In future works, we aim at using larger syntactic contexts as well as additional semantic information. Despite the successful results, we plan to also examine the context between x and y in order to further strengthen the technique. Then, the problem of finding meaningful noun modifiers as part of the entities involved in hypernym relations needs to be studied carefully, starting from a task-specific annotated corpus. Finally, we aim at evaluating our approach on the construction of entire taxonomies relying on domain-specific text corpora, as in [20,27].

References

1. Berland, M., Charniak, E.: Finding parts in very large corpora. In: Annual Meeting Association for Computational Linguistics, vol. 37, pp. 57–64. Association for Computational Linguistics (1999)
2. Biemann, C.: Ontology learning from text: A survey of methods. LDV Forum 20, 75–93 (2005)
3. Borg, C., Rosner, M., Pace, G.: Evolutionary algorithms for definition extraction. In: Proceedings of the 1st Workshop on Definition Extraction, pp. 26–32. Association for Computational Linguistics (2009)
4. Buitelaar, P., Cimiano, P., Magnini, B.: Ontology learning from text: An overview. Ontology Learning from Text: Methods, Evaluation and Applications 123, 3–12 (2005)
5. Candan, K., Di Caro, L., Sapino, M.: Creating tag hierarchies for effective navigation in social media. In: Proceedings of the 2008 ACM Workshop on Search in Social Media, pp. 75–82. ACM (2008)
6. Cataldi, M., Schifanella, C., Candan, K.S., Sapino, M.L., Di Caro, L.: Cosena: a context-based search and navigation system. In: Proceedings of the International Conference on Management of Emergent Digital EcoSystems, p. 33. ACM (2009)
7. Cortes, C., Vapnik, V.: Support-vector networks. Machine Learning 20(3), 273–297 (1995)
8. Cui, H., Kan, M.Y., Chua, T.S.: Soft pattern matching models for definitional question answering. ACM Trans. Inf. Syst. 25(2) (April 2007),
http://doi.acm.org/10.1145/1229179.1229182

9. Del Gaudio, R., Branco, A.: Automatic extraction of definitions in portuguese: A rule-based approach. In: Neves, J., Santos, M.F., Machado, J.M. (eds.) EPIA 2007. LNCS (LNAI), vol. 4874, pp. 659–670. Springer, Heidelberg (2007)

10. Fahmi, I., Bouma, G.: Learning to identify definitions using syntactic features. In: Proceedings of the EACL 2006 Workshop on Learning Structured Information in Natural Language Applications, pp. 64–71 (2006)

11. Ferraresi, A., Zanchetta, E., Baroni, M., Bernardini, S.: Introducing and evaluating ukwac, a very large web-derived corpus of english. In: Proceedings of the 4th Web as Corpus Workshop (WAC-4) Can We Beat Google, pp. 47–54 (2008)

12. Fortuna, B., Mladenič, D., Grobelnik, M.: Semi-automatic construction of topic ontologies. In: Ackermann, M., et al. (eds.) EWMF 2005 and KDO 2005. LNCS (LNAI), vol. 4289, pp. 121–131. Springer, Heidelberg (2006)

13. Gangemi, A., Navigli, R., Velardi, P.: The ontowordnet project: Extension and axiomatization of conceptual relations in wordnet. In: Meersman, R., Schmidt, D.C. (eds.) CoopIS 2003, DOA 2003, and ODBASE 2003. LNCS, vol. 2888, pp. 820–838. Springer, Heidelberg (2003),
http://dx.doi.org/10.1007/978-3-540-39964-3_52

14. Hall, M., Frank, E., Holmes, G., Pfahringer, B., Reutemann, P., Witten, I.H.: The weka data mining software: an update. ACM SIGKDD Explorations Newsletter 11(1), 10–18 (2009)

15. Harris, Z.: Distributional structure. Word 10(23), 146–162 (1954)

16. Hearst, M.: Automatic acquisition of hyponyms from large text corpora. In: Proceedings of the 14th Conference on Computational Linguistics, vol. 2, pp. 539–545. Association for Computational Linguistics (1992)

17. Hovy, E., Philpot, A., Klavans, J., Germann, U., Davis, P., Popper, S.: Extending metadata definitions by automatically extracting and organizing glossary definitions. In: Proceedings of the 2003 Annual National Conference on Digital Government Research, pp. 1–6. Digital Government Society of North America (2003)

18. Klavans, J., Muresan, S.: Evaluation of the definder system for fully automatic glossary construction. In: Proceedings of the AMIA Symposium, p. 324. American Medical Informatics Association (2001)

19. Navigli, R.: Using cycles and quasi-cycles to disambiguate dictionary glosses. In: Proceedings of the 12th Conference of the European Chapter of the Association for Computational Linguistics, pp. 594–602. Association for Computational Linguistics (2009)

20. Navigli, R., Velardi, P., Faralli, S.: A graph-based algorithm for inducing lexical taxonomies from scratch. In: Proceedings of the Twenty-Second International Joint Conference on Artificial Intelligence, vol. 3, pp. 1872–1877. AAAI Press (2011)

21. Navigli, R., Velardi, P.: Learning word-class lattices for definition and hypernym extraction. In: Proceedings of the 48th Annual Meeting of the Association for Computational Linguistics, pp. 1318–1327. Association for Computational Linguistics, Uppsala (2010), http://www.aclweb.org/anthology/P10-1134

22. Navigli, R., Velardi, P., Ruiz-Martnez, J.M.: An annotated dataset for extracting definitions and hypernyms from the web. In: Proceedings of the Seventh International Conference on Language Resources and Evaluation (LREC 2010). European Language Resources Association (ELRA), Valletta (2010)

23. Ponzetto, S., Strube, M.: Deriving a large scale taxonomy from wikipedia. In: Proceedings of the National Conference on Artificial Intelligence, vol. 22, p. 1440. AAAI Press, MIT Press, Menlo Park, Cambridge (2007)

24. Salton, G., Wong, A., Yang, C.S.: A vector space model for automatic indexing. Commun. ACM 18(11), 613–620 (1975),
http://doi.acm.org/10.1145/361219.361220
25. Snow, R., Jurafsky, D., Ng, A.: Learning syntactic patterns for automatic hypernym discovery. In: Advances in Neural Information Processing Systems 17 (2004)
26. Storrer, A., Wellinghoff, S.: Automated detection and annotation of term definitions in german text corpora. In: Proceedings of LREC, vol. 2006 (2006)
27. Velardi, P., Faralli, S., Navigli, R.: Ontolearn reloaded: A graph-based algorithm for taxonomy induction. Computational Linguistics, 1–72 (2012)
28. Westerhout, E.: Definition extraction using linguistic and structural features. In: Proceedings of the 1st Workshop on Definition Extraction, WDE 2009, pp. 61–67. Association for Computational Linguistics, Stroudsburg (2009),
http://dl.acm.org/citation.cfm?id=1859765.1859775
29. Yamada, I., Torisawa, K., Kazama, J., Kuroda, K., Murata, M., De Saeger, S., Bond, F., Sumida, A.: Hypernym discovery based on distributional similarity and hierarchical structures. In: Proceedings of the 2009 Conference on Empirical Methods in Natural Language Processing, vol. 2, pp. 929–937. Association for Computational Linguistics (2009)
30. Yang, H., Callan, J.: Ontology generation for large email collections. In: Proceedings of the 2008 International Conference on Digital Government Research, pp. 254–261. Digital Government Society of North America (2008)
31. Zhang, C., Jiang, P.: Automatic extraction of definitions. In: 2nd IEEE International Conference on Computer Science and Information Technology, ICCSIT 2009, pp. 364–368 (August 2009)

Error Prediction with Partial Feedback

William Darling, Cédric Archambeau,
Shachar Mirkin, and Guillaume Bouchard

Xerox Research Centre Europe
Meylan, France
`firstname.lastname@xrce.xerox.com`

Abstract. In this paper, we propose a probabilistic framework for predicting the root causes of errors in data processing pipelines made up of several components when we only have access to partial feedback; that is, we are aware when *some* error has occurred in one or more of the components, but we do not know which one. The proposed error model enables us to direct the user feedback to the correct components in the pipeline to either automatically correct errors as they occur, retrain the component with assimilated training examples, or take other corrective action. We present the model and describe an Expectation Maximization (EM)-based algorithm to learn the model parameters and predict the error configuration. We demonstrate the accuracy and usefulness of our method first on synthetic data, and then on two distinct tasks: error correction in a 2-component opinion summarization system, and phrase error detection in statistical machine translation.

Keywords: error modeling, user feedback, binary classification, EM.

1 Introduction and Motivation

In this work we are interested in predicting the root cause of errors for data that have been processed through a pipeline of components when we only have access to partial feedback. That is, an input X goes through a series of components that ultimately results in an output Y. Each component in the processing pipeline performs some action on X, and each of the components might result in an error. However, the user often only has access to the final output, and so it is unclear which of the components was at fault when an error is observed in the final output. In cases where the user is aware of the intermediate results, it is also typically more complex to have to specify the exact component that was at fault when providing error feedback. Therefore, given only the fact that an error has occurred or not, we would like to predict the root causes of the error.

Pipeline processing, or Pipes and Filters, has been a stalwart of computing since the concept was invented and integrated into the UNIX operating system in the 1970's [1]. The simple idea is that complex processing and powerful results can be achieved by running an input through a series of more basic components to in turn produce a more intricate output than would have been possible with a single method. This approach has seen increasing use in recent years as the

H. Blockeel et al. (Eds.): ECML PKDD 2013, Part II, LNAI 8189, pp. 80–94, 2013.

Fig. 1. Typical data processing pipeline with two components resulting in a marked-up output. The components are often black boxes that the user is unaware of which can render providing user feedback complicated.

outputs desired by users have become more complex. It is seen especially in Natural Language Processing (NLP) applications such as named entity recognition [2], text summarization [3], and recognizing textual entailment [4].

For example, the majority of comment or opinion summarization systems described in the literature make use of a collection of diverse techniques in a pipeline-like architecture [5,6]. A first component might filter out spam comments and then a second could categorize the comments into aspects. This fits a typical data processing pipeline consisting of two components as is shown in Figure 1. It is also a common technique for applications such as identifying *evaluative* sentences [7] and MacCartney et al.'s three stage approach to textual inference: linguistic analysis (which consists of a pipeline itself), followed by graph alignment, ending with determining an entailment [8]. More generally, GATE provides a software architecture for building NLP pipelines [9]. Recent work has also shown that running two binary classifiers in a series can result in improved results over a more complex multi-class classification approach providing a further reason to consider problems associated with errors in processing pipelines [10].

We begin with a simple motivating toy example. Figure 2 visualizes a set of input data $\mathcal{X} = (x_1, x_2) \in \mathbb{R}^2$ (left) running through two affine transformation components in a pipeline. The first component translates the data by $\mathcal{T}_1 = x_1 + 6$, and the second component translates the output of the first component by $\mathcal{T}_2 = x_2 - 6$. However, each of the components have some region where they commit errors and an error causes the translation to be distorted by scaling it by uniform noise. Since the user only observes the final output (right), there are two classes of partial feedback: $f_i = 0$ means that for point $\mathcal{X}_i$, the data transformations were successful and no errors occurred; $f_i = 1$, on the other hand, means that there was some error, though it is not clear whether it was the result of the first component committing an error, the second component committing an error, or both.

The user is generally able to identify an overall error much more efficiently than having to specify its source. With K components, full feedback would require selecting from $2^K - 1$ distinct configurations of error. In many cases, it will be impossible to identify the source of error. In a Named Entity Recognition pipeline that consists first of a part-of-speech (POS) tagging component, followed by a chunker, followed by a named entity classifier, an incorrectly identified named entity can easily be spotted, but it may very well be impossible to identify if it was the POS tagger, the chunker, or the classifier that was the root

Fig. 2. Input data $\mathcal{X} = (x_1, x_2) \in \mathbb{R}^2$ (left) is first translated by $\mathcal{T}_1$ (centre) and then $\mathcal{T}_2$ (right). When either translation component commits an error, the translation fails and instead results in a translation distorted by uniform random error.

cause of the error. In Figure 3, the incorrectly translated output datapoints have been identified as red circles and the error-free datapoints as blue squares (right). We also show the same colour-coding on the input and intermediate results (left) and (centre). In these latter plots, we have also plotted the component-specific linear error predictor. Knowing this relationship, we can then predict the prior probability of a component committing an error given some input, and the posterior probability of error configuration given that an error has been observed. One could then take corrective measure by directing training data to the component at fault, or automatically attempting to rectify the error through a component wrapper.

In this paper, we propose a probabilistic framework that aims to uncover the predictors of error for each of the arbitrary number of components in a data processing pipeline, and predict the configuration of error for each data observation. After discussing some related work, we present our probabilistic model that is based on binary classification of error through logistic regression. We then present an Expectation Maximization (EM)-based algorithm to learn component-specific error model parameters and to estimate the configuration of error. We demonstrate the accuracy of our approach first on synthetic data, and then on two real-world tasks: a two-component opinion summarization pipeline and a phrase error prediction task for post-editing in machine translation. We conclude with some discussion and thoughts on future work.

2   Related Work

While our probabilistic error model has some connections to sigmoid belief networks (SBN) [11], the most closely related work with respect to improving performance in pipelines when errors occur comes from the NLP domain. In [12], Marciniak and Strube explain how NLP problems can generally be cast as a set of several classification tasks, some of which are mutually related. A discrete optimization model is presented that is shown to deliver higher accuracy in a language generation task than the equivalent task implemented as solving

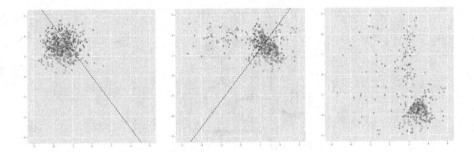

Fig. 3. Colour-coded data $\mathcal{X}$ where blue represents the feedback $f = 0$ (no error) and red represents the partial feedback $f = 1$ (error in one or more of the components). The linear predictors of error for $\mathcal{T}_1$ (left) and $\mathcal{T}_2$ (centre) are also plotted.

classification tasks sequentially. However, they do not address either a general approach to improve the accuracy of each of the classifiers, nor do they consider how user feedback might be taken into consideration.

Finkel et al. show that modeling a pipeline as a Bayesian network where each component is seen as a random variable and then performing approximate inference to determine the best output can outperform a greedy pipeline architecture where a best decision is made at each node [4]. While a general method to solve any kind of multi-stage algorithm is proposed, a principal requirement is that each component must be able to generate samples from a posterior. The authors note:

> If ... all NLP researchers wrote packages which can generate samples from the posterior, then the entire NLP community could use this method as easily as they can use the greedy methods that are common today... [4]

Our proposed method is also a Bayesian network [13]. However, its aim is to predict the root causes of errors in a pipeline and it requires no changes to any of the underlying methods. Depending on the composition of the underlying components, knowing the cause of an output error could allow us to dynamically correct it by asking for a training label in an active learning setting, or flipping the erroneous prediction if the component consists of a binary classifier.

3   Probabilistic Model

For each component n in a pipeline processing system, we model the probability that it will commit an error e_n as a Bernoulli random variable modeled using binary logistic regression: $p(e_n = 1 | x, \beta) = \sigma(\phi_n(x)^\top \beta)$ where $\sigma(\cdot)$ is the logistic function and $\phi_n(\cdot)$ is a function that extracts the features required for component n. In this setting we address the case where the system only has access to *partial* feedback; that is, the only error observation, f, is with respect to the aggregate error. In this case, a user provides feedback only pertaining to whether *some*

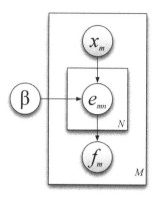

Fig. 4. Graphical model of error prediction framework. There are M observations and N components in the pipeline; β is a vector parameter of the component error models.

error occurred at an indeterminate set of components ($f = 1$), or that the output contains no errors at all ($f = 0$).

We let $e = (e_1, \ldots, e_N)$ be the collection of error random variables for each component, such that

$$p(f, e|x, \beta) = p(e|x, \beta)p(f|e), \tag{1}$$

where the first term $p(e|x, \beta)$ contains the probability of a given error configuration e, and the second term $p(f|e)$ encodes how the user feedback f relates to the error configuration. In the general case of the former, we have

$$p(e|x, \beta) = \prod_{i=1}^{N} p(e_i|x, \beta) \tag{2}$$

For the latter term in the standard case where 1 or more errors committed in the components leads to an observed final error $f = 1$ we have $p(f = 1|e) = \delta(\sum_i e_i)$ and $p(f = 0|e) = 1 - p(f = 1|e) = 1 - \delta(\sum_i e_i)$ where $\delta(z) = 1$ if $z > 0$ and $\delta(z) = 0$ otherwise. Note that this term could be modeled more intricately by allowing a user to specify a *degree* of error or by leading the model in the general direction of error(s) without having to explicitly report them.

All errors are assumed to be conditionally independent given the features x: $p(e|x, \beta) = p(e_1, \ldots, e_N|x, \beta) = \prod_{i=1}^{N} p(e_i|x, \beta)$, and the posterior probabilities of error are then given by

$$p(e|f, x, \beta) = \begin{cases} \delta_{\{e_1=0,\ldots,e_n=0\}} & \text{if } f = 0, \\ \dfrac{\prod_{i=1}^{N} \sigma((2e_i-1)\phi_i(x)^{\top}\beta)}{1 - \prod_{i=1}^{N} \sigma(-\phi_i(x)^{\top}\beta)}. & \text{if } f = 1. \end{cases} \tag{3}$$

A graphical model depiction of the error model framework is shown in Figure 4.

4   Parameter Estimation

We can learn the component-specific error model parameters β by maximizing the likelihood which is obtained by integrating out the latent error variables e_i. The likelihood and its derivative can be computed in closed form and the parameters then optimized using gradient descent [14]. However, as the number of components grows, the terms in the gradient and the likelihood grow unwieldy. For simplicity, we therefore decompose the error estimation and parameter learning by turning to a stochastic EM-based approach [15].

Where there are M observations and N components, the log likelihood is:

$$\ln \mathcal{L} = \ell(\beta) = \sum_{m=1}^{M} \ln \sum_{e_1} \cdots \sum_{e_N} p(f_m, e_1, \ldots, e_N | x_m, \beta)$$

$$= \sum_{m=1}^{M} \ln \sum_{e} p(f_m, e | x_m, \beta) \qquad (4)$$

which includes the log of a sum. By the Jensen inequality, however,

$$\ell(\beta) = \sum_{m=1}^{M} \ln \sum_{e \in e} p(f_m, e | x_m, \beta)$$

$$\geq \sum_{m=1}^{M} \sum_{e \in e} w_{m,e} \ln p(f_m, e | x_m, \beta) + \mathcal{H}(\mathbf{w_m})$$

$$= g(\mathbf{w}, \beta) \qquad (5)$$

where $\mathbf{w_m}$ contains a non-negative weight for each configuration of error (size $2^N - 1$), $\sum_{e \in \{e \backslash e_{0\ldots 0}\}} w_{m,e} = 1$ and $\forall e, w_{m,e} \geq 0$. $g(\mathbf{w}, \beta)$ is then a lower bound for the log likelihood.

Because $g(\mathbf{w}, \beta)$ is a lower bound for the log likelihood, maximizing $g(\mathbf{w}, \beta)$ will also maximize $\ell(\beta)$. However, we now have the latent parameters $\mathbf{w}$ so we iteratively maximize $\mathbf{w}$ (E-step) and β (M-step).

E-step. Where $e \in e$ is one of the $2^N - 1$ permutations of $e_1 e_2 \ldots e_N$ where there is at least one error we have, for each observation m:

$$w_{m,e} = \frac{p(f_m, e = e | x_m, \beta)}{\sum_{e' \in e} p(f_m, e = e' | x_m, \beta)} \qquad (6)$$

Therefore, for the example where there are $N = 3$ components in an observation, there will be $2^3 - 1 = 7$ w's for each configuration of error (e_1, e_2, e_3): w_{001}, w_{010}, w_{100}, w_{110}, w_{101}, w_{011}, and w_{111}. Each w is a weight in the sense that it represents the probability of the given configuration (for observations where there is no error, $f = 0$, the weight $w_{0\ldots 0} = 1$). This exponential explosion of error combinations can be managed for medium numbers of components, which is reasonable for many applications. For large numbers of components, an approximate E-step could be derived using a variational EM algorithm.

M-step. The M-step is a weighted maximum likelihood of the following:

$$g(\mathbf{w}, \beta) = \sum_{m=1}^{M} \sum_{e \in e} w_{m,e} \ln p(f_m, e | x_m, \beta)$$

$$= \sum_{m=1}^{M} \sum_{e \in e} w_{m,e} \sum_{i=1}^{N} [e_i \ln \sigma(\phi_i(x_m)^\top \beta) + (1 - e_i) \ln(1 - \sigma(\phi_i(x_m)^\top \beta))]$$

$$(7)$$

where each e_i takes on its value assigned by the permutation indexed by e. For example, if $N = 2$, then $e = (e_1, e_2) = \{1 : (0, 1), 2 : (1, 0), 3 : (1, 1), 4 : (0, 0)\}$. Therefore, each observation m with $f_m = 1$ requires 3 w calculations, and contributes 3 weighted samples to the maximum likelihood. Things are further complicated by the fact that β will generally be different for each component. We get around this issue by having each feature vector $\phi_i(x)$ be of size $D \times N$ where there are D features and place zeros in the components that align with β values not considered by this component. A dot product between a sparse feature vector and the parameters that pertain to the given component can be efficiently computed. For the M-step we run a small number of iterations of SGD or batch gradient descent (depending on the application) at each step.

It is well known that EM algorithms are often highly sensitive to how the parameters are initialized [16]. Our algorithm is no different and we empirically observed falling into local minima for certain initializations. We overcome this problem by initializing the model parameters to those obtained by running an independent logistic regression with the observed labels being the overall feedback for the entire pipeline. In other words, for observation $\mathcal{X}$ with 2 components, we learn β_i for component i with features $\phi_i(\mathcal{X})$ and label f, even though $f = 1$ is partial as it could imply any of the following configurations: $(e_1 = 1, e_2 = 0), (e_1 = 0, e_2 = 1), (e_1 = 1, e_2 = 1)$. This initialization seems to discourage local minima that could trap the algorithm with a random initialization.

5   Experiments

We demonstrate the viability of our method on three separate tasks. First, we show that our model and inference algorithm are sound by learning the error configuration and model parameters on synthetic data for different lengths of pipelines and numbers of feedback observations. We then show results on improving a 2-stage opinion summarization system by learning the probability of two static components committing an error given partial feedback. Finally, we describe the results of a semi-synthetic experiment on phrase error prediction for post-editing in machine translation where we predict the phrases most likely to contain translation errors given that we know there is some error in the translation.

5.1 Synthetic Data

To demonstrate how our model is able to learn the probability of a component committing an error with access only to partial feedback, we revisit the motivating $N = 2$ component example that we presented in the introduction. Here, we draw up to $M = 500$ datapoints with $d = 2$ features from a multivariate normal distribution with $\mu = (-3, 3)$ and $\Sigma = \mathbb{I}_2$. We randomly select a true β parameter for each component which corresponds to what we hope to learn. An error matrix $E \in \{0, 1\}^{M \times N}$ is generated where $e_{m,n} = 1$ implies that $\phi_n(x_m)$ would result in an error for component n. Each element $e_{m,n}$ is computed by drawing a random Bernoulli variable with parameter modeled by a binary logistic regression resulting in data with some added noise. This tends to result in a dataset that is roughly balanced between $f_m = 0$ and $f_m = 1$ observations. The translation $\mathcal{T}_n$ is applied for point x_m if $e_{m,n} = \text{Bernoulli}(\sigma(\phi_n(x_m)^\top \beta)) = 0$, otherwise the translation is scaled by some noise and the data gets translated randomly. The observations are then $(x_m, f_m)_{m=1}^M$ where $f_m = 1$ if any of $e_{m,n} = 1$ and $f_m = 0$ otherwise. Before proceeding all $e_{m,n}$ are removed and the algorithm learns β (and eventually $e_{m,n}$) given only $x_{m,n}$ and f_m. This is synthetic data experiment 1.

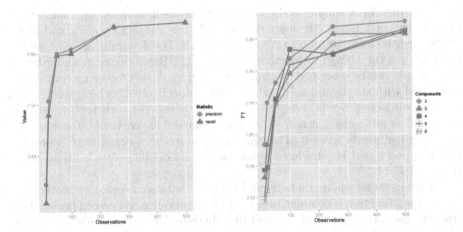

Fig. 5. Precision and Recall values for synthetic data experiment 1 (left) and 2 (right). Both experiments demonstrate that the precision and recall increase with the number of observations, as expected, but also that we approach perfect prediction with only very few labels.

We learn the parameters with varying number of error observations and then test the precision and recall of predicted prior probability of error on a separate test set of 500 observations drawn from the same distribution. For each number of observations (10 to 500), we run 5 trials and report the average precision and recall. Figure 5 (left) shows that we do very well even from very few observations and we predict essentially perfectly from 250 observations on.

Next, we examine the precision and recall statistics for another synthetic experiment that considers how our model performs as the number of components varies. This is synthetic data experiment 2. Here, to simplify things, independent features are drawn for each component from a standard multivariate normal distribution. We randomly select error parameters and generate ground truth labels. Our algorithm then observes the features and only the partial feedback f for each observation. We are then interested in how many observations are required for different lengths of processing pipelines. We consider between $N = 2$ and $N = 6$ components. For testing, we again draw data from the same distribution but with $M = 100$ observations; this will amount to MN values of e_i to be predicted. We show F1-scores for different lengths of pipelines as the number of observations grows in Figure 5 (right). This shows that even with up to 6 components, we can learn the error model parameters very well with few observation examples. Also, the number of required observations for good predictive performance does not seem to heavily depend on the number of components at least for medium numbers of components as we tested.

5.2   Opinion Summarization

Next, we present a simple 2-component deterministic opinion summarization system that first filters out comments that do not contain opinion, and then labels the comments with up to K category labels. For determining opinionated texts, we use the MPQA Subjectivity Lexicon [17]. Here, among other designations, words can be described as *strong_subj* and *weak_subj* for being commonly associated with strong and weak subjectivity, respectively. Our intuition is that strongly subjective words result in opinionated texts. For each text, if a word is marked as *strong_subj* it scores 1.0, if it is marked as *weak_subj* it scores 0.5, and all other words score 0. The opinion score is the average word score for the text, and a text is considered opinionated if its opinion score is above some threshold Γ_O.

For determining whether a text can be labeled with some category marker c_k, we use a method that is common in text summarization: average word probability [18]. We use LDA [19] to learn word distributions for each category and then consider a text's average word probability under a word distribution for each category. Again, we consider a text to be a positive example for a category if its average word probability for that category is above some threshold Γ_C. The underlying methods are relatively basic but our aim is to demonstrate how we can predict when each of the components has commit an error given that the final observation resulted in an error. Because each component is made up of binary classifiers, we can improve the system in the light of user feedback without modifying the underlying components. We wrap each of the components in an error model *wrapper* such that when the error model predicts that the current input would result in an error, we flip the prediction.

Our data to summarize consists of a subset of public comments on the US Department of Agriculture's (USDA) proposed National Organic Program (NOP)

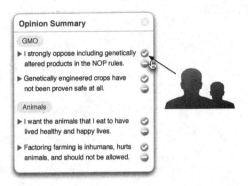

Fig. 6. The base components produce a summary and the user gives partial feedback by stating whether a given sentence either contains opinion and is in the correct category ($f = 0$), or that one or both of these is incorrect ($f = 1$)

("USDA-TMD-94-00-2").[1] These are comments by concerned citizens with respect to a proposed rule on what standards would apply to allow products to be designated as *organic*. This data fits our problem nicely because a sizable portion of the data consists of no opinion, and most of the texts can be sensibly placed into different categories given what aspect of the proposed legislation a citizen was referring to (animal well-being, genetically modified organisms, human health, etc.). 650 texts were manually labeled as either containing opinion or not, and for membership in up to 6 categories. We randomly select 100, 300, and 500 texts for training and leave the rest aside for testing. In this experiment, the feedback is whether a comment is correctly identified as containing opinion and labeled with the correct category ($f = 0$), or some labeling error exists. Figure 6 visualizes the setting, and the features are simple bag-of-words. We are interested in the accuracy of all predicted labels for the "wrapped" system. That is, we use the base system described above, run the testing data through the pipeline, and at each component if our error model predicts an error, we flip that prediction. We run each experiment 5 times with different random permutations of training and testing data and report the average accuracy. Figure 7 shows the opinion component accuracy (left) and the aspect component accuracy (right) as the number of feedback examples varies.

With 100 and more partial feedback examples, the error model-wrapped opinion component does better than the base component. For the aspect labeling component, 100 examples was not enough to provide adequate predictive accuracy to do better than the base component. However, with 300 labels and more, it handily beats the base component. The reason for this discrepancy is data sparsity; each feedback example is only with respect to one aspect label and with 6 labels, a training set perfectly balanced amongst the 4 different error combinations would only include $100 \times \frac{1}{6} \times \frac{1}{4} \approx 4$ training examples per

---

[1] http://erulemaking.cs.cmu.edu/Data/USDA

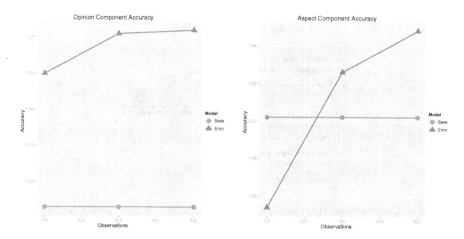

Fig. 7. Accuracy results for the opinion (left) and aspect (right) components in our augmented summarization system. The "base" curve shows the accuracy of the base components and the "error" curve shows the accuracy when our error model is applied.

context. Of course, in practice we never achieve this perfect balance and certain contexts will be over-represented while others will have no training examples at all. Nevertheless, even with a relatively small amount of feedback, we can see that the system is able to predict the error configuration and therefore improve the accuracy of the overall system. In practice, when the error model is used as a wrapper in such an experiment, it would only be activated once an appropriate amount of training data was obtained.

5.3   Error Detection in Machine Translation

For our final experiment, we describe a semi-synthetic experiment in that the features are true data, but the labels are partially generated. Machine Translation (MT) quality has yet to reach a state where translations can be used reliably without user supervision. Therefore, when a high quality translation is required, a *post-editing* stage is typically conducted. In post-editing, professional translators review and correct a translation before it is used. Error detection is therefore an important subject in machine translation [20,21]. It is a useful means of reducing post-editing effort by directing the translator to specific segments in the translation which are estimated to be erroneous. This could also be used within the MT system itself, by avoiding erroneous translations and reverting to the next best alternatives proposed by the system in the light of a predicted error.

Here, we use our error model framework to predict the phrases in a translated sentence that are most likely to contain errors. Connecting the setting to the previous examples, each phrase is a component, and each sentence is a pipeline. Feedback consists of either a perfectly translated sentence ($f = 0$) or a sentence that contains at least one error ($f = 1$). There are simply 4 features for this

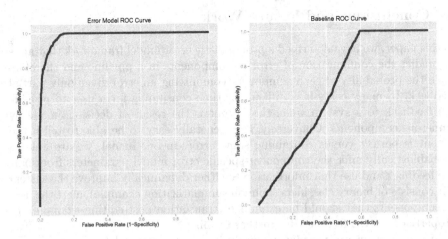

Fig. 8. ROC Curves for predicting prior probability of an incorrect phrase translation learned with the partial feedback error model (left) vs. the baseline which assigns an error to all phrases in a sentence when $f = 1$ (right)

experiment: the probability of the source phrase given the target phrase; the lexical weighting of the source phrase given the target phrase; the probability of the target phrase given the source phrase; and the lexical weighting of the target phrase given the source phrase. Each of these features is computed automatically using the Moses phrase-based SMT system [22].

Because we need phrase-specific error labels for testing, we take a synthetic approach to labeling. We manually labeled ~ 400 translated phrases as either containing or not containing an error and then learned an independent binary classifier on this fully-labeled data. Using this classifier, we then generated labels for a set of 5000 sentences that are segmented into phrases. We then took all of the sentences that contained 6 phrases or less to end up with 1002 training sentences. Each of these sentences receives a label $f = 1$ if any of its phrases contain errors, and $f = 0$ otherwise. We learn the error model and then predict the prior probability of each phrase-pair containing an error.

We compare our model to a simple baseline. The baseline learns a binary logistic regression classifier on phrases where the labels are simply the partial feedback f. That is, when $f = 0$, each phrase is an independent example with the (correct) label 0. When $f = 1$, each phrase is also an independent example but now the label will only sometimes be correct. In fact, it will rarely be correct because most translated sentence errors are confined to 1 or 2 phrases. The behavior of the baseline is best understood by showing its ROC curve. The ROC curves for each method are shown in Figure 8 and demonstrate that there is little problem with choosing a discrimination threshold in our method (left) and that the baseline (right) is a poor method especially when the density of errors in a pipeline is low.

6   Conclusions and Future Work

In this paper, we have described a probabilistic error model framework that aims to predict the configuration of error in components in a pipeline and therefore learn the probability of each component committing an error given only partial feedback. In many cases it is difficult or time consuming for a user to provide full feedback to a system when the output is the result of decisions made by numerous components. Conversely, it is generally easy to be able to tell if the output is perfect versus containing some error. In our model, we are able to probabilistically infer the component-specific error model parameters from partial feedback and use that information to either dynamically improve the system if it consists of binary classifiers (opinion summarization example), alert the user to components that should be examined (translation post-editing example), or properly direct the feedback for further training.

In the near future, we plan to apply it to other natural language processing tasks, such as named entity recognition, which could benefit a great deal from human feedback, and which would be very difficult for a human to pinpoint the exact cause of error. We also plan to combine it with active learning techniques to choose which examples to show to the user. At the moment, this is done randomly, irrespective of the quality of the individual components and the predictions we make. This strategy is clearly suboptimal. In principle, we could reduce the amount of feedback to be provided by the user for a same level of accuracy if we choose the examples to provide feedback on in a sensible way. Finally, we are also interested in exploring non-linear versions of the error model.

Acknowledgments. The research leading to these results has received funding from the European Commission Seventh Framework Programme (FP/2007-2013) through the projects Fupol and Fusepool.

References

1. Ritchie, D.M.: The evolution of the unix time-sharing system. Communications of the ACM 17, 365–375 (1984)
2. Ritter, A., Clark, S., Mausam, Etzioni, O.: Named entity recognition in tweets: An experimental study. In: Proceedings of the 2011 Conference on Empirical Methods in Natural Language Processing, pp. 1524–1534. Association for Computational Linguistics, Edinburgh (2011)
3. Ly, D.K., Sugiyama, K., Lin, Z., Kan, M.Y.: Product review summarization from a deeper perspective. In: Proceedings of the 11th Annual International ACM/IEEE Joint Conference on Digital Libraries, JCDL 2011, pp. 311–314. ACM, New York (2011)
4. Finkel, J.R., Manning, C.D., Ng, A.Y.: Solving the problem of cascading errors: approximate bayesian inference for linguistic annotation pipelines. In: Proceedings of the 2006 Conference on Empirical Methods in Natural Language Processing, EMNLP 2006, pp. 618–626. Association for Computational Linguistics, Stroudsburg (2006)

5. Blair-goldensohn, S., Neylon, T., Hannan, K., Reis, G.A., Mcdonald, R., Reynar, J.: Building a sentiment summarizer for local service reviews. In: NLP in the Information Explosion Era (2008)
6. Lu, Y., Zhai, C., Sundaresan, N.: Rated aspect summarization of short comments. In: Proceedings of the 18th International Conference on World Wide Web, WWW 2009, pp. 131–140. ACM, New York (2009)
7. Zhai, Z., Liu, B., Zhang, L., Xu, H., Jia, P.: Identifying evaluative sentences in online discussions. In: AAAI (2011)
8. MacCartney, B., Grenager, T., de Marneffe, M.C., Cer, D., Manning, C.D.: Learning to recognize features of valid textual entailments. In: Proceedings of the Main Conference on Human Language Technology Conference of the North American Chapter of the Association of Computational Linguistics, HLT-NAACL 2006, pp. 41–48. Association for Computational Linguistics, Stroudsburg (2006)
9. Cunningham, H., Humphreys, K., Gaizauskas, R., Wilks, Y.: Software infrastructure for natural language processing. In: Proceedings of the Fifth Conference on Applied Natural Language Processing, ANLC 1997, pp. 237–244. Association for Computational Linguistics, Stroudsburg (1997)
10. Lamb, A., Paul, M.J., Dredze, M.: Separating fact from fear: Tracking flu infections on twitter. In: Proceedings of NAACL-HLT, pp. 789–795 (2013)
11. Neal, R.M.: Connectionist learning of belief networks. Artificial Intelligence 56(1), 71–113 (1992)
12. Marciniak, T., Strube, M.: Beyond the pipeline: discrete optimization in nlp. In: Proceedings of the Ninth Conference on Computational Natural Language Learning, CONLL 2005, pp. 136–143. Association for Computational Linguistics, Stroudsburg (2005)
13. Koller, D., Friedman, N.: Probabilistic Graphical Models: Principles and Techniques. MIT Press (2009)
14. Bottou, L.: Large-scale machine learning with stochastic gradient descent. In: Lechevallier, Y., Saporta, G. (eds.) Proceedings of the 19th International Conference on Computational Statistics (COMPSTAT 2010), Paris, France, pp. 177–187. Springer (August 2010)
15. Liang, P., Klein, D.: Online em for unsupervised models. In: Proceedings of Human Language Technologies: The 2009 Annual Conference of the North American Chapter of the Association for Computational Linguistics, NAACL 2009, pp. 611–619. Association for Computational Linguistics, Stroudsburg (2009)
16. Fayyad, U., Reina, C., Bradley, P.S.: Initialization of iterative refinement clustering algorithms. In: Proceedings of the Fourth International Conference on Knowledge Discovery and Data Mining, pp. 194–198 (1998)
17. Wilson, T., Wiebe, J., Hoffmann, P.: Recognizing contextual polarity in phrase-level sentiment analysis. In: Proceedings of the Conference on Human Language Technology and Empirical Methods in Natural Language Processing, HLT 2005, pp. 347–354. Association for Computational Linguistics, Stroudsburg (2005)
18. Nenkova, A., Vanderwende, L., McKeown, K.: A compositional context sensitive multi-document summarizer: exploring the factors that influence summarization. In: SIGIR 2006: Proceedings of the 29th Annual International ACM SIGIR Conference on Research and Development in Information Retrieval, pp. 573–580. ACM, New York (2006)
19. Blei, D.M., Ng, A.Y., Jordan, M.I.: Latent dirichlet allocation. J. Mach. Learn. Res. 3, 993–1022 (2003)

20. Xiong, D., Zhang, M., Li, H.: Error detection for statistical machine translation using linguistic features. In: Proceedings of the 48th Annual Meeting of the Association for Computational Linguistics, ACL 2010, pp. 604–611. Association for Computational Linguistics, Stroudsburg (2010)

21. Ueffing, N., Ney, H.: Word-level confidence estimation for machine translation. Comput. Linguist. 33(1), 9–40 (2007)

22. Koehn, P., Hoang, H., Birch, A., Callison-Burch, C., Federico, M., Bertoldi, N., Cowan, B., Shen, W., Moran, C., Zens, R., Dyer, C., Bojar, O., Constantin, A., Herbst, E.: Moses: Open source toolkit for statistical machine translation. In: Proceedings of ACL, Demo and Poster Sessions (2007)

Boot-Strapping Language Identifiers
for Short Colloquial Postings

Moises Goldszmidt, Marc Najork, and Stelios Paparizos

Microsoft Research, Mountain View, CA 94043, USA
{moises,najork,steliosp}@microsoft.com

Abstract. There is tremendous interest in mining the abundant user generated content on the web. Many analysis techniques are language dependent and rely on accurate language identification as a building block. Even though there is already research on language identification, it focused on very 'clean' editorially managed corpora, on a limited number of languages, and on relatively large-sized documents. These are not the characteristics of the content to be found in say, Twitter or Facebook postings, which are short and riddled with vernacular.

In this paper, we propose an automated, unsupervised, scalable solution based on publicly available data. To this end we thoroughly evaluate the use of Wikipedia to build language identifiers for a large number of languages (52) and a large corpus and conduct a large scale study of the best-known algorithms for automated language identification, quantifying how accuracy varies in correlation to document size, language (model) profile size and number of languages tested. Then, we show the value in using Wikipedia to train a language identifier directly applicable to Twitter. Finally, we augment the language models and customize them to Twitter by combining our Wikipedia models with location information from tweets. This method provides massive amount of automatically labeled data that act as a bootstrapping mechanism which we empirically show boosts the accuracy of the models.

With this work we provide a guide and a publicly available tool [1] to the mining community for language identification on web and social data.

Keywords: Language Identification, Wikipedia, Twitter.

1 Introduction

The last decade has seen the exponential rise of user-generated content such as contributions to web forums, Facebook posts and Twitter messages. There is a tremendous interest in mining this content to extract trends, to perform sentiment analysis [14], for automatic machine translation [10], and for different types of social analytics [2]. There is consequently an entire industry providing infrastructure, tools, and platform support to address these problems. Many of the techniques are either language dependent (i.e., affective words for sentiment analysis) or can benefit dramatically from knowing the language to apply certain

H. Blockeel et al. (Eds.): ECML PKDD 2013, Part II, LNAI 8189, pp. 95–111, 2013.

type of rules and knowledge. Thus, reliable automatic language identification is a basic requirement for providing higher level analytic capabilities.

Even though there is a large body of research on language identification, labeled as "solved" by some [13], the conventional wisdom is no longer valid. Computers and smartphones have become internationalized, making it trivial to publish postings in one's native language. Constraints in the number of characters have resulted in abbreviations (such as OMG, LOL), and in sentiment and emphasis being expressed using repetition of letters. In addition a new vernacular, particular to each language, using misspellings and incorporating other languages into the mix has emerged. These characteristics are different from clean and editorially managed corpora used to train language identifiers in the past. These trends combined with the rise of microblogging has caused renewed interest on research in language identification [3,4].

An big obstacle to adapting established models and techniques for language identification is the generation of a sizable corpus of labeled data for all languages in the world that keeps up with the trends described above. In this paper we propose and evaluate such a methodology. The first step is to use Wikipedia as training material for producing an initial set of language identifiers. Wikipedia provides a good source of user generated content, covering a wide variety of topics in 280 languages. We restrict ourselves to the 52 languages with at least 50,000 documents each. Besides being user-generated, these documents have interesting characteristics such as incorporating words from non-primary languages (e.g., Latin and Greek word definitions, pronunciation guidance, etc.), making classifiers more robust to multilingual content. Using language identifiers trained this way, we first characterize the tradeoff of the various parameters for a popular set of language classifiers. We then directly apply the classifiers trained on Wikipedia to two sets of labeled Tweets with different characteristics. Our results indicate that this procedure already yield acceptable identifiers. In order to further improve performance in the context of Twitter, we combine our language prediction with the country information found in tweets to assemble a more appropriate training corpus – each tweet that is labeled as being in language "L" comes from a region where the native language is indeed "L". We retrain our language identifiers using this corpus and repeat our experiment on tweets, resulting in a significant accuracy increase.

All our experiments are performed on well established language identification models and algorithms. The contributions of this paper are:

1. A careful empirical quantification of the different tradeoffs in the selection of the free parameters of existing language identifiers to be used as a guide on selecting the most efficient solution for training and testing depending on the task and the scale.
2. A methodology for automated and unsupervised labeled data generation for current social and colloquial postings by taking advantage of side information such as location. This can be generalized for any additional feature besides location and provides a boot-strapping mechanism for other social data sets.

This paper is organized as follows: Section 2 provides a taxonomic overview of statistical language classifiers and tokenization methodology; Section 3 defines a principled way to evaluate language identification, Section 4 describes the setup and detailed results of our experiments; Section 5 surveys related work, and Section 6 offers concluding thoughts.

2   Statistical Language Classifiers

Automatic language identification is a classification task consisting of finding a mapping from a *document* to the language it is written in. In this paper we consider statistical classifiers, namely those that model both languages and documents using token frequencies.

In order to construct language models, we use a corpus of documents labeled with their respective languages, extract tokens from each document, compile their frequencies, and use the token frequencies of all documents in a given language to build a *language profile*, a function from tokens to frequencies. The document to be classified is represented by a *document profile* using the same techniques. The classification process consists of computing a similarity score between a document profile and each language profile, and reporting the language whose profile is most similar to the document.

Such a statistical approach is language-agnostic and presents the advantages that (i) models can be constructed without any knowledge of the morphology or grammar of a language; and (ii) there is no need for stemming, part-of-speech tagging, and the like. Still, there are several design choices to be made regarding text normalization, tokenization, profile size, and the choice of profile similarity metric, all of which we explore in this paper.

Text normalization refers to any cleanup performed on the training corpus or the document instance prior to tokenization, e.g. the removal of spurious whitespace characters and punctuation. It may also include case folding: whether to convert upper-case characters into their lower-case equivalents or not. For some languages, there are firm grammatical rules governing capitalization; for example, in German, all nouns are capitalized. Other languages, including English, only capitalize the first word of a sentence and certain parts of speech.

In virtually all previous work, tokens are either character n-grams or word n-grams[1]. Language profiles may contain n-grams for multiple values of n. We write $\{n_1, n_2, n_3\}$ -grams to denote token sets that include n_1-grams, n_2-grams and n_3-grams. There are various trade-offs to consider when choosing between character n-grams and word n-grams and when deciding on the value(s) of n. The alphabet of Western languages is quite limited (52 lower- and upper-case letters in English), so the set of possible n-grams remain manageable for small values of n. By contrast, the set of valid words in a language is very large.

---

[1] In the context of this paper, a character is a Unicode symbol. We break words on whitespace characters; ignoring the different word-breaking rules of Asian languages.

We consider three different families of profile similarity measures:

1. Rank-based approaches (RANK). In these approaches the tokens in the language and document models are ranked (according to their observed frequencies in the training corpus and document respectively). In this paper we investigated two rank correlation measures in order to compute a similarity score between these two rankings: Spearman's footrule (SF) [16] and Spearman's rho (SR) [15].
2. Vector-based approaches (VECTOR). These approaches assume that each language model defines a vector in the (very high-dimensional) space of tokens, and the similarity between the vector of the language model and the vector of the document model is given by the cosine between the vectors. Vector-based approaches are used widely in Information Retrieval systems to quantify the similarity between a document and a query.
3. Likelihood-based approaches (LIKELY). These approaches use the frequencies in the models to estimate the probability of seeing a token. Given a document to be classified we can estimate the likelihood that the document was generated from the language model. Likelihood-based approaches differ from the rank and vector approaches in that the document under consideration does not need to be converted into a document profile.

The remainder of this section provides a more formal definition of the profile similarity measures we consider in this paper. We use D to denote a document and L to denote a language profile. All of the measures described below have been adjusted such that higher values indicate greater similarity between D and L. Let T denote the universe of tokens. We write $F_\Psi(t)$ to denote the frequency of token t in profile Ψ, and we write $P_\Psi(t)$ to denote the normalized frequency of token t in profile Ψ, $P_\Psi(t) = \frac{F_\Psi(t)}{\sum_{t' \in T} F_\Psi(t')}$. We will write $t \in \Psi$ to denote the tokens in Ψ, that is, those $t \in T$ where $F_\Psi(t) > 0$.

Rank-Based Approaches (RANK): All rank-based similarity measures rank-order the tokens in D and L and then compute a similarity measure on the rank-ordering. It is worth pointing out that rank-based approaches use token frequency information only to rank-order tokens (and discards the difference in magnitude). Let $R_\Psi(t_i) = i$ denote the rank of the token in the sorted order. If there are z ranks, we define $R_\Psi(t) = z + 1$ for all $t \notin T$. Our variant of **Spearman's footrule** is defined as $SF(D, L) = -\sum_{t \in D} |R_L(t) - R_D(t)|$, where $|x|$ denotes the absolute value of x. Our variant of **Spearman's rho** is defined as $SR(D, L) = -\sum_{t \in D} (R_L(t) - R_D(t))^2$. In both these measures a value of 0 indicates perfect similarity between D and L.

Vector-Based Approaches (VECTOR): The prototypical similarity measure between two vectors in a high-dimensional space is their **cosine similarity**, that is, their dot product normalized by their Euclidean lengths: $CS(D, L) = \frac{\sum_{t \in D} F_D(t) F_L(t)}{\sqrt{\sum_{t \in D} F_D(t)^2} \sqrt{\sum_{t \in L} F_L(t)^2}}$.

Likelihood-Based Approaches (LIKELY): Likelihood-based approaches use the statistical information captured in each language profile to estimate the **probability** that document d was generated from a given language model. Assume that d consists of symbols (characters or words) $(s_1 \ldots s_z) \in S$, S the universe of symbols, and that the language models capture n-gram frequency information. We estimate the probability that d is generated by language model L by directly looking up $P_L(s_1 \ldots s_n)$, and then sliding an n-symbol window over d, advancing one symbol at a time, and computing $P_L(s_{1+i} \ldots s_{n+i}|s_{1+i} \ldots s_{n+i-1})$, the probability that symbol s_{n+i} is preceded by $s_{1+i} \ldots s_{n+i-1}$. Thus $P_L(s_1 \ldots s_z)$ $\approx P_L(s_1 \ldots s_n) \prod_{i=1}^{z-n} P_L(s_{1+i} \ldots s_{n+i} | s_{1+i} \ldots s_{n+i-1})$, where $P_L(s_{1+i} \ldots s_{n+i} |$ $s_{1+i} \ldots s_{n+i-1}) = \frac{P_L(s_{1+i}\ldots s_{n+i})}{\sum_{s\in S} P_L(s_{1+i}\ldots s_{n+i-1}s)}$. It is possible for L not to contain an entry for some n-gram in d. As we are performing a maximum likelihood estimation of the parameters this will be a problem as we will divide by zero. One standard approach to this problem is to smooth the maximum likelihood estimate by using information from a larger (and related) population that has a non-zero values. There are many choices including the amount of smoothing (usually a linear combination of values) as well as the selection of a suitable population. After experimentation, we settled for an approach that assumes a small value such as the min $P_L'(t')$ amongst all tokens t' in all languages L'.

3   Performance Measures

The accuracy A of an identifier is estimated as follows [6]: Let $True(\ell)$ be the number of documents in language ℓ in $\mathcal{L}$, and $TP(\ell)$ be the number of true positives – that is, documents in ℓ that were correctly classified, then $A = \frac{\sum_{\ell \in \mathcal{L}} TP(\ell)}{\sum_{\ell \in \mathcal{L}} True(\ell)}$. Given the fact that different languages have very different numbers of documents for testing (and training), we would like our estimate to reflect that our confidence in a classifier performance depends on the number of samples. To this end, we use *weighted accuracy*, where each language is weighted using the standard deviation of the estimate of the accuracy as a measure of the uncertainty on that estimate. Thus we think of $TP(\ell)$ as coming from a Binomial distribution with parameters $(A(\ell), True(\ell))$, where $A(\ell)$ is the maximum likelihood estimate of the accuracy: $A(\ell) = \frac{TP(\ell)}{True(\ell)}$. Now, we take the weight the inverse of the standard deviation of this estimate: $W(\ell) = \sqrt{\frac{True(\ell)}{A(\ell)(1-A(\ell))}}$. So the weighted accuracy of an identifier is: $WA = \frac{\sum_{\ell \in \mathcal{L}} A(\ell)W(\ell)}{\sum_{\ell \in \mathcal{L}} W(\ell)}$.

To compare the statistical significance between the reported accuracies WA_1 and WA_2 of identifiers 1 and 2, we can use a Wald test [19] directed at rejecting the null hypothesis (H_0) that $WA_1 - WA_2 = 0$ as advocated in [19] when comparing predictors. Again assuming that the accuracies come from a Binomial distribution, and letting n_1 and n_2 denote the number of samples used in classifiers 1 and 2 respectively, we have that

$$\frac{WA_1 - WA_2}{\sqrt{\frac{WA_1*(1-WA_1)}{n_1} + \frac{WA_2*(1-WA_2)}{n_2}}} \tag{1}$$

essentially computes the number of standard deviations providing a degree of confidence necessary to reject H_0. Regarding this expression as approximating a standard Normal distribution,[2] we can reject H_0 and declare the accuracies as different if this expression is bigger than 2 with approximately 95% confidence.

For example, assume $n_1 = n_2 = 20$ million samples, and one classifier has a weighted accuracy of 0.901 and another has a weighted accuracy of 0.90. The difference is 0.001 – is this significant? An application of Eq. 1 yields a results of 10.5, which is bigger than 2 and provides us with confidence that the difference is significant.[3] This should not be surprising with this number of samples. Indeed, taking the weighted accuracy as coming from a Binomial with around 20 million samples, only differences in the fourth digit should start concerning us.

4   Experiments

We used Wikipedia to build our language profiles and evaluate the various language identifier design choices. We downloaded the XML files for all languages. We selected the languages that have over 50000 documents as the popular and most representative ones. We concatenated the text of all documents of a language into a single document and build the language profile. Table 1 shows a summary of information for the languages we considered.

In our experiments, we probed a variety of design choices for language identifiers. For character n-grams, we considered {3}-grams (a popular choice in the literature) as well as {1, 2, 3, 4, 5}-grams (as suggested by Cavnar & Trenkle [6]). Going above the 5 character grams did not produce any benefits and increased the language profiles dramatically. For word n-grams, we considered only 1-grams – in other words, a language profile is simply the lexicon of that language. Multi-gram words could be used, but this would be more suitable for phrase prediction than language prediction, and the space required for even bi-grams words made it computationally infeasible for us. For language and document profile sizes, we explored retaining all tokens, and for performance reasons we also explored using the 10k or 500 most frequent ones. Table 2 summarizes our choices. In each case, we explored both case folding and leaving the capitalization unchanged. In other words, in each of the experiments described below we compare the performance of 42 different classifiers.

4.1   Language Identification Design Alternatives

In-Model Accuracy: Our first experiment compares the performance of the 21 design choices described in Table 2. We used the uncleaned Wikipedia abstracts of the 52 languages shown in Table 1 for both training and testing; we did not perform any case folding. We conducted in-model testing using every abstract in the collection. The results are summarized in Table 3. We observe the following:

---

[2] This will certainly be true for all tests sets involving Wikipedia data, given that even the smallest test set contains millions of samples.

[3] When the samples are the same we are violating an assumption of independence that will impact the degrees of freedom; yet with this many samples this has no effect.

Table 1. Basic statistics for the 52 languages we trained and tested on

Language	# docs	# words	# chars	Language	# docs	# words	# chars
English	3,841,701	69,995,728	420,822,801	Arabic	154,105	3,638,910	21,077,370
German	1,334,677	21,476,871	153,119,580	Serbian	151,409	1,850,106	12,006,824
French	1,175,638	19,664,258	116,384,106	Lithuanian	142,468	1,314,498	10,254,168
Italian	874,827	14,892,786	92,885,963	Slovak	130,348	1,703,213	11,452,875
Dutch	871,310	16,666,710	108,570,923	Malay	130,170	1,878,252	12,846,493
Polish	852,219	11,107,886	79,515,315	Hebrew	124,884	2,408,802	13,677,123
Spanish	851,369	18,239,492	109,873,261	Bulgarian	124,665	2,192,227	13,911,993
Russian	800,527	8,972,267	66,586,091	Kazakh	122,442	1,879,736	14,817,377
Japanese	791,350	5,390,566	40,296,474	Slovene	121,968	1,496,767	10,103,516
Portuguese	706,771	13,518,871	81,500,258	Volapük	118,923	1,757,761	9,813,402
Swedish	417,092	7,671,785	49,951,384	Croatian	109,103	1,499,300	10,120,238
Chinese	385,528	2,526,386	19,823,371	Basque	106,846	1,463,312	11,361,999
Catalan	359,848	7,860,184	44,345,082	Hindi	92,371	4,337,272	10,367,220
Ukrainian	330,559	4,148,304	29,714,043	Estonian	90,333	1,061,293	8,128,345
Norwegian	320,318	5,557,555	34,993,643	Azerbaijani	84,265	714,444	5,041,032
Finnish	284,303	3,093,762	25,700,474	Galician	78,419	2,089,220	12,265,696
Vietnamese	247,286	5,141,075	25,607,852	Nynorsk	75,399	1,385,501	8,466,704
Czech	214,219	3,040,486	20,456,549	Thai	70,863	2,085,809	9,704,446
Hungarian	206,518	2,613,681	19,410,794	Greek	67,634	1,676,886	11,126,017
Korean	186,746	2,751,819	11,251,750	Latin	62,985	1,149,511	8,383,645
Indonesian	182,026	3,174,474	22,359,174	Occitan	55,520	830,662	4,685,239
Romanian	170,328	2,372,032	15,006,843	Tagalog	54,796	809,647	4,856,032
Persian	170,137	3,397,007	16,845,518	Georgian	53,736	731,235	5,847,744
Turkish	164,263	2,114,148	15,257,393	Haitian	53,575	509,151	2,650,222
Danish	158,497	2,711,688	17,229,078	Slavomacedonian	53,185	857,418	5,538,833
Esperanto	158,152	2,761,440	17,518,543	Serbo-Croatian	52,922	826,400	5,460,844

Table 2. Design alternatives explored in this paper

	$\{1,\cdots,5\}$-char			$\{3\}$-char			$\{1\}$-word				$\{3\}$-char	$\{1\}$-word
	all	10k	500	all	10k	500	all	10k	500		all	all
RANK SF	×	×	×	×	×	×	×	×	×	VECTOR	×	×
RANK SR	×	×	×	×	×	×	×	×	×	LIKELY	×	

Table 3. Weighted accuracy % of in-model testing

	$\{1,\cdots,5\}$-char			$\{3\}$-char			$\{1\}$-word				$\{3\}$-char	$\{1\}$-word
	all	10k	500	all	10k	500	all	10k	500		all	all
RANK SF	93.35	89.17	78.56	89.45	88.56	80.92	90.97	88.66	83.39	VECTOR	80.19	79.88
RANK SR	93.35	89.64	79.08	87.41	87.57	81.14	83.03	81.17	77.96	LIKELY	88.68	

- $\{1,\cdots,5\}$-char RANK(SF, all) and RANK(SR, all) have the highest weighted accuracy.
- $\{1,\cdots,5\}$-char RANK is preferable.
- Using the full language profile (all) is better than restricting to smaller sizes.
- The vector-based approaches are not competitive.

Table 4. Weighted accuracy change for ten-fold cross validation vs in-model testing from Table 3. Negative values indicate over-fitting.

	$\{1,\cdots,5\}$-char			$\{3\}$-char			$\{1\}$-word				$\{3\}$-char	$\{1\}$-word
	all	10k	500	all	10k	500	all	10k	500		all	all
RANK SF	-0.51	0.01	0.00	-2.87	-0.04	0.00	-2.90	-0.09	0.03	VECTOR	-0.02	-0.51
RANK SR	-0.57	0.00	0.00	-2.87	-0.08	-0.02	-3.65	-0.12	0.01	LIKELY	-0.90	

Table 5. Weighted accuracy change for case folding. Negative values indicated that case folding hurts accuracy (compared to Table 4).

	$\{1,\cdots,5\}$-char			$\{3\}$-char			$\{1\}$-word				$\{3\}$-char	$\{1\}$-word
	all	10k	500	all	10k	500	all	10k	500		all	all
RANK SF	-0.82	0.02	0.21	0.00	-0.56	-0.41	-0.96	-0.14	0.31	VECTOR	-0.42	-0.61
RANK SR	-0.96	-0.02	0.28	0.13	-1.03	-0.35	-1.06	-0.19	0.19	LIKELY	-0.61	

Cross-Validation: In order to quantify the impact of over-fitting we repeated the same experiment using ten-fold cross validation. Table 4 shows the difference between the weighted accuracies of ten-fold and in-model experiments. A negative value indicates the ten-fold accuracy is lower, i.e. over-fitting occurred. We observe the following:

- $\{1,\cdots,5\}$-char, RANK(SF, all) and RANK(SR, all) still have the highest weighted accuracy.
- Over-fitting is a bigger issue when using the full language profile (all). This makes sense: the truncated language profiles omit low frequency tokens.
- Amongst rank-based approaches, $\{1,\cdots,5\}$-char are less affected.
- Rank-based classifiers are more affected than Vector or Likelihood-based for the same tokenization scheme and profile limit.

Case Folding: We lowercased both training and test data and repeated the same ten-fold cross validation experiment with all other choices unmodified. Table 5 shows the difference between the weighted accuracies of using case folding vs leaving the capitalization as is. A negative value indicates that case folding lowers accuracy, i.e. lower casing is a bad idea. We observe the following:

- $\{1,\cdots,5\}$-char, RANK(SF, all) and RANK(SR, all) still have the highest weighted accuracy.
- By and large, case folding not only does not help much, but in many cases produces statistically significant worse results. We attribute this to the fact that in some languages, such as German, capitalization is governed by strict grammatical rules.

Language Specific Results: Next, we tested the accuracy of classifiers with respect to the 52 languages in our corpus. The results are shown in Figure 1. The solid black line shows the weighted accuracy of the classifier from Table 2 that performed best for each given language; the dotted red line shows the weighted

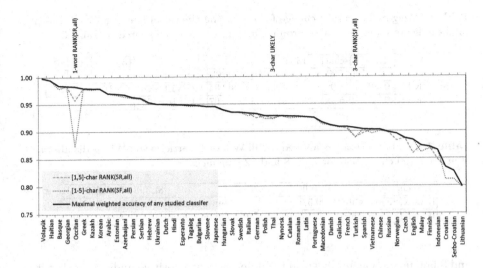

Fig. 1. Classification accuracy broken down by language. The solid black line shows the accuracy of the classifier from Table 2 that performed best for each given language; the dotted red line shows the accuracy of $\{1,\cdots,5\}$-char RANK(SF, all), and the dashed green line shows the accuracy of $\{1,\cdots,5\}$-char RANK(SR, all). For the three languages (Occitan, Thai, and Turkish) where these two classifiers did not perform best, the best-performing classifier is listed along the upper horizontal axis.

accuracy of $\{1,\cdots,5\}$-char RANK(SF, all), and the dashed green line shows the weighted accuracy of $\{1,\cdots,5\}$-char RANK(SR, all). There were only three languages where neither of these two classifiers performed best: Occitan, Thai, and Turkish. In the case of Thai, the accuracy gap was minor, in the case of Turkish and more so Occitan it was substantial. For Occitan, 1-word RANK(SR,all) performed best; for Turkish, it was 3-char RANK(SF,all), and in the case of Thai, the 3-char LIKELY classifier slightly outperformed the overall leaders. We speculate that Occitan is sufficiently close to Catalan that their character n-gram profiles are very similar, while their lexica are different enough to allow differentiation, giving an edge to word-based approached. We are not sure why 3-character tokens work better than one- to five-character tokens for Turkish. Languages with languages-specific characters sets (such as Hebrew and Greek) provide a strong signal to any character-based classifier, and thus can be recognized with high accuracy. By contrast, our classifiers performed relatively poorly on Chinese. This is to be expected because it has a large alphabet of information-rich symbols, meaning the space of character n-grams is both large and sparse.

Number of Languages: We hypothesized that language detection becomes harder as the set of languages increases. In order to test that hypothesis we repeated the experiment described in Table 3 using only the 8 languages listed in [6] (namely, English, German, Dutch, Polish, Portuguese, Spanish, Italian and French). Table 6 shows the difference between the weighted accuracies of 52

Table 6. Weighted accuracy change for restricting the set of languages to 8. Positive values indicate fewer languages produce better accuracy (compared to Table 3).

	$\{1,\cdots,5\}$-char			$\{3\}$-char			$\{1\}$-word				$\{3\}$-char	$\{1\}$-word
	all	10k	500	all	10k	500	all	10k	500		all	all
RANK SF	-0.39	2.10	5.99	0.04	1.49	6.14	1.18	2.84	4.22	VECTOR	6.31	9.02
RANK SR	-0.71	1.90	5.58	-1.25	1.37	5.72	3.35	5.07	7.41	LIKELY	2.65	

Table 7. Two best classifiers applied to full Wikipedia articles and Wikipedia abstracts of at least 150 characters for both 8 and 52 languages

[1, 5]-char, all	8L full	8L abs	52L abs		8L full	8L abs	52L abs
RANK(SF) weighted	99.52	99.96	99.95	un-weighted	98.91	99.90	99.74
RANK(SR) weighted	99.46	99.97	99.98	un-weighted	97.47	99.82	99.71

and 8 languages. Generally speaking differences are positive, indicating that it is indeed easier to classify languages when presented with fewer choices. However, the two classifiers that perform the best ($\{1,\cdots,5\}$-char RANK(SF, all) and RANK(SR, all)) perform slightly worse for fewer languages. Moreover, none of the accuracy values approaches the accuracy scores reported by [6].

Limiting Minimum Document Size: In order to understand the difference between our performance and that of [6] we explored three alternative hypotheses: (i) the quality of Wikipedia abstracts might be low; full Wikipedia documents might be better, (ii) some abstracts are very short, and these tend to be of low quality, (iii) the weighted accuracy measure we use might produce different results. We tested this by taking the two best performing classifiers from our previous experiments (($\{1,\cdots,5\}$-char RANK(SF, all) and RANK(SR, all))) and applied them to (i) full Wikipedia articles, (ii) only abstracts with at least 150 characters, and (iii) computed both weighted and un-weighted accuracy measures. Besides testing on only the 8 languages described in [6], we also computed weighted and un-weighted accuracies for the length-restricted abstracts of the 52 languages from Table 1. Restricting to abstracts of minimum length eliminated about 71.5% of the Wikipedia abstracts. The results are shown in Table 7. We observed that both the weighted and un-weighted accuracies are in the same range to those from [6]. This is a dramatic improvement over the previous results we reported in Tables 3 and 6. The false negative rate was reduced by 99%. We attribute this to the longer average size of each document tested. Using the full Wikipedia articles does not provide an improvement over the length-restricted abstracts. Furthermore, there is no statistically significant difference between the 8 and 52 languages. Finally, the un-weighted accuracy is lower but inline with [6]. So considering only documents with at least 150 characters when building the language profile shows significant improvement.

Length of Test Documents: To examine whether classifier accuracy is indeed affected by the length of test documents, we constructed a synthetic collection

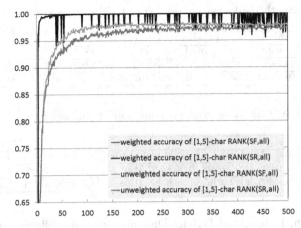

Fig. 2. Accuracy as a function of test document size in characters

of test cases where we control the size of the documents. For a language ℓ with n_ℓ Wikipedia abstracts, we generated $0.1n_\ell$ documents consisting of consecutive words drawn from the concatenated wikipedia abstracts of ℓ. The lengths of these synthetic documents is uniformly distributed in the range of 1 to 500 characters per language. We tested the performance of the two best classifiers – $\{1, \cdots, 5\}$-char RANK(SF, all) and RANK(SR, all) – on this collection, using the 52 languages from Table 1. Figure 2 shows the weighted and un-weighted accuracies for the classifiers. We observe that the weighted accuracy of both classifiers reaches 99% at around 9 characters average document length. The un-weighted accuracy grows more gradually, approaching 98% at the 500 character limit. The curves show some noise which is due to the limited sample size – for example, less popular languages have just 10 test cases per bucket.

4.2   Short Colloquial Postings and Tweets

Synthetic Tweets: We assembled a synthetic collection with a word count distribution that follows that of Twitter. We used all 52 languages and again $0.1n_\ell$ documents per language. The Twitter word count distribution was based on one month worth of real tweets, with Twitter commands, usernames, emoticons and URLs removed. 12.5% of cleaned tweets contained a single word and 64.5% contained at most ten words. Table 8 shows the weighted accuracy for all our classifiers applied to this synthetic data set. We observe that the two best classifiers – $\{1, \cdots, 5\}$-char RANK(SF, all) and RANK(SR, all) – still perform best. However, when comparing with the results of Table 3, RANK(SR, all) improves slightly whereas RANK(SF, all) deteriorates slightly.

Real Tweets from Tromp and Pechenizkiy [17]: The experiment uses the 9066 labeled tweets made available by [17]. They extracted tweets from accounts known to only contain messages in a specific language, namely German, English,

Table 8. Synthetic documents using Twitter's word-length distribution

	{1,···,5}-char			{3}-char			{1}-word				{3}-char	{1}-word
	all	10k	500	all	10k	500	all	10k	500		all	all
RANK SF	94.05	89.78	81.40	87.55	86.27	78.39	82.56	79.38	72.96	VECTOR	76.87	66.06
RANK SR	93.03	89.05	80.47	84.03	83.68	77.45	79.07	76.60	71.23	LIKELY	88.80	

Table 9. Trained on Wikipedia and applied on tweets from [17]

	{1,···,5}-char			{3}-char			{1}-word				{3}-char	{1}-word
	all	10k	500	all	10k	500	all	10k	500		all	all
RANK SF	98.00	99.10	95.34	95.83	98.19	96.84	97.51	98.35	94.89	VECTOR	91.51	85.15
RANK SR	89.22	99.12	95.08	89.21	95.51	96.67	96.41	98.03	94.78	LIKELY	98.75	

French, Spanish, Italian, and Dutch. They also used six different accounts per language, to provide diversity (tweets from different accounts) and consistency (several tweets from the same account). The tweets in the corpus are all lowercase, thus for this experiment we only applied our classifiers with models that had case-folding. We limited the language profiles used by our classifier to the six languages found in the data set. The results are displayed in Table 9. As in our previous experiments the {1,···,5}-char RANK classifiers perform best. We observed the tweets in the test set are longer and relatively cleaner than average tweets, in fact the length is even higher than the median length of Wikipedia abstracts. Therefore, the numbers produced are inline with our findings on minimum 150 character length. Furthermore, we highlight that we are able to beat the results from [17]. Even though we used Wikipedia as the training data and our rank-based classifier, in contrast to their specialized Machine Learning approach trained and tested on this very same data set.

Real Tweets – Uniform Random Sample: We sampled 2000 random tweets from the Twitter stream using uniform distribution. We labeled them ourselves to the correct language resulting in 869 tweets that we could process confidently. We run the same set of language identifiers on the labeled tweets and report the results in Table 10. The two best-performing classifiers are {1,···,5}-char RANK(SF,10k) and {1,···,5}-char RANK(SR,10k). Compared to Table 9, the performance is worse overall. We attribute this to the tweets in this sample being shorter and much less pristine than those provided by [17]. The best way to illustrate this is with some examples: (a) *"u nah commmming ?"*, (b) *"alreadyyyy!!!!!!!!"*, (c) *"omg im bussin up"*. Note the use of letter repetition for emotion, single letter such as '*u*' for 'you', and what is becoming standard abbreviations such as '*omg*' for '*oh my god*'. These should be contrasted with the tweets from [17], which look like: *"egypts supreme military council to sack cabinet suspend both houses of parliament and govern with head of supreme court from reuters"*. The performance difference is expected once we consider such differences in the sets of tweets. We feel the random sample of our tweets is representative of what to expect from average tweets. The results indicate that our language identifiers trained on Wikipedia do a good job in processing tweets.

Table 10. Trained on Wikipedia and applied on a random sample of real tweets

	{1,···,5}-char			{3}-char			{1}-word				{3}-char	{1}-word
	all	10k	500	all	10k	500	all	10k	500		all	all
RANK SF	81.73	87.90	72.57	81.20	82.47	70.55	79.40	80.28	67.81	VECTOR	53.61	37.11
RANK SR	74.58	87.89	74.24	67.44	72.51	69.61	74.13	77.77	66.00	LIKELY	79.60	

Tweets from an Uncommon Language with Latin Alphabet: To test the strength of our methodology, we chose to explore the performance of a classifier trained on Wikipedia and applied on Twitter, but this time on a language that had the following characteristics: (i) is written mostly in Latin characters (so the character set will not help the classification), (ii) was not on the top list in terms of data available for training (i.e., relatively obscure), and (iii) we have access to a native speaker capable of labeling the results. One language that met all these conditions was Romanian (with 170,000 abstracts – see Table 1). For this experiment we sampled 250 tweets that our classifier labeled as Romanian and that had more than 50 characters (to avoid too much ambiguity) and asked our native speaker to label them. We found that 85.2% (213 of 250) were indeed Romanian. While this result is based on a moderately sized sample, in combination with the previous experiments it provides evidence that classifiers trained on Wikipedia is generally helpful in automatically classifying tweets.

Boot-strapping Twitter Labels: As our experiments have demonstrated, Wikipedia is a great resource for building language profiles, however its corpus is missing idioms that are particular to social postings such as tweets (e.g., the examples given above). One way to address this problem is to train models using large amounts of labeled tweets which is an expensive proposition (when human provide the labels). We automatically (and inexpensively) created this labeled set using the following methodology: by selecting tweets containing a country code in their metadata, labeling each tweet with our best language prediction as "L", and verifying it comes from a country where the native language is indeed "L". To avoid over-representation of popular languages, we put a cap of 10 million tweets at most per language. We were able to generate 88.3 million labeled documents – pairs of tweet and corresponding language. The set of 52 languages from Table 1 was reduced to 26, as we can only capture the actively used languages for which we could extract location information from the tweets. Using these documents we constructed new language profiles and repeated the set of experiments on real tweets.

Table 11 summarizes the results for the tweets from Tromp & Pechenizkiy [17]. In comparison to Table 9, accuracy is higher across the set of models tested, and even the (previously) weaker performing algorithms perform great. This is truly an indication of the value of our method for generating boot-strapped labels – the large amount of automatically training data generated by our method boosts the accuracy of our relatively simple classifiers, perform better than the specialized approach in [17] and the relatively more complex approach in [11] which were tested on the same dataset.

Table 11. Boot-strapped Twitter labels, applied on tweets as in Table 9

	{1,···,5}-char			{3}-char			{1}-word				{3}-char	{1}-word
	all	10k	500	all	10k	500	all	10k	500		all	all
RANK SF	99.11	99.52	96.40	99.02	99.06	97.97	99.31	99.44	97.03	VECTOR	95.41	93.98
RANK SR	95.05	99.53	96.42	96.53	97.44	97.93	98.83	99.27	96.99	LIKELY	99.64	

Table 12. Boot-strapped Twitter labels, applied on tweets as in Table 10

	{1,···,5}-char			{3}-char			{1}-word				{3}-char	{1}-word
	all	10k	500	all	10k	500	all	10k	500		all	all
RANK SF	96.23	95.37	89.07	93.85	95.05	90.22	92.80	94.25	90.06	VECTOR	84.55	79.16
RANK SR	93.95	95.44	88.78	87.99	90.27	90.83	90.25	92.02	88.73	LIKELY	95.09	

Table 12 summarizes the results for the random sample of tweets we obtained directly from the Twitter stream. As we discussed earlier, we believe this data set to be a closer representation of the average tweet. The results show a very high degree of accuracy. We note that our language identifiers are triggered on all tweets – we will always return our top language guess. Consequently, the overall best results have precision above 96% with 100% recall.

5   Related Work

Our study is inspired by the work on using n-grams for text classification and statistical language identification, like [5,7,8], and in particular Cavnar and Trenkle [6]. This work proposed to use character n-grams for values of n ranging from 1 to 5, to limit the number of n-gram/frequency pairs retained for each language to 400, to compare languages and documents using a rank-based approach, and specifically to use Spearman's footrule [16] to estimate rank correlation. Cavnar and Trenkle recognized 8 languages and trained on a total of 3713 samples (documents) from a Usenet group. They reported accuracy of 92.9% for language profiles of 100 tokens and documents of less than 300 characters, and accuracy of 99.8% for language profiles of 400 tokens documents over 400 characters in length. We achieved similar accuracy (99.82; see Table 7) using the same 8 languages but training and testing on Wikipedia. Furthermore, we did a more exhaustive study: we experimented on 52 languages, trained on over 18 million Wikipedia documents, and tested on both Wikipedia and Twitter documents with different characteristics. We found that language profiles should be one to two orders of magnitude larger than suggested by [6].

Our work compares a larger number of models and classifiers including those based on using words as tokens, and those that take the frequencies of the tokens to estimate probabilities, and use likelihood for discrimination. In that sense our work is closer in spirit to Grothe et al. [9]. Once again, our study differs from theirs in terms of scale: we recognize almost three times as many languages (52 vs. 18), use a much larger corpus for training and testing, and also expand on the number of classifiers, the use of capitalization, and the length of the documents to be classified.[4] We are also able to explain their observed phenomena of restricting the number of tokens in the profile based on word models. Using all the words (as opposed to the top ones) will result in over-fitting.

More recently Majlis [12] proposed a study using n-grams with W2C (WebToCorpus) training corpora and various methods like SVM, naive Bayes, regression tree and their own YALI algorithm. The results show high precision for large documents but much lower precision for document sizes of 30 to 140 characters. In some cases the results are in the order of 70% precision for the smaller documents. The YALI method proposed in the paper produces the best results when using 4-grams and shows an in-model testing accuracy of around 93% for small documents. Conceptually, there are similarities between YALI and our approach in that we both use a variation of n-gram retrieval to determine the language of documents. Another important difference is in the testing methodology. We trained and tested on Wikipedia and also tested on real tweets. The authors of YALI trained and tested on a carefully curated W2C corpus. By comparison, we achieved significantly better accuracies on "dirtier" data.

Dunning in [8] considered using probabilistic methods to identify the language of a document including Markov chains and Bayesian inference. Our likelihood classifier is similar in spirit and has similar performance characteristics.

Tromp and Pechenizkly [17] explore a supervised machine learning approach to language identification in the context of Twitter. They made their corpus available, enabling us to benchmark our implementations on their test corpus. In comparing the tweets contained in their collection to a uniform random sample of tweets we collected and labeled, we found the tweets in their collection to be much "cleaner" than the average tweet: They are fairly long, tend to be grammatically well formed, and contain fewer Twitter-specific acronyms than we observe "in the wild". Tromp and Pechenizkly used the same corpus for training and for testing. Our classifiers are unsupervised, start with Wikipedia and can be extended with automatically generated Twitter labels, and produce better results on their test set.

Another approach that is similar to ours is the work described in [11]. Their approach is based on a combination of a naive Bayes classifier using n-grams as features, with separate feature selection for each use case. By contrast our approach relies on side information to automatically obtain labeled data, allowing us to use over 80 million Tweets for training – three orders of magnitude larger than the corpus used by [11]. Moreover, the corpus can be generated dynamically and therefore our approach adapts to changes in style and usage patterns. Finally, [11] report 94% accuracy on the dataset from [17] while our method yields accuracies above 99% on the same dataset.

The work in [18] explores the task of identification of very short text segments on the order of 5 to 21 characters, training on the Universal Declaration of Human Rights as a corpus. They explore this space at full scale in terms of the

---

[4] We note that in one of their evaluation they use 135 documents from Wikipedia.

number of languages, using 281 (we restricted to 52 because of limitations on the labeled data of the corpus). They also explore n-grams-based models, but because of their restrictions to very short text they have to explore smoothing techniques (which do not seem to be necessary for identification of longer objects). We did not find that expanding the set of languages has a deleterious effect, but we did find that accuracy is sensitive to document length.

Finally, there is recent work in identifying the language in coping with short idiomatic text, such as tweets. The authors in [4] propose to use both endogenous and exogenous information to improve classification accuracy. In addition to the text of a tweet, they use the posting history of the author and other users mentioned in a tweet and web page content referred in the message. The underlying idea is to increase the size of each document by leveraging the structure of Twitter messages. The results reported are in line with our findings in that an increase in document length will yield higher accuracy. The authors in [3] focus on tweets from low-resource languages. Their approach is to collect such tweets, label them via the use of Mechanical Turk and use a supervised learning algorithm to train a classifier. They compare against three readily available language identifier systems, including an implementation of Cavnar and Trenkle [6]. Like [4], they also incorporate some meta information, such as tweet authorship. In contrast, our classifiers are unsupervised, using Wikipedia and tweet location to boot-strap mass labeling, and perform in a higher accuracy range.

6   Conclusions

In this paper we focused on two aspects of language identification. (i) study the various algorithms and free parameters and offer a guide on what works well and what not, and (ii) determine a methodology to provide high quality language identification for short colloquial postings, e.g. Twitter.

For language identification in general we learned that: (a) rank based classifiers are both effective and efficient, (b) if memory-limited, we can obtain good results with the top 10k tokens in the language model, (c) case folding does not matter, (d) the number of languages makes little difference, if enough training data exists, and (e) the length of test documents make a big difference, with larger being easier to classify.

Specifically to Twitter postings we learned that (a) Wikipedia works great for a solid baseline language model, and (b) generating labeled data using a combination of Wikipedia classification and a Twitter specific signal, like location, allows us to boot-strap superior language models.

According to our knowledge, this work describes the best overall methodology for an automated, unsupervised, and scalable technique in language identification on short colloquial postings. Our implementation is available [1] as a building block for further research in the area of social analytics.

References

1. Automatic language identification tool, http://research.microsoft.com/lid/
2. Aggarwal, C.C. (ed.): Social Network Data Analytics. Springer (2011)
3. Bergsma, S., McNamee, P., Bagdouri, M., Fink, C., Wilson, T.: Language identification for creating language-specific twitter collections. In: Proc. Second Workshop on Language in Social Media, pp. 65–74 (2012)
4. Carter, S., Weerkamp, W., Tsagkias, E.: Microblog language identification: Overcoming the limitations of short, unedited and idiomatic text. Language Resources and Evaluation Journal (2013)
5. Cavnar, W.: Using an n-gram-based document representation with a vector processing retrieval model, pp. 269–269. NIST SPECIAL PUBLICATION SP (1995)
6. Cavnar, W., Trenkle, J.: N-gram-based text categorization. In: SIDAIR (1994)
7. Damashek, M.: Gauging similarity with n-grams: Language-independent categorization of text. Science 267(5199), 843–848 (1995)
8. Dunning, T.: Statistical identification of language. Technical Report MCCS-94-273, New Mexico State University (1994)
9. Grothe, L., Luca, W.D., Nurnberger, A.: A comparative study on language identification methods. In: Proc. of LREC (2008)
10. Lopez, A.: Statistical machine translation. ACM Comput. Surv. 40(3) (2008)
11. Lui, M., Baldwin, T.: landid.py: An off-the-shelf language identification tool. In: Proc. of ACL (2012)
12. Majliš, M.: Yet another language identifier. In: EACL 2012, p. 46 (2012)
13. McNamee, P.: Language identification: A solved problem suitable for undergraduate instruction. Journal of Computing Sciences in Colleges 20(3) (2005)
14. Pang, B., Lee, L.: Opinion mining and sentiment analysis. Foundations and Trends in Information Retrieval 2(1-2), 1–135 (2007)
15. Spearman, C.: The proof and measurement of association between two things. The American Journal of Psychology 15(1), 72–101 (1904)
16. Spearman, C.: Footrule for measuring correlation. The British Journal of Psychiatry 2(1), 89–108 (1906)
17. Tromp, E., Pechenizkly, M.: Graph-based n-gram language identification on short texts. In: Proc. of BENELEARN (2011)
18. Vatanen, T., Vayrynen, J., Virpioja, S.: Language identification of short text segments with n-gram models. In: Proc. of LREC (2010)
19. Wasserman, L.: All of statistics. Springer (2004)

A Pairwise Label Ranking Method with Imprecise Scores and Partial Predictions

Sebastien Destercke

Université de Technologie de Compiegne U.M.R. C.N.R.S. 7253 Heudiasyc Centre de
recherches de Royallieu F-60205 Compiegne Cedex France
sebastien.destercke@hds.utc.fr

Abstract. In this paper, we are interested in the label ranking problem. We are
more specifically interested in the recent trend consisting in predicting partial but
more accurate (i.e., making less incorrect statements) orders rather than complete
ones. To do so, we propose a ranking method based on pairwise imprecise scores
obtained from likelihood functions. We discuss how such imprecise scores can
be aggregated to produce interval orders, which are specific types of partial orders. We then analyse the performances of the method as well as its sensitivity to
missing data and parameter values.

Keywords: Label ranking, imprecise probabilities, Pairwise voting.

1 Introduction

In recent years, learning problems with structured outputs have received a growing
interest. Such problems appear in a variety of applications fields requiring to deal with
complex data: natural language treatment [6], biological data [32], image analysis...

In this paper, we are concerned with the problem of *label ranking*, where one has
to learn a mapping from instances to rankings (complete orders) defined over a finite
number of labels. Different methods have been proposed to perform this task. Ranking
by pairwise comparison (RPC) [25] transforms the problem of label ranking into binary
classification problems, combining all results to obtain the final ranking. Constraint
classification and log-linear models [23,15] intend to learn, for each label, a (linear)
utility function from which the ranking is deduced. Other approaches propose to fit a
probabilistic ranking model (Mallows, Placket-Luce [28]) using different approaches
(instance-based, linear models, etc. [29,10]).

Recently, some authors [13] have discussed the interest, in label ranking and more
generally in preference learning problems, to predict partial orders rather than complete
rankings. Such an approach can be seen as an extension of the reject option implemented in learning problems [3] or of the fact of making partial predictions [14]. Such
cautious predictions can prevent harmful decisions based on incorrect predictions. In
practice, current methods [13] consist in thresholding a pairwise comparison matrix
containing probabilistic estimates. More recently, it was shown [12] that probabilities
issued from Placket-Luce and Mallows models are particularly interesting in such a
thresholding approach, as they are guaranteed to produce consistent orders (i.e., without cycles) that belong to the family of semi-orders.

H. Blockeel et al. (Eds.): ECML PKDD 2013, Part II, LNAI 8189, pp. 112–127, 2013.

In this paper, we adopt a different approach in which we propose to use imprecise probabilities and non-parametric estimations to predict partial orderings. As making partial predictions is one central feature of imprecise probabilistic approaches [14], it seems interesting to investigate how one can use such approaches to predict partial orders. In addition, these approaches are also well-designed to cope with the problem of missing or incomplete data [33].

More precisely, we extend the proposal of [25] to imprecise estimates, and propose to get these estimates from a method based on instance-based learning and likelihood functions. The paper is organized as follows: Section 2 discusses the basics of label ranking and of label ranking evaluation when predicting complete and partial orders. Section 3 then presents the method. It first provides a means to obtain imprecise estimates from a likelihood-based approach, before discussing how such estimates can be aggregated to produce interval orders (a sub-family of partial orders including semi-orders) as predictions. Finally, Section 4 ends up with experiments performed on various data sets.

2 Preliminaries

This section introduces the necessary elements concerning label ranking problems.

2.1 Label Ranking Problem

The usual goal of classification problems is to associate an instance $\mathbf{x}$ coming from an instance space $\mathscr{X}$ to a single (preferred) label of the space $\Lambda = \{\lambda_1, \ldots, \lambda_m\}$ of possible classes. Label ranking problems correspond to the case where an instance $\mathbf{x}$ is no longer associated to a single label of $\mathscr{X}$ but to a total order over the labels, that is a complete, transitive, and asymmetric relation $\succ_\mathbf{x}$ over $\Lambda \times \Lambda$, or equivalently to a complete ranking over the labels $\Lambda = \{\lambda_1, \ldots, \lambda_m\}$. Hence, the space of prediction is now the whole set $\mathscr{L}(\Lambda)$ of complete rankings of Λ. It is equivalent to the set of permutations of Λ and contains $|\mathscr{L}(\Lambda)| = m!$ elements. We can identify a ranking $\succ_\mathbf{x}$ with a permutation $\sigma_\mathbf{x}$ on $\{1, \ldots, m\}$ such that $\sigma_\mathbf{x}(i) < \sigma_\mathbf{x}(j)$ iff $\lambda_i \succ_\mathbf{x} \lambda_j$, as they are in one-to-one correspondence. In the following, we will use the terms rankings and permutations interchangeably.

The task in label ranking is the same as the task in usual classification: to use the training instances $(\mathbf{x}_i, y_i)$, $i = 1, \ldots, n$ to estimate the theoretical conditional probability measure $P_\mathbf{x} : 2^{\mathscr{L}(\Lambda)} \to [0,1]$ associated to an instance $\mathbf{x} \in \mathscr{X}$. Ideally, observed outputs y_i should be complete orders over Λ, however this is seldom the case and in this paper we allow training instance outputs y_i to be incomplete (i.e., partial orders over Λ).

In label ranking problems the size of the prediction space quickly increases, even when Λ is of limited size (for instance, $|\mathscr{L}(\Lambda)| = 3628800$ for $m = 10$). This makes the estimation of $P_\mathbf{x}$ difficult and potentially quite inaccurate if only little data is available, hence an increased interest in providing accurate yet possibly partial predictions. This rapid increase in $|\mathscr{L}(\Lambda)|$ size also means that estimating directly the whole measure $P_\mathbf{x}$ is in general untractable except for very small problems. The most usual means to solve this issue is either to decompose the problem into many simpler ones or to assume that $P_\mathbf{x}$ follows some parametric law. In this paper, we shall focus on the pairwise decomposition approach, recalled and extended in Section 3. To simplify notations, we will drop the subscript x in the following when there is no possible ambiguity.

2.2 Evaluating Partial Predictions

The classical task of label ranking is to predict a complete ranking $\hat{y} \in \mathscr{L}(\Lambda)$ of labels as close as possible to an observed complete ranking y. When the observed and predicted rankings y and $\hat{y}$ are complete, various accuracy measures [25] (0/1 accuracy, Spearman's rank, ...) have been proposed to measure how close $\hat{y}$ is to y. In this paper, we retain Kendall's Tau, as it will be generalized to measure the quality of partial predictions. Given y and $\hat{y}$, Kendall's Tau is

$$A_\tau(y, \hat{y}) = \frac{C - D}{m(m-1)/2} \tag{1}$$

where $C = |\{(\lambda_i, \lambda_j) | (\sigma(i) < \sigma(j)) \wedge (\hat{\sigma}(i) < \hat{\sigma}(j))\}|$ is the number of concording pairs of labels in the two rankings, and $D = |\{(\lambda_i, \lambda_j) | (\sigma(j) < \sigma(i)) \wedge (\hat{\sigma}(i) < \hat{\sigma}(j))\}|$ the number of discording pairs of labels. $A_\tau(y, \hat{y})$ has value 1 when $y = \hat{y}$ and -1 when $\hat{y}$ and y are reversed rankings.

$A_\tau(y, \hat{y})$ assumes that the prediction $\hat{y}$ is a complete ranking and that the model can compare each pair of labels in a reliable way. Such an assumption is quite strong, especially if the information in the training samples is not complete (e.g., incomplete rankings). When we allow the prediction $\hat{y}$ to be a partial order, $A_\tau(y, \hat{y})$ needs to be adapted.

[13] propose to decompose the usual accuracy measures into two components: the correctness (CR) measuring the quality of the predicted comparisons; and the completeness (CP) measuring the completeness of the prediction. They are defined as

$$CR(y, \hat{y}) = \frac{C - D}{C + D} \quad \text{and} \quad CP(y, \hat{y}) = \frac{C + D}{m(m-1)/2}, \tag{2}$$

where $\hat{y}$ is a partial order and where C and D have the same definitions as in Eq. (1). When the predicted order $\hat{y}$ is complete ($C + D = m(m-1)/2$), $CR(y, \hat{y}) = A_\tau(y, \hat{y})$ and $CP(y, \hat{y}) = 1$, while $CP(y, \hat{y}) = 0$ and by convention $CR(y, \hat{y}) = 1$ if no comparison is done (as all orders are then considered possible).

To summarize, the following assumptions are made in this paper:

- the theoretical model we seek to estimate is a probability measure P defined on the space $\mathscr{L}(\Lambda)$ of complete rankings;
- training instance outputs $y_i, i = 1, \ldots, n$ are allowed to be incompletely observed (i.e., partial orders), while test instances are assumed to be fully observed (i.e. complete rankings);
- the predictions $\hat{y}$ are allowed to be partial orders.

3   Partial Orders Prediction Method

This section describes our likelihood pairwise comparison (LPC) method. It first recalls the principle of pairwise decomposition (Section 3.1). It then details the proposed likelihood-based method used to obtain imprecise estimates (Section 3.2) before discussing how such estimates can be aggregated to obtain partial orders (Section 3.3).

3.1 Pairwise Decomposition

Pairwise decomposition is a well-known procedure used in classification to simplify the initial problem [27,20], which is divided into several binary problems then combined into a final prediction. A similar approach can also be used in preference learning and label ranking problems: it consists in estimating, for each pair of labels λ_i, λ_j, the probabilities $P(\{\lambda_i \succ \lambda_j\})$ or $P(\lambda_j \succ \lambda_i)$ and then to predict an (partial) order on Λ from such estimates. In practice, this can be done by decomposing the data set into $(m-1)m/2$ data sets, one for each pair. This decomposition is illustrated in Figure 1 on an imaginary label ranking training data set with four input attributes.

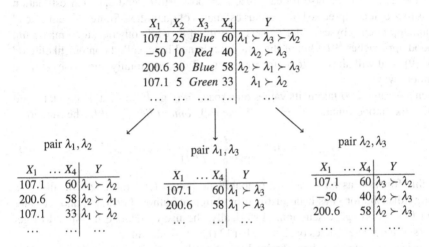

Fig. 1. Pairwise decomposition of rankings

From a data perspective, working with pairwise comparisons is not very restrictive, as almost all models working with preferences can be decomposed into such pairwise preferences: complete rankings, top-t preferences where only the first t labels $\sigma(1),\ldots,\sigma(t)$ are ordered [8], rankings of subsets A of Λ where only labels in A are ordered [22], rankings over a partition of Λ (or bucket orders) [21]. Pairwise preferences are even more general, as most of the previous models cannot model a unique preference $\lambda_i \succ \lambda_j$ [7].

For each pair λ_i, λ_j, the data $(\mathbf{x}_k, y_k)$ is retained if it contains the information $\lambda_i \succ \lambda_j$ or $\lambda_j \succ \lambda_i$, and is forgotten otherwise. Using all data for which λ_i, λ_j are compared, the goal is then to estimate the probability

$$P(\{\lambda_i \succ \lambda_j\}) = 1 - P(\{\lambda_j \succ \lambda_i\}). \tag{3}$$

Any probabilistic binary classifier can be used to estimate this probability once the data set has been decomposed (possibly by mapping $\lambda_i \succ \lambda_j$ to value 1 and $\lambda_i \prec \lambda_j$ to 0 or -1). We will denote by $\hat{P}(\{\lambda_i \succ \lambda_j\})$ the obtained estimate of the theoretical probability $P(\{\lambda_i \succ \lambda_j\})$.

Rather than using only a precise estimate $\hat{P}(\{\lambda_i \succ \lambda_j\})$ as score, we propose in this paper to learn an imprecise estimate of $P(\{\lambda_i \succ \lambda_j\})$ in the form of an interval $[\hat{P}(\{\lambda_i \succ \lambda_j\})] = [\underline{\hat{P}}(\{\lambda_i \succ \lambda_j\}), \overline{\hat{P}}(\{\lambda_i \succ \lambda_j\})]$ and to use such imprecise estimates to predict partial orders. The next section introduces a method inspired from imprecise probabilistic approaches that provides a continuous range of nested imprecise estimates, going from a precise one $(\underline{\hat{P}}(\{\lambda_i \succ \lambda_j\}) = \overline{\hat{P}}(\{\lambda_i \succ \lambda_j\}))$ to a completely imprecise one $([\hat{P}(\{\lambda_i \succ \lambda_j\})] = [0, 1])$.

3.2 Pairwise Imprecise Estimates by Contour Likelihood

One of the goal of imprecise probabilistic methods is to extend classical estimation methods to provide imprecise but cautious estimates of quantities. Some of them extend Bayesian approaches by considering sets of priors [4], while others extend maximum likelihood approaches [9]. One of the latter is retained here, called contour likelihood method [9], as it will allow us to go from a totally precise to a totally imprecise estimate in a smooth way.

Given a parameter θ taking its values on a space Θ (e.g., $P(\{\lambda_i \succ \lambda_j\})$ on $[0, 1]$) and a positive likelihood function $L : \Theta \to \mathbb{R}^+$, we call contour likelihood L^* the function

$$L^*(\theta) = \frac{L(\theta)}{\max_{\theta \in \Theta} L(\theta)}. \qquad (4)$$

Using this function as an imprecise probabilistic model (and more specifically as a possibility distribution) has been justified by different authors [17,31,9], and we refer to them for a thorough discussion. Historically, the use of relative likelihood to get estimates of parameters dates back to Fisher [24] and Birnbaum [5].

Imprecise estimates are then obtained by using the notion of β-cut. Given a value $\beta \in [0, 1]$, the β-cut L^*_β of L^* is the set such that

$$L^*_\beta = \{\theta \in \Theta | L^*(\theta) \geq \beta\}. \qquad (5)$$

Given Eq. (4), we have $L^*_1 = \arg\max_{\theta \in \Theta} L(\theta)$ (the precise maximum likelihood estimator) and $L^*_0 = \Theta$ (the whole set of possible parameter values). In between, we have that $L^*_{\beta_1} \subseteq L^*_{\beta_2}$ for any values $\beta_1 > \beta_2$, that is the lower the value of β, the more imprecise and cautious is our estimate L^*_β. Such estimates are usually simple to obtain and have the advantage (e.g., over frequentist confidence intervals) to follow the likelihood principle, that is to say they depend on the sampling model and data only through the likelihood function (they do not require extra information such as prior probabilities).

In a binary space where $\theta \in [0, 1]$ is the probability of success, Eq. (4) becomes

$$L^*(\theta) = \frac{\theta^s (1 - \theta)^{n-s}}{(s/n)^s (1 - s/n)^{n-s}} \qquad (6)$$

with n the number of observations, s the number of success and $\arg\max_{\theta \in \Theta} L(\theta) = s/n$.

Once they are decomposed into pairwise preferences, we can use training examples $(\mathbf{x}_k, y_k)$, $k = 1, \ldots, n$ and Eq. (6) to estimate $P(\{\lambda_i \succ \lambda_j\})$ for any pair λ_i, λ_j and for a new instance $\mathbf{x}$. To do so, we assume that a metric d is defined (or can be defined) on $\mathcal{X}$ and we propose a simple instance-based strategy.

For a given upper distance $\overline{d}$, let $\mathcal{N}_{\overline{d},\mathbf{x}} = \{\mathbf{x}_i : d(\mathbf{x},\mathbf{x}_i) \leq \overline{d}\}$ the set of training instances whose distance from $\mathbf{x}$ is lower than some (upper) distance $\overline{d}$. In this set of training examples, let us denote by

- $\mathcal{N}_{\overline{d},\mathbf{x}}(i,j) = \{\mathbf{x}_k : \mathbf{x}_k \in \mathcal{N}_{\overline{d},\mathbf{x}}, (\lambda_i \succ \lambda_j) \vee (\lambda_j \succ \lambda_i) \in y_k\}$ the set of all instances that provides a comparison for the labels λ_i and λ_j;
- $\mathcal{N}_{\overline{d},\mathbf{x}}(i > j) = \{\mathbf{x}_k : \mathbf{x}_k \in \mathcal{N}_{\overline{d},\mathbf{x}}(i,j) \wedge (\lambda_i \succ \lambda_j) \in y_k\}$ the set of items where λ_i is preferred to λ_j.

Using these information, interval estimates $[\hat{P}(\{\lambda_i \succ \lambda_j\})]_\beta$ are then simply obtained using Eq. (6) with $|\mathcal{N}_{\overline{d},\mathbf{x}}(i,j)|$ the number of observations, $|\mathcal{N}_{\overline{d},\mathbf{x}}(i > j)|$ the number of successes and β a fixed level of confidence.

Figure 2 pictures two contour likelihoods together with estimates obtained for a given β. As can be seen from the picture, for a given β the imprecision of the estimate $[\hat{P}(\{\lambda_i \succ \lambda_j\})]_\beta$ will depend on the amount of data used to compute $L^*(\theta)$.

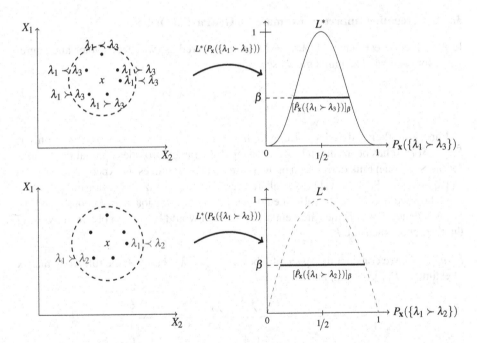

Fig. 2. Imprecise estimates through β-cut of relative likelihood: illustration

As other instance-based methods (e.g., k-nearest neighbour), this approach is based on the assumption that $P_\mathbf{x}$ is *constant* around the instance $\mathbf{x}$. In practice, this means that $\overline{d}$ should not be too high, but also not too small (otherwise there may be no data in the neighbourhood). As we fix $\overline{d}$ rather than the number of neighbours, the presence of missing data will lead to a lower value of $|\mathcal{N}_{\bar{d},\mathbf{x}}(i,j)|$ and to a more imprecise estimate (for a given β). The amount of missing data is therefore automatically considered by the method.

Once precise or imprecise estimates $[\hat{P}(\{\lambda_i \succ \lambda_j\})]$ are obtained, the next step is to combine them to obtain a predicted (partial) order $\hat{y}$. In this paper, we extend the voting approach detailed in [25]. Note that the values that (theoretical) probabilities

$$P(\{\lambda_i \succ \lambda_j\}) = \sum_{y \in \mathcal{L}(\Lambda), \lambda_i \succ \lambda_j} P(\{y\})$$

can assume are constrained, as they are linked by the following weak transitivity relation for any three labels $\lambda_i, \lambda_j, \lambda_k$ [25]:

$$P(\{\lambda_i \succ \lambda_j\}) \geq P(\{\lambda_i \succ \lambda_k\}) + P(\{\lambda_k \succ \lambda_j\}) - 1. \tag{7}$$

This relation may not be satisfied by all estimates $[\hat{P}(\{\lambda_i \succ \lambda_j\})]$. Luckily, post-processing methods of estimates $[\hat{P}(\{\lambda_i \succ \lambda_j\})]$ (such as the voting approach) usually do not require this relation to be satisfied to predict a consistent order.

3.3 Aggregating Imprecise Estimates to Get Partial Orders

In [25], precise estimates $\hat{P}(\{\lambda_i \succ \lambda_j\})$ are considered as (weighted) votes and aggregated for each label λ_i into a global score

$$\hat{S}(i) = \sum_{j \in \{1,\ldots,m\}\setminus i} \hat{P}(\{\lambda_i \succ \lambda_j\}). \tag{8}$$

Labels are then ordered according to their scores $\hat{S}(i)$, that is $\lambda_i \succ \lambda_j$ if and only if $\hat{S}(i) \geq \hat{S}(j)$. It has been shown [25] that using this strategy provides optimal predictions for the Spearman rank correlation in the sense that it maximizes its expected accuracy if $\hat{P}(\{\lambda_i \succ \lambda_j\}) = P(\{\lambda_i \succ \lambda_j\})$ (Kendall Tau can also be optimized by using only $P(\{\lambda_i \succ \lambda_j\})$), however it requires to solve the NP-hard minimum feedback arc set problem [1]). We will denote by $S(j)$ the (theoretical) scores that would have been obtained by using the theoretical measure P.

Example 1. We consider the space of labels $\Lambda = \{\lambda_1, \lambda_2, \lambda_3\}$ with the following matrix of estimates $\hat{P}(\{\lambda_i \succ \lambda_j\})$ and scores $\hat{S}(i)$

$\hat{P}(\{\lambda_i \succ \lambda_j\})$	λ_1	λ_2	λ_3	$\hat{S}(i)$
λ_1		0.6	0.6	1.2
λ_2	0.4		0.3	0.7
λ_3	0.4	0.7		1.1

The obtained prediction is $\lambda_2 \prec \lambda_3 \prec \lambda_1$ (Note that estimates $\hat{P}(\{\lambda_i \succ \lambda_j\})$ satisfy constraints (7) and could originate from a theoretical model P).

Let us now deal with the case where estimates $[\hat{P}(\{\lambda_i \succ \lambda_j\})]$ are imprecise (here, originating from the method presented in Section 3.2). It is straightforward to extend Eq. (8) to imprecise estimates by defining the imprecise score $[\hat{S}(i)]$ as

$$[\hat{S}(i)] = \sum_{j \in \{1,\ldots,m\} \setminus i} [\hat{P}(\{\lambda_i \succ \lambda_j\})] \tag{9}$$

$$= [\sum_{j \in \{1,\ldots,m\} \setminus i} \underline{\hat{P}}(\{\lambda_i \succ \lambda_j\}), \sum_{j \in \{1,\ldots,m\} \setminus i} \overline{\hat{P}}(\{\lambda_i \succ \lambda_j\})].$$

There are multiple ways to compare intervals $[\hat{S}(i)]$ to get a partial or a complete order. Let us denote $\hat{y}_D$ the prediction obtained by a decision rule D and intervals $[\hat{S}(i)]$. We think that decision rules producing partial orders from interval-valued scores should obey at least the two following properties:

Definition 1 (Imprecision monotonicity). *Consider two assessments $[\hat{P}]$ and $[\hat{P}^*]$ such that, for every pair $i, j \in \{1,\ldots,m\}$ we have $[\hat{P}(\{\lambda_i \succ \lambda_j\})] \subseteq [\hat{P}^*(\{\lambda_i \succ \lambda_j\})]$. A decision rule D is said* imprecision monotonic *if*

$$\lambda_i \succ \lambda_j \in \hat{y}_D^* \Rightarrow \lambda_i \succ \lambda_j \in \hat{y}_D$$

for any pair i, j and with $\hat{y}_D, \hat{y}_D^*$ the predictions produced using estimates $[\hat{P}]$ and $[\hat{P}^*]$, respectively.

This property basically says that getting less information cannot make our prediction more precise, in the sense that every label pair comparable according to estimates $[\hat{P}^*]$ should also be comparable according to $[\hat{P}]$ under decision rule D. If we denote by $\mathscr{C}(\hat{y})$ the set of linear extensions (i.e., of completions into complete orders) of the partial order $\hat{y}$, another way to formalise imprecision monotonicity is to ask $\mathscr{C}(\hat{y}_D) \subseteq \mathscr{C}(\hat{y}_D^*)$, that is to require every possible completion of $\hat{y}_D$ to be also a completion of $\hat{y}_D^*$. The second property is not related to imprecision, but to the coherence between the predicted partial order and the complete order that would be obtained using the theoretical model P.

Definition 2 (Model coherence). *Let $S(j)$ be the theoretical scores, y the associated complete order and $[\hat{P}]$ an assessment with associated scores $[\hat{S}(j)]$. Then, a decision rule D is said* model coherent *if*

$$S(j) \in [\hat{S}(j)] \forall j \in \{1,\ldots,m\} \Rightarrow y \in \mathscr{C}(\hat{y}_D)$$

This property requires that if our estimates are consistent with the theoretical model (i.e., include the true value), then the optimal complete ranking is an extension of our prediction. That is, our prediction is totally consistent with the optimal solution, but is possibly incomplete. In particular, satisfying model coherence ensures that the prediction optimizing Spearman rank correlation is in $\mathscr{C}(\hat{y}_D)$, provided $P \in [\hat{P}]$.

To produce a partial ranking from intervals $[\hat{S}(j)]$, we propose to use the following decision rule, that we call *strict dominance* and denote $\mathscr{I}$:

$$\lambda_i \succ_{\mathscr{I}} \lambda_j \Leftrightarrow \underline{S}(i) \geq \overline{S}(j).$$

That is, label λ_i is ranked before label λ_j only when we are certain that the score of λ_j is lower than the one of λ_i. Partial orders obtained following this rule correspond to so-called interval-orders [18], that have been widely studied in the literature. The next proposition shows that such a procedure satisfies the properties we find appealing.

Proposition 1. *The ordering $\succ_{\mathscr{I}}$ is imprecision monotonic and model coherent.*

Proof. Let $\succ_{\mathscr{I}}, \succ_{\mathscr{I}}^*$ be the interval orders obtained by estimates $[\hat{P}]$ and $[\hat{P}^*]$ with rule $\mathscr{I}$.

Imprecision Monotonic: if for every pair $i, j \in \{1, \ldots, m\}$ we have $[\hat{P}(\{\lambda_i \succ \lambda_j\})] \subseteq [\hat{P}^*(\{\lambda_i \succ \lambda_j\})]$, then $[\hat{S}(i)] \subseteq [\hat{S}^*(i)]$ for any label λ_i and $\lambda_i \succ_{\mathscr{I}}^* \lambda_j$ implies $\lambda_i \succ_{\mathscr{I}} \lambda_j$, since the inequalities

$$\overline{\hat{S}}(j) \leq \overline{\hat{S}^*}(j) \leq \underline{\hat{S}^*}(i) \leq \underline{\hat{S}}(i)$$

hold. The second is due to $\lambda_i \succ_{\mathscr{I}}^* \lambda_j$, while the first and third are due to $[\hat{S}(i)] \subseteq [\hat{S}^*(i)]$ for any label λ_i. This is sufficient to show imprecision monotonicity.

Model Coherence: assume that $P(\{\lambda_i \succ \lambda_j\}) \in [\hat{P}(\{\lambda_i \succ \lambda_j\})]$. Then we can show that $\lambda_i \succ_{\mathscr{I}} \lambda_j$ implies $\lambda_i \succ \lambda_j$, where $\succ$ is the ordering obtained from P. Simply observe that inequalities

$$S(j) \leq \overline{\hat{S}}(j) \leq \underline{\hat{S}}(i) \leq S(i)$$

hold as $\lambda_i \succ \lambda_j$ and $S(j) \in [\hat{S}(j)] \forall j \in \{1, \ldots, m\}$ by definition. This is sufficient to show model coherence.

Example 2. Consider the following matrix of imprecise scores that include the matrix of Example 1

$\hat{P}(\{\lambda_i \succ \lambda_j\})$	λ_1	λ_2	λ_3	$S(i)$
λ_1		$[0.4, 0.6]$	0.6	$[1, 1.2]$
λ_2	$[0.4, 0.6]$		$[0.1, 0.3]$	$[0.5, 0.9]$
λ_3	0.4	$[0.7, 0.9]$		$[1.1, 1.3]$

Applying the $\mathscr{I}$ rule results in $\lambda_2 \prec_{\mathscr{I}} \lambda_3$ and $\lambda_2 \prec_{\mathscr{I}} \lambda_1$, without being able to compare λ_3 and λ_1. This prediction is more cautious but coherent with the ordering obtained in Example 1, which was $\lambda_2 \prec \lambda_3 \prec \lambda_1$.

In summary, the likelihood pairwise comparison (LPC) method consists in the following steps:

1. decompose the training data set $(\mathbf{x}_k, y_k)$, $k = 1, \ldots, n$ into pairwise data sets;
2. pick a distance $\overline{d}$ and a level $\beta \in [0, 1]$;
3. for each pair λ_i, λ_j, take as estimate the interval $[\hat{P}(\{\lambda_i \succ \lambda_j\})]_\beta$ from L^*;
4. compute $[\hat{S}(j)]$ and use strict dominance to predict an interval order $\succ_{\mathscr{I}}$.

By varying β, we can go smoothly from a precise ordering $\succ_{\mathscr{I}}$ ($\beta = 1$) to an ordering making no comparison at all ($\beta = 0$), similarly to what is done in [13] by varying the threshold. Note, however, that this approach is quite different from [13,12]:

- we rely on aggregation of imprecise estimates to predict partial orders rather than thresholding a precise model (in [13,12], the prediction $\lambda_i \succ \lambda_j$ is made if $\hat{P}_{\mathbf{x}}(\{\lambda_i \succ \lambda_j\}) \geq \alpha$ with $\alpha \in [0.5, 1]$);
- we always predict an interval order (a family of partial orders that includes semi-orders, the type of orders predicted in [12]) and do not have to face issues related to the presence of cycles [13];
- we use a non-parametric estimation method rather than parametric probabilities [12], which makes our approach computationally more demanding. As the aggregation methods presented in Section 3.3 applies to any imprecise estimates, it would be interesting to study how confidence intervals can be extracted from estimated parametric models, or to which extent are the results affected by considering other imprecise estimates (e.g., confidence intervals).

4 Experiments

In this section, we first compare the performances of our approach with two other techniques in the case of complete order predictions. We use the WEKA-LR [2] implementation. We also discuss the behaviour of our approach with respect to missing data.

The datasets used in the experiments come from the UCI machine repository [19] and the Statlog collection [26]. They are synthetic label ranking data sets built either from classification or regression problems. From each original data set, a transformed data set $(\mathbf{x}_i, y_i)$ with complete rankings was obtained by following the procedure described in [11]. A summary of the data sets used in the experiments is given in Table 1.

Table 1. Experimental data sets

Data set	Type	#Inst	#Attributes	#Labels
authorship	classification	841	70	4
bodyfat	regression	252	7	7
calhousing	regression	20640	4	4
cpu-small	regression	8192	6	5
elevators	regression	16599	9	9
fried	regression	40768	9	5
glass	classification	214	9	6
housing	regression	506	6	6
iris	classification	150	4	3
pendigits	classification	10992	16	10
segment	classification	2310	18	7
stock	regression	950	5	5
vehicle	classification	846	18	4
vowel	classification	528	10	11
wine	classification	178	13	3
wisconsin	regression	194	16	16

4.1   Comparative Experiments with Precise Predictions

To show that our approach performs satisfyingly, we apply our method to complete data sets of Table 1 and compare its results with other label ranking approaches in the case where predictions are complete ($\beta = 1$ in Eq. (5)). More precisely, we compare the results of the proposed approach with the Ranking by Pairwise Comparison method (RPC) using a logistic regression as base classifier [25] and the Label Ranking Tree (LRT) method [11]. Note that if $\beta = 1$ in Eq. (5), LPC is equivalent to adopt a ranking by pairwise comparison approach (RPC) with another base classifier.

Kendall tau is used to assess the accuracy of the classifiers, and reported results are averages over a 10-fold cross validation. Concerning the LPC method, the Euclidean distance d was used, with a maximum radius $\overline{d} = a\mathbb{E}_d$ where $a > 0$ multiplies the average distance

$$\mathbb{E}_d = n(n-1)/2 \sum_{\substack{\mathbf{x}_i, \mathbf{x}_j \in \mathbf{x}_1, \ldots, \mathbf{x}_n \\ j \neq i}} d(\mathbf{x}_i, \mathbf{x}_j)$$

between all training instances. As the goal of these experiments is to assess whether our method provides satisfying results, we set $a = 1.0$ (the effect of modifying a when preferences are missing is studied in the next section).

Table 2. Results on precise case

Data set	LPC accuracy	LPC rank	RPC accuracy	RPC rank	LRT accuracy	LRT rank
authorship	0.910	1	0.908	2	0.887	3
bodyfat	0.216	2	0.282	1	0.11	3
calhousing	0.273	2	0.244	3	0.357	1
cpu-small	0.421	2	0.449	1	0.423	3
elevators	0.701	3	0.749	2	0.758	1
fried	0.789	3	0.999	1	0.888	2
glass	0.853	3	0.887	2	0.893	1
housing	0.699	2	0.674	3	0.799	1
iris	0.92	2	0.893	3	0.947	1
pendigits	0.879	1	0.932	3	0.942	2
segment	0.880	3	0.934	2	0.956	1
stock	0.792	2	0.779	3	0.892	1
vehicle	0.843	2	0.857	1	0.833	3
vowel	0.805	1	0.652	3	0.795	2
wine	0.947	1	0.914	2	0.88	3
wisconsin	0.451	2	0.634	1	0.328	3
Average rank	1.8		2		2.2	

To compare the different results, Demsar [16] approach is used on the results of Table 2. Friedman test was used on the ranks of algorithm performances for each data-set, finding a value 1.13 for the Chi-square test with 2 degree of freedom and a corresponding p-value of 0.57, hence the null-hypothesis (no significant differences between algorithms) cannot be rejected. The algorithms therefore display comparable performances. It should be noted, however, that no optimisation was performed for the proposed method (either on the value $\overline{d}$ or on the shape of the neighbourhood region).

4.2 Accuracy of Partial Predictions

In the previous section, we have shown that our method is competitive with other label ranking methods in situations where complete orders are predicted ($\beta = 1$ in LPC). In this section, we study the behaviour of LPC with respect to completeness and correctness (2) when we allow for partial predictions, that is $\beta \in [0, 1]$ in Eq. (5) and we use the strict dominance rule $\mathscr{I}$ to produce predictions. As in [11,12], we span a whole set of partial orders by going from precise orders ($\beta = 1$) to completely imprecise ones ($\beta = 0$). However we span a richer family of partial orders, namely interval orders.

To study how LPC behaves when some pairwise preferences are missing, we also consider incomplete rankings in training instances. Missing preferences in the training data sets are induced with the following strategy [11]: for a given training instance y_k, each label is removed with a probability γ (here, either 30 or 60%).

Intuitively, we may expect the predictions to be more accurate (i.e., predicted comparisons to be more often correct) as they become more partial. That is, as β decreases, the average completeness CP decreases, with the hope that this decrease is counterbalanced by an increase in correctness CR. To verify this intuition, we have compared our approach with the following base-line: for a given β, we have considered the complete ordering obtained with $\beta^* = 1$ in (5), and have randomly removed each pairwise comparison induced by this ordering with a probability $1 - \beta$.

Figure 3 shows the evolution of completeness and correctness for two data sets (a classification one, vowel, and a regression one, wisconsin) as β decreases for various choices of $\overline{d}$ and for different percentages of missing data. As expected, the (average) correctness is increasing as completeness decreases for our method, while the baseline that performs random suppression of preferences does not show a significant increase of correctness as completeness decreases. This confirms that our method provides cautious yet more accurate predictions as β decreases.

There are other facts that we may notice from the graphs in Figure 3:

- the higher is the distance $\overline{d}$, the more stable is the evolution of correctness/completeness, showing that LPC with higher distances is less affected by missing preferences. In particular, correctness for a level $\beta = 1$ ($CP = 1$) does not change significantly when $\overline{d}$ is high, whether preferences are missing or not. On the contrary, the effect of missing preferences is quite noticeable for lower values of $\overline{d}$, particularly when β is low. This is not surprising, as a higher $\overline{d}$ means using more training instances to assess the model;
- when there are no missing preference, taking a lower $\overline{d}$ usually provides better correctness than a higher one. This can be explained by the fact that the instance-based assumption (i.e. assuming P_x constant around x) becomes less and less supported when $\overline{d}$ increases. However, when the number of missing preferences is significant, correctness is usually better for large $\overline{d}$, as the model is then less sensible to such missing preferences.

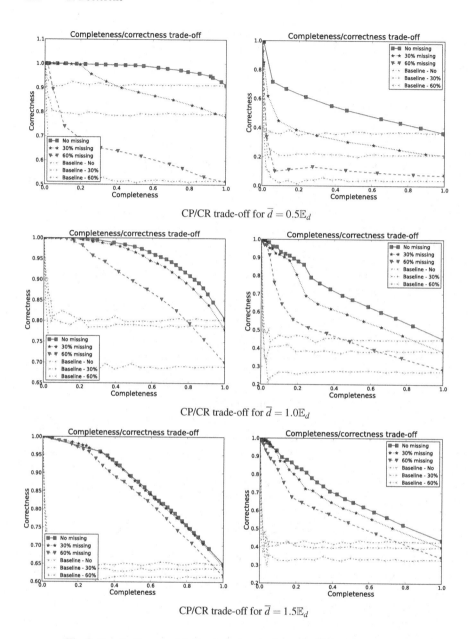

CP/CR trade-off for $\bar{d} = 0.5\mathbb{E}_d$

CP/CR trade-off for $\bar{d} = 1.0\mathbb{E}_d$

CP/CR trade-off for $\bar{d} = 1.5\mathbb{E}_d$

Fig. 3. Results for vowel (left column) and wisconsin (right column) data set

These experiments suggest that the choice of a good $\bar{d}$ heavily depends on the data: if full rankings are given for each training instance, then $\bar{d}$ should be kept low, while if the information is poor (many partial rankings with few preferences), a higher $\bar{d}$ should be preferred.

5 Conclusion

In this paper, we have introduced the likelihood pairwise comparison (LPC) approach to achieve label ranking by pairwise comparisons, which can be seen as an instance-based non-parametric approach.

Although the method can produce complete rankings as predictions, one of its main interest lies in the ability to produce partial but more accurate orders as predictions. This is done by using the interpretation [17,9] seeing the normalised likelihood function as an imprecise probabilistic model, and more precisely as a possibility distribution from which are derived imprecise estimates. To build this normalised likelihood, we have proposed a simple instance-based approach using the neighbours that are within a given radius of the instance.

Our results indicate that the choice of the distance (i.e., radius) used to estimate our model can be important: a higher distance will usually produce less accurate predictions when preferences are complete and more accurate predictions when preferences are missing, while a lower distance will produce more accurate predictions when preferences are complete, but will be more sensible to missing data.

Compared to [12], our method also guarantees the consistency of predicted partial orders while being potentially more expressive, as predicted partial orders are interval orders (that include semi-orders). First experimental results show a good increase of correctness when partial predictions are considered. In the future, it would be interesting to compare the obtained results to other methods, or to study the problem of predicting partial orders from imprecisely specified parametric models (as non-parametric instance-based methods are computationally costly), possibly combining them with other decision rules of imprecise probabilistic approaches [30].

References

1. Alon, N.: Ranking tournaments. SIAM Journal on Discrete Mathematics 20, 137–142 (2006)
2. Balz, A., Senge, R.: Weka-lr: A label ranking extension for weka (July 2011),
   http://www.uni-marburg.de/fb12/kebi/research/software/WEKA-LR-PAGE
3. Bartlett, P., Wegkamp, M.: Classification with a reject option using a hinge loss. The Journal of Machine Learning Research 9, 1823–1840 (2008)
4. Bernard, J.: An introduction to the imprecise dirichlet model for multinomial data. International Journal of Approximate Reasoning 39(2), 123–150 (2005)
5. Birnbaum, A.: On the foundations of statistical inference. Journal of the American Statistical Association 57(298), 269–306 (1962)
6. Bordes, A., Glorot, X., Weston, J., Bengio, Y.: Joint learning of words and meaning representations for open-text semantic parsing. Journal of Machine Learning Research - Proceedings Track 22, 127–135 (2012)
7. Boutilier, C., Lu, T.: Learning mallows models with pairwise preferences. In: Proceedings of the 28th Annual International Conference on Machine Learning - ICML, pp. 145–152 (2011)
8. Busse, L., Orbanz, P., Buhmann, J.: Cluster analysis of heterogeneous rank data. In: ACM International Conference Proceeding Series, vol. 227, pp. 113–120 (2007)
9. Cattaneo, M.: Statistical Decisions Based Directly on the Likelihood Function. Ph.D. thesis, ETH Zurich (2007)

10. Cheng, W., Dembczynski, K., Hüllermeier, E.: Label ranking methods based on the plackett-luce model. In: Proceedings of the 27th Annual International Conference on Machine Learning - ICML, pp. 215–222 (2010)
11. Cheng, W., Hühn, J., Hüllermeier, E.: Decision tree and instance-based learning for label ranking. In: Proceedings of the 26th Annual International Conference on Machine Learning - ICML 2009 (2009)
12. Cheng, W., Hüllermeier, E., Waegeman, W., Welker, V.: Label ranking with partial abstention based on thresholded probabilistic models. In: Advances in Neural Information Processing Systems 25 (NIPS 2012), pp. 2510–2518 (2012)
13. Cheng, W., Rademaker, M., De Baets, B., Hüllermeier, E.: Predicting partial orders: Ranking with abstention. In: Balcázar, J.L., Bonchi, F., Gionis, A., Sebag, M. (eds.) ECML PKDD 2010, Part I. LNCS, vol. 6321, pp. 215–230. Springer, Heidelberg (2010)
14. Corani, G., Antonucci, A., Zaffalon, M.: Bayesian networks with imprecise probabilities: Theory and application to classification. In: Holmes, D.E., Jain, L.C. (eds.) Data Mining: Foundations and Intelligent Paradigms. ISRL, vol. 23, pp. 49–93. Springer, Heidelberg (2012)
15. Dekel, O., Manning, C.D., Singer, Y.: Log-linear models for label ranking. In: Advances in Neural Information Processing Systems (2003)
16. Demšar, J.: Statistical comparisons of classifiers over multiple data sets. The Journal of Machine Learning Research 9(7), 1–30 (2006)
17. Dubois, D., Moral, S., Prade, H.: A semantics for possibility theory based on likelihoods. Journal of Mathematical Analysis and Applications 205, 359–380 (1997)
18. Fishburn, P.: Interval Orderings. Wiley (1987)
19. Frank, A., Asuncion, A.: UCI machine learning repository (2010), http://archive.ics.uci.edu/ml
20. Fürnkranz, J.: Pairwise classification as an ensemble technique. In: Elomaa, T., Mannila, H., Toivonen, H. (eds.) ECML 2002. LNCS (LNAI), vol. 2430, pp. 97–110. Springer, Heidelberg (2002)
21. Gionis, A., Mannila, H., Puolamäki, K., Ukkonen, A.: Algorithms for discovering bucket orders from data. In: Proceedings of the 12th ACM SIGKDD International Conference on Knowledge Discovery and Data Mining, pp. 561–566. ACM (2006)
22. Guiver, J., Snelson, E.: Bayesian inference for plackett-luce ranking models. In: Proceedings of the 26th Annual International Conference on Machine Learning, pp. 377–384. ACM (2009)
23. Har-peled, S., Roth, D., Zimak, D.: Constraint classification: A new approach to multiclass classification and ranking. In: Advances in Neural Information Processing Systems, pp. 785–792 (2002)
24. Hudson, D.: Interval estimation from the likelihood function. Journal of the Royal Statistical Society. Series B (Methodological), 256–262 (1971)
25. Hüllermeier, E., Fürnkranz, J., Cheng, W., Brinker, K.: Label ranking by learning pairwise preferences. Artificial Intelligence 172, 1897–1916 (2008)
26. King, R., Feng, C., Sutherland, A.: Statlog: Comparison of classification algorithms on large real-world problems. Applied Artificial Intelligence 9(3), 289–333 (1995)
27. Lorena, A., de Carvalho, A., Gama, J.: A review on the combination of binary classifiers in multiclass problems. Artificial Intelligence Review 30(1), 19–37 (2008)
28. Marden, J.: Analyzing and modeling rank data, vol. 64. Chapman & Hall/CRC (1996)
29. Meila, M., Chen, H.: Dirichlet process mixtures of generalized mallows models. In: Proceedings of the 26th Conference in Uncertainty in Artificial Intelligence, pp. 358–367 (2010)
30. Troffaes, M.: Decision making under uncertainty using imprecise probabilities. Int. J. of Approximate Reasoning 45, 17–29 (2007)

31. Walley, P., Moral, S.: Upper probabilities based only on the likelihood functions. Journal of the Royal Statistical Society: Series B (Statistical Methodology) 61, 831–847 (1999)
32. Weskamp, N., Hüllermeier, E., Kuhn, D., Klebe, G.: Multiple graph alignment for the structural analysis of protein active sites. IEEE/ACM Transactions on Computational Biology and Bioinformatics 4(2), 310–320 (2007)
33. Zaffalon, M.: Exact credal treatment of missing data. Journal of Statistical Planning and Inference 105(1), 105–122 (2002)

Learning Socially Optimal Information Systems from Egoistic Users

Karthik Raman and Thorsten Joachims

Department of Computer Science,
Cornell University, Ithaca, NY, USA
{karthik,tj}@cs.cornell.edu
http://www.cs.cornell.edu

Abstract. Many information systems aim to present results that maximize the collective satisfaction of the user population. The product search of an online store, for example, needs to present an appropriately diverse set of products to best satisfy the different tastes and needs of its user population. To address this problem, we propose two algorithms that can exploit observable user actions (e.g. clicks) to learn how to compose diverse sets (and rankings) that optimize expected utility over a distribution of utility functions. A key challenge is that individual users evaluate and act according to their own utility function, but that the system aims to optimize collective satisfaction. We characterize the behavior of our algorithms by providing upper bounds on the social regret for a class of submodular utility functions in the coactive learning model. Furthermore, we empirically demonstrate the efficacy and robustness of the proposed algorithms for the problem of search result diversification.

Keywords: Online Learning, Coactive Learning, Implicit Feedback, Diversified Retrieval.

1 Introduction

Many information systems serve a diverse population of users who have conflicting preferences. This poses the challenge of maximizing collective user satisfaction over a distribution of conflicting needs. A typical example is the problem of search result diversification [1]. For an ambiguous query such as "jaguar", a diversified set of results should ideally provide some relevant results for each of the query intents. Similar challenges also arise in an online store that wants to appeal to a range of customers with different tastes, or in a movie recommendation system where even a single user may have different preferences (e.g. moods, viewing companions) on different days. More generally, "diversification" describes the problem of hedging against uncertainty about a user's preferences.

Much prior work on this problem has focused on manually-tuned methods for generating diverse results [2–6]. Some learning approaches exist as well and have been shown to outperform manually tuned methods [7–10]. Unfortunately, the practical use of those learning methods is limited, since most require expert

H. Blockeel et al. (Eds.): ECML PKDD 2013, Part II, LNAI 8189, pp. 128–144, 2013.

annotated training data that explicitly lists all facets of an information need (e.g. the different moods a user can be in).

The use of implicit feedback from user interaction (e.g. clicks) has the potential to overcome this data bottleneck. Not only is it available in abundance, but it also directly reflects the users' – not the experts' – preferences. The challenge, however, is that the learning algorithm no longer gets (expert constructed) examples of socially optimal results, but needs to construct a socially optimal compromise from the egoistic actions of the users. Some learning methods for this problem already exist, but these methods either cannot generalize across queries [11] or are specific to a particular notion of user utility that cannot be adjusted through learning [12].

In this paper we consider the problem of learning socially optimal rankings from egoistic user feedback in the following online learning model. In each iteration a user, drawn from an unknown but fixed distribution, presents a context (e.g., query) to the system and receives a ranking in response. The user is represented by a utility function that determines the actions (e.g. clicks) and therefore the feedback to the learning algorithm. The same utility function also determines the value of the presented ranking. We focus on utility functions that are submodular in the elements of the ranking, since those naturally lead to diverse result sets. The goal is to learn a ranking function that has high social utility, which is the expected utility over the user distribution.

For this model, we present two coactive learning algorithms that learn to compose rankings that optimize social utility. Note that this setup is fundamentally different from previous coactive learning problems [13–15], which assume that user actions always come from the same utility function. After characterizing the informativeness and noisiness of the implicit feedback, we give theoretical results bounding the regret of our algorithms in terms of the social utility. Furthermore, we empirically show that the algorithms perform well for both single query and cross-query diversification tasks. In particular, we show that the algorithms can robustly learn, using only implicit feedback, to compose rankings with an appropriate amount of diversity.

2 Related Work

Coactive learning [13] is a framework for modeling the interaction between users and a learning algorithm, where the user feedback is interpreted as a revealed preference from a boundedly rational user. Recently, coactive learning [14] has been applied to the problem of *intrinsic diversity*. As opposed to our problem (*i.e.,* extrinsic diversity [1]) intrinsic diversity is diversity required by a single individual among their various different interests. More specifically, in their problem there is only a single user utility, based on which feedback is received. However in our problem, users have different utilities, which may conflict, and the goal of the system is finding a socially optimal solution.

Yue and Guestrin [16] also proposed online learning algorithms for the problem of intrinsic diversity. While they too maximize submodular utilities, their model relies on observing cardinal utilities which can be substantially less reliable than preference feedback, as shown in user studies [17]. El-Arini and Guestrin [18] also propose submodularity-based techniques to optimize for both diversity and relevance in the context of scientific literature discovery. However, their model is motivated by *exploration* diversity that hedges against uncertainty about a single utility, while we optimize social utility over a distribution of utility functions.

Our work also relates to a large body of work in the game theory literature on finding social optimally solutions, such as work on maximizing welfare in congestion games [19, 20], auctions [21, 22] and social choices [23]. However, to the best of our knowledge, there has been no work on learning socially optimal rankings from noisy user feedback. While coactive learning is related to partial monitoring games [24], here the loss and feedback matrices are not revealed to the learning algorithm. Furthermore partial monitoring games have no explicit notion of context that is available at the beginning of each round.

3   Learning Problem and Model

Let's start with an example to motivate the formalization of the learning problem considered in this paper. Suppose we have a search engine that receives an ambiguous query (e.g. "jaguar"). In particular, there are three user populations that consider different documents relevant to the query as detailed in Fig. 1. The user populations have different sizes, and Fig. 1 lists the probability of each type. Note that the search engine has no way of identifying which type of user issued the query (*i.e.*, the search engine does not know whether "jaguar" refers to the cat or the car for this user). Suppose the utility of a ranking R to users of type i is $U_i(R) = \sqrt{\#\text{of rel docs in top 4 of R}}$. This means it is beneficial to show at least one relevant document, and that the marginal utility of showing additional relevant documents is sub-linear.

User Type	Prob.	Relevant docs
1	0.5	$a_1, a_2, a_3, \ldots$
2	0.25	$b_1, b_2, b_3, \ldots$
3	0.25	$c_1, c_2, c_3, \ldots$

Fig. 1. Illustrative example showing different user preferences

Now consider the following two rankings that the search engine could show.

- $R_1 = (a_1, a_2, a_3, a_4)$: While ideal for the predominant users (*i.e.*, type 1 users get utility $U_1 = 2$), it provides no value for the other users (utility $U_2 = U_3 = 0$). Thus in expectation, this ranking has expected utility of $\mathbf{E}[U] = 1$.
- $R_2 = (a_1, b_1, c_1, a_2)$: This ranking provides some relevant documents for all user types ($U_1 \sim 1.4; U_2 = 1; U_3 = 1$), maximizing the collective user satisfaction with $\mathbf{E}[U] \sim 1.2$.

Our goal is to find rankings of the latter type, which we call *socially optimal* since they maximize expected utility (*i.e.*, *social utility*).

In this paper we use *implicit feedback* from the users for learning these rankings. Consider, for example, a user of type 1 that chooses to click/read relevant documents a_1, a_2 from the *presented ranking* $\mathbf{y}_t = (b_1, c_1, b_2, a_1, c_2, a_2)$. These actions reveal information about the user's utility functions which we can exploit to construct a *feedback ranking* $\bar{\mathbf{y}}_t$, such as $(b_1, c_1, a_1, b_2, a_2, c_2)$, that has higher utility for that user (or at least not worse utility) *i.e.*, $U_1(\bar{\mathbf{y}}_t) \geq U_1(\mathbf{y}_t)$.

The key challenge in learning socially optimal rankings from the feedback of individual users lies in resolving the contradicting feedback from different user types. Each user's feedback reflects only their own utility, not social utility. For example, even if presented with the socially optimal ranking R_2, users may provide feedback indicating a preference for a different ranking (e.g. type 1 users may indicate their preference for R_1). Thus, a successful learning algorithm for this problem should be able to reconcile such differences in preference and display stability despite the *egoistic* feedback, especially when the presented solution approaches the optimal.

3.1 Learning Problem

We now define the learning problem and user-interaction model more formally. We assume there are N types of users, each associated with a probability p_i according to which individual users accessing the system are sampled. Given a context $\mathbf{x}_t$ (e.g., query), the **personal utility** of an object (*e.g.*, ranking) $\mathbf{y}_t$ for users of type i is $U_i(\mathbf{x}_t, \mathbf{y}_t)$. The **social utility** $U(\mathbf{x}_t, \mathbf{y}_t)$ is defined as the expected utility over the user distribution.

$$U(\mathbf{x}_t, \mathbf{y}_t) = \mathbf{E}[U_i(\mathbf{x}_t, \mathbf{y}_t)] = \sum_{i=1}^{N} p_i U_i(\mathbf{x}_t, \mathbf{y}_t) \tag{1}$$

The optimal object for context $\mathbf{x}_t$ and user type i is denoted as

$$\mathbf{y}_t^{*,i} := \arg \max_{\mathbf{y}_t \in \mathcal{Y}} U_i(\mathbf{x}_t, \mathbf{y}_t). \tag{2}$$

The socially optimal object for context $\mathbf{x}_t$ is denoted as

$$\mathbf{y}_t^* := \arg \max_{\mathbf{y}_t \in \mathcal{Y}} U(\mathbf{x}_t, \mathbf{y}_t). \tag{3}$$

Users interact with the system like in the standard coactive learning model [13], but it is no longer assumed that all users act according to a single utility function. Specifically, at each timestep t the system receives a context $\mathbf{x}_t$ and a user type i is sampled from the user distribution. In response, the system presents the user with an object $\mathbf{y}_t$ and the user draws utility $U_i(\mathbf{x}_t, \mathbf{y}_t)$. The algorithm then observes (implicit) feedback from the user (who acts according to U_i), updates its model, and repeats. The goal of the algorithm is to present objects as close to the social optimal $\mathbf{y}_t^*$, as measured by the following notion of regret over time steps t of the learning process:

$$REG_T := \frac{1}{T} \sum_{t=0}^{T-1} \left(U(\mathbf{x}_t, \mathbf{y}_t^*) - U(\mathbf{x}_t, \mathbf{y}_t) \right). \tag{4}$$

Thus the lower the regret, the better the performance of the algorithm. Note that the social optimal $\mathbf{y}_t^*$ is never given to the learning algorithms, but nevertheless used to measure predictive performance.

To be able to prove anything about the regret of any learning algorithm, we need to make an assumption about the quality of the user feedback. Coactive learning assumes that the user feedback reveals an object $\bar{\mathbf{y}}_t$ that has improved utility compared to the presented object $\mathbf{y}_t$. In a ranked retrieval system, for example, $\bar{\mathbf{y}}_t$ can be constructed from $\mathbf{y}_t$ by moving the clicked documents to the top. Unlike in the traditional coactive learning model, this paper studies the case where users do not provide feedback from a single global utility function that directly reflects social utility. Instead, users valuate and provide feedback according to their own personal utility. Thus, we characterize feedback quality through the following definition.

Definition 1. *User feedback is* **expected** α_i, δ_i**-informative** *for a presented object* $\mathbf{y}_t$ *under context* $\mathbf{x}_t$ *for a user with personal utility function* U_i, *if* $\bar{\xi}_t \in \Re$ *is chosen such that for some given* $\alpha_i \in]0,1]$ *and* $\delta_i > 0$ *it holds that*

$$\mathbf{E}_{\bar{\mathbf{y}}_t}[U_i(\mathbf{x}_t, \bar{\mathbf{y}}_t)] \geq (1 + \delta_i)U_i(\mathbf{x}_t, \mathbf{y}_t) + \alpha_i \left(U_i(\mathbf{x}_t, \mathbf{y}_t^{*,i}) - U_i(\mathbf{x}_t, \mathbf{y}_t) \right) - \bar{\xi}_t. \quad (5)$$

Note that the expectation is over the user feedback.

The expected α_i, δ_i-informative criterion states that the user's feedback object $\bar{\mathbf{y}}_t$ has better personal utility than the presented object $\mathbf{y}_t$ on average. More precisely, the first term on the right-hand side implies that the improvement should be at least by a factor of $(1 + \delta_i)$. Note, though, that this condition is based only on the personal utility of the specific user, not the social utility. The second term on the right-hand side further prescribes that personal utility increases proportional to how far $\mathbf{y}_t$ is away from the optimal object $\mathbf{y}_t^{*,i}$, and the factor $\alpha_i \in [0,1]$ describes the informativeness of the feedback. This second term captures that it is easier to make large improvements in utility when the presented $\mathbf{y}_t$ is far from optimal for this user. Finally, since it would be unreasonable to assume that user feedback is always strictly α_i, δ_i-informative, the $\bar{\xi}_t$ captures the amount of violation.

3.2   Submodular Utility Model

The following defines the class of utility function we consider for modeling users. As done in previous work [14], we assume that the utility functions $U_i(\mathbf{x}_t, \mathbf{y}_t)$ is linear in its parameters $\mathbf{v}_i \in \mathbf{R}^m$.

$$U_i(\mathbf{x}_t, \mathbf{y}_t) = \mathbf{v}_i^\top \phi_F(\mathbf{x}_t, \mathbf{y}_t) \quad (6)$$

$\phi_F(\mathbf{x}_t, \mathbf{y}_t)$ is a feature vector representation of the context-object pair and F is a submodular function as further elaborated on below. We require that all $\mathbf{v}_i$'s and $\phi_F(\mathbf{x}_t, \mathbf{y}_t)$'s are component-wise non-negative. The linear model implies that one can write the *social utility* as

$$U(\mathbf{x}_t, \mathbf{y}_t) = \mathbf{w}_*^\top \phi_F(\mathbf{x}_t, \mathbf{y}_t), \text{ where } \mathbf{w}_* = \sum_{i=1}^N p_i \mathbf{v}_i. \quad (7)$$

Algorithm 1. GreedyRanking($\mathbf{w}, \mathbf{x}$)

1: $\mathbf{y} \leftarrow 0$
2: **for** $i = 1$ **to** k **do**
3:     $bestU \leftarrow -\infty$
4:     **for all** $d \in \mathbf{x}/\mathbf{y}$ **do**
5:         **if** $\mathbf{w}^{\top}(\mathbf{x}, \mathbf{y} \oplus d) > bestU$ **then**
6:             $bestU \leftarrow \mathbf{w}^{\top}\phi(\mathbf{x}, \mathbf{y} \oplus d)$
7:             $best \leftarrow d$
8:     $\mathbf{y} \leftarrow \mathbf{y} \oplus best$ {Append document to ranking $\mathbf{y}$}
9: **return** $\mathbf{y}$

We model $\phi_F(\mathbf{x}_t, \mathbf{y}_t)$ using a submodular aggregation of its components, which is a well accepted method for modeling diversity [9, 14]. To simplify the exposition, we focus on rankings as objects $\mathbf{y}$, but analogous constructions also work for other types of objects. Given context $\mathbf{x}$, each document in ranking $\mathbf{y} = (d_{i_1}, d_{i_2}, \ldots, d_{i_n})$ has a feature representation given by $\phi(\mathbf{x}, d_{i_j}) \in \mathbf{R}^m$. We then obtain the overall feature vector $\phi_F(\mathbf{x}, \mathbf{y})$ as

$$\phi_F^j(\mathbf{x}, \mathbf{y}) = F(\gamma_1 \phi^j(\mathbf{x}, d_{i_1}), \gamma_2 \phi^j(\mathbf{x}, d_{i_2}), \ldots \ldots, \gamma_n \phi^j(\mathbf{x}, d_{i_n})) \qquad (8)$$

where $\phi^j(\mathbf{x}, d)$ and $\phi_F^j(\mathbf{x}, \mathbf{y})$ represent the jth feature in the vectors $\phi(\mathbf{x}, d)$ and $\phi_F(\mathbf{x}, \mathbf{y})$ respectively. We also introduce position-discounting factors $\gamma_1 \geq \ldots \geq \gamma_j \geq \ldots \geq \gamma_n \geq 0$, which determine how important each position in the ranking is. The choice of aggregation function F determines the diminishing returns profile of the users utility. For example, using a coverage-like aggregation function $F(A) = \max_{a \in A} a$, strongly promotes diversity, since a single document can already maximize utility. On the other extreme lies the additive aggregation function $F(A) = \sum_{a \in A} a$, which leads to a diversity-agnostic (i.e., modular) feature vector. More generally, any monotone increasing and concave function of $\sum_{a \in A} a$ can be used. It was shown [9, 14] that this allows for a broad class of performance measures to be modeled, including many common IR performance metrics (e.g. NDCG, Precision, Coverage).

For a component-wise non-negative vector $\mathbf{w}$, we can compute a ranking that approximately maximizes the utility function, *i.e.*, $\mathbf{y} := \arg\max_{\mathbf{y} \in \mathcal{Y}} \mathbf{w}^{\top}\phi_F(\mathbf{x}, \mathbf{y})$, using the Greedy Algorithm 1. The algorithm iteratively picks the document with the highest marginal utility to be added to the ranking. Despite its simplicity, Algorithm 1 has good approximation properties for this NP-hard problem.

Lemma 1. *For $\mathbf{w} \geq 0$ and monotone, concave $F : \mathbf{R}_{\geq 0}^n \to \mathbf{R}_{\geq 0}$ that commutes in all arguments, Algorithm 1 produces a ranking that is a β_{gr}-approximate solution, with $\beta_{gr} = \left(1 - \frac{1}{e}\right)$ if $\gamma_1 = \cdots = \gamma_k$ or $\beta_{gr} = 1/2$ otherwise.*

Proof. For $\gamma_1 = \ldots = \gamma_k$ this is a straightforward reduction to monotone submodular maximization with a cardinality constraint for which the greedy algorithm is $(1 - \frac{1}{e})$-approximate [25]. For the more general case we reduce it to submodular maximization over a partition matroid. Suppose we have documents $\{d_1, \ldots, d_N\}$ and want to find a ranking of length k. Let the new ground set A contain k copies $d_{i,j} : j \in \{1, k\}$ of each document d_i, one for each position. The matroid only permits sets containing at most one document per position. Define set function H over A: For set $B(\subseteq A)$, let $C = \{\ldots d_{i_j, j} \ldots\}$ be the set obtained by removing all duplicates from B (*i.e.*, keep only the highest ranked occurrence of a document). Define $H(B) = F(\ldots, \gamma_j \phi(\mathbf{x}, d_{i_j}), \ldots)$. The lemma follows from observing that Algorithm 1 is equivalent to the greedy algorithm for maximizing H over A under a matroid constraint, which is known to provide a $\frac{1}{2}$-approximate solution [25].

4   Social Learning Algorithms

In this section, we present two coactive learning algorithms for predicting rankings that optimize social utility. The first considers rankings with discount factors for each rank while the second considers the special case of evaluating the top k results as a set. For both algorithms, we characterize their regret by providing upper bounds.

4.1   Social Perceptron for Rankings (SoPer-R)

Following the utility model introduced in Section 3.2, we now present an algorithm for learning rankings $\mathbf{y} = (d_{i_1}, d_{i_2}, \ldots, d_{i_n})$ that aim to optimize social utility where personal user utility can be represented as

$$U_i(\mathbf{x}_t, \mathbf{y}_t) = \mathbf{v}_i^\top \phi_F(\mathbf{x}_t, \mathbf{y}_t), \tag{9}$$

$$\phi_F^j(\mathbf{x}, \mathbf{y}) = F(\gamma_1 \phi^j(\mathbf{x}, d_{i_1}), \gamma_2 \phi^j(\mathbf{x}, d_{i_2}), \ldots \ldots, \gamma_n \phi^j(\mathbf{x}, d_{i_n})), \tag{10}$$

with $\gamma_1 \geq \gamma_2 \geq \ldots \geq \gamma_n \geq 0$. The submodular DCG metric proposed in [9], where the discount factors are $\gamma_i = \frac{1}{\log_2(1+i)}$, is an example of such a utility function.

The Social Perceptron for Rankings (SoPer-R) is detailed in Algorithm 2. It applies to any F that satisfies the conditions of Lemma 1. The algorithm maintains a weight vector $\mathbf{w}_t$, which is its estimate of $\mathbf{w}_*$. For the given context $\mathbf{x}_t$, the algorithm first computes ranking $\mathbf{y}_t$ using the greedy Algorithm 1, which is then presented to the user. The user actions (e.g., clicks) are observed and used to construct the feedback as follows. The ranking is first partitioned into adjacent pairs by randomly selecting an odd or even grouping. The feedback ranking $\bar{\mathbf{y}}_t$ is constructed by swapping the documents whenever the user clicks on the lower element of the pair. This relates to the idea of FairPairs [26], which is used to help de-bias click data. Note that feedback is only generated whenever the lower elements was clicked but not the upper, otherwise $\bar{\mathbf{y}}_t := \mathbf{y}_t$. After the feedback $\bar{\mathbf{y}}_t$ is received, the algorithm performs a perceptron-style update to the weight vector. To to ensure that the weight vector contains only non-negative weights, any negative weights are *clipped* to zero.

Given function g and constant $0 \leq \lambda \leq 1$ define $\tau_g(\lambda)$ as:

$$\tau_g(\lambda) = \lim_{x \to 0} \frac{g(\lambda \cdot x, 0, \ldots, 0)}{g(x, 0, \ldots, 0)} \tag{11}$$

Algorithm 2. Social Perceptron for Ranking (SoPer-R)

1: Initialize $\mathbf{w}_0 \leftarrow \mathbf{0}$
2: **for** $t = 0$ **to** $T - 1$ **do**
3:     Observe $\mathbf{x}_t$
4:     Present $\mathbf{y}_t \leftarrow GreedyRanking(\mathbf{w}_t, \mathbf{x}_t)$ {Present argmax ranking}
5:     Observe user clicks $\mathcal{D}$ {Get User Feedback}
6:     Construct feedback $\bar{\mathbf{y}}_t \leftarrow ListFeedback(\mathbf{y}_t, \mathcal{D})$ {Create Feedback Object}
7:     Update: $\bar{\mathbf{w}}_{t+1} \leftarrow \mathbf{w}_t + \phi(\mathbf{x}_t, \bar{\mathbf{y}}_t) - \phi(\mathbf{x}_t, \mathbf{y}_t)$ {Perceptron Update}
8:     Clip: $\mathbf{w}_{t+1}^j \leftarrow \max(\bar{\mathbf{w}}_{t+1}^j, 0)$ $\quad \forall 1 \leq j \leq m$.
9:
10: **Function ListFeedback$(\mathbf{y}, \mathcal{D})$**   {$\mathbf{y}$: Presented Ranking; $\mathcal{D}$: User clicks }
11: $\bar{\mathbf{y}} \leftarrow \mathbf{y}$ {Initialize with presented object}
12: With probability 0.5: $\mathcal{PR} \leftarrow (\{1, 2\}, \{3, 4\}, \{5, 6\} \cdots)$
13: else: $\mathcal{PR} \leftarrow (\{1\}, \{2, 3\}, \{4, 5\}, \{6, 7\} \cdots)$
14: **for** $i = 0 \cdots len(\mathcal{PR})$ **do**
15:     $\{j_{upper}, j_{lower}\} \leftarrow \mathcal{PR}[i]$ {Get Pair}
16:     **if** $\mathbf{y}[j_{lower}] \in \mathcal{D}$ AND $\mathbf{y}[j_{upper}] \notin \mathcal{D}$ **then**
17:         Swap($\bar{\mathbf{y}}[j_{upper}], \bar{\mathbf{y}}[j_{lower}]$)   {Place clicked doc above the other doc}
18: **return** $\bar{\mathbf{y}}$

The below lemma bounds the change in a concave function on scaling arguments.

Lemma 2. *For any function g (satisfying the conditions of Lemma 1), constant $0 \leq \lambda \leq 1$ and values $v_1, v_2, \ldots, v_n \geq 0$, we can bound the change in value of g on scaling the values v_i by λ as follows:*

$$g(v_1, \ldots, v_i, \ldots, v_n) \geq \tau_g(\lambda) \cdot g(\lambda \cdot v_1, \ldots, \lambda \cdot v_i, \ldots, \lambda \cdot v_n) \tag{12}$$

We use this to characterize the sequence of position discounts and their smoothness, which is a key parameter of the main theorem. Thus for a utility measure with function F and γ_i discount factors, we define:

$$\Gamma_F = 1 - \min_i \tau_F\left(\frac{\gamma_{i+1}}{\gamma_i}\right) \tag{13}$$

We can now characterize the regret suffered by the SoPer-R algorithm for list-based utilities, as shown below in Theorem 1.

Theorem 1. *For any $\mathbf{w}_* \in \mathbf{R}^m$ and $\|\phi(\mathbf{x}, \mathbf{y})\|_{\ell_2} \leq R$ the average regret of the SoPer-R algorithm can be upper bounded as:*

$$REG_T \leq \frac{1}{\eta T} \sum_{t=0}^{T-1} \mathbf{E}_i[p_i \bar{\xi}_t] + \frac{\beta R \|\mathbf{w}_*\|}{\eta} + \frac{\sqrt{2}\sqrt{4 - \beta^2} R \|\mathbf{w}_*\|}{\eta \sqrt{T}}. \tag{14}$$

with: $\delta_i \geq \left(\Gamma_F \cdot \frac{1-p_i}{p_i}\right)$, $\eta = \min_i p_i \alpha_i$ *and* $\beta = (1 - \beta_{gr}) = \frac{1}{2}$.

Before presenting the proof of the theorem, we first analyze the structure of the regret bound. The first term on the right-hand side characterizes in how far the user feedback violates the desired α_i, δ_i-informative feedback assumption due to model misspecification and bias/noise in the user feedback. This term implies that the regret does not necessarily converge to zero in such cases.

The second term results from the fact that we can only guarantee a β_{gr}-approximate solution for greedy Algorithm 1. In practice, however, the solutions computed by greedy Algorithm 1 tend to be much better, making the second term much smaller than in the worst case.

The third and final term converges to zero at a rate of $T^{0.5}$. Note that none of the terms in the bound depend explicitly on the number of features, but that that it scales only in terms of margin $R\|\mathbf{w}_*\|$.

Proof. From Lemma 1, we get that:

$$\mathbf{w}_t^\top \phi(\mathbf{x}_t, \mathbf{y}_t) \geq \beta_{gr} \mathbf{w}_t^\top \phi(\mathbf{x}_t, \bar{\mathbf{y}}_t)$$
$$\mathbf{w}_t^\top (\phi(\mathbf{x}_t, \bar{\mathbf{y}}_t) - \phi(\mathbf{x}_t, \mathbf{y}_t)) \leq (1 - \beta_{gr})\mathbf{w}_t^\top \phi(\mathbf{x}_t, \bar{\mathbf{y}}_t) \leq \beta R \|\mathbf{w}_t\| \tag{15}$$

Next, we bound the ℓ_2 norm of $\mathbf{w}_T$:

$$\begin{aligned}
\|\mathbf{w}_T\|^2 &= \|\mathbf{w}_{T-1}\|^2 + 2\mathbf{w}_{T-1}^\top(\phi(\mathbf{x}_{T-1}, \bar{\mathbf{y}}_{T-1}) - \phi(\mathbf{x}_{T-1}, \mathbf{y}_{T-1})) \\
&\quad + \|\phi(\mathbf{x}_{T-1}, \bar{\mathbf{y}}_{T-1}) - \phi(\mathbf{x}_{T-1}, \mathbf{y}_{T-1})\|^2 \\
&\leq \|\mathbf{w}_{T-1}\|^2 + 2\beta\|\mathbf{w}_{T-1}\|R + 4R^2 \\
&\leq (\beta T + \sqrt{4 - \beta^2}\sqrt{2T})^2 R^2
\end{aligned} \tag{16}$$

Eq. (15) is used for the second inequality. The last line is obtained using the inductive argument made in [14]. Similarly we bound $\mathbf{E}[\mathbf{w}_T^\top \mathbf{w}_*]$ using Cauchy-Schwartz and concavity:

$$\|\mathbf{w}_*\|\mathbf{E}[\|\mathbf{w}_{T+1}\|] \geq \mathbf{E}[\mathbf{w}_T^\top \mathbf{w}_*] = \sum_{t=0}^{T-1} \mathbf{E}[U(\mathbf{x}_t, \bar{\mathbf{y}}_t) - U(\mathbf{x}_t, \mathbf{y}_t)] \tag{17}$$

Now we use the α_i, δ_i-informativeness condition:

$$\begin{aligned}
\mathbf{E}[U_i(\mathbf{x}_t, \bar{\mathbf{y}}_t) - U_i(\mathbf{x}_t, \mathbf{y}_t)] &\geq \alpha_i \left(U_i(\mathbf{x}_t, \mathbf{y}_t^{*,i}) - U_i(\mathbf{x}_t, \mathbf{y}_t) \right) + \delta_i U_i(\mathbf{x}_t, \mathbf{y}_t) - \bar{\xi}_t \\
&\geq \frac{\eta}{p_i}\left(U_i(\mathbf{x}_t, \mathbf{y}_t^{*,i}) - U_i(\mathbf{x}_t, \mathbf{y}_t) \right) + \delta_i U_i(\mathbf{x}_t, \mathbf{y}_t) - \bar{\xi}_t
\end{aligned} \tag{18}$$

Next we bound the expected difference in the social utility between $\bar{\mathbf{y}}_t$ and $\mathbf{y}_t$ IF a user of type i provided feedback at iteration t:

$$\begin{aligned}
\Delta_i &= \mathbf{E}[U(\mathbf{x}_t, \bar{\mathbf{y}}_t) - U(\mathbf{x}_t, \mathbf{y}_t)] \geq -\Gamma_F \sum_{j \neq i} p_j U_j(\mathbf{x}_t, \mathbf{y}_t) + p_i \mathbf{E}[U_i(\mathbf{x}_t, \bar{\mathbf{y}}_t) - U_i(\mathbf{x}_t, \mathbf{y}_t)] \\
&= -\Gamma_F(U(\mathbf{x}_t, \mathbf{y}_t) - p_i U_i(\mathbf{x}_t, \mathbf{y}_t)) + p_i \mathbf{E}[U_i(\mathbf{x}_t, \bar{\mathbf{y}}_t) - U_i(\mathbf{x}_t, \mathbf{y}_t)] \\
&\geq -\Gamma_F U(\mathbf{x}_t, \mathbf{y}_t) + p_i \Gamma_F U_i(\mathbf{x}_t, \mathbf{y}_t) + \eta\left(U_i(\mathbf{x}_t, \mathbf{y}_t^{*,i}) - U_i(\mathbf{x}_t, \mathbf{y}_t) \right) + p_i \delta_i U_i(\mathbf{x}_t, \mathbf{y}_t) - p_i \bar{\xi}_t \\
&\geq \eta\left(U_i(\mathbf{x}_t, \mathbf{y}_t^{*,i}) - U_i(\mathbf{x}_t, \mathbf{y}_t) \right) + \Gamma_F\left(U_i(\mathbf{x}_t, \mathbf{y}_t) - U(\mathbf{x}_t, \mathbf{y}_t) \right) - p_i \bar{\xi}_t
\end{aligned} \tag{19}$$

Algorithm 3. Social-Set-Based-Perceptron(C, M, p)

1: **Function SetFeedback$(\mathbf{y}, \mathcal{D})$**
2: $\bar{\mathbf{y}} \leftarrow \mathbf{y}$ {Initialize with presented object}
3: $\mathcal{D}_{\mathcal{O}} \leftarrow \mathcal{D}/\mathbf{y}[1:M]$ {Clicks on docs outside top M}
4: **for** $i = 1 \cdots \min(C, |\mathcal{D}_{\mathcal{O}}|)$ **do**
5:    $c \leftarrow \mathcal{D}_{\mathcal{O}}[i]$ {Clicked document}
6:    $u \leftarrow$ Random (unclicked) document from $\mathbf{y}[1:M]$ {Non-clicked document}
7:    Swap$(\bar{\mathbf{y}}[j_u], \bar{\mathbf{y}}[j_c])$
8: **return** $\bar{\mathbf{y}}$

The first line is obtained by using Lemma 2 and definition of Γ_F (Eq. 13). The second line uses the definition of the social utility (Eq. 1). The third line uses Eq. 18. The fourth step uses the condition on δ_i and rearranging of terms. Note that the expectations in the above lines are w.r.t. the user feedback (and the feedback construction process).

We next consider the expected value of Δ_i (over the user distribution):

$$\mathbf{E}_i[\Delta_i] = \mathbf{E}[U(\mathbf{x}_t, \bar{\mathbf{y}}_t) - U(\mathbf{x}_t, \mathbf{y}_t)] \geq \eta\Big(\mathbf{E}_i[U_i(\mathbf{x}_t, \mathbf{y}_t^{*,i})] - U(\mathbf{x}_t, \mathbf{y}_t)\Big) - \mathbf{E}_i[p_i \bar{\xi}_t]$$

$$\geq \eta\Big(U(\mathbf{x}_t, \mathbf{y}_t^*) - U(\mathbf{x}_t, \mathbf{y}_t)\Big) - \mathbf{E}_i[p_i \bar{\xi}_t] \quad (20)$$

where the second line uses the fact that $\mathbf{E}_i[U_i(\mathbf{x}_t, \mathbf{y}_t^{*,i})] \geq U(\mathbf{x}_t, \mathbf{y}_t^*)$. We can put together Eqns. 16, 17 and 20 to give us the required bound.

4.2   Social Perceptron for Sets (SoPer-S)

While DCG-style position discounts γ_i that decay smoothly are often appropriate, other models of utility require more discrete changes in the rank discounts. The *coverage* metric is an example of such a metric, which measures what fraction of the users will find atleast 1 document relevant to them in the set of M documents [11, 8, 12]. We call these metrics *set-based*, since they consider the first M documents in a ranking as a set (*i.e.*, position within the top-M positions does not matter). Clearly, we can model such metrics by setting the γ_i in the aggregation step (defined in Eq. 8) as

$$\gamma_i = \begin{cases} 1 & \text{if } i \leq M \\ 0 & \text{if } i > M. \end{cases}$$

However, the bound in Theorem 1 can be rather loose for this case, and the pairwise feedback construction model "wastes" information. In particular, since utility is invariant to reordering in the top M or below the top M, only pairwise feedback between position M and $M+1$ provides information. To overcome this problem, we now present an alternate algorithm that is more appropriate for set-based utility functions.

The *Social Perceptron for Sets* (SoPer-S), shown in Algorithm 3, uses the same basic algorithm, but replaces the feedback mechanism. Now, clicked documents outside the top M are swapped with a random non-clicked document in the top M. This leads to a feedback set $\bar{\mathbf{y}}_t$ (of size M), that contains more (or at least as many) of the user's preferred documents than the top M elements of the presented ranking. Note that during the feedback creation, we only consider the first C clicks outside the top M. This parameter C is used to restrict the difference between the feedback set and the presented set. We now state a lemma we will use to bound the regret of the SoPer-S algorithm.

Lemma 3. *For any non-negative, submodular function g and set X with $|X| = n$, we can lower bound the function value of a random subset of size k as:*

$$\mathbf{E}_{Y:Y\subseteq X,|Y|=k}[g(Y)] \geq \frac{k}{n}g(X) \tag{21}$$

Using Lemma 3, we can now characterize the regret suffered by the SoPer-S algorithm for set-based utilities, as shown below in Theorem 2.

Theorem 2. *For any $\mathbf{w}_* \in \mathbf{R}^m$ and $\|\phi(\mathbf{x},\mathbf{y})\|_{\ell_2} \leq R$ the average regret of the SoPer-S algorithm can be upper bounded as:*

$$REG_T \leq \frac{1}{\eta T}\sum_{t=0}^{T-1}\mathbf{E}_i[p_i\bar{\xi}_t] + \frac{\beta R\|\mathbf{w}_*\|}{\eta} + \frac{\sqrt{2}\sqrt{4-\beta^2}R\|\mathbf{w}_*\|}{\eta\sqrt{T}}. \tag{22}$$

with: $\delta_i \geq \left(\frac{C}{M}\cdot\frac{1-p_i}{p_i}\right)$, $\eta = \min_i\ p_i\alpha_i$ *and* $\beta = (1-\beta_{gr}) = \frac{1}{e}$.

Note that the proposed algorithms are efficient (due to the online updates) and scalable with the greedy algorithm requiring $O(nk)$ time to find a length k ranking over n documents. This can be further improved using lazy evaluation.

5   Empirical Evaluation

In this section, we empirically analyze the proposed learning algorithms for the task of *extrinsic* [1] search result diversification. In particular, we (a) explore how well the algorithms perform compared to existing algorithms that do single-query learning; we (b) compare how close our algorithms get to the performance of algorithms that require expert annotated examples of socially optimal ranking for cross-query learning; and (c) we explore the robustness of our algorithm to noise and misspecification of the utility model.

5.1   Experiment Setup

We performed experiments using the standard diversification dataset from the *TREC 6-8 Interactive Track*. The dataset contains 17 queries, each with binary relevance judgments for 7 to 56 different user types, which we translate into binary utility values. Similar to previous work [9], we consider the probability of a user type to be proportional to the number of documents relevant to that

Table 1. Summary of key properties of the TREC dataset

Statistic	Value
Average number of documents per query	46.3
Average number of user types	20.8
Fraction of docs. relevant to > 1 user	0.21
Average number of users a document is relevant for	1.33
Fraction of docs. relevant to most popular user	0.38
Average probability of most popular user	0.29

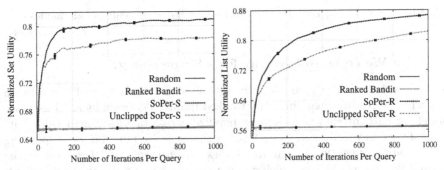

Fig. 2. Performance of different methods for single-query learning to diversify. Performance is averaged over all queries, separately considering Set Utility (Left) and List Utility (Right). Standard error bars are shown in black.

user type. Also following [9], we only consider documents that are relevant to at least 1 user type to focus the experiments on learning to diversify, not learning to determine relevance. Table 1 summarizes some key properties of the data.

To simulate user behavior, we use the following model. Users scan the documents of a ranking in order and click on the first document they consider relevant. Each (binary) decision of relevance is made incorrectly with a small probability of error. This error probability was set to zero for most experiments but later varied when studying the effect of user noise.

Unless mentioned otherwise, we used the coverage function ($F(x_1, \ldots, x_n) = \max_i x_i$) to define the submodular function for utility aggregation. We measured performance of the different methods in terms of the utility being optimized - *i.e.*, Set Utility (of size 5 sets) for the Set-Based methods and List Utility (up to rank 5) with DCG discounting factors, for the List-Based methods. Additionally we normalize the maximum scores per query to 1 (i.e., $\forall x : U(\mathbf{x}, \mathbf{y}^*) = 1$), so as to get comparable scores across queries. We report the performance of each algorithm in terms of its running average of these scores (i.e., $1 - REG_T$).

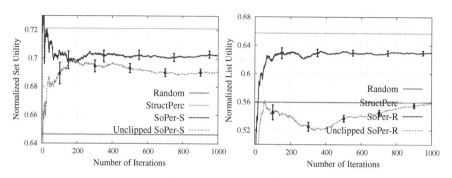

Fig. 3. Set (L) and List (R) Utilities for learning to diversify across queries

5.2   Can We Learn to Diversify for a Single Query?

We first evaluate our algorithms in the setting of the Ranked Bandits algorithm [11], which serves as a baseline. The Ranked Bandit algorithm learns a separate model for each query and cannot generalize across queries. Furthermore, its original version was limited to optimizing the coverage function, corresponding to the *max* aggregation in our framework. We use the *UCB1* variant of the Ranked Bandits algorithm, which was empirically found to be the best variant.

As a second baseline we report randomly ordering the results. Note that this is a competitive baseline, since (a) all documents are relevant to at least 1 user, and (b) the probability of users is proportional to the number of documents relevant to them.

For the SoPer-R and SoPer-S algorithms, documents were represented as unit-norm TFIDF word vectors. All learning algorithms were run twice for each of the 17 queries (with different random seeds) and the results are averaged across all 34 runs. As seen from Figure 2, the proposed algorithms perform much better than either of the two baselines. The Ranked Bandits algorithm converges extremely slowly, and is barely better than the random baseline after 1000 iterations. Both the SoPer-R and SoPer-S algorithm are able to learn substantially faster. Already within 200 iterations, the SoPer-S method is able to provide at least 1 relevant document to 80% of the user population, while random and Ranked Bandits perform at around 65%. Thus both proposed methods are clearly able to learn the diversity required in such rankings from individual user feedback.

We also explore variants of the SoPer-S and SoPer-R algorithms where we omit the final step of clipping negative weights to 0. While the unclipped versions of both algorithms still perform better than random, they fall short of the corresponding clipped versions as seen from Figure 2. Thus we can conclude that ensuring non-negative weights not only guarantees theoretical results, but is important for empirical performance.

5.3   Can We Learn a Cross-Query Model for Diversification?

While the previous experiments indicate that the new algorithms can learn to diversify for a single query, such single-query learning is restricted to frequent queries that are issued hundreds of times. Instead, it is more desirable for diversification models to be trained across a distribution of queries.

To get a suitable representation that allows cross-query learning, we use the same *word-importance* feature vectors that were used in previous work on learning from expert-annotated feedback [8, 9]. These features capture both the overall importance of a word (e.g., "Does the word appears in at least $x\%$ of the documents?"), as well as the importance in the documents of the ranking (e.g., "Does the word appear with frequency of atleast $y\%$ in the document?"). Using different such values of x and y along with other similar features, we get a total of 1197 features.

To produce the following results, all methods were run for 1000 iterations with 5 random seeds. The values reported are averaged across these 5 runs.

In this cross-query setting, we cannot apply Ranked-Bandits as it only works for a single query. Thus we again use the Random baseline in this experiment. Existing supervised learning algorithms for diversification are also not applicable here, as they require explicit training data of socially optimal rankings (i.e., knowledge of all document-user relevance labels). However, we would like to estimate how well our algorithms can learn from (far weaker) implicit feedback data, in relation to conventional methods trained in such a *full information* setting. Thus we trained a structural perceptron, which internally uses the greedy algorithm for prediction. This uses the same feature vector representation as our algorithm, but is provided the *social optimal* at every iteration.

Fig. 3 shows the average utility for the SoPer-S and SoPer-R algorithms, as well as the random baseline and the Structured Perceptron after 1000 iterations. Both SoPer-S and SoPer-R substantially outperform the random baseline, indicating that the proposed algorithms can learn to diversify for this cross-query setting. Both methods get close to the performance of the supervised method despite learning from far weaker feedback. For example, the SoPer-S method is able to satisfy 70% of the user population, as compared to the 64% of the baseline and 72% of the Structured Perceptron. We also again evaluate the unclipped versions of the algorithms. For the the unclipped SoPer-R, performance never rises above random, indicating the practical importance of maintaining a positive weight vector to ensure good performance of the greedy algorithm.

5.4   How Robust Are the Algorithms to Misspecification of the Model?

While the previous experiments showed that the algorithms can learn efficiently when the submodular function of the user population (as used in computing the personal and social utilities) and the algorithm match, we now study what happens when there is a mismatch. More specifically, for the *cross-query diversification* setting, we ran the algorithms with three different submodular functions

Table 2. Set and List Utilities (with standard error) when the two submodular functions *i.e.*, of the population (fixed for row) and the algorithm (fixed for column) are mismatched

UserF	SET			
	Max	Sqrt	Lin	Rand
Max	.699 ±.005	.695 ±.005	.683 ±.005	.646 ±.006
Sqrt	.675 ±.006	.686 ±.006	.706 ±.006	.634 ±.006
Lin	.509 ±.006	.532 ±.006	.574 ±.007	.492 ±.006

UserF	LIST			
	Max	Sqrt	Lin	Rand
Max	.630 ±.007	.620 ±.006	.618 ±.006	.557 ±.006
Sqrt	.656 ±.007	.654 ±.007	.684 ±.006	.610 ±.007
Lin	.500 ±.006	.504 ±.006	.566 ±.007	.474 ±.007

Table 3. Ranking performance in the presence of feedback noise

Utility	Random	No Noise	Noise
Set	.646 ±.006	.699 ±.005	.694 ±.006
List	.557 ±.006	.630 ±.007	.631 ±.007

as defined by the concave function F: a) **Max**: $F(x_1, \ldots, x_n) = \max_i x_i$; b) **Lin**: $F(x_1, \ldots, x_n) = \sum_i x_i$; c) **Sqrt**: $F(x_1, \ldots, x_n) = \sqrt{\sum_i x_i}$. We also varied the population utility to each of these three functions, and obtained the average utility value (after 200 iterations) for all 9 combinations of functions. Note that we still ensured that SoPer-R was used to optimize the List based utilities, while SoPer-S was used for set-based ones.

The results (averaged over 5 runs) are shown in Table 2. We find that for both methods and all three population utility functions, the utility value is always better than the random baseline, regardless of the algorithm and function used. While the values may be highest when the functions align, we still find significant improvements over the baselines even when there is a mismatch. In fact, for some situations we find that the utility is highest when there is a mismatch: The case of a linear algorithm utility but SQRT population utility is one such example. We conjecture that is due to the relatively small set/list size of 5. On short rankings LIN and SQRT do not differ as much as on longer rankings. Additionally LIN does not suffer any approximation degradation as the greedy algorithm always provides an optimal solution for LIN.

5.5   Is the Method Robust to Noise in the Feedback?

In the real world, users make errors in judging the relevance of documents. To model this, we simulated users who make an error in each binary relevance judgment with 0.1 probability. This means that, as users go down the ranking, they may flip the true relevance label. Users now return as feedback the first document they *perceive* as relevant, which contains significant noise. We ran both our algorithms and measured the average utility after 200 iterations in the cross-query setting, with matching algorithm and population utilities using the Max function.

Table 3 shows the results (averaged over 5 runs) comparing the performance of the algorithms in both the noise-free and noisy settings. We see that the performance for both SoPer-S and SoPer-R is almost the same, with the gap to the baseline still being significant. The robustness to noise is also supported by the theoretical results. In particular, note that the definition of α_i, δ_i-informative feedback only requires that feedback be informative in expectations, such that the slack terms $\bar{\xi}_t$ may be zero even for noisy feedback. In general, we conclude that the algorithms are robust and applicable in noisy settings.

6   Conclusions

We proposed two online-learning algorithms in the coactive setting for aggregating the conflicting preferences of a diverse user population into a ranking that aims to optimize social utility. Formalizing the learning problem and model as learning an aggregate utility function that is submodular in the elements of the ranking and linear in the parameters, we were able to provide regret bounds that characterize the worst-case behavior of the algorithm. In an empirical evaluation, the algorithms learned substantially faster than existing algorithms for single-query diversification. For learning cross-query diversification models, the algorithms are robust and the first that can be trained using implicit feedback. This work was supported in part by NSF Awards IIS-1217686, IIS-1247696, IIS-0905467, the Cornell-Technion Joint Research Fund, and a Google Fellowship.

References

1. Radlinski, F., Bennett, P.N., Carterette, B., Joachims, T.: Redundancy, diversity and interdependent document relevance. SIGIR Forum 43(2), 46–52 (2009)
2. Carbonell, J., Goldstein, J.: The use of MMR, diversity-based reranking for reordering documents and producing summaries. In: SIGIR, pp. 335–336 (1998)
3. Zhai, C.X., Cohen, W.W., Lafferty, J.: Beyond independent relevance: methods and evaluation metrics for subtopic retrieval. In: SIGIR, pp. 10–17 (2003)
4. Chen, H., Karger, D.R.: Less is more: probabilistic models for retrieving fewer relevant documents. In: SIGIR, pp. 429–436 (2006)
5. Swaminathan, A., Mathew, C.V., Kirovski, D.: Essential pages. In: Web Intelligence, pp. 173–182 (2009)
6. Clarke, C.L.A., Kolla, M., Vechtomova, O.: An effectiveness measure for ambiguous and underspecified queries. In: Azzopardi, L., Kazai, G., Robertson, S., Rüger, S., Shokouhi, M., Song, D., Yilmaz, E. (eds.) ICTIR 2009. LNCS, vol. 5766, pp. 188–199. Springer, Heidelberg (2009)
7. Santos, R.L., Macdonald, C., Ounis, I.: Selectively diversifying web search results. In: CIKM, pp. 1179–1188 (2010)

8. Yue, Y., Joachims, T.: Predicting diverse subsets using structural SVMs. In: ICML, pp. 1224–1231 (2008)
9. Raman, K., Joachims, T., Shivaswamy, P.: Structured learning of two-level dynamic rankings. In: CIKM, pp. 291–296 (2011)
10. Kulesza, A., Taskar, B.: Learning determinantal point processes. In: UAI, pp. 419–427 (2011)
11. Radlinski, F., Kleinberg, R., Joachims, T.: Learning diverse rankings with multi-armed bandits. In: ICML, pp. 784–791 (2008)
12. Slivkins, A., Radlinski, F., Gollapudi, S.: Ranked bandits in metric spaces: learning optimally diverse rankings over large document collections. JMLR 14, 399–436 (2013)
13. Shivaswamy, P., Joachims, T.: Online structured prediction via coactive learning. In: ICML (2012)
14. Raman, K., Shivaswamy, P., Joachims, T.: Online learning to diversify from implicit feedback. In: KDD, pp. 705–713 (2012)
15. Raman, K., Joachims, T., Shivaswamy, P., Schnabel, T.: Stable coactive learning via perturbation. In: ICML (2013)
16. Yue, Y., Guestrin, C.: Linear submodular bandits and their application to diversified retrieval. In: NIPS, pp. 2483–2491 (2012)
17. Joachims, T., Granka, L., Pan, B., Hembrooke, H., Radlinski, F., Gay, G.: Evaluating the accuracy of implicit feedback from clicks and query reformulations in web search. TOIS 25(2) (April 2007)
18. El-Arini, K., Guestrin, C.: Beyond keyword search: discovering relevant scientific literature. In: KDD, pp. 439–447 (2011)
19. Blumrosen, L., Dobzinski, S.: Welfare maximization in congestion games. In: EC, pp. 52–61 (2006)
20. Meyers, C.A., Schulz, A.S.: The complexity of welfare maximization in congestion games. Netw. 59(2), 252–260 (2012)
21. Dobzinski, S., Schapira, M.: An improved approximation algorithm for combinatorial auctions with submodular bidders. In: SODA, pp. 1064–1073 (2006)
22. Feige, U.: On maximizing welfare when utility functions are subadditive. In: STOC, pp. 41–50 (2006)
23. Boutilier, C., Caragiannis, I., Haber, S., Lu, T., Procaccia, A.D., Sheffet, O.: Optimal social choice functions: a utilitarian view. In: EC, pp. 197–214 (2012)
24. Bartók, G., Pál, D., Szepesvári, C.: Toward a classification of finite partial-monitoring games. In: Hutter, M., Stephan, F., Vovk, V., Zeugmann, T. (eds.) Algorithmic Learning Theory. LNCS, vol. 6331, pp. 224–238. Springer, Heidelberg (2010)
25. Nemhauser, G.L., Wolsey, L.A., Fisher, M.L.: An analysis of approximations for maximizing submodular set functions. Mathematical Programming 14, 265–294 (1978)
26. Radlinski, F., Joachims, T.: Minimally invasive randomization for collecting unbiased preferences from clickthrough logs. In: AAAI, pp. 1406–1412 (2006)

Socially Enabled Preference Learning from Implicit Feedback Data

Julien Delporte[1], Alexandros Karatzoglou[2],
Tomasz Matuszczyk[3], and Stéphane Canu[1]

[1] INSA de Rouen, LITIS, Rouen, France
`firstname.lastname@insa-rouen.fr`
[2] Telefonica Research, Barcelona, Spain
`alexk@tid.es`
[3] Tuenti, Barcelona, Spain
`tomasz@tuenti.com`

Abstract. In the age of information overload, collaborative filtering and recommender systems have become essential tools for content discovery. The advent of online social networks has added another approach to recommendation whereby the social network itself is used as a source for recommendations i.e. users are recommended items that are preferred by their friends.

In this paper we develop a new model-based recommendation method that merges collaborative and social approaches and utilizes implicit feedback and the social graph data. Employing factor models, we represent each user profile as a mixture of his own and his friends' profiles. This assumes and exploits "homophily" in the social network, a phenomenon that has been studied in the social sciences. We test our model on the *Epinions* data and on the *Tuenti* Places Recommendation data, a large-scale industry dataset, where it outperforms several state-of-the-art methods.

1 Introduction

Online social networks (OSN) provide users with new forms of interaction that currently shape the social lives of millions of people. The main ingredient of the success of OSN's is the ease with which friendships, groups and communities arise. These groups often arise among like-minded users, i.e. users that share the same interests. To explain our inexorable tendency to link up with one another in ways that reinforce rather than test our preferences sociologists in the 1950s, coined the term "homophily" a Greek word meaning love of the same.

Fundamental to online social networks and their commercial success is the commercial exploitation of this phenomena. The principle of homophily is used to recommend products and services through the social graph, i.e. if your friends like an item it will be recommended to you. In effect, the social graph is used as the recommendation engine. Leveraging the social graph to serve the user with potentially useful services (e.g. places, videos, coupons, etc.) can improve the

H. Blockeel et al. (Eds.): ECML PKDD 2013, Part II, LNAI 8189, pp. 145–160, 2013.

satisfaction the involvement and the time the user spends on the network. Most recommendation algorithms work by modeling the bipartite graph of user-item preferences. In effect, an implicit social network among users who share the same taste is built and exploited. Methods based on this principle are often referred to as Collaborative Filtering (CF) methods.

Notation. Before going any further with CF some notations have to be introduced. The data from which recommendations can be produced is typically derived from interactions between users $i \in \mathcal{U}$ and items $j \in \mathcal{M}$ with a response $Y_{ij} \in \mathcal{Y}$. The data for n total users interacting with a collection of m items can be thought of as a sparse $n \times m$ so-called user-item matrix $Y \in \mathcal{Y}^{|\mathcal{U}| \times |\mathcal{M}|}$ where $|\mathcal{U}|$ denotes the cardinal of set $\mathcal{U}$. In this context, $Y_{ij} = 1$ indicates the existence of an interaction (purchase, rating, etc.) between user i and item j. In this sense, $Y_{ij} = 0$ is special since it does *not* indicate that a user dislikes an item but rather that data is missing. We thus only have implicit information on which items a user likes. To thus avoid an estimator that is overly optimistic with regards to user preferences we need to take into account unobserved entries ($Y_{ij} = 0$) as some form of negative feedback. Moreover from the social network graph we know the set of friends $\mathcal{F}_i \subset \mathcal{U}$ of user i.

A class of CF methods often used in recommender systems is memory or similarity based methods [1] that work by computing similarity measures (*e.g.* Pearson correlation) between users. Another common approach to collaborative filtering and recommendation is to fit a factor model (e.g. [2]) to the data. For example by extracting a feature vector U_i, M_j for each user and item in the data set such that the inner product of these features minimizes an explicit or implicit loss functional following a probabilistic approach). The underlying idea behind these methods is that both user preferences and item properties can be modeled by a number of latent factors.

The basic idea in matrix factorization approaches is to fit the original Y matrix with a low rank approximation $F = UM$ where matrix U contains the user features and M the item features. More specifically, the goal is to find such an approximation that minimizes the sum of the squared distances $\sum_{ij}(Y_{ij} - F_{ij})^2$ between the known entries in Y and their predictions in F. Combining the two approaches, i.e. direct recommendation over the social graph and recommendations using a collaborative filtering method can yield significant advantages both in terms of the quality of the recommendations but also in terms of computational efficiency and speed.

In most recommendation domains the data come in the form of implicit feedback (purchases, clicks, etc.) in contrast to explicit feedback such as ratings where a user explicitly expresses his positive, neutral or negative attitude towards an item. A key challenge in modeling implicit feedback data is defining negative feedback, since in this case the observed data (user-item interactions) can only be considered as a form of positive feedback. Moreover for non-observed *user − item* interactions we cannot be certain if the user did not consider the items or if the user considered the items and simply chose not to interact with the items (reflecting a negative feedback). Hence we cannot ignore these entries as this would lead to a model that would be overly optimistic with regard to user preferences.

The Socially Enabled Collaborative Filtering model denoted as *SECoFi* introduced here has several novel aspects:

- We develop a collaborative filtering model that also directly models the social interactions and quantifies the influence/trust between each users based on the implicit feedback data from the user and his friends.
- We develop a way to quantify and use this influence in the proposed collaborative filtering model, to our knowledge this is the first model to do so without precomputing any affinity or similarity measures among users.
- *SECoFi* scales linearly to the number of user-item interactions and is tested on a large-scale industry dataset with over 10M users where it outperforms state-of-the-art socially enabled collaborative filtering methods.
- We extensively test *SECoFi* on two datasets (*Tuenti, Epinions*) and compare it to three state-of-the-art socially-enabled collaborative filtering methods and a matrix factorization method.

2 Socially Enabled Collaborative Filtering

The main idea behind factor models is to fit a model of a d dimensional latent user $U \in \mathbb{R}^{|\mathcal{U}| \times d}$ and item factors $M \in \mathbb{R}^{d \times |\mathcal{M}|}$ so that the scores between a user and an item calculated by the inner product between the corresponding rows of the user i and item j latent factor matrices $F_{ij} = U_i M_j$ can be used to provide recommendations typically by displaying the top N scoring items to the user. The latent factors U and M are typically computed by minimizing some objective functions that either stem from regularized loss functions [3,4,5] or are derived from probabilistic models [6]. In both cases the objectives are of the form:

$$L(F, Y) + \Omega(F) \tag{1}$$

where $L(F, Y)$ is typically a loss function such as Frobenius norm of the error $\|F - Y\|_F^2$ and $\Omega(F)$ is a regularization term preventing from overfitting. A typical choice is the Frobenius norm of the factors $\|M\|_F^2 + \|U\|_F^2$ [7].

2.1 Friends Influence

The key challenge of this work is to include the influence of the social graph in a matrix factorization model. We choose to model the users preferences as a mixture of his own and those of his friends. To this end we change the score function to include the influence of the friendship network, and thus the score function becomes:

$$F_{ij} = U_i M_j + \sum_{k \in \mathcal{F}_i} \frac{\alpha_{ik}}{|\mathcal{F}_i|} U_k M_j \tag{2}$$

where α_{ik} is a weight parameter that encodes how much friend k influences user i. This weight α takes value between 0 and 1.

As we presume "homophily" in the social network it is reasonable to assume that some of the users latent preferences might not have been expressed in the user-item data but could instead be encoded in the users friendship network.

Moreover the score function in Equation 2 encodes the fact that the user is "influenced" by his friendship network and the weight α_{ik} quantifies the amount of influence each individual friend k has on the user i. OSN users tend to have dozens of friends and we can expect that the user might have similar preferences to only a fraction of his friends. Moreover it should be noted that the influence is not necessarily symmetric as a user might be "influenced" by a friend but might not be exerting influence on his friend in the same manner.

Given this score function and the objective function in Equation 1, we can devise an objective function with respect to the U, M factors and the influence weights α_{ik}. We define the matrix A such that $A_{ik} = \alpha_{ik}, \forall i, \forall k \in \mathcal{F}_i$, 0 otherwise.

$$\min_{U,M,A} J = \sum_{(i,j)\in\mathcal{Y}} c_{ij} \left(U_i M_j + \sum_{k\in\mathcal{F}_i} \frac{\alpha_{ik} U_k M_j}{|\mathcal{F}_i|} - Y_{ij} \right)^2 + \Omega_{U,M,A} \qquad (3)$$

where $\Omega_{U,M,A} = \lambda_1 \|U\|_F^2 + \lambda_2 \|M\|_F^2 + \lambda_3 \|A\|_F^2$ is a regularizer term and c_{ij} is a constant defined to give more weight to the loss function when dealing with observed entries $Y_{ij} = 1$ than when $Y_{ij} = 0$.

2.2 Optimization

Although Equation 3 is not *jointly convex* in U, M, and A, it is still convex in each of this factors whenever the remaining two are kept fixed. Since we are dealing with implicit feedback data, we cannot give the same importance to information we know to be true, (i.e. the user clicked/purchased an item represented as a 1 in the Y matrix and thus showed and interest in it), and to information we do not know the real meaning (i.e. the user had no interaction with the item thus a 0 in the Y matrix and thus we are unsure about the potential interest). Note that in contrast to factor models for explicit data (i.e. ratings) where learning is performed only over the observed ratings in this case we perform the optimization over the whole matrix Y including the unobserved entries as a form of weak negative feedback.

We optimize the objective function in Equation 3 using the following block Gauss-Seidel process: fixing alternatively two of the three parameters (U, M or A), we update the third parameter. When two out of three parameters are fixed the remaining problem is a basic and convex quadratic least-square minimization that can be efficiently solved. So the optimization process consists in efficiently updating, at each iteration, alternatively the user matrix U, the item matrix M and the weight matrix A.

To get the proper updates for each of the three parameters (U_i, M_j and $\alpha_{ii'}$), we need to calculate the partial derivative of the objective in Formula 3 according to the corresponding factor matrices.

Update U. To compute the update for the factor vector of a single user i, U_i, we calculate $\frac{\partial J}{\partial U_i}$ the derivative of the objective with respect to the users factors and set it to 0. We can then analytically solve this expression with respect to U_i. To formulate the update it is convenient to write the equations in a matrix form. To this end we define a diagonal matrix $C^i \in \mathbb{R}^{|\mathcal{M}| \times |\mathcal{M}|}$ such that $C_{jj}^i = c_{ij}$. c_{ij} encodes the confidence we have in each entry y_{ij} in the Y matrix, i.e. observed entries clicks/purchases etc. get high confidence and thus a higher weight $c_{ij} = 1 + \beta y_{ij}$ where e.g. $\beta = 20$ while when $y_{ij} = 0$ i.e. no action has been taken by user i on item j, $y_{ij} = 0$ and thus $c_{ij} = 1$.

$$U_i = \left(Y_{i\bullet} C^i M^T - \frac{A_i U M C^i M^T}{|\mathcal{F}_i|} \right) (MC^i M^T + \lambda_1 I)^{-1} \qquad (4)$$

In this update rule, the real problem is not the inversion of the $d \times d$ (which is in $O(d^3)$) matrix but the computation of $MC^i M^T$ (which seems to be at first glance $O(|\mathcal{M}| \times d^2)$). Note that $MC^i M^T$ is an operation quadratic in $|\mathcal{M}|$ the number of items. Computing this product is too expensive even for the smallest datasets since it has to be done for each user. In the spirit of [8] we can replace $MC^i M^T$ by $MM^T + M(C^i - I)M^T$. Computing MM^T is independent of the user i and thus can be calculated once before each iteration (and not for each user i), and by cleverly choosing c_{ij}, the product $M(C^i - I)M^T$ can be computed efficiently. Since $c_{ij} = 1 + \beta y_{ij}$, the diagonal terms of $C^i - I$ will be zero for each j where $y_{ij} = 0$. we can thus just compute $M_{y_i}(C^i - I)_{y_i} M_{y_i}^T$, where $\mathcal{Y}_i$ is the set of items of user i. Because matrix Y is by it's nature very sparse, we have $|\mathcal{Y}_i| \ll |\mathcal{M}|$. This leads to a computational complexity of $O(|\mathcal{Y}_i| \times d^2)$ which is linear in the number of items user i had interactions.

Update M. To update matrix M, we use a matrix U' defined by $U_i' = U_i + \sum_{k \in \mathcal{F}_i} \frac{\alpha_{ik} U_k}{|\mathcal{F}_i|}$ for each user i. Using U' the loss function becomes:

$$L(U, M, A) = \sum_{i,j} c_{ij} (U_i' M_j - Y_{ij})^2$$

The partial derivative calculation is pretty much straightforward and can be easily written in a matrix notation if we use a diagonal matrix C^j, defined by $C_{ii}^j = c_{ij}$[1]. The update rule of M_j is as follows:

$$M_j = (U'^T C^j U' + \lambda_2 I)^{-1} U'^T C^j Y_{\bullet j} \qquad (5)$$

To compute the expensive product, we propose to reuse the trick described above by writing it $U'^T C^j U' = U'^T U' + U_{\mathcal{Y}_j}'^T (C^j - I)_{y_j} U_{\mathcal{Y}_j}'$, where $\mathcal{Y}_j$ is the set of the users that have purchased/consumed item j . Just like in the paragraph concerning the update of U, we compute $U'^T U'$ once before the iteration over all items. The computational complexity of the update is $U_{\mathcal{Y}_j}'^T (C^j - I)_{y_j} U_{\mathcal{Y}_j}'$ is $O(|\mathcal{Y}_j| \times d^2)$.

---

[1] Note that $C^j \in \mathbb{R}^{|\mathcal{U}| \times |\mathcal{U}|}$ while $C^i \in \mathbb{R}^{|\mathcal{M}| \times |\mathcal{M}|}$.

Update A. One approach for updating A consists in working row by row, i.e. update $A_{i\bullet}$ for each user i. Since $A_{i\bullet}$ has the same sparsity structure as the adjacency matrix of the social graph we only need to compute the values $A_{i\mathcal{F}_i}$. By using the same procedure as above and setting the partial derivative of the objective to 0 we get:

$$A_{i\mathcal{F}_i} = (Y_{i\bullet} C^i M^T U_{\mathcal{F}_i}^T - U_i M C^i M^T U_{\mathcal{F}_i}^T) \left(\frac{U_{\mathcal{F}_i} M C^i M^T U_{\mathcal{F}_i}^T}{|\mathcal{F}_i|} + \lambda_3 \right)^{-1} \tag{6}$$

Note again that the computational cost for calculating the product $U_i M C^i M^T U_{\mathcal{F}_i}^T$, is limited since we can employ here the same trick we used in the update rules for U and M. The main computational bottleneck is in the computation of the inverse of the matrix which is of size $|\mathcal{F}_i| \times |\mathcal{F}_i|$, implying a complexity in $O(|\mathcal{F}_i|^3)$ i.e. the computation scales cubically to the number of friends per user. Depending on the social network, if we have $d \ll |\mathcal{F}_i|$ for a significant fraction of users, this update rule could be problematic.

Another approach for the update of α, is to compute them not in a user-by-user fashion but relationship-by-relationship, i.e. update $\alpha_{ii'}$ for given user i and friend i'. By calculating the gradient and setting it to zero, we reach the following update rule:

$$\alpha_{ii'} = \left(Y_{i\bullet} - U_i M - \sum_{k \in \mathcal{F}_i k \neq i'} \frac{\alpha_{ik} U_k M}{|\mathcal{F}_i|} \right) C^i M^T U_{i'}^T \left(\frac{U_{i'} M C^i M^T U_{i'}^T}{|\mathcal{F}_i|} + \lambda_3 \right)^{-1} \tag{7}$$

In this case, we just have to invert a scalar. And we can use the same trick as in the update of U to compute the product $M C^i M^T$. This can indeed be rewritten as $M C^i M^T = M M^T + M_{\mathcal{Y}_i} (C^i - I)_{\mathcal{Y}_i} M_{\mathcal{Y}_i}^T$, where $\mathcal{Y}_i$ is the set of the items liked/purchased by the user i.

Given that the complexity of computing Equation 7 is linear to the number of friends of i, while the complexity of Equation 6 is polynomial to the number of friends of i we choose to use 7. Finally note that the α parameters provide a relative measure of the influence (or trust) of a given user on his friends.

Given the optimization procedures for U, M and A we iterativly update each of the factor matrices by keeping the other two factor matrices fixed. We repeat this procedure until convergence.

Prediction. Using Equation 2 at prediction time can be slow since it requires extensive memory access due to the need to retrieve the friends from the social graph. To speedup the computation of the scores at prediction time we can simply precompute the mixed user factors $U_i' = U_i + \sum_{k \in \mathcal{F}_i} \frac{\alpha_{ik}}{|\mathcal{F}_i|} U_k$. The score computation then becomes $F_{ij} = U_i' M_j$.

3 Related Work

Much of the current work on OSN data and collaborative filtering models utilize the social graph data in order to impose additional constrains on the modeling process. Some methods [9,10,11] leverage the OSN graph information in factor models by adding an additional term to the objective function of the matrix factorization that penalizes the distance $|U_i - \frac{1}{|\mathcal{F}_i|} \sum_{k \in \mathcal{F}_i} U_k|^2$ between the factors of friends. This forces profiles among users that are friends to be similar. In [9] a refinement to this approach was presented whereby the penalization of the distance between friends was proportional to a Pearson correlation similarity measure $|U_i - \frac{\sum_{k \in \mathcal{F}_i} sim_{ik} U_k}{\sum_{k \in \mathcal{F}_i} sim_{ik}}|^2$ computed on the items the users had consumed. This enforced even greater similarity among friends that have consumed the same items.

Another approach [12,13] that builds on [14] adds the OSN information by minimizing a second binary loss function $\sum_{k \in \mathcal{F}_i} L(S_{ik}, U_i U_k)$, where S the adjacency matrix of the graph, in the objective function that penalizes mistakes in predicting friendship. These models also leverage side information (i.e. user, item features) in the model. A similar method utilizing both a social regularization and a social loss function approach was introduced in [15].

In [16] a trust ensemble model is introduced, the user is modeled as an ensemble of his own and his friends preferences. While the functional form of this model has similarities with the approach introduced in the current work there are two crucial differences: 1) their method only deals with explicit feedback data (ratings) while we focus on implicit feedback data which is the norm in industry applications, 2) they precompute the weight of the influence or trust of friends on the users based on the ratings, while in *SECoFi* the interaction weights are computed in the model. This allows us to compute the interaction weights even when the users do not actually share a common subset of items. We demonstrate in the **Experiments** section that these are essential components for the performance of the model.

The matrix factorization approach for implicit feedback data introduced in [8] relies on using a least squares loss function and uses a trick that exploits the sparse structure of the data (dominated by non-observed entries) to speed up the optimization process. This approach though does not include any OSN information. An approach that leverages the social network for apps recommendation was introduced in [17]. Approaches such as [18] and [19] exploit geolocation information and context to recommend places to user. The focus of the current work is on the OSN integration for place recommandation.

4 Experiments

Tuenti Places. In the experiment section we use data from the places service of the Tuenti OSN. Tuenti is Spain's leading online Social Network in terms of traffic. Over 80% of Spaniards aged 14-27 actively use the service and today counts more than 14M users and over a billion daily page views. Early 2010,

Table 1. Summary of the data used for the experiments

	Users	Places/Items	Edges in SN
Tuenti 10M		100K	700M
Epinions 50K		140K	500K

a feature was added to the Tuenti web platform whereby users could tell their friends where they were, and which places they particularly enjoyed. These places where added to the users profile.

The Data. To test *SECoFi*, we used the *Tuenti* place-user interaction matrix, as the matrix Y, that contains all the places the users have added to their profile. We also used the social network $\mathcal{F}$, i.e. the friendship matrix of *Tuenti* users. The data contains about 10 million users and approximately 100,000 places. Both of the matrices are very sparse, as each user has on average 4 places in his profile and 60 friends. The social graph among the *Tuenti* users contains approximately 700M edges that is each user has on average 70 friends. Note that this is an industry-scale dataset where the user/places graph takes up 2GB of storage space and the social graph data 22GB.

The *Epinions* data contains about 50k users and approximately 140,000 articles. Here users form a social graph (500k edges) based on the trust they show on each others reviews/ratings. Unlike the *Tuenti* data, the *Epinions* data is in the form of ratings with values between 1 and 5. We replace the rating values by 1 to convert the data to implicit feedback (just like in the KDD cup challenge 2007 *who-rated-what?*).

In contrast to the *Tuenti* data the relationships of the users are much more well defined in the *Epinions* data in that they reflect trust in another users opinion. Social relationships as the ones in the *Tuenti* data capture a much wider range of relationships between users e.g. family relationships, neighbours, classmates etc. which might not always translate into trust/influence.

Evaluation Protocol. For the evaluation procedure we adopt a similar strategy to [20]. We split the dataset into two parts, a training set to learn our model and a test set for evaluation. The test set contains the last 25% of places or items added to each users profile, and the training set contains all the remaining places/items that where added in the users profile. For each user we draw randomly some unobserved entries $Y_{ij} = 0$ assuming that these places/items are irrelevant to the user. We use these randomly chosen unobserved entries for training some of the methods in comparison (see Section 4). We used this protocol for both datasets.

We trained the model to compute a score F_{ij} for each user i place j in the test set along with the randomly drawn irrelevant items and rank the items for each user according to their scores. In recommendation algorithms we ultimately care about the ranking of the items, we thus use ranking metrics for the evaluation.

A popular list-wise ranking measure for this type of data is the Mean Average Precision metric (MAP) which is particularly well suited to recommendations ranking since it puts an emphasis in getting the first items in the ranked list right. MAP can be written as in equation 8.

$$\text{MAP} = \frac{1}{|\mathcal{U}|} \sum_{i=1}^{|\mathcal{U}|} \frac{\sum_{k=1}^{|\mathcal{M}|} P(k) Y_{ik}}{|\mathcal{Y}_i|} \tag{8}$$

Where $P(k)$ is the precision at the cut-off k. We also compute the RANK metric described in [8] to evaluate the performance of the different models. In contrast to the MAP metric, here smaller values indicate better performance. As we have no rating data, the RANK metric can be written, as follows:

$$\text{RANK} = \frac{\sum_{i,j} Y_{ij} rank_{ij}}{|\mathcal{Y}|} \tag{9}$$

Where $rank_{ij}$ is the percentile-ranking of the item j for a given user i.

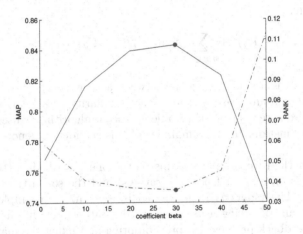

Fig. 1. The MAP and RANK metrics with respect to the value of the coefficient β

Methods in Comparison. The first method we compare against is a matrix factorization method based on alternating least squares optimization described in [8]. This method is tailored to implicit feedback data, but does not take the social graph into account. We can gauge based on the comparison with this method how much the use of the social data improves the recommendation performance. In the remaining we denote this method as *iMF*.

The second method we compare against is of [12], which takes advantage of the social graph along with contextual information to perform their recommendation. The resulting model is used to predict both items and friends for a given user. As the focus here is on the social aspect we do not use any contextual information but only the social graph. Thus adapting their objective function to our evaluation environment we end up optimizing:

$$\min_{U,M} \sum_{(i,j)\in\mathcal{Y}} L(U_i M_j, Y_{ij}) + \sum_{i,i'\in\mathcal{F}_i} L(U_i U_{i'}^T, S_{ij}) + \Omega_{U,M} \qquad (10)$$

Where S represents the social graph (in which $S_{ii'}$ is 1 if the users i and i' are friends, 0 otherwise), and where L and Ω are respectively the loss function and the regularizer. The method was tested with several different loss function, we picked the one that gave the best results, the logistic loss function and used a simple $l2$-norm for the regularization term. Following [12] a stochastic gradient descent algorithm was used to optimize this objective. For the rest of the experiment section, we will denote this method as *LLA*.

The third method we compare against was introduced in [9] and takes the social data into account by penalizing the $l2$ distance between friends in the objective function. Two ways are proposed to penalize the distance between friends, we choose the one that gave them the best performance, i.e the one denoted individual-based regularization. The objective function minimized in [9] is the following:

$$\min_{U,M} \sum_{(i,j)\in\mathcal{Y}} (U_i M_j - Y_{ij})^2 + \sum_{i,i'\in\mathcal{F}_i} \mathrm{sim}(i,i')\|U_i - U_{i'}\|_F^2 + \lambda_1\|U\|^2 + \lambda_2\|M\|^2$$

Where $\mathrm{sim}(i,i')$ is a similarity score between a user i and a user i'. This similarity can be computed using vector space similarity or a Pearson correlation coefficient. Also here a stochastic gradient descent algorithm is used to optimize the objective function. In the remainder of this section we denote this method as *RSR*.

The last method we compare against is the one described in [16]. The focus of that work is on explicit feedback (ratings) and the social trust matrix A is precomputed. The model is then trained by optimizing a simple loss among the factors U and V, using a user-item rating dataset. We fit their method to the implicit feedback problem by, precomputing and fixing the matrix A at the beginning, and using the ponderation trick on the objective (with the use of the coefficients c_{ij}) to make implicit feedback learning possible. We will denote this method as *Trust Ensemble*. We also compare *SECoFi* to a baseline : the average predictor, which will recommend the overall most popular places to each user.

Computational Complexity. We first validate the efficiency of *SECoFi* by measuring the time needed to execute one iteration of the *SECoFi* method using a varying portion of the training data. We expect the method to show linear scalability in terms of the users and the observed entries in the user/item dataset.

To this end we use the *Epinions* data and run one iteration of the algorithm for each random data split. Those tests has been performed using a single Intel i5 core. The resulting timing information is displayed in Figure 2. Note the linear growth in the running-time of the method given the different data splits.

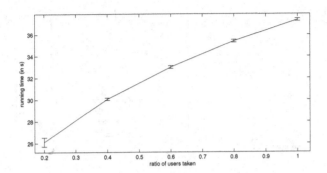

Fig. 2. The running time of one iteration of *SECoFi* given different random data splits 20%, 40% etc. over the *Epinions data*

Results on the *Tuenti* Data. We performed cross-validation for model selection. We randomly initialize U and M drawing from a uniform distribution between 0 and 1. For the initialization of the friendship weight α_{ij}, we found out empirically that the best performance is achieved by initializing with $\alpha_{ij} = 1$. We also estimate that the optimal value of the parameter β (used in the coefficient c_{ij}) is $\beta = 30$, according to the MAP and the RANK metrics (see Figure 1). We used this value of β for all the experiments.

We validate the performance of *SECoFi* also over a range of values of the number of factors d parameter (1, 5, 10, 15 and 20) on *Tuenti*. We repeated the experiments several (10) times for each method and report the mean values of the runs along with the standard deviations. We run experiments for different values of d for each method, results are shown in Figure 3.

We observe that even for a small number of factors, *SECoFi* outperforms the alternative social *LLA* and *RSR* enabled methods both in terms of MAP and RANK (over 17% improvement for the MAP and over 14% for the RANK). Moreover *SECoFi* is significantly better than *iMF* in terms of MAP, and for higher values of d our method becomes statistically equivalent to *iMF* in terms of RANK. Note that for recommendations where only a small number of items k is shown to the user the importance of MAP is bigger then RANK since MAP is a top-biased evaluation measure, i.e. placing items at the top of the list is more important then lowering the overall ranking of the all the items. *SECoFi* clearly outperforms in terms of MAP and RANK the *Trust Ensemble* method. Surprisingly *iMF* seems to outperform the alternative socially-enabled LLA and RSR methods in the comparison. One of the reason for this might be the strong

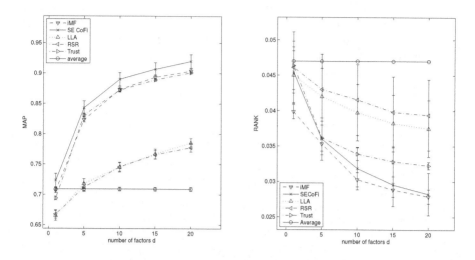

Fig. 3. The MAP and RANK metrics of the various methods on the *Tuenti* data depending on the number of factors d

sparsity of the data which bodes well with methods that take all the non-observed entries into account.

The relative performance between the methods does not depend strongly on the number of latent variables used. Except for *Trust Ensemble* method which was statistically equivalent to our method for small dimension, but we clearly see the difference for bigger dimensions. Indeed, *SECoFi* outperforms *Trust Ensemble* as well in terms of MAP as of RANK for a number of factors $d \geq 10$. *SECoFi* outperforms the other methods for all the values of d we tested with.

We thus confirm that the relative performance of our model does not depend on d for most of the alternative methods, we also observe that the relative performance *SECoFi* method with regards to *Trust Ensemble* is enhanced with higher numbers of factors. We also observe that the optimal regularization parameters for *SECoFi* were always the same, independent of the value of d. This eases the model selection process particularly compared to SGD based methods where both a learning rate and a regularizer need to be tuned. Moreover it seems that methods that are based on alternated least-square (ALS) optimization perform better predictions than those that use SGD. Note that the SGD-based methods subsample the unobserved entries to avoid biasing the estimator.

Experiments on the *Epinions* Data. We repeat the experimental evaluation of *SECoFi* on the publicly available *Epinions*[2] dataset. We follow the same procedure as described for the *Tuenti* data, the experiment results for the different methods on the *Epinions* data are shown in Figure 4.

---

[2] http://snap.stanford.edu/data/soc-Epinions1.html

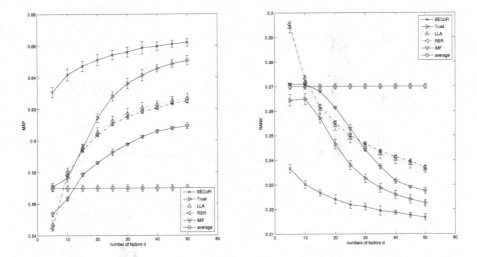

Fig. 4. The MAP and RANK metrics of *SECoFi* and the remaining methods on the *Epinions* dataset

From the results we can draw similar conclusions to the experiments with the *Tuenti* data: learning the friendship weights A during the optimization process significantly improves the performance over methods that just use the social network information as proposed by [16] without quantifying these relationships. Note that *SECoFi* outperforms the second best method *Trust Ensemble* by 2.4% in terms of MAP and by 4.1% in terms of RANK, while *SECoFi* outperforms the remaining methods in comparison by more then 6% both in terms of MAP and RANK. We observe that ALS based methods that take all the "unobserved entries" of the data into account perform better then SGD-based approaches that sample the space of "unobserved entries". Moreover *SECoFi* performs relatively well even utilizing a smaller numbers of factors d. This can be particularly useful in recommendation engines that need to be compact in terms of memory usage e.g. on a smartphone.

In Figure 5 we plot the distribution of the values of α for the two datasets. Recall that the values of alpha encode the degree of influence or trust among users. We observe that for both of the datasets there is a bimodal distribution. For the *Epinions* dataset, most of the values are between 0 and 1 (99%) and 70% of the values are around 1, signalling strong trust relationships among users. For Tuenti, fewer values of α are around 1, and most of the values are close to 0. While there is still some significant influence/trust among users it is less prevalent than in the *Epinions* dataset. This reflects the nature of the data: in the *Epinions* dataset the social network of the users is based on the trust that the users put on each others opinions/ratings while the social relationships on the *Tuenti* network are of much broader scope and can range from close friendships to simple acquaintances, thus we also expect that a smaller fraction of these

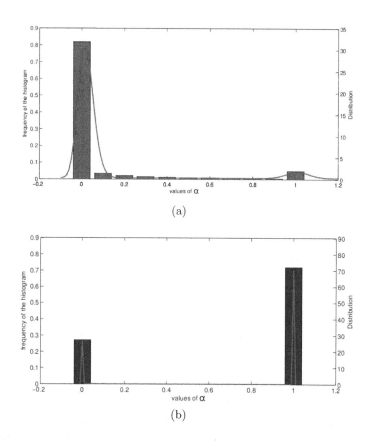

(a)

(b)

Fig. 5. Distribution of the values of α on *Tuenti* (a) and *Epinions* (b) datasets. The distributions are bimodal.

relationships will reflect trust/influence. Note also that *SECoFi* outperforms the competing methods to a higher degree on the *Epinions* data, another indication that the social information in this dataset provides more information on the preferences of the users.

Another important point is that *SECoFi* depends less on the "quality" of the users Social Network. In fact *iMF*, which does not utilize OSN information, is the best runner up in the experiments on the *Tuenti* dataset. This can be attributed to the more relaxed definition of friends in a general purpose social network such as *Tuenti* where we can expect that not all friends share the same taste and preferences with the user. Alternative approches relying on a non-adaptive contribution of friends (*RSR*, *LLA*) suffer more in this context, while learning the weights α helps *SECoFI* to keep only the useful part of the users social network with respect to the recommendations.

5   Conclusions

We presented a method minimizing a novel objective function that takes advantage of the social graph data to perform personalized item recommendation on implicit feedback data. *SECoFi* outperforms alternative state-of-the-art methods in terms of ranking measures and also provides the added benefit of quantifying the influence/trust relationships among users. The latter can be particularly helpful when providing group recommendations e.g. when inviting a group of users to an event etc. We can moreover use the computed α_{ij} values to perform friend recommendation, by leveraging the fact that these values represent a measure of shared interest and taste among users which quantifies the "homophily" effect.

Acknowledgements. This work is funded as part of a Marie Curie Intra European Fellowship for Career Development (IEF) awards held by Alexandros Karatzoglou (CARS, PIEF-GA-2010-273739).

References

1. Deshpande, M., Karypis, G.: Item-based top-n recommendation algorithms. ACM Transactions on Information Systems (TOIS) 22(1), 143–177 (2004)
2. Basilico, J., Hofmann, T.: Unifying collaborative and content-based filtering. In: Proc. Intl. Conf. Machine Learning, pp. 65–72. ACM Press, New York (2004)
3. Srebro, N., Rennie, J., Jaakkola, T.: Maximum-margin matrix factorization. In: Saul, L.K., Weiss, Y., Bottou, L. (eds.) Advances in Neural Information Processing Systems 17. MIT Press, Cambridge (2005)
4. Takacs, G., Pilaszy, I., Nemeth, B., Tikk, D.: Scalable collaborative filtering approaches for large recommender systems. Journal of Machine Learning Research 10, 623–656 (2009)
5. Weimer, M., Karatzoglou, A., Le, Q., Smola, A.: Cofirank - maximum margin matrix factorization for collaborative ranking. In: Advances in Neural Information Processing Systems 20 (NIPS). MIT Press, Cambridge (2008)
6. Salakhutdinov, R., Mnih, A.: Bayesian probabilistic matrix factorization using markov chain monte carlo. In: Proceedings of the 25th International Conference on Machine learning, ICML 2008, pp. 880–887. ACM, New York (2008)
7. Srebro, N., Shraibman, A.: Rank, trace-norm and max-norm. In: Auer, P., Meir, R. (eds.) COLT 2005. LNCS (LNAI), vol. 3559, pp. 545–560. Springer, Heidelberg (2005)
8. Hu, Y., Koren, Y., Volinsky, C.: Collaborative filtering for implicit feedback datasets. In: Proc. of ICDM 2008, pp. 263–272. IEEE Computer Society, Washington, DC (2008)
9. Ma, H., Zhou, D., Liu, C., Lyu, M.R., King, I.: Recommender systems with social regularization. In: Proceedings of the Fourth ACM International Conference on Web Search and Data Mining, WSDM 2011, pp. 287–296. ACM, New York (2011)
10. Jamali, M., Ester, M.: A matrix factorization technique with trust propagation for recommendation in social networks. In: Proceedings of the Fourth ACM Conference on Recommender Systems, RecSys 2010, pp. 135–142. ACM, New York (2010)

11. Li, W.-J., Yeung, D.-Y.: Relation regularized matrix factorization. In: Proceedings of the 21st International Joint Conference on Artifical Intelligence, IJCAI 2009, pp. 1126–1131. Morgan Kaufmann Publishers Inc., San Francisco (2009)

12. Yang, S.-H., Long, B., Smola, A., Sadagopan, N., Zheng, Z., Zha, H.: Like like alike: joint friendship and interest propagation in social networks. In: Proceedings of the 20th International Conference on World Wide Web, WWW 2011, pp. 537–546. ACM, New York (2011)

13. Cui, P., Wang, F., Liu, S., Ou, M., Yang, S., Sun, L.: Who should share what?: item-level social influence prediction for users and posts ranking. In: Proceedings of the 34th International ACM SIGIR Conference on Research and Development in Information Retrieval, SIGIR 2011, pp. 185–194. ACM, New York (2011)

14. Agarwal, D., Chen, B.-C.: Regression-based latent factor models. In: Proceedings of the 15th ACM SIGKDD International Conference on Knowledge Discovery and Data Mining, KDD 2009, pp. 19–28. ACM, New York (2009)

15. Noel, J., Sanner, S., Tran, K.-N., Christen, P., Xie, L., Bonilla, E.V., Abbasnejad, E., Penna, N.D.: New objective functions for social collaborative filtering. In: Proceedings of the 21st International Conference on World Wide Web, WWW 2012, pp. 859–868. ACM, New York (2012)

16. Ma, H., King, I., Lyu, M.R.: Learning to recommend with social trust ensemble. In: Proceedings of the 32nd International ACM SIGIR Conference on Research and Development in Information Retrieval, SIGIR 2009, pp. 203–210. ACM, New York (2009)

17. Pan, W., Aharony, N., Pentland, A.: Composite social network for predicting mobile apps installation. In: Proceedings of the 25th Conference on Artificial Intelligence (AAAI 2011), San Francisco, CA (2011)

18. Zheng, Y., Zhang, L., Ma, Z., Xie, X., Ma, W.Y.: Recommending friends and locations based on individual location history. ACM Transactions on the Web (TWEB) 5(1), 5 (2011)

19. Scellato, S., Noulas, A., Mascolo, C.: Exploiting place features in link prediction on location-based social networks. In: Proceedings of the 17th ACM SIGKDD International Conference on Knowledge Discovery and Data Mining, pp. 1046–1054. ACM (2011)

20. Cremonesi, P., Koren, Y., Turrin, R.: Performance of recommender algorithms on top-n recommendation tasks. In: Proc. of the 4th ACM Conference on Recommender Systems (RecSys 2010), pp. 39–46 (2010)

Cross-Domain Recommendation via Cluster-Level Latent Factor Model

Sheng Gao[1], Hao Luo[1], Da Chen[1], Shantao Li[1],
Patrick Gallinari[2], and Jun Guo[1]

[1] PRIS - Beijing University of Posts and Telecommunications, China
{gaosheng,guojun}@bupt.edu.cn, {legand1989,chenda104}@gmail.com,
buptlishantao@163.com
[2] LIP6 - Université Pierre et Marie Curie, France
patrick.gallinari@lip6.fr

Abstract. Recommender systems always aim to provide recommendations for a user based on historical ratings collected from a single domain (e.g., movies or books) only, which may suffer from the data sparsity problem. Recently, several recommendation models have been proposed to transfer knowledge by pooling together the rating data from multiple domains to alleviate the sparsity problem, which typically assume that multiple domains share a latent common rating pattern based on the user-item co-clustering. In practice, however, the related domains do not necessarily share such a common rating pattern, and diversity among the related domains might outweigh the advantages of such common pattern, which may result in performance degradations. In this paper, we propose a novel cluster-level based latent factor model to enhance the cross-domain recommendation, which can not only learn the common rating pattern shared across domains with the flexibility in controlling the optimal level of sharing, but also learn the domain-specific rating patterns of users in each domain that involve the discriminative information propitious to performance improvement. To this end, the proposed model is formulated as an optimization problem based on joint nonnegative matrix tri-factorization and an efficient alternating minimization algorithm is developed with convergence guarantee. Extensive experiments on several real world datasets suggest that our proposed model outperforms the state-of-the-art methods for the cross-domain recommendation task.

1 Introduction

Most recommender systems based on collaborative filtering aim to provide recommendations or rating predictions of an active user on a set of items belonging to only a single domain (e.g., movies or books) based on the historical user-item preference records [1]. However, in many cases, users rate only a limited number of items, even the item space is often very large. Then the available rating data can be extremely sparse, which may cause the recommendation models suffer from the overfitting problem and result in low-quality predictions as well.

H. Blockeel et al. (Eds.): ECML PKDD 2013, Part II, LNAI 8189, pp. 161–176, 2013.

In fact, there exists a considerable number of publicly available user-item rating datasets from multiple domains, which could have dependencies and correlations among the domains. Taking `Amazon` as an example, since the products in *Book* domain and *Music* domain may have correspondence in genre, and the respective customers can be considered to belong to similar groups sampled from the same population with alike social aspects [2], it would be useful to exploit a user's preferences on best-sellers in *Book* domain to help generate recommendations for that user on MP3 albums from the *Music* domain. Thus, instead of treating items from each single domain independently, users' preferences knowledge acquired in a single domain could be transferred and shared in other related domains, which has been referred to as *Cross-Domain Recommendation* [3]. Recently, several cross-domain recommendation models [2] [4] [5] have been proposed to transfer a common user-item rating pattern from a dense auxiliary rating dataset in other domains to a sparse rating dataset in the target domain of interest, which typically assume that multiple domains share the latent common rating pattern based on the user-item co-clustering. Thus, knowledge transfer and sharing among the related domains can be beneficial to alleviate the data sparsity problem.

However, the limitation of the existing methods is two-fold.

1. The existing models usually construct a latent space to represent the common latent structure shared across domains, which captures the rating pattern of user groups provided on item clusters. But in practice, the rating datasets from multiple domains may rarely contain exactly the same items or users, some domains are more closely related to the target domain of interest than others, simply forcing the subspaces in each domain to be identical is highly unrealistic. For example, books are more closely related to movies than to electronic gadgets, thus the different level of relatedness among multiple domains can not be captured by the identical rating patterns, which implies the existing methods are inflexible under the "all shared" latent factor assumption.

2. In practice, some related domains do not necessarily share such a common rating pattern, which has the intuition of *"Harmony in Diversity"* from the rating patterns in multiple domains. Moreover, the diversity among the related domains might outweigh the advantages of the common rating pattern, which may result in performance degradations. That is, the existing models cannot consider the domain-specific knowledge about the rating patterns to improve the mutual strengths in cross-domain recommendation.

To this end, in this paper, we propose a novel cluster-level based latent factor model to enhance the cross-domain recommendation. By deriving latent user-cluster factor and latent item-cluster factor from the available rating data, our proposed model can construct a latent space to represent the rating patterns of user groups on the item clusters. Based on a subspace learning of the latent space, the model can learn the common cluster-level user-item rating pattern that is shared across domains, especially, with the flexibility in controlling the optimal level of sharing the relatedness among multiple domains, while the existing models do not provide this function for fine analysis of intrinsic cluster structure in rating data. Meanwhile, our proposed model can simultaneously learn the domain-specific cluster-level rating pattern from each domain, which contains the discriminative information propitious to improve across recommendation accuracy. The proposed model thus can exploit the mutual strengths of related domains by the shared common rating patterns as well as the domain-specific rating patterns immune to the discernable knowledge from each domain.

Moreover, our proposed model can be formulated as an optimization problem based on joint nonnegative matrix tri-factorization, and an efficient alternating method is developed to optimize the objective function with convergence guarantee. Extensive experiments on several real world datasets suggest that our proposed model outperforms the state-of-the-art methods for cross-domain recommendation task.

The paper is structured as follows. In Section 2 we briefly introduce the background and problem definition. In Section 3 the proposed framework based on the cluster-level latent factor model and the model specification are presented, followed by the efficient optimization algorithm. Then we describe experiments on several real world datasets, and provide comparisons with state-of-the-art methods in Section 4. The related work is discussed in Section 5. In Section 6 we present conclusions.

2 Background

2.1 Basic Model

Our proposed latent factor model is based on the orthogonal non-negative matrix tri-factorization (ONMTF) clustering algorithm [6], which is an effective framework for data mining. In this section, we introduce the background paradigm behind ONMTF that motivates our model.

In ONMTF model, a data matrix $\mathbf{X} \in \mathbb{R}^{M \times N}$ from a rating dataset is factorized into a product of three nonnegative factors $\mathbf{U} \in \mathbb{R}^{M \times K}$, $\mathbf{S} \in \mathbb{R}^{K \times L}$ and $\mathbf{V} \in \mathbb{R}^{N \times L}$, such that $\mathbf{X} \approx \mathbf{U}\mathbf{S}\mathbf{V}^T$. This approximation can be achieved by the following matrix norm optimization:

$$\min_{\mathbf{U},\mathbf{S},\mathbf{V} \geq 0} \mathcal{J}_{\text{ONMTF}} = \|\mathbf{X} - \mathbf{U}\mathbf{S}\mathbf{V}^T\| \tag{1}$$

where $\|\cdot\|$ is the Frobenious norm of matrix. $\mathbf{X} = [\mathbf{x}_1, ..., \mathbf{x}_N]$ is an $M \times N$ rating matrix containing M users and N items. From the co-clustering perspective, the three nonnegative factors decomposed from ONMTF can be interpreted in the following way:

- $\mathbf{U} = [\mathbf{u}_1, ..., \mathbf{u}_K]$ represent latent user factors, where each $\mathbf{u}_k$ is an $M \times 1$ vector indicating a probability distribution over M users and referred to as a *user-cluster latent factor*. Here $\arg\max_k (\mathbf{U})_{ik} = k^*$ means the ith user belongs to the k^*th user cluster (i.e., user group).

- $\mathbf{V} = [\mathbf{v}_1, ..., \mathbf{v}_L]$ represent latent item factors, where each $\mathbf{v}_l$ is an $N \times 1$ vector indicating a probability distribution over N items and referred to as a *item-cluster latent factor*. Here $\arg\max_l(\mathbf{V})_{il} = l^*$ means the ith item belongs to the l^*th item cluster (i.e., item topic).
- $\mathbf{S} = [\mathbf{s}_1, ..., \mathbf{s}_L]$ is an $K \times L$ weight matrix representing the rating patterns from K user clusters provided on L item clusters. $\mathbf{S}_{ij}$ can be considered as the probability that the kth user group rates the lth item cluster.

By clustering both sides of the data matrix simultaneously, ONMTF makes use of the interrelatedness between users and items, leading to better performance than other clustering methods. In [4] the authors proposed a latent factor model based on ONMTF clustering algorithm (i.e., Equation (1)) to provide recommendations for a sparse target domain (e.g., $\mathbf{X}_{tgt}$) by sharing the latent common rating pattern knowledge in a latent space from the related dense domain (e.g., $\mathbf{X}_{src}$), which was referred to as a *codebook* as $\mathbf{S}$ (i.e., $\mathbf{X}_{src} \approx \mathbf{U}_{src}\mathbf{S}\mathbf{V}_{src}^T$). Thus, the codebook $\mathbf{S}$ was constructed by simultaneously clustering the users (rows) and items (columns) of $\mathbf{X}_{src}$, indicating the rating that a user belonging to a specific user cluster $\mathbf{u}_{src}$ provides on an item belonging to a specific item cluster $\mathbf{v}_{src}$. Then the missing values in the target domain $\mathbf{X}_{tgt}$ could be learned by duplicating the rows and columns of the codebook using $\mathbf{U}_{tgt}\mathbf{S}\mathbf{V}_{tgt}^T$, which was called CodeBook Transfer (CBT) recommendation model. Experimental results have shown that latent common information from a related domain can be derived to improve performance in the target domain. In the next section, we will discuss the way how to derive the latent common rating pattern and domain-specific rating pattern to enhance the cross-domain recommendation.

2.2 Problem Definition

Suppose that we are given multiple rating matrices from related domains for personalized item recommendation. Let τ be the domain index, and $\tau \in [1, t]$. In the τ-th domain rating matrix $\mathbf{D}_\tau$ there are a set of users $X_\tau = \{x_1^\tau, ..., x_{M_\tau}^\tau\}$ to rate a set of items $Y_\tau = \{y_1^\tau, ..., y_{N_\tau}^\tau\}$, where M_τ and N_τ represent the numbers of rows (users) and columns (items) respectively. Here the set of users and items across multiple domains may overlap or be isolated with each other. In this work we consider the more difficult case that neither the users or the items in the multiple rating matrices are overlapping. Moreover, each rating matrix contains a few observed ratings and some missing values to predict. We thus employ a binary weighting matrix $\mathbf{W}_\tau$ of the same size as $\mathbf{D}_\tau$ to mask the missing entries, where $[\mathbf{W}_\tau]_{ij} = 1$ if $[\mathbf{D}_\tau]_{ij}$ is observed and $[\mathbf{W}_\tau]_{ij} = 0$ otherwise. For easy understanding, we call the rating matrix of interest as the *target domain* and other related rating matrices the *source domains*.

In this paper, we consider how to predict the missing ratings in the target domain of interest by transferring correlated knowledge from the source domains as well as to learn the relatedness among multiple domains.

3   Our Proposed Model

Existing cross-domain recommendation models [2] [4] assume that the cluster-level structures hidden across domains can be extracted to learn the rating-pattern of user groups on the item clusters for knowledge transfer and sharing, and to clearly demonstrate the co-clusters of users and items. In this paper, we follow the framework proposed in [4] to extract the co-clustering of users and items as well as the shared common rating pattern. Thus, the initial co-clustering of the data matrix in domain τ can be performed by using ONMTF model as follows:

$$\min_{\mathbf{U}_\tau, \mathbf{S}_\tau, \mathbf{V}_\tau \geq 0} \mathcal{J}_\tau = \left\| [\mathbf{D}_\tau - \mathbf{U}_\tau \mathbf{S}_\tau^* \mathbf{V}_\tau^T] \circ \mathbf{W}_\tau \right\|^2 \tag{2}$$

where $\mathbf{U}_\tau \in \mathbb{R}^{M_\tau \times K_\tau}$ denotes the K_τ user clusters in the τth domain, and $\mathbf{V}_\tau \in \mathbb{R}^{N_\tau \times L_\tau}$ denotes the L_τ item clusters in the τth domain. $\mathbf{S}_\tau^* \in \mathbb{R}^{K_\tau \times L_\tau}$ represents the rating pattern of the kth user cluster made on the lth item cluster in the τth domain, where each entry $[\mathbf{S}_\tau^*]_{kl}$ is considered to be the average rating of the corresponding user-item co-cluster. $\mathbf{W}_\tau$ is the binary mask matrix, and $\circ$ denotes the entry-wise product. In the case of multiple related domains involving different sets of users or items, the assumption that users from different domains have a *similar rating pattern* on similar item clusters or topics can be held due to the harmony of users and items across the related domains, that denotes the same clustering topics in items (i.e., $L_\tau = L$) and the same cluster distributions over user groups (i.e., $K_\tau = K$) in each domain [4].

However, as we have introduced, the assumption does not hold in many real-world applications, where the items from multiple domains can not always find their explicit correspondence in the cluster level. Taking movie-rating and book-rating web sites for example, the movies and books can be considered to have similar clusters or topics based on their genre information (e.g., the categories of comedy or tragedy), but various customer groups from different websites may keep some domain-specific knowledge about the items of their respective domains in mind, showing different rating patterns and cognitive styles, such as the rating information about some Oscar-winning movies can not necessarily help to discover the clustering of the books on the topic of Oscar history. Inspired by this observation, we relax the unrealistic assumption in [4], and consider that the users from different domains should have similar explicit cluster-level correspondence while the items in each domain may hold their domain-specific knowledge.

3.1   Model Formulation

We propose a latent factor model based on the ONMTF framework to cluster the users and items in τth domain simultaneously and then learn a latent space to construct the cluster-level rating pattern of user-item co-clusters. Specifically, we partition the latent rating pattern *across domains* into a common part and a domain-specific part by the subspace learning of the latent space, that is $\mathbf{S}_\tau^* = [\mathbf{S}_0, \mathbf{S}_\tau]$, where $\mathbf{S}_0 \in \mathbb{R}^{K_\tau \times T}$ and $\mathbf{S}_\tau \in \mathbb{R}^{K_\tau \times (L_\tau - T)}$, T is the dimensionality of

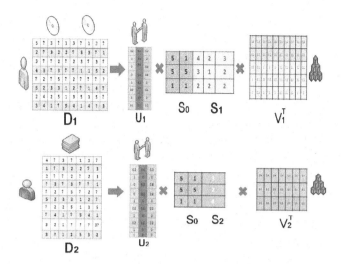

Fig. 1. Illustration of our proposed CLFM model in the context of two related domains. $\mathbf{U}_1$ and $\mathbf{U}_2$ are the respective user-cluster latent factors, $\mathbf{V}_1$ and $\mathbf{V}_2$ are the respective item-cluster latent factors, $\mathbf{S}_0$ denotes the common latent rating pattern across domains, $\mathbf{S}_1$ and $\mathbf{S}_2$ are the domain-specific rating patterns in each domain.

shared common rating pattern and $(L_\tau - T)$ the dimensionality of τth domain-specific rating pattern.

Here the common part of rating pattern $\mathbf{S}_0$ captures the similar behavior styles of user clusters when they face the shared T clusters of related items from different domains, which can be captured to help alleviate the sparsity problem in each rating matrix. While the domain-specific part of rating pattern $\mathbf{S}_\tau$ can be learned to denote the discriminative aspect of user groups providing ratings on $(L_\tau - T)$ item clusters, which can be used to reveal the relatedness among multiple domains and improve the accuracy of recommendation.

Accordingly, in each domain our proposed Cluster-Level Latent Factor Model (CLFM) can learn the user-cluster latent factor $\mathbf{U}_\tau \in \mathbb{R}^{M_\tau \times K_\tau}$ where $K_\tau = K$, and item-cluster latent factor $\mathbf{V}_\tau = [\mathbf{V}_{\tau 0}^T, \mathbf{V}_{\tau 1}^T] \in \mathbb{R}^{N_\tau \times L_\tau}$ where $\mathbf{V}_{\tau 0} \in \mathbb{R}^{T \times N_\tau}$ corresponds to shared topics of item clusters and $\mathbf{V}_{\tau 1} \in \mathbb{R}^{(L_\tau - T) \times N_\tau}$ corresponds to domain-specific topics of item clusters in τth domain. The illustration of our proposed CLFM model can be found in Figure 1.

Then the learning of our proposed model can be derived in a unified subspace learning paradigm as the following objective function:

$$\min_{\mathbf{U}_\tau, \mathbf{S}_0, \mathbf{S}_\tau, \mathbf{V}_\tau \geq 0} \mathcal{J} = \sum_\tau \left\| [\mathbf{D}_\tau - \mathbf{U}_\tau [\mathbf{S}_0, \mathbf{S}_\tau] \mathbf{V}_\tau^T] \circ \mathbf{W}_\tau \right\|^2 \tag{3}$$

Specifically, to make the latent factors more accurate, we can impose some prior knowledge on the latent factors during the model learning. For example, the ℓ_1 normalization constraint can be imposed on each row of $\mathbf{U}_\tau$ and $\mathbf{V}_\tau$, i.e., $\mathbf{U}_\tau \mathbf{1} = \mathbf{1}$ and $\mathbf{V}_\tau \mathbf{1} = \mathbf{1}$.

Note that, from the construction of our proposed CLFM model, we can easily find that the recently proposed cross-domain recommendation model CBT is a special case of CLFM model with $L_\tau = T$, which means that there is no domain-specific rating pattern for each domain but only the shared common rating pattern across all the domains with the dimensionality of T. Therefore, our proposed CLFM model can not only exploit the optimal level of sharing information across multiple domains, but also reveal the individual differences in each domain.

3.2 Optimization

The optimization of our proposed model can be performed by an alternating minimization algorithm until convergence. For ease of understanding and without loss of generality, we set $\tau = 2$ [1]. The general objective function in Equation (3) can be rewritten as follows:

$$\min_{\mathbf{U},\mathbf{S}_0,\mathbf{S}_1,\mathbf{S}_2,\mathbf{V}\geq 0} \mathcal{J} = \left\| [\mathbf{D}_1 - \mathbf{U}_1[\mathbf{S}_0,\mathbf{S}_1]\mathbf{V}_1^T] \circ \mathbf{W}_1 \right\|^2$$
$$+ \left\| [\mathbf{D}_2 - \mathbf{U}_2[\mathbf{S}_0,\mathbf{S}_2]\mathbf{V}_2^T] \circ \mathbf{W}_2 \right\|^2 \qquad (4)$$
$$s.t. \quad \mathbf{U}_1\mathbf{1} = 1, \mathbf{U}_2\mathbf{1} = 1, \mathbf{V}_1\mathbf{1} = 1, \mathbf{V}_2\mathbf{1} = 1$$

where $\mathbf{U}_1 \in \mathbb{R}^{M_1 \times K}$, $\mathbf{U}_2 \in \mathbb{R}^{M_2 \times K}$, $\mathbf{V}_1 = [\mathbf{V}_{10}^T, \mathbf{V}_{11}^T] \in \mathbb{R}^{N_1 \times L_1}$, $\mathbf{V}_2 = [\mathbf{V}_{20}^T, \mathbf{V}_{21}^T] \in \mathbb{R}^{N_2 \times L_2}$, $\mathbf{S}_0 \in \mathbb{R}^{K \times T}$, $\mathbf{S}_1 \in \mathbb{R}^{K \times (L_1-T)}$, $\mathbf{S}_2 \in \mathbb{R}^{K \times (L_2-T)}$.

To optimize the proposed model, we employ the alternating multiplicative updating algorithm [7], which warrants the nonnegativity of latent factors and provides an automatic step parameter selection. Since the objective function $\mathcal{J}$ in Equation (4) is not convex in $\mathbf{U},\mathbf{S}_0,\mathbf{S}_1,\mathbf{S}_2$ and $\mathbf{V}$ together, the alternating updating algorithm optimizes the objective function with respect to one set of parameters while fixing the others, and then exchanges the roles of the parameter sets alternatively. This procedure will be repeated for several iterations until convergence.

Learning $\mathbf{S}_1$. Taking the learning of $\mathbf{S}_1$ as an example, we will show how to optimize $\mathbf{S}_1$ by deriving its updating rule while fixing the other factors. For that we can rewrite the objective function in Equation (4) as follows:

$$\min_{\mathbf{S}_1} \mathcal{J}(\mathbf{S}_1) = \left\| [\mathbf{D}_1 - \mathbf{U}_1\mathbf{S}_0\mathbf{V}_{10} - \mathbf{U}_1\mathbf{S}_1\mathbf{V}_{11}] \circ \mathbf{W}_1 \right\|^2$$
$$+ \left\| [\mathbf{D}_2 - \mathbf{U}_2\mathbf{S}_0\mathbf{V}_{20} - \mathbf{U}_2\mathbf{S}_2\mathbf{V}_{21}] \circ \mathbf{W}_2 \right\|^2 \qquad (5)$$

Then the derivative of $\mathcal{J}(\mathbf{S}_1)$ with respect to $\mathbf{S}_1$ is as follows:

$$\frac{\partial \mathcal{J}(\mathbf{S}_1)}{\partial \mathbf{S}_1} = 2(\mathbf{U}_1^T([\mathbf{U}_1\mathbf{S}_0\mathbf{V}_{10}] \circ \mathbf{W}_1)\mathbf{V}_{11}^T - \mathbf{U}_1^T(\mathbf{D}_1 \circ \mathbf{W}_1)\mathbf{V}_{11}^T)$$
$$+ 2\mathbf{U}_1^T([\mathbf{U}_1\mathbf{S}_1\mathbf{V}_{11}] \circ \mathbf{W}_1)\mathbf{V}_{11}^T$$

---

[1] Following the definition in [4], domain $\mathbf{D}_1$ can be considered as the source domain, and $\mathbf{D}_2$ the target domain of interest.

Using the Karush-Kuhn-Tucker complementary condition for the nonnegativity of $\mathbf{S}_1$ and let $\frac{\partial \mathcal{J}(\mathbf{S}_1)}{\partial \mathbf{S}_1} = 0$, we can get the following updating rule for $\mathbf{S}_1$:

$$\mathbf{S}_1 \longleftarrow \mathbf{S}_1 \sqrt{\frac{\mathbf{U}_1^T(\mathbf{D}_1 \circ \mathbf{W}_1)\mathbf{V}_{11}^T}{\mathbf{U}_1^T([\mathbf{U}_1\mathbf{S}_0\mathbf{V}_{10}] \circ \mathbf{W}_1)\mathbf{V}_{11}^T + \mathbf{U}_1^T([\mathbf{U}_1\mathbf{S}_1\mathbf{V}_{11}] \circ \mathbf{W}_1)\mathbf{V}_{11}^T}} \tag{6}$$

Learning $\mathbf{S}_2$. The latent factor $\mathbf{S}_2$ can be learned in a similar way. Here we can derive the updating rules for learning $\mathbf{S}_2$ as follows:

$$\mathbf{S}_2 \longleftarrow \mathbf{S}_2 \sqrt{\frac{\mathbf{U}_2^T(\mathbf{D}_2 \circ \mathbf{W}_2)\mathbf{V}_{21}^T}{\mathbf{U}_2^T([\mathbf{U}_2\mathbf{S}_0\mathbf{V}_{20}] \circ \mathbf{W}_2)\mathbf{V}_{21}^T + \mathbf{U}_2^T([\mathbf{U}_2\mathbf{S}_2\mathbf{V}_{21}] \circ \mathbf{W}_2)\mathbf{V}_{21}^T}} \tag{7}$$

Learning $\mathbf{S}_0$. The updating rules for learning latent factor $\mathbf{S}_0$ can be derived as follows:

$$\begin{aligned}
\mathbf{S}_0 &\longleftarrow \mathbf{S}_0 \sqrt{\frac{\mathbf{U}_1^T(\mathbf{D}_1 \circ \mathbf{W}_1)\mathbf{V}_{10}^T + \mathbf{U}_2^T(\mathbf{D}_2 \circ \mathbf{W}_2)\mathbf{V}_{20}^T}{\mathbf{A} + \mathbf{B}}} \\
\mathbf{A} &= \mathbf{U}_1^T([\mathbf{U}_1\mathbf{S}_0\mathbf{V}_{10}] \circ \mathbf{W}_1)\mathbf{V}_{10}^T + \mathbf{U}_1^T([\mathbf{U}_1\mathbf{S}_1\mathbf{V}_{11}] \circ \mathbf{W}_1)\mathbf{V}_{10}^T \\
\mathbf{B} &= \mathbf{U}_2^T([\mathbf{U}_2\mathbf{S}_0\mathbf{V}_{20}] \circ \mathbf{W}_2)\mathbf{V}_{20}^T + \mathbf{U}_2^T([\mathbf{U}_2\mathbf{S}_2\mathbf{V}_{21}] \circ \mathbf{W}_2)\mathbf{V}_{20}^T
\end{aligned} \tag{8}$$

Learning $\mathbf{U}_1$. The latent factor $\mathbf{U}_1$ can be learned in the similar way. Here we can derive the updating rules for learning $\mathbf{U}_1$ as follows:

$$\mathbf{U}_1 \longleftarrow \mathbf{U}_1 \sqrt{\frac{(\mathbf{D}_1 \circ \mathbf{W}_1)\mathbf{V}_1[\mathbf{S}_0, \mathbf{S}_1]^T}{([\mathbf{U}_1[\mathbf{S}_0, \mathbf{S}_1]\mathbf{V}_1^T] \circ \mathbf{W}_1)\mathbf{V}_1[\mathbf{S}_0, \mathbf{S}_1]^T}} \tag{9}$$

Note that during the learning of $\mathbf{U}_1$, we formulate the Lagrange function for the optimization with normalization constraint on the latent factor.

Learning $\mathbf{U}_2$. The updating rules for learning latent factor $\mathbf{U}_2$ can be derived as follows:

$$\mathbf{U}_2 \longleftarrow \mathbf{U}_2 \sqrt{\frac{(\mathbf{D}_2 \circ \mathbf{W}_2)\mathbf{V}_2[\mathbf{S}_0, \mathbf{S}_2]^T}{([\mathbf{U}_2[\mathbf{S}_0, \mathbf{S}_2]\mathbf{V}_2^T] \circ \mathbf{W}_2)\mathbf{V}_2[\mathbf{S}_0, \mathbf{S}_2]^T}} \tag{10}$$

Here we also formulate the Lagrange function for the optimization with normalization constraint in learning $\mathbf{U}_2$.

Learning $\mathbf{V}_1$. The latent factor $\mathbf{V}_1$ can be learned in the similar way as for constrained optimization. Here we can derive the updating rules for learning $\mathbf{V}_1$ as follows:

$$\mathbf{V}_1 \longleftarrow \mathbf{V}_1 \sqrt{\frac{[\mathbf{S}_0, \mathbf{S}_1]^T\mathbf{U}_1^T(\mathbf{D}_1 \circ \mathbf{W}_1)}{[\mathbf{S}_0, \mathbf{S}_1]^T\mathbf{U}_1^T([\mathbf{U}_1[\mathbf{S}_0, \mathbf{S}_1]\mathbf{V}_1^T] \circ \mathbf{W}_1)}} \tag{11}$$

Note that $\mathbf{V}_{10}^T = \mathbf{V}_1(:, 1 : T)$ and $\mathbf{V}_{11}^T = \mathbf{V}_1(:, (T+1) : L_1)$.

Learning V_2. The updating rules for learning latent factor V_2 as for constrained optimization can be derived as follows:

$$V_2 \leftarrow V_2 \sqrt{\frac{[S_0, S_1]^T U_2^T (D_2 \circ W_2)}{[S_0, S_2]^T U_2^T ([U_2[S_0, S_2] V_2^T] \circ W_2)}} \qquad (12)$$

Note that $V_{20}^T = V_2(:, 1 : T)$ and $V_{21}^T = V_2(:, (T + 1) : L_2)$.

Convergence Analysis. Based on the above updating rules for learning different latent factors, we can prove that the learning algorithm is convergent.

Theorem 1. *Using the updating rules for S_0 in Equation (8), S_1 in Equation (6), S_2 in Equation (7), U_1 in Equation (9), U_2 in Equation (10), V_1 in Equation (11) and V_2 in Equation (12), the objective function in Equation (4) will monotonically decrease, thus the learning algorithm converges.*

The proof could be refereed to [8] [9] for more details.

4   Experiments

In the experiments, we examine how our proposed model behaves on real-world rating datasets and compare it with several state-of-the-art single-domain recommendation models and cross-domain recommendation models:

- NMF (Nonnegative Matrix Factorization) based model [7]: a single-domain model which employs nonnegative matrix factorization method to learn the latent factors in each domain and provide the prediction performance separately.
- FMM (Flexible Mixture Model) based model [10]: a single-domain model which uses probabilistic mixture model to learn latent cluster structure in each single domain and then provide the single domain performance separately.
- CBT (CodeBook Transfer) based model [4]: a cross-domain model which can only transfer and share the common rating pattern by the codebook information across multiple domains.
- CLFM model: our proposed model.

In terms of the cross-domain recommendation task, we evaluate these methods in terms of two ways: one is the impact of different level of sharing common information across domains, the other is to check the effectiveness of the cross-domain models to alleviate the sparsity problem.

4.1   Datasets

For the experiments we have used three benchmark real-world datasets for performance evaluation:

- *MovieLens* dataset[2]: contains more than 100,000 movie ratings with the scales from 1 to 5 provided by 943 users on 1,682 movies. Following [2] we randomly choose 500 users with more than 16 ratings and 1000 movies for experiments.
- *EachMovie* dataset[3]: contains 2.8 million movie ratings with the scales from 1 to 6 provided by 72,916 users on 1,628 movies. Following [2] we also randomly choose 500 users with more than 20 ratings and 1000 movies for experiments.
- *Book-Crossing* dataset[4]: contains more than 1.1 million ratings with the scales from 0 to 9 provided by 278,858 users on 271,379 books. We still randomly select 500 users and 1000 books with more than 16 ratings for each item in the experiments.

Note that for all the datasets we have normalized the rating scales from 1 to 5 in the average style for fair comparison. Our proposed CLFM model can handle various types of users' rating information, including the explicit rating (e.g., from 1 to 5) as well as the implicit preferences of users (e.g., visit, click or comment), which are based on the flexible function of the inner product of the latent factors learned from the observations in different styles.

4.2   Experimental Setup

Following the work in [4], we examine the compared models for the cross-domain recommendation task. For that, 300 users with their ratings in each dataset are randomly selected as the training data, and the remaining 200 users for testing. For each test user, we consider to keep different sizes of the observed ratings as the initialization of each user in the experiments, i.e., 5 or 10 ratings of each test user are given to avoid cold-start problem (e.g., ML − Given5 or ML − Given10 in the *MovieLens* dataset as illustrated in Table 1) and the remaining ratings are used for evaluation.

We choose *MovieLens* vs *EachMovie*, *EachMovie* vs *Book-Crossing* and *Movie-Lens* vs *Book-Crossing* as three kinds of related domains (the former is source domain and the latter the target domain for CBT model in the experiments) to discover the relatedness among different domains. To check the performances of different methods, we use MAE (Mean Absolute Error)[5] as the evaluation metric. In the experiments we conduct the methods by repeating the process 10 times and report the average results.

4.3   Experimental Results

We compare the performances of different models under different configurations. The parameters of different models have been manually tuned and we report here

---

[2] http://www.grouplens.org/node/73

[3] http://www.cs.cmu.edu/~lebanon/IR-lab.htm

[4] http://www.informatik.uni-freiburg.de/~cziegler/BX/

[5] MAE is computed as $MAE = \sum_{i \in O} |R_i - R_i^*|/|O|$, where $|O|$ denotes the number of test ratings, R_i is the true value and R_i^* is the predicted rating. The smaller the value of MAE is, the better the model performs.

Table 1. MAE performances of the compared models on *MovieLens* vs *EachMovie* related domains under different configurations. ML-Given5 means 5 ratings of each test user in *MovieLens* dataset are given while the remaining ratings are used for evaluation. The combined settings *ML-Given5* vs *EM-Given5* and *ML-Given10* vs *EM-Given10* are conducted. Best results are in bold.

	NMF	FMM	CBT	CLFM
ML-Given5	0.9652	0.9338	0.9242	**0.9121**
ML-Given10	0.9411	0.9203	0.9101	**0.8815**
EM-Given5	0.9803	0.9569	0.9333	**0.9209**
EM-Given10	0.9425	0.9396	0.9185	**0.8907**

the best results obtained based on the optimal combination of many parameter settings. The number of users clusters K and item clusters L in each rating dataset have been chosen in the range of $[10, 100]$. For NMF model, the dimension of factorization is set to be 50. For CBT model, we set $K = 30$ and $L = 80$. For all the matrix factorization based models, the number of iterations has been set at 50.

Table 1 shows the MAE performance of the compared models in the *Movie-Lens* vs *EachMovie* related domains under different configurations, where in the both domains the number of users and item clusters $K = 30$ and $L = 50$ respectively, the parameter $T = 40$ denoting the dimensionality of shared common subspace. In the experiments, we have 5 and 10 ratings of each test user in the *MovieLens* and *EachMovie* datasets that are given for training while the remaining ratings are used for test, and the compared models are evaluated on the different combined settings as *ML-Given5* vs *EM-Given5* and *ML-Given10* vs *EM-Given10*. From the results we can observe that the best performing method among all the models is our proposed CLFM model. The FMM model performs slightly better than the NMF model, which implies that the cluster-level based methods can gain meaningful knowledge from user and item clusters due to the co-clustering property in the FMM model.

Moreover, the cross-domain based models (i.e., CBT and CLFM) clearly outperforms the single domain based models (i.e., NMF and FMM), which shows that the latent cross-domain common rating pattern can indeed aggregate more useful knowledge than the single-domain methods do individually. Specifically, our proposed CLFM model provides even better results than the state-of-the-art cross-domain recommendation model CBT, which proves the benefits of the CLFM model with the ability of extracting the common rating pattern and the domain-specific knowledge to enhance the cross-domain recommendation accuracy.

Again, Table 2 shows the MAE performances of the compared models on *EachMovie* vs *Book-Crossing* related domains under different configurations, where in the both domains the parameters $K = 30$, $L = 80$ and $T = 40$ as the optimal values. The combined settings *EM-Given5* vs *BC-Given5* and *EM-Given10* vs *BC-Given10* are conducted in the experiments. From the results we can draw the similar conclusion, that is, our proposed CLFM model performs

Table 2. MAE performances of the compared models on *EachMovie* vs *Book-Crossing* related domains under different configurations. The combined settings *EM-Given5* vs *BC-Given5* and *EM-Given10* vs *BC-Given10* are conducted. Best results are in bold.

	NMF	FMM	CBT	CLFM
EM-Given5	0.9803	0.9569	0.9541	**0.9334**
EM-Given10	0.9425	0.9366	0.9225	**0.9091**
BC-Given5	0.7326	0.7192	0.6978	**0.6757**
BC-Given10	0.7198	0.6924	0.6805	**0.6514**

better than the other related methods. The results about *MovieLens* vs *Book-Crossing* domains have similar characteristics and are omitted here.

Meanwhile, from Table 1 and Table 2 we can also discover that the performances for the item recommendation in the *EachMovie* dataset are not identical even in terms of the same users and items when combined with different related domains in the experiments. The results show that different domains may have various levels of shared information, which are hidden across domains. The remarkable advantage of our proposed CLFM model is to capture and control the level of sharing the relatedness by the shared subspace dimensionality T. In the CLFM model the value of the parameter T is limited between 0 and $\min(L_1, L_2)$, i.e., ranges from *no sharing* to *full sharing*. Figure 2 provides the performances of the compared models in *EachMovie* domain as a function of the parameter T under the configuration of $K = 30$, $L = 80$ given 10 ratings observed for each test user. From the figure, we can find that T increases with the level of sharing pattern between the two domains until it reaches the optimal value $T = 40$.

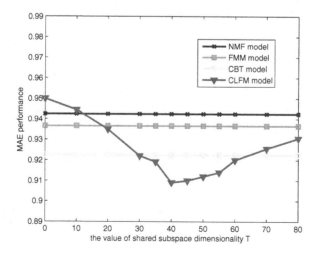

Fig. 2. MAE performance of the compared models with respect to the value of shared subspace dimensionality T in *EachMovie* domain

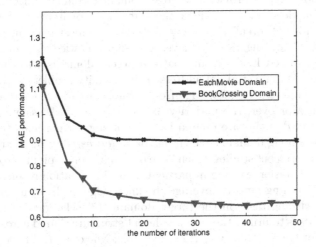

Fig. 3. Convergence curve of running CLFM model in *MovieLens* vs *Book-Crossing* domains with respect to nb. of iterations

This observation confirms that our proposed CLFM model has more flexible and efficient ability to capture latent common rating pattern than the other methods.

Figure 3 demonstrates the convergence curve of running the proposed CLFM model in *MovieLens* vs *Book-Crossing* domains under the configuration of $K = 30$, $L = 80$ and $T = 40$ given 10 ratings available for each test user. From the results we can observe that our proposed model can converge quickly after about 20 iterations, which proves the efficiency and scalability of the CLFM model in the cross-domain recommendation task.

5   Related Work

Cross-domain recommendation is an emerging research topic, which considers to incorporate relevant data sources from different domains and combine with the original target data to improve the recommendation [11]. For example, it is possible to merge multiple rating matrices to leverage rating behavior similarity in one domain to predict ratings in the other domain. Kuflik et al. [12] first proposed cross domain mediation problem and introduced several techniques for importing relevant data. Pan et al. [13] proposed the model to transform knowledge from domains which have heterogeneous forms of user feedback. Cremonesi et al. [14] considered to model the classical similarity relationships as a direct graph and explore all possible paths connecting users or items in order to find new cross-domain relationships. Tang et al. [15] proposed the cross-domain topic learning model to predict the cross-domain collaborations through topic layers instead of at author layers, which alleviated the sparseness issue. Winoto et al. [16] proposed to uncover the association between user preferences on related items across domains to provide the cross-domain recommendation. However, all the above models always consider the same users or items across different domains, which is not a realistic setting. Actually, the most relevant works with ours are [2] [4]. They suggested to leverage useful knowledge from a different domain and extracted latent rating pattern across domains with non-overlap users and items. Moreno et al. [5] addressed the sparsity problem by integrating the appropriate amount of knowledge from each domain in order to enrich the target domain. Moreover, the majority of the existing work assumes that the source and target domains are related but do not suggest methods to calculate the relatedness among domains, which has been addressed in our work.

There are other recent studies which have been made on applying cross domain techniques, and transfer learning in particular into information recommendation task. Transfer learning aims to leverage the knowledge in the source domain to facilitate the learning tasks in the target domain [17]. The general idea of the existing methods is to utilize the common latent structure shared across domains as the bridge for knowledge transfer. For example, Xue et al. [18] have addressed the problem of using the auxiliary labeled data to help classify the unlabeled data in the target domain. Shi et al. [19] proposed a generalized tag-induced cross-domain collaborative filtering model to exploit user-contributed tags as common characteristics to link different domains together and transfer the knowledge between different domains. However, transferring knowledge across domains is a challenging task since it cannot be guaranteed that the knowledge of other domains is useful for the target domain. In this paper, we not only consider the common latent rating pattern across domains but also extract the discriminative domain-specific information to improve the mutual strengths in each domain.

6   Conclusion

In this paper, we proposed a novel Cluster-Level based Latent Factor Model (CLFM) based on the framework of joint nonnegative matrix tri-factorization. The CLFM model can construct a latent space to represent the rating patterns of user groups on the item clusters from each domain, then based on a subspace learning of the latent space, CLFM model not only learn shared common rating pattern across multiple rating matrices to alleviate the sparsity problems in individual domain, with the flexibility in controlling the optimal level of sharing the relatedness among domains, but also learn the domain-specific cluster-level rating pattern from each domain that contains the discriminative information propitious to improve across recommendation accuracy. The proposed model thus could exploit the mutual strengths of related domains by the shared common rating patterns as well as the domain-specific rating patterns from each domain. The experimental results have validated that our proposed CLFM model indeed can benefit from the cluster-level rating patterns and outperforms the state-of-the-art methods for cross-domain recommendation task.

There are still several extensions to improve our work. Firstly, it is necessary to compare our proposed model against the two more recent methods [5] to explore a deeper understanding in the cross-domain recommendation task. Second, our proposed CLFM model should be evaluated on large scale rating dataset to exploit its scalable computational ability. Third, a probabilistic version would be the natural extension of our proposed CLFM model, which may exhibit better interpretable properties.

Acknowledgment. This research was supported by the National Natural Science Foundation of China under grant No. 61273217, the Fundamental Research Funds for the Central Universities of China under grant No. 2013RC0119, the Chinese 111 program of 'Advanced Intelligence and Network Service' under grant No. B08004 and key project of Ministry of Science and Technology of China under grant No. 2011ZX03002-005-01. The authors are also supported by the Key Laboratory of Advanced Information Science and Network Technology of Beijing under grant No. XDXX1304.

References

1. Koren, Y.: Factorization meets the neighborhood: a multifaceted collaborative filtering model. In: Proceedings of the 14th ACM SIGKDD International Conference on Knowledge Discovery and Data Mining, KDD 2008, pp. 426–434 (2008)
2. Li, B., Yang, Q., Xue, X.: Transfer learning for collaborative filtering via a rating-matrix generative model. In: Proceedings of the 26th Annual International Conference on Machine Learning, ICML 2009, pp. 617–624 (2009)
3. Li, B.: Cross-domain collaborative filtering: a brief survey. In: 23rd IEEE International Conference on Tools with Artificial Intelligence, pp. 1085–1086 (2011)
4. Li, B., Yang, Q., Xue, X.: Can movies and books collaborate? cross-domain collaborative filtering for sparsity reduction. In: Proceedings of the 21st International Joint Conference on Artifical Intelligence, IJCAI 2009, pp. 2052–2057 (2009)
5. Moreno, O., Shapira, B., Rokach, L., Shani, G.: Talmud: transfer learning for multiple domains. In: 21st ACM International Conference on Information and Knowledge Management, CIKM 2012, pp. 425–434 (2012)
6. Ding, C., Li, T., Peng, W., Park, H.: Orthogonal nonnegative matrix t-factorizations for clustering. In: SIGKDD, pp. 126–135 (2006)
7. Lee, D.D., Seung, H.S.: Algorithms for non-negative matrix factorization. In: NIPS, pp. 556–562 (2000)
8. Gao, S., Denoyer, L., Gallinari, P.: Temporal link prediction by integrating content and structure information. In: CIKM 2011, pp. 1169–1174 (2011)
9. Ding, C., Li, T., Jordan, M.I.: Convex and semi-nonnegative matrix factorizations. IEEE Transactions on Pattern Analysis and Machine Intelligence, 2052–2067 (2008)
10. Si, L., Jin, R.: Flexible mixture model for collaborative filtering. In: ICML 2003, pp. 704–711 (2003)
11. Fernadez-Tobis, I., Cantador, I., Kaminskas, M., Ricci, F.: Cross-domain recommender systems: A survey of the state of the art. In: Proceedings of the 2nd Spanish Conference on Information Retrieval (2012)
12. Berkovsky, S., Kuflik, T., Ricci, F.: Cross-domain mediation in collaborative filtering. In: Conati, C., McCoy, K., Paliouras, G. (eds.) UM 2007. LNCS (LNAI), vol. 4511, pp. 355–359. Springer, Heidelberg (2007)

13. Pan, W., Xiang, E., Liu, N., Yang, Q.: Transfer learning in collaborative filtering for sparsity reduction. In: Proceedings of the 24rd AAAI Conference on Artificial Intelligence, pp. 425–434 (2010)
14. Cremonesi, P., Tripodi, A., Turrin, R.: Cross-domain recommender systems. In: Proceedings of the 2011 IEEE 11th International Conference on Data Mining Workshops, ICDMW 2011, pp. 496–503 (2011)
15. Tang, J., Wu, S., Sun, J., Su, H.: Cross-domain collaboration recommendation. In: Proceedings of the 18th ACM SIGKDD International Conference on Knowledge Discovery and Data Mining, pp. 1285–1293 (2012)
16. Winoto, P., Tang, T.Y.: If you like the devil wears prada the book, will you also enjoy the devil wears prada the movie? a study of cross-domain recommendations. New Generation Comput., 209–225 (2008)
17. Pan, S., Yang, Q.: A survey on transfer learning. IEEE Transactions on Knowledge and Data Engineering 22(10), 1345–1359 (2010)
18. Xue, G., Dai, W., Yang, Q., Yu, Y.: Topic-bridged plsa for cross-domain text classification. In: SIGIR, pp. 627–634 (2008)
19. Shi, Y., Larson, M., Hanjalic, A.: Generalized tag-induced cross-domain collaborative filtering. CoRR, abs/1302.4888 (2013)

Minimal Shrinkage for Noisy Data Recovery Using Schatten-p Norm Objective

Deguang Kong, Miao Zhang, and Chris Ding

Dept. of Computer Science & Engineering, University of Texas at Arlington, TX, 76013
doogkong@gmail.com, miao.zhang@mavs.uta.edu, chqding@uta.edu

Abstract. Noisy data recovery is an important problem in machine learning field, which has widely applications for collaborative prediction, recommendation systems, etc. One popular model is to use trace norm model for noisy data recovery. However, it is ignored that the reconstructed data could be shrank (i.e., singular values could be greatly suppressed). In this paper, we present novel noisy data recovery models, which replaces the standard rank constraint (i.e., trace norm) using Schatten-p Norm. The proposed model is attractive due to its suppression on the shrinkage of singular values at smaller parameter p. We analyze the optimal solution of proposed models, and characterize the rank of optimal solution. Efficient algorithms are presented, the convergences of which are rigorously proved. Extensive experiment results on 6 noisy datasets demonstrate the good performance of proposed minimum shrinkage models.

1 Introduction

In big-data era, data is always noisy, development of robust noise tolerant algorithm for data recovery, is always useful and highly demanded. On the other hand, the available of large amount of data makes it more difficult to control the quality the data. The chances of the damaged data or noisy data are increasing. Given input noisy data $\mathbf{X}$, the goal of low rank data recovery problem [1,2,3], is to find a low rank approximation $\mathbf{Z}$. Recovered data $\mathbf{Z}$ is expected to be low rank, and retain minimum reconstruction errors (such as least square error) as compared to input data matrix $\mathbf{X}$. In practice, input data can be noisy and also has missing values. This problem has attracted a lot of attentions due to its widely applications in recommendation systems [4], collaborative prediction [5], image/video completion [6], etc.

Data recovery problem has close relations with dimension reduction or low dimension subspace recovery, since for most of high-dimensional data, they may have low-dimensional subspace. Many efforts have been devoted along the direction of principal component analysis (PCA) [7], compressive sensing [8], affine rank minimization [3], etc. For example, Principal component analysis (PCA) seeks for a low-dimensional subspace given data matrix, which can be efficiently computed using singular value decomposition (SVD). However, a major drawback of classical PCA [9] is that, it breaks down under grossly corrupted or noisy observations, such as noises/corruptions in images, and dis-measurement in bio-informatics, etc. In Regularized PCA model (*e.g.*, [10,11]), it aims at reducing the rank of the data without explicitly reducing the dimension. However, they do not return the clear representation of subspace and low-dimensional data explicitly.

H. Blockeel et al. (Eds.): ECML PKDD 2013, Part II, LNAI 8189, pp. 177–193, 2013.

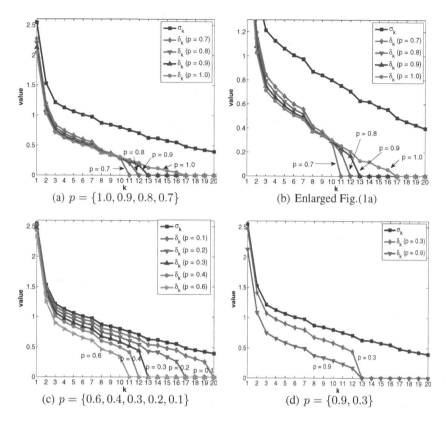

(a) $p = \{1.0, 0.9, 0.8, 0.7\}$

(b) Enlarged Fig.(1a)

(c) $p = \{0.6, 0.4, 0.3, 0.2, 0.1\}$

(d) $p = \{0.9, 0.3\}$

Fig. 1. Optimal solution δ_k given singular value σ_k of input data $\mathbf{X}$, at different $p = \{1, 0.9, 0.8, \cdots, 0.1\}$ values with fixed $\beta = 0.5$, on dataset Mnist with 20 images, i.e., $\mathbf{X} = \{\mathbf{x}_1, \mathbf{x}_2, \cdots, \mathbf{x}_{20}\}$. To avoid clutter, part of Fig.1a is zoomed in and shown in Fig.1b. In Fig.1d, the solution at $p = 0.3$ is a *faithful* low-rank solution, and the solution at $p = 0.9$ is a *suppressed* low-rank solution.

It is well known that it is a NP-hard problem to directly minimizing the rank of data for recovering input data. Since trace norm can be viewed as a convex envelope of rank function [12], different methods (*e.g.*, [13,14,1,15,16,17]), have been proposed by minimizing the trace norm. In this paper, we point out that, standard trace norm model suffers from a serious problem: *shrinkage of reconstructed data and suppression of singular values* (see more details in Figs.(1-2) and §3). We find that the trace norm relaxation may deviate the solution away from the real solution of original rank minimization problem.

The goal of this paper is to develop new methods to solve the approximation of the rank minimization problem. In this paper, we reformulate the noisy data recovery problem using schatten p norm, where efficient algorithms are presented. To summarize, the main contribution of this paper is listed as follows.

- From model construction point of view, we present new models for noisy data recovery, which minimize both data recovery error and rank of recovery data. The proposed models give the minimum shrinkage of recovered data.
- From algorithmic development point of view, we present a complete analysis for proposed model, where the rank of optimal solution is characterized by Theorem 1. Efficient algorithms are developed.
- Extensive experiments on noisy datasets indicate better noisy data recovery performance at smaller p values (p is parameter of our model).

2 Proposed Data Recovery Models

Notation. Let $\mathbf{X} = (x_1 \cdots x_n) \in \Re^{d \times n}$ be input n data, each of dimension d. For standard Schatten p norm of matrix $\mathbf{Z}$,

$$\|\mathbf{Z}\|_{sp} = \left(\sum_{k=1}^{r} \sigma_k^p\right)^{\frac{1}{p}} = \left(\text{Tr}[(\mathbf{Z}^T\mathbf{Z})^{\frac{p}{2}}]\right)^{\frac{1}{p}}, \tag{1}$$

where σ_k is the singular value of $\mathbf{Z}$, $r = rank(\mathbf{Z})$.

Given a data matrix $\mathbf{X}$, it is often of interest to compute a matrix $\mathbf{Z}$ that is "close" to $\mathbf{X}$ and satisfies the constraint $rank(\mathbf{Z}) < rank(\mathbf{X})$. Singular value decomposition [18] is the most popular method for such approximations. There are alternative methods that replace this constraint with a more friendly constraint, like, for example, the trace norm. In this paper, we present two models:

Model 1: Schatten p Model
We wish to solve the data recovery problem, i.e.,

$$\min_{\mathbf{Z}} \frac{1}{2}\|\mathbf{Z} - \mathbf{X}\|_F^2 + \beta\text{Tr}[(\mathbf{Z}^T\mathbf{Z})^{\frac{p}{2}}], \tag{2}$$

where $\text{Tr}(\mathbf{Z}^T\mathbf{Z})^{\frac{p}{2}} = \sum_{k=1}^{r} \sigma_k^p$, and σ_k is the singular value of $\mathbf{Z}$, β is a parameter to control the scale of schatten p term.

The fact is that the approximation has the same eigen-vectors as the original matrix, and that only eigen-values are shrinked in standard matrix linear algebra. The particular shrinkage of p Schatten norm is better than trace norm ($p = 1$, see Fig. 1), which is corresponding to soft thresholding. At $p = 0$, this is corresponding to hard thresholding (exactly the rank).

Model 2: Robust Schatten p Model
We wish to find low-rank data recovery $\mathbf{Z}$ given $\mathbf{X}$, i.e.,

$$\min_{\mathbf{Z}} \|\mathbf{Z} - \mathbf{X}\|_1 + \beta\text{Tr}[(\mathbf{Z}^T\mathbf{Z})^{\frac{p}{2}}]. \tag{3}$$

This is used for noisy data recovery purpose, which can be viewed as an extension of robust PCA [10].

Motivation

The goal of proposed models is to provide minimum shrinkage of reconstructed data and suppression of singular values. This is the reason, why we replace the trace norm regularization with schatten p regularization. More detailed analysis is provided in §3-4. Our experiment results indicate that proposed models at smaller p values give better recovery performance.

As p becomes small, it is closer to the desired rank constraint:

$$\lim_{p \to 0} \mathrm{Tr}(\mathbf{Z}^T \mathbf{Z})^{\frac{p}{2}} = \lim_{p \to 0} \sum_k \sigma_k^p = rank(\mathbf{Z}).$$

This indicates that the lower p, the better that Schatten norm resembles the rank. Since we wish to do reconstruction with low rank, thus parameter p is usually set to $0 \le p \le 1$. In general $p > 1$ case is un-interesting.

Differences of Two Models. The difference of above two models of Eqs.(2, 3) lies in the first term. In Model 1 of Eq.(2), Frobenius norm or the least square error is used to minimize the reconstruction error. In Model 2 of Eq.(3), the L_1-norm is used to minimize the reconstruction error. As is known to us, L_1 error is more robust to noises and outliers, because $||\mathbf{X} - \mathbf{Z}||_1 = \sum_{ij} |\mathbf{X} - \mathbf{Z}|_{ij}$, where residue term is *not* squared. In real world, the observations (like images, text features, etc) can be contaminated by noises or outliers. Model of Eq.(2) is for the data recovery problem polluted by Gaussian noise, while model of Eq.(3) is for data contaminated by Laplacian noises. Both models can be used to solve noisy data recovery, matrix completion problem, etc. For second term, for computational purpose, we add p power to standard term $||\mathbf{Z}||_{sp}$, which plays the same role as standard schatten term for low rank approximation purpose.

Relations with Previous Methods. At $p = 1$, Eq.(3) is equivalent to standard trace-norm model, which optimizes

$$\min_{\mathbf{Z}} ||\mathbf{Z} - \mathbf{X}||_1 + \beta ||\mathbf{Z}||_*, \tag{4}$$

where $||\mathbf{Z}||_* = \mathrm{Tr}(\mathbf{Z}^T \mathbf{Z})^{\frac{1}{2}}$ is the trace norm, and σ_k is the singular value of $\mathbf{Z}$. This study is a special case of our model. Note in [10], Schatten p-Norm model at $p = 1$ is called as Robust PCA, because it can correctly recover underlying low-rank structure $\mathbf{Z}$ from the data $\mathbf{X}$ in the presence of gross errors and outlying observations.

3   Illustration of Model 1 and Model 2

Due to the non-smoothness of Schatten norm at $p < 1$, the computational algorithm is challenging. We provide detailed analysis and efficient algorithms of both models in §4, §5 and §6. Here we discuss the general features of the optimal solutions to these two models. The key conclusion is that the solutions at small p are much better than the solution at $p = 1$, which is a previously studied model.

3.1 Illustration of Model 1

To illustrate results of Model 1, we use 20 images from real-world dataset mnist (more details of this dataset is in §7). Let δ_k be the singular values of the optimal solution $\mathbf{Z}^*$. Let σ_k be the singular values of input data $\mathbf{X}$. We show solution δ_k in Fig.1 along with σ_k. We fix $\beta = 0.5$, but let p vary from $p = 1$ to $p = 0.1$. From Fig.1, we see that at $p = 1$, the optimal solution $\mathbf{Z}^*_{p=1}$, which is represented by $(\delta_1, \delta_2, \cdots, \delta_{20})$, is a simple downshift of $(\sigma_1, \sigma_2, \cdots, \sigma_{20})$. The high rank part $(k = 17 - 20)$ is zero. As p decreases, more high rank part of the solution $\{\delta_k\}$ becomes zero, while the lower rank part of $\{\delta_k\}$ moves closer to $\{\sigma_k\}$ of the input data. For example, in Fig.1a, Fig.1b, in optimal solution $\mathbf{Z}^*_{p=0.9}$, the high rank part $(k = 13 - 20)$ becomes zero, while the low-rank part $(k = 1 - 7)$ is higher than that of $\mathbf{Z}^*_{p=1}$, i.e., this part moves towards corresponding $\{\sigma_k\}$.

In general in low-rank data recovery, we wish the low-rank part of $\mathbf{Z}^*$ is close to those of the input data, while the high-rank part is cut-off (close to zero). Looking in Fig.1d, the solution at $p = 0.3$ is a "faithful" low-rank solution, because the low-rank part is more close or *faithful* to the original data. The solution at $p = 0.9$ is a "suppressed" low-rank solution because the low-rank part is far below the original data, i.e., they are *suppressed*. Clearly, the solution at $p = 0.3$ is more desirable than solution at $p = 0.9$, even though both solutions are low-rank: rank$(\mathbf{Z}^*_{p=0.9})$= rank$(\mathbf{Z}^*_{p=0.3}) = 12$.

The Schatten p norm model at small p provides the desirable "faithful" low-rank solution, while the previous work using $p = 1$ also provides a low-rank solution, but the low-rank part is more *suppressed*.

3.2 Illustration of Model 2

Model 2 of Eq.(3) differs from Model 1 by using the L_1 norm in error function. This enables the model to do robust data recovery (e.g., moving outliers back to the correct subspace). However, this model does not change the observed *suppression* in Model 1 at p close to 1 (see Fig.1d). The suppression of singular values leads to the *shrinkage* effect in reconstructed data.

We demonstrate the robust data recovery and the shrinkage effects for Model 2 at different p values on a simple toy data in Fig.(2a). The original data $\mathbf{X}$ are shown as black circles. Reconstructed data z_i are shown as red-squares. We show the reconstructed results at $p = 0.2$ Fig.(2b, 2e, 2f), $p = 0.5$ (Fig.2c, 2g), $p = 1$ (Fig.2d, 2h). We have two observations.

First, at $0 \leq p \leq 1$, outliers $(\mathbf{x}_{13}, \mathbf{x}_{14}, \mathbf{x}_{15})$ all move towards the correct subspace, indicating the desired denoising data recovery effects.

Second, for non-outlier data, the reconstructed data shrink strongly at $p = 1$, but they shrink much less at $p = \{0.2, 0.5\}$. This shrinkage is result of the singular value suppression in computed $\mathbf{Z}$. At $p = \{0.2, 0.5, 1\}$, the largest singular value are $\{5.35, 4.49, 2.93\}$, while the second singular values are very small, i.e., $\{1.7e\text{-}8, 1.7e\text{-}16, 9.8e\text{-}9\}$, respectively.

In summary, the Schatten model at small p enables us to do robust data recovery but without significant shrinkage in previous models which use $p = 1$.

To our knowledge the singular value suppression and shrinkage (both at $p = 1$ and smaller p values) have not been studied previously.

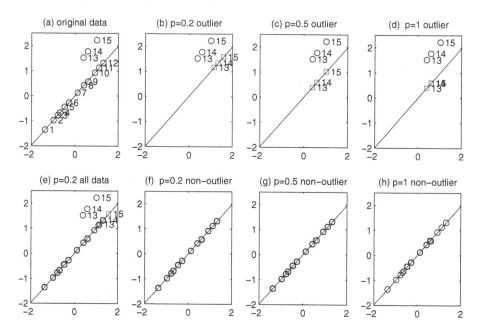

Fig. 2. Demonstration of robust Schatten-p model of Eq.(3) on a toy data shown in panel (a): original data shown as black circles. $(\mathbf{x}_1 \cdots \mathbf{x}_{12})$ are non-outliers and $(\mathbf{x}_{13} \cdots \mathbf{x}_{15})$ are outliers. Reconstructed data $\mathbf{z}_i$ are shown as red-diamonds. Blue line indicates the subspace computed from standard PCA on non-outlier data. Results of Schatten model at $p = 0.2$ are shown in (e). This $p = 0.2$ results are split to outliers and non-outliers as shown in (b) and (f). Similarly, results for $p = 0.5$ shown in (c) and (g); results for $p = 1$ shown in (d) and (h). At $p = 1$, non-outliers shrink towards coordinate (0,0). At smaller p, non-outliers shrink far less.

4 Analysis and Algorithm of Model 1

We show how to solve Model 1 of Eq.(2) at different p values. This also serves as the basic step in solving Model 2 of Eq.(3) using the ALM of §5. To our knowledge, this problem has not been studied before.

Property 1. The global optimal solution for Eq.(2) at all $0 \le p \le 1$, can be efficiently computed, even though it is non-convex at $p < 1$.

Property 2. Rank of the optimal solution $\mathbf{Z}^*$ has a closed form solution:

Theorem 1. *Let the singular value decomposition (SVD) of $\mathbf{X}$ be $\mathbf{X} = \sum_k \sigma_k \mathbf{u}_k \mathbf{v}_k^T$. Then rank of optimal solution $\mathbf{Z}^*$: $rank(\mathbf{Z}^*) = largest\ k$, such that*

$$\sigma_k \le \left(\frac{\beta p (2-p)^{(2-p)}}{(1-p)^{(1-p)}} \right)^{\frac{1}{2-p}}, \quad 0 < p \le 1. \tag{5}$$

In particular, $p = 1, \sigma_k \le \beta; p = \frac{1}{2}, \sigma_k \le (\sqrt{\frac{27}{16}}\beta)^{\frac{2}{3}}$.

Property 3. Optimal solution $\mathbf{Z}^*$ has a closed form solution at $p = \frac{1}{2}$.

Property 4. Optimal solution $\mathbf{Z}^*$ at $0 < p < 1$ can be obtained using Newton's method.

To prove above 4 properties for Model 1 of Eq.(2), we need the following useful lemma.

Lemma 1. *Let the singular value decomposition (SVD) of $\mathbf{X}$ be $\mathbf{X} = \sum_k \sigma_k \mathbf{u}_k \mathbf{v}_k^T$. The global optimal $\mathbf{Z}$ for Eq.(2) is given by $\mathbf{Z} = \sum_k \delta_k \mathbf{u}_k \mathbf{v}_k^T$, where δ_k is given by solving,*

$$\min_{\delta_1, \cdots, \delta_r} \sum_{k=1}^{r} \left[\frac{1}{2}(\delta_k - \sigma_k)^2 + \beta \delta_k^p \right], \quad s.t. \ \delta_k \geq 0, k = 1 \cdots r. \tag{6}$$

4.1 Proof of Lemma 1

Proof. Let the optimal solution of $\mathbf{Z}$ have the SVD $\mathbf{Z} = \mathbf{F} \Delta \mathbf{G}^T$ where $\mathbf{F} = (\mathbf{f}_1 \cdots \mathbf{f}_r)$ and $\mathbf{G} = (\mathbf{g}_1 \cdots \mathbf{g}_r)$ are the singular vectors of $\mathbf{Z}$, and $\Delta = \text{diag}(\delta_1 \cdots \delta_r)$ be their singular values. The key is to prove that the singular vectors of $\mathbf{Z}^*$ are the same as those of the input data $\mathbf{X}$. Using von Neumann's trace inequality

$$|\text{Tr}(\mathbf{Z}^T \mathbf{X})| \leq \text{Tr}\Delta\Sigma = \sum_{k=1}^{r} \delta_k \sigma_k. \tag{7}$$

From this, we have

$$\text{Tr}(\mathbf{U}\Delta\mathbf{V}^T)^T \mathbf{X} = \text{Tr}\Delta\Sigma \geq \text{Tr}(\mathbf{Z}^T \mathbf{X}) = \text{Tr}(\mathbf{F}\Delta\mathbf{G}^T)^T \mathbf{X}, \tag{8}$$

where the inequality comes from Eq.(7). The inequality

$$\text{Tr}(\mathbf{U}\Delta\mathbf{V}^T)^T \mathbf{X} \geq \text{Tr}(\mathbf{F}\Delta\mathbf{G}^T)^T \mathbf{X}$$

implies

$$\frac{1}{2}\|\mathbf{U}\Delta\mathbf{V}^T - \mathbf{X}\|^2 + \beta \text{Tr}\Delta^p \leq \frac{1}{2}\|\mathbf{F}\Delta\mathbf{G}^T - \mathbf{X}\|^2 + \beta \text{Tr}\Delta^p.$$

This indicates $(\mathbf{U}, \mathbf{V})$ are better singular vectors for $\mathbf{Z}$ than $(\mathbf{F}, \mathbf{G})$. This proves that the optimal singular vectors for $\mathbf{Z}$ must be the same singular vectors of $\mathbf{X}$. Setting $\mathbf{Z} = \mathbf{U}\Delta\mathbf{V}^T$ in Eq.(2), we obtain Eq.(6).

4.2 Analysis of Property 1

Due to Lemma 1, we now solve the simpler problem of Eq.(6) instead of the original harder problem of Eq.(2). Clearly the optimization of Eq.(6) decouples into r independent subproblems, each for one δ_k:

$$\min_{\delta_k} \frac{1}{2}(\delta_k - \sigma_k)^2 + \beta \delta_k^p, \quad s.t. \ \delta_k \geq 0. \tag{9}$$

KKT complementarity slackness condition for $\delta_k \geq 0$ leads to $\left[(\delta_k - \sigma_k) + p\beta\delta_k^{p-1} \right] \delta_k = 0$. The optimization of Eq.(9) decouples into r independent subproblems, and each of them is of the type:

$$\min_{x \geq 0} J(x) = \frac{1}{2}(x - a)^2 + \beta x^p, \tag{10}$$

where $x, a \in \Re$. Here the correspondence between Eq.(10) and Eq.(9) is $a = \sigma_k$, $x = \delta_k$. $J(x)$ is a weight sum over two functions: $J(x) = f_1(x) + \beta f_2(x)$, where $f_1(x) = \frac{1}{2}(x - a)^2, f_2(x) = x^p$. $f_1(x)$ has a local minima at $x_1 = a$. $f_2(x)$ is a singular function, $p \leq 1$ with singularity at $x_2 = 0$, which is also a local minima.

Therefore, $J(x)$ in general has two local minima (x_1^*, x_2^*). Because $f_2(x)$ is singular at x_2, for Eq.(10), the singular point (local minima) does not change with different weight β. Thus $x_2^* = 0$ is always a local minima.

When β is small, $x_1^* = a$. As β increases, x_1^* moves towards 0. At certain (β, p), this local minima disappears, $J(x)$ has only one local minima $x_2^* = 0$. This condition is determined by the same condition as in Theorem 1 or Eq.(12) with $\sigma_k = a$. x_1^* is easily computed using Property 4.

In summary, the optimal solution of Eq.(10) is either the trivial one $x_2^* = 0$ or $\min(x_1^*, x_2^*)$, when x_1^* exits. This means Eq.(9) can be easily solved. Thus Eq.(6) can be easily solved for each rank one at a time.

4.3 Proof of Theorem 1

Proof. First, optimization of Eq.(2) is equivalent to optimizing Eq.(10), which can be further written as,

$$\min_{z \geq 0} g(z) = \frac{1}{2}(z - 1)^2 + \hat{\beta} z^p, \tag{11}$$

where $z = x/a, \hat{\beta} = \beta a^{(p-2)}$. First, we note a key quantity, the zero crossing point z_0 exists, where the second derivative $g''(z)$ changes its sign, i.e., $g''(z_0) = 0$. We need two lemmas.

Lemma 2. *This cross point z_0 always exists at any β.*

Lemma 3. *If the slope of cost function of Eq.(11) at the crossing point z_0 is negative, i.e., $g'(z_0) < 0$, there exists two distinct local minima: $z_2 = 0$ and $z_1 > 0$. If $g'(z_0) \geq 0$, $z_2 = 0$ is the global optimal solution.*

Lemmas 2 and 3 give the key properties of optimization of Eq.(11). Set $g''(z_0) = 0$, we obtain $z_0 = [\hat{\beta} p(1 - p)]^{\frac{1}{2-p}}$. Lemma 2 states that $z_2 = 0$ is the global solution, $g'(z_0) = z_0 - 1 + \hat{\beta} p z_0^{p-1} \geq 0$, i.e., $[\hat{\beta} p(1 - p)]^{\frac{1}{2-p}} - 1 + \hat{\beta} p[\hat{\beta} p(1 - p)]^{\frac{p-1}{2-p}} \geq 0$. Solving for β, we have,

$$\beta \geq \frac{1(1 - p)^{(1-p)}}{p(2 - p)^{(2-p)}} \cdot \sigma_k^{(2-p)}, \quad 0 < p \leq 1. \tag{12}$$

This indicates that the optimal solution δ_k of Eq.(11) is zero (i.e., $\delta_k = 0$), if Eq.(12) holds. This completes the proof.

4.4 Analysis of Property 3

Clearly, at $p = \frac{1}{2}$, from Eq.(9), we need to solve $\delta_k - \sigma_k + (\beta/2)\delta_k^{-1/2} = 0$, s.t. $\delta_k \geq 0$. Let $\rho_k = \left(\frac{\delta_k}{\sigma_k}\right)^{1/2}, \mu = \frac{\beta}{2\sigma_k^{\frac{3}{2}}}$, this becomes $\rho_k^3 - \rho_k + \mu = 0$, where $\rho_k \geq 0$. The analytic solution of this cubic equation can be solved in closed form.

Algorithm 1. ALM algorithm to solve Eq.(3)

Input: data matrix $\mathbf{X}$, parameter $\rho > 1$.
Output: low rank approximation $\mathbf{Z}$.
Procedure:

1: Initialize $\mathbf{E}, \mathbf{Z}, \Omega, \mu > 0, t = 0,$; $\rho = 1.1$
2: **while** Not converge **do**
3:     Updating $\mathbf{E}$ according to Eq.(16)
4:     Updating $\mathbf{Z}$ according to Eq.(17)
5:     Updating Ω: $\Omega := \Omega + \mu(\mathbf{Z} - \mathbf{X} - \mathbf{E})$
6:     Updating μ: $\mu := \rho\mu$
7: **end while**

4.5 Analysis of Property 4

From analysis of property 1, the optimization of Eq.(10) has two local optima: $x_1^* > 0, x_2^* = 0$. Our algorithm is: (b1) to use Newton's method to compute x_1^*; (b2) compare $J(x_1^*), J(x_2^*)$, and pick the smaller one. It is easy to see $J'(x) = x - a + \beta p x^{p-1}$, $J''(x) = 1 + \beta p(p-1)x^{p-2}$. Using standard Newton's method, we can update x through $x \leftarrow x - \frac{J'(x)}{J''(x)}$. This algorithm has quadratic convergence. In practical applications, we found this Newton's algorithm typically converges to local minima within a few iterations.

5 ALM Algorithm to Solve Model 2

Augmented lagrange multipliers(ALM) have been widely used to solve different kinds of optimization problems ([10], [19]). Here we adapt standard ALM method [20,19] to solve Schatten-p model of Eq.(3). It is worth noting that it is not trivial to solve Eq.(3) using ALM method. One challenging step is to solve the associated Schatten-p term shown in §4.

According to ALM algorithm, by imposing constraint variable $\mathbf{E} = \mathbf{Z} - \mathbf{X}$, the problem of Eq.(3) is equivalent to solve,

$$\min_{\mathbf{E}, \mathbf{Z}} \|\mathbf{E}\|_1 + \beta \text{Tr}(\mathbf{Z}^T \mathbf{Z})^{\frac{p}{2}}, \qquad s.t. \qquad \mathbf{Z} - \mathbf{X} - \mathbf{E} = 0. \tag{13}$$

According to ALM algorithm, we need to solve,

$$\min_{\mathbf{E}, \mathbf{Z}} \|\mathbf{E}\|_1 + \langle \Omega, \mathbf{Z} - \mathbf{X} - \mathbf{E} \rangle + \frac{\mu}{2}\|\mathbf{Z} - \mathbf{X} - \mathbf{E}\|_F^2 + \beta \text{Tr}(\mathbf{Z}^T \mathbf{Z})^{\frac{p}{2}}, \tag{14}$$

where Lagrange multiplier is Ω and μ is penalty constant. For this problem, Ω and μ updated in a specified pattern:

$$\Omega \leftarrow \Omega + \mu(\dot{\mathbf{Z}} - \mathbf{X} - \mathbf{E}), \quad \mu \leftarrow \rho\mu.$$

We need to search for optimal $\mathbf{E}, \mathbf{Z}$ iteratively until the algorithm converges. Now we discuss how we solve $\mathbf{E}, \mathbf{Z}$ in each step.

Update E. To update the error matrix $\mathbf{E}$, we derive Eq.(15) with fixed $\mathbf{Z}$ and obtain the following form:

$$\min_{\mathbf{E}} \frac{\mu}{2}||\mathbf{E} - \mathbf{A}||_F^2 + ||\mathbf{E}||_1 \tag{15}$$

where $\mathbf{A} = \mathbf{X} - \mathbf{Z} + \frac{\Omega}{\mu}$. It is well-known that the solution to the above LASSO type problem [21] is given by,

$$\mathbf{E}_{ij} = sign(\mathbf{A}_{ij}) \max(|\mathbf{A}_{ij}| - \frac{1}{\mu}, 0). \tag{16}$$

Update Z. To update $\mathbf{Z}$ while fixing $\mathbf{E}$, we minimize the relevant part of Eq.(14), which is

$$\min_{\mathbf{Z}} \beta \mathrm{Tr}(\mathbf{Z}\mathbf{Z}^T)^{\frac{p}{2}} + \frac{\mu}{2}||\mathbf{Z} - \mathbf{X} - \mathbf{E} + \frac{\Omega}{\mu}||_F^2. \tag{17}$$

Setting $\mathbf{B} = \mathbf{X} + \mathbf{E} - \frac{\Omega}{\mu}, \hat{\beta} = \frac{\beta}{\mu}$, this optimization becomes Eq.(2), which has been solved in §4.

6   Iterative Algorithm to Solve Model 2

We present another efficient iterative algorithm to solve Eq.(3), where the variable matrix $\mathbf{Z}$ is updated iteratively. Suppose $\mathbf{Z}_t$ is the value of $\mathbf{Z}$ at t-th step. At step t, the key step of our algorithm is to iteratively update j-th column ($\mathbf{z}_j$) of $\mathbf{Z}$ one at a time, according to

$$\mathbf{z}_j = \mathbf{A}^{-1}(\mathbf{A}^{-1} + p\lambda\mathbf{D}_j^{-1})^{-1}\mathbf{x}_j, \tag{18}$$

where $\mathbf{A} = (\mathbf{Z}_t\mathbf{Z}_t^T)^{p/2-1}, \mathbf{W}_{ij} = 1/|(\mathbf{Z}_t - \mathbf{X})_{ij}|, \mathbf{D}_j = \mathrm{diag}(\mathbf{w}_j), \mathbf{w}_j$ is the j-th column of $\mathbf{W}$. This process is iteratively done for $1 \leq j \leq n$. Then $\mathbf{Z}$ is updated until the algorithm converges. More detailed algorithm is summarized in Algorithm 2. In computing $\mathbf{z}_j$ of Eq.(18), we first use conjugate gradient method to compute $\tilde{\mathbf{z}}_j$, where $(\mathbf{A}^{-1} + p\lambda\mathbf{D}_j^{-1})\tilde{\mathbf{z}}^j = \mathbf{x}_j$, and then $\mathbf{z}_j = \mathbf{A}^{-1}\tilde{\mathbf{z}}_j$.

Algorithm 2. An iterative algorithm to solve Eq.(3)

Input: $\mathbf{X}, \lambda$
Output: $\mathbf{Z}$

1: **while** *not converge* **do**
2:     compute $\mathbf{A}^{-1}$
3:     **for** $j = 1 : n$ **do**
4:         compute $\mathbf{D}_j^{-1}$, solve $\mathbf{z}_j$ according to Eq.(18)
5:     **end for**
6: **end while**

6.1 Convergence of Algorithm

Let $J(\mathbf{Z}) = \|\mathbf{Z} - \mathbf{X}\|_1 + \beta \mathrm{Tr}(\mathbf{Z}^T \mathbf{Z})^{\frac{p}{2}}$, we have

Theorem 2. *Updating $\mathbf{Z}$ using Eq.(18), $J(\mathbf{Z})$ decreases monotonically.*

The proof requires the following two Lemmas.

Lemma 4. *Define the objective function*

$$J_2(\mathbf{Z}) = \|\mathbf{Z} - \mathbf{X}\|_{\mathbf{W}}^2 + p\beta \mathrm{Tr}(\mathbf{Z}^T \mathbf{A}\mathbf{Z}). \tag{19}$$

where $\|\mathbf{A}\|_{\mathbf{W}}^2 = \sum_{ij} A_{ij}^2 W_{ij}$. The updated $\mathbf{Z}_{t+1}$ using Eq.(18) satisfies

$$J_2(\mathbf{Z}_{t+1}) \le J_2(\mathbf{Z}_t) \tag{20}$$

Lemma 5. *The updated $\mathbf{Z}_{t+1}$ using Eq.(18) satisfies*

$$J(\mathbf{Z}_{t+1}) - J(\mathbf{Z}_t) \le \frac{1}{2}\big[J_2(\mathbf{Z}_{t+1}) - J_2(\mathbf{Z}_t)\big] \tag{21}$$

6.2 Proof of Theorem 2

Proof. From Eq.(20), clearly, LHS of Eq.(21) is LHS ≤ 0.

6.3 Proof of Lemma 4

Proof. Setting $\partial J_2(\mathbf{Z})/\partial Z_{ij} = 0$, we have $(\mathbf{Z} - \mathbf{X})_{ij} W_{ij} + p\lambda(\mathbf{A}\mathbf{Z})_{ij} = 0$. This can be written as $Z_{ij}W_{ij} + p\lambda(\mathbf{A}\mathbf{Z})_{ij} = X_{ij}W_{ij}$. In matrix form, $\mathbf{D}_j\mathbf{z}_j + p\lambda\mathbf{A}\mathbf{z}_j = \mathbf{D}_j\mathbf{x}_j$. Thus we have

$$\mathbf{z}_j = (\mathbf{D}_j + p\lambda\mathbf{A})^{-1}\mathbf{D}_j\mathbf{x}_j = [\mathbf{D}_j(\mathbf{A}^{-1} + p\lambda\mathbf{D}_j^{-1})\mathbf{A}]^{-1}\mathbf{D}_j\mathbf{x}_j, \tag{22}$$

which gives Eq.(18).

6.4 Proof of Lemma 5

Proof. Let $\Delta = $ LHS - RHS of Eq.(21). We have $\Delta = \alpha + \beta$ where

$$\alpha = \sum_{ij}\left[|(\mathbf{Z}_{t+1} - \mathbf{X})_{ij}| - |(\mathbf{Z}_t - \mathbf{X})_{ij}| - \frac{(\mathbf{Z}_{t+1} - \mathbf{X})_{ij}^2}{2|(\mathbf{Z}_t - \mathbf{X})_{ij}|} \frac{(\mathbf{Z}_t - \mathbf{X})_{ij}^2}{2|(\mathbf{Z}_t - \mathbf{X})_{ij}|}\right]$$

$$= \sum_{ij} \frac{-1}{2|(\mathbf{Z}_t - \mathbf{X})_{ij}|}\Big[|(\mathbf{Z}_{t+1} - \mathbf{X})_{ij}| - |(\mathbf{Z}_t - \mathbf{X})_{ij}|\Big]^2 \le 0.$$

and

$$\beta = \lambda\Big[\mathrm{Tr}(\mathbf{Z}_{t+1}\mathbf{Z}_{t+1}^T)^{\frac{p}{2}} - \mathrm{Tr}(\mathbf{Z}_t\mathbf{Z}_t^T)^{\frac{p}{2}}\Big] - \frac{p}{2}\lambda\Big[\mathrm{Tr}\mathbf{Z}_{t+1}^T(\mathbf{Z}_t\mathbf{Z}_t^T)^{\frac{p}{2}}\mathbf{Z}_{t+1} - \mathrm{Tr}\mathbf{Z}_t^T(\mathbf{Z}_t\mathbf{Z}_t^T)^{\frac{p}{2}}\mathbf{Z}_t\Big]$$

$$= \lambda\Big[\mathrm{Tr}(\mathbf{Z}_{t+1}\mathbf{Z}_{t+1}^T)^{\frac{p}{2}} - \mathrm{Tr}(\mathbf{Z}_t\mathbf{Z}_t^T)^{\frac{p}{2}}\Big] - \frac{p}{2}\lambda\mathrm{Tr}\Big[(\mathbf{Z}_{t+1}\mathbf{Z}_{t+1}^T - \mathbf{Z}_t\mathbf{Z}_t^T)(\mathbf{Z}_t\mathbf{Z}_t^T)^{\frac{p}{2}}\Big]$$

$$\le 0, \tag{23}$$

where in the last inequality, we set $\mathbf{A} = \mathbf{Z}_{t+1}\mathbf{Z}_{t+1}^T$, $\mathbf{B} = \mathbf{Z}_t\mathbf{Z}_t^T$ and used Lemma 6 below. Clearly $\Delta = \alpha + \beta \le 0$.

Table 1. Description of Data sets

Dataset	#data	#dimension	#class
AT&T_oc	400	2576	40
Binalpha_oc	1404	320	36
Umist_oc	360	644	20
YaleB	256	2016	4
CMUPIE_oc	680	1024	68
Mnist_oc	150	784	10

Lemma 6. *[24] For any two symmetric positive definite matrices* $\mathbf{F}, \mathbf{G}$ *and* $0 < p \le 2$,

$$Tr\left[\mathbf{F}^{p/2} - \mathbf{G}^{p/2}\right] \le \frac{p}{2} Tr\left[(\mathbf{F} - \mathbf{G})\mathbf{G}^{p/2-1}\right] \tag{24}$$

Due to space limit, we omit the proofs of Lemma 6 here.

7 Connection to Related Works

We note [22] proposes an algorithm to solve squared schatten p model, i.e., $\min_{\mathbf{Z}} f(\mathbf{Z}) + \beta\left(\text{Tr}(\mathbf{Z}^T\mathbf{Z})^{\frac{p}{2}}\right)^{\frac{2}{p}}$, which cannot be directly applied here. [23] proposes an iterative reweighted algorithm for trace norm minimization problem, in the similar vein as what has been proposed for adaptive lasso. However, it cannot be directly applied to solve Eq.(3). As compared to [24,25,26,27], our goal is for noisy data recovery problem raised in computer vision, instead of for matrix completion problems with missing values.

8 Experiments

We use six widely used image data sets, including four face datasets: AT&T Umist, YaleB [28] and CMUPIE; and two digit datasets: Mnist [29] and Binalpha[1]. We generate **occluded** image datasets corresponding to 5 original data sets (except YaleB). For YaleB dataset, the images are taken under different poses with different illumination conditions. The shading parts of the images play the similar role of occlusion (noises). Thus we use the original YaleB data with first 4 persons in our experiments. For the other 5 datasets, half of the images are selected from each category for occlusion with block size of $w{\times}w$ pixels (e.g., $w = 10$). The locations of occlusions are random generated without overlaps among the images from the same category. *Occluded* images (with occlusion size 7×7) generated from Umist data sets are shown in Fig. 4. Table 1 summarizes the characteristics of these occluded data sets.

We did all experiments using Eq.(3). At $p < 1$, objective function in Eq.(3) is not convex any more, and we cannot get global minima. We initialize $\mathbf{Z}$ using trace norm minimization solution, i.e., set $p = 1$ in Eq.(3). In the following experiments, we did both algorithms proposed in §5-6, and reported the results using the one achieving smaller objectives.

---

[1] http://www.kyb.tuebingen.mpg.de/ssl-book/benchmarks.html

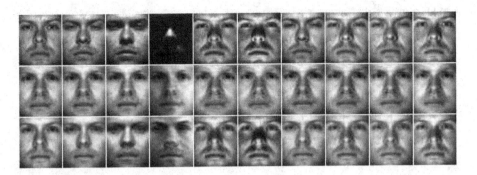

Fig. 3. Reconstructed images ($\mathbf{Z}$) of YaleB dataset using Model 2 of Eq.(3) shown in 1 panel. First line: original images of one person, Second line: reconstructed images $\mathbf{Z}$ at $p = 1$, Third line: reconstructed images at $p = 0.2$. One can see $p = 1$ images are very similar to each other (most fine details are lost), while $p = 0.2$ images retain some fine details and are closer to original images.

Fig. 4. Occluded image dataset Umist

Illustrative Examples. To visualize the denoising effect of proposed method, we apply our model on YaleB dataset. YaleB contains images with different shading which plays similar role of occlusion (noises). Thus we did not add occlusion and use the original data. In this demonstration and following experiment, each data (image) is linearized into a vector each $\mathbf{x}_i$, and the input matrix $\mathbf{X}$ is constructed as $\mathbf{X} = (\mathbf{x}_1, \mathbf{x}_2, \cdots, \mathbf{x}_n)$. We typically set the rank k equal to the number of classes in the dataset. Due to space limit, computed $\mathbf{Z}$ at different p values for the two persons are shown. In Fig.(3), 20 images are shown as 2 panels, each panel for one person. On each panel, the first line images are original images $\mathbf{X}$, the 2nd line are computed $\mathbf{Z}$ at $p = 1$, and 3rd line are computed $\mathbf{Z}$ at 0.2.

Clearly, at different p values (such as $p = \{1, 0.2\}$), Schatten p-Norm model can effectively recover the original data by removing the shadings. See 2nd line on each panel in Fig.(3), almost every person is recovered to same template, not any difference any more. In contrast, we have much better visualization results (with more details) when $p = 0.2$ (see 3rd line on each panel). Moreover, these fine details are expected to be helpful for classification on images from different persons.

True Data Recovery: True Signal Reconstruction Error. Given noisy data $\mathbf{X}$, $\mathbf{X} = \mathbf{X}_0 + \mathbf{X}_E$, where $\mathbf{X}_0$ is the true signal and $\mathbf{X}_E$ is the noise. Our goal is to recover $\mathbf{X}_0$ using Eq.(3). We did experiments on above 6 datasets. To evaluate the performance,

Table 2. True data recovery: True signal reconstruction error at different p on six datasets

dataset	$\frac{\|\mathbf{X}_E\|_F}{\|\mathbf{X}_0\|_F}$	Noise-free reconstruction error at different p				
		$p = 1$	$p = 0.75$	$p = 0.5$	$p = 0.2$	$p = 0$
AT&T	0.3657	0.2672	0.2240	0.2199	0.2159	**0.2132**
Binalpha	0.2359	0.2023	0.1974	0.1845	**0.1594**	0.1729
Umist	0.3123	0.2816	0.2290	0.2199	0.2153	**0.2151**
YaleB	N/A	0.2304	0.2264	0.2174	**0.1912**	0.2126
CMUPIE	0.2542	0.2012	0.1925	0.1845	**0.1594**	0.1729
Mnist	0.5574	0.5123	0.4993	0.4814	**0.4542**	0.4553

we define the true signal data recovery error, $E_{\text{true-signal}} = \frac{\|\mathbf{Z}-\mathbf{X}_0\|_F}{\|\mathbf{X}_0\|_F}$. Clearly, smaller $E_{\text{true-signal}}$ values indicate better recovery. Computed true signal reconstruction error are shown in Table.2. The experiment results indicate that true signal reconstruction errors are *smaller* at smaller p values. We also list $\frac{\|\mathbf{X}_E\|_F}{\|\mathbf{X}_0\|_F}$ values in Table.2 to indicate the level of occlusions. Interestingly, $E_{\text{true-signal}} < \frac{\|\mathbf{X}_E\|_F}{\|\mathbf{X}_0\|_F}$ on all datasets at different p values. This further confirms "de-noisy" effects of proposed data recovery model.

Table 3. Loss of fine-details: variance of reconstructed $\mathbf{Z}$ on six datasets, original images: $\mathbf{X}_0$, occluded images: $\mathbf{X}$

dataset	$\mathbf{X}_0$	$\mathbf{X}$	Variance of $\mathbf{Z}$ at different p				
			$p = 1$	$p = 0.75$	$p = 0.5$	$p = 0.2$	$p = 0$
AT&T	8.89	9.03	5.83	7.13	7.45	8.11	7.80
Binalpha	27.90	31.13	13.40	22.89	25.38	26.89	26.73
Umist	7.01	7.42	3.87	5.31	5.71	6.38	6.01
YaleB	9.75	9.75	7.28	8.22	8.59	9.19	8.76
CMUPIE	12.09	13.16	8.12	10.07	10.54	11.30	10.87
Mnist	9.24	10.26	0.49	4.41	5.45	7.04	5.85

Loss of Fine Details in Recovered Data and Its Measure. Due to suppression of higher order/frequency terms associated with smaller singular values, fine details of original data $\mathbf{X}$ are lost in the recovered $\mathbf{Z}$. As a consequence, recovered individual images are very similar to each other. One numeric measure is the variance of reconstructed images. We therefore define $var(\mathbf{Z}) = \sum_{i=1}^{n} \|\mathbf{z}_i - \bar{\mathbf{z}}\|^2$, $\bar{\mathbf{z}} = \frac{1}{n} \sum_{i=1}^{n} \mathbf{z}_i$, where $\mathbf{z}_i \in \Re^{d \times 1}$ is the reconstructed image corresponding to each original image $\mathbf{x}_i$. Larger variance values indicate more fine details are preserved in the solution $\mathbf{Z}$. Computed variance of $\mathbf{Z}$ are shown in Table.3. Clearly, reconstructed images preserve more detailed information at small p (say $p = 0.2$). One demonstrating example is shown in Fig.3, where fine details of individual images are mostly suppressed at $p = 1$, but are generally preserved/reetained at $p = 0.2$.

Table 4. Classification accuracy(shown as percentage) on six occluded datasets using input corrupted data **X** and reconstructed **Z** at different p values

dataset	method	X	Reconstructed **Z** at different p				
			$p = 1$	$p = 0.75$	$p = 0.5$	$p = 0.2$	$p = 0$
AT&T	SVM	29.75	30.52	33.75	34.25	**36.78**	35.53
	KNN	25.75	28.63	**30.25**	28.75	29.31	28.33
Binalpha	SVM	38.35	44.78	42.74	43.84	**48.53**	47.43
	KNN	52.09	56.78	55.10	54.65	**58.23**	57.87
Umist	SVM	59.83	65.89	63.17	64.35	**68.33**	67.67
	KNN	89.12	93.89	92.67	93.75	**94.23**	93.01
YaleB	SVM	46.11	52.12	51.89	53.78	54.67	**54.96**
	KNN	85.43	90.89	90.36	91.15	**91.76**	91.40
CMUPIE	SVM	29.24	33.57	**36.74**	34.21	35.39	34.98
	KNN	58.12	64.03	65.38	64.27	**66.39**	65.64
Mnist	SVM	49.38	51.93	53.24	**57.18**	56.79	54.67
	KNN	76.63	81.35	80.75	81.56	**82.47**	82.34

Classification Results Using Recovered Z. So far we have discussed low rank recovery capability of computed **Z**. Reconstructed low rank **Z** is expected to have much clear structure after removing noises and outliers. As a by-product of solving low-rank data recovery problem, computed **Z** can be used for classification tasks. We compare the classification results by using the occluded images **X** and recovered data **Z** at different p. The experiments are done on two widely used classifiers: k nearest neighbor (kNN) and support vector machine[2] using 5 fold cross validation. Since the regularization coefficient is also a hyper-parameter, the performance of each Schatten-p norm model is evaluated at an optimal value of β (which is determined by cross validation). The experiment results are shown in Table.4. We have two important observations from experiment results. **(1)** Performances for image categorization tasks are improved by using computed **Z** at different p values; **(2)** Classification accuracy is consistently better at smaller p values on both SVM and kNN classifiers, as compared to that at large p values. All above results suggest us to use Schatten p-Norm at small p values.

9 Conclusion

We present novel models for low-rank data recovery, where efficient algorithms are proposed. Extensive experiment results indicate schatten p model gives relatively better reconstructed results at small p values. In the next step, we will further explore how to scale our model for large-size problems.

Acknowledgement. This work is partially supported by NSF-CCF-0939187, NSF-DMS-0915228, NSF-CCF-0917274.

---

[2] http://www.csie.ntu.edu.tw/~cjlin/libsvm/

References

1. Cai, J., Candès, E., Shen, Z.: A singular value thresholding algorithm for matrix completion. SIAM Journal on Optimization 20, 1956–1982 (2010)
2. Candès, E., Recht, B.: Exact matrix completion via convex optimization. Communications of the ACM 55, 111–119 (2012)
3. Recht, B., Fazel, M., Parrilo, P.: Guaranteed minimum-rank solutions of linear matrix equations via nuclear norm minimization. SIAM Review 52, 471–501 (2010)
4. Rennie, J.M., Srebro, N.: Fast maximum margin matrix factorization for collaborative prediction. In: ICML, pp. 713–719 (2005)
5. Abernethy, J., Bach, F., Evgeniou, T., Paristech, M., Jaakkola, T.: A new approach to collaborative filtering: Operator estimation with spectral regularization. Journal of Machine Learning Research 10, 803–826 (2009)
6. Liu, J., Musialski, P., Wonka, P., Ye, J.: Tensor completion for estimating missing values in visual data. In: ICCV, pp. 2114–2121 (2009)
7. Jolliffe, I.: Principal Component Analysis. Springer (1986)
8. Donoho, D.: Compressed sensing. IEEE Transactions on Information Theory 52, 1289–1306 (2006)
9. Ding, C.H.Q., Zhou, D., He, X., Zha, H.: R_1-pca: rotational invariant l_1-norm principal component analysis for robust subspace factorization. In: ICML, pp. 281–288 (2006)
10. Wright, J., Ganesh, A., Rao, S., Peng, Y., Ma, Y.: Robust principal component analysis: Exact recovery of corrupted low-rank matrices via convex optimization. In: NIPS (2009)
11. Peng, Y., Ganesh, A., Wright, J., Xu, W., Ma, Y.: Rasl: Robust alignment by sparse and low-rank decomposition for linearly correlated images. In: CVPR, pp. 763–770 (2010)
12. Fazel, M.: Matrix rank minimization with applications. PhD thesis, Stanford University, pp. 1–130 (2002)
13. Candès, E., Tao, T.: The power of convex relaxation: near-optimal matrix completion. IEEE Transactions on Information Theory 56, 2053–2080 (2010)
14. Chandrasekaran, V., Sanghavi, S., Parrilo, P., Willsky, A.: Sparse and low-rank matrix decompositions. In: Allerton 2009 Proceedings of the 47th Annual Allerton Conference on Communication, Control, and Computing, pp. 962–967 (2009)
15. Ma, S., Goldfarb, D., Chen, L.: Fixed point and bregman iterative methods for matrix rank minimization. Mathematical Programming: Series A and B 128, 321–353 (2011)
16. Toh, K., Yun, S.: An accelerated proximal gradient algorithm for nuclear norm regularized linear least squares problems. Pacific Journal of Optimization (2010)
17. Pong, T., Tseng, P., Ji, S., Ye, J.: Trace norm regularization: Reformulations, algorithms, and multi-task learning. SIAM Journal on Optimization 20, 3465–3489 (2010)
18. Golub, G.H., Reinsch, C.: Singular value decomposition and least squares solutions. Numerische Mathematik 14, 403–420 (1970)
19. Lin, Z., Chen, M., Ma, Y.: The augmented lagrange multiplier method for exact recovery of corrupted low-rank matrices. In: UIUC Tech. Rep., UILU-ENG-09-2214 (2010)
20. Bertsekas, D.: Constrained Optimization and Lagrange Multiplier Methods. Athena Scientific (1996)
21. Tibshirani, R.: Regression shrinkage and selection via the lasso. Journal of the Royal Statistical Society, Series B 58, 267–288 (1994)
22. Argyriou, A., Micchelli, C., Pontil, M., Ying, Y.: A spectral regularization framework for multi-task structure learning. In: NIPS (2007)
23. Jain, P., Meka, R., Dhillon, I.: Guaranteed rank minimization via singular value projection. In: NIPS (2010)

24. Nie, F., Huang, H., Ding, C.: Low-rank matrix recovery via efficient schatten p-norm minimization. In: Twenty-Sixth AAAI Conference on Artificial Intelligence (2012)
25. Mohan, K., Fazel, M.: Iterative reweighted least squares for matrix rank minimization. In: 2010 48th Annual Allerton Conference on Communication, Control, and Computing (Allerton), pp. 653–661. IEEE (2010)
26. Keshavan, R., Montanari, A., Oh, S.: Matrix completion from noisy entries. The Journal of Machine Learning Research 11, 2057–2078 (2010)
27. Wipf, D.: Non-convex rank minimization via an empirical bayesian approach. In: UAI, pp. 914–923 (2012)
28. Georghiades, A., Belhumeur, P., Kriegman, D.: From few to many: Illumination cone models for face recognition under variable lighting and pose. PAMI, 643–660 (2001)
29. Lecun, Y., Bottou, L., Bengio, Y., Haffner, P.: Gradient-based learning applied to document recognition. Proceedings of the IEEE, 2278–2324 (1998)

Noisy Matrix Completion Using Alternating Minimization

Suriya Gunasekar, Ayan Acharya, Neeraj Gaur, and Joydeep Ghosh

Department of ECE, University of Texas at Austin, USA
{suriya,aacharya,neeraj.gaur}@utexas.edu, ghosh@ece.utexas.edu

Abstract. The task of matrix completion involves estimating the entries of a matrix, $M \in \mathbb{R}^{m \times n}$, when a subset, $\Omega \subset \{(i,j) : 1 \leq i \leq m, 1 \leq j \leq n\}$ of the entries are observed. A particular set of low rank models for this task approximate the matrix as a product of two low rank matrices, $\widehat{M} = UV^T$, where $U \in \mathbb{R}^{m \times k}$ and $V \in \mathbb{R}^{n \times k}$ and $k \ll \min\{m, n\}$. A popular algorithm of choice in practice for recovering M from the partially observed matrix using the low rank assumption is alternating least square (ALS) minimization, which involves optimizing over U and V in an alternating manner to minimize the squared error over observed entries while keeping the other factor fixed. Despite being widely experimented in practice, only recently were theoretical guarantees established bounding the error of the matrix estimated from ALS to that of the original matrix M. In this work we extend the results for a noiseless setting and provide *the first guarantees for recovery under noise for alternating minimization*. We specifically show that for well conditioned matrices corrupted by random noise of bounded Frobenius norm, if the number of observed entries is $\mathcal{O}\left(k^7 n \log n\right)$, then the ALS algorithm recovers the original matrix within an error bound that depends on the norm of the noise matrix. The sample complexity is the same as derived in [7] for the noise–free matrix completion using ALS.

1 Introduction

The problem of matrix completion has found application in a number of research areas such as in recommender systems [10], multi-task learning [15], remote sensing[12] and image inpainting [1]. In a typical setting for matrix completion, a matrix $M \in \mathbb{R}^{m \times n}$ is observed on a subset of entries $\Omega \subset \{(i,j) : 1 \leq i \leq m, 1 \leq j \leq n\}$, while a large number of entries are missing. The task is then to fill in the missing entries of the matrix yielding an estimate $\widehat{M}$ of the complete matrix that is consistent with the original matrix M.

Among the many models that try and tackle the matrix completion problem, low rank models have enjoyed a great deal of success in practice and have proven to be very popular and effective for the matrix completion task on real life datasets [3,8,10,13,11]. Low rank models with numerous variations have been heavily used in practice for matrix completion specially towards the application of collaborative filtering [10,13]. Though it is one of the most widely used techniques to model incomplete matrix data, there are only a few algorithms for which theoretical guarantees have been established, most notably the nuclear norm minimization [3,4] and OptSpace [8]. However, these algorithms are computationally expensive and hence not scalable.

H. Blockeel et al. (Eds.): ECML PKDD 2013, Part II, LNAI 8189, pp. 194–209, 2013.

A popular algorithm that is heavily used in practice for recovering M from the entries observed on Ω under the low rank assumption is the alternating least squares minimization (ALS) [16,10]. The algorithm makes the assumption that the matrix M is of a fixed low rank that has a latent factor representation $M = UV^T$, where $U \in \mathbb{R}^{m \times k}$, $V \in \mathbb{R}^{n \times k}$ and $k \ll n, m$. Hence, one is interested in solving the following:

$$\min_{U,V} \|P_\Omega(M) - P_\Omega(UV^T)\|_F^2$$

Where Ω is the set of observed entries and $P_\Omega(M)$, also denoted by M^Ω, is the projection of the matrix M onto the observed set Ω, given by, $M_{ij}^\Omega = \begin{cases} M_{ij} & \text{if } (i,j) \in \Omega \\ 0 & \text{otherwise} \end{cases}$

The above problem as described is jointly non–convex in U and V. Alternating minimization proceeds by alternatively fixing one of the latent factors and optimizing the other. Once one of the factors (say U) is fixed, solving for the other (V) is a convex problem. In fact, it is a simple least squares problem. This simplicity of the alternating minimization has made it a popular approach for low rank matrix factorization in practice. Recent results [7,6,14] give recovery guarantees for ALS in a noiseless setting. However theoretical guarantees for ALS when the observed entries are corrupted by noise are still lacking. On the other hand, in real life applications, the matrix entries are often corrupted by various means including the noise in the matrix generation process, outliers and inaccurate measurements. In this work we present *the first guarantees for recovery under noise for alternating least squares minimization*. We rely heavily on the analysis of [7,6] and also borrow results from [9].

The paper is organized as follows. After explaining the notations and defining a few quantities in Section 1.1, we briefly review relevant work in Section 2. In Section 3, we describe the algorithm and state the main result of the paper and compare the results with the existing results. Our primary contribution in this paper is the proof of the result stated in Section 3. We build the proof in Section 4. As the proof is fairly involved, the proof of various lemmata in this section are deferred to the Appendix. We conclude with an analysis of the results and possible future directions in Section 5.

1.1 Notations and Preliminaries

Unless stated otherwise, we use the following notation in the rest of the paper. Matrices are represented by uppercase letters. For a matrix M, M_i represents the ith column, $M^{(i)}$ represents the vector corresponding to the ith row, (all the vectors are column vectors *i.e* they are or dimension $d \times 1$, where d is the length of the vector) and M_{ij} is the $(i, j)^{\text{th}}$ entry. The spectral norm and Frobenius norm of a matrix M are denoted by $\|M\|_2$ and $\|M\|_F$, respectively. The max norm of M, denoted by M_{max}, is the maximum of the absolute values of the entries of M. The transpose of a matrix M is denoted by $M^\dagger$. Vectors are denoted by lowercase letters. For a vector u, u_i is the ith component of u. The p-norm of a vector is given by $\|u\|_p = (\sum_i |u_i|^p)^{1/p}$, $p \geq 1$. Finally, set of integers from 1 to m is denoted by $[m] = \{1, 2, \ldots, m\}$.

Definition 1 (SVD (or truncated SVD)). *The singular value decomposition (SVD) of a matrix* $M \in \mathbb{R}^{m \times n}$ *of rank* k *is given by* $M = U \Sigma V^{\dagger}$, *where* $U \in \mathbb{R}^{m \times k}$ *and* $V \in \mathbb{R}^{n \times k}$ *have orthonormal columns, i.e.* $U^{\dagger}U = V^{\dagger}V = I$ *and* $\Sigma \in \mathbb{R}^{k \times k}$ *is a diagonal matrix whose entries are* $(\sigma_1, \sigma_2, \ldots, \sigma_k)$. *Here, the columns of* U *and* V *are called the **left and right singular vectors** of* M *respectively and* $\sigma_1 \geq \sigma_2, \ldots, \sigma_k > 0$ *are the **singular values**.*

Definition 2 (Condition number). *Consider a matrix* M *of rank* k, *with singular values,* $\sigma_1 \geq \sigma_2, \ldots, \sigma_k > 0$. *The condition number of the matrix* M, *denoted by* κ_M *is defined as* $\kappa_M = \frac{\sigma_1}{\sigma_k}$

Definition 3 (Reduced–QR factorization (or simply QR factorization)). *The Reduced–QR factorization, which is often overloaded as QR factorization, of a matrix* $X \in \mathbb{R}^{m \times k}$, $m \geq k$, *is given by* $X = QR$, *where* $Q \in \mathbb{R}^{m \times k}$ *has orthonormal columns and* $R \in \mathbb{R}^{k \times k}$ *is an upper triangular matrix. The columns of the matrix* Q *is an orthonormal basis for the subspace spanned by the columns of* X.

Definition 4 (Distance between two matrices [5]). *Given two matrices* $\widehat{U}, \widehat{W} \in \mathbb{R}^{m \times k}$, *the distance between the subspaces spanned by the columns of* $\widehat{U}$ *and* $\widehat{W}$ *is given by* $dist(\widehat{U}, \widehat{W}) = \|U_{\perp}^{\dagger} W\|_2 = \|U W_{\perp}^{\dagger}\|_2$ *where* U *and* W *are orthonormal bases of the spaces* $span(\widehat{U})$ *and* $span(\widehat{W})$, *respectively. Similarly,* $U_{\perp}$ *and* $W_{\perp}$ *are orthonormal bases of the spaces* $span(\widehat{U}_{\perp})$ *and* $span(\widehat{W}_{\perp})$.

Definition 5 (Incoherence of a matrix). *A matrix* $M \in \mathbb{R}^{m \times n}$ *is incoherent with parameter* μ *if* $\|U^{(i)}\|_2 \leq \mu \frac{\sqrt{k}}{\sqrt{m}} \ \forall i \in [m]$ *and* $\|V^{(j)}\|_2 \leq \mu \frac{\sqrt{k}}{\sqrt{n}} \ \forall j \in [n]$ *where* $M = U \Sigma V^{\dagger}$ *is the SVD of* M. *We remind that* $X^{(i)}$ *is the* i^{th} *row of matrix* X.

Definition 6 (Vector to matrix conversion). *The operator* vec2mat() *converts a vector to matrix in column–order, i.e.* $\forall \ x \in \mathbb{R}^{nk}, vec2mat(x) =$
$$\begin{bmatrix} \uparrow & \uparrow & \cdots & \uparrow \\ x_{1:n} & x_{n+1:2n} & \cdots & x_{(k-1)n+1:kn} \\ \downarrow & \downarrow & \cdots & \downarrow \end{bmatrix}$$

2 Related Work

Candès and Recht [3] first demonstrated that under the assumptions of random sampling and incoherence conditions $O(kn^{1.2} \log n)$ samples allow for exact recovery of the true underlying matrix via convex nuclear–norm based minimization. The sample complexity result was further improved to $O(kn \log n)$ by Candès and Tao [4]. Later on, Candès and Plan [2] analyzed the recovery guarantees for nuclear–norm based optimization algorithm under bounded noise added to the true underlying matrix. However, one should note that nuclear–norm based minimization approach is computationally expensive and infeasible in practice for large scale matrices.

In the OptSpace algorithm [8], Keshavan *et al.* adopted a different approach for the matrix completion problem where they first took the SVD of the matrix M^{Ω}. Their analysis showed that such a SVD provides a reasonably good initial estimate for the spanning subspace, which can further be refined by gradient descent on a Grassmanian manifold. They show asymptotic recovery guarantees of original matrix if the number of samples is $O(nk \, (\sigma_1^*/\sigma_k^*)^2 \log n)$. In a later paper, Keshavan $et \, al.$ [9] also examined the reconstruction guarantee of OptSpace under two noise models. The analysis (for both noiseless and noisy recovery) of the algorithm only guarantees asymptotic convergence and the convergence might take exponential time in the problem size in the worst case.

In practice, however, alternating minimization based approach produces good optimal solution. Though the underlying optimization problem is non–convex, each step is convex, computationally cheaper and solutions close to global optimal are often reported in experiments [11]. The algorithm and its variations have been practically deployed in many real life collaborative filtering datasets and have shown good performance [10,13]. Wang and Xu [14] first showed that given a factorization algorithm attains a global optimum, the space of the factors, U and V, and the estimated matrix $\widehat{M}$ are robust against corruption of the observed entries by bounded noise. Jain $et \, al.$ [7,6], however, were the first to formulate the conditions for recovery of the underlying matrix using alternating minimization. They showed that the true underlying matrix M can be recovered within an error of ϵ in $O(\log(\|M\|_F/\epsilon))$ steps and this requires $O((\sigma_1^*/\sigma_k^*)^4 k^7 n \log n \log \|M\|_F/\epsilon)$ number of samples. We build on the results of Jain $et \, al.$ [7] and provide recovery guarantees of noisy matrix completion problem with alternating minimization.

3   Main Result

In the rest of the paper, the underlying true rank–k matrix to be completed is denoted by $M \in \mathbb{R}^{m \times n}$. With a slight abuse of notation, the truncated SVD of M is given by $M = U^* \Sigma^* V^{*\dagger}$ with $U^* \in \mathbb{R}^{m \times k}$, $V^* \in \mathbb{R}^{n \times k}$ and $\Sigma^* = \mathrm{diag}(\sigma_1^*, \sigma_2^*, \ldots, \sigma_k^*)$. Without loss of generality, it is assumed that $m \leq n$ and $\alpha = n/m \geq 1$ is a constant (independent of n). The noisy matrix which is partially observed is given by $\widetilde{M} = M + N$, where $N \in \mathbb{R}^{m \times n}$ is the noise matrix. Further, let $N = U_N \Sigma_N V_N^\dagger$ be the SVD of the noise matrix with $U_N \in \mathbb{R}^{m \times m}$, $V_N \in \mathbb{R}^{n \times m}$ and $\Sigma_N = \mathrm{diag}(\sigma_1^N, \sigma_2^N, \ldots, \sigma_m^N)$. Each entry of the matrix $\widetilde{M}$ is independently observed with probability p. Let Ω be the set of indices where the matrix $\widetilde{M}$ is observed. The task is to estimate M given $\widetilde{M}^\Omega$ and Ω.

3.1   Noise Model

We consider a fairly general, $worst \, case \, model$ for the noise matrix N, also used in [9]. In this model N is distributed arbitrarily but bounded as $|N_{ij}| \leq N_{\mathrm{max}}$. This is a generic setting, and any noise distribution with sub Gaussian tails can be approximated by this model with high probability. However, tighter bounds can be obtained for individual cases. Our bounds primarily depend on N_{max} and the fractional operator norm of N^Ω, $\|N^\Omega\|_2/p$. We use the following result from [9]:

Theorem 1 ([9]). $If \, N \, is \, a \, matrix \, from \, the \, worst \, case \, model, \, then \, for \, any \, realization$ $of \, N, \, \|N^\Omega\|_2 \leq \frac{2|\Omega|}{m\sqrt{\alpha}} N_{max}.$

Using, $|\Omega| \approx pmn$ in Theorem 1, we have the following bound:

$$\frac{\|N^\Omega\|_2}{p} \leq 2\sqrt{mn}N_{max},$$ (1)

3.2   Algorithm

The algorithm analyzed in this paper is presented below [7]:

Algorithm 1. ALSM

1: **Input:** observed set Ω, values $P_\Omega(\widetilde{M})$
2: Create $(2T + 1)$ subsets from Ω — $\Omega_1, \Omega_2 \cdots , \Omega_{2T}$, each of size $|\Omega|$, with the elements of Ω belonging to one of the Ω_t's with equal probability and sampled independently
3: Set $\widehat{U}^0 = \text{SVD}(P_{\Omega_0}(\widetilde{M})/p, k)$ i.e., top-k left singular vectors of $P_{\Omega_0}(\widetilde{M})/p$
4: Clipping step: Set all elements of $\widehat{U}^0$ that have magnitude greater than $\frac{2\mu\sqrt{k}}{\sqrt{n}}$ to zero and orthonormalize the columns of $\widehat{U}^0$ (using QR decomposition)
5: **for** $t = 0, \cdots , (T - 1)$ **do**

$$\widehat{V}^{(t+1)} \leftarrow \underset{V \in \mathbb{R}^{n \times k}}{\text{argmin}} \|P_{\Omega^{(t+1)}}(\widehat{U}^t V^\dagger - \widetilde{M})\|_F$$ (2)

$$\widehat{U}^{(t+1)} \leftarrow \underset{U \in \mathbb{R}^{m \times k}}{\text{argmin}} \|P_{\Omega^{(T+t+1)}}\left(U(\widehat{V}^{(t+1)})^\dagger - \widetilde{M}\right)\|_F$$ (3)

end
6: **Output:** $X = \widehat{U}^T(\widehat{V}^T)^\dagger$

For ease of analysis, we have modified the standard ALS algorithm. In Step 2 of the algorithm, independently sampled subsets of Ω are generated that are further used in the rest of the algorithm. This modification was introduced purely for the ease of theoretical analysis and is not required in practice. In the above algorithm, in each iteration, t, the observed set $\Omega^{(t)}$ is independent of the other iterations and hence, each iteration could be analyzed independently. In the proof of our main result, while analyzing iteration, t, we overload Ω to represent $\Omega^{(t)}$ to avoid cluttering of symbols. Thus, the final sample complexity for recovery would be $2T$ times the sample complexity requirements in each iteration, where T is the total number of iterations required for convergence.

3.3   Result

Theorem 2. *Let $M = U^*\Sigma^*(V^*)^\dagger \in \mathbb{R}^{m \times n}$ be a rank–k, incoherent matrix with both U^* and V^* being μ incoherent. Further, it is assumed that, $N_{max} \leq C_3 \frac{\sigma_k^*}{n\sqrt{k}}$ and $\frac{\|N^\Omega\|_2}{p} \leq C_2 \frac{\sigma_k^*}{\kappa_M k}$. Additionally, let each entry of $\widetilde{M} = M + N$ be observed uniformly and independently with probability*

$$p > C\frac{\kappa_M^4 \mu^4 k^7 \log n \log \frac{\|M\|_F}{\epsilon}}{m\delta_{2k}^2}$$ (4)

where, $\kappa_M = \frac{\sigma_1^*}{\sigma_k^*}$ is the condition number of the M, $\delta_{2k} \leq \frac{\sigma_k^*}{64\sigma_1^*}$ and $C > 0$ is a global constant. Then with high probability, for $T \geq C' \log \frac{\|M\|_F}{\epsilon}$, the outputs $\widehat{U}^T$ and $\widehat{V}^T$ of Algorithm 1 with input $(\Omega, P_\Omega(\widetilde{M}))$ satisfy

$$\frac{1}{\sqrt{mn}}\|M - \widehat{U}^T(\widehat{V}^T)^\dagger\|_F \leq \epsilon + 20\mu\kappa_M^2 k^{1.5}\left(\frac{\|N^\Omega\|_2}{|\Omega|}\right) \leq \epsilon + 40\mu\kappa_M^2 k^{1.5}N_{max} \quad (5)$$

Worst Case Noise Model Requirements. The theorem requires that $N_{max} \leq C_3 \frac{\sigma_k^*}{n\sqrt{k}}$ and $\frac{\|N^\Omega\|_2}{p} \leq C_2 \frac{\sigma_k^*}{\kappa_M k}$. For the *worst case noise model*, if $N_{max} \leq C_2 \frac{\sigma_k^*}{2\kappa_M nk} \implies \frac{\|N^\Omega\|_2}{p} \leq 2\sqrt{mn}N_{max} \leq C_2\frac{\sigma_k^*}{\kappa_M k}$ Further, $N_{max} \leq C_3\frac{\sigma_k^*}{\kappa_M nk} \implies N_{max} \leq C_3\frac{\sigma_k^*}{nk} \leq C_3\frac{\sigma_k^*}{n\sqrt{k}}$. Thus, choosing $C = \min\{C_2/2, C_3\}$, and $N_{max} \leq C\frac{\sigma_k^*}{\kappa_M nk}$, both the conditions on noise matrix for Theorem 2 are satisfied.

For a well conditioned matrix M of condition number close to 1, the above requirement is approximately equivalent to $N_{max} \leq C'k^{-1.5}\frac{\|M\|_F}{\sqrt{mn}}$, which is $k^{-1.5}$ fraction of root mean square value of the entries of matrix M. This is a fairly reasonable assumption on the noise matrix for recovery guarantees.

3.4 Comparison with Similar Results

The most relevant work for our analysis is the analysis of low rank matrix completion under alternating minimization approach proposed by Jain et. al. [7]. They have the following result for ALS under noiseless setting, $N = 0$:

Theorem 3 ([7]). *Let $M = U^*\Sigma^*(V^*)^\dagger \in \mathbb{R}^{m \times n}$ be a rank–k, incoherent matrix with both U^* and V^* being μ incoherent. Let each entry of M be observed uniformly and independently with probability, $p > C\frac{\kappa_M^4 \mu^4 k^7 \log n \log \frac{\sqrt{k}\|M\|_2}{\epsilon}}{m\delta_{2k}^2}$ where, $\delta_{2k} \leq \frac{\sigma_k^*}{64\sigma_1^*}$ and $C > 0$ is a global constant. Then with high probability, for $T \geq C' \log \frac{\|M\|_F}{\epsilon}$, the outputs $\widehat{U}^T$ and $\widehat{V}^T$ of Algorithm 1 with input $(\Omega, P_\Omega(\widetilde{M}))$ satisfy $\|M - \widehat{U}^T(\widehat{V}^T)^\dagger\|_F \leq \epsilon$*

Even for a very general noise model, the sample complexity required for our analysis is the same as that required by the noise–free analysis.

Next, we compare our bounds with the bounds obtained for noisy matrix completion by Keshavan et. al [9]. The algorithm suggested by Keshavan et. al., OptSpace, involves optimizing the initial estimate from SVD of $P_\Omega(\widetilde{M})$ over a Grassmann manifold. The main result in their paper is stated below:

Theorem 4 ([9]). *Let $\widetilde{M} = M + N$, where M is a rank–k, μ incoherent matrix. A subset, $\Omega \subset [m] \times [n]$, of entries of $\widetilde{M}$ are revealed. Let $\widehat{M}$ be the output of OptSpace on the input of $(\widetilde{M}, \Omega)$. Then, there exists numerical constants, C and C' such that, if $|\Omega| \geq Cn\sqrt{\alpha}\kappa_M^2 \max\{\mu k\sqrt{\alpha}\log n; \mu^2 k^2 \alpha\kappa_M^4\}$ and $\frac{\|N^\Omega\|_2}{p} \leq C'\frac{\sigma_k^*}{\kappa_M^2\sqrt{k}}$ then, with probability atleast $1 - 1/n^3$, $\frac{1}{\sqrt{mn}}\|M - \widehat{M}\|_F \leq C'\kappa_M^2 k^{0.5}\left(\frac{\|N^\Omega\|_2}{|\Omega|}\right)$*

The requirements on the noise matrix for recovery guarantees by OptSpace is close to that derived in our results for Alternating minimization. Also, the error in the recovered matrix in our analysis is off by a small factor of k as compared to the analysis in [9]. However, the sample complexity required by ALS as evaluated by our analysis is much higher than that of Keshavan et. al.

4   Proof of Theorem 2

In this section we present the proof of Theorem 2. The outline of the proof is as follows. In Section 4.1, Theorem 5 states that the initialization step of the Algorithm described in 1 provides a good starting point. In Section 4.2, we first propose a modification to the ALSM algorithm and prove that the modification in practice is equivalent to the original ALSM algorithm, while the modified algorithm is easier to analyze. Theorem 6 is then stated without proof. This theorem establishes that the space spanned by ALSM estimates of $\widehat{U}$ and $\widehat{V}$ converge towards U^* and V^* respectively. Finally, we combine the results on initialization and above mentioned theorem to prove the main result. The proof of Theorem 6 is deferred to Section 4.3. In each subsection, the relevant lemmata are first presented and then the main theorems are proved. The proofs of the lemmata are provided in the Áppendix.

4.1   Initialization

Lemma 1 (Theorem 1.1 **of [8]).** *Let* $\widetilde{M} = M + N$ *be such that* M *is rank–k and* μ*–incoherent and* $|\Omega| \geq Cnk \max\{\log n, k\}$. *Further, from the SVD of* $\frac{\widetilde{M}^{\Omega}}{p}$, *we get a rank–k approximation as,* $\widetilde{M}_k^{\Omega} = \widetilde{U}^0 \widetilde{\Sigma}^0 \widetilde{V}^{0\dagger}$, *where* $\widetilde{U}^0 \in \mathbb{R}^{m \times k}$ *and* $\widetilde{V}^0 \in \mathbb{R}^{n \times k}$. *Let* $\alpha = n/m \geq 1$. *Then the following is true with probability greater than* $(1 - 1/n^3)$,

$$\frac{1}{\sqrt{mn}}\|M - \widetilde{M}_k^{\Omega}\|_2 \leq CM_{max}\left(\frac{m\alpha^{3/2}}{|\Omega|}\right)^{1/2} + \frac{2m\sqrt{\alpha}}{|\Omega|}\|N^{\Omega}\|_2. \tag{6}$$

Lemma 2. *Let* $\widetilde{U}^0$ *be defined as in Lemma 1. Further, under the conditions of Theorem 2, the following is true with probability greater than* $(1 - 1/n^3)$,

$$dist(\widetilde{U}^0, U^*) \leq \frac{1}{64k}.$$

The proof of Lemma 2 is presented in Appendix B.1.

Theorem 5 (ALSM has a good initial point). *Let* U^c *be obtained from* $\widetilde{U}^0$ *defined above, by setting all the entries greater than* $\frac{2\mu\sqrt{k}}{\sqrt{m}}$ *to zero. Let* U^0 *be the orthonormal basis of* U^c. *Then under the conditions of Lemma 2, w.h.p. we have*

- $dist(U^0, U^*) \leq 1/2.$
- U^0 *is incoherent with parameter* $\mu_1 = \frac{32\sigma_1^*\mu\sqrt{k}}{\sigma_k^*}.$

The proof follows directly from Lemma C.2 in [6] and Lemma 2.   □

Note that the U^0 defined above is the same as the the initial estimate, $\widehat{U}^0$ from the initialization step of the Algorithm 1.

4.2 Convergence of ALS Minimization

Consider the following modification to Equation 2 and 3 of Algorithm 1:

$$\widehat{V}^{(t+1)} \leftarrow \underset{\widehat{V} \in \mathbb{R}^{n \times k}}{\operatorname{argmin}} \| P_{\Omega^{(t+1)}} (U^t \widehat{V}^\dagger - \widetilde{M}) \|_F$$

$$V^{(t+1)} R_V^{(t+1)} = \widehat{V}^{(t+1)} \quad \text{(QR decomposition)}$$

$$\widehat{U}^{(t+1)} \leftarrow \underset{\widehat{U} \in \mathbb{R}^{m \times k}}{\operatorname{argmin}} \| P_{\Omega^{(T+t+1)}} \left(\widehat{U} V^{(t+1)\dagger} - \widetilde{M} \right) \|_F$$

$$U^{(t+1)} R_U^{(t+1)} = \widehat{U}^{(t+1)} \quad \text{(QR decomposition)} \tag{7}$$

Lemma 3 (Lemma 4.4 of [7]). *Let $\widehat{U}^{(t)}$ be the t-th step iterate of ALSM Algorithm 1, and $\widetilde{U}^{(t)} = U^{(t)} R_U^{(t)}$ be that of the modified algorithm presented above. Suppose that both $\widetilde{U}^{(t)}$ and $\widehat{U}^{(t)}$ are full rank and span the same space, then the same will be true for subsequent iterates. i.e $\operatorname{span}(\widehat{V}^{(t+1)}) = \operatorname{span}(\widetilde{V}^{(t+1)})$ and $\operatorname{span}(\widehat{U}^{(t+1)}) = \operatorname{span}(\widetilde{U}^{(t+1)})$ and all the matrices at iterate $t+1$ are full rank.*

Proof. As both $\widetilde{U}^{(t)}, \widehat{U}^{(t)} \in \mathbb{R}^{m \times k}$ have full rank and span same subspace, there exists a $k \times k$ full rank matrix R such that $\widehat{U}^{(t)} = \widetilde{U}^{(t)} R = U^{(t)} R_U^{(t)} R$. Thus,

$$\min_{V \in \mathbb{R}^{n \times k}} \| P_{\Omega^{(t+1)}} \left(\widehat{U}^{(t)} V^\dagger - \widetilde{M} \right) \|_2 \quad = \| P_{\Omega^{(t+1)}} \left(\widehat{U}^{(t)} \widehat{V}^{(t+1)\dagger} - \widetilde{M} \right) \|_2$$

$$= \| P_{\Omega^{(t+1)}} \left(U^{(t)} (\widehat{V}^{(t+1)} (R_U^{(t)} R)^\dagger)^\dagger - \widetilde{M} \right) \|_2 \quad \geq \min_{V \in \mathbb{R}^{n \times k}} \| P_{\Omega^{(t+1)}} \left(U^{(t)} V^\dagger - \widetilde{M} \right) \|_2$$

$$= \| P_{\Omega^{(t+1)}} \left(U^{(t)} \widetilde{V}^{(t+1)\dagger} - \widetilde{M} \right) \|_2$$

The above equation holds with equality for $\widehat{V}^{(t+1)} = \widetilde{V}^{(t+1)} \left((R_U^t R)^\dagger \right)^{-1}$. Further Theorem 6 shows that $\widetilde{V}^{(t+1)}$ is full rank (as $\operatorname{dist}(\widetilde{V}^{(t+1)}, V^*) < 1$) and hence, $\widehat{V}^{(t+1)} = \widetilde{V}^{(t+1)} \left((R_U^t R)^\dagger \right)^{-1}$ is full rank and their columns span the same subspace. Similar arguments can be used to show that $\widehat{U}^{(t+1)}$ and $\widetilde{U}^{(t+1)}$ are both full rank and span the same subspace. $\square$

Further, as the initial estimate, $\widehat{U}^0$ satisfies the conditions of the above lemma, in the rest of the proof it is assumed that the distances $\operatorname{dist}(\widehat{U}^t, U^*)$ and $\operatorname{dist}(\widehat{V}^t, V^*)$ are the same for the updates from both ALSM and its modified version presented above.

Theorem 6 (Each step of ALSM is good). *Under the assumptions of Theorem 2, the $(t+1)^{th}$ iterates, $\widehat{U}^{t+1}$ and $\widehat{V}^{t+1}$ satisfy the following w.h.p:*

$$\operatorname{dist}\left(\widehat{V}^{t+1}, V^* \right) \leq \frac{1}{4} \operatorname{dist}\left(\widehat{U}^t, U^* \right) + 10 \frac{\mu \kappa_M \| N^\Omega \|_2 k}{\sigma_k^* p}$$

$$\operatorname{dist}\left(\widehat{U}^{t+1}, U^* \right) \leq \frac{1}{4} \operatorname{dist}\left(\widehat{V}^{t+1}, U^* \right) + 10 \frac{\mu \kappa_M \| N^\Omega \|_2 k}{\sigma_k^* p} \tag{8}$$

where, $\kappa_M = \sigma_1^* / \sigma_k^*$ is the condition number of the matrix M.

The proof of Theorem 6, involves few other lemmata and is deferred to Section 4.3. The main theorem is now proved using the results from Theorem 5 and 6. $\square$

Proof of Main Result, Theorem 2. From Theorem 6, after $T = \mathcal{O}(\log \frac{\sqrt{k}\|M\|_2}{\epsilon})$ steps, we have:

$$\text{dist}\left(\widehat{U}^T, U^*\right) \leq \frac{\epsilon}{2\sqrt{k}\|M\|_2} + 10\frac{\mu\kappa_M\|N^\Omega\|_2 k}{\sigma_k^* p} \tag{9}$$

Using Lemma 4, we have that:

$$\|M - \widehat{U}^T\widehat{V}^{(T+1)\dagger}\|_F \leq \|(I - \widehat{U}^T\widehat{U}^{T\dagger})U^*\Sigma^*\|_F + \|F\|_F + \|N_{res}\|_F \tag{10}$$

Note that the bounds on $\|F\|_2$ and $\|N_{res}\|_2$ from Lemma 5 and Equation 16, also hold for both $\|F\|_F$ and $\|N_{res}\|_F$ respectively. This can be seen from the proofs of Lemmata 5 and 6. Using these bounds we have the following:

$$\begin{aligned}
\|M - \widehat{U}^T\widehat{V}^{(T+1)\dagger}\|_F &\leq \sqrt{k}\sigma_1^*\text{dist}(\widehat{U}^T, U^*) + \frac{\delta_{2k}\sigma_1^*}{1-\delta_{2k}}\text{dist}(\widehat{U}^T, U^*) + C\sigma_k^*\frac{10\mu\kappa_M\|N^\Omega\|_2 k}{\sigma_k^* p} \\
&\leq \epsilon + \frac{20\mu\kappa_M^2\|N^\Omega\|_2 k^{1.5}}{p}
\end{aligned} \tag{11}$$

Further, in order that each of the $2T + 1$ sub-sampled indices Ω_t has $\mathcal{O}\left(\frac{\mu^4\kappa_M^4 k^7\log n}{m\delta_{2k}^2}\right)$ samples, the total sample complexity required is $\mathcal{O}\left(\frac{\mu^4\kappa_M^4 k^7\log n\log\frac{\sqrt{k}\|M\|_2}{\epsilon}}{m\delta_{2k}^2}\right)$. □

4.3   Proof of Theorem 6

To avoid cluttering of notations we define a few quantities first. In the following definitions, we recall that U^* and U_N are the left singular vectors of M and N respectively, and U^t is the tth step iterate of the modified algorithm ($U^t = \widehat{U}^t(R_U^t)^{-1}$). Further $U^{(i)}$ and U_i represent the ith row and column vectors of U respectively and U_{ij} is the $(i, j)^{\text{th}}$ entry of U. For $1 \leq p \leq k$ and $1 \leq q \leq k$ we define diagonal matrices $B_{pq}, C_{pq}, D_{pq} \in \mathbb{R}^{n\times n}$, where, $D_{pq} = \langle U_p^t, U_q^*\rangle\mathbb{I}_{n\times n}$ and the, j^{th} diagonal entries of B_{pq} and C_{pq} are given by:

$$(B_{pq})_{jj} = \left[\frac{1}{p}\sum_{i:(i,j)\in\Omega} U_{ip}^t U_{iq}^t\right], (C_{pq})_{jj} = \left[\frac{1}{p}\sum_{i:(i,j)\in\Omega} U_{ip}^t U_{iq}^*\right].$$

Using the above matrices, we define the following matrices of dimension $nk \times nk$:

$$B \triangleq \begin{bmatrix} B_{11} & \cdots & B_{1k} \\ \vdots & \ddots & \vdots \\ B_{k1} & \cdots & B_{kk} \end{bmatrix}, C \triangleq \begin{bmatrix} C_{11} & \cdots & C_{1k} \\ \vdots & \ddots & \vdots \\ C_{k1} & \cdots & C_{kk} \end{bmatrix}, D \triangleq \begin{bmatrix} D_{11} & \cdots & D_{1k} \\ \vdots & \ddots & \vdots \\ D_{k1} & \cdots & D_{kk} \end{bmatrix}, S \triangleq \begin{bmatrix} \sigma_1^*\mathbb{I}_n & \cdots & 0 \\ \vdots & \ddots & \vdots \\ 0 & \cdots & \sigma_k^*\mathbb{I}_n \end{bmatrix}.$$

Analogously, we define matrices, $C^N \in \mathbb{R}^{nk\times nm}$ and $S^N \in \mathbb{R}^{nm\times nm}$ as follows:

$$C^N \triangleq \begin{bmatrix} C_{11}^N & \cdots & C_{1m}^N \\ \vdots & \ddots & \vdots \\ C_{k1}^N & \cdots & C_{km}^N \end{bmatrix}, S^N \triangleq \begin{bmatrix} \sigma_1^N\mathbb{I}_n & \cdots & 0 \\ \vdots & \ddots & \vdots \\ 0 & \cdots & \sigma_m^N\mathbb{I}_n \end{bmatrix} \tag{12}$$

where, $\forall \, 1 \leq p \leq k$ and $1 \leq q \leq m$, diagonal matrices $C_{pq}^N \in \mathbb{R}^{n \times n}$ are defined as

$$(C_{pq}^N)_{jj} = \left[\frac{1}{p} \sum_{i:(i,j) \in \Omega} U_{ip}^t U_{iq}^N \right].$$ Additionally, we define the following vectors:

$$v^* = [V_1^\dagger, V_2^\dagger, \cdots, V_k^\dagger]^\dagger \in \mathbb{R}^{nk}, \quad v^N = [V_1^{N\dagger}, V_2^{N\dagger}, \cdots, V_m^{N\dagger}]^\dagger \in \mathbb{R}^{nm}.$$

Finally, we define the matrices, $F = \text{vec2mat}\left(B^{-1}(BD - C)Sv^*\right) \in \mathbb{R}^{n \times k}$ and $N_{res} = \text{vec2mat}\left(B^{-1}C^N S^N v^N\right) \in \mathbb{R}^{n \times k}$.

Lemma 4. *Let $\widehat{U}^t$ be the tth step iterate of the above algorithm and let U^t, $\widehat{V}^{t+1}$ and V^{t+1} be obtained by Updates in 7. Then, using the matrices defined above, we have:*

$$\widehat{V}^{t+1} = V^* \Sigma^* U^{*\dagger} U^t - F + N_{res} \tag{13}$$

The proof of the above lemma is provided in Appendix B.2.

Lemma 5. *Let F be the error matrix defined above and let U^t be a μ_1 incoherent orthonormal matrix obtained from the tth update. Under the conditions of Theorem 2, with probability at least $1 - 1/n^3$, $\|F\|_2 \leq \frac{\delta_{2k}\sigma_1^*}{1 - \delta_{2k}} \, dist(U^t, U^*)$.*

This is the same as Lemma 5.6 of [7] and the proof follows exactly for the noisy case.

Lemma 6. *Let N_{res} be the matrix defined above. Under the conditions of the Theorem 2 with probability at least $1 - 1/n^3$*

$$\|N_{res}\|_2 \leq \frac{\mu_1\sqrt{k}}{(1 - \delta_{2k})} \left(\frac{\|N^\Omega\|_2}{p} \right) \tag{14}$$

Lemma 7. *Let $R_V^{(t+1)}$ be the upper triangular matrix obtained by QR decomposition of $\widehat{V}^{t+1}$ an. Let F, N_{res} and U^t be defined as above. Then,*

$$\left\| \left(R_V^{(t+1)} \right)^{-1} \right\|_2 \leq \frac{1}{\left[\sigma_k^* \sqrt{1 - dist^2(U^t, U^*)} - \|F\|_2 - \|N_{res}\|_2 \right]} \tag{15}$$

The proof of Lemma 6 and 7 are provided in Appendix B.3 and B.4 respectively. We now use the above lemmata to prove Theorem 6.

If $\delta_{2k} \leq \frac{\sigma_k^*}{C\sigma_1^*}$ for appropriate $C > 1$, then $\frac{1}{1 - \delta_{2k}} \leq C/(C - 1) = C_1$. Further as $dist(U^{(t)}, U^*) \leq dist(U^{(0)}, U^*) \leq 1/2$, we have $\sqrt{1 - dist^2(U^{(t)}, U^*)} \geq \sqrt{3}/2$. Finally, from Lemma 8, we have $\mu_1 = \frac{32\sigma_1^* \mu \sqrt{k}}{\sigma_k^*}$. This implies that $\|N_{res}\|_2 \leq \frac{32\mu\kappa_M k}{1 - \delta_{2k}} \left(\frac{\|N^\Omega\|_2}{p} \right)$. If further we have that $\frac{\|N^\Omega\|_2}{p} \leq C_2 \frac{\sigma_k^*}{\kappa_M k}$, then for small enough C_2, we have

$$\|N_{res}\|_2 \leq C_4 \mu\kappa_M k \frac{\|N^\Omega\|_2}{p} \leq C'\sigma_k^* \tag{16}$$

Using Lemma 4, we have:

$$\text{dist}\left(V^*, V^{(t+1)}\right) = \left\| \left[V_\perp^{*\dagger} V^* \Sigma^* U^{*\dagger} - V_\perp^{*\dagger} F + V_\perp^{*\dagger} N_{res} \right] (R_V^{(t+1)})^{-1} \right\|_2$$

$$\leq \left(\|F\|_2 + \|N_{res}\|_2 \right) \left\| \left(R_V^{(t+1)} \right)^{-1} \right\|_2 \qquad (17)$$

For appropriate choice of $C > 1$ and small enough $C' < 1$, we have From Lemma 5, 6 and 7 and Equation 17, we have the following:

$$\text{dist}(V^*, V^{(t+1)}) \leq \frac{1}{4} dist(U^{(t)}, U^*) + 10 \frac{\mu \kappa_M \|N^\Omega\|_2 k}{\sigma_k^* p}$$

Incoherence of Solutions in Each Iteration

Lemma 8. *Under the conditions of Theorem 2, let U^t be the tth step iterate obtained by Eq. 3. If U^t is $\mu_1 = \frac{32\sigma_1^* \mu \sqrt{k}}{\sigma_k^*}$ incoherent then with probability at least $(1 - 1/n^3)$, the solution $V^{(t+1)}$ obtained from Eq. 7 is also μ_1 incoherent.*

The proof of the above lemma can be found in Appendix B.5. As for $t = 0$, U^0 is μ_1 incoherent, the theorem can be used for inductively proving that U^t and V^t are μ_1 incoherent for all t.   □

5   Conclusion

We have established the first theoretical guaranties for recovery of a low rank matrix perturbed by bounded noise, using alternating least squares minimization algorithm. The algorithm is computationally more scalable than the algorithms that have previously established error bounds under noisy observations. We use the *worst case noise model* and it is observed that for well conditioned matrices, the main result requires a reasonable bound on the maximum noise entry. The results establish that under the conditions of incoherence of the underlying matrix M and bounded noise, with sufficient samples, the Frobenius norm of the deviation of the recovered matrix, $\widehat{M}$, from the original matrix M, $\frac{\|M - \widehat{M}\|_F}{\sqrt{mn}}$ can be made arbitrarily close to $Ck^{1.5} N_{max}$. Finally, for well conditioned matrices, the sample complexity is $\mathcal{O}(k^7 n \log n)$. This is the same complexity as that required by the current proof of recovery guaranties of ALSM under noiseless setting. However, this is looser compared to the established bounds of other algorithms like nuclear norm minimization and OptSpace and tightening the sample complexity will be considered in the future work. Another direction for future work would include bounding the ALSM algorithm with cost function modified to include regularization on the factors U and V.

Acknowledgments. This research was supported by ONR ATL Grant N00014-11-1-0105, NSF Grants (IIS-0713142 and IIS-1016614). We also thank Praneeth Netrapalli for his valuable comments.

References

1. Bertalmio, M., Vese, L., Sapiro, G., Osher, S.: Simultaneous structure and texture image inpainting. IEEE Transactions on Image Processing (2003)
2. Candès, E.J., Plan, Y.: Matrix completion with noise. CoRR (2009)
3. Candès, E.J., Recht, B.: Exact matrix completion via convex optimization. Foundations of Computational Mathematics (2009)
4. Candès, E.J., Tao, T.: The power of convex relaxation: near-optimal matrix completion. IEEE Transactions on Information Theory (2010)
5. Golub, G.H., van Van Loan, C.F.: Matrix Computations (Johns Hopkins Studies in Mathematical Sciences), 3rd edn. The Johns Hopkins University Press (1996)
6. Jain, P., Netrapalli, P., Sanghavi, S.: Low-rank matrix completion using alternating minimization. ArXiv e-prints (December 2012)
7. Jain, P., Netrapalli, P., Sanghavi, S.: Low-rank matrix completion using alternating minimization. In: STOC (2013)
8. Keshavan, R.H., Montanari, A., Oh, S.: Matrix completion from a few entries. IEEE Transactions on Information Theory (2010)
9. Keshavan, R.H., Montanari, A., Oh, S.: Matrix completion from noisy entries. JMLR (2010)
10. Koren, Y., Bell, R., Volinsky, C.: Matrix factorization techniques for recommender systems. IEEE Computer (2009)
11. Mitra, K., Sheorey, S., Chellappa, R.: Large-scale matrix factorization with missing data under additional constraints. In: NIPS (2010)
12. So, A.M.C., Ye, Y.: Theory of semidefinite programming for sensor network localization. In: ACM-SIAM Symposium on Discrete Algorithms (2005)
13. Takács, G., Pilászy, I., Németh, B., Tikk, D.: Investigation of various matrix factorization methods for large recommender systems. In: KDD Workshop on Large-Scale Recommender Systems and the Netflix Prize Competition (2008)
14. Wang, Y., Xu, H.: Stability of matrix factorization for collaborative filtering. In: ICML (2012)
15. Yu, K., Tresp, V.: Learning to learn and collaborative filtering. In: NIPS Workshop on Inductive Transfer: 10 Years Later (2005)
16. Zhou, Y., Wilkinson, D., Schreiber, R., Pan, R.: Large-scale parallel collaborative filtering for the netflix prize. In: Fleischer, R., Xu, J. (eds.) AAIM 2008. LNCS, vol. 5034, pp. 337–348. Springer, Heidelberg (2008)

Appendix A

- $\|M\|_2 \leq \|M\|_F \leq \sqrt{k}\|M\|_2$
- If a matrix M is μ-incoherent, then,

$$M_{max} \leq \frac{\mu^2\sqrt{k}}{\sqrt{mn}}\|M\|_F \leq \frac{\mu^2 k}{\sqrt{mn}}\|M\|_2 \tag{18}$$

- *Bernstein's Inequality*: Let X_i, $i = \{1, 2, \ldots, n\}$ be independent random numbers. Let $|X_i| \leq L \; \forall \; i$ w.p. 1. Then we have the following inequalities:

$$P\left[\sum_{i=1}^n X_i - \sum_{i=1}^n \mathbb{E}[X_i] > t\right] \leq \exp\left(\frac{-t^2/2}{\sum_{i=1}^n Var(X_i)+Lt/3}\right)$$

$$P\left[\sum_{i=1}^n X_i - \sum_{i=1}^n \mathbb{E}[X_i] < -t\right] \leq \exp\left(\frac{-t^2/2}{\sum_{i=1}^n Var(X_i)+Lt/3}\right) \tag{19}$$

Appendix B

B.1   Initialization Proofs

Proof (Proof of Lemma 2)

$$
\begin{aligned}
\|M - \widetilde{M}_k^\Omega\|_2^2 &= \|U^*\Sigma^*V^{*\dagger} - \widetilde{U}^0\widetilde{\Sigma}^0\widetilde{V}^{0\dagger}\|_2^2 \\
&= \|(I - \widetilde{U}^0\widetilde{U}^{0\dagger})U^*\Sigma^*V^{*\dagger} + \widetilde{U}^0(\widetilde{U}^{0\dagger}U^*\Sigma^*V^{*\dagger} - \widetilde{\Sigma}^0\widetilde{V}^{0\dagger})\|_2^2 \\
&\overset{(1)}{=} \|(I - \widetilde{U}^0\widetilde{U}^{0\dagger})U^*\Sigma^*V^{*\dagger}\|_2^2 + \|\widetilde{U}^0(\widetilde{U}^{0\dagger}U^*\Sigma^*V^{*\dagger} - \Sigma V^\dagger)\|_2^2 \\
&\geq \|(I - \widetilde{U}^0\widetilde{U}^{0\dagger})U^*\Sigma^*V^{*\dagger}\|_2^2 = \|\widetilde{U}_\perp^{0\dagger}U^*\Sigma^*\|_2^2 \geq \sigma_k^{*2}\|\widetilde{U}_\perp^{0\dagger}U^*\|_2^2
\end{aligned}
$$

where, (1) follows as the two terms span orthogonal spaces. Hence,

$$
\begin{aligned}
\operatorname{dist}(\widetilde{U}^0, U^*) &\leq \frac{1}{\sigma_k^*}\|M - \widetilde{M}_k^\Omega\|_2 \overset{(2)}{\leq} \frac{1}{\sigma_k^*}\left(CM_{max}\sqrt{\frac{m\alpha^{3/2}}{p}} + 2\frac{\|N^\Omega\|_2}{p}\right) \\
&\overset{(3)}{\leq} C\mu^2 k\frac{\sigma_1^*}{\sigma_k^*}\sqrt{\frac{m\alpha^{3/2}}{pmn}} + \frac{2\|N^\Omega\|_2}{p\sigma_k^*} \\
&\leq \frac{1}{64k}, \text{ if } p > \frac{C'k^4\mu^4\sigma_1^{*2}}{m\sigma_k^{*2}} \text{ and } \frac{\|N^\Omega\|_2}{p} \leq C''\frac{\sigma_k^*}{k}.
\end{aligned}
$$

where, (2) follows from Lemma 1 and (3) follows from Equation 18

B.2   Proof of Lemma 4

We recall that $M^{(i)}$ is the ith row of the matrix M. Given, U^t, the tth step iterate. The update of $\widehat{V}^{(t+1)}$ is guided by the following equation from 7.

$$
\begin{aligned}
\widehat{V}^{(t+1)} &= \underset{V\in\mathbb{R}^{n\times k}}{\operatorname{argmin}} \|P_\Omega(U^tV^\dagger) - P_\Omega(\widetilde{M})\|_F^2 \\
&= \underset{V\in\mathbb{R}^{n\times k}}{\operatorname{argmin}} \sum_{(i,j)\in\Omega}\left(U^{t(i)\dagger}V^{(j)} - U^{*(i)\dagger}\Sigma^*V^{*(j)} - U_N^{(i)\dagger}\Sigma_N V_N^{(j)}\right)^2
\end{aligned}
$$

Taking the gradient with respect to each $V^{(j)}$ and setting it to 0 for the optimum $V = \widehat{V}^{(t+1)}$, we have the following $\forall\, j \in [n]$:

$$
\sum_{i:(i,j)\in\Omega} U^{t(i)}\left(U^{t(i)\dagger}\left(\widehat{V}^{(t+1)}\right)^{(j)} - U^{*(i)\dagger}\Sigma^*V^{*(j)} - U_N^{(i)\dagger}\Sigma_N V_N^{(j)}\right) = 0 \quad (20)
$$

We further define matrices $B^j, C^j, D^j \in \mathbb{R}^{k\times k}$ and $C_N^j \in \mathbb{R}^{k\times m}$ for $1 \leq j \leq n$ as follows:

$$
B^j = \frac{1}{p}\sum_{i:(i,j)\in\Omega} U^{t(i)}U^{t(i)\dagger}, \quad C^j = \frac{1}{p}\sum_{i:(i,j)\in\Omega} U^{t(i)}U^{*(i)\dagger},
$$

$$
D^j = U^{t\dagger}U^*, \quad C_N^j = \frac{1}{p}\sum_{i:(i,j)\in\Omega} U^{t(i)}U^{N(i)\dagger}. \quad (21)
$$

It is useful to note that $B^j \in \mathbb{R}^{k \times k}$ is obtained by taking the jth diagonal elements of B_{pq} (defined in Equation 12), for $1 \leq p, q \leq k$, i.e. $(B^j)_{pq} = (B_{pq})_{jj}$. The other matrices are defined similarly. Using the above matrices, we have from the previous equation:

$$
\begin{aligned}
\left(\widehat{V}^{(t+1)}\right)^{(j)} &= D^j \Sigma^* V^{*(j)} - (B^j)^{-1} \left(B^j D^j - C^j\right) \Sigma^* V^{*(j)} + (B^j)^{-1} C_N^j \Sigma_N V_N^{(j)} \\
\widehat{V}^{(t+1)} &= V^* \Sigma^* U^{*\dagger} U^t - F + N_{res}
\end{aligned}
$$

$$(22)$$

The last equation above can be easily seen by writing the structure of matrices defined above.

B.3   Proof of Lemma 6

From Lemma C.6 of [6], under the assumptions on p and M specified in the Lemma, we have, $\|B^{-1}\|_2 \leq \frac{1}{1-\delta_{2k}}$. Further from the structure of the matrices C^N, S^N and C_N^j, it can be verified that $\|C^N S^N v^N\|_2^2 = \sum_{j=1}^{n} \|C_N^{(j)} \Sigma_N V_N^{(j)}\|_2^2$. Recall that $V_N^{(j)} \in \mathbb{R}^{m \times 1}$ is the jth row of $V_N \in \mathbb{R}^{n \times m}$ (a similar decomposition is used in Equation 22). Thus we have:

$$
\begin{aligned}
\|C^N S^N v^N\|_2^2 &= \sum_{j=1}^{n} \left\| C^{(j)} \Sigma_N V_N^{(j)} \right\|_2^2 = \sum_{j=1}^{n} \left\| \frac{1}{p} \sum_{i:(i,j)\in\Omega} U^{t(i)} U_N^{(i)\dagger} \Sigma_N V_N^{(j)} \right\|_2^2 \\
&\leq \frac{1}{p^2} \sum_{j=1}^{n} \sum_{i:(i,j)\in\Omega} \|U^{t(i)} N_{ij}\|_2^2 \leq \frac{1}{p^2} \sum_{(i,j)\in\Omega} \|U^{t(i)}\|_2^2 |N_{ij}|_2^2 \\
&\leq \mu_1^2 k \left(\frac{\|N^\Omega\|_F}{\sqrt{mp}}\right)^2 \leq \mu_1^2 k \left(\frac{\|N^\Omega\|_2}{p}\right)^2
\end{aligned}
$$

$$(23)$$

This implies that

$$
\|N_{res}\|_2 \leq \|N_{res}\|_F = \|B^{-1} C^N S^N v^N\|_2 \leq \|B^{-1}\|_2 \|C^N S^N v^N\|_2 \quad (24)
$$

$$
\leq \frac{\mu_1 \sqrt{k}}{(1-\delta_{2k})} \left(\frac{\|N^\Omega\|_2}{p}\right).
$$

B.4   Proof of Lemma 7

$$
\begin{aligned}
\frac{1}{\|(R_V^{(t+1)})^{-1}\|_2} &= \sigma_{min}(R_V^{(t+1)}) = \min_{z:\|z\|_2=1} \|R_V^{(t+1)} z\|_2 = \min_{z:\|z\|_2=1} \|V^{(t+1)} R_V^{(t+1)} z\|_2 \\
&= \min_{z:\|z\|_2=1} \|\widehat{V}^{(t+1)} z\|_2 = \min_{z:\|z\|_2=1} \| \left(V^* \Sigma^* (U^*)^\dagger U^t - F + N_{res}\right) z\|_2 \\
&\geq \min_{z:\|z\|_2=1} \left[\|V^* \Sigma^* (U^*)^\dagger U^t z\|_2 - \|F\|_2 - \|N_{res}\|_2 \right] \\
&\geq \sigma_k^* \min_{z:\|z\|_2=1} \|(U^*)^\dagger U^t z\|_2 - \|F\|_2 - \|N_{res}\|_2 \\
&= \sigma_k^* \sqrt{1 - \text{dist}(U^t, U^*)^2} - \|F\|_2 - \|N_{res}\|_2
\end{aligned}
$$

Thus, $\|(R_V^{(t+1)})^{-1}\|_2 \leq \dfrac{1}{\sigma_k^* \sqrt{1-\mathrm{dist}(U^t,U^*)^2}-\|F\|_2-\|N_{res}\|_2}$

B.5 Proof of Lemma 8

In this proof, we use the following set of inequalities:

$$\|(B^j)^{-1}\|_2 \leq \frac{1}{1+\delta_{2k}}$$

$$\|B^j\|_2 \leq 1+\delta_{2k}, \ \|C^j\|_2 \leq 1+\delta_{2k}, \ \|D^j\| \leq \|U^*\|_2\|U^t\|_2 = 1. \tag{25}$$

The above set of equations involve terms that does not depend on the noise and hence are incorporated from Appendix C.3 of [7]. It can be verified that the proof does not change for the noisy case. We omit the derivation here to avoid redundancy.

Lemma 9. *Under the conditions of Theorem 2, w.p. greater that* $1 - 1/n^3$

$$\|C_N^j \Sigma_N V_N^{(j)}\|_2 \leq N_{max}\mu_1\sqrt{km}(1 + \delta_{2k})$$

We prove the above lemma at the end of this section. Now, from Equation 22, we have:

$$\left(\widehat{V}^{(t+1)}\right)^{(j)} = D^j \Sigma^* V^{*(j)} - (B^j)^{-1}\left(B^j D^j - C^j\right)\Sigma^* V^{*(j)} + (B^j)^{-1}C_N^j \Sigma_N V_N^{*(j)}$$

Thus,

$$\left\|\left(V^{(t+1)}\right)^{(j)}\right\|_2 \leq \ \|(R^{(t+1)})^{-1}\|_2\left[\left(\|D^j\|_2 + \|(B^j)^{-1}\|_2(\|B^j D^j\|_2 + \|C^j\|_2)\right)\|\Sigma^*\|\|V^{*(j)}\|_2\right.$$
$$\left. + \ \|(B^j)^{-1}\|_2\|C_N^j \Sigma_N V_N^{*(j)}\|_2\right] \tag{26}$$

Using equations from 25, Lemma 9 and 7, and $\delta_{2k} \leq \frac{1}{C}, C > 1$ we have the following:

$$\left\|\left(V^{(t+1)}\right)^{(j)}\right\|_2 \leq \ \|(R^{(t+1)})^{-1}\|_2\left[\frac{\sigma_1^*\mu\sqrt{k}}{\sqrt{n}}\left(1 + \frac{(2(1+\delta_{2k}))}{1-\delta_{2k}}\right) + \frac{N_{max}\mu_1\sqrt{km}(1+\delta_{2k})}{1-\delta_{2k}}\right]$$
$$= \ \|(R^{(t+1)})^{-1}\|_2\left[\frac{4\sigma_1^*\mu\sqrt{k}}{\sqrt{n}} + 2N_{max}\mu_1\sqrt{km}\right] \tag{27}$$

We now use that for, $\mu_1 = \frac{32\mu\sigma_1^*\sqrt{k}}{\sigma_k^*}$ and further, $N_{max} \leq C_3\frac{\sigma_k^*}{n\sqrt{k}}$. Choosing C_3 appropriately, we have $N_{max}\mu_1\sqrt{km} \leq \frac{2\sigma_1^*\mu\sqrt{k}}{\sqrt{n}}$. Finally, using Lemmata 5 and 6, the fact that $\mathrm{dist}(U^*, U^t) \leq \mathrm{dist}(U^*, U^0) \leq 0.5$ and using the conditions on $\|N^\Omega\|_2$ from Theorem 2, we have, $\|(R^{(t+1)})^{-1}\|_2 \leq \frac{4}{\sigma_k^*}$. Using these we have:

$$\left\|\left(V^{(t+1)}\right)^{(j)}\right\|_2 \leq \|(R^{(t+1)})^{-1}\|_2\frac{8\sigma_1^*\mu\sqrt{k}}{\sqrt{n}} \leq \frac{32\sigma_1^*\mu\sqrt{k}}{\sigma_k^*\sqrt{n}} \tag{28}$$

Thus we have, $\mu(V^{(t+1)}) \leq \frac{32\sigma_1^*\mu}{\sigma_k^*} \leq \frac{32\sigma_1^*\mu\sqrt{k}}{\sigma_k^*}$.

Proof of Lemma 9. $\|C_N^j \Sigma_N V_N^{(j)}\|_2 = \max_{x:\|x\|_2=1} x^\dagger C_N^j \Sigma_N V_N^{(j)}$. Given any x such that $\|x\|_2 = 1$, then

$$x^\dagger C_N^j \Sigma_N V_N^{(j)} = \frac{1}{p} \sum_{i:(i,j)\in\Omega} x^\dagger U^{t(i)} U_N^{(i)\dagger} \Sigma_N V_N^{(j)} = \frac{1}{p} \sum_{i:(i,j)\in\Omega} x^\dagger U^{t(i)} N_{ij} \quad (29)$$

We define $\delta_{ij} = \begin{cases} 1 & \text{if } (i,j) \in \Omega \\ 0 & \text{otherwise} \end{cases}$, $Z_i = \frac{1}{p}\delta_{ij} x^\dagger U^{t(i)} N_{ij}$ and $Z = \sum_{i=1}^m Z_i$

$$E[Z] = \sum_{i=1}^m E[Z_i] = \sum_{i=1}^m x^\dagger U^{t(i)} N_{ij} \le N_{max} \sum_{i=1}^m \|U^{t(i)}\|_2 = N_{max}\mu_1\sqrt{km} \quad (30)$$

$$\begin{aligned}
var(Z) &= \sum_{i=1}^m E[Z_i^2] - (E[Z_i])^2 = \frac{1-p}{p} \sum_{i=1}^m \left(x^\dagger U^{t(i)}\right)^2 N_{ij}^2 \\
&\le \frac{1}{p} N_{max}^2 \sum_{i=1}^m \|U^{t(i)}\|_2^2 = \frac{1}{p} N_{max}^2 \|U^t\|_F^2 = \frac{N_{max}^2 k}{p}
\end{aligned} \quad (31)$$

$$\max_i Z_i = \frac{1}{p} \max_i x^t U^{t(i)} N_{ij} \le \frac{N_{max}\mu_1\sqrt{k}}{p\sqrt{m}} \quad (32)$$

From Equations 29, 30, 31, 32 and using Bernstein's inequality in Equation 19, we have the following:

$$P\left(Z \ge N_{max}\mu_1\sqrt{km}(1+\delta_{2k})\right) \le \exp\left(\frac{-\delta_{2k}^2 N_{max}^2 \mu_1^2 km/2}{\frac{N_{max}^2 k}{p} + \frac{N_{max}^2 \mu_1^2 k\delta_{2k}}{3p}}\right) = \exp\left(\frac{-\delta_{2k}^2 \mu_1^2 mp}{2\left(1+\frac{\mu_1^2\delta_{2k}}{3}\right)}\right)$$

From the conditions of Theorem 2, $\delta_{2k} \le \frac{\sigma_1^*}{C\sigma_k^*}$, $p > 12\frac{\log n}{m\delta_{2k}}$, using $(1+\mu_1^2\delta_{2k}/3) \le (1+\mu_1^2) \le 2\mu_1^2$, we have:

$$P\left(Z \ge N_{max}\mu_1\sqrt{km}(1+\delta_{2k})\right) \le \exp\left(\frac{-12\mu_1^2 \log n}{4\mu_1}\right) = \frac{1}{n^3}.$$

Thus, we have with probability grater that $1 - 1/n^3$, $\forall\ x : \|x\|_2 = 1$, including the maximizing x, we have $x^\dagger C_N^j \Sigma_N V_N^{(j)} \le N_{max}\mu_1\sqrt{km}(1+\delta_{2k})$. Thus, $\|C_N^j \Sigma_N V_N^{(j)}\|_2 \le N_{max}\mu_1\sqrt{km}(1+\delta_{2k})$. $\square$

A Nearly Unbiased Matrix Completion Approach

Dehua Liu, Tengfei Zhou, Hui Qian*, Congfu Xu, and Zhihua Zhang

College of Computer Science and Technology
Zhejiang University, Hangzhou 310007, China
{dehua,zhoutengfei_zju,qianhui,xucongfu,zhzhang}@zju.edu.cn

Abstract. Low-rank matrix completion is an important theme both theoretically and practically. However, the state-of-the-art methods based on convex optimization usually lead to a certain amount of deviation from the original matrix. To perfectly recover a data matrix from a sampling of its entries, we consider a non-convex alternative to approximate the matrix rank. In particular, we minimize a matrix γ-norm under a set of linear constraints. Accordingly, we derive a shrinkage operator, which is nearly unbiased in comparison with the well-known soft shrinkage operator. Furthermore, we devise two algorithms, non-convex soft imputation (NCSI) and non-convex alternative direction method of multipliers (NCADMM), to fulfil the numerical estimation. Experimental results show that these algorithms outperform existing matrix completion methods in accuracy. Moreover, the NCADMM is as efficient as the current state-of-the-art algorithms.

1 Introduction

Applications of low-rank matrix completion become increasingly popular in machine learning and data mining. For instance, in the system of collaborative filtering, we aim to predict the unknown preference of a user on a set of unrated items, only according to a few submitted rating. In image inpainting problems, large amount of missing pixels should be estimated by exploiting the known content.

Typically, matrix completion is formed as minimizing the rank of matrix when given a few known entries. However, the rank minimization problem is often numerically prohibitive. Thus, many approximation strategies are encouraged. One principled approach is to replace the matrix rank by the nuclear norm, because the nuclear norm is the best convex relaxation of the matrix rank. In the literature [1–3], the authors proved that under certain assumptions on the proportion of the missing entries and locations, most low-rank matrices can be completed exactly by minimizing the nuclear norm under the linear constraints (the completed matrix must be consistent with the observed matrix for the few known entries).

Based on the nuclear norm, Cai *et al.* [4] devised a singular value thresholding (SVT) algorithm for this convex optimization problem. Mazumder *et al.* [5] formed a unconstrained convex optimization problem and developed a soft-impute algorithm. Ma *et al.* [6] devised a fixed point iterative algorithm inspired from the work of Hale *et al.* [7] in the ℓ_1 regularization problem. Lin *et al.* [8, 9] proposed an alternative direction algorithm based on the augmented Lagrangian multipliers. Other efficient algorithms based

---

* Corresponding author.

H. Blockeel et al. (Eds.): ECML PKDD 2013, Part II, LNAI 8189, pp. 210–225, 2013.

on convex relaxation include [3, 10, 11]. There are also some other matrix norms that have been considered, e.g., the max-norm which is also a convex approximation of the matrix rank [12].

However, convex relaxation often makes the resulting solution deviate from the original matrix [13]. To address this problem, non-convex approximation to the matrix norm has been also exploited recently. Such treatments include the Schatten ℓ_p-norm ($0 < p < 1$) used by Nie et al. [13], the truncated nuclear norm proposed by Hu et al. [14] and a so-called matrix γ-norm studied by Wang et al. [15]. Note that matrix γ-norm is not really a norm, since it does not satisfy triangle inequality.

The matrix γ-norm is a matrix extension of the MC+ function studied by Mazumder et al. [16] and Zhang [17] for variable selection. The γ-norm is characterized by a positive factor γ, and is tighter than the nuclear norm to the matrix rank. Wang et al. [15] employed the γ-norm, giving a non-convex approach to robust principle component analysis (RPCA). In this paper we introduce the γ-norm into the matrix completion problem. We develop a shrinkage operator which is nearly unbiased from non-convex rank approximation and put forward two effective algorithms called NCSI and NCADMM.

The remaining parts of the paper are organized as follows. Section 2 reviews the preliminaries for matrix completion. Section 3 presents the NCSI and NCADMM algorithms. Section 4 gives the convergence analysis of our NCSI algorithm. Section 5 conducts the experimental analysis. Finally, we conclude our work in Section 6.

2 Preliminaries

We are given a matrix $M = [m_{ij}] \in \mathbb{R}^{n \times m}$ with missing entries. Without loss of generality, we assume $m \leq n$. Let $X = [x_{ij}] \in \mathbb{R}^{n \times m}$ be an unknown low-rank matrix. The matrix completion problem is to address the following rank minimization problem:

$$\min_{X} \quad \text{rank}(X)$$
$$\text{s.t.} \quad x_{ij} = m_{ij}, \quad \forall (i, j) \in \Omega,$$

in which $\Omega \subset \{1, \ldots, n\} \times \{1, \ldots, m\}$ is the set of indices of observation entries of M. We denote the indices of the missing entries by $\bar{\Omega} = \{1, \ldots, n\} \times \{1, \ldots, m\} \setminus \Omega$.

This rank minimization problem is generally NP-hard. However, it can be relaxed to a feasible optimization problem via rank approximation. That is, we consider the following alternative:

$$\min_{X} \quad P(X; \theta) \tag{1}$$
$$\text{s.t.} \quad x_{ij} = m_{ij}, \quad \forall (i, j) \in \Omega, \tag{2}$$

where $P(X; \theta)$ represents the approximation of rank(X).

It is well known that the nuclear norm, the sum of singular values, is the tightest convex relaxation of the matrix rank. Candès and Tao [2] proved that most low rank matrices can be completed from (1) with $P(X; \theta)$ as the nuclear norm $\|X\|_*$ if the number of given entries is greater than $nr\mathrm{polylog}(n)$ up to a positive constant C. Meanwhile, some researchers developed efficient algorithms to solve the above problem such as singular value thresholding SVT [4].

In order to accommodate the small noise in observation, it is better to relax the constraints in (1) by adding a square loss to the objective function, forming an unconstrained problem [5]:

$$\min_{\mathbf{X}} \tfrac{1}{2}\|P_{\Omega}(\mathbf{X} - \mathbf{M})\|_F^2 + \lambda P(\mathbf{X}; \theta), \tag{3}$$

where $\|\mathbf{A}\|_F = \sqrt{\sum_{i,j} a_{ij}^2} = \sqrt{\mathrm{tr}(\mathbf{A}\mathbf{A}^\top)} = \sqrt{\sum_i \sigma_i^2(\mathbf{A})}$ is the Frobenius norm of $\mathbf{A} = [a_{ij}]$, and $P_{\Omega}(\mathbf{A})$ is such an $n \times m$ matrix that its (i, j)th entry is a_{ij} if $(i, j) \in \Omega$ and zero otherwise. $P(\mathbf{X}; \theta)$ is usually called a regularization or penalty term.

In order to solve problem (3), a key step is to solve a subproblem of the form:

$$\min_{\mathbf{X}} \left\{ \frac{1}{2}\|\mathbf{X} - \boldsymbol{\Phi}\|_F^2 + \lambda P(\mathbf{X}; \theta) \right\}. \tag{4}$$

First of all, we introduce a so-called shrinkage operator.

Definition 1 (Shrinkage Operator). $S_{\lambda,\theta}(\boldsymbol{\Phi}) = \mathrm{argmin}_{\mathbf{X}}\{\frac{1}{2}\|\mathbf{X} - \boldsymbol{\Phi}\|_F^2 + \lambda P(\mathbf{X}; \theta)\}$ *is a shrinkage operator if it shrinks the small singular value of $\boldsymbol{\Phi}$ to 0.*

In this paper, we would like to consider the special penalty $P(\mathbf{X}; \theta)$ which is constructed from a single variable function. Suppose $p(x; \theta)$ is a function of single variable function with domain $\mathbb{R}_+$, then $P(\mathbf{X}; \theta) = \sum_i p(\sigma_i(\mathbf{X}))$. We can construct many penalty on matrix by this way. In this case we define the overloading of shrinkage operator on $\mathbb{R}_+$ as $S_{\lambda,\theta}(z) = \mathrm{argmin}_{x \geq 0}\{\frac{1}{2}(x - z)^2 + \lambda p(x; \theta)\}$.

For example, the popular used nuclear norm $P(\mathbf{X}; \theta) = \|\mathbf{X}\|_*$ is derived from function $p(x) = x$. And $S_{\lambda}(z) = \mathrm{argmin}_{x \geq 0}\{\frac{1}{2}(x - z)^2 + \lambda x\}$ for $z \geq 0$.

Let $\boldsymbol{\Phi} = \mathbf{U}\boldsymbol{\Sigma}\mathbf{V}^\top$ be the thin SVD of $\boldsymbol{\Phi}$. It has been proved that in the case of $P(\mathbf{X}; \theta) = \|\mathbf{X}\|_*$ the shrinkage operator has a simple form which is given by $S_{\lambda}(\boldsymbol{\Phi}) = \mathbf{U}\boldsymbol{\Sigma}_{\lambda}\mathbf{V}^\top$ with $\boldsymbol{\Sigma}_{\lambda} = \mathrm{diag}(S_{\lambda}(\sigma_1), \ldots, S_{\lambda}(\sigma_m))$ [4–6], where $S_{\lambda}(\cdot)$ defined for a single variable is the overloading operator which is given by

$$S_{\lambda}(z) = [z - \lambda]_+ = \begin{cases} z - \lambda & \text{if } z > \lambda, \\ 0 & \text{if } z \leq \lambda. \end{cases}$$

We call $S_{\lambda}(\cdot)$ the soft shrinkage operator.

Observe that problem (4) can be viewed as the extreme case of (3), when Ω is the set of subscript indices of all entries of matrix. We find that soft shrinkage operator derived from the nuclear norm may lead to deviation for large λ, since a same positive number is subtracted from all the singular values of a matrix. This encourages us to use a non-convex penalty which results in a shrinkage operator keeping the large singular values unchanged while shrinking the small ones to zero. We establish this thought the following definition.

Definition 2 (Nearly Unbiasedness). *We say that the operator $S_{\lambda,\theta}(\boldsymbol{\Phi})$ is nearly unbiased, if it keeps the sufficiently large singular value of $\boldsymbol{\Phi}$ unchanged.*

The non-convex penalty has been mentioned in [5] and further explored by some researchers in matrix recovery as well as matrix completion problems [14, 15]. Here we employ the treatment of Wang *et al.* [15] who devised a so-called γ-norm in their work of matrix recovery problem. We will show that this non-convex penalty can make a nearly unbiased estimator for the matrix completion problem.

3 Methodology

For a matrix $\mathbf{X} \in \mathbb{R}^{n \times m}$ with $m \leq n$, $\mathbf{X} = \mathbf{U}\boldsymbol{\Sigma}\mathbf{V}^{\top}$ is the SVD factorization with $\boldsymbol{\Sigma} = \text{diag}\{\sigma_1, ..., \sigma_m\}$. According to Wang *et al.* [15], γ-norm is defined as

$$\|\mathbf{X}\|_{\gamma} = \sum_{i=1}^{m} p(\sigma_i; \gamma),$$

where $p(x; \gamma) = \int_0^x (1 - \frac{u}{\gamma})_+ du = (x - \frac{x^2}{2\gamma})I(x < \gamma) + \frac{\gamma}{2}I(x \geq \gamma)$. A key step to construct an algorithm is an optimization problem whose solution is summarized as below.

Theorem 1. *Suppose $\boldsymbol{\Phi} = \mathbf{U}\boldsymbol{\Sigma}\mathbf{V}^{\top}$ is the SVD factorization. The minimizer of $\phi(\mathbf{X}) = \frac{1}{2}\|\boldsymbol{\Phi} - \mathbf{X}\|_F^2 + \lambda\|\mathbf{X}\|_{\gamma}$ with $\gamma > \lambda$ is $S_{\lambda,\gamma}(\boldsymbol{\Phi}) = \mathbf{U}\boldsymbol{\Sigma}_{\lambda,\gamma}\mathbf{V}^{\top}$, where $\boldsymbol{\Sigma}_{\lambda,\gamma} = \text{diag}(S_{\lambda,\gamma}(\sigma_1), \cdots, S_{\lambda,\gamma}(\sigma_m))$ is a diagonal matrix with the diagonal elements*

$$S_{\lambda,\gamma}(\sigma_i) = \underset{x \geq 0}{\text{argmin}}\{\frac{1}{2}(x - \sigma_i)^2 + \lambda p(x; \gamma)\} = \begin{cases} \sigma_i & \text{if } \sigma_i \geq \gamma, \\ \frac{\sigma_i - \lambda}{1 - \frac{\lambda}{\gamma}} & \text{if } \lambda \leq \sigma_i < \gamma, \\ 0 & \text{if } \sigma_i < \lambda. \end{cases}$$

Proof. Let the thin SVD of $\mathbf{X}$ be of $\mathbf{X} = \mathbf{U}\boldsymbol{\Lambda}\mathbf{V}^T$, where $\mathbf{U} = [\mathbf{u}_1, \ldots, \mathbf{u}_m]$ has orthonormal columns, $\mathbf{V} = [\mathbf{v}_1, \ldots, \mathbf{v}_m]$ is orthogonal, and $\boldsymbol{\Lambda} = \text{diag}(\lambda_1, \ldots, \lambda_m)$ is arranged as $\lambda_1 \geq \lambda_2 \geq \cdots \geq \lambda_m \geq 0$.

$$\phi(\mathbf{X}) = \frac{1}{2}\|\boldsymbol{\Phi} - \mathbf{X}\|_F^2 + \lambda\sum_{i=1}^{m}\int_0^{\lambda_i}(1 - \frac{u}{\gamma})_+ du$$

$$= \frac{1}{2}\|\boldsymbol{\Phi} - \mathbf{U}\boldsymbol{\Sigma}\mathbf{V}^T\|_F^2 + \lambda\sum_{i=1}^{m}\int_0^{\lambda_i}(1 - \frac{u}{\gamma})_+ du$$

$$= \frac{1}{2}\|\boldsymbol{\Phi}\|_F^2 + \frac{1}{2}(\sum_{i=1}^{m}\lambda_i^2 - 2\sum_{i=1}^{m}\lambda_i\mathbf{u}_i^T\boldsymbol{\Phi}\mathbf{v}_i) + \lambda\sum_{i=1}^{m}\int_0^{\lambda_i}(1 - \frac{u}{\gamma})_+ du.$$

Fixing $\mathbf{u}_i$ and $\mathbf{v}_i$, and then differentiating $\phi(\mathbf{X})$ with respect to λ_i yields

$$\lambda_i - \mathbf{u}_i^T\boldsymbol{\Phi}\mathbf{v}_i + \lambda(1 - \frac{\lambda_i}{\gamma})_+ = 0.$$

Denoting $\xi_i = \mathbf{u}_i^T \boldsymbol{\Phi} \mathbf{v}_i$, we obtain

$$\hat{\lambda}_i = S_{\lambda,\gamma}(\xi_i) = \begin{cases} \xi_i & \text{if } \gamma < \xi_i, \\ \frac{\xi_i - \lambda}{1 - \frac{\lambda}{\gamma}} & \text{if } \lambda < \xi_i \leq \gamma, \\ 0 & \text{if } \xi_i \leq \lambda. \end{cases}$$

Substituting the $\hat{\lambda}_i$ back into $g(\mathbf{X})$ yields

$$\phi(\mathbf{X}) = \frac{1}{2}\|\boldsymbol{\Phi}\|_F^2 + \frac{1}{2}\sum_{i=1}^m \hat{\lambda}_i^2 - \sum_{i=1}^m \hat{\lambda}_i \xi_i + \lambda \sum_{i=1}^m \left\{ (\hat{\lambda}_i - \frac{\hat{\lambda}_i^2}{2\gamma}) I(\hat{\lambda}_i \leq \gamma) + \frac{\gamma}{2} I(\hat{\lambda}_i > \gamma) \right\}.$$

We now see that minimizing $\phi(\mathbf{X})$ w.r.t. $\mathbf{X}$ is equivalent to the minimization of ψ w.r.t. the ξ_i. Here

$$\begin{aligned} \psi &= \frac{1}{2}\sum_{i=1}^m \hat{\lambda}_i^2 - \sum_{i=1}^m \hat{\lambda}_i \xi_i + \lambda \sum_{i=1}^m \left\{ \left(\hat{\lambda}_i - \frac{\hat{\lambda}_i^2}{2\gamma}\right) I(\hat{\lambda}_i \leq \gamma) + \frac{\lambda\gamma}{2} I(\hat{\lambda}_i > \gamma) \right\} \\ &= \frac{1}{2}\sum_{\lambda < \xi_i \leq \gamma} \mu^2(\xi_i - \lambda)^2 + \frac{1}{2}\sum_{\xi_i > \gamma} \xi_i^2 - \sum_{\lambda < \xi_i \leq \gamma} \mu(\xi_i - \lambda)\xi_i - \sum_{\xi_i > \gamma} \xi_i^2 \\ &\quad + \lambda \sum_{\lambda < \xi_i \leq \gamma} \left\{ \mu(\xi_i - \lambda) - \frac{\mu^2(\xi_i - \lambda)^2}{2\gamma} \right\} + \sum_{\xi_i > \gamma} \frac{\lambda\gamma}{2}, \end{aligned}$$

where $\mu = \frac{1}{1 - \frac{\lambda}{\gamma}} = \frac{\gamma}{\gamma - \lambda}$. Since the ξ_i are partitioned into the three parts, we consider the corresponding terms of ψ separately. The term of ψ corresponding to $\lambda < \xi_i \leq \gamma$ is

$$\begin{aligned} \psi_1 &= \sum_{\lambda < \xi_i \leq \gamma} \left\{ \frac{1}{2}\mu^2(\xi_i - \lambda)^2 - \mu(\xi_i - \lambda)\xi_i + \lambda\mu(\xi_i - \lambda) - \frac{\lambda\mu^2(\xi_i - \lambda)^2}{2\gamma} \right\} \\ &= -\sum_i \frac{\mu}{2}(\xi_i - \lambda)^2. \end{aligned}$$

The term of ψ corresponding to $\xi_i > \gamma$ is

$$\psi_2 = \sum_{\xi_i > \gamma} \left\{ -\frac{1}{2}\xi_i^2 + \frac{\lambda\gamma}{2} \right\}.$$

Recall that $\xi_i = \mathbf{u}_i^T \boldsymbol{\Phi} \mathbf{v}_i$. In order to minimize ψ w.r.t. ξ_i, $\mathbf{u}_i$ and $\mathbf{v}_i$ should be the singular vectors corresponding the singular values of $\boldsymbol{\Phi}$ and $\xi_i = \sigma_i$. Thus, it is necessary that the optimal solution to $\min \phi(\mathbf{X})$ is $\mathbf{X} = S_{\lambda,\gamma}(\boldsymbol{\Phi})$. $\square$

Now we have introduced a so-called non-convex soft shrinkage operator $S_{\lambda,\gamma}(\cdot)$. Compared to the popular soft shrinkage operator, it has an advantage of nearly unbiasedness, since it keeps large singular values of a matrix unchanged, see Figure 1. It is expected that our algorithms have higher accuracy compared to the state-of-the-art algorithms.

Based on this non-convex soft shrinkage operator, we develop two algorithms. One is to directly solve an unconstrained problem and the other is to solve an equivalent form but with explicit constraints.

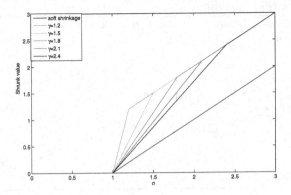

Fig. 1. Non-convex soft shrinkage operator vs. soft shrinkage operator ($\lambda = 1$)

3.1 The Non-convex Soft Imputation Algorithm

The γ-norm regularization problem is

$$\min_{\mathbf{X}}\{J(\mathbf{X}) = \frac{1}{2}\|P_\Omega(\mathbf{X} - \mathbf{M})\|_F^2 + \lambda\|\mathbf{X}\|_\gamma\}. \tag{5}$$

We now derive an iterative process to minimize $J(\mathbf{X})$. Suppose we have obtained $\mathbf{X}^k$ at iteration k for a fixed λ, we bound $J(\mathbf{X})$ from above by

$$Q(\mathbf{X}|\mathbf{X}^k) = \frac{1}{2}\|P_\Omega(\mathbf{X}^k - \mathbf{M})\|_F^2 + \langle P_\Omega(\mathbf{X}^k - \mathbf{M}), \mathbf{X} - \mathbf{X}^k \rangle + \lambda\|\mathbf{X}\|_\gamma + \frac{1}{2\alpha}\|\mathbf{X} - \mathbf{X}^k\|_F^2,$$

where $0 < \alpha < 1$. It is obvious that $J(\mathbf{X}) \leq Q(\mathbf{X}|\mathbf{X}^k)$, and the equality hold only when $\mathbf{X} = \mathbf{X}^k$. Then we set

$$\begin{aligned}
\mathbf{X}^{k+1} &= \underset{\mathbf{X}}{\arg\min}\, Q(\mathbf{X}|\mathbf{X}^k) \\
&= \underset{\mathbf{X}}{\arg\min}\{\frac{1}{2}\|\mathbf{X} - (\mathbf{X}^k + \alpha P_\Omega(\mathbf{M} - \mathbf{X}^k))\|_F^2 + \alpha\lambda\|\mathbf{X}\|_\gamma\} \\
&= S_{\alpha\lambda,\gamma}(\mathbf{X}^k + \alpha P_\Omega(\mathbf{M} - \mathbf{X}^k)).
\end{aligned}$$

The above described iterations constitute inner loop for a fixed λ. In the outer loop, we decrease λ every time and use the previous solution as warm start for next iteration. We call this algorithm as non-convex soft imputation (NCSI), see Algorithm 1.

3.2 The Non-convex Alternating Direction Method of Multipliers

The above NCSI algorithm is designed to iteratively minimize an unconstrained problem (5). We can equivalently reform it as a optimization problem with linear constraints:

$$\min_{\mathbf{X},\mathbf{E}} \|\mathbf{X}\|_\gamma + \frac{\tau}{2}\|P_\Omega(\mathbf{E})\|_F^2 \tag{6}$$

$$\text{s.t.} \quad P_\Omega(\mathbf{M}) = \mathbf{X} + \mathbf{E}.$$

Algorithm 1. The NCSI algorithm

---

input: ν, $P_\Omega(\mathbf{M})$ and tolerance ϵ, $0 < \alpha < 1$, $0 < \rho < 1$, γ
Initialize: $\mathbf{Z}_0 = \mathbf{0}$, λ_0
while $\lambda_i > \nu$ **do**
    $\mathbf{X}^0 = \mathbf{Z}_i$
    repeat
        $\mathbf{X}^{k+1} = S_{\alpha\lambda_i,\gamma}(\mathbf{X}^k + \alpha P_\Omega(\mathbf{M} - \mathbf{X}^k))$
    until $\frac{\|P_\Omega(\mathbf{X}^k - \mathbf{M})\|_F}{\|P_\Omega(\mathbf{M})\|_F} < \epsilon$
    $\mathbf{Z}_{i+1} = \mathbf{X}^k$
    $\lambda_{k+1} = \rho\lambda_k$
end while
output $\mathbf{X}_{sol} = \mathbf{Z}_i$

---

where $\tau = \frac{1}{\lambda}$. We employ alternating direction method of multipliers (ADMM) to solve the optimization problem. ADMM was originally proposed in [18], and has been applied to a number of convex optimization problems [19]. Recently ADMM algorithm have been used in the minimization of a non-convex function [13, 20]. Here we use ADMM to solve the non-convex problem (6). The derived algorithm is called non-convex alternating direction method of multipliers (NCADMM). This algorithm is similar to [8], while it has an advantage of taking noise into consideration. Thus it is expected to have a higher accuracy.

The augmented Lagrangian function of problem (6) is

$$L(\mathbf{X}, \mathbf{E}, \mathbf{Y}, \mu) = \|\mathbf{X}\|_\gamma + \frac{\tau}{2}\|P_\Omega\mathbf{E}\|_F^2 + \langle \mathbf{Y}, P_\Omega(\mathbf{M}) - \mathbf{X} - \mathbf{E} \rangle + \frac{\mu}{2}\|P_\Omega(\mathbf{M}) - \mathbf{X} - \mathbf{E}\|_F^2.$$

The NCADMM optimize w.r.t. one variable while keeping the others fixed. Specifically the optimization problem can be solved efficiently by the following iterations.

$$\begin{aligned}
\mathbf{X}^{k+1} &= \arg\min_{\mathbf{X}} L(\mathbf{X}, \mathbf{E}^k, \mathbf{Y}^k, \mu) \\
&= \arg\min_{\mathbf{X}} \|\mathbf{X}\|_\gamma + \frac{\mu}{2}\|\mathbf{X} - (P_\Omega(\mathbf{M}) - \mathbf{E}^k + \frac{1}{\mu}\mathbf{Y}^k)\|_F^2 \\
&= S_{\frac{1}{\mu},\gamma}(P_\Omega(\mathbf{M}) - \mathbf{E}^k + \frac{1}{\mu}\mathbf{Y}^k),
\end{aligned}$$

$$\begin{aligned}
\mathbf{E}^{k+1} &= \arg\min_{\mathbf{E}} L(\mathbf{X}^{k+1}, \mathbf{E}, \mathbf{Y}^k, \mu) \\
&= \arg\min_{P_\Omega(\mathbf{E})} \frac{\tau + \mu}{2}\|P_\Omega(\mathbf{E})\|_F^2 - \langle P_\Omega(\mathbf{Y}^k + \mu(P_\Omega(\mathbf{M}) - \mathbf{X}^{k+1})), P_\Omega(\mathbf{E}) \rangle \\
&\quad + \arg\min_{P_{\bar{\Omega}}(\mathbf{E})} \frac{\mu}{2}\|P_{\bar{\Omega}}(\mathbf{E})\|_F^2 - \langle P_{\bar{\Omega}}(\mathbf{Y}^k + \mu(P_\Omega(\mathbf{M}) - \mathbf{X}^{k+1})), P_{\bar{\Omega}}(\mathbf{E}) \rangle \\
&= P_\Omega(\mathbf{E}^{k+1}) + P_{\bar{\Omega}}(\mathbf{E}^{k+1}),
\end{aligned}$$

and

$$\mathbf{Y}^{k+1} = \mathbf{Y}^k + \mu(P_\Omega(\mathbf{M}) - \mathbf{X}^{k+1} - \mathbf{E}^{k+1}),$$

where

$$P_\Omega(\mathbf{E}^{k+1}) = \frac{\mu}{\tau + \mu} P_\Omega(\mathbf{M} - \mathbf{X}^{k+1}) + \frac{1}{\mu + \tau} P_\Omega(\mathbf{Y}^k),$$

$$P_{\bar{\Omega}}(\mathbf{E}^{k+1}) = P_{\bar{\Omega}}(-\mathbf{X}^{k+1}) + \frac{1}{\mu} P_{\bar{\Omega}}(\mathbf{Y}^k).$$

Note that if we set $\mathbf{Y}^0 = \mathbf{0}$, then during the iterations $P_{\bar{\Omega}}(\mathbf{Y}^k) = \mathbf{0}$ for all k. Using this property and eliminating the variable E leads to the iteration as follows

$$\mathbf{X}^{k+1} = S_{\frac{1}{\mu}, \gamma} \left(\mathbf{X}^k + \frac{\tau}{\tau + \mu} P_\Omega(\mathbf{M} - \mathbf{X}^k) - \frac{1}{\mu + \tau} P_\Omega(\mathbf{Y}^{k-1}) + \frac{1}{\mu} \mathbf{Y}^k \right), \quad (7)$$

$$\mathbf{Y}^{k+1} = \mathbf{Y}^k + \mu \left(\frac{\tau}{\tau + \mu} P_\Omega(\mathbf{M} - \mathbf{X}^{k+1}) - \frac{1}{\mu + \tau} P_\Omega(\mathbf{Y}^k) \right). \quad (8)$$

In previously described process, the penalty parameter μ is fixed. It is found that a small constant μ may lead to slow convergence, while large μ may make the algorithm ill-conditioned. Thus a dynamic μ is preferred in practice. Inspired by [9] we use the following update rule for μ.

$$\mu_{k+1} = \min(\mu_{max}, \rho\mu_k), \quad (9)$$

where μ_{max} is the upper bound on the penalty parameter μ. The value of ρ is determined by

$$\rho = \begin{cases} \rho_0, & \mu_k \frac{\|\mathbf{X}^{k+1} - \mathbf{X}^k\|}{\|P_\Omega(\mathbf{M})\|_F^2} < \zeta, \\ 1, & \text{otherwise,} \end{cases} \quad (10)$$

where $\rho_0 > 1$ and $\zeta > 0$ is a threshold fixed in advance. We summarize the entire procedure in Algorithm 2.

4   Convergence Analysis of NCSI Algorithm

We need to further explore the γ-norm before proving the convergence property of NCSI algorihtm. First we need a definition called absolutely symmetric function [21].

Definition 3 (Absolutely Symmetric). *Suppose f is a mapping from $\mathbf{R}^m$ to $\mathbf{R}$. We say that f is absolutely symmetric if $f(x_1, x_2, \ldots, x_m) = f(|x_{\pi(1)}|, |x_{\pi(2)}|, \ldots, |x_{\pi(m)}|)$ for any permutation π.*

Lemma 1. *The gamma-norm $\|\mathbf{X}\|_\gamma$ of a $n \times m$ matrix $\mathbf{X}$ can be decomposed as the difference of two convex functions $f(\boldsymbol{\sigma}(\mathbf{X}))$ and $g(\boldsymbol{\sigma}(\mathbf{X}))$ of matrix $\mathbf{X}$, where*

$$f(\boldsymbol{\sigma}(\mathbf{X})) = \sum_{i=1}^{m} \sigma_i(\mathbf{X}), \quad (11)$$

Algorithm 2. The NCADMM algorithm

input: $\tau = 1/\lambda$, $P_\Omega(\mathbf{M})$, tolerance ϵ, threshold ζ, μ_{max}, $\rho_0 > 1$, γ
Initialize: $\mathbf{X}_1^0 = 0, \mathbf{Y}^0 = 0$,
repeat

$$\mathbf{X}^{k+1} = S_{\frac{1}{\mu},\gamma}\left(\mathbf{X}^k + \frac{\tau}{\tau + \mu}P_\Omega(\mathbf{M} - \mathbf{X}^k) - \frac{1}{\mu + \tau}P_\Omega(\mathbf{Y}^{k-1}) + \frac{1}{\mu}\mathbf{Y}^k\right)$$

$$\mathbf{Y}^{k+1} = \mathbf{Y}^k + \mu\left(\frac{\tau}{\tau + \mu}P_\Omega(\mathbf{M} - \mathbf{X}^{k+1}) - \frac{1}{\mu + \tau}P_\Omega(\mathbf{Y}^k)\right)$$

Update μ_{k+1} according to (9) and (10)
until $\frac{\|P_\Omega(\mathbf{X}^{k+1} - \mathbf{M})\|_F}{\|P_\Omega(\mathbf{M})\|_F} < \epsilon$
Output $\mathbf{X}_{sol} = \mathbf{X}^k$

$$g(\boldsymbol{\sigma}(\mathbf{X})) = \sum_{i=1}^m \frac{\sigma_i^2(\mathbf{X})}{2\gamma}\mathbb{I}\{\sigma_i(\mathbf{X}) < \gamma\} + (\sigma_i(\mathbf{X}) - \frac{\gamma}{2})\mathbb{I}\{\sigma_i(\mathbf{X}) \geq \gamma\}. \qquad (12)$$

The above Lemma can be inferred from [22]: If a mapping f is absolutely symmetric and convex on $\mathbb{R}^m$, then $f(\boldsymbol{\sigma}(\mathbf{X}))$ is convex w.r.t. matrix $\mathbf{X}$. In Lemma 1, both $f(\boldsymbol{\sigma})$ and $g(\boldsymbol{\sigma})$ are absolutely symmetric and convex, so $f(\boldsymbol{\sigma}(\mathbf{X}))$ and $g(\boldsymbol{\sigma}(\mathbf{X}))$ are convex functions on matrix $\mathbf{X}$.

Definition 4. *We say that two matrices $\mathbf{X}$ and $\mathbf{Y}$ in $\mathbf{R}^{n \times m}$ have a simultaneous ordered singular value decomposition if there exist two orthonormal matrices $\mathbf{U} \in \mathbf{R}^{n \times m}$ and $\mathbf{V} \in \mathbf{R}^{m \times m}$ such that $\mathbf{X} = \mathbf{U}\mathrm{diag}(\boldsymbol{\sigma}(\mathbf{X}))\mathbf{V}^\top, \mathbf{Y} = \mathbf{U}\mathrm{diag}(\boldsymbol{\sigma}(\mathbf{Y}))\mathbf{V}^\top$.*

Theorem 2. *let a function f be absolutely symmetric and convex. Consider the corresponding convex function $f(\boldsymbol{\sigma}(\mathbf{X}))$. The matrix Y is a subgradient of $f(\boldsymbol{\sigma}(\mathbf{X}))$ at $\mathbf{X}$ if and only if $\boldsymbol{\sigma}(\mathbf{Y})$ is a subgradient of f at $\boldsymbol{\sigma}(X)$ and the two matrices $\mathbf{X}$ and $\mathbf{Y}$ admit simultaneous ordered singular value decomposition.*

Detailed proof of this theorem can be found in [21]. We can compute the subgradient of function $f(\boldsymbol{\sigma}(\mathbf{X}))$ and $g(\boldsymbol{\sigma}(\mathbf{X}))$ w.r.t. $\mathbf{X}$ by applying this theorem directly.

Corollary 1. *(1) Let $f(\boldsymbol{\sigma}(\mathbf{X}))$ and $g(\boldsymbol{\sigma}(\mathbf{X}))$ be defined as Eqn. (11) and (12). Suppose $\mathbf{X} \in \mathbb{R}^{n \times m}$. The matrix $\mathbf{Y}_f$ is a subgradient of $f(\boldsymbol{\sigma}(\mathbf{X}))$ if and only if $\sigma_i(\mathbf{Y}_f) =$*
$$\begin{cases} 1 & \text{if } \sigma_i(\mathbf{X}) > 0 \\ \alpha & \text{if } \sigma_i(\mathbf{X}) = 0 \end{cases} \text{ where } 0 \leq \alpha \leq 1, \text{ and the two matrices admit simultaneous ordered}$$
singular value decomposition.

(2) Suppose $h_i(\boldsymbol{\sigma}(\mathbf{X})) = \frac{\sigma_i(\mathbf{X})}{\gamma}\mathbb{I}\{\sigma_i(\mathbf{X}) < \gamma\} + \mathbb{I}\{\sigma_i(\mathbf{X}) \geq \gamma\}$. $\mathbf{Y}_g$ is a subgradient of $g(\boldsymbol{\sigma}(\mathbf{X}))$ if and only if $\sigma_i(\mathbf{Y}_g) = h_i(\boldsymbol{\sigma}(\mathbf{X}))$ and the two matrices $\mathbf{X}$ and $\mathbf{Y}_g$ admit simultaneous ordered singular value decomposition.

The following theorem shows that our algorithm NSCI decreases the objective function at every iteration.

Theorem 3. *For every fixed* $0 < \alpha < 1$ *and* $\lambda > 0$, *define a sequence* $\mathbf{X}^k$

$$\mathbf{X}^{k+1} = S_{\alpha\lambda,\gamma}(\mathbf{X}^k + \alpha P_\Omega(\mathbf{M} - \mathbf{X}^k)) \tag{13}$$

with a starting point $\mathbf{X}^0$. *The sequence* $\mathbf{X}^k$ *satisfies*

$$J(\mathbf{X}^{k+1}) \leq J(\mathbf{X}^k) - \frac{1-\alpha}{2\alpha}\|\mathbf{X}^{k+1} - \mathbf{X}^k\|_F^2. \tag{14}$$

Proof. Suppose

$$L(\mathbf{X}, \mathbf{Z}) = \frac{1}{2}\|P_\Omega(\mathbf{Z} - \mathbf{M})\|_F^2 + \langle P_\Omega(\mathbf{Z} - \mathbf{M}), \mathbf{X} - \mathbf{Z}\rangle + \lambda\|\mathbf{X}\|_\gamma.$$

Since

$$\mathbf{X}^{k+1} = \underset{\mathbf{X}}{\text{argmin}}\, L(\mathbf{X}, \mathbf{X}^k) + \frac{1}{2\alpha}\|\mathbf{X} - \mathbf{X}^k\|_F^2, \tag{15}$$

then we have

$$\begin{aligned}
J(\mathbf{X}^{k+1}) &\leq L(\mathbf{X}^{k+1}, \mathbf{X}^k) + \frac{1}{2}\|\mathbf{X}^{k+1} - \mathbf{X}^k\|_F^2 \\
&= L(\mathbf{X}^{k+1}, \mathbf{X}^k) + \frac{1}{2\alpha}\|\mathbf{X}^{k+1} - \mathbf{X}^k\|_F^2 - \frac{1-\alpha}{2\alpha}\|\mathbf{X}^{k+1} - \mathbf{X}^k\|_F^2 \\
&\leq L(\mathbf{X}^k, \mathbf{X}^k) + \frac{1}{2\alpha}\|\mathbf{X}^k - \mathbf{X}^k\|_F^2 - \frac{1-\alpha}{2\alpha}\|\mathbf{X}^{k+1} - \mathbf{X}^k\|_F^2 \\
&= J(\mathbf{X}^k) - \frac{1-\alpha}{2\alpha}\|\mathbf{X}^{k+1} - \mathbf{X}^k\|_F^2.
\end{aligned}$$

$\square$

The inequality (14) tells us that $J(\mathbf{X}^k)$ monotonously decrease to its limit point since $J(\mathbf{X}) \geq 0$. Meanwhile the sequence $\{\|\mathbf{X}^{k+1} - \mathbf{X}^k\|_F^2\}$ converges to 0.

The next theorem states that any limit point generated by Algorithm 1 is a critical point of objective function (5).

Theorem 4. *For every fixed* $0 < \alpha < 1$, $\lambda > 0$ *and* $\gamma > \lambda$. *Each limit point of* $\mathbf{X}^k$ *generated by Eqn. (13) is a critical point of* $J(\mathbf{X})$.

Proof. Suppose there is a subsequence $\{\mathbf{X}^k\}_{k\in\mathbb{K}}$ converging to $\mathbf{X}^\infty$. According to the minimization problem (15) and Lemma 1 we have

$$0 \in P_\Omega(\mathbf{X}^k - \mathbf{M}) + \frac{1}{\alpha}(\mathbf{X}^{k+1} - \mathbf{X}^k) + \lambda(\partial f(\sigma(\mathbf{X}^{k+1})) - \partial g(\sigma(\mathbf{X}^{k+1}))).$$

Suppose $\mathbf{S}^{k+1} \in \partial f(\sigma(\mathbf{X}^{k+1}))$, $\mathbf{T}^{k+1} \in \partial g(\sigma(\mathbf{X}^{k+1}))$ satisfying

$$P_\Omega(\mathbf{X}^k - \mathbf{M}) + \frac{1}{\alpha}(\mathbf{X}^{k+1} - \mathbf{X}^k) + \lambda(\mathbf{S}^{k+1} - \mathbf{T}^{k+1}) = 0. \tag{16}$$

Since $\mathbf{S}^{k+1}$ and $\mathbf{T}^{k+1}$ are subgradient of $f(\boldsymbol{\sigma}(\mathbf{X}))$ and $g(\boldsymbol{\sigma}(\mathbf{X}))$ at $\mathbf{X}^{k+1}$, according to Corollary 1, there exist a $n \times m$ orthonormal matrix $\mathbf{U}^{k+1}$ and a $m \times m$ orthogonal matrix $\mathbf{V}^{k+1}$ such that

$$\mathbf{S}^{k+1} = \mathbf{U}^{k+1}\mathrm{diag}\{\boldsymbol{\sigma}(\mathbf{S}^{k+1})\}\mathbf{V}^{(k+1)\top}, \tag{17}$$

$$\mathbf{T}^{k+1} = \mathbf{U}^{k+1}\mathrm{diag}\{\mathbf{h}(\boldsymbol{\sigma}(X^{k+1}))\}\mathbf{V}^{(k+1)\top}, \tag{18}$$

$$\mathbf{X}^{k+1} = \mathbf{U}^{k+1}\mathrm{diag}\{\boldsymbol{\sigma}(X^{k+1})\}\mathbf{V}^{(k+1)\top}. \tag{19}$$

Since orthogonal matrices $\mathbf{U}^{k+1}$, $\mathbf{V}^{k+1}$ and the singular values $\boldsymbol{\sigma}(S^{k+1})$ are bounded, without loss of generality we suppose they converging to $\mathbf{U}^{\infty}$, $\mathbf{V}^{\infty}$ and $\boldsymbol{\sigma}(S^{\infty})$. According to (17), (18) and (19) we infer that $\mathbf{S}^{\infty}$, the limit point of $\mathbf{S}^{k+1}$ is subgradient of $f(\boldsymbol{\sigma}(\mathbf{X}^{\infty}))$ and $\mathbf{T}^{\infty}$, the limit point of $\mathbf{T}^{k+1}$ is subgradient of $g(\boldsymbol{\sigma}(\mathbf{X}^{\infty}))$.

Make $k \to \infty$, $k \in \mathbb{K}$ and use $\lim_{k \to \infty} \mathbf{X}^{k+1} - \mathbf{X}^k = 0$, the Eqn. (16) transfers to

$$P_{\Omega}(\mathbf{X}^{\infty} - \mathbf{M}) + \lambda(\mathbf{S}^{\infty} - \mathbf{T}^{\infty}) = 0.$$

So $\mathbf{X}^{\infty}$ is a critical point of $J(\mathbf{X})$.   □

5 Experiments

In this section, we conduct experiments on synthetic data, image data and three standard collaborative filtering datasets. To show the effectiveness of NCSI and NCADMM, we compare them with the following matrix completion solvers: ALM [8], SVT [4], Soft-Impute [5], and OptSpace [3]. Particularly, ALM, SVT, and SoftImpute are based on the nuclear norm, while OptSpace represent matrix as its factors and optimize a non-convex objective function. Besides, in collaborative filtering experiment, we also add PMF [23] and GECO [24] into our comparison list.

5.1 Synthetic Data

We generate synthetic data $\mathbf{X}$ by $\mathbf{X} = \mathbf{M} + \sigma\mathbf{Z}$, where $\mathbf{X}, \mathbf{M}, \mathbf{Z} = [z_{ij}] \in \mathbb{R}^{m \times n}$. $z_{i,j}$ is Gaussian white noise with zero mean and standard deviation of one. And $\mathbf{M}$ is a matrix with rank of r produced by $\mathbf{M} = \mathbf{L}\mathbf{R}^{\top}$, in which both $\mathbf{L} \in \mathbb{R}^{m \times r}$ and $\mathbf{R} \in \mathbb{R}^{n \times r}$ have i.i.d. Guassian entries. The set of observed entries Ω is uniformly sampled among the $m \times n$ indices. Suppose that the degree of freedom of matrix with rank r is d_r. We fixed the number of observed entries to $5d_r$ and σ to 10^{-6}.

We only compare our methods with ALM since none of the algorithms mentioned before claimed to outperform ALM in terms of accuracy or efficiency on large synthetic matrices. Additionally, We set the parameter γ to 4 in both NCSI and NCADMM. And the same stop criterion is adopted for all algorithms:

$$\frac{\|P_{\Omega}(\mathbf{X} - \mathbf{M})\|_F}{\|P_{\Omega}(\mathbf{M})\|_F} < \epsilon,$$

in which ϵ is set to 0.3σ. We evaluate the accuracy of the solution $\mathbf{X}_{sol}$ of our algorithm by the relative error (RE), which is a widely used metric in matrix completion, defined by

$$RE = \frac{\|\mathbf{X}_{sol} - \mathbf{M}\|_{\mathbf{F}}}{\|\mathbf{M}\|_{\mathbf{F}}}.$$

We report the RE and #SVD (number of doing SVD) in Table 1. Experimental results demonstrate that NCSI and NCADMM consistently outperform ALM in accuracy; NCADMM achieve higher accuracy with nearly the same time cost as ALM; and NCSI and NCADMM have almost the same accuracy (since they solve matrix completion problem using same γ-norm based scheme).

Table 1. Comparisons among NCSI, NCADMM, ALM on the synthetic data

(rank ratio)	$RE(10^{-7})$			#SVD			CPU-time(minutes)		
$(r, \frac{|\Omega|}{(m \times n)})$	NCSI	NCADMM	ALM	NCSI	NCADMM	ALM	NCSI	NCADMM	ALM
$m = 10000, n = 10000$									
(10,0.012)	5.412	5.412	6.230	1818	322	325	10.45	4.36	4.7
(20,0.024)	3.853	3.790	4.040	850	179	180	13.78	5.73	6.18
(30,0.035)	3.026	3.020	4.106	522	149	150	17.73	8.58	8.68
(50,0.057)	2.794	2.794	3.864	313	122	115	26.58	11.32	14.78
$m = 20000, n = 20000$									
(10,0.006)	5.976	5.963	6.410	3755	800	683	48.41	37.02	32.28
(20,0.012)	4.428	4.416	5.246	1606	283	246	76.85	30.21	28.95
(30,0.018)	3.797	3.801	3.952	1034	199	201	76.85	28.24	28.61
(50,0.030)	2.839	2.839	3.957	613	173	164	114.3	60.13	57.41

5.2 Experiment on Image Data

In the image inpainting experiment, we aim to estimate missing (or masked) pixels by exploiting the known content. As colored image is commonly represented as three matrices(containing red, green and blue components respectively) we simply deal with each of three matrix and combine them together to obtain the final results.

Performance of different algorithms are evaluated by the $PSNR$ (Peak Signal-to-Noise Ratio) metric. Suppose that the total number of missing pixel is T and the total squared error TSE is defined by $TSE = error_r^2 + error_g^2 + error_b^2$, then the total mean squared error MSE is defined by $MSE = TSE/3T$. And the $PSNR$ can be evaluated as $PSNR = 10 \log_{10} 255^2 / MSE$.

In our experiments, the parameters of ALM, SVT, Soft-Impute, and OptSpace are carefully chosen to achieve the best performance. For NCSI and NCADMM we fix $\gamma = 100$ and empirically set $\lambda = 0.001$. Since large μ in NCADMM will make the minimization problem ill-conditioned, we set $\mu_{max} = 10^{10}$.

Two experiments using different image masks are reported. The first is a relatively easy matrix completion problem with random mask. We report the results in Fig. 2 and Fig. 3. We see that the γ-Norm minimization scheme always achieve larger $PSNR$ compared with other five methods from Fig. 2. Second experiment uses text mask. It is generally agreed that image inpainting with text mask is more difficult since the observed pixels are not randomly sampled and text mask may result in loss of important image information. We report our results in Fig. 4. The results of NCSI and NCADMM are also encouraging.

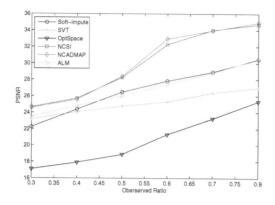

Fig. 2. Comparison of matrix completion algorithms for recovery of a image under different observed ratios

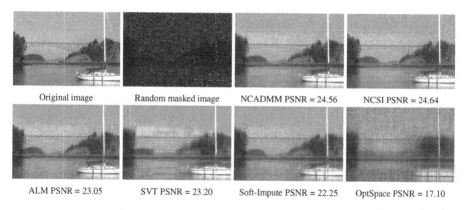

Fig. 3. Comparison of matrix completion algorithms for recovery of a image only 30% of its pixels are observed

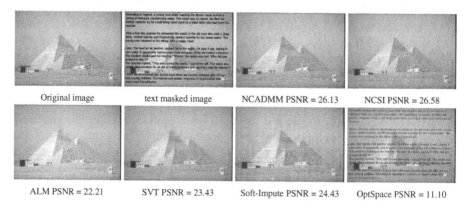

Fig. 4. Comparison of matrix completion algorithms for recovery of a image masked by text

5.3 Collaborative Filtering

Collaborative filtering (CF) is a technique used by some recommender systems. One of the CF's main purposes is to predict the unknown preference of a user on a set of unrated items, according to other similar users or similar items. In order to validate the performance of our methods, we compare our NCSI and NCADMM with three algorithms using nuclear norm: ALM, SVT and Soft-Impute, and three other non-nuclear-norm algorithms: OptSpace, GECO and PMF.

We use three standard MovieLens Data Sets[1]:

MovieLens-100K contains 100,000 ratings for 1682 movies by 943 users;
MovieLens-1M contains 1million ratings for 3,900 movies by 6,040 users;
MovieLens-10M contains 10 million ratings for 10,681 movies by 71,567 users.

For each data set, we randomly select 70% ratings as known samples, and use the rest ratings to test the performance of the methods. Then, we run 5 repeats for each data set and each method, and report the average results in table 2.

In our experiment, We fix $\gamma = \|P_\Omega(\mathbf{M})\|_F$ and use the commonly accepted CF metric $RMSE$ (Root Mean Square Error) to evaluate the eight methods. $RMSE$ is defined by

$$RMSE = \frac{1}{|T|} \sqrt{\sum_{(i,j) \in T}^{|T|} (X_{ij} - M_{ij})^2},$$

where T is the test set.

Our results in Table 2 show that γ-norm based algorithms outperform other matrix completion algorithms and are competitive to the state-of-the-art collaborative filtering method PMF.

Table 2. Performance of deference matrix completion methods on real collaborative filtering data sets

Data set	NCSI	NCADMM	ALM	SVT	Soft-Impute	OptSpace	GECO	PMF
MovieLens-100k	0.9710	0.9710	1.083	1.536	1.071	1.583	0.9810	0.9790
MovieLens-1M	0.8670	0.8670	0.9037	0.9498	0.9185	1.007	0.8808	0.8683
MovieLens-10M	0.8250	0.8250	0.8843	0.9731	0.8854	too long	0.8402	0.8247

6 Conclusion

In this paper we have employed the matrix γ-norm as a non-convex relaxation to the matrix rank and devised two algorithms: non-convex soft imputation (NCSI) and non-convex alternative direction method of multipliers algorithm (NCADMM), to solve the

---

[1] http://www.grouplens.org

matrix completion problem. The algorithms are effective, because they can achieve high accuracy in the simulated datasets and real world datasets. Moreover, the NCADMM is quite efficient as its running CPU-time is comparable with the current state-of-the-art algorithms.

Acknowledgments. This work is partially supported by the National Natural Science Foundation of china (No. 61070239 and No. 61272303) and the National Program on Key Basic Research Project of China (973 Program, No. 2010CB327903).

References

1. Candès, E.J., Recht, B.: Exact matrix completion via convex optimization. Foundations of Computational Mathematics 9(6), 717–772 (2009)
2. Candès, E.J., Tao, T.: The power of convex relaxation: Near-optimal matrix completion. IEEE Transactions on Information Theory 56(5), 2053–2080 (2010)
3. Keshavan, R., Montanari, A., Sewoong, O.: Matrix Completion from a Few Entries. IEEE Transactions on Information Theory 56(6), 2980–2998 (2010)
4. Cai, J., Candès, E.J., Shen, Z.: A singular value thresholding algorithm for matrix completion. SIAM Journal on Optimization 20(4), 1956–1982 (2010)
5. Mazumder, R., Hastie, T., Tibshirani, R.: Spectral Regularization Algorithms for Learning Large Incomplete Matrices. The Journal of Machine Learning Research 99, 2287–2322 (2010)
6. Ma, S., Goldfarb, D., Chen, L.: Fixed point and Bregman iterative methods for matrix rank minimization. Mathematical Programming 128(1-2), 321–353 (2011)
7. Hale, E.T., Yin, W., Zhang, Y.: A fixed-point continuation method for ℓ_1-regularized minimization with applications to compressed sensing. echnical Report, CAAM TR07-07 (2007)
8. Lin, Z., Chen, M., Wu, L., Ma, Y.: The augmented Lagrange multiplier method for Exact Recovery of Corrupted Low-Rank Matrices. arXiv preprint arXiv:1009.5055 (2010)
9. Lin, Z., Liu, R., Su, Z.: Linearized alternating direction method with adaptive penalty for low-rank representation. In: Advances in Neural Information Processing Systems (2011)
10. Hu, Y., Zhang, D., Liu, J., Ye, J., He, X.: Accelerated singular value thresholding for matrix completion. In: Proceedings of the 18th ACM SIGKDD International Conference on Knowledge Discovery and Data Mining (2012)
11. Recht, B., Fazel, M., Parrilo, P.A.: Guaranteed minimum-rank solutions of linear matrix equations via nuclear norm minimization. SIAM Review 52(3), 471–501 (2010)
12. Cai, T.T., Zhou, W.: Matrix completion via max-norm constrained optimization. arXiv preprint arXiv:1303.0341 (2013)
13. Nie, F., Wang, H., Cai, X., Huang, H., Ding, C.: Robust Matrix Completion via Joint Schatten p-Norm and ℓ_p-Norm Minimization. In: 2012 IEEE 12th International Conference on Data Mining (ICDM), pp. 566–574 (2012)
14. Hu, Y., Zhang, D., Ye, J., Li, X., He, X.: Fast and Accurate Matrix Completion via Truncated Nuclear Norm Regularization. EEE Transactions on Pattern Analysis and Machine Intelligence (2012)
15. Wang, S., Liu, D., Zhang, Z.: Nonconvex Relaxation Approaches to Robust Matrix Recovery. In: International Joint Conference on Artificial Intelligence (2013)
16. Mazumder, R., Friedman, J.H., Hastie, T.: SparseNet: Coordinate descent with nonconvex penalties. Journal of the American Statistical Association 106(495) (2011)

17. Zhang, C.H.: Nearly Unbiased Variable Selection Under Minimax Concave Penalty. The Annals of Statistics 38(2), 894–942 (2010)
18. Gabay, D., Bertrand, M.: A dual algorithm for the solution of nonlinear variational problems via finite element approximation. Computers & Mathematics with Applications 2(1), 1–40 (1976)
19. Boyd, S., Parikh, N., Chu, E., Peleato, B., Eckstein, J.: Distributed optimization and statistical learning via the alternating direction method of multipliers. Foundations and Trends in Machine Learning 3(1), 1–122 (2011)
20. Xiang, S., Shen, X., Ye, J.: Efficient Sparse Group Feature Selection via Nonconvex Optimization. arXiv preprint arXiv:1205.5075 (2012)
21. Lewis, A.S., Hristo, S.S.: Nonsmooth analysis of singular values. Part I: Theory. Set-Valued Analysis 13(3), 213–241 (2005)
22. Lewis, A.S.: Group invariance and convex matrix analysis. SIAM Journal on Matrix Analysis and Applications 17(4), 927–949 (1996)
23. Salakhutdinov, R., Mnih, A.: Probabilistic matrix factorization. In: Advances in Neural Information Processing Systems, pp. 1257–1264 (2008)
24. Shalev-Shwartz, S., Gonen, A., Shamir, O.: Large-scale convex minimization with a low-rank constraint. In: International Conference on Machine Learning (2011)

A Counterexample for the Validity of Using Nuclear Norm as a Convex Surrogate of Rank

Hongyang Zhang, Zhouchen Lin*, and Chao Zhang

Key Lab. of Machine Perception (MOE), School of EECS
Peking University, Beijing, China
{hy_zh,zlin}@pku.edu.cn, chzhang@cis.pku.edu.cn

Abstract. Rank minimization has attracted a lot of attention due to its robustness in data recovery. To overcome the computational difficulty, rank is often replaced with nuclear norm. For several rank minimization problems, such a replacement has been theoretically proven to be valid, i.e., the solution to nuclear norm minimization problem is also the solution to rank minimization problem. Although it is easy to *believe* that such a replacement may not always be valid, no concrete example has ever been found. We argue that such a validity checking cannot be done by numerical computation and show, by analyzing the noiseless latent low rank representation (LatLRR) model, that even for very simple rank minimization problems the validity may still break down. As a by-product, we find that the solution to the nuclear norm minimization formulation of LatLRR is *non-unique*. Hence the results of LatLRR reported in the literature may be questionable.

1 Introduction

We are now in an era of big data as well as high dimensional data. Fortunately, high dimensional data are not unstructured. Usually, they lie near low dimensional manifolds. This is the basis of linear and nonlinear dimensionality reduction [1]. As a simple yet effective approximation, linear subspaces are usually adopted to model the data distribution. Because low dimensional subspaces correspond to low rank data matrices, rank minimization problem, which models the real problem into an optimization by minimizing the rank in the objective function (cf. models (1), (3) and (4)), is now widely used in machine learning and data recovery [2–5]. Actually, rank is regarded as a sparsity measure for matrices [3]. So low rank recovery problems are studied [6–9] in parallel with the compressed sensing theories for sparse vector recovery. Typical rank minimization problems include matrix completion [2, 4], which aims at completing the entire matrix from a small sample of its entries, robust principal component analysis [3], which recovers the ground truth data from sparsely corrupted elements, and low rank representation [10, 11], which finds an affinity matrix of subspaces that has the lowest rank. All of these techniques have found wide applications,

---

* Corresponding author.

H. Blockeel et al. (Eds.): ECML PKDD 2013, Part II, LNAI 8189, pp. 226–241, 2013.
© Springer-Verlag Berlin Heidelberg 2013

such as background modeling [3], image repairing [12], image alignment [12], image rectification [13], motion segmentation [10, 11], image segmentation [14], and saliency detection [15].

Since the rank of a matrix is discrete, rank minimization problems are usually hard to solve. They can even be NP hard [3]. To overcome the computational obstacle, as a common practice people usually replace rank in the objective function with nuclear norm, which is the sum of singular values and is the convex envelope of rank on the unit ball of matrix operator norm [5], to transform rank minimization problems into nuclear norm minimization problems (cf. models (2) and (5)). Such a strategy is widely adopted in most rank minimization problems [2–4, 10–15]. However, this naturally brings a *replacement validity problem* which is defined as follows.

Definition 1 (Replacement Validity Problem). *Given a rank minimization problem together with its corresponding nuclear norm formulation, the replacement validity problem investigates whether the solution to the nuclear norm minimization problem is also a solution to the rank minimization one.*

In this paper, we focus on the replacement validity problem. There is a related problem, called *exact recovery problem*, that is more widely studied by scholars. It is defined as follows.

Definition 2 (Exact Recovery Problem). *Given a nuclear norm minimization problem, the exact recovery problem investigates the sufficient conditions under which the nuclear norm minimization problem could exactly recover the real structure of the data.*

As an example of the exact recovery problem, Candès et al. proved that when the rank of optimal solution is sufficiently low and the missing data is sufficiently few or the corruption is sufficiently sparse, solving nuclear norm minimization problems of matrix completion [2] or robust PCA problems [3] can exactly recover the ground truth low rank solution with an overwhelming probability. As another example, Liu et al. [10, 16] proved that when the rank of optimal solution is sufficiently low and the percentage of corruption does not exceed a threshold, solving the nuclear norm minimization problem of low rank representation (LRR) [10, 11] can exactly recover the ground truth subspaces of the data.

We want to highlight the difference between our replacement validity problem and the exact recovery problem that scholars have considered before. The replacement validity problem is to compare the solutions between two optimization problems, while the exact recovery problem is to study whether solving a nuclear norm minimization problem can exactly recover a ground truth low rank matrix. As a result, in all the existing exact recovery problems, the scholars have to assume that the rank of the ground truth solution is sufficiently low. In contrast, the replacement validity problem does not rely on this assumption: even if the ground truth low rank solution cannot be recovered, we can still investigate whether the solution to a nuclear norm minimization problem is also the solution to the corresponding rank minimization problem.

For replacement validity problems, it is easy to *believe* that the replacement of rank with nuclear norm will break down for complex rank minimization problems. While for exact recovery problems, the existing analysis all focuses on relatively simple rank minimization problems, such as matrix completion [2], robust PCA problems [3], and LRR [10, 11], and has achieved affirmative results under some conditions. So it is also easy to *believe* that for simple rank minimization problems the replacement of rank with nuclear norm will work. This paper aims at breaking such an illusion. Here, we have to point out that replacement validity problem *cannot* be studied by numerical experiments. This is because: 1. rank is sensitive to numerical errors. Without prior knowledge, one may not correctly determine the rank of a given matrix, even if there is a clear drop in its singular values; 2. it is hard to verify whether a given solution to nuclear norm minimization problem is a *global* minimizer to a rank minimization problem, whose objective function is discrete and non-convex. So we should study replacement validity problem by purely theoretical analysis. We analyze a simple rank minimization problem, noiseless latent LRR (LatLRR) [17], to show that solutions to a nuclear norm minimization problem may not be solutions of the corresponding rank minimization problem.

The contributions of this paper include:

1. We use a simple rank minimization problem, noiseless LatLRR, to prove that solutions to a nuclear norm minimization problem may not be solutions of the corresponding rank minimization problem, even for very simple rank minimization problems.
2. As a by-product, we find that LatLRR is not a good mathematical model because the solution to its nuclear norm minimization formulation is *non-unique*. So the results of LatLRR reported in the literature, e.g., [10, 17], may be questionable.

2   Latent Low Rank Representation

In this section, we first explain the notations that will be used in this paper and then introduce latent low rank representation which we will analyze its closed form solutions.

2.1   Summary of Main Notations

A large amount of matrix related symbols will be used in this paper. Capital letters are used to represent matrices. Especially, I denotes the identity matrix and 0 is the all-zero matrix. The entry at the ith row and the jth column of a matrix is denoted by $[\cdot]_{ij}$. Nuclear norm, the sum of all the singular values of a matrix, is denoted by $||\cdot||_*$. Operator norm, the maximum singular value, is denoted by $||\cdot||_2$. Trace(A) represents the sum of the diagonal entries of A and $A^\dagger$ is the Moore-Penrose pseudo-inverse of A. For simplicity, we use the same letter to present the subspace spanned by the columns of a matrix. The dimension of a space V is presented by $\dim(V)$. The orthogonal complement of V is denoted by $V_\perp$. Range(A) indicates the linear space spanned by all the columns of matrix A, while Null(A) represents the null space of A. They are closely related: $(\text{Range}(A))_\perp = \text{Null}(A^T)$. Finally, we always use $U_X \Sigma_X V_X^T$ to represent the *skinny* SVD of the data matrix X. Namely, the numbers of columns in U_X and V_X are both rank(X) and Σ_X consists of all the non-zero singular values of X, making Σ_X invertible.

2.2 Low Rank Subspace Clustering Models

Low rankness based subspace clustering stems from low rank representation (LRR) [10, 11]. An interested reader may refer to an excellent review on subspace clustering approaches provided by Vidal [18]. The mathematical model of the original LRR is

$$\min_Z \text{rank}(Z), \quad \text{s.t.} \quad X = XZ, \tag{1}$$

where X is the data matrix we observe. LRR extends sparse subspace clustering [19] by generalizing the sparsity from 1D to 2D. When there is noise or corruption, a noise term can be added to the model [10, 11]. Since this paper considers closed form solutions for noiseless models, to save space we omit the noisy model. The corresponding nuclear norm minimization formulation of (1) is

$$\min_Z \|Z\|_*, \quad \text{s.t.} \quad X = XZ, \tag{2}$$

which we call the heuristic LRR. LRR has been very successful in clustering data into subspaces robustly [20]. It is proven that when the underlying subspaces are independent, the optimal representation matrix is block diagonal, each block corresponding to a subspace [10, 11].

LRR works well only when the samples are sufficient. This condition may not be fulfilled in practice, particularly when the dimension of samples is large. To resolve this issue, Liu et al. [17] proposed latent low rank representation (LatLRR). Another model to overcome this drawback of LRR is fixed rank representation [21]. LatLRR assumes that the observed samples can be expressed as the linear combinations of themselves together with the unobserved data:

$$\min_Z \text{rank}(Z), \quad \text{s.t.} \quad X = [X, X_H]Z, \tag{3}$$

where X_H is the unobserved samples for supplementing the shortage of the observed ones. Since X_H is unobserved and problem (3) cannot be solved directly, by some deduction and mathematical approximation, LatLRR [17] is modeled as follows:

$$\min_{Z,L} \text{rank}(Z) + \text{rank}(L), \quad \text{s.t.} \quad X = XZ + LX. \tag{4}$$

Both the optimal Z and L can be utilized for learning tasks: Z can be used for subspace clustering, while L is for feature extraction, thus providing us with the possibility for integrating two tasks into a unified framework. We call (4) the original LatLRR. Similarly, it has a nuclear norm minimization formulation

$$\min_{Z,L} ||Z||_* + ||L||_*, \quad \text{s.t. } X = XZ + LX, \tag{5}$$

which we call the heuristic LatLRR. LatLRR has been reported to have better performance than LRR [10, 17].

In this paper, we focus on studying the solutions to problems (1), (2), (4) and (5), in order to investigate the replacement validity problem.

3   Analysis on LatLRR

This section provides surprising results: both the original and heuristic LatLRR have closed form solutions! We are able to write down *all* their solutions, as presented in the following theorems.

Theorem 1. *The complete solutions to the original LatLRR problem* (4) *are as follows*

$$Z^* = V_X \tilde{W} V_X^T + S_1 \tilde{W} V_X^T \text{ and } L^* = U_X \Sigma_X (I - \tilde{W}) \Sigma_X^{-1} U_X^T + U_X \Sigma_X (I - \tilde{W}) S_2, \tag{6}$$

where $\tilde{W}$ is any idempotent matrix and S_1 and S_2 are any matrices satisfying:
1. $V_X^T S_1 = 0$ and $S_2 U_X = 0$; and 2. $\text{rank}(S_1) \leq \text{rank}(\tilde{W})$ and $\text{rank}(S_2) \leq \text{rank}(I - \tilde{W})$.

Theorem 2. *The complete solutions to the heuristic LatLRR problem* (5) *are as follows*

$$Z^* = V_X \widehat{W} V_X^T \text{ and } L^* = U_X (I - \widehat{W}) U_X^T, \tag{7}$$

where $\widehat{W}$ is any block diagonal matrix satisfying: 1. its blocks are compatible with Σ_X, i.e., if $[\Sigma_X]_{ii} \neq [\Sigma_X]_{jj}$ then $[\widehat{W}]_{ij} = 0$; and 2. both $\widehat{W}$ and $I - \widehat{W}$ are positive semi-definite.

By Theorems 1 and 2, we can conclude that if the $\widehat{W}$ in Theorem 2 is not idempotent, then the corresponding (Z^*, L^*) is not the solution to the original LatLRR, due to the following proposition:

Proposition 1. *If the $\widehat{W}$ in Theorem 2 is not idempotent, then $Z^* = V_X \widehat{W} V_X^T$ cannot be written as $Z^* = V_X \tilde{W} V_X^T + S_1 \tilde{W} V_X^T$, where $\tilde{W}$ and S_1 satisfy the conditions stated in Theorem 1.*

The above results show that for noiseless LatLRR, nuclear norm is not a valid replacement of rank. As a by-product, since the solution to the heuristic LatLRR is *non-unique*, the results of LatLRR reported in [11, 17] may be questionable.

We provide detailed proofs of the above theorems and proposition in the following section.

4 Proofs

4.1 Proof of Theorem 1

We first provide the complete closed form solutions to the original LRR in a more general form

$$\min_Z \operatorname{rank}(Z), \quad \text{s.t. } A = XZ, \tag{8}$$

where $A \in \operatorname{Range}(X)$ so that the constraint is feasible. We call (8) the generalized original LRR. Then we have the following proposition.

Proposition 2. *Suppose $U_A \Sigma_A V_A^T$ is the skinny SVD of A. Then the minimum objective function value of the generalized original LRR problem (8) is $\operatorname{rank}(A)$ and the complete solutions to (8) are as follows*

$$Z^* = X^\dagger A + S V_A^T, \tag{9}$$

where S is any matrix such that $V_X^T S = 0$.

Proof. Suppose Z^* is an optimal solution to problem (8). First, we have

$$\operatorname{rank}(A) = \operatorname{rank}(XZ^*) \le \operatorname{rank}(Z^*). \tag{10}$$

On the other hand, because $A = XZ$ is feasible, there exists Z_1 such that $A = XZ_1$. Then $Z_0 = X^\dagger A$ is feasible: $XZ_0 = XX^\dagger A = XX^\dagger XZ_1 = XZ_1 = A$, where we have utilized a property of Moore-Penrose pseudo-inverse $XX^\dagger X = X$. So we obtain

$$\operatorname{rank}(Z^*) \le \operatorname{rank}(Z_0) \le \operatorname{rank}(A). \tag{11}$$

Combining (10) with (11), we conclude that $\operatorname{rank}(A)$ is the minimum objective function value of problem (8).

Next, let $Z^* = PQ^T$ be the full rank decomposition of the optimal Z^*, where both P and Q have $\operatorname{rank}(A)$ columns. From $U_A \Sigma_A V_A^T = XPQ^T$, we have $V_A^T = (\Sigma_A^{-1} U_A^T XP)Q^T$. Since both V_A and Q are full column rank and $Y = \Sigma_A^{-1} U_A^T XP$ is square, Y must be invertible. So V_A and Q represent the same subspace. Because P and Q are unique up to an invertible matrix, we may simply choose $Q = V_A$. Thus $U_A \Sigma_A V_A^T = XPQ^T$ reduces to $U_A \Sigma_A = U_X \Sigma_X V_X^T P$, i.e., $V_X^T P = \Sigma_X^{-1} U_X^T U_A \Sigma_A$, and we conclude that the complete choices of P are given by $P = V_X \Sigma_X^{-1} U_X^T U_A \Sigma_A + S$, where S is any matrix such that $V_X^T S = 0$. Multiplying P with $Q^T = V_A^T$, we obtain that the entire solutions to problem (8) can be written as $Z^* = X^\dagger A + S V_A^T$, where S is any matrix satisfying $V_X^T S = 0$. $\square$

Remark 1. Friedland and Torokhti [22] studied a similar model as (8), which is

$$\min_Z \|X - AZ\|_F, \quad \text{s.t. } \operatorname{rank}(Z) \le k. \tag{12}$$

However, (8) is different from (12) in two aspects. First, (8) requires the data matrix X to be strictly expressed as linear combinations of the columns in A. Second, (8) does not impose an upper bound for the rank of Z. Rather, (8) solves for the Z with the lowest rank. As a result, (8) has infinitely many solutions, as shown by Proposition 2, while (12) has a unique solution when k fulfills some conditions. So the results in [22] do not apply to (8).

Similar to Proposition 2, we can have the complete closed form solution to the following problem

$$\min_{Z} \operatorname{rank}(L), \quad \text{s.t.} \quad A = LX, \tag{13}$$

which will be used in the proof of Theorem 1.

Proposition 3. *Suppose $U_A \Sigma_A V_A^T$ is the skinny SVD of A. Then the minimum objective function value of problem* (13) *is $\operatorname{rank}(A)$ and the complete solutions to problem* (13) *are as follows*

$$L^* = AX^\dagger + U_A S, \tag{14}$$

where S is any matrix such that $SU_X = 0$.

Next, we provide the following propositions.

Proposition 4. *$\operatorname{rank}(X)$ is the minimum objective function value of the original LatLRR problem* (4).

Proof. Suppose (Z^*, L^*) is an optimal solution to problem (4). By Proposition 2 and fixing Z^*, we have $\operatorname{rank}(L^*) = \operatorname{rank}(X - XZ^*)$. Thus

$$\operatorname{rank}(Z^*) + \operatorname{rank}(L^*) \geq \operatorname{rank}(XZ^*) + \operatorname{rank}(X - XZ^*) \geq \operatorname{rank}(X). \tag{15}$$

On the other hand, if Z^* and L^* are adopted as $X^\dagger X$ and 0, respectively, the lower bound is achieved and the constraint is fulfilled as well. So we conclude that $\operatorname{rank}(X)$ is the minimum objective function value of the original LatLRR problem (4). □

Proposition 5. *Suppose (Z^*, L^*) is one of the solutions to problem* (4). *Then there must exist another solution $(\widetilde{Z}, \widetilde{L})$, such that $XZ^* = X\widetilde{Z}$ and $\widetilde{Z} = V_X \widetilde{W} V_X^T$ for some matrix $\widetilde{W}$.* .

Proof. According to the constraint of problem (4), we have $XZ = (I - L)X$, i.e., $(XZ)^T \in \operatorname{Range}(X^T)$. Since $V_X V_X^T$ is the projection matrix onto $\operatorname{Range}(X^T)$, we have

$$XZ^* V_X V_X^T = XZ^*. \tag{16}$$

On the other hand, given the optimal Z^*, L^* is the optimal solution to

$$\min_{L} \operatorname{rank}(L) \quad \text{s.t.} \quad X(I - Z^*) = LX. \tag{17}$$

So by Proposition 2 we get

$$\operatorname{rank}(L^*) = \operatorname{rank}(X(I - Z^*)X^\dagger). \tag{18}$$

As a result,

$$\begin{aligned}
\operatorname{rank}(X) &= \operatorname{rank}(Z^*) + \operatorname{rank}(L^*) \\
&= \operatorname{rank}(Z^*) + \operatorname{rank}(X(I - Z^*)X^\dagger) \\
&= \operatorname{rank}(Z^*) + \operatorname{rank}(X(I - V_X V_X^T Z^* V_X V_X^T)X^\dagger) \\
&\geq \operatorname{rank}(V_X V_X^T Z^* V_X V_X^T) + \operatorname{rank}(X(I - V_X V_X^T Z^* V_X V_X^T)X^\dagger) \\
&\geq \operatorname{rank}(X),
\end{aligned} \tag{19}$$

where the last inequality holds since $(V_X V_X^T Z^* V_X V_X^T, X(I - V_X V_X^T Z^* V_X V_X^T) X^\dagger)$ is a feasible solution to problem (4) and $\text{rank}(X)$ is the minimum objective according to Proposition 4. (19) shows that $(V_X V_X^T Z^* V_X V_X^T, X(I - V_X V_X^T Z^* V_X V_X^T) X^\dagger)$ is an optimal solution. So we may take $\tilde{Z} = V_X V_X^T Z^* V_X V_X^T$ and write it as $\tilde{Z} = V_X \tilde{W} V_X^T$, where $\tilde{W} = V_X^T Z^* V_X$.

Finally, combining with equation (16), we conclude that

$$X\tilde{Z} = U_X \Sigma_X V_X^T V_X V_X^T Z^* V_X V_X^T = XZ^* V_X V_X^T = XZ^*. \tag{20}$$

$\square$

Proposition 5 provides us with a great insight into the structure of problem (4): we may break (4) into two subproblems

$$\min_Z \text{rank}(Z), \quad \text{s.t.} \quad X V_X \tilde{W} V_X^T = XZ, \tag{21}$$

and

$$\min_L \text{rank}(L), \quad \text{s.t.} \quad X - X V_X \tilde{W} V_X^T = LX, \tag{22}$$

and then apply Propositions 2 and 3 to find the complete solutions to problem (4).

For investigating the properties of $\tilde{W}$ in (21) and (22), the following lemma is critical.

Lemma 1. *For $A, B \in \mathbb{R}^{n \times n}$, if $AB = BA$, then the following inequality holds*

$$\text{rank}(A + B) \leq \text{rank}(A) + \text{rank}(B) - \text{rank}(AB). \tag{23}$$

Proof. On the basis of $AB = BA$, it is easy to check that

$$\text{Null}(A) + \text{Null}(B) \subset \text{Null}(AB), \tag{24}$$

and

$$\text{Null}(A) \cap \text{Null}(B) \subset \text{Null}(A + B). \tag{25}$$

On the other hand, according to the well-known dimension formula

$$\dim(\text{Null}(A)) + \dim(\text{Null}(B)) = \dim(\text{Null}(A) + \text{Null}(B)) + \dim(\text{Null}(A) \cap \text{Null}(B)), \tag{26}$$

by combining (26) with (24) and (25), we get

$$\dim(\text{Null}(A)) + \dim(\text{Null}(B)) \leq \dim(\text{Null}(AB)) + \dim(\text{Null}(A + B)). \tag{27}$$

Then by the relationship $\text{rank}(S) = n - \dim(\text{Null}(S))$ for any $S \in \mathbb{R}^{n \times n}$, we arrive at the inequality (23). $\square$

Based on the above lemma, the following proposition presents the sufficient and necessary condition on $\tilde{W}$.

Proposition 6. *Let L^* be any optimal solution to subproblem (22), then $(V_X \tilde{W} V_X^T, L^*)$ is optimal to problem (4) if and only if the square matrix $\tilde{W}$ is idempotent.*

Proof. Obviously, $(V_X \tilde{W} V_X^T, L^*)$ is feasible based on the constraint in problem (22). By considering the optimality of L^* for (22) and replacing Z^* with $V_X \tilde{W} V_X^T$ in equation (18), we have

$$\text{rank}(L^*) = \text{rank}(X(I - V_X \tilde{W} V_X^T)X^\dagger). \tag{28}$$

First, we prove the sufficiency. According to the property of idempotent matrices, we have

$$\text{rank}(\tilde{W}) = \text{trace}(\tilde{W}) \text{ and } \text{rank}(I - \tilde{W}) = \text{trace}(I - \tilde{W}). \tag{29}$$

By substituting $(V_X \tilde{W} V_X^T, L^*)$ into the objective function, the following equalities hold

$$\begin{aligned}
\text{rank}(V_X \tilde{W} V_X^T) + \text{rank}(L^*) &= \text{rank}(\tilde{W}) + \text{rank}(X(I - V_X \tilde{W} V_X^T)X^\dagger) \\
&= \text{rank}(\tilde{W}) + \text{rank}(U_X \Sigma_X (I - \tilde{W}) \Sigma_X^{-1} U_X^T) \\
&= \text{rank}(\tilde{W}) + \text{rank}(I - \tilde{W}) \\
&= \text{trace}(\tilde{W}) + \text{trace}(I - \tilde{W}) \\
&= \text{rank}(X).
\end{aligned} \tag{30}$$

So $(V_X \tilde{W} V_X^T, L^*)$ is optimal since it achieves the minimum objective function value of problem (4).

Second, we prove the necessity. Suppose $(V_X \tilde{W} V_X^T, L^*)$ is optimal to problem (4). Substituting it into the objective follows

$$\begin{aligned}
\text{rank}(X) &= \text{rank}(V_X \tilde{W} V_X^T) + \text{rank}(X(I - V_X \tilde{W} V_X)X^\dagger) \\
&= \text{rank}(\tilde{W}) + \text{rank}(I - \tilde{W}) \\
&\geq \text{rank}(X).
\end{aligned} \tag{31}$$

Hence $\text{rank}(\tilde{W}) + \text{rank}(I - \tilde{W}) = \text{rank}(X)$. On the other hand, as $\tilde{W}$ and $I - \tilde{W}$ are commutative, by Lemma 1 we have $\text{rank}(X) \leq \text{rank}(\tilde{W}) + \text{rank}(I - \tilde{W}) - \text{rank}(\tilde{W} - \tilde{W}^2)$. So $\text{rank}(\tilde{W} - \tilde{W}^2) = 0$ and thus $\tilde{W} = \tilde{W}^2$. $\square$

We are now ready to prove Theorem 1.

Proof. Solving problems (21) and (22) by using Propositions 2 and 3, where $\tilde{W}$ is idempotent as Proposition 6 shows, we directly get

$$Z^* = V_X \tilde{W} V_X^T + \tilde{S}_1 V_A^T \text{ and } L^* = U_X \Sigma_X (I - \tilde{W}) \Sigma_X^{-1} U_X^T + U_B \tilde{S}_2, \tag{32}$$

where $U_A \Sigma_A V_A^T$ and $U_B \Sigma_B V_B^T$ are the skinny SVDs of $U_X \Sigma_X \tilde{W} V_X^T$ and $U_X \Sigma_X (I - \tilde{W}) V_X^T$, respectively, and $\tilde{S}_1$ and $\tilde{S}_2$ are matrices such that $V_X^T \tilde{S}_1 = 0$

and $\widetilde{S}_2 U_X = 0$. Since we have $\text{Range}((\tilde{W}V_X^T)^T) = \text{Range}(V_A)$ and $\text{Range}(U_X \Sigma_X (I - \tilde{W})) = \text{Range}(U_B)$, there exist full column rank matrices M_1 and M_2 satisfying $V_A = (\tilde{W}V_X^T)^T M_1$ and $U_B = U_X \Sigma_X (I - \tilde{W}) M_2$, respectively. The sizes of M_1 and M_2 are $\text{rank}(X) \times \text{rank}(\tilde{W})$ and $\text{rank}(X) \times \text{rank}(I - \tilde{W})$, respectively. We can easily see that a matrix S_1 can be decomposed into $S_1 = \widetilde{S}_1 M_1^T$, such that $V_X^T \widetilde{S}_1 = 0$ and M_1 is full column rank, if and only if $V_X^T S_1 = 0$ and $\text{rank}(S_1) \leq \text{rank}(\tilde{W})$. Similarly, a matrix S_2 can be decomposed into $S_2 = M_2 \widetilde{S}_2$, such that $\widetilde{S}_2 U_X = 0$ and M_2 is full column rank, if and only if $S_2 U_X = 0$ and $\text{rank}(S_2) \leq \text{rank}(I - \tilde{W})$. By substituting $V_A = (\tilde{W}V_X^T)^T M_1$, $U_B = U_X \Sigma_X (I - \tilde{W}) M_2$, $S_1 = \widetilde{S}_1 M_1^T$, and $S_2 = M_2 \widetilde{S}_2$ into (32), we obtain the conclusion of Theorem 1. $\qquad \square$

4.2   Proof of Theorem 2

We first quote two results from [10].

Lemma 2. *Assume $X \neq 0$ and $A = XZ$ have feasible solution(s), i.e., $A \in \text{Range}(X)$. Then*

$$Z^* = X^\dagger A \tag{33}$$

is the unique minimizer to the generalized heuristic LRR problem:

$$\min_Z ||Z||_*, \quad s.t. \quad A = XZ. \tag{34}$$

Lemma 3. *For any four matrices B, C, D and F of compatible dimensions, we have the inequalities*

$$\left\| \begin{bmatrix} B & C \\ D & F \end{bmatrix} \right\|_* \geq ||B||_* + ||F||_* \quad and \quad \left\| \begin{bmatrix} B & C \\ D & F \end{bmatrix} \right\|_* \geq ||B||_*, \tag{35}$$

where the second equality holds if and only if $C = 0$, $D = 0$, and $F = 0$.

Then we prove the following lemma.

Lemma 4. *For any square matrix $Y \in \mathbb{R}^{n \times n}$, we have $||Y||_* \geq trace(Y)$, where the equality holds if and only if Y is positive semi-definite.*

Proof. We prove by mathematical induction. When $n = 1$, the conclusion is clearly true. When $n = 2$, we may simply write down the singular values of Y to prove.

Now suppose for any square matrix $\tilde{Y}$, whose size does not exceed $n - 1$, the inequality holds. Then for any matrix $Y \in \mathbb{R}^{n \times n}$, using Lemma 3, we get

$$\begin{aligned}
||Y||_* &= \left\| \begin{bmatrix} Y_{11} & Y_{12} \\ Y_{21} & Y_{22} \end{bmatrix} \right\|_* \\
&\geq ||Y_{11}||_* + ||Y_{22}||_* \\
&\geq trace(Y_{11}) + trace(Y_{22}) \\
&= trace(Y),
\end{aligned} \tag{36}$$

where the second inequality holds due to the inductive assumption on the matrices Y_{11} and Y_{22}. So we always have $||Y||_* \geq \text{trace}(Y)$.

It is easy to check that any positive semi-definite matrix Y, it satisfies $||Y||_* = \text{trace}(Y)$. On the other hand, just following the above proof by choosing Y_{22} as 2×2 submatrices, we can easily get that $||Y||_* > \text{trace}(Y)$ strictly holds if $Y \in \mathbb{R}^{n \times n}$ is asymmetric. So if $||Y||_* = \text{trace}(Y)$, then Y must be symmetric. Then the singular values of Y are simply the absolute values of its eigenvalues. As $\text{trace}(Y)$ equals the sum of all eigenvalues of Y, $||Y||_* = \text{trace}(Y)$ holds only if all the eigenvalues of Y are non-negative. $\qquad\square$

Using Lemma 2, we may consider the following unconstrained problem

$$\min_Z f(Z) \triangleq ||Z||_* + ||X(I - Z)X^\dagger||_*, \tag{37}$$

which is transformed from (5) be eliminating L therein. Then we have the following result.

Proposition 7. *Unconstrained optimization problem* (37) *has a minimum objective function value rank*(X).

Proof. Recall that the sub-differential of the nuclear norm of a matrix Z is [23]

$$\partial_Z ||Z||_* = \{U_Z V_Z^T + R | U_Z^T R = 0, RV_Z = 0, ||R||_2 \leq 1\}, \tag{38}$$

where $U_Z \Sigma_Z V_Z^T$ is the skinny SVD of the matrix Z. We prove that $Z^* = 1/2 X^\dagger X$ is an optimal solution to (37). It is sufficient to show that

$$0 \in \partial_Z f(Z^*) = \partial_Z ||Z^*||_* + \partial_Z ||X(I - Z^*)X^\dagger||_*$$
$$= \partial_Z ||Z^*||_* - X^T \partial_{X(I-Z)X^\dagger} ||X(I - Z^*)X^\dagger||_* (X^\dagger)^T. \tag{39}$$

Notice that $X(I - Z^*)X^\dagger = U_X(1/2I)U_X^T$ is the skinny SVD of $X(I - Z^*)X^\dagger$ and $Z^* = V_X(1/2I)V_X^T$ is the skinny SVD of Z^*. So $\partial_Z f(Z^*)$ contains

$$V_X V_X^T - X^T(U_X U_X^T)(X^\dagger)^T = V_X V_X^T - V_X \Sigma_X U_X^T U_X U_X^T U_X \Sigma_X^{-1} V_X^T = 0. \tag{40}$$

Substituting $Z^* = 1/2 X^\dagger X$ into (37), we get the minimum objective function value rank(X). $\qquad\square$

Next, we have the form of the optimal solutions to (37) as follows.

Proposition 8. *The optimal solutions to the unconstrained optimization problem* (37) *can be written as* $Z^* = V_X \widehat{W} V_X^T$.

Proof. Let $(V_X)_\perp$ be the orthogonal complement of V_X. According to Proposition 7, rank(X) is the minimum objective function value of (37). Thus we get

$$
\begin{aligned}
\mathrm{rank}(X) &= ||Z^*||_* + ||X(I - Z^*)X^\dagger||_* \\
&= \left\| \begin{bmatrix} V_X^T \\ (V_X)_\perp^T \end{bmatrix} Z^* \left[V_X, (V_X)_\perp \right] \right\|_* + ||X(I - Z^*)X^\dagger||_* \\
&= \left\| \begin{bmatrix} V_X^T Z^* V_X & V_X^T Z^* (V_X)_\perp \\ (V_X)_\perp^T Z^* V_X & (V_X)_\perp^T Z^* (V_X)_\perp \end{bmatrix} \right\|_* + ||X(I - Z^*)X^\dagger||_* \\
&\geq ||V_X^T Z^* V_X||_* + ||U_X \Sigma_X V_X^T (I - Z^*) V_X \Sigma_X^{-1} U_X^T||_* \\
&= ||V_X V_X^T Z^* V_X V_X^T||_* + ||U_X \Sigma_X V_X^T (I - V_X V_X^T Z^* V_X V_X^T) V_X \Sigma_X^{-1} U_X^T||_* \\
&= ||V_X V_X^T Z^* V_X V_X^T||_* + ||X(I - V_X V_X^T Z^* V_X V_X^T)X^\dagger||_* \\
&\geq \mathrm{rank}(X),
\end{aligned}
\tag{41}
$$

where the second inequality holds by viewing $Z = V_X V_X^T Z^* V_X V_X^T$ as a feasible solution to (37). Then all the inequalities in (41) must be equalities. By Lemma 3 we have

$$
V_X^T Z^* (V_X)_\perp = (V_X)_\perp^T Z^* V_X = (V_X)_\perp^T Z^* (V_X)_\perp = 0.
\tag{42}
$$

That is to say

$$
\begin{bmatrix} V_X^T \\ (V_X)_\perp^T \end{bmatrix} Z^* \left[V_X, (V_X)_\perp \right] = \begin{bmatrix} \widehat{W} & 0 \\ 0 & 0 \end{bmatrix},
\tag{43}
$$

where $\widehat{W} = V_X^T Z^* V_X$. Hence the equality

$$
Z^* = \left[V_X, (V_X)_\perp \right] \begin{bmatrix} \widehat{W} & 0 \\ 0 & 0 \end{bmatrix} \begin{bmatrix} V_X^T \\ (V_X)_\perp^T \end{bmatrix} = V_X \widehat{W} V_X^T
\tag{44}
$$

holds.   □

Based on all the above lemmas and propositions, the following proposition gives the whole closed form solutions to the unconstrained optimization problem (37). So the solution to problem (37) is non-unique.

Proposition 9. *The solutions to the unconstrained optimization problem* (37) *are* $Z^* = V_X \widehat{W} V_X^T$, *where* $\widehat{W}$ *satisfies: 1. it is block diagonal and its blocks are compatible with* Σ_X[1]; *2. both* $\widehat{W}$ *and* $I - \widehat{W}$ *are positive semi-definite.*

Proof. First, we prove the sufficiency. Suppose $Z^* = V_X \widehat{W} V_X^T$ satisfies all the conditions in the theorem. Substitute it into the objective function, we have

$$
\begin{aligned}
||Z^*||_* + ||X(I - Z^*)X^\dagger||_* &= ||\widehat{W}||_* + ||\Sigma_X(I - \widehat{W})\Sigma_X^{-1}||_* \\
&= ||\widehat{W}||_* + \mathrm{trace}(\Sigma_X(I - \widehat{W})\Sigma_X^{-1}) \\
&= ||\widehat{W}||_* + \mathrm{trace}(I - \widehat{W}) \\
&= ||\widehat{W}||_* + \mathrm{rank}(X) - \mathrm{trace}(\widehat{W}) \\
&= \mathrm{rank}(X) \\
&= \min_Z ||Z||_* + ||X(I - Z)X^\dagger||_*,
\end{aligned}
\tag{45}
$$

---
[1] Please refer to Theorem 2 for the meaning of "compatible with Σ_X."

where based on Lemma 4 the second and the fifth equalities hold since $I - \widehat{W} = \Sigma_X(I - \widehat{W})\Sigma_X^{-1}$ as $\widehat{W}$ is block diagonal and both $I - \widehat{W}$ and $\widehat{W}$ are positive semi-definite.

Next, we give the proof of the necessity. Let Z^* represent a minimizer. According to Proposition 8, Z^* could be written as $Z^* = V_X \widehat{W} V_X^T$. We will show that $\widehat{W}$ satisfies the stated conditions. Based on Lemma 4, we have

$$
\begin{aligned}
\mathrm{rank}(X) &= ||Z^*||_* + ||X(I - Z^*)X^\dagger||_* \\
&= ||\widehat{W}||_* + ||\Sigma_X(I - \widehat{W})\Sigma_X^{-1}||_* \\
&\geq ||\widehat{W}||_* + \mathrm{trace}(\Sigma_X(I - \widehat{W})\Sigma_X^{-1}) \\
&= ||\widehat{W}||_* + \mathrm{trace}(I - \widehat{W}) \\
&= ||\widehat{W}||_* + \mathrm{rank}(X) - \mathrm{trace}(\widehat{W}) \\
&\geq \mathrm{rank}(X).
\end{aligned}
\tag{46}
$$

Thus all the inequalities above must be equalities. From the last equality and Lemma 4, we directly get that $\widehat{W}$ is positive semi-definite. By the first inequality and Lemma 4, we know that $\Sigma_X(I - \widehat{W})\Sigma_X^{-1}$ is symmetric, i.e.,

$$
\frac{\sigma_i}{\sigma_j}[I - \widehat{W}]_{ij} = \frac{\sigma_j}{\sigma_i}[I - \widehat{W}]_{ij},
\tag{47}
$$

where σ_i represents the ith entry on the diagonal of Σ_X. Thus if $\sigma_i \neq \sigma_j$, then $[I - \widehat{W}]_{ij} = 0$, i.e., $\widehat{W}$ is block diagonal and its blocks are compatible with Σ_X. Notice that $I - \widehat{W} = \Sigma_X(I - \widehat{W})\Sigma_X^{-1}$. By Lemma 4, we get that $I - \widehat{W}$ is also positive semi-definite. Hence the proof is completed. □

Now we can prove Theorem 2.

Proof. Let $\widehat{W}$ satisfy all the conditions in the theorem. According to Proposition 8, since the row space of $Z^* = V_X \widehat{W} V_X^T$ belongs to that of X, it is obvious that $(Z^*, X(I - Z^*)X^\dagger)$ is feasible to problem (5). Now suppose that (5) has a better solution $(\widetilde{Z}, \widetilde{L})$ than (Z^*, L^*), i.e.,

$$
X = X\widetilde{Z} + \widetilde{L}X,
\tag{48}
$$

and

$$
||\widetilde{Z}||_* + ||\widetilde{L}||_* < ||Z^*||_* + ||L^*||_*.
\tag{49}
$$

Fixing Z in (5) and by Lemma 2, we have

$$
||\widetilde{Z}||_* + ||(X - X\widetilde{Z})X^\dagger||_* \leq ||\widetilde{Z}||_* + ||\widetilde{L}||.
\tag{50}
$$

Thus

$$
||\widetilde{Z}||_* + ||(X - X\widetilde{Z})X^\dagger||_* < ||Z^*||_* + ||X(I - Z^*)X^\dagger||_*.
\tag{51}
$$

So we obtain a contradiction with respect to the optimality of Z^* in Proposition 9, hence proving the theorem. □

4.3 Proof of Proposition 1

Proof. Suppose the optimal formulation $Z^* = V_X \widehat{W} V_X^T$ in Theorem 2 could be written as $Z^* = V_X \tilde{W} V_X^T + S_1 \tilde{W} V_X^T$, where $\tilde{W}$ is idempotent and S_1 satisfies $\tilde{W} V_X^T S_1 = 0$. Then we have

$$V_X \widehat{W} V_X^T = V_X \tilde{W} V_X^T + S_1 \tilde{W} V_X^T. \tag{52}$$

By multiplying both sides with V_X^T and V_X on the left and right, respectively, we get

$$\widehat{W} = \tilde{W} + V_X^T S_1 \tilde{W}. \tag{53}$$

As a result, $\widehat{W}$ is idempotent:

$$
\begin{aligned}
\widehat{W}^2 &= (\tilde{W} + V_X^T S_1 \tilde{W})(\tilde{W} + V_X^T S_1 \tilde{W}) \\
&= \tilde{W}^2 + V_X^T S_1 \tilde{W}^2 + \tilde{W} V_X^T S_1 \tilde{W} + V_X^T S_1 \tilde{W} V_X^T S_1 \tilde{W} \\
&= \tilde{W} + V_X^T S_1 \tilde{W} + \tilde{W} V_X^T S_1 \tilde{W} + V_X^T S_1 \tilde{W} V_X^T S_1 \tilde{W} \\
&= \tilde{W} + V_X^T S_1 \tilde{W} = \widehat{W},
\end{aligned}
\tag{54}
$$

which is contradictory to the assumption. $\square$

5 Conclusions

Based on the expositions in Section 3 and the proofs in Section 4, we conclude that even for rank minimization problems as simple as noiseless LatLRR, replacing rank with nuclear norm is not valid. We have also found that LatLRR is actually problematic because the solution to its nuclear norm minimization formation is not unique. We can also have the following interesting connections between LRR and LatLRR. Namely, LatLRR is indeed an extension of LRR because its solution set strictly includes that of LRR, no matter for the rank minimization problem or the nuclear norm minimization formulation. So we can summarize their relationship as Figure 1.

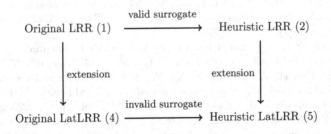

Fig. 1. The detailed relationship among the original LRR (1), the heuristic LRR (2), the original LatLRR (4), and the heuristic LatLRR (5) in the sense of their solution sets

Although the existing formulation of LatLRR is imperfect, since some scholars have demonstrated its effectiveness in subspace clustering by using a solution which is randomly chosen in some sense, in the future we will consider how to choose the best solution in the solution set in order to further improve the performance of LatLRR.

Acknowledgments. Hongyang Zhang and Chao Zhang are supported by National Key Basic Research Project of China (973 Program) 2011CB302400 and National Nature Science Foundation of China (NSFC Grant, no. 61071156). Zhouchen Lin is supported by National Nature Science Foundation of China (Grant nos. 61272341, 61231002, and 61121002).

References

1. Wang, J.: Geometric Structure of High-Dimensional Data and Dimensionality Reduction. Springer (2012)
2. Candès, E., Recht, B.: Exact matrix completion via convex optimization. Foundations of Computational Mathematics 9(6), 717–772 (2009)
3. Candès, E., Li, X., Ma, Y., Wright, J.: Robust principal component analysis? Journal of the ACM 58(3), 1–37 (2009)
4. Candès, E.: Matrix completion with noise. Proceedings of the IEEE 98(6), 925–936 (2010)
5. Fazel, M.: Matrix Rank Minimization with Applications. PhD thesis, Standford University (2002), http://search.proquest.com/docview/305537461
6. Gross, D.: Recovering low-rank matrices from few coefficients in any basis. IEEE Transactions on Information Theory 57(3), 1548–1566 (2011)
7. Wright, J., Ganesh, A., Min, K., Ma, Y.: Compressive principal component pursuit. In: IEEE International Symposium on Information Theory Proceedings, pp. 1276–1280 (2012)
8. Waters, A.E., Sankaranarayanan, A.C., Baraniuk, R.G.: SpaRCS: Recovering low-rank and sparse matrices from compressive measurements. In: Advances in Neural Information Processing Systems, pp. 1089–1097 (2011)
9. Liu, Y.K.: Universal low-rank matrix recovery from pauli measurements. In: Advances in Neural Information Processing Systems, pp. 1638–1646 (2011)
10. Liu, G., Lin, Z., Yan, S., Sun, J., Ma, Y.: Robust recovery of subspace structures by low-rank representation. IEEE Transactions on Pattern Analysis and Machine Intelligence 35(1), 171–184 (2013)
11. Liu, G., Lin, Z., Yu, Y.: Robust subspace segmentation by low-rank representation. In: International Conference on Machine Learning, vol. 3, pp. 663–670 (2010)
12. Peng, Y., Ganesh, A., Wright, J., Xu, W., Ma, Y.: RASL: Robust alignment by sparse and low-rank decomposition for linearly correlated images. In: IEEE Conference on Computer Vision and Pattern Recognition, vol. 34, pp. 2233–2246 (2010)
13. Zhang, Z., Ganesh, A., Liang, X., Ma, Y.: TILT: Transform invariant low-rank textures. International Journal of Computer Vision 99(1), 1–24 (2012)
14. Cheng, B., Liu, G., Huang, Z., Yan, S.: Multi-task low-rank affinities pursuit for image segmentation. In: IEEE International Conference on Computer Vision, pp. 2439–2446 (2011)

15. Lang, C., Liu, G., Yu, J., Yan, S.: Saliency detection by multi-task sparsity pursuit. IEEE Transactions on Image Processing 21(3), 1327–1338 (2012)
16. Liu, G., Xu, H., Yan, S.: Exact subspace segmentation and outlier detection by low-rank representation. In: International Conference on Artificial Intelligence and Statistics (2012)
17. Liu, G., Yan, S.: Latent low-rank representation for subspace segmentation and feature extraction. In: IEEE International Conference on Computer Vision, pp. 1615–1622 (2011)
18. Vidal, R.: Subspace clustering. IEEE Signal Processing Magazine 28(2), 52–68 (2011)
19. Elhamifar, E., Vidal, R.: Sparse subspace clustering. In: IEEE Conference on Computer Vision and Pattern Recognition, vol. 2, pp. 2790–2797 (2009)
20. Adler, A., Elad, M., Hel-Or, Y.: Probabilistic subspace clustering via sparse representations. IEEE Signal Processing Letters 20(1), 63–66 (2013)
21. Liu, R., Lin, Z., Torre, F.D.L., Su, Z.: Fixed-rank representation for unsupervised visual learning. In: IEEE Conference on Computer Vision and Pattern Recognition, pp. 598–605 (2012)
22. Friedland, S., Torokhti, A.: Generalized rank-constrained matrix approximations. SIAM Journal on Matrix Analysis and Applications 29(2), 656–659 (2007)
23. Cai, J., Candès, E., Shen, Z.: A singular value thresholding algorithm for matrix completion. SIAM Journal of Optimization 20(4), 1956–1982 (2010)

Efficient Rank-one Residue Approximation Method for Graph Regularized Non-negative Matrix Factorization

Qing Liao and Qian Zhang

Department of Computer Science and Engineering
The Hong Kong University of Science and Technology
{qnature,qianzh}@cse.ust.hk

Abstract. Nonnegative matrix factorization (NMF) aims to decompose a given data matrix X into the product of two lower-rank nonnegative factor matrices UV^T. Graph regularized NMF (GNMF) is a recently proposed NMF method that preserves the geometric structure of X during such decomposition. Although GNMF has been widely used in computer vision and data mining, its multiplicative update rule (MUR) based solver suffers from both slow convergence and non-stationarity problems. In this paper, we propose a new efficient GNMF solver called rank-one residue approximation (RRA). Different from MUR, which updates both factor matrices (U and V) as a whole in each iteration round, RRA updates each of their columns by approximating the residue matrix by their outer product. Since each column of both factor matrices is updated optimally in an analytic formulation, RRA is theoretical and empirically proven to converge rapidly to a stationary point. Moreover, since RRA needs neither extra computational cost nor parametric tuning, it enjoys a similar simplicity to MUR but performs much faster. Experimental results on real-world datasets show that RRA is much more efficient than MUR for GNMF. To confirm the stationarity of the solution obtained by RRA, we conduct clustering experiments on real-world image datasets by comparing with the representative solvers such as MUR and NeNMF for GNMF. The experimental results confirm the effectiveness of RRA.

Keywords: Nonnegative matrix factorization, Manifold regularization, Rank-one residue iteration, Block coordinate descent.

1 Introduction

Given n data points arranged in a nonnegative matrix $X \in \mathbb{R}_+^{m \times n}$ and m stands for the dimension of the data, nonnegative matrix factorization (NMF) decomposes X into the product of two lower-rank nonnegative factor matrices, that is, UV^T, where $U \in \mathbb{R}_+^{m \times r}$ and $V \in \mathbb{R}_+^{n \times r}$ signify the basis of the low-dimensional space and the coefficient, respectively. Although NMF performs well in several tasks, it completely ignores the property where many datasets, for example, human faces or hand-written digits, reside in a manifold that lies in a low-dimensional space. Recently, Cai et al. [3] proposed graph regularized NMF

H. Blockeel et al. (Eds.): ECML PKDD 2013, Part II, LNAI 8189, pp. 242–255, 2013.
© Springer-Verlag Berlin Heidelberg 2013

(GNMF) to address this problem. GNMF assumes that the neighborhoods of one data point in high-dimensional space should be as close as possible to the images of that point in low-dimensional space. Since GNMF preserves the geometric structure of the data set, any label information can be propagated along the surface of the manifold to its neighbourhoods. Such an advantage greatly enhances the clustering performance of NMF and makes GNMF a powerful tool in data mining [2][18].

Recently, many GNMF variants have been proposed for various applications. Zhang et al. [18] proposed a topology preserving NMF (TPNMF) for face recognition. TPNMF considers the manifold structure in face datasets and preserves the local topology while face images are transformed to a subspace. For the same purpose, Gu et al. [6] proposed neighborhood preserving NMF (NPNMF). NPNMF represents each data point by a linear combination of its neighbours in high-dimensional space and keeps such relationships in low-dimensional space with the same combination coefficient. Yang et al. [17] developed non-negative graph embedding (NGE) to integrate both intrinsic graph and penalty graph in NMF in the spirit of marginal fisher analysis [16]. Guan et al. [7] proposed a manifold regularized discriminative NMF (MD-NMF) for classification tasks. MD-NMF preserves both local geometry and label information of data points simultaneously by expecting data points in the same class close to be each other and data points in different classes far from each other. Shen and Si [13] proposed an NMF on multiple manifolds method (MM-NMF) to model the intrinsic geometrical structure of data on multiple manifolds. MM-NMF assumes the data points are drawn from multiple manifolds, if one data point can be reconstructed by several neighbourhoods on the same manifold in high-dimensional space, it can also be reconstructed in a similar way in low-dimensional space.

Although GNMF and its variants perform well in many fields, the multiplicative update rules (MUR) based algorithm suffers two drawbacks: 1) MUR converges slowly because it is intrinsically a first-order gradient descent method, and 2) MUR does not guarantee convergence to any stationary point [10]. These two deficiencies seriously prohibit GNMF from practical applications. To remedy these problems, Guan et al. [8] has recently proposed a NeNMF method for optimizing GNMF. NeNMF applies Nesterovs's method to alternatively update each factor matrix. Since Nesterovs's method updates each factor matrix in a second-order convergence rate, it converges rapidly for optimizing GNMF. However, NeNMF needs a stop condition to check when to stop Nesterovs's method, and it is non-trivial to determine such tolerance on many datasets.

In this paper, we propose an efficient rank-one residue approximation (RRA) method for GNMF and its variants. RRA decomposes the reconstruction UV^T into a summation of r rank-one matrices, i.e., $X \approx \sum_{i=1}^{r} U_{\cdot i} V_{\cdot i}^T$. In contrast to MUR which recursively updates the whole factor matrix, RRA recursively updates each column of each factor matrix with the remaining variables fixed, that is, $U_k V_k^T \approx X - \sum_{l \neq k}^{r} U_l V_l^T$ for each $1 \leq k \leq r$. It is obvious that each column of U can be updated in an analytic formulation based on non-negative least squares, but it is difficult to update columns of V due to the incorporated manifold regularization term. In this paper, we show that each column of V can also be updated in an analytic formulation. Since the objective function is continuously differentiable on a Cartesian product of $2r$ closed convex sets, RRA is proved to converge to a stationary point. However, an inverse of the Hessian matrix is needed in updating each column of V and such matrix inverse operator costs too much computational time. To overcome such deficiency, we introduce the Sherman-Morrison-Woodbury (SMW) formula to approximate the inverse of Hessian matrix. Since the approximation can be efficiently calculated in advance, the SMW-based formula greatly reduces the time cost of RRA. Experimental results on real-world datasets show the efficiency of RRA compared with representative GNMF solvers. To evaluate the effectiveness of RRA for GNMF, we conduct clustering experiments on two popular image datasets including COIL-20 [12] and CMU PIE [14], the experimental results show that RRA is effective.

The rest of the paper is organized as follows. Section 2 briefly reviews the graph regularized NMF (GNMF) method and its state-of-the-art solvers. Section 3 presents the RRA method for GNMF. Section 4 evaluate the efficiency and effectiveness and gives the experimental results, while Section 5 summaries this paper.

2   Related Works

Nonnegative matrix factorization (NMF) is a popular dimension reduction method which has been widely used in pattern recognition and data mining. Given a dataset $X = [x_1, x_2, \ldots, x_n] \in \mathbb{R}_+^{m \times n}$, where each column of X presents an data point, NMF aims to find two lower-rank nonnegative matrices $U \in \mathbb{R}_+^{m \times r}$ and $V \in \mathbb{R}_+^{n \times r}$, where $r \leq \min\{m, n\}$ is the reduced dimensionality, such that their product UV^T approximates the original matrix X:

$$X = UV^T + E \tag{1}$$

where E denotes the residual error. Assuming the entries in E to be I.I.D. Gaussian distributed, we get the objective function of NMF:

$$\min_{U \geq 0, V \geq 0} \frac{1}{2} \|X - UV^T\|_F^2 \tag{2}$$

where $\|\cdot\|_F$ signifies the Frobenius norm. The smaller the cost function, the better the approximation of UV^T.

Since NMF does not consider the intrinsic geometrical structure of dataset, it does not always perform well in some real-world datasets, for example, face images and hand-written digits. To remedy this problem, Cai et al. [3] proposed graph regularized NMF (GNMF), which considers the geometrical structure of the dataset in NMF. The basic assumption is that data points reside on the surface of a manifold that lies in a low-dimensional space, that is, if two data points are close enough in high-dimensional space they are still close in low-dimensional space. To this end, GNMF constructs an adjacent graph G on a scatter of data points to represent the local geometric structure. In graph G, each node associates an data point and an edge is established between two nodes once one node belongs to the k nearest neighbourhoods of another. Based on G, we can build an adjacent matrix W as follows:

$$W_{ij} = \begin{cases} 1, & x_j \in N_k(x_i) \mid x_i \in N_k(x_j) \\ 0, & \text{otherwise} \end{cases} \tag{3}$$

where $N_k(x_i)$ denotes the k nearest neighbourhoods of x_i. We are now ready to preserve the geometrical structure of X in the low-dimensional space, the objective is to minimize

$$\sum_{i=1}^{n}\sum_{j=1}^{n}\|v_i - v_j\|_2^2 W_{ij} = tr(V^T L V) \tag{4}$$

where $L = D - W$ is the Laplacian matrix of G, where D is a diagonal matrix and $D_{jj} = \sum_l W_{jl}$, and $tr(\cdot)$ signifies the trace operator over a symmetric matrix. Combing (2) and (4) together, we arrive at the objective of GNMF:

$$\min_{U \geq 0, V \geq 0} f(U, V) = \frac{1}{2}\|X - UV^T\|_F^2 + \frac{\beta}{2}tr(V^T L V) \tag{5}$$

where $\beta > 0$ is a trade-off parameter over the manifold regularization term.

Although $f(U, V)$ is jointly non-convex w.r.t. both U and V, it is convex w.r.t. either U or V. Therefore, we can apply block coordinate descent method [1] to solve (5). Cai et al. [3] proposed a multiplicative update rule (MUR) to recursively update matrices U and V, respectively. At the $t-th$ iteration round, U_{t+1} and V_{t+1} are updated as follows:

$$U_{t+1} = U_t \circ \frac{XV_t}{U_t V_t^T V_t} \tag{6}$$

and

$$V_{t+1} = V_t \circ \frac{X^T U_{t+1} + \beta W V_t}{V_t U_{t+1}^T U_{t+1} + \beta D V_t} \tag{7}$$

where $\circ$ signifies the element-wise multiplication.

Although MUR is proved to reduce the objective function $f(U, V)$, it converges slowly in terms of number of iteration because it is intrinsically a first-order gradient descent method. In addition, MUR suffers from non-stationarity problems like NMF [10].

Recently, to remedy the problem of MUR, Guan et al. [8] proposed a NeNMF algorithm to update both factor matrices by solving the following two subproblems

$$\min_{U \geq 0} \frac{1}{2}\|X - UV^T\|_F^2 \tag{8}$$

and

$$\min_{V \geq 0} \frac{1}{2}\|X - UV^T\|_F^2 \tag{9}$$

Since NeNMF solves both (8) and (9) with Nesterovs's method [8], it converges rapidly because Nesterovs's method is intrinsically a second-order gradient descent method. However, NeNMF needs a tolerance to check the stopping of the Nesterovs's method. If the tolerance is set too small, NeNMF costs too much extra iterations for solving (8) and (9). If the tolerance is set too large, NeNMF gets a low-quality solution for either (8) or (9). Therefore, it is not easy to use NeNMF in practical applications because it is non-trivial to determine a suitable tolerance on many datasets.

3   Rank-one Residue Approximation for GNMF

In this section, we proposed an efficient rank-one residue approximation (RRA) algorithm to overcome the deficiencies of both MUR and NeNMF by recursively updating columns of U and V with an analytic formulation.

The main idea is inspired by the well-known rank-one residue iteration (RRI, [9]) method and hierarchical alternating least squares (HALS, [5]) method for NMF. In contrast to MUR and NeNMF which alternately updates the whole U and V, RRA recursively updates columns of them with the remaining variables fixed. For the $k-th$ column of U and V, the sub-problems are

$$\min_{U_{\cdot k} \geq 0} \frac{1}{2}\|R_k - U_{\cdot k}V_{\cdot k}^T\|_F^2 \tag{10}$$

and

$$\min_{V_{\cdot k} \geq 0} \frac{1}{2}\|R_k - U_{\cdot k}V_{\cdot k}^T\|_F^2 + \frac{\beta}{2}V_{\cdot k}^T L V_{\cdot k} \tag{11}$$

where R_k denotes the residue of X after eliminating the $k-th$ column of U and V, i.e., $R_k = X - \sum_{l \neq k} U_{\cdot l}V_{\cdot l}^T$. The formula (11) is derived from the following equation: $tr(V^T L V) = \sum_{k=1}^{r} V_{\cdot k}^T L V_{\cdot k}$.

The problem (10) should be solved in two cases, that is, $V_{\cdot k} = 0$ and $V_{\cdot k} \neq 0$. If $V_{\cdot k} = 0$, eq. (10) has an infinite number of solutions. Therefore, the $k-th$ column of both U and V should be taken off in the remaining computation. If $V_{\cdot k} \neq 0$, according to [3], the sub-problem (10) has a closed-form solution

$$U_{\cdot k} = \frac{\prod_{+}(R_k V_{\cdot k})}{\|V_{\cdot k}\|_2^2} \tag{12}$$

where $\prod_{+}(x) = \max(0, x)$ is an element-wise projection that shrinks negative entries of x to zero. Similar to (10), the problem (11) should be considered in two cases, that is, $U_{\cdot k} = 0$ and $U_{\cdot k} \neq 0$. If $U_{\cdot k} = 0$, the $k-th$ column of both U and V does not take part in the remaining computation and should be taken off. If $U_{\cdot k} \neq 0$, below we show how to solve (11) in an analytic formulation though it is not as direct as (12).

We solved the constrained optimization (11) by using the Lagrangian multiplier method [1]. The Lagrangian function of (11) is

$$\mathcal{L} = \frac{1}{2}\|R_k - U_{\cdot k}V_{\cdot k}^T\|_F^2 + \frac{\beta}{2}V_{\cdot k}^T L V_{\cdot k} - \langle V_{\cdot k}, \lambda \rangle \tag{13}$$

where λ is the Lagrangian multiplier for the constraint $V_{\cdot k} \geq 0$. Based on the K.K.T. conditions, the solution of (11) satisfies

$$\begin{cases} V_{\cdot k} \geq 0, \lambda \geq 0, \\ \frac{\partial \mathcal{L}}{\partial V_{\cdot k}} = -R_k^T U_{\cdot k} + (\|U_{\cdot k}\|_2^2 I + \beta L) V_{\cdot k} - \lambda = 0 \\ \lambda \circ V_{\cdot k} = 0 \end{cases} \tag{14}$$

where $I \in \mathbb{R}^{n \times n}$ is an identity matrix. With simple algebra, based on (14), we update columns of V as follows:

$$V_{\cdot k} = \prod_+ ((\|U_{\cdot k}\|_2^2 I + \beta L)^{-1} R_k^T U_{\cdot k}) \tag{15}$$

by updating columns of U and V alternatively with (12) and (15), respectively, until convergence, RRA solves GNMF. The following **Proposition 1** shows that alternating (12) and (15) reaches a stationary point. The proof is similar to [9], we only include it here for completeness.

Proposition 1. *Every limited point generated by alternating (12) and (15) is a stationary point.*

Proof. From (10) and (11), we know that the feasible sets of $U_{\cdot k}$ and $V_{\cdot k}$ are $\Omega_k^U \subset \mathbb{R}_+^m$ and $\Omega_k^V \subset \mathbb{R}_+^n$. According to [11], since X is bounded, we can set an upper bound for Ω_k^U and Ω_k^V and thus can consider them as closed convex sets.

Therefore, the GNMF problem can be written as a bound-constrained optimization problem

$$\min_{[U,V] \in \Omega} \frac{1}{2} \|X - \sum_{k=1}^r U_{\cdot k} V_{\cdot k}^T\|_F^2 + \frac{\beta}{2} \sum_{k=1}^r V_{\cdot k}^T L V_{\cdot k} \tag{16}$$

where $\Omega = \prod_{k=1}^r \Omega_k^U \times \prod_{k=1}^r \Omega_k^V$ is a Cartesian product of closed convex sets. Since the objective function of (16) is continuously differentiable over Ω and RRA updates the $k-th$ column of U and V with the optimal solutions of (12) and (15), according to **Proposition 2.7.1** in [1], every limit point generated by alternating (12) and (15) is a stationary point.

For completeness, we must consider cases when either $U_{\cdot k}$ or $V_{\cdot k}$ is zero. In these cases, eq. (12) and (15) do not give unique solutions, and **Proposition 2.7.1** in [1] cannot be applied. As mentioned above, such columns should be taken off without changing the value of the objective function of (16). Therefore, these columns do not destroy the theoretic analysis. It completes the proof.

Since L is a positive semi-definite matrix, the matrix inverse operator $(\|U_{\cdot k}\|_2^2 I + \beta L)^{-1}$ in (15) is well-defined if $U_{\cdot k} \neq 0$. However, the matrix inverse operator is inefficient because its time cost is $O(n^3)$. Fortunately, the matrix $\|U_{\cdot k}\|_2^2 I + \beta L$ has a nice property, that is, it is composed of a symmetric positive semi-definite matrix and a diagonal matrix. This property motivates us to apply the well-known Sherman-Morrison-Woodbury (SMW, [15]) formula to approximate the matrix inverse operator efficiently. That is why our method is called rank-one

---

Algorithm 1. Rank-one Residue Approximation for GNMF

Input: $X \in \mathbb{R}_+^{m \times n}, L \in \mathbb{R}^{n \times n}, 1 \leq r \leq \min(m,n), \beta$

Output: $U \in \mathbb{R}_+^{m \times r}, V \in \mathbb{R}_+^{n \times r}$

1: Initialize: $U^1 \in R_+^{m \times r}, V^1 \in R_+^{n \times r}, t = 1$
2: Calculate: $L = \Theta \Sigma \Theta^T \approx \tilde{\Theta}\tilde{\Sigma}\tilde{\Theta}^T, R^1 = X - U^1 V^{1T}$
3: **repeat**
4:　　**for** $k = 1 \ldots r$ **do**
5:　　　　Calculate: $R_k^t = R^t + U_{\cdot k}^t V_{\cdot k}^{t}{}^T$
6:　　　　Update: $U_{\cdot k}^{t+1} = \prod_+ (R_k^t V_{\cdot k}^t))/\|V_{\cdot k}^t\|_2^2$
7:　　　　Calculate: $A_k^t \approx (\|U_{\cdot k}^{t+1}\|_2^2 I + \beta L)^{-1}$
8:　　　　Update: $V_{\cdot k}^{t+1} = \prod_+ (A_k^t R_k^{t}{}^T U_{\cdot k}^{t+1})$
9:　　　　Update: $R^t = R_k^t - U_{\cdot k}^{t+1} V_{\cdot k}^{t+1}{}^T$
10:　　**end for**
11:　　Update: $R^{t+1} = R^t$
12:　　Update: $t \leftarrow t + 1$
13: **until** The Stopping Condition (18) is Satisfied.
14: $U = U^t, V = V^t$

---

residue approximation (RRA). In particular, we can approximate the matrix inverse in (15) as

$$(\|U_{\cdot k}\|_2^2 I + \beta L)^{-1} = \beta^{-1}\left(\frac{\|U_{\cdot k}\|_2^2}{\beta}I + \Theta \Sigma \Theta_T\right)^{-1}$$

$$\approx \beta^{-1}\left(\frac{\|U_{\cdot k}\|_2^2}{\beta}I + \tilde{\Theta}\tilde{\Sigma}\tilde{\Theta}^T\right)^{-1}$$

$$= \left(\frac{1}{\|U_{\cdot k}\|_2^2}I - \frac{\beta}{\|U_{\cdot k}\|_2^4}\tilde{\Theta}(\tilde{\Sigma}^{-1} + \frac{\beta}{\|U_{\cdot k}\|_2^2})^{-1}\tilde{\Theta}^T\right) \tag{17}$$

where $L = \Theta \Sigma \Theta^T$ is a SVD of L, and $\tilde{\Theta}\tilde{\Sigma}\tilde{\Theta}^T$ is its approximation by taking the first p most important components. The value p is usually determined by keeping 95% of the energy, i.e., $\sum_{i=1}^p \sigma_i^2 / \sum_{i=1}^n \sigma_i^2 \leq 95\%$, where σ_i^2 is the $i-th$ largest singular value. The following section will discuss how to choose the percentage of energy kept. The main time cost of (17) is spent on the calculation of the second term, whose time complexity is $O(n^2 p)$ because the inverse operator is performed over a diagonal matrix. Since $p \ll n$, the SMW formula greatly reduces the time cost of (16) from $O(n^3 + mn)$ to $O(n^2 p + mn)$.

By recursively updating columns of U and V with (12) and (15), respectively, we can solve GNMF efficiently without tuning extra parameters. The total procedure is summarized in **Algorithm 1**, where the stopping condition is given as follows:

$$|f(U^t, V^t) - f(U^{t+1}, V^{t+1})| \leq \varepsilon |f(U^1, V^1) - f(U^2, V^2)| \tag{18}$$

where ε is the tolerance, for example, $\varepsilon = 10^{-3}$.

The proposed RRA algorithm avoids parameter tuning thus it is more flexible and convenient in practical applications. In **Algorithm 1**, since the SVD of L

can be calculated beforehand and the residue R_k^t can be updated recursively as lines 5 and 9 in $O(mn)$ time, the time complexity of RRA is mainly spent on line 7 because the matrix-vector multiplication in lines 6 and 8 cost $O(mn)$ time. In summary, the time complexity of one iteration round of RRA is $O(mnr + prn^2)$. According to [8], the time complexity of one iteration round of NeNMF is $O(mnr + mr^2 + nr^2) + K \times O(mr^2 + nr^2)$, where K is the number of iterations performed by Nesterovs's method. An unsuitable tolerance leads to a rather large K for NeNMF and pulls down its efficiency. RRA overcomes such deficiency and performs more efficiently than NeNMF without any parameter tuning. Although the time complexity of one iteration round of MUR is $O(mnr + rn^2)$ [3], RRA costs less time in total than MUR because it converges in far less iteration rounds.

4 Experiments

In this section, we evaluate the efficiency of our RRA method by comparing it with state-of-the-art GNMF solvers including MUR and NeNMF in terms of CPU seconds. In addition, we evaluate the clustering performance of the RRA based GNMF on popular image datasets, such as, COIL-20 [12] and CMU PIE [14] to confirm its effectiveness.

4.1 Preliminaries

We followed [3] to evaluate RRA on two popular image datasets including COIL-20 [12] and CMU PIE [14]. The COIL-20[1] image library contains images of 20 objects viewed from different angles. Totally 72 images were taken for each object and each image was cropped to 32×32 pixels and rescaled to an 1024-dimensional long vector. The CMU PIE[2] face image database contains face images of 68 individuals. There are totally 42 facial images for each individual taken under different lighting and illumination conditions. Similarly, each image was cropped to 32×32 pixels and rescaled to a 1024-dimensional long vector. In summary, the COIL-20 is composed of a 1024×1440 matrix and CMU PIE is composed of a 1024×2856 matrix.

4.2 Efficiency Evaluation

We evaluated the efficiency of RRA by comparing its CPU seconds with those spent by both MUR and NeNMF on whole COIL-20 and CMU PIE datasets. For GNMF, we set the number of neighborhoods to $k = 5$, the trade-off parameter $\beta = 1$, and $r = 10, 50, 100, 200$ to study the scalability of RRA. To keep the fairness of comparison, all GNMF solvers start from an identical point and stop when they reach an identical objective value. Then the CPU seconds it costs are compared for the purpose of evaluation.

---

[1] http://www1.cs.columbia.edu/CAVE/software/softlib/coil-20.php
[2] http://www.ri.cmu.edu/projects/project_418.html

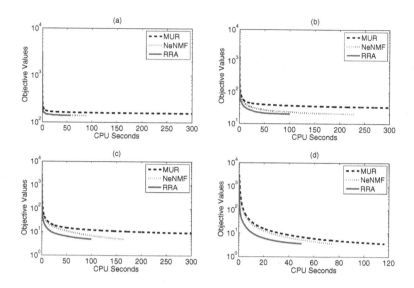

Fig. 1. Objective values versus CPU seconds of MUR, NeNMF, and RRA on COIL-20 dataset. (a) r=20, (b) r=50, (c) r=100, (d) r=200

Figure 1 compares the objective values versus CPU seconds of MUR, NeNMF and RRA on COIL-20 dataset. It shows that RRA performs much more rapidly than MUR and NeNMF because it reaches lower objective values in the same number of CPU seconds. In other words, to get a similar solution RRA costs far less CPU seconds. When the reduced dimensionality $r = 20$, NeNMF performs comparably with RRA because its time complexity $O(mnr + mr^2 + nr^2) + K \times O(mr^2 + nr^2 + rn^2)$ is dominated by $O(mnr) + K \times O(rn^2)$ which is comparable with the time complexity of RRA, that is, $O(mnr + prn^2)$. From subfigures (a) to (d), we can see that RRA always performs better and better than MUR, but it performs closely to NeNMF when the reduced dimensionality is 200.

Figure 2 compares the objective values versus CPU seconds of MUR, NeNMF and RRA on the CMU PIE dataset. From Figure 2, we have the same observation as Figure 1. It confirms that RRA is much more efficient than MUR and NeNMF. To study the speedup rate of RRA versus MUR and NeNMF, we repeated the experiments on the PIE dataset with r varying from 10 to 100. MUR, NeNMF, and RRA start from the identical initial point and stop when the same objective value is reached. Then, we calculated the speedup rate as the ratio of time costs of MUR (or NeNMF) dividing those of RRA. Such trial was repeated ten times with different randomly generated initial point.

Figure 3 gives the speedup rates of RRA versus NeNMF and MUR. It shows that RRA is much faster than both MUR especially when the reduced dimensionality is 10. RRA is also faster than NeNMF when the reduced dimensionality is 10, and costs comparable CPU time in other cases. This observation shows that RRA is much efficient when the reduced dimensionality is relatively small.

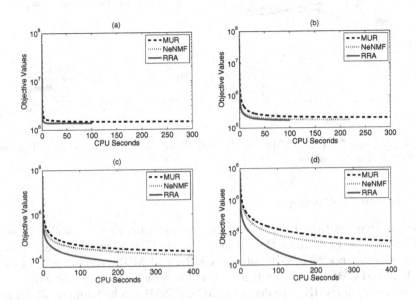

Fig. 2. Objective values versus CPU seconds of MUR, NeNMF, and RRA on PIE dataset. (a) r=20, (b) r=50, (c) r=100, (d) r=200.

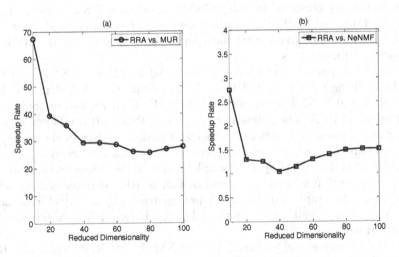

Fig. 3. Speedup rate versus reduced dimensionality: (a) RRA versus MUR and (b) RRA versus NeNMF

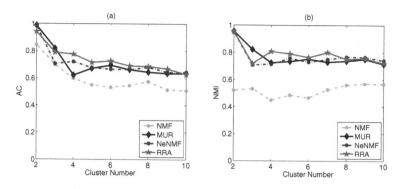

Fig. 4. AC and NMI of GNMF by MUR, NeNMF, and RRA on COIL-20 dataset. NMF is used as a baseline.

4.3   Image Clustering

Based on **Proposition 1**, RRA converges to a stationary point of GNMF while MUR does not. To evaluate the stationary level of RRA, we compared the clustering performance of GNMF solved by RRA and MUR, respectively, on both COIL-20 and CMU PIE datasets in terms of accuracy (AC) and normalized mutual information (NMI). The details of AC and NMI can be found in [3]. To keep the fairness of comparison, all GNMF solvers start from an identical initial point and stop when the same stopping condition (18) is satisfied. In this experiment, we randomly selected $r = 2, \ldots, 10$ individuals from both datasets and the selected images are clustered by using GNMF based on different solvers, that is, RRA, NeNMF, and MUR. Such trails were repeated ten times and the average AC and NMI are used to compare their performance. The standard NMF is also compared as a baseline.

Figure 4 compares the averaged AC and NMI by MUR, NeNMF, and RRA based GNMF on the COIL-20 dataset. It shows that the RRA slightly outperforms MUR and NeNMF in terms of AC and NMI. There are two reasons for this observation: 1) RRA gets a stationary point (see **Proposition 1**) which better approximates the data points and thus might perform better in clustering, and 2) RRA approximates the graph Laplacian L in (4) with the largest eigenvectors (see (17)), according to the spectral graph theory [4], these eigenvectors associate with most smooth functions over graph G, that is, RRA eliminates some outlier functions on the graph and thus RRA propagates the geometrical information better than the original GNMF method. For this reason, RRA clusters the data points better.

Figure 5 compares the averaged AC and NMI by MUR, NeNMF, and RRA based GNMF on the PIE dataset. Figure 5 confirms our analysis Figure 3. In summary, RRA is effective for optimizing GNMF.

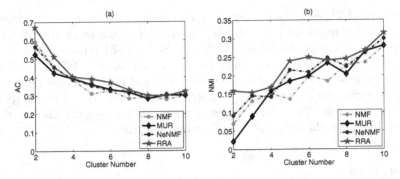

Fig. 5. AC and NMI of GNMF by MUR, NeNMF, and RRA on PIE dataset. NMF is used as a baseline.

Table 1. CPU seconds versus the percentage of energy kept for RRA on the PIE dataset

the percentage of energy kept	65%	75%	85%	95%	100%
CPU Seconds	199.18	184.59	183.92	177.76	209.47

4.4 Parameter Selection

In RRA, the percentage of energy kept in (17) controls the approximation quality of the SMW formula to the Hessian inverse as well as the time complexity because one iteration round of RRA costs $O(mnr + prn^2)$ time. The more energy kept, the larger the value p. It implies the higher the approximation quality and the less iteration rounds RRA spent. However, a large p means that each iteration round consumes much CPU time. So the percentage of energy kept must be carefully selected to balance the two facts. To study this point, we test RRA on PIE dataset when setting the reduced dimensionality to 10 and varying the percentage of energy kept from 65% to 95% with a step-size 10%. Table 1 compares the CPU time spent. To keep the fairness of comparison, all trials of RRA start from identical initial point and stop when same objective values are reached.

Table 1 shows that the time cost decreases with increasing of the percentage of energy kept. That is because the approximation quality is improved and iteration number is reduced in this case. However, the percentage of energy kept cannot be chosen too large such as 100% because that will increase the time complexity of each iteration round. In our experiment, 95% is a good choice.

5 Conclusion

In this paper, we proposed a novel efficient rank-one residue approximation (RRA) solver for graph regularized non-negative matrix factorization (GNMF). Unlike the existing GNMF solvers which recursively update each factor matrix as a whole, RRA recursively updates each column of both factor matrices in an analytic formulation. Although RRA needs a time-consuming matrix inverse operator, it can be approximated by using the Sherman-Morrison-Woodbury formula. Since the objective function of GNMF is continuously differentiable over a Cartesian product of several closed convex sets and RRA finds the optimal solution for each column of both factor matrices, RRA theoretically converges to a stationary point of GNMF. Experimental results and real-world image datasets confirm both the efficiency and the effectiveness of RRA.

References

1. Bertsekas, D.P.: Nonlinear programming. Athena Scientific (1999)
2. Cai, D., He, X., Han, J., Huang, T.S.: Graph regularized nonnegative matrix factorization for data representation. IEEE Transactions on Pattern Analysis and Machine Intelligence 33(8), 1548–1560 (2011)
3. Cai, D., He, X., Wu, X., Han, J.: Non-negative matrix factorization on manifold. In: Eighth IEEE International Conference on Data Mining (ICDM 2008), pp. 63–72. IEEE (2008)
4. Chung, F.R.: Spectral graph theory. In: CBMS Regional Conference Series in Mathematics, vol. 92 (1997)
5. Cichocki, A., Zdunek, R., Amari, S.-I.: Hierarchical als algorithms for nonnegative matrix and 3d tensor factorization. In: Davies, M.E., James, C.J., Abdallah, S.A., Plumbley, M.D. (eds.) ICA 2007. LNCS, vol. 4666, pp. 169–176. Springer, Heidelberg (2007)
6. Gu, Q., Zhou, J.: Neighborhood preserving nonnegative matrix factorization. In: Proc. 20th British Machine Vision Conf. (2009)
7. Guan, N., Tao, D., Luo, Z., Yuan, B.: Manifold regularized discriminative nonnegative matrix factorization with fast gradient descent. IEEE Transactions on Image Processing 20(7), 2030–2048 (2011)
8. Guan, N., Tao, D., Luo, Z., Yuan, B.: Nenmf: an optimal gradient method for nonnegative matrix factorization. IEEE Transactions on Signal Processing 60(6), 2882–2898 (2012)
9. Ho, N.D., Van Dooren, P., Blondel, V.D.: Descent methods for nonnegative matrix factorization. In: Numerical Linear Algebra in Signals, Systems and Control, pp. 251–293. Springer (2011)
10. Lin, C.J.: On the convergence of multiplicative update algorithms for nonnegative matrix factorization. IEEE Transactions on Neural Networks 18(6), 1589–1596 (2007)
11. Lin, C.J.: Projected gradient methods for nonnegative matrix factorization. Neural Computation 19(10), 2756–2779 (2007)
12. Nene, S.A., Nayar, S.K., Murase, H.: Columbia object image library (coil-20). Technical Report CUS-005-96, Columbia Univ. USA (1996)
13. Shen, B., Si, L.: Nonnegative matrix factorization clustering on multiple manifolds. In: Proceedings of the 24th AAAI Conference on Artificial Intelligence, pp. 575–580 (2010)
14. Sim, T., Baker, S., Bsat, M.: The cmu pose, illumination, and expression database. IEEE Transactions on Pattern Analysis and Machine Intelligence 25(12), 1615–1618 (2003)

15. Woodbury, M.A.: Inverting modified matrices. Memorandum Report 42, 106 (1950)
16. Xu, D., Yan, S., Tao, D., Lin, S., Zhang, H.-J.: Marginal fisher analysis and its variants for human gait recognition and content-based image retrieval. IEEE Transactions on Image Processing 16(11), 2811–2821 (2007)
17. Yang, J., Yang, S., Fu, Y., Li, X., Huang, T.: Non-negative graph embedding. In: IEEE Conference on Computer Vision and Pattern Recognition (CVPR), pp. 1–8. IEEE (2008)
18. Zhang, T., Fang, B., Tang, Y.Y., He, G., Wen, J.: Topology preserving non-negative matrix factorization for face recognition. IEEE Transactions on Image Processing 17(4), 574–584 (2008)

Maximum Entropy Models for Iteratively Identifying Subjectively Interesting Structure in Real-Valued Data

Kleanthis-Nikolaos Kontonasios[1], Jilles Vreeken[2], and Tijl De Bie[1]

[1] Intelligent Systems Laboratory, University of Bristol, Bristol, United Kingdom
{kk8232,Tijl.DeBie}@bristol.ac.uk
[2] Advanced Database Research and Modelling, University of Antwerp, Antwerp, Belgium
Jilles.Vreeken@ua.ac.be

Abstract. In exploratory data mining it is important to assess the significance of results. Given that analysts have only limited time, it is important that we can measure this *with regard to what we already know*. That is, we want to be able to measure whether a result is interesting from a subjective point of view.

With this as our goal, we formalise how to probabilistically model real-valued data by the Maximum Entropy principle, where we allow statistics on *arbitrary* sets of cells as background knowledge. As statistics, we consider means and variances, as well as histograms. The resulting models allow us to assess the likelihood of values, and can be used to verify the significance of (possibly overlapping) structures discovered in the data. As we can feed those structures back in, our model enables iterative identification of subjectively interesting structures.

To show the flexibility of our model, we propose a subjective informativeness measure for tiles, i.e. rectangular sub-matrices, in real-valued data. The *Information Ratio* quantifies how strongly the knowledge of a structure reduces our uncertainty about the data with the amount of effort it would cost to consider it.

Empirical evaluation shows that iterative scoring effectively reduces redundancy in ranking candidate tiles—showing the applicability of our model for a range of data mining fields aimed at discovering structure in real-valued data.

1 Introduction

When analysing a database, what we already know will strongly determine which results we will find interesting. Rather than analysing results representing knowledge we already have, we want to discover knowledge that is novel from *our* perspective. That is, the interestingness of a data mining result is highly subjective. Hence, in order to identify and mine such results we need theory and methods by which we can measure interestingness, significance, or surprise, from a *subjective* point of view.

With this as our goal, we formalise how to probabilistically model real-valued data using the Maximum Entropy principle [8], where we allow statistics over arbitrary sets of cells—such as, but not limited to, rectangular tiles—as background information. Using our model, we can measure the likelihood of values and value configurations from the perspective of what we already know about the data. As possible background knowledge, we develop theory for two easily considered sets of statistics for characterising distributions, namely means and variances, and histograms.

H. Blockeel et al. (Eds.): ECML PKDD 2013, Part II, LNAI 8189, pp. 256–271, 2013.

While being able to measure the significance of data mining results is important in general, it is particularly so for the sub-fields of data mining that discover local structures, such as in (frequent) pattern mining, subspace clustering, subgroup discovery and bi-clustering—each of which areas where we are typically faced with overly large and highly redundant result sets. As our model allows us to incorporate information on any area of the data, we are able to *iteratively* identify the most surprising pattern out of a given collection of candidate patterns; as by subsequently updating our model with this new information, all variations of the same theme will become predictable and hence onward be considered uninformative.

To quantify the subjective informativeness of a local structure we propose an *Information Ratio* measure for submatrices, or tiles, in real-valued data. It trades off how strongly the knowledge of the structure reduces our uncertainty about the data, with the amount of effort it would cost the analyst to consider it. For binary data, De Bie [5], and Kontonasios and De Bie [10] successfully used an information ratio based measure for identifying surprisingly dense tiles in binary data. Here, we generalise this notion to more rich statistics on expected values of tiles in real-valued data.

The topic of measuring the significance of data mining results was first discussed by Gionis et al. [6], who gave a method for testing significance by swap randomizing binary data. Ojala et al. [15] extended swap randomization to real-valued data. With swap randomization, however, only empirical p-values can be determined. Recently, we gave theory for modelling real-valued data by the Maximum Entropy principle [11]. One key advantage, besides speed, of MaxEnt modelling over swap randomization is that analytical modelling allows us to calculate *exact* probabilities.

Unlike the model we propose here, all of the above provide only relatively weak and *static* null-hypotheses. That is, they can not incorporate information beyond row and column margins, and hence can not identify redundancy in light of previous discoveries. Our model, on the other hand, allows for much stronger statistical testing as rich background knowledge can be incorporated, as well as for iterative use.

Empirical evaluation of our model shows that the information ratio measure reliably identifies highly informative tiles, and that by iteratively including the highest ranked tiles in our model, we correctly and non-redundantly identify the subjectively most informative tiles, subgroups, and bi-clusters on both synthetic and real data.

In sum, the main contributions of this paper are as follows:

- We give the first Maximum Entropy model for real-valued data that can incorporate knowledge of certain useful statistics (mean/variance, and histograms) over arbitrary areas of the data, and hence allows for *iterative* identification of the subjectively most informative data mining result;
- We formalise *Information Ratio* scores for contrasting the subjective informativeness of tiles in real-valued data against their descriptive complexity;

Finally, it is important to stress the main goal of this paper is theoretical. It is explicitly *not* our goal to define practical algorithms for the efficient search of non-redundant selection of tiles/subgroups/biclusters/subspace clusters/etc., all of which will likely need specialised solutions. Instead, our aim here is to provide the theoretical foundations that can be used as building blocks to those ends. As such, the information ratio we here propose for tiles in real-valued data is a proof-of-concept, not a general solution.

2   Related Work

We discuss related work along the following lines: measuring significance, iterative data mining, and identifying informative submatrices in real-valued data.

Assessing the significance of data mining results was first discussed by Gionis et al. [6]. The general idea is to check how likely a result is in data that shares basic properties, but is fully random otherwise. Swap-randomisation, essentially a Markov-chain of many random, local, value-swaps, for generating random data that exactly maintains the row and column margins of the original data. By sampling many random datasets (10 000s), we can subsequently obtain empirical p-values for the result at hand.

Ojala et al. generalised this to maintaining means and variances over rows and columns of real-valued data [15,14]. Recently, De Bie [5], and Kontonasios et al. [11] gave theory to instead model real-valued data analytically by the Maximum Entropy (MaxEnt) principle [8], maintaining properties by expectation instead of exactly. The formalisation by De Bie only allowed to maintain row and column means [5], Kontonasios et al. [11] generalised towards means and variances as well as histograms over rows and columns. Key advantages over swap-randomisation include the speed at which random data can be sampled, and that *exact* probabilities and p-values can be calculated.

None of these can incorporate information on arbitrary *submatrices* as background knowledge. As such, they are essentially static null hypotheses, and hence not applicable for identifying which result is most significant in light of previous discoveries.

The general concept of iterative data mining was first proposed by Hanhijärvi et al. [7]. A key advantage of the iterative approach is that it naturally eliminates redundancy. Based on MaxEnt, to the end that we can make an informed decision which result we should analyse next, Tatti and Vreeken [17] gave a general framework for measuring the difference of results of arbitrary methods in terms of the information they provide about the data at hand, and gave a proof-of-concept for binary data.

This paper broadly follows the lines of the framework for iterative data mining based on Maximum Entropy modelling of prior beliefs [4]. While that work was mostly abstract, we here bring these ideas closer to practice for a broad class of data and patterns. The Information Ratio was introduced in De Bie [5] (there called Compression Ratio) and Kontonasios and De Bie [10], respectively in the context of exact and noisy tile patterns in binary databases. In the current paper it is extended to real-valued data.

Bi-clustering [13] and sub-space clustering [12] are also concerned with identifying sub-matrices in real-valued data that exhibit structure different from the background. As in pattern mining, these approaches typically result in overly large, and highly redundant result sets [19]. Existing proposals to identify significant results either employ simple null-hypotheses that do not allow for iterative updating, require strong assumptions on the distribution of the data, or cannot deal with overlapping tiles [12].

3   Preliminaries

3.1   Notation and Basic Concepts

As data we consider rectangular real-valued matrices $\mathbf{D} \in \mathbb{R}^{n \times m}$. We denote the set of row indices as $\mathcal{I} = \{1, \ldots, n\}$ and the set of column indices as $\mathcal{J} = \{1, \ldots, m\}$.

A matrix element can be referred to using an element e from the Cartesian product of $\mathcal{I}$ and $\mathcal{J}$, i.e.: $e \in \mathcal{I} \times \mathcal{J}$. For $e = (i,j)$, $\mathbf{D}_e$ denotes the matrix element on row i and column j, sometimes also explicitly denoted as $\mathbf{D}_{ij}$. W.l.o.g. we assume the attributes to be normalised between 0 and 1.

Since the patterns we will consider in this paper are local patterns, we need to establish a notation for referring to such subsets. A subset of database entries can be indexed using an *index set*, here defined as a subset of the Cartesian product of $\mathcal{I}$ and $\mathcal{J}$, i.e.: $\mathcal{E} \subseteq \mathcal{I} \times \mathcal{J}$. We will use $\mathbf{D}_{\mathcal{E}}$ to refer to the (multi-)set of matrix values indexed by $\mathcal{E}$.

Special cases of index sets that will turn out to be of particular importance in this paper are index sets referring to: the elements within one particular row i (i.e. $\mathcal{E} = \{(i,j) \mid j \in \mathcal{J}\}$); the elements within one particular column j (i.e. $\mathcal{E} = \{(i,j) \mid i \in \mathcal{I}\}$); and the elements within a tile τ, which is defined as the set of elements in the intersection between a set of rows $\mathcal{I}_\tau \subseteq \mathcal{I}$ and a set of columns $\mathcal{J}_\tau \subseteq \mathcal{J}$.

Central in this paper is the notion of a pattern p, which we define as a triple $p = (\mathcal{E}, \mathbf{s}, \hat{\mathbf{s}})$. Here, $\mathcal{E}$ is an index set; $\mathbf{s}$ is a vector-valued function defined over sets of real-valued elements, called a *statistic*; and $\hat{\mathbf{s}}$ is the value the data miner beliefs to hold over the part of the data matrix indexed by $\mathcal{E}$. In general, this will be the empirical value, i.e.: $\hat{\mathbf{s}} = \mathbf{s}(\mathbf{D}_{\mathcal{E}})$, yet our theory below allows $\hat{\mathbf{s}}$ to take any valid value for $\mathbf{s}$.

We will focus on two statistics in particular. We define these by specifying the components of this vector-valued function.

- The first statistic we will consider has two components $s^{(1)}$ and $s^{(2)}$: resp. the function computing the sum of the set of values it is evaluated on, and the function computing the sum of their squares: $s^{(1)}(\mathbf{D}_{\mathcal{E}}) = \sum_{(i,j) \in \mathcal{E}} \mathbf{D}_{ij}$, and $s^{(2)}(\mathbf{D}_{\mathcal{E}}) = \sum_{(i,j) \in \mathcal{E}} \mathbf{D}_{ij}^2$. These two values (along with the cardinality of $\mathbf{D}_{\mathcal{E}}$) are sufficient to compute the mean and the variance of $\mathbf{D}_{\mathcal{E}}$, so patterns defined in these terms inform a user also on the mean and the standard deviation of a set of database elements.

- The second statistic has $d_{\mathcal{E}}$ components $s_{\mathcal{E}}^{(k)}$, specifying a $d_{\mathcal{E}}$-dimensional histogram for the elements indexed by $\mathcal{E}$. More specifically, given a set of $d_{\mathcal{E}} + 1$ real values $b_{0,\mathcal{E}} < b_{1,\mathcal{E}} < \ldots < b_{d_{\mathcal{E}},\mathcal{E}} \in \mathbb{R}$ specifying the boundary values of the histogram bins used for the set $\mathcal{E}$ (typically with $b_{0,\mathcal{E}} = -\infty$ and $b_{d_{\mathcal{E}},\mathcal{E}} = \infty$), we write $k_{\mathcal{E}}$ for the bin index for any $x \in \mathbb{R}$, i.e. $k_{\mathcal{E}}(x) = \max\{k \mid x < b_{k,\mathcal{E}}\}$. Then the k'th statistic $s_{\mathcal{E}}^{(k)}(\mathbf{D}_{\mathcal{E}})$ in this set is equal to the number of elements from $\mathbf{D}_{\mathcal{E}}$ that fall between $b_{k,\mathcal{E}}$ and $b_{k-1,\mathcal{E}}$, i.e. $s_{\mathcal{E}}^{(k)}(\mathbf{D}_{\mathcal{E}}) = \sum_{(i,j) \in \mathcal{E}} I(k_{\mathcal{E}}(\mathbf{D}_{ij}) = k)$, with I the indicator function. A further piece of notation that will prove useful is $w_{\mathcal{E}}(k) = b_{k,\mathcal{E}} - b_{k-1,\mathcal{E}}$, denoting the width of the k'th bin.

Given our broad definition of a pattern, we can assume that also background knowledge can be specified in terms of patterns, as formalised above, the data miner expects to be present in the data. The set of all triples specifying such background knowledge will be denoted as $\mathcal{B}$. We will show below how we will use such background knowledge to obtain a probability distribution P for the data, representing the data miner's background knowledge on $\mathbf{D}$. Note that the analogy between patterns and the background knowledge will prove useful in an iterative data mining setting: it allows us to naturally incorporate previously shown patterns into the background knowledge.

All logarithms are base 2, and by convention, $0 \log 0 = 0$.

3.2   The Maximum Entropy Principle, A Brief Primer

In our approach we make use of maximum entropy models, a class of probabilistic models that are identified by the Maximum Entropy principle [8]. This principle states that the best probabilistic model is the model that makes optimal use of the provided information, and that is fully random otherwise. This makes these models very suited for identifying informative patterns: when measuring quality, we know our measurement is based on our background information only, not on undue bias of the distribution.

Entropy maximisation problems are particularly well-suited to deal with background information on *expected* values of certain properties of the data. For example, the data miner may have an expectation about the value $\hat{f}$ of a certain function f when evaluated on the data. We embed this information in the background distribution by requiring that the expected value of f evaluated on $\mathbf{D}$ is equal to $\hat{f}$:

$$\int P(\mathbf{D})f(\mathbf{D})\, d\mathbf{D} = \hat{f}$$

Thus, inference of background model P is done by solving the following problem:

$$P^* = \max_P - \int P(\mathbf{D}) \log \left(P(\mathbf{D}) \right)\, d\mathbf{D} \,, \tag{1}$$

$$\text{s.t.} \quad \int P(\mathbf{D})f(\mathbf{D})\, d\mathbf{D} = \hat{f}\,, \quad \forall f \,, \tag{2}$$

$$\int P(\mathbf{D})\, d\mathbf{D} = 1 \,. \tag{3}$$

where each function f computes a statistic of which the data miner expects the value to be $\hat{f}$. It is clear that in the current context, these functions are determined by the triples $(\mathcal{E}, \mathbf{s}, \hat{\mathbf{s}})$ in the background knowledge. More specifically, for each pattern $(\mathcal{E}, \mathbf{s}, \hat{\mathbf{s}})$ there would be a corresponding set of functions f defined as $f(\mathbf{D}) = s^{(k)}(\mathbf{D}_{\mathcal{E}})$, and $\hat{f} = \hat{s}^{(k)}$, and this for each component $s^{(k)}$ of $\mathbf{s}$.

Entropy maximisation subject to such constraints is a convex problem, which can often be solved efficiently. Furthermore, the resulting distributions are known to belong to the exponential family of distributions, the properties of which are very well understood [18]. In particular, the maximum entropy distribution is of the form:

$$P(\mathbf{D}) = \frac{1}{Z} \exp \left\{ \sum_i \lambda_f f(\mathbf{D}) \right\} \,,$$

where Z is a normalisation factor known as the partition function and λ_f is a Lagrange multiplier corresponding to the constraint involving f, the value of which can be found by solving the dual optimisation problem.

4   Maximum Entropy Modeling

In this section we discuss the details of the maximum entropy modelling of the statistics discussed in Sec. 3. More in particular, we discuss how the constraint functions f in

Eq. (2) are to be instantiated, and we will provide the resulting probability distribution as the solution to the *MaxEnt* optimisation problem.

Here, we focus on how to *obtain* the MaxEnt distribution; in the next section, we will show how to *use* this model for identifying subjectively interesting tiles.

4.1 Encoding Means and Variances as Prior Knowledge

As we argued in Sec. 3, the information on the empirical mean and variance of a set of values is equivalent with information on their sum and sum of squares, computed by the functions $s^{(1)}$ and $s^{(2)}$ respectively. Thus, to consider background knowledge on the empirical mean and variance for a set of elements $\mathcal{E}$, we need the following two functions as constraint functions in Eq. (2):

$$f_{\mathcal{E}}^{(1)}(\mathbf{D}) \triangleq s^{(1)}(\mathbf{D}_{\mathcal{E}}) = \sum_{(i,j)\in\mathcal{E}} \mathbf{D}_{ij} \quad , \quad f_{\mathcal{E}}^{(2)}(\mathbf{D}) \triangleq s^{(2)}(\mathbf{D}_{\mathcal{E}}) = \sum_{(i,j)\in\mathcal{E}} \mathbf{D}_{ij}^2 \quad .$$

We will denote the corresponding Lagrange multipliers as $\lambda_{\mathcal{E}}^{(1)}$ and $\lambda_{\mathcal{E}}^{(2)}$ respectively. Since we need these functions for each $\mathcal{E}$ for which background knowledge is available, the total number of constraints of the type of Eq. (2) is equal to twice the number of index sets $\mathcal{E}$ for which the sum of the elements and the sum of the squares of the elements are part of the background knowledge.

Shape of the Solution. By solving the MaxEnt optimization problem using Lagrange Multiplier theory we obtain the following probability distribution:

$$P(\mathbf{D}) = \prod_{i,j} \sqrt{\frac{\beta_{ij}}{\pi}} \cdot \exp\left\{ -\frac{(\mathbf{D}_{ij} + \frac{1}{2} \cdot \frac{\alpha_{ij}}{\beta_{ij}})^2}{\frac{1}{\beta_{ij}}} \right\} , \tag{4}$$

where the variables α_{ij} and β_{ij} can be expressed as follows in terms of the Lagrange multipliers for the individual constraints:

$$\alpha_{ij} = \sum_{\mathcal{E}:(i,j)\in\mathcal{E}} \lambda_{\mathcal{E}}^{(1)} \quad , \quad \beta_{ij} = \sum_{\mathcal{E}:(i,j)\in\mathcal{E}} \lambda_{\mathcal{E}}^{(2)} \quad .$$

Thus the resulting probability distribution is a product of $n \times m$ independent *Normal* distributions, with means equal to $\mu_{ij} = -\frac{\alpha_{ij}}{2\beta_{ij}}$ and variance $\sigma_{ij}^2 = \frac{1}{2\beta_{ij}}$. Each factor in the product of Eq. (4) corresponds to the probability distribution for one value $\mathbf{D}_{ij}$ in the database. Sampling a full database from P hence comes down to $n \times m$ independent univariate sampling operations. Also, note that the distribution of $\mathbf{D}_{ij}$ depends *only* on the Lagrange multipliers of the subsets $\mathcal{E}$ in which the cell participates, i.e. for which $(i,j) \in \mathcal{E}$, which makes the sampling easier to conduct.

4.2 Encoding Histograms as Prior Knowledge

In order to encode background knowledge on the histogram within an index set $\mathcal{E}$, we need $d_{\mathcal{E}}$ constraint functions (as explained in Sec. 3, we assume that the histogram for subset $\mathcal{E}$ contains $d_\mathcal{E}$ bins, defined by the bin boundaries $b_{0,\mathcal{E}}, b_{1,\mathcal{E}}, \ldots, b_{d_\mathcal{E},\mathcal{E}} \in \mathbb{R}$, i.e. we allow number and values of the bin boundaries to be different for different $\mathcal{E}$). These functions are defined as: $f_\mathcal{E}^{(k)}(\mathbf{D}) \triangleq s_\mathcal{E}^{(k)}(\mathbf{D}_\mathcal{E}) = \sum_{(i,j)\in\mathcal{E}} I(k_\mathcal{E}(\mathbf{D}_{ij}) = k)$ where $k_\mathcal{E}(\mathbf{D}_{ij})$ denotes the bin index for the value $\mathbf{D}_{ij}$ (see Sec. 3). The functions $f_\mathcal{E}^{(k)}$ calculate the number of elements from $\mathbf{D}_\mathcal{E}$ that fall in the k'th bin of the histogram. We will denote the corresponding Lagrange multiplier as $\lambda_\mathcal{E}^{(k)}$.

Shape of the Solution. Using basic Lagrange multiplier theory one can show that the resulting probability distribution decomposes to a product of $n \times m$ independent probability distributions, i.e. $P(\mathbf{D}) = \prod_{i,j} P_{ij}(\mathbf{D}_{ij})$. Each component P_{ij} has the form

$$P_{ij}(\mathbf{D}_{ij}) = \frac{1}{Z_{ij}} \cdot \exp\left\{ \sum_{\mathcal{E}:(i,j)\in\mathcal{E}} \sum_{k=1}^{d_\mathcal{E}} \lambda_\mathcal{E}^{(k)} I\left(k_\mathcal{E}(\mathbf{D}_{ij}) = k\right) \right\} \quad,$$

where Z_{ij} is the decomposed partition function, formally defined as

$$Z_{ij} = \int_0^1 \exp\left\{ \sum_{\mathcal{E}:(i,j)\in\mathcal{E}} \sum_{k=1}^{d_\mathcal{E}} \lambda_\mathcal{E}^{(k)} I\left(k_\mathcal{E}(\mathbf{D}_{ij}) = k\right) \right\} d\mathbf{D}_{ij} \quad.$$

Each component refers to one data cell and it is affected only by the Lagrange Multipliers assigned to the sets $\mathcal{E}$ in which the entry participates.

4.3 Inferring the Model

The values of the λ parameters, which uniquely determine the MaxEnt model, are inferred by solving the Lagrange duals of the respective MaxEnt optimisation problems. These dual optimisation problems are convex, such that they can be solved efficiently using simple and well-known optimisation methods such as Conjugate Gradient (CG, the method of our choice in this paper) providing the gradient can be computed efficiently. Due to lack of space details cannot be shown here, but the theoretical complexity of each CG step is $O(\#\lambda^2)$ (with $\#\lambda$ the number of Lagrange multipliers), dominated by the cost of computing the gradient vector. In practice, however, we observe that run time develops linearly with the number of λs.

5 Measuring Subjective Interestingness

In this section we discuss how to use a MaxEnt model for measuring the subjective interestingness of a pattern. From 5.4 onward we focus on the specific case of tiles, a well-known and intuitive pattern type that lends itself for description in simple terms.

5.1 Quantifying Subjective Interestingness

The main goal of this section is defining a measure for the *subjective* interestingness of a pattern from the user's point of view. That is, how strongly a pattern contrasts to what the user already knows or beliefs about the data.

Loosely speaking, for a given pattern p, we have to define two important properties; the first we refer to as the *Information Content* of a pattern, or, the amount of information the pattern can convey to the user. The second property we refer to as the *Description Length* of a pattern, or, the cost for transmitting this information.

We can assume that a user, given his limited capacity for information processing, will find patterns with larger *Information Content* for a constant *Description Length* more interesting. In other words, the interestingness of a pattern can be formalised as a trade-off between these two quantities. This motivates the *Information Ratio* as a suitable measure of a pattern's interestingness.

Definition. The Information Ratio of a pattern p is the ratio of its *Information Content* and *Description Length*:

$$InfRatio(p) = \frac{InfContent(p)}{DescrLength(p)} \quad,$$

The Information Ratio was originally introduced in the context of exact tiles in binary data [5]. There it was shown that iteratively mining the tile with the highest *InfRatio* amounts to searching for a set of tiles with maximal total self-information given a limited budget on the overall description length. This was abstracted in [4] by relating iterative data mining in general to the budgeted set coverage problem, further justifying the use of this measure in iterative data mining.

5.2   Information Content of a Pattern

Our goal here is to quantify the information a user can extract from a pattern. We define the *InfContent* for a pattern p as the number of bits we gain when using p in addition to our current background knowledge when describing $\mathbf{D}$. More formally,

$$InfContent(p) = L(\mathbf{D} \mid \mathcal{B}) - L(\mathbf{D} \mid \mathcal{B} \cup \{p\}) \quad,$$

where $L(\mathbf{D} \mid \mathcal{B})$ is the number of bits required to describe $\mathbf{D}$ using only $\mathcal{B}$, and $L(\mathbf{D} \mid \mathcal{B} \cup \{p\})$ is the number of bits to describe $\mathbf{D}$ using both $\mathcal{B}$ and p.

We have, for the number of bits to encode $\mathbf{D}$ given background knowledge $\mathcal{B}$,

$$L(\mathbf{D} \mid \mathcal{B}) = \sum_{(i,j) \in \mathcal{I} \times \mathcal{J}} L(\mathbf{D}_{ij} \mid \mathcal{B}) \quad,$$

where we transmit, in a fixed order, the value of each cell $\mathbf{D}_{ij} \in \mathbf{D}$. To encode a specific value, we use optimal prefix codes [3]. We obtain the probabilities from our maximum entropy model P built using the information in $\mathcal{B}$. We hence have

$$L(\mathbf{D}_{ij} \mid \mathcal{B}) = -\log P_{ij}(\mathbf{D}_{ij}) \quad,$$

Intuitively, the better our model predicts the value of cell, i.e. the more expected a value is, the fewer bits are needed to encode it.

5.3   Estimating *InfContent*

With the above, we can measure the subjective interestingness of a pattern p, and so identify the most informative pattern from a large collection. To do so, however, the above formalisation requires us to infer the maximum entropy model for $\mathcal{B} \cup p$, and this for each candidate pattern p. This is clearly bound to be computationally prohibitive for large collections of candidate patterns. We therefore take a different approach, and instead *estimate* the gain of adding a pattern $p = (\mathcal{E}, \mathbf{s}, \hat{\mathbf{s}})$ to our model by focusing on the information gained regarding the database entries in $\mathcal{E}$. That is,

$$InfContent(p) = L(\mathbf{D}_\mathcal{E} \mid \mathcal{B}) - L(\mathbf{D}_\mathcal{E} \mid p) \quad,$$

which is the difference between the number of bits we need under the current model to describe the values of $\mathbf{D}_\mathcal{E}$ corresponding to the pattern, and the number of bits we would need to encode this area by solely using the information the pattern p provides. This approximation will be good as long as the pattern contains significantly more information about the database entries $\mathcal{E}$ concerned than the background information—a reasonable assumption given our focus on identifying the most informative patterns.

As discussed in Sec. 5.2, the first term $L(\mathbf{D}_\mathcal{E} \mid \mathcal{B})$ can be computed directly. For calculating $L(\mathbf{D}_\mathcal{E} \mid p)$ we consider two approaches, corresponding to the statistics discussed in Sec. 3.1. Each results in a different way of calculating the overall *InfContent*:

- Considering the first statistic, a pattern specifies the mean μ and variance σ^2 of the values in $\mathbf{D}_\mathcal{E}$. We know [3] that in this case the maximum entropy model for $\mathbf{D}_\mathcal{E}$ reduces to the normal distribution $\mathcal{N}(\mu, \sigma^2)$. Hence, with ϕ_{μ,σ^2} the normal density function, we have (with the subscript m to denote *mean and variance*):

$$L_m(\mathbf{D}_\mathcal{E} \mid p) = \sum_{(i,j) \in \mathcal{E}} - \log \phi_{\mu,\sigma^2}(\mathbf{D}_{ij}) \quad,$$

by which we encode the values in $\mathbf{D}_\mathcal{E}$ by an optimal prefix code proportional to the probability of the entry under this normal distribution.

- Patterns defined in terms of the second type of statistic specify a histogram for the values in $\mathcal{E}$. The maximum entropy model subject to this information reduces to a piecewise-constant density function, uniform with value $\frac{\hat{s}_\mathcal{E}^{(k)}}{|\mathcal{E}|} \times \frac{1}{w(k)}$ within the k'th bin of the histogram, where $\frac{\hat{s}_\mathcal{E}^{(k)}}{|\mathcal{E}|}$ represents the fraction of values from $\mathbf{D}_\mathcal{E}$ specified to belong to the k'th bin. (This follows from the fact that the MaxEnt distribution with a bounded domain is uniform [3].) Hence, we have (with the subscript h to denote *histogram*):

$$L_h(\mathbf{D}_\mathcal{E} \mid p) = \sum_{(i,j) \in \mathcal{E}} L_h(\mathbf{D}_{ij} \mid p) \quad,$$

where

$$L_h(\mathbf{D}_{ij} \mid p) = - \log \left(\frac{\hat{s}_\mathcal{E}^{(k_\mathcal{E}(\mathbf{D}_{ij}))}}{|\mathcal{E}|} \right) - \log \left(\frac{1}{w(k_\mathcal{E}(\mathbf{D}_{ij}))} \right) \quad,$$

in which, per entry $\mathbf{D}_{ij}$, the first term represents the description length for the bin index the entry falls in, and the second term for the actual value within that bin.

5.4   Description Length of a Pattern

So far our discussion has been fully general with respect to the choice of $\mathcal{E}$. From now on our exposition will be focused on patterns defined by a tile. Recall that a tile, denoted as τ, is defined as a sub-matrix of the original data matrix. The reason for focusing on tiles is that a tile τ can be described conveniently and compactly by specifying their defining row set $\mathcal{I}_\tau$ and column set $\mathcal{J}_\tau$.

Thus, for the *Description Length* of a tile pattern $p = (\mathcal{E}, \mathbf{s}, \hat{\mathbf{s}})$ where $\mathcal{E} = \{(i,j) \mid i \in \mathcal{I}_\tau, j \in \mathcal{J}_\tau\}$, we have $DescrLength(p) = L(\mathcal{E}) + L(\mathbf{s}, \hat{\mathbf{s}}) = L(\mathcal{I}_\tau) + L(\mathcal{J}_\tau) + L(\mathbf{s}, \hat{\mathbf{s}})$, where $L(\mathcal{I}_\tau) + L(\mathcal{J}_\tau)$ is the number of bits we need to identify the position of the tile. $L(\mathcal{I}_\tau)$ can be computed as:

$$L(\mathcal{I}_\tau) = \log(n) + \log \binom{n}{|\mathcal{I}_\tau|}$$

where the first term accounts for the transmission of the height $|\mathcal{I}_\tau|$ of tile τ. With this information, we can now identify which rows of $\mathbf{D}$ are part of the tile using an index over a binomial. We calculate $L(\mathcal{J}_\tau)$ analogously. Note that as we know the exact counts, encoding through an index over a binomial is at least as efficient as using individual prefix codes [3] as it makes optimal use of the available knowledge.

The quantity $L(\mathbf{s}, \hat{\mathbf{s}})$ scores the cost for transmitting the remaining information conveyed by the pattern. It is straightforward that the way *InfContent* is calculated dictates a certain approach for the *DescrLength* as well. Let us discuss these here in turn.

- For mean and variance of a tile as background knowledge we have

$$L_m(\mathbf{s}, \hat{\mathbf{s}}) = 2 \log(10^{acc})$$

  where we encode $mean(\mathbf{D}_\tau)$ and $var(\mathbf{D}_\tau)$ using a uniform prior.
- In the case of histograms as background information, we have

$$L_h(\mathbf{s}, \hat{\mathbf{s}}) = L_\mathbb{N}(d_\mathcal{E}) + \log \binom{10^{acc}}{d_\mathcal{E} - 1} + \log \binom{|\mathcal{E}| + d_\mathcal{E} - 1}{d_\mathcal{E} - 1}$$

where we first encode the number of bins $d_\mathcal{E}$ in the histogram, using the MDL optimal code $L_\mathbb{N}$ for integers ≥ 1 [16]. This encoding requires progressively more bits the higher the value—by which we explicitly reward simple histograms. In the next term, $\log \binom{10^{acc}}{d_\mathcal{E}-1}$, we encode the split points between the bins. These terms account for specifying the histogram, i.e. the statistic s used.

The last term, $\log \binom{|\mathcal{E}|+d_\mathcal{E}-1}{d_\mathcal{E}-1}$, encodes $\hat{\mathbf{s}}$: how many observations fall within each bin. We have to partition $|\mathcal{E}|$ entries over $d_\mathcal{E}$ possibly empty bins. This is known as a weak composition. The number of weak compositions of k non-negative terms summing up to n is given by $\binom{n+k-1}{k-1}$. Assuming an ordered enumeration, we need $\log \binom{|\mathcal{E}|+d_\mathcal{E}-1}{d_\mathcal{E}-1}$ bits to identify our composition. Note that $\log \binom{n}{k} = \log \Gamma(n+1) - \log \Gamma(k+1) - \log \Gamma(n-k+1)$ and hence is calculable even for large n and k.

Each of these methods require an accuracy level acc to be specified. Ultimately to be decided by the user, a natural choice is the number of significant digits of the data [9].

5.5   Instantiating *InfRatio*

The discrimination between transmitting mean/variance and histograms allows us to instantiate the *InfRatio* of a tile τ in two different ways per scheme.

$$InfRatio_s(p) = \frac{L(\mathbf{D}_{\mathcal{E}} \mid \mathcal{B}) - L_s(\mathbf{D}_{\mathcal{E}} \mid p)}{L(\mathcal{E}) + L_s(\mathbf{s}, \hat{\mathbf{s}})}$$

where s is the statistic used as prior knowledge in the modelling of the database.

6   Iteratively Identifying Subjectively Interesting Structure

In Section 4 we discussed how to obtain a Maximum Entropy model for a real-valued database under background information of statistics $\mathbf{s}$ over arbitrary sets of cells $\mathcal{E}$ of the data, and Section 5 proposed *InfRatio* as a measure for the informativeness of patterns in the data based on their unexpectedness under this background knowledge.

In practice, to discover novel knowledge, we propose to iteratively find the most informative pattern with regard to background knowledge; then present this pattern to the user, and continue our search after incorporating the pattern in the background knowledge and updating the MaxEnt model accordingly.

What background knowledge to start the search from is up to the user; it may be empty, or can consist of already known patterns. In our experiments, we choose to compute the initial Maximum Entropy model with statistics on the row and column distributions as prior knowledge. We then use *InfRatio* to rank a collection of candidate patterns $\mathcal{F}$, and identify the top-ranked pattern as most interesting. This pattern is henceforth considered prior knowledge. From this point the algorithm iterates with

1. a Maximum Entropy modelling step, using all accumulated background knowledge
2. an *InfRatio* ranking for the tiles in $\mathcal{F}$ in each iteration.

We terminate when user dependent criteria are met. These criteria can be objective (e.g. top-k, log-likelihood of the data, or a model selection criterion such as MDL, BIC or AIC), subjective (e.g. human evaluation of patterns) or a combination of both.

7   Experiments

In this section we empirically evaluate our MaxEnt model for real-valued data. We stress, however, that the contribution of this paper is theoretical in nature—this section should be regarded as proof-of-concept, not a final application. Though our modelling theory is general for $\mathcal{E}$, for the practical reasons of defining meaningful *InfRatio*, as well as for mining candidate patterns $\mathcal{F}$ we here restrict ourselves to tiles.

7.1   Setup

We evaluate whether iterative ranking helps to correctly identify the most informative tiles on data with known ground truth. Second, for synthetic and real data, we investigate the *InfRatio* and log-likelihood curves of the top-k identified patterns.

We implemented our model in C++, and make our code available for research purposes.[1] All experiments were performed on a 2.66GHz Windows 7 machine.

As initial background knowledge to our model, we always include statistics over the row and column distributions.

Data. In order to evaluate performance with a known ground truth, we use synthetic data. We generate a 500-by-500 dataset, in which we plant four complexes of five overlapping tiles, for which the values are distributed significantly different from the background. We refer to this dataset as *Synthetic*.

Per complex, we plant three tiles of 15-by-15, one of 10-by-10, and one of 8-by-8. Values of cells not associated with a complex are drawn from a Normal distribution with mean $\mu = 0.3$ and variance $\sigma = 0.15$. For cells associated with complex A we use $\mu = 0.15$ and $\sigma = 0.05$; for complex B we use $\mu = 0.2$ and $\sigma = 0.05$; for complex C, $\mu = 0.3$ and $\sigma = 0.05$; and for complex D, $\mu = 0.4$ and $\sigma = 0.05$.

For the set of candidate tiles to be ranked, in addition to the 20 true tiles, we randomly generate 230 tiles of 15-by-15. Per random tile we uniformly randomly select columns and rows. Note that by generating these at random, they may or may not overlap with the complexes, and hence may or may not identify a local distribution significantly different from the background.

We also evaluate on real data. The *Alon* dataset is a 62-by-2000 gene expression dataset [1]. To obtain tiles for this data, using CORTANA[2] at default settings, we mined the top-1000 subgroups of up to 3 attributes. Subgroups naturally translate into tiles; simply consider the features of the rule as columns, and the transactions satisfying the rule as rows. The Arabidopsis thaliana, or *Thalia*, is a 734-by-69 gene expression dataset.[3] For this data we mined 1 600 biclusters using BIVISU at default settings [2].

7.2 Ranking Tiles

In our first experiment we evaluate iteratively identifying subjectively informative patterns. In particular, we use *InfRatio* to rank the 250 candidate tiles of the *Synthetic* dataset, starting with background knowledge $\mathcal{B}$ containing only statistics on the row and column distributions. At every iteration, we incorporate the top-ranked tile into the model, and re-rank the candidates.

We evaluate modelling with means-variances, and with histograms information. For both, we report both the ranking of the first five iterations, and the final top-10 iterative ranking. We depict the id of the top-ranked tile in an iteration in bold. We give the results in Table 1. On the left, we give the *InfRatio* rankings for transmitting means-variances, i.e. using L_m, while on the right we give the rankings for when transmitting histograms information, i.e. using L_h.

Table 1 shows that *InfRatio* consistently ranks the largest and least-overlapping tiles at the top. In the first iteration the top-10 tiles include the largest tiles from each of

---

[1] http://www.tijldebie.net/software
[2] CORTANA: http://datamining.liacs.nl/cortana.html
[3] http://www.tik.ee.ethz.ch/~sop/bimax/

Table 1. Iterative Ranking. The top-10 ranked tiles for the first five iterations, plus the final ranking of the top-10 most informative tiles, from 250 candidate tiles. We give the results for the *Synthetic* dataset both for mean-variance modelling and transmission using L_m (left), resp. for histogram modelling with transmission using L_h (right). Tiles prefixed with a letter denote planted tiles, those denoted with a plain number are random.

	Mean-Variances						Histograms					
Rank	It 1	It 2	It 3	It 4	It 5	Final	It 1	It 2	It 3	It 4	It 5	Final
1.	**A2**	**B3**	**A3**	**B2**	**C3**	**A2**	A1	C2	**B1**	C1	D1	**A1**
2.	A4	B4	B2	C3	C4	**B3**	C2	B1	B2	D1	D3	**C2**
3.	A3	B2	C3	C4	C2	**A3**	B1	C1	B3	D3	B2	**B1**
4.	B3	A3	C4	C2	D2	**B2**	C1	C3	C1	B2	D2	**C1**
5.	B4	C3	C2	B4	D4	**C3**	C3	B2	D1	D2	A2	**D1**
6.	B2	C4	B4	D2	D3	**C2**	B2	B3	D3	A2	84	**B2**
7.	C3	C2	D2	D4	D1	**D2**	B3	D1	D2	C3	25	**A2**
8.	C4	D2	D4	D3	A5	**D3**	A3	D3	A2	25	228	**D2**
9.	C2	D4	D3	D1	21	**A5**	A2	D2	C3	84	43	**228**
10.	D2	D3	B1	A5	B5	**B5**	D1	A2	25	228	33	**43**

the four complexes. The final ranking shows that for each complex large and little-overlapping tiles are selected. Note that due to the random value generation some tiles within a complex may stand out more from the rest of the data than others.

In general, modelling with histograms is more powerful than the means-variances case. Clearly, this does rely on the quality of the used histograms. Here, we use histograms that balance complexity with the amount of data by the MDL principle [9]. As we here mainly consider small tiles, the obtained histograms are likely to underfit the underlying Normal distribution of the data. Nevertheless, Table 1 shows that the largest tiles of each complex are again top-ranked. As an artifact of the above effect, the final ranking does include a few randomly generated tiles—for which we find that by chance their values indeed differ strongly from the background distribution, and hence identify potentially interesting areas.

From this experiment we conclude that our *InfRatio* for real-valued tiles coupled with iterative modelling leads to the correct identification of subjectively interesting patterns, while strongly reducing redundancy.

7.3   *InfRatio* and Log-likelihood Curves

Next, we examine *InfRatio* and the iteratively obtained rankings on both artificial and real-world data. Evaluating the interestingness of patterns found in real-world data is highly subjective, however. The negative log-likelihood of the data is often considered as a measure of the surprise of the data as a whole. Finding and encoding informative patterns will provide more insight in the data, and so decrease surprise; the negative log-likelihood scores. Since we have a probabilistic model computing these scores is straightforward. In our experiments we expect to see significant decreases of the score after the discovery of a surprising pattern.

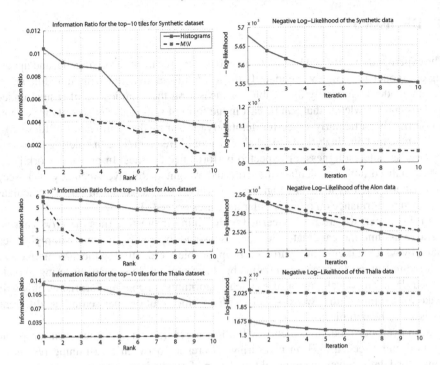

Fig. 1. *InfRatio* values for the top-10 ranked tiles (left column) and negative log-likelihood of the data (right) for the *Synthetic* (top row), *Alon* (middle row), and *Thalia* (bottom row) datasets. The solid blue line corresponds to histograms, the dashed black line to means variance modelling. Note the downward trend for both plot types: the most informative tiles are ranked first.

Figure 1 presents the *InfRatio* and negative log-likelihood scores for the first 10 iterations of our algorithm for resp. the *Synthetic, Alon,* and *Thalia* datasets.

We observe the *InfRatio* values are non-increasing over the whole range of the ten iterations, for all datasets and coding schemes. An immediate explanation is that in successive iterations more (correct) information is encoded into the model, and hence patterns become at most as surprising as they were in previous iterations. We further observe decreasing negative log-likelihood curves over the ten iterations. The fact that the data gets less surprising in every iteration is an indication that the patterns discovered and encoded in the model are significant. Loosely speaking, we can also say that the more the decrease per iteration the more interesting the corresponding pattern is.

We see that for *Synth* means-variance modelling obtains the best scores (note the very low negative log-likelihood scores). This follows as the data was generated by Normal distributions. For our real datasets, on the other hand, histogram modelling is in its element as these datasets are less well explained by Gaussians.

The modelling time for these datasets, for the first ten iterations range between seconds (*Synthetic*), tens of seconds (*Thalia*, and Histograms on *Alon*), up to hundreds of seconds for *Alon* when modelling means/variances. Scalability experiments (not shown due to lack of space) show run time scales linearly with the number of added tiles.

8 Discussion

The experiments verify that by iteratively identifying the most surprising tile, and updating our model accordingly, we identify the top-k informative tiles without redundancy.

Our model is generally applicable for measuring the significance of results under background knowledge that can be cast in the form of statistics over sets of entries, yet it is currently particularly suited for iteratively measuring the significance of tiles.

We note that our analytical model allows a lot of freedom for constructing measures of surprise, interestingness, or expected performance for tiles. In previous work [11], we formalised how to calculate the expected Weighted Relative Accuracy (WRAcc) of a pattern, such as used in subgroup discovery. More research on how to calculate expected performance on other popular measures is needed.

In this paper, we are not concerned with mining interesting tiles *directly*; That is, we here assume the candidate tiles are discovered externally, and we 'simply' order them; by which it is currently particularly applicable for identifying the informative and non-redundant results in pattern mining, bi-clustering, and sub-space clustering of real-valued data. The development of efficient algorithms for mining surprising tiles will make for exciting future work—likely with specific solutions per setting, as different goals are to be optimised.

Moreover, our modelling theory will allow a generalisation of the recent proposal by Tatti and Vreeken [17] on measuring differences between data mining results to real-valued data—given meaningful translation of results into patterns $p = (\mathcal{E}, \mathbf{s}, \hat{\mathbf{s}})$.

Perhaps most interesting from our perspective is to extend our modelling theory further towards richer types of structure. A straightforward step would be to incorporate into the model not just a statistic over the whole tile, but instead do so per row and/or column. Beyond means-variances, and histograms, there are other important notions of structure such as similarities between rows, as well as correlations between attributes—each of which will further extend the applicability towards the above applications.

In this light it is important to note that our models, and in particular the histogram variant, are already directly applicable for evaluating significances and measuring subjective interestingness on ordinal and integer valued data.

9 Conclusion

We formalised how to probabilistically model a real-valued dataset by the Maximum Entropy principle, such that we can iteratively feed in background information on the distributions of arbitrary subsets of elements in the database. To show the flexibility of our model, we proposed the *InfRatio* measure to quantify the subjective interestingness of tiles, trading off how strongly the knowledge of the pattern reduces our uncertainty about the data with how much effort would it costs the analyst to consider it.

Empirical evaluation showed that by iteratively scoring candidate tiles, and subsequently updating the model with the most informative tile, we can effectively reduce redundancy in the set of candidate tiles—showing the applicability of our model for a range of data mining fields dealing with real-valued data.

Acknowledgements. Kleanthis-Nikolaos Kontonasios and Tijl De Bie are supported by EPSRC grant EP/ G056447/1, the European Commission through the PASCAL2 Network of Excellence (FP7-216866) and a UoB Scholarship. Jilles Vreeken is supported by a Post-Doctoral Fellowship of the Research Foundation – Flanders (FWO).

References

1. Alon, U., Barkai, N., Notterman, D.A., Gish, K., Mack, D., Ybarra, S., Levine, A.J.: Broad patterns of gene expression revealed by clustering of tumor and normal colon tissues probed by oligonucleotide arrays. PNAS 96(12), 6745–6750 (1999)
2. Cheng, K.O., Law, N.F., Siu, W.C., Lau, T.H.: Bivisu: software tool for bicluster detection and visualization. Bioinformatics 23(17), 23–42 (2007)
3. Cover, T.M., Thomas, J.A.: Elements of Information Theory. Wiley-Interscience, New York (2006)
4. De Bie, T.: An information theoretic framework for data mining. In: KDD, pp. 564–572. ACM (2011)
5. De Bie, T.: Maximum entropy models and subjective interestingness: an application to tiles in binary databases. Data Min. Knowl. Disc. 23(3), 407–446 (2011)
6. Gionis, A., Mannila, H., Mielikäinen, T., Tsaparas, P.: Assessing data mining results via swap randomization. TKDD 1(3), 167–176 (2007)
7. Hanhijärvi, S., Ojala, M., Vuokko, N., Puolamäki, K., Tatti, N., Mannila, H.: Tell me something I don't know: randomization strategies for iterative data mining. In: KDD 2009, pp. 379–388. ACM (2009)
8. Jaynes, E.T.: On the rationale of maximum-entropy methods. Proc. IEEE 70(9), 939–952 (1982)
9. Kontkanen, P., Myllymäki, P.: MDL histogram density estimation. In: AISTATS (2007)
10. Kontonasios, K.-N., De Bie, T.: An information-theoretic approach to finding noisy tiles in binary databases. In: SDM, pp. 153–164. SIAM (2010)
11. Kontonasios, K.-N., Vreeken, J., De Bie, T.: Maximum entropy modelling for assessing results on real-valued data. In: ICDM, pp. 350–359. IEEE (2011)
12. Kriegel, H.-P., Kröger, P., Zimek, A.: Clustering high-dimensional data: A survey on subspace clustering, pattern-based clustering, and correlation clustering. ACM TKDD 3(1), 1–58 (2009)
13. Madeira, S.C., Oliveira, A.L.: Biclustering algorithms for biological data analysis: A survey. IEEE/ACM TCBB 1(1), 24–45 (2004)
14. Ojala, M.: Assessing data mining results on matrices with randomization. In: ICDM, pp. 959–964 (2010)
15. Ojala, M., Vuokko, N., Kallio, A., Haiminen, N., Mannila, H.: Randomization methods for assessing data analysis results on real-valued matrices. Stat. Anal. Data Min. 2(4), 209–230 (2009)
16. Rissanen, J.: Modeling by shortest data description. Annals Stat. 11(2), 416–431 (1983)
17. Tatti, N., Vreeken, J.: Comparing apples and oranges - measuring differences between exploratory data mining results. Data Min. Knowl. Disc. 25(2), 173–207 (2012)
18. Wainwright, M., Jordan, M.I.: Graphical models, exponential families, and variational inference. Foun. Trends Mach. Learn. 1(1-2), 1–305 (2008)
19. Zimek, A., Vreeken, J.: The blind men and the elephant: on meeting the problem of multiple truths in data from clustering and pattern mining perspectives. Mach. Learn. (in Press, 2013)

An Analysis of Tensor Models
for Learning on Structured Data

Maximilian Nickel[1] and Volker Tresp[2]

[1] Ludwig Maximilian University, Oettingenstr. 67, Munich, Germany
nickel@dbs.ifi.lmu.de
[2] Siemens AG, Corporate Technology, Otto-Hahn-Ring 6, Munich, Germany
volker.tresp@siemens.com

Abstract. While tensor factorizations have become increasingly popular for learning on various forms of structured data, only very few theoretical results exist on the generalization abilities of these methods. Here, we discuss the tensor product as a principled way to represent structured data in vector spaces for machine learning tasks. By extending known bounds for matrix factorizations, we are able to derive generalization error bounds for the tensor case. Furthermore, we analyze analytically and experimentally how tensor factorization behaves when applied to over- and understructured representations, for instance, when two-way tensor factorization, i.e. matrix factorization, is applied to three-way tensor data.

Keywords: Tensor Factorization, Structured Data, Generalization Error Bounds.

1 Introduction

Learning from structured data is a very active line of research in a variety of fields, including social network analysis, natural language processing, bioinformatics, and artificial intelligence. While tensor factorizations have a long tradition in psycho- and chemometrics, only more recently have they been applied to various tasks on structured data in machine learning. Examples include link prediction and entity resolution on multi-relational data [18,13] and large knowledge bases [3,19], item recommendation on sequential data [20,21], or the analysis of time varying social networks [2]; only to name a few examples. A reason for the success of tensor methods in these tasks is their very appealing property to efficiently impose structure on the vector space representation of data. Moreover, tensor factorizations can be related to multilinear models, which overcome some limitations of linear models, such as their limited expressiveness, but at the same time remain more scalable and easier to handle than non-linear approaches. However, despite their increasing popularity and their appealing properties, only very few theoretical results exist on the generalization abilities of tensor factorizations. Furthermore, an important open question is what kind of generalization improvements over simpler, less structured models can be expected. For instance,

H. Blockeel et al. (Eds.): ECML PKDD 2013, Part II, LNAI 8189, pp. 272–287, 2013.
© Springer-Verlag Berlin Heidelberg 2013

propositionalization, which transforms relational data into feature-based representations, has been considered as a mean for relational learning [15,12]. In terms of tensor factorization, propositionalization would be equivalent to transforming a tensor into a matrix representation prior to computing the factorization. While it has been shown empirically that tensor methods usually scale better with the amount of missing data than their matrix counterparts [26,16,25,22] and that they can yield significantly improved results over "flat" methods which ignore a large part of the data structure [18], no theoretical justification of this behavior is known in terms of generalization bounds.

In this paper, we approach several of these open questions. First, we will briefly discuss the tensor product as a principled way to derive vector space representations of structured data. Subsequently, we will present the first generalization error bounds of tensor factorizations for classification tasks. We will analyze experimentally the effect of imposing structure on vector space representations via the tensor product as well as the effect of constraints that are applied to popular tensor decompositions. Based on the newly derived bounds we discuss how these results can be interpreted analytically.

2   Theory and Methods

In this section we will briefly review concepts related to tensor factorization, as far as they are important for the course of this paper. Furthermore, we will discuss how structured data can be modeled as weighted sets of n-tuples, which enables a closer analysis of the relations between tensor factorizations and structured data.

In the following, scalars will be denoted by lowercase letters x; vectors will be denoted by bold lowercase letters $\boldsymbol{x}, \boldsymbol{y}$ with elements x_i, y_j. Vectors are assumed to be column vectors. Matrices will be denoted by uppercase letters X, Y with elements x_{ij}. Tensors will be indicated by upright bold uppercase letters $\mathbf{X}, \mathbf{Y}$ with elements $x_{i_1,\dots,i_n}$. For notational convenience, we will often group tensor indices into a vector $\boldsymbol{i} = [i_1, \cdots, i_n]^T$ and write $x_{\boldsymbol{i}}$ instead of $x_{i_1,\dots,i_n}$. Sets will be denoted by calligraphic letters $\mathcal{S}$ and their cardinality will be denoted by $|\mathcal{S}|$.

2.1   Tensor Product

First, we will review basic properties of the tensor product. The review closely follows the discussions in [4] and [14].

Definition 1 (Tensor Product of Vectors). *The tensor product of vectors* $\boldsymbol{x} \in \mathbb{R}^{n_1}$ *and* $\boldsymbol{y} \in \mathbb{R}^{n_2}$, *denoted by* $\boldsymbol{x} \otimes \boldsymbol{y}$, *is an array with* $n_1 n_2$ *entries, where*

$$(\boldsymbol{x} \otimes \boldsymbol{y})_{ij} = x_i y_j$$

The defining property of the tensor product of vectors is that $(\boldsymbol{x} \otimes \boldsymbol{y})_{ij} = x_i y_j$. However, since the "shape" of $\boldsymbol{x} \otimes \boldsymbol{y}$ is not defined, there exists a deliberate ambiguity in how to compute the tensor product of vectors. In particular, for two vectors $\boldsymbol{x}$, $\boldsymbol{y}$, we might obtain one- or two-dimensional arrays with

$$\boldsymbol{x} \otimes \boldsymbol{y} = \left[x_1 \boldsymbol{y}^T \; x_2 \boldsymbol{y}^T \; \ldots \; x_n \boldsymbol{y}^T \right]^T \in \mathbb{R}^{mn} \tag{1}$$

$$\boldsymbol{x} \otimes \boldsymbol{y} = \boldsymbol{x} \boldsymbol{y}^T \in \mathbb{R}^{m \times n} \tag{2}$$

We will refer to eq. (1) as a vectorized representation of the tensor product, as its results is again a vector, while eq. (2) will be called a *structured* representation. Usually, it will be clear from context which representation is used. The tensor product of vectors is easily extended to more than two vectors, e.g. $(\boldsymbol{x} \otimes \boldsymbol{y} \otimes \boldsymbol{z})_{ijk} = x_i y_j z_k$. In the following, we will denote the tensor product of n vectors also by $\bigotimes_n \boldsymbol{v}_n$. In the structured representation, the tensor product of n vectors corresponds to an n-dimensional array. Furthermore, the tensor product of vectors preserves their linear independence: if the vectors $\{\boldsymbol{x}_1, \ldots, \boldsymbol{x}_n\}$ and $\{\boldsymbol{y}_1, \ldots, \boldsymbol{y}_m\}$ are, respectively, linearly independent, then the vectors $\{\boldsymbol{x}_i \otimes \boldsymbol{y}_j \mid i = 1 \ldots n, \; j = 1 \ldots m\}$ are also linearly independent.

Definition 2 (Tensor Product of Vector Spaces). *The tensor product of vector spaces V and W, denoted by $V \otimes W$, is the vector space consisting of all linear combinations $\sum_i a_i \boldsymbol{v}_i \otimes \boldsymbol{w}_i$, where $\boldsymbol{v}_i \in V$ and $\boldsymbol{w}_i \in W$.*

Similarly to the tensor product of vectors, the tensor product of vector spaces is easily extended to more than two vector spaces. In the following, $\bigotimes_n V_n$ will denote the tensor product of n vector spaces. We will refer to a vector space that is the result of tensor products of vector spaces also as a *tensor product space*.

Definition 3 (Tensor). *Let $V = \bigotimes_n W_n$ be a tensor product space with $n \geq 1$. The elements of V are called n-th order tensors.*

Following definition 1 and definition 3, tensors can be interpreted in different ways. One way is as a *vector in a structured vector space*, what corresponds to the vectorized representation in eq. (1). However, according to the structured representation in eq. (2), tensors can also be viewed as *multidimensional arrays*, which is the more commonly used interpretation. Here, we will use both interpretations interchangeably. It also follows immediately that any vector is a first-order tensor and each matrix is a second-order tensor. In the following, $\text{ord}(\mathbf{X})$ will denote the order of a tensor $\mathbf{X}$. For notational convenience, we will also write $\mathbf{X} \in \mathbb{R}^{\prod_i n_i}$ instead of $\mathbf{X} \in \mathbb{R}^{n_1 \times \cdots \times n_k}$.

2.2 Structured Data, the Cartesian, and the Tensor Product

To analyze the relation between the order of a tensor and the "structuredness" of data representation we introduce the concept of the *order of structured data*. The general framework in which we will describe structured data is in form of sets of weighted m-tuples, which are defined as follows:

Definition 4 (Set of Weighted m-Tuples). *Let $V = V^{(1)} \times \cdots \times V^{(m)}$ be the Cartesian product over m sets $V^{(1)}, \ldots, V^{(m)}$ and let $\phi : \mathcal{E} \mapsto \mathbb{R}$ be a real-valued function that assigns a weight to each m-tuple in $\mathcal{E} \subseteq V$. A set of weighted m-tuples $\mathcal{T}$ is then defined as a 4-tuple $(V, \mathcal{E}, \phi, m)$. The order of $\mathcal{T}$ is defined as the length of its tuples m. For conciseness, we will refer to sets of weighted m-tuples also as* weighted tuple-sets.

Weighted tuple-sets can be interpreted in the following way: The elements of the sets $V^{(1)}, \ldots, V^{(m)}$ correspond to the constituents of the structured data. The set $\mathcal{E}$ corresponds to the observed m-tuples, while V corresponds to all possible m-tuples. For a tuple $t \in \mathcal{E}$, the pair $(t, \phi(t))$ corresponds to an observed data point. This is a very general form of data representation that allows us to consider many forms of structured data. For instance, *dyadic multi-relational* data – as it arises in the Semantic Web or Linked Data – has a natural representation as a weighted tuple-set, where $V^{(e)}$ is the set of all entities in the data, $V^{(p)}$ is the set of all predicates, and the weight function $\phi : V^{(p)} \times V^{(e)} \times V^{(e)} \mapsto \{\pm 1\}$ is defined as

$$\phi(p_i, e_j, e_k) = \begin{cases} +1, & \text{if the relationship } p_i(e_j, e_k) \text{ exits} \\ -1, & \text{otherwise} \end{cases}.$$

Similarly, *sequential* or *time-varying* data can be modeled via m-tuples such as (`user`, `item`, `last item`) triples for item recommendation [20] or (`person`, `person`, `month`) triples in time-varying social networks [2]. In these cases, the function ϕ could model the rating of a product or the interaction of persons. Furthermore, traditional *attribute-value data*, as it is common in many machine learning applications, can be modeled via (`object`, `attribute`) pairs, which are weighted by the respective attribute values, e.g. $\phi(\text{Anne}, \text{age}) = 36$.

Tuple-sets can be modeled very naturally using tensors in the following way: Let $\mathcal{T} = (V, \mathcal{E}, \phi, m)$ be a weighted tuple-set and let $I^{(i)}$ be the standard basis of dimension $|V^{(i)}|$, such that it indexes all elements of $V^{(i)}$. $\mathcal{T}$ can then be modeled as a tensor $\mathbf{Y} \in \bigotimes_{i=1}^{m} I^{(i)}$ with entries $y_{i_1, \ldots, i_m} = \phi(v_{i_1}, \ldots, v_{i_m})$ for all observed tuples $(v_{i_1}, \ldots, v_{i_m}) \in \mathcal{E}$. For unobserved tuples $(v_{i_1}, \ldots, v_{i_m}) \in V \setminus \mathcal{E}$, the corresponding entries in $\mathbf{Y}$ are modeled as missing. Using this construction, each set of objects $V^{(i)}$ is indexed separately by a mode of the tensor $\mathbf{Y}$. Therefore, it holds that the order of the tensor $\mathbf{Y}$ is identical to the order of the weighted tuple-set $\mathcal{T}$. This enables us to rephrase the question how the structuring of a vector space representation affects the generalization ability of a factorization in terms of the order of weighted tuple-sets and the order of tensors. In particular we are interested in how the generalization ability changes for a tensor representation that has *not* the same order as the underlying weighted tuple-set; compared to a tensor representation that has the identical order.

In this work, we will only consider the problem of learning from sets of binary-weighted tuples, i.e. tuple-sets with weight functions of the form $\phi : \mathcal{E} \mapsto \{\pm 1\}$. This corresponds to a classification setting on binary tensors where $y_i \in \{\pm 1\}$ indicates the presence or absence of an m-tuple.

2.3   Tensor Factorizations

Learning via tensor factorizations is based on the idea of explaining an observed tensor $\mathbf{Y}$ through a set of latent factors. The Tucker decomposition is a very general form of factorizing a tensor and allows us to consider different factorization methods within this framework through additional constraints. The Tucker decomposition is defined as

Definition 5 (Tucker Decomposition). *Let* $\mathbf{Y} \in \mathbb{R}^{\prod_i n_i}$ *be an observed tensor with* $\operatorname{ord}(\mathbf{Y}) = m$. *The Tucker decomposition with n-rank* $(r_1, \ldots, r_m)$ *factorizes* $\mathbf{Y}$ *such that each entry of* $\mathbf{Y}$ *is described by the multilinear polynomial*

$$y_{i_1,\ldots,i_m} \approx \sum_{j_1=1}^{r_1} \sum_{j_2=1}^{r_2} \cdots \sum_{j_m=1}^{r_m} w_{j_1,\ldots,j_m} \prod_{k=1}^{m} u^{(k)}_{i_k,j_k} \tag{3}$$

We can now make the connection between the Tucker decomposition of a tensor and weighted tuple-sets as defined in definition 4: the factorization eq. (3) can be interpreted as learning a multilinear function $\gamma : \mathcal{V}^{(1)} \times \cdots \times \mathcal{V}^{(m)} \mapsto \mathbb{R}$ which maps m-tuples to the entries of $\mathbf{Y}$. In contrast to the weight function ϕ of a tuple set, γ is defined over the whole Cartesian product $\mathcal{V}^{(1)} \times \cdots \times \mathcal{V}^{(m)}$.

In the following, it will prove convenient to state eq. (3) in different notations. In tensor notation, eq. (3) is equivalent to

$$\mathbf{Y} \approx \mathbf{X} = \mathbf{W} \times_1 U^{(1)} \times_2 \cdots \times_m U^{(m)} \tag{4}$$

where $\times_k$ denotes the n-mode product of a tensor and a matrix in mode k, while $U^{(k)} \in \mathbb{R}^{n_k \times r_k}$ is the latent factor matrix for mode k and $\mathbf{W} \in \mathbb{R}^{r_1 \times \cdots \times r_m}$ is the core tensor of the factorization. Furthermore, via the *unfolding* operation on tensors and the Kronecker product, eq. (4) can be stated in matrix notation as

$$Y_{(k)} \approx U^{(k)} W_{(k)} \left(U^{(m)} \otimes \cdots \otimes U^{(k+1)} \otimes U^{(k-1)} \otimes \cdots \otimes U^{(1)} \right)^T \tag{5}$$

We will also shorten n-rank$(\mathbf{Y}) = (r_1, \ldots, r_m)$ to n-rank$(\mathbf{Y}) = \boldsymbol{r}$. Furthermore, we define some quantities associated with the Tucker decomposition that will prove convenient for the rest of this paper.

Definition 6. *Let* $\mathbf{X} = \mathbf{W} \times_1 U^{(1)} \times_2 \cdots \times_m U^{(m)}$ *with* n-rank$(\mathbf{X}) = \boldsymbol{r}$, $m = \operatorname{ord}(\mathbf{X})$ *and* $\mathbf{X} \in \mathbb{R}^{\prod_i n_i}$. *The number of variables of a Tucker decomposition, i.e. the number of entries in the latent factors, is then given by*

$$\operatorname{var}(\mathbf{X}) = \prod_{i=1}^{m} r_i + \sum_{i=1}^{m} n_i r_i$$

The number of polynomials associated with $\mathbf{X}$, *i.e. the number of entries in* $\mathbf{X}$, *is denoted by*

$$\operatorname{pol}(\mathbf{X}) = \prod_{i=1}^{m} n_i$$

By applying specific constraints on the core tensor or the latent factors, various important factorization methods can be expressed as special cases within the Tucker decomposition framework. One focus of this work is to analyze how these constraints affect the generalization ability of a factorization. In the following, we will briefly discuss some important models to illustrate these constraints: Most matrix factorization methods, can be considered a Tucker decomposition of a second-order tensor. For instance, the *singular value decomposition* can be expressed as a Tucker decomposition of a second order tensor with orthogonal factor matrices. Furthermore, *Candecomp / Parafac* (CP) [10,7] can be described as a Tucker decomposition with the additional constraints that the core tensor $\mathbf{W}$ is superdiagonal and $r_1 = r_2 = \cdots = r_m$. Similarly, the *Block-Term decomposition* (BTD) [8] can be viewed as imposing the constraint that the core tensor $\mathbf{W}$ is block-diagonal. While CP and BTD are decompositions that put special constraints on the core tensor, RESCAL [18] is a factorization that constrains the number of different vector spaces under consideration and is particularly useful for modeling knowledge representations [19]. Specifically, it requires that some of the latent factors are identical, which corresponds to the fact that for some sets $\mathcal{V}^{(i)}$, $\mathcal{V}^{(j)}$ of the underlying tuple-set, it holds that $\mathcal{V}^{(i)} = \mathcal{V}^{(j)}$. Due to space constraints we refer the interested reader to [14] for further details on tensor factorization and the Tucker decomposition on particular.

3   Generalization Bounds for Low-Rank Factorizations

To get deeper theoretical insight into the generalization ability of tensor factorizations, we will now present generalization error bounds. In section 3.1 and section 3.2 we will derive generalization error bounds for the zero-one loss and real-valued loss functions, based on the number of sign patterns that a factorization can express. In these sections, we will closely follow the theory developed in [24,23] and extend it to the general multilinear setting. The actual upper and lower bounds on the number of sign patterns that a tensor factorization can express are then given in section 3.3. To derive these bounds, we will employ properties of the tensor product as discussed in section 2.

Consider the following setting: Let $\mathbf{Y}$ be the tensor representation of structured data $\mathcal{T}$, where a subset of entries y_i has been observed and let the set $\Omega = \{i \mid y_i \text{ observed}\}$ hold the indices of these observed entries. Then, we seek to predict the missing entries in $\mathbf{Y}$, by computing a factorization such that

$$\mathbf{Y} \approx \mathbf{X} = \mathbf{W} \times_1 U^{(1)} \times_2 \cdots \times_m U^{(m)}.$$

Similar to the matrix case [23], we now seek to bound the true discrepancy between the predicted tensor $\mathbf{X}$ and the target tensor $\mathbf{Y}$ as a function of the discrepancy of the observed entries Ω of $\mathbf{Y}$. The discrepancy of tensors is defined relative to a specific loss function $\Delta(\cdot, \cdot)$. The *true discrepancy* of a predicted tensor $\mathbf{X}$ and a target tensor $\mathbf{Y}$ with $\text{ord}(\mathbf{X}) = \text{ord}(\mathbf{Y}) = m$ is defined as

$$\mathcal{D}(\mathbf{X}, \mathbf{Y}) = \frac{1}{\prod_{i=1}^{m} n_i} \sum_{i_1=1}^{n_1} \sum_{i_2=1}^{n_2} \cdots \sum_{i_m=1}^{n_m} \Delta(x_{i_1,\ldots i_m}, y_{i_1,\ldots,i_m})$$

while the *empirical discrepancy* is given as

$$\mathcal{D}_\Omega(\mathbf{X}, \mathbf{Y}) = \frac{1}{|\Omega|} \sum_{i \in \Omega} \Delta(x_i, y_i)$$

We restrict the latent tensor $\mathbf{X}$ to the class of *fixed n-rank* tensors of a given order, which will be denoted by

$$\mathcal{X}_r := \{\mathbf{X} \mid \text{n-rank}(\mathbf{X}) \leq r\}$$

Please note that by restricting the factorization to a Tucker-type decomposition and by fixing n-rank$(\mathbf{X}) = r$, we also fix the quantity var$(\mathbf{X})$, while ord$(\mathbf{X})$ and pol$(\mathbf{X})$ are already determined by the target tensor $\mathbf{Y}$. We now seek to derive PAC-type error bounds of the form

$$\forall \mathbf{Y} \in \mathbb{R}^{\prod n} : \Pr_\Omega \left(\forall \mathbf{X} \in \mathcal{X}_r : \mathcal{D}(\mathbf{X}, \mathbf{Y}) \leq \mathcal{D}_\Omega(\mathbf{X}, \mathbf{Y}) + \varepsilon \right) > 1 - \delta \qquad (6)$$

such that the true discrepancy for all tensors in $\mathcal{X}_r$ is bounded by their discrepancy on the observed entries Ω plus a second term ε. An important assumption that will be made is that the set of observed entries Ω is chosen uniformly at random.

3.1   Bounds for Zero-One Sign Agreement Loss

A reasonable choice for $\Delta(\cdot, \cdot)$ in a classification setting is the zero-one loss, i.e.

$$\Delta(a, b) = \begin{cases} 0, & \text{if } \text{sgn}(a) = \text{sgn}(b) \\ 1, & \text{otherwise.} \end{cases}$$

For target entries $y_i \in \{\pm 1\}$, the zero-one loss $\Delta(x_i, y_i)$ is independent of the magnitude of the predictions x_i and only depends on their sign. A central concept in the following discussion will therefore be the equivalence classes of tensors with identical sign patterns, i.e. the elements of the set

$$\mathcal{S}_{n,r} = \left\{ \text{sgn}(\mathbf{X}) \in \{-1, 0, +1\}^{\prod n} \mid \mathbf{X} \in \mathbb{R}^{\prod n}, \text{n-rank}(\mathbf{X}) \leq r \right\}.$$

The cardinality $|\mathcal{S}_{n,r}|$ specifies therefore, how many different sign patterns can be expressed by factorizations with n-rank$(\mathbf{X}) \leq r$ and pol$(\mathbf{X}) = \prod n$.

Lemma 1. *Let $\mathbf{Y} \in \{\pm 1\}^{\prod n}$ be any binary tensor with $n_i > 2$. Furthermore, let Ω be a set of $|\Omega|$ uniformly chosen entries of $\mathbf{Y}$, let $\delta > 0$, and let $r \in \mathbb{N}_+^{\text{ord}(\mathbf{Y})}$. Then, it holds with probability at least $1 - \delta$ that*

$$\forall \mathbf{X} \in \mathcal{X}_r : \mathcal{D}(\mathbf{X}, \mathbf{Y}) < \mathcal{D}_\Omega(\mathbf{X}, \mathbf{Y}) + \sqrt{\frac{\log |\mathcal{S}_{n,r}| - \log \delta}{2|\Omega|}}$$

where $|\mathcal{S}_{n,r}| \leq \left(\frac{4e \ (\text{ord}(\mathbf{X})+1) \ \text{pol}(\mathbf{X})}{\text{var}(\mathbf{X})} \right)^{\text{var}(\mathbf{X})}$

Proof. The following proof is analogue to the matrix case [24], hence we will only provide a brief outline. First, we fix $\mathbf{Y}$ and $\mathbf{X}$. For an index i, chosen uniformly at random, it holds that $\Delta(x_i, y_i) \sim Bernoulli(\mathcal{D}(\mathbf{X}, \mathbf{Y}))$. Consequently, for independently and uniformly chosen observed entries, the sum of Bernoulli distributed random variables $|\Omega|\mathcal{D}_\Omega(\mathbf{X}, \mathbf{Y})$ follows a binomial distribution with mean $|\Omega|\mathcal{D}(\mathbf{X}, \mathbf{Y})$. It follows from Chernoff's inequality that

$$\Pr_\Omega \left(\mathcal{D}(\mathbf{X}, \mathbf{Y}) \geq \mathcal{D}_\Omega(\mathbf{X}, \mathbf{Y}) + \varepsilon \right) \leq \exp \left(-2|\Omega|\varepsilon^2 \right)$$

Furthermore, since $\Delta(x_i, y_i)$ depends only on the sign of x_i, the random variable $\mathcal{D}_\Omega(\mathbf{X}, \mathbf{Y})$ is identical for all tensors $\mathbf{X}$ in the same equivalence class of sign patterns. Since there exist $|\mathcal{S}_{n,r}|$ different equivalence classes, lemma 1 follows by taking a union bound of the events $\mathcal{D}(\mathbf{X}, \mathbf{Y}) \geq \mathcal{D}_\Omega(\mathbf{X}, \mathbf{Y}) + \varepsilon$ for these random variables. The actual bound on $|\mathcal{S}_{n,r}|$ is deferred until section 3.3.   □

3.2   Bounds for Real-Valued Loss Functions

Before deriving upper and lower bounds for the number of sign patterns, we also provide a bound for real-valued loss functions, which is the more commonly used setting for tensor factorizations. However, these loss functions, and therefore also their associated discrepancies, are not only determined by the sign of an entry x_i but are also determined by the value of this entry. We will therefore derive bounds for the pseudodimension of low-rank tensors.

Lemma 2. *Let* $\mathbf{Y} \in \{\pm 1\}^n$ *be any binary tensor with* $n_i > 2$. *Furthermore, let* $|\Delta(\cdot, \cdot)| \leq b$ *be a bounded monotone loss function, let* Ω *be a set of* $|\Omega|$ *uniformly chosen entries of* $\mathbf{Y}$, *let* $\delta > 0$, *and let* $\mathbf{r} \in \mathbb{N}_+^{\mathrm{ord}(\mathbf{Y})}$. *Then, it holds with probability at least* $1 - \delta$

$$\forall \mathbf{X} \in \mathcal{X}_r : \mathcal{D}(\mathbf{X}, \mathbf{Y}) < \mathcal{D}_\Omega(\mathbf{X}, \mathbf{Y}) + \sqrt{32 \frac{\log |\mathcal{S}_{n,r,\mathbf{T}}| \log \frac{b|\Omega|}{\mathrm{var}(\mathbf{X})} - \log \delta}{|\Omega|}}$$

Proof. Again, the following proof is analogue to the matrix case [24], hence we will outline it only briefly. As mentioned in section 2.3, tensor factorizations can be interpreted as real-valued functions, which map from tuples of indices to entries of the tensor, i.e. a multilinear function $\gamma : \mathcal{I}^{(1)} \times \cdots \times \mathcal{I}^{(n)} \mapsto \mathbb{R}$, where $\mathcal{I}^{(i)}$ indexes the i-th mode. This allows to use the pseudodimension of classes of real-valued functions to obtain similar generalization error bounds as for matrices. The difference to the matrix case is that for tensors the domain of the function ϕ ranges of tuples of fixed length n, while for matrices it ranges over ordered pairs. Therefore, we first bound the pseudodimension of n-rank tensors via the number of sign patterns *relative to a threshold tensor* $\mathbf{T} \in \mathbb{R}^{\Pi n}$. The equivalence classes for these relative sign patterns are given by the set

$$\mathcal{S}_{n,r,\mathbf{T}} = \left\{ \mathrm{sgn}(\mathbf{X} - \mathbf{T}) \in \{-1, 0, +1\}^{\Pi n} \,\middle|\, \mathbf{X} \in \mathbb{R}^{\Pi n}, \text{n-rank}(\mathbf{X}) \leq r \right\}.$$

The concrete bound for $|\mathcal{S}_{n,r,\mathbf{T}}|$ will be given in section 3.3. Using [23, Theorem 44] we can then obtain the desired bound.   □

3.3   Bounds on the Number of Sign Patterns

Following the discussion in section 3.1 and section 3.2, we now seek to bound the number of possible sign patterns $|\mathcal{S}_{n,r}|$ and the number of relative sign patterns $|\mathcal{S}_{n,r,T}|$ for tensors $\mathbf{X} \in \mathcal{X}_r$. For this purpose, consider the polynomial form of the Tucker decompositions as given in eq. (3). Due to the multilinearity of tensor factorizations, the degree of the polynomial in eq. (3) is equal to $\operatorname{ord}(\mathbf{X}) + 1$. Furthermore, for tensors of fixed size and n-rank, the quantities $\operatorname{pol}(\mathbf{X})$ and $\operatorname{var}(\mathbf{X})$ are also fixed. Using this property of multilinear factorizations, we can bound the number of possible sign patterns of tensors with n-rank$(\mathbf{X}) = r$ by using their polynomial representation. Following [27] it has been shown, that the number of possible sign patterns for polynomials are bounded by

Theorem 1 ([23, Theorem 34, 35]). *The number of sign patterns of m polynomials, each of degree at most d, over q variables is at most*

$$\left(\frac{4edm}{q} \right)^q$$

for all $m > q > 2$.

By combining the polynomial form of tensor factorizations eq. (3) and theorem 1, we can immediately derive the following lemma which bounds the number of possible sign patterns for n-rank tensors.

Lemma 3 (Upper Bound for Sign Patterns). *The number of possible sign patterns of a m-th order tensor $\mathbf{X} \in \mathbb{R}^{\prod n} = \mathbf{W} \times_1 U^{(1)} \times_2 \cdots \times_m U^{(m)}$ with n-rank$(\mathbf{X}) = r$ is at most*

$$|\mathcal{S}_{n,r}| \leq \left(\frac{4e\,(\operatorname{ord}(\mathbf{X}) + 1)\ \operatorname{pol}(\mathbf{X})}{\operatorname{var}(\mathbf{X})} \right)^{\operatorname{var}(\mathbf{X})}$$

for $\operatorname{pol}(\mathbf{X}) > \operatorname{var}(\mathbf{X}) > 2$.

Furthermore, the number of relative sign patterns, i.e. $|\mathcal{S}_{n,r,T}|$, can be bounded in the same way, since for

$$y_{i_1,\dots,i_m} - t_{i_1\dots,i_m} = \sum_{j_1=1}^{r_1} \sum_{j_2=1}^{r_2} \cdots \sum_{j_m=1}^{r_m} w_{j_1,\dots,j_m} \prod_{k=1}^{m} u_{i_k j_k}^{(k)} - t_{i_1\dots,i_m}$$

we have again $\operatorname{pol}(\mathbf{X})$ polynomials of degree $\operatorname{ord}(\mathbf{X}) + 1$ over $\operatorname{var}(\mathbf{X})$ variables.

Next, we provide a lower bound on the number of sign patterns, by interpreting tensor factorization as multiple simultaneous linear classifications.

Lemma 4 (Lower Bound for Sign Patterns). *The number of possible sign patterns of a m-th order tensor $\mathbf{X} \in \mathbb{R}^{\prod n} = \mathbf{W} \times_1 U^{(1)} \times_2 \cdots \times_m U^{(m)}$ with n-rank$(\mathbf{X}) = r$ is at least*

$$|\mathcal{S}_{n,r}| \geq \left(\frac{n_i}{r_i - 1} \right)^{\frac{1}{n_i}(r_i - 1)\,\operatorname{pol}(\mathbf{X})}$$

Proof. First, consider the Tucker decomposition in its unfolded variant, i.e.

$$X_{(i)} = U^{(i)} W_{(i)} \left(U^{(m)} \otimes \cdots \otimes U^{(i+1)} \otimes U^{(i-1)} \otimes \cdots \otimes U^{(1)} \right)^T$$

Let $B = U^{(m)} \otimes \cdots \otimes U^{(i+1)} \otimes U^{(i-1)} \otimes \cdots \otimes U^{(1)} \in \mathbb{R}^{\prod n/n_i \times \prod r/r_i}$, and fix $U^{(k)} \in \mathbb{R}^{n_k \times r_k}$ with rows in general position for all $k = 1 \ldots m$. We now consider the number of possible sign patterns of matrices $U^{(i)} W_{(i)} B^T$. It follows from the rows being in general position that $\text{rank}\left(U^{(k)}\right) = r_k$ for all $k = 1 \ldots m$ [11, Sec. 1.3.2]. Furthermore, since the tensor product preserves the linear independence of vectors, it follows that $\text{span}(B) = \mathbb{R}^{\prod r/r_i}$ [1, Sec. 6.1.4]. Although B is highly structured, it follows that the matrix product $W_{(i)} B^T$ varies over all possible $r_i \times \prod n/n_i$ matrices. Therefore, each column of $\text{sgn}(U^{(i)} W_{(i)} B^T)$ can be considered an independent homogeneous linear classification of n_i vectors in $\mathbb{R}^{r_i}$, for which exactly

$$2 \sum_{k=0}^{r_i-1} \binom{n_i}{k} > \left(\frac{n_i}{r_i - 1} \right)^{r_i - 1}$$

such classifications exists. Consequently, this many sign patterns exist for each of the $\prod n/n_i = \text{pol}(\mathbf{X})/n_i$ columns of $U^{(i)} W_{(i)} B^T$.   □

Next we analyze the tightness of bounds in lemma 3 and lemma 4. Let $m = \text{ord}(\mathbf{X})$, let $\alpha = 4e(m+1)$, let $\forall i : r_{min} \leq r_i$, and similarly let $\forall i : n_{max} \geq n_i$. Then, for $r_{min} \geq \sqrt[m]{\alpha}$ it follows from lemma 3 that

$$|\mathcal{S}_{\mathbf{n},\mathbf{r}}| \leq \left(\frac{\alpha n_{max}^m}{r_{min}^m} \right)^{\text{var}(\mathbf{X})} \leq \left(\frac{\sqrt[m]{\alpha} n_{max}}{r_{min}} \right)^{m \, \text{var}(\mathbf{X})} \leq n_{max}^{m \, \text{var}(\mathbf{X})}$$

Furthermore, for low-rank factorizations with $n_i > r_i^2$ and $\text{pol}(\mathbf{X}) > \frac{m}{r_i-1} \text{var}(\mathbf{X})$ it follows from lemma 4 that

$$|\mathcal{S}_{\mathbf{n},\mathbf{r}}| \geq \left(\frac{n_i}{r_i - 1} \right)^{\frac{1}{n_i}(r_i-1)\text{pol}(\mathbf{X})} \geq \sqrt{n_i}^{\frac{1}{n_i}(r_i-1)\text{pol}(\mathbf{X})} \geq n_i^{\frac{1}{2n_i} m \, \text{var}(\mathbf{X})}$$

Hence, the bound is tight up to a multiplicative factor in the exponent.

4   The Effect of Structure and Constraints

In section 3 we derived bounds on the generalization error of tensor factorizations. In this section we discuss what conclusions can be drawn from the derived bounds. In particular, we are interested in how additional structure or constraints affect the generalization ability of tensor factorizations. For this purpose, we will first present a setting in which it is reasonable to compare tensor factorizations of different order. Furthermore, we will evaluate experimentally how the generalization ability of tensor factorizations behaves with the change of structure and constraints. At last, we will discuss how these results can be interpreted with respect to the derived generalization bounds.

4.1   Comparable Tensors

Since it is not reasonable to compare arbitrary tensor factorizations, consider the following setting: Let $\mathcal{T} = (\mathcal{V}, \mathcal{E}, \phi, m)$ be a weighted tuple-set of order m and let $\mathbf{Y}$ be the tensor representation of $\mathcal{T}$. Furthermore, let $\mathbf{Y}^-$ be a tensor representation of $\mathcal{T}$ such that the k-th mode of $\mathbf{Y}^-$ is indexed by the set

$$\mathcal{U}^{(k)} = \begin{cases} \mathcal{V}^{(k)} & , k \neq i \neq j \\ \mathcal{V}^{(i)} \times \mathcal{V}^{(j)} & , k = i. \end{cases}$$

This means that for two index sets $\mathcal{V}^{(i)}$, $\mathcal{V}^{(j)}$ of $\mathcal{T}$ only a single vector space representation is used in $\mathbf{Y}^-$. Consequently, it holds that $\mathrm{ord}(\mathbf{Y}^-) = \mathrm{ord}(\mathbf{Y}) - 1$. This setting corresponds, for example, to propositionalization in multi-relational learning. We will refer to $\mathbf{Y}^-$ as an *understructured representation* of $\mathcal{T}$. The opposite setting would be an *overstructured representation* where the tensor $\mathbf{Y}^-$ is the correct representation of $\mathcal{T}$, while $\mathbf{Y}$ represents one index set $\mathcal{V}^{(i)}$ of $\mathcal{T}$ by two modes, i.e.

$$\mathcal{V}^{(k)} = \begin{cases} \mathcal{U}^{(k)} & , k \neq i \neq j \\ \mathcal{U}^{(i)} \times \mathcal{U}^{(j)} & , k = i \end{cases}$$

For both, the under- and the overstructured case, we are interested to see how the generalization ability of a tensor factorization changes by factorizing $\mathbf{Y}$ compared to $\mathbf{Y}^-$. Without loss of generalization, let $i = m - 1$, $j = m$ where $m = \mathrm{ord}(\mathbf{Y})$ and $\ell = \mathrm{ord}(\mathbf{Y}^-) = m - 1$. Furthermore, let $\mathbf{X} = \mathbf{W} \times_1 U^{(1)} \times_2 \cdots \times_m U^{(m)} \in \mathbb{R}^n$ and $\mathbf{X}^- = \mathbf{W}^- \times_1 U^{(1)-} \times_2 \cdots \times_\ell U^{(\ell)-} \in \mathbb{R}^{n^-}$ be factorizations of $\mathbf{Y}$ and $\mathbf{Y}^-$. Since we are only interested in the effect that the order of data representation has on the generalization ability, we want to exclude the effect of different ranks. Analogously to section 3, we restrict therefore $\mathbf{X}$ and $\mathbf{X}^-$ to be of similar n-rank, in order to get comparable models. Since it holds for the Kronecker product that $\mathrm{rank}(V \otimes W) = \mathrm{rank}(V)\,\mathrm{rank}(W)$, we require that

$$r_k^- = \begin{cases} r_k & , k \neq m \neq \ell \\ r_m r_\ell & , k = \ell \end{cases}$$

It also follows immediately from the construction of $\mathbf{Y}$ and $\mathbf{Y}^-$ and the properties of the Cartesian product that

$$n_k^- = \begin{cases} n_k & , k \neq m \neq \ell \\ n_m n_\ell & , k = \ell \end{cases}$$

In the following, we will refer to tensors $\mathbf{X}$, $\mathbf{X}^-$ who have these properties as *comparable tensors*. Please note that for comparable tensors, it holds that $\mathrm{var}(\mathbf{X}^-) > \mathrm{var}(\mathbf{X})$, since $n_m n_\ell r_m r_\ell > n_m r_m + n_\ell r_\ell$. Furthermore, it holds that $\mathrm{ord}(\mathbf{X}^-) + 1 = \mathrm{ord}(\mathbf{X})$ and $\mathrm{pol}(\mathbf{X}^-) = \mathrm{pol}(\mathbf{X})$.

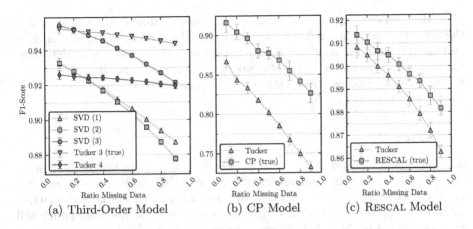

Fig. 1. Mean and standard error of the F1-Score over 100 iterations per percentage of missing data. SVD (i) denotes the singular value decomposition of $Y_{(i)}$, i.e. the unfolding of the i-th mode of $\mathbf{Y}$.

4.2 Experimental Results

Given comparable tensors, we evaluated experimentally how tensor factorization behaves under the change of structure and constraints. The experiments were carried out on synthetic data with different amounts of missing data. To evaluate the *effects of structure*, we created a third-order tensor $\mathbf{X} = \mathbf{W} \times_1 A \times_2 B \times_3 C$, where $\mathbf{W} \in \mathbb{R}^{5\times10\times2}$, $A \in \mathbb{R}^{50\times5}$, $B \in \mathbb{R}^{100\times10}$, $C \in \mathbb{R}^{20\times2}$ and where all entries of the core tensor and the factor matrices had been drawn from the standard normal distribution $\mathcal{N}(0,1)$. From $\mathbf{X}$ we created the target tensor $\mathbf{Y}$ by setting $y_{ijk} = \mathrm{sgn}(x_{ijk})$. Furthermore, the set of observed entries Ω has been drawn uniformly at random, where we increased the ratio of missing entries from $[0.1, 0.9]$. To evaluate the effects of under- and overstructuring, we compared three models: a Tucker-3 decomposition, which is the correct model, the SVD which is an understructured model and a Tucker-4 decomposition, which is an overstructured model. Moreover, the SVD has been computed on all possible unfoldings $Y_{(i)}$, where $i \in \{1, 2, 3\}$. For the Tucker-4 decomposition, we split the second mode of $\mathbf{Y}$ into two size-10 modes, such that $\mathbf{Y}_4 \in \mathbb{R}^{50\times10\times10\times20}$. For each model and each ratio of missing entries we computed 100 factorizations and recorded the F1-score for the classification of the missing entries compared to the ground truth. fig. 1(a) shows the results of these experiments. As expected, the true model provides the best overall performance. One understructured model, i.e. SVD (3), shows comparable results to the true model for low amounts of missing entries but scales significantly worse as the missing data increases. The overstructured model displays the opposite behaviour; it shows reduced overall generalization ability compared to the true model but is more stable with the amount of missing data.

In similar experiments we also evaluated the *effects of constraints*. For this purpose, we created synthetic CP and RESCAL models under similar conditions as in the previous experiment. However, in this experiment we evaluated how the correct model compared to an unconstrained Tucker model. Figures fig. 1(b) and fig. 1(c) show the results of these experiments. Again, the true models show the best overall performance in both experiments. Furthermore, in both settings, the constrained models scale better with the amount of missing data than the unconstrained tucker model.

4.3   Discussion

The previously derived generalization bounds can provide insight in how to interpret these experimental results. First, note that both terms in eq. (6), i.e. $\mathcal{D}_\Omega(\mathbf{X}, \mathbf{Y})$ and ε, are influenced by the number of sign patterns that a factorization can express. For $\mathcal{D}_\Omega(\mathbf{X}, \mathbf{Y})$ this is the case because the discrepancy will increase when a model $\mathbf{X}$ is not expressive enough to model the sign patterns of a target tensor $\mathbf{Y}$. Furthermore, it has been shown in section 3 that the term ε grows with the number of sign patterns. Since it has also been shown that the upper bound on the number of sign configurations in lemma 3 is tight at least up to a multiplicative factor in the exponent, we consider how this bound changes with the order of the data representation; to see what possible effects the change of structure can have in terms of the generalization ability.

Corollary 1. *For comparable tensors* $\mathbf{X} \in \mathbb{R}^n$, $\mathbf{X}^- \in \mathbb{R}^{n^-}$ *with* $\mathrm{ord}(\mathbf{X}) = \mathrm{ord}(\mathbf{X})^- + 1$, n-rank$(\mathbf{X}) = \mathbf{r}$ *and* n-rank$(\mathbf{X})^- = \mathbf{r}^-$, *the ratio of upper bounds on then number of possible sign patterns is at most*

$$1 < \frac{\mathcal{O}(|\mathcal{S}_{n,r}^-|)}{\mathcal{O}(|\mathcal{S}_{n,r}|)} < \left(\frac{4e \ (\mathrm{ord}(\mathbf{X}^-) + 1) \ \mathrm{pol}(\mathbf{X})}{\mathrm{var}(\mathbf{X}^-)} \right)^v$$

where $v = n_m n_\ell r_m r_\ell - (n_\ell r_\ell + n_m r_m) > 0$

Proof. It follows straight from the definition of comparable tensors that $\mathrm{var}(\mathbf{X}^-)$ can be rewritten as $\mathrm{var}(\mathbf{X}^-) = \mathrm{var}(\mathbf{X}) + v$. Furthermore, let

$$\alpha = 4e \ (\mathrm{ord}(\mathbf{X}^-) + 1) \ \mathrm{pol}(\mathbf{X})$$
$$\beta = 4e \ (\mathrm{ord}(\mathbf{X}) + 1) \ \mathrm{pol}(\mathbf{X}) = \alpha + 4e \ \mathrm{pol}(\mathbf{X})$$

Then, it holds that

$$\frac{\mathcal{O}(|\mathcal{S}_{n,r}^-|)}{\mathcal{O}(|\mathcal{S}_{n,r}|)} = \frac{\alpha^{\mathrm{var}(\mathbf{X})+v}}{\mathrm{var}(\mathbf{X}^-)^{\mathrm{var}(\mathbf{X})+v}} \frac{\mathrm{var}(\mathbf{X})^{\mathrm{var}(\mathbf{X})}}{\beta^{\mathrm{var}(\mathbf{X})}}$$

$$= \left(\frac{\alpha}{\mathrm{var}(\mathbf{X}^-)} \right)^v \frac{\alpha^{\mathrm{var}(\mathbf{X})}}{\beta^{\mathrm{var}(\mathbf{X})}} \frac{\mathrm{var}(\mathbf{X})^{\mathrm{var}(\mathbf{X})}}{(\mathrm{var}(\mathbf{X}) + v)^{\mathrm{var}(\mathbf{X})}} \leq \left(\frac{\alpha}{\mathrm{var}(\mathbf{X}^-)} \right)^v$$

$\square$

The main result of corollary 1 for this discussion is that the bound increases as we decrease the order of the tensor. This suggests that as we increase the order of the data representation, we will reduce the term ε in eq. (6). As the amount of missing data increases, it is therefore likely to see increasingly severe overfitting for $\mathbf{X}^-$ compared to $\mathbf{X}$. However, when $\mathbf{X}^-$ is the correct and $\mathbf{X}$ is an understructured representation, $\mathcal{O}(|\mathcal{S}_{n,r}^-|) > \mathcal{O}(|\mathcal{S}_n, r|)$ also suggests that the model $\mathbf{X}$ might not be expressive enough to model the sign patterns of $\mathbf{Y}^-$. This corresponds nicely to the experimental results shown in fig. 1(a). The understructured models are expressive enough to model the sign patterns of $\mathbf{Y}$, as seen in the case of SVD (3). However, they also scale significantly worse than the correct model with the amount of missing data. The overstructured Tucker-4 model scales even better with missing data than the true model, but at the same time gives significantly worse overall results, what suggests that it might not be expressive enough. A possible interpretation is therefore, that the ratio between expressiveness and overfitting is superior for a correct model specification. Since the correct model $\mathbf{X}$ has a much smaller number of variables, it should also be noted that the memory complexity of $\mathbf{X}$ is significantly reduced compared to $\mathbf{X}^-$.

Similar arguments apply for the effect of constraints. Here, the key insight is that both CP-type and RESCAL-type constraints decrease the number of variables in a model. Models like CP or the Block-Term Decompostion, require that $\mathbf{W}$ is superdiagonal or block-superdiagonal and therefore set most entries in the core tensor to $w_i = 0$. Models like RESCAL on the other hand, decrease the number of variables through the constraint that some factor matrices $U^{(i)}$, $U^{(j)}$ have to be identical. Since $\mathcal{O}(|\mathcal{S}_{n,r}|)$ depends exponentially on var$(\mathbf{X})$, conclusions similar to the effects of structure can be drawn with regard to the effects of constraints. It suggests that a model with a larger number of variables, i.e. fewer constraints, has more capacity to model sign patterns, but at the same time is more likely to overfit as the amount of missing data increases. Again, this corresponds nicely to the experimental results in fig. 1(b) and fig. 1(c).

5 Related Work

We are not aware of any previous generalization error bounds for tensor factorizations or of any theoretical results that relate the order of a tensor and the order of structured data to the generalization ability of factorizations. Our derivation of error bounds for the tensor case builds strongly on the work of [24,23], which provided error bounds for matrix factorizations with zero-one loss and general loss functions. [28] derived similar bounds in the context of rank-k SVMs. For general matrices, [6,5] show that under suitable conditions a low-rank matrix can be recovered from a minimal set of entries via convex optimization and also provide theoretical bounds. [9,26] extends these methods to tensor completion, although without providing error bounds. It has also been shown experimentally that by adding structure to the vector space representations via the tensor product, the amount of data needed for exact recovery can be greatly reduced [26,25].

6 Conclusion

To obtain a deeper understanding of the generalization ability of tensor factorizations, we derived generalization error bounds based on the number of sign patterns that a tensor factorization can model. Using a general framework to describe structured data based on weighted tuple-sets, we analyzed how tensor factorizations behave when their order does not match the true order of the data. We showed experimentally that structuring vector space representations via the tensor product, up to the true order of the data, adds important information such that tensor models often scale better with sparsity or missing data than their understructured counterparts. We also discussed analytically how this behaviour can be explained in the light of the newly derived generalization bounds. In this work, we only considered binary values for the target tensor $\mathbf{Y}$, which corresponds to a classification setting. For future work, it would prove very valuable to also derive error bounds for the more general case of real-valued weight functions. Since the current error bounds are based on the assumption that the observed entries are independently and identically distributed what – especially on structured data – might not hold, it might also be useful to consider techniques as in [17], to overcome this limitation.

References

1. Anthony, M., Harvey, M.: Linear Algebra: Concepts and Methods. Cambridge University Press (2012)
2. Bader, B.W., Harshman, R.A., Kolda, T.G.: Temporal analysis of semantic graphs using ASALSAN. In: Seventh IEEE International Conference on Data Mining (ICDM 2007), Omaha, NE, USA, pp. 33–42 (2007)
3. Bordes, A., Weston, J., Collobert, R., Bengio, Y.: Learning structured embeddings of knowledge bases. In: Proceedings of the 25th Conference on Artificial Intelligence, San Francisco, USA (2011)
4. Burdick, D.S.: An introduction to tensor products with applications to multiway data analysis. Chemometrics and Intelligent Laboratory Systems 28(2), 229–237 (1995)
5. Candes, E.J., Plan, Y.: Matrix completion with noise. Proceedings of the IEEE 98(6), 925–936 (2010)
6. Candès, E.J., Recht, B.: Exact matrix completion via convex optimization. Foundations of Computational Mathematics 9(6), 717–772 (2009)
7. Carroll, J.D., Chang, J.J.: Analysis of individual differences in multidimensional scaling via an n-way generalization of Eckart-Young decomposition. Psychometrika 35(3), 283–319 (1970)
8. De Lathauwer, L.: Decompositions of a higher-order tensor in block termsPart II: definitions and uniqueness. SIAM Journal on Matrix Analysis and Applications 30(3), 1033–1066 (2008)
9. Gandy, S., Recht, B., Yamada, I.: Tensor completion and low-n-rank tensor recovery via convex optimization. Inverse Problems 27(2), 025010 (2011)
10. Harshman, R.A., Lundy, M.E.: PARAFAC: parallel factor analysis. Computational Statistics & Data Analysis 18(1), 39–72 (1994)

11. Hassoun, M.H.: Fundamentals of Artificial Neural Networks. MIT Press (1995)
12. Huang, Y., Tresp, V., Bundschus, M., Rettinger, A., Kriegel, H.-P.: Multivariate prediction for learning on the semantic web. In: Frasconi, P., Lisi, F.A. (eds.) ILP 2010. LNCS, vol. 6489, pp. 92–104. Springer, Heidelberg (2011)
13. Jenatton, R., Le Roux, N., Bordes, A., Obozinski, G.: A latent factor model for highly multi-relational data. In: Advances in Neural Information Processing Systems, vol. 25, pp. 3176–3184. MIT Press, Lake Tahoe (2012)
14. Kolda, T.G., Bader, B.W.: Tensor decompositions and applications. SIAM Review 51(3), 455–500 (2009)
15. Kramer, S., Lavrac, N., Flach, P.: Propositionalization approaches to relational data mining. Springer-Verlag New York, Inc. (2001)
16. Liu, J., Musialski, P., Wonka, P., Ye, J.: Tensor completion for estimating missing values in visual data. In: 2009 IEEE 12th International Conference on Computer Vision, pp. 2114–2121 (2009)
17. Mohri, M., Rostamizadeh, A.: Rademacher complexity bounds for non-iid processes. In: Advances in Neural Information Processing Systems, vol. 21, pp. 1097–1104. MIT Press, Cambridge (2009)
18. Nickel, M., Tresp, V., Kriegel, H.P.: A three-way model for collective learning on multi-relational data. In: Proceedings of the 28th International Conference on Machine Learning, pp. 809–816. ACM, Bellevue (2011)
19. Nickel, M., Tresp, V., Kriegel, H.P.: Factorizing YAGO: scalable machine learning for linked data. In: Proceedings of the 21st International Conference on World Wide Web, pp. 271–280. ACM, New York (2012)
20. Rendle, S., Freudenthaler, C., Schmidt-Thieme, L.: Factorizing personalized markov chains for next-basket recommendation. In: Proceedings of the 19th International Conference on World Wide Web, pp. 811–820. ACM (2010)
21. Rettinger, A., Wermser, H., Huang, Y., Tresp, V.: Context-aware tensor decomposition for relation prediction in social networks. Social Network Analysis and Mining 2(4), 373–385 (2012)
22. Signoretto, M., Van de Plas, R., De Moor, B., Suykens, J.A.: Tensor versus matrix completion: a comparison with application to spectral data. IEEE Signal Processing Letters 18(7), 403–406 (2011)
23. Srebro, N.: Learning with matrix factorizations. Ph.D. thesis, Massachusetts Institute of Technology, Cambridge, MA, USA (2004)
24. Srebro, N., Alon, N., Jaakkola, T.S.: Generalization error bounds for collaborative prediction with low-rank matrices. In: Advances in Neural Information Processing Systems, vol. 17, pp. 1321–1328. MIT Press, Cambridge (2005)
25. Tomioka, R., Hayashi, K., Kashima, H.: Estimation of low-rank tensors via convex optimization. arXiv preprint arXiv:1010.0789 (2010)
26. Tomioka, R., Suzuki, T., Hayashi, K., Kashima, H.: Statistical performance of convex tensor decomposition. In: Advances in Neural Information Processing Systems, vol. 24, pp. 972–980 (2012)
27. Warren, H.E.: Lower bounds for approximation by nonlinear manifolds. Transactions of the American Mathematical Society 133(1), 167–178 (1968)
28. Wolf, L., Jhuang, H., Hazan, T.: Modeling appearances with low-rank SVM. In: IEEE Conference on Computer Vision and Pattern Recognition, pp. 1–6 (2007)

Learning Modewise Independent Components from Tensor Data Using Multilinear Mixing Model

Haiping Lu*

Institute for Infocomm Research
Singapore
hplu@ieee.org

Abstract. Independent component analysis (ICA) is a popular unsupervised learning method. This paper extends it to multilinear modewise ICA (MMICA) for tensors and explores two architectures in learning and recognition. MMICA models tensor data as mixtures generated from modewise source matrices that encode statistically independent information. Its sources have more compact representations than the sources in ICA. We embed ICA into the multilinear principal component analysis framework to solve for each source matrix alternatively with a few iterations. Then we obtain mixing tensors through regularized inverses of the source matrices. Simulations on synthetic data show that MMICA can estimate hidden sources accurately from structured tensor data. Moreover, in face recognition experiments, it outperforms competing solutions with both architectures.

Keywords: independent component analysis, mixing model, tensor, multilinear subspace learning, unsupervised learning.

1 Introduction

Independent component analysis (ICA) is an important *unsupervised learning* method for finding representational components of data with maximum statistical independence [1]. While principal component analysis (PCA) [2] gives *independent components* (ICs) only for *Gaussian* data, ICA finds ICs for the general case of *non-Gaussian* data [3]. ICA can be performed under two different architectures for image representation and recognition [4]. *Architecture I* treats images as random variables and pixels as random trials to find spatially local basis images that are statistically independent. *Architecture II* treats pixels as random variables and images as random trials to find factorial code that reflects global properties.

Real-world data are often specified in a high-dimensional space while they are highly constrained to a *subspace* [5]. Thus, *dimensionality reduction* is frequently

---

* The author (Haiping Lu) is currently affiliated with Hong Kong Baptist University (email: haiping@hkbu.edu.hk)

H. Blockeel et al. (Eds.): ECML PKDD 2013, Part II, LNAI 8189, pp. 288–303, 2013.
© Springer-Verlag Berlin Heidelberg 2013

employed to transform a high-dimensional data set into a low-dimensional subspace while retaining most of the underlying structure. As the number of ICs found by ICA typically corresponds to the dimension of the input, dimensionality reduction in ICA is usually achieved through PCA [4].

Recently, there has been a surge of interest in learning subspace of multidimensional data, i.e., *tensor* data, from their natural multidimensional representations. Examples are 2-D/3-D images, videos, and multi-way social networks [6, 7]. *Multilinear subspace learning* of tensor data operates on natural tensor representations without reshaping into vectors. Thus, it can obtain simpler and more compact representations, and handle big data more efficiently [6]. There have been many multilinear extensions of PCA [8–12], linear discriminant analysis [13–17], and canonical correlation analysis [18–20]. In contrast, multilinear extensions of ICA have not been well addressed, though several works deal with the ICA problem using tensor-based approaches.

In [21], ICA mixing matrix is identified by decomposing the higher-order cumulant tensor for vector-valued observation data. In [22], a tensor probabilistic ICA algorithm was formulated for fMRI analysis, with selected voxels represented as very-high-dimensional vectors. The *multilinear ICA* (MICA) model in [23] analyzes multiple factors for image ensembles organized into a tensor according to different image formation factors such as people, views, and illuminations. It requires a large number of samples for training, e.g., 36 well-selected samples per class in [23]. Furthermore, MICA represents images as vectors and needs to know the forming factors. In this sense, MICA is a *supervised learning* method requiring data to be labeled with such information. In *unsupervised learning* without labels, it *degenerates* to classical ICA. Another work with the same name MICA in [24] uses a multilinear expansion of the probability density function of source statistics but represents data as vectors too. To the best of our knowledge, the only multilinear ICA formulation based on tensor input data is the *directional tensor ICA* (DTICA) in [25, 26], which estimates *two* mixing matrices for images. It forms row and column directional images by shifting the rows/columns and applies FastICA [27] to row/column vectors. As in [4], PCA is used for dimensionality reduction in DTICA. Neither MICA in [23] nor DTICA has demonstrated **blind source separation** capability, a classical application of ICA.

This paper aims to develop a multilinear extension of ICA that can do blind source separation for tensors with appropriate modeling. For example, Fig. 1(a) shows ten mixtures generated from two simple binary patterns in Fig. 1(b) with a *multilinear mixing model* similar to classical ICA. We propose a *multilinear modewise ICA* (MMICA) model for tensor data with *modewise ICs* that can model this generation process like ICA. We then develop an MMICA algorithm to estimate these modewise ICs, i.e., to estimate the sources in Fig. 1(b) from the observed mixtures in Fig. 1(a). As an ICA extension to tensors, MMICA can be applied to domains where ICA has been applied in the past.

This work is inspired by previous attempts in multilinear extensions of ICA and motivated by the compact representations of multilinear subspace learning methods. The MMICA model, to be presented in Section 2, assumes that tensor

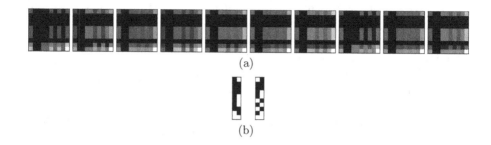

(a)

(b)

Fig. 1. The structured data in (a) are all mixtures generated from the source data in (b) with a *multilinear mixing model*. MMICA can recover the sources in (b) from observed mixtures in (a).

observation data have rich structures and they are mixtures generated from simple *modewise sources*, as illustrated in Fig. 1. We formulate the MMICA algorithm in Section 3 as an extension of *multilinear PCA* (MPCA) [10] to the non-Gaussian case by embedding ICA into MPCA to deal with non-Gaussianity of data in each mode. Also, we explore two architectures suggested in [4] for MMICA. Next, we discuss the differences with related works in Section 4. Finally, in Section 5, we show the *blind source separation* capability of MMICA through simulations and its recognition capability on real face data.

Note: for convenience of discussion, the acronym MICA below refers to the method in [23] rather than that in [24].

2 Multilinear Mixing Model for Tensors

2.1 Notations and Fundamentals

We briefly introduce some notations and operations needed. For more details, please refer to [6, 28–30].

Vectors are denoted by lowercase boldface letters, e.g., $\mathbf{x}$; matrices by uppercase boldface, e.g., $\mathbf{U}$; and tensors by calligraphic letters, e.g., $\mathcal{A}$. Their elements are denoted with indices in parentheses. *Indices* are denoted by lowercase letters and span the range from 1 to the uppercase letter of the index whenever appropriate, e.g., $n = 1, 2, ..., N$.

Multidimensional arrays are referred to as *tensors* in mathematics. The number of dimensions N defines the *order* of a tensor. Tensor is a generalization of vector and matrix. Vectors are first-order tensors, and matrices are second-order tensors. An Nth-order tensor is denoted as $\mathcal{A} \in \mathbb{R}^{I_1 \times I_2 \times ... \times I_N}$. It is addressed by N indices i_n, $n = 1, ..., N$, and each i_n addresses the n-mode of $\mathcal{A}$.

The *n-mode product* of a tensor $\mathcal{A}$ by a matrix $\mathbf{U} \in \mathbb{R}^{J_n \times I_n}$, denoted by $\mathcal{A} \times_n \mathbf{U}$, is a tensor with entries [29]:

$$(\mathcal{A} \times_n \mathbf{U})(i_1, ..., i_{n-1}, j_n, i_{n+1}, ..., i_N) = \sum_{i_n} \mathcal{A}(i_1, ..., i_N) \cdot \mathbf{U}(j_n, i_n). \quad (1)$$

The n-*mode vectors* of $\mathcal{A}$ are the I_n-dimensional vectors obtained by varying i_n while keeping all the other indices fixed. The n-*mode unfolded matrix* of $\mathcal{A}$, denoted as $\mathbf{A}_{(n)} \in \mathbb{R}^{I_n \times (I_1 \times ... \times I_{n-1} \times I_{n+1} \times ... \times I_N)}$, is formed with the n-mode vectors of $\mathcal{A}$ as its column vectors. An n-mode matrix or vector is denoted as $\mathbf{A}^{(n)}$ or $\mathbf{a}^{(n)}$, respectively. A *rank-one tensor* $\mathcal{A}$ equals to the outer product (denoted by 'o') of N vectors [29]:

$$\mathcal{A} = \mathbf{u}^{(1)} \circ \mathbf{u}^{(2)} \circ ... \circ \mathbf{u}^{(N)}, \tag{2}$$

which means that

$$\mathcal{A}(i_1, i_2, ..., i_N) = \mathbf{u}^{(1)}(i_1) \cdot \mathbf{u}^{(2)}(i_2) \cdot ... \cdot \mathbf{u}^{(N)}(i_N) \tag{3}$$

for all values of indices.

2.2   MMICA Model for Tensor Mixtures

The simplified noise-free ICA model [3] assumes that we observe M linear mixtures $\{x_m\}$ $(m = 1, ..., M)$ of P sources $\{s_p\}$ (the latent variables):

$$x_m = a_{m_1} s_1 + a_{m_2} s_2 + ... + a_{m_P} s_P, \tag{4}$$

where each mixture x_m and each IC (source) s_p are random scalar variables. The P sources $\{s_p\}$ are assumed to be independent. In ICA for random vector variables $\{\mathbf{x}_m\}$, each $\mathbf{x}_m$ is a mixture of P independent vector sources $\{\mathbf{s}_p\}$:

$$\mathbf{x}_m = a_{m_1} \mathbf{s}_1 + a_{m_2} \mathbf{s}_2 + ... + a_{m_P} \mathbf{s}_P. \tag{5}$$

For random Nth-order tensor variables $\{\mathcal{X}_m\}$ of dimension $I_1 \times ... \times I_N$, we propose a mixing model similar to (4) and (5) assuming P tensor variables $\{\mathcal{S}_p\}$ as the sources:

$$\mathcal{X}_m = a_{m_1} \mathcal{S}_1 + a_{m_2} \mathcal{S}_2 + ... + a_{m_P} \mathcal{S}_P. \tag{6}$$

Real-world tensor data often have rich structures. Therefore, we assume that the source tensors have compact representation as rank-one tensors (see (2)). For a simpler model, we further assume that these simple rank-one tensors are formed by $P_1 \times P_2 \times ... \times P_N = P$ vectors with one set in each mode, where the n-mode set has P_n *independent column vectors*: $\{\mathbf{s}_{p_n}^{(n)}, p_n = 1, ..., P_n\}$, and each source tensor is the outer product of N vectors, one from each mode, i.e.,

$$\mathcal{S}_p = \mathbf{s}_{p_1}^{(1)} \circ \mathbf{s}_{p_2}^{(2)} \circ ... \circ \mathbf{s}_{p_N}^{(N)}. \tag{7}$$

Next, we form an Nth-order *mixing tensor* $\mathcal{A}_m \in \mathbb{R}^{P_1 \times P_2 \times ... \times P_N}$ by stacking all the P mixing parameters $\{a_{m_1}, a_{m_2}, ..., a_{m_P}\}$ in (6) into an Nth-order tensor so its size $P_1 \times P_2 \times ... \times P_N = P$. Correspondingly, we form the n-*mode source matrix* $\mathbf{S}^{(n)} \in \mathbb{R}^{I_n \times P_n}$ with *independent columns* $\{\mathbf{s}_{p_n}^{(n)}, p_n = 1, ..., P_n\}$. We can then write the multilinear mixing model (6) in a form of the tensor-to-tensor-projection [6], an adaption of the Tucker decomposition model [31] to subspace learning, as

$$\mathcal{X}_m = \mathcal{A}_m \times_1 \mathbf{S}^{(1)} \times_2 \mathbf{S}^{(2)} \times ... \times_N \mathbf{S}^{(N)}. \tag{8}$$

We name this model as the **MMICA** model.

2.3   Regularized Estimation of Mixing Tensor

When applying MMICA to learning and recognition, we estimate the source matrices $\{\mathbf{S}^{(n)}\}$ from M observed mixtures $\{\mathcal{X}_m\}$ (to be described in Sec. 3). To get the mixing tensor $\mathcal{A}_m$ from an observation tensor $\mathcal{X}_m$ based on $\{\mathbf{S}^{(n)}\}$, we use (8) to get

$$\mathcal{A}_m = \mathcal{X}_m \times_1 \mathbf{S}^{(1)^+} \times_2 \mathbf{S}^{(2)^+} \times ... \times_N \mathbf{S}^{(N)^+}, \tag{9}$$

where $\mathbf{S}^{(n)^+} = (\mathbf{S}^{(n)^T}\mathbf{S}^{(n)})^{-1}\mathbf{S}^{(n)^T}$ is the left inverse of $\mathbf{S}^{(n)}$. The superscript 'T' denotes the transpose of a matrix[1]. As $\mathbf{S}^{(n)^T}\mathbf{S}^{(n)}$ can be poorly conditioned in practice, we introduce a *regularized* left inverse of $\mathbf{S}^{(n)}$ to reduce the estimation variance by adding some small bias as [32, 33]

$$\mathbf{S}_r^{(n)^+} = (\mathbf{S}^{(n)^T}\mathbf{S}^{(n)} + \eta \mathbf{I}_{P_n})^{-1}\mathbf{S}^{(n)^T}, \tag{10}$$

where η is a small *regularization* parameter and $\mathbf{I}_{P_n}$ is an identity matrix of size $P_n \times P_n$. Thus, the mixing tensor is approximated as

$$\hat{\mathcal{A}}_m = \mathcal{X}_m \times_1 \mathbf{S}_r^{(1)^+} \times_2 \mathbf{S}_r^{(2)^+} \times ... \times_N \mathbf{S}_r^{(N)^+}. \tag{11}$$

3   MMICA Algorithm

3.1   MMICA by Embedding ICA into MPCA

We solve the MMICA problem by embedding ICA into the MPCA framework [10], following the PCA+ICA in [4]. The procedures are *centering, initialization of source matrices, partial multilinear projection, modewise PCA, and modewise ICA*. Modewise ICA can be carried out in two architectures as in [4], where Architecture I is commonly used for traditional blind source separation task of ICA and Architecture II is for estimation of ICs for images[2]. The MMICA algorithm is summarized in Algorithm 1, with details described below.

The input to MMICA is a set of M tensor data samples $\{\mathcal{X}_m \in \mathbb{R}^{I_1 \times ... \times I_N}, m = 1, ..., M\}$. We need to specify two parameters: one is Q, the percentage of energy to be kept in PCA, and the other is K, the maximum number of iterations. Input data are centered first as in ICA or MPCA by subtracting the sample mean

$$\bar{\mathcal{X}} = \frac{1}{M} \sum_{m=1}^{M} \mathcal{X}_m. \tag{12}$$

There is no other data manipulation involved, such as data *re-sampling* and *re-arrangement* in DTICA [26].

---

[1] Only real-valued data are considered in this paper.

[2] We refer to the code at: http://mplab.ucsd.edu/~marni/icaFacesCode.tar

---

Algorithm 1. Multilinear Modewise ICA (MMICA)

Input: M tensor samples $\{\mathcal{X}_m \in \mathbb{R}^{I_1 \times \cdots \times I_N}, m = 1, ..., M\}$, the percentage of energy to be kept in PCA Q, the maximum number of iterations K.

$\diamond$ Center the input samples by subtracting the mean $\bar{\mathcal{X}}$.

$\diamond$ Initialize source matrices $\mathbf{S}^{(n)} = \mathbf{I}_{I_n}$ and $P_n = I_n$ for $n = 1, ..., N$.

for $k = 1$ **to** K **do**

  for $n = 1$ **to** N **do**

    • Calculate partial multilinear projection $\tilde{\mathcal{A}}_m$ according to (13) for $m = 1, ..., M$.

    • Form $\tilde{\mathbf{A}}$ with columns consisting of n-mode vectors from $\{\tilde{\mathcal{A}}_m, m = 1, ..., M\}$.

    • Perform PCA on $\tilde{\mathbf{A}}$ and keep $Q\%$ of the total energy. Obtain $\mathbf{U}$ with the first R eigenvectors as its columns. Update $P_n = R$.

    • **Architecture I:** Perform FastICA on $\mathbf{U}^T$ to get $\mathbf{A}$ and $\mathbf{W}$. Set $\mathbf{S}^{(n)} = \mathbf{U}\mathbf{A}$.

    • **Architecture II:** Get $\mathbf{V} = \mathbf{U}^T \tilde{\mathbf{A}}$ and perform FastICA on $\mathbf{V}^T$ to get $\mathbf{A}$ and $\mathbf{W}$. Set $\mathbf{S}^{(n)} = \mathbf{U}\mathbf{W}^T$.

  end for

end for

Output: $\{\mathbf{S}^{(n)}, n = 1, ..., N\}$

3.2 Iterative Alternating Estimation

In the MMICA model (8), data are generated from all N source matrices $\{\mathbf{S}^{(n)}, n = 1, ..., N\}$ rather than any one of them individually. Unfortunately, we can not determine these N matrices simultaneously, except when $N = 1$ where it is degenerated to the classical ICA. Estimating $\mathbf{S}^{(n)}$ in a particular mode n needs the knowledge of other source matrices $\{\mathbf{S}^{(j)}, j \neq n\}$. Therefore, to solve MMICA, we follow the *iterative alternating projection* method [6]. We estimate $\mathbf{S}^{(n)}$ conditioned on all the other source matrices $\{\mathbf{S}^{(j)}, j \neq n\}$, alternating between modes. This is *significantly different* from DTICA [26], which is non-iterative.

Initialization: Since all source matrices depend on each other in estimation, we need to initialize them before proceeding. We adopt a simple strategy to initialize the n-mode source matrix $\mathbf{S}^{(n)}$ to an identity matrix $\mathbf{I}_{I_n}$ of size $I_n \times I_n$. Thus, the n-mode source dimension P_n is initialized to I_n.

Modewise Processing: In each iteration, we process modewise from 1-mode to N-mode, a simple mode ordering used by many other algorithms [6]. For a particular mode n, we have all other source matrices $\{\mathbf{S}^{(j)}, j \neq n\}$ fixed and estimate $\mathbf{S}^{(n)}$ by first calculating the *partial multilinear projection* based on (11) as

$$\tilde{\mathcal{A}}_m = \mathcal{X}_m \times_1 \mathbf{S}_r^{(1)^+} \cdots \times_{n-1} \mathbf{S}_r^{(n-1)^+} \times_{n+1} \mathbf{S}_r^{(n+1)^+} \cdots \times_N \mathbf{S}_r^{(N)^+}. \quad (13)$$

Next, we form a matrix $\tilde{\mathbf{A}} \in \mathbb{R}^{I_n \times (M \times \prod_{j=1, j \neq n}^{N} P_j)}$ by concatenating $\{\tilde{\mathbf{A}}_{m(n)}, m = 1, ..., M\}$, the n-mode unfolded matrix of $\{\tilde{\mathcal{A}}_m\}$, so that the columns of $\tilde{\mathbf{A}}$ consist of n-mode vectors from $\{\tilde{\mathcal{A}}_m\}$. We then perform PCA on $\tilde{\mathbf{A}}$ and keep Q percent of the total energy/variations, resulting a PCA basis matrix $\mathbf{U} \in \mathbb{R}^{I_n \times R}$ with R leading eigenvectors. Subsequently, we update the n-mode source dimension as $P_n = R$.

3.3   Two Architectures

ICA can be performed under two architectures [4] and so can MMICA. We use the popular FastICA [27] to maximize the modewise non-Gaussianity for modewise IC estimation. FastICA takes a data matrix in and returns a mixing matrix $\mathbf{A}$ and a separating matrix $\mathbf{W}$. They can be used under two architectures in MMICA in the following ways:

– **Architecture I:** FastICA on $\mathbf{U}^T$ gives mixing matrix $\mathbf{A}$ and separating matrix $\mathbf{W}$. Thus, we set the n-mode source matrix as

$$\mathbf{S}^{(n)} = \mathbf{UA}. \tag{14}$$

– **Architecture II:** We first obtain the PCA projection as $\mathbf{V} = \mathbf{U}^T \tilde{\mathbf{A}}$, and FastICA on $\mathbf{V}^T$ gives $\mathbf{A}$ and $\mathbf{W}$. Hence, we set the n-mode source matrix as

$$\mathbf{S}^{(n)} = \mathbf{UW}^T. \tag{15}$$

3.4   Discussion on MMICA

Identifiability and Number of ICs: Following ICA [1], the independent column vectors of modewise source matrices in MMICA are identifiable up to permutation and scaling if they (except one at most) have non-Gaussian distributions and the number of mixtures is no smaller than the number of ICs to be estimated. However, MMICA can not estimate the number of modewise ICs, as in the general case of ICA. When this number is unknown, we determine it by specifying Q in PCA, as described above.

Convergence and Termination: The convergence problem is difficult in ICA. To the best of our knowledge, for FastICA, local convergence analysis is only available for the so-called one-unit case, which considers only one row of the separating matrix [34]. Here, we provide empirical results on the convergence properties of MMICA in Sec. 5, where it converges in one iteration in studies on synthetic data while its classification accuracy stabilizes in just a few iterations in face recognition experiments. Thus, we terminate the iteration by setting K, the maximum number of iterations, to a small number for efficiency.

3.5   Feature Selection for Classification

After obtaining the separated source matrices $\{\mathbf{S}^{(n)}\}$, we have the MMICA representation (coordinates in the mixing tensor) $\hat{\mathcal{A}}$ of a sample $\mathcal{X}$ from (11). Though we can use $\hat{\mathcal{A}}$ directly for classification tasks, we can select a subspace for the convenience of comparison with linear learning algorithms and also for better classification accuracy, as pointed out in [4]. Thus, we further select and sort MMICA features through the same class discriminability as in [4, 25, 26] to study its classification performance.

We can view the MMICA representation $\hat{\mathcal{A}} \in \mathbb{R}^{P_1 \times \cdots \times P_N}$ as being projected through $\prod_{n=1}^N P_n$ *elementary multilinear projections* (EMPs) $\{\mathbf{s}_{p_n}^{(n)}, n = 1, ..., N\}$

[12], where $\mathbf{s}_{p_n}^{(n)}$ is the p_nth column of $\mathbf{S}^{(n)}$. For each component $\hat{A}(p_1, ..., p_N)$, extracted by EMP $\{\mathbf{s}_{p_n}^{(n)}, n = 1, ..., N\}$, we define a class discriminability $\gamma_{p_1 p_2 ... p_N}$ as the ratio of between-class variability to within-class variability, measured by scatters calculated from the training samples:

$$\gamma_{p_1 ... p_N} = \frac{\sum_{c=1}^{C} N_c \cdot \left[\bar{\hat{A}}_c(p_1, ..., p_N) - \bar{\hat{A}}(p_1, ..., p_N)\right]^2}{\sum_{m=1}^{M} \left[\hat{A}_m(p_1, ..., p_N) - \bar{\hat{A}}_{c_m}(p_1, ..., p_N)\right]^2}, \tag{16}$$

where C is the number of classes, M is the number of training samples, N_c is the number of samples for class c and c_m is the class label for the mth training sample $\mathcal{X}_m$. $\hat{A}_m$ is the mixing tensor for $\mathcal{X}_m$. The mean feature tensor

$$\bar{\hat{A}} = \frac{1}{M} \sum_m \hat{A}_m \tag{17}$$

and the class mean feature tensor

$$\bar{\hat{A}}_c = \frac{1}{N_c} \sum_{m, c_m = c} \hat{A}_m. \tag{18}$$

We arrange the entries in $\hat{A}$ into a feature vector $\hat{\mathbf{a}}$ according to the magnitude of $\gamma_{p_1 ... p_N}$ in descending order. The first P entries of $\hat{\mathbf{a}}$, i.e., the P most discriminable components, are selected for classification tasks.

4 Differences with Related Works

MMICA is different from MICA, DTICA and ICA in several aspects, as discussed in the following.

4.1 MMICA vs. MICA

The origin of MMICA in (8) traces back to the higher-order singular value decomposition (HOSVD) [29] and Tucker decomposition [31]. Therefore, it shares mathematical similarity with MICA [23] and DTICA [25, 26], which are both based on HOSVD. However, the MMICA model represents multidimensional data as *tensors* while the MICA model represents them as *vectors* (e.g., 2D faces are represented as 8560×1 vectors in [23]). Thus, 'N' in MMICA represents the order (*number of dimensions*) of a single tensor sample, while 'N' in MICA represents the *number of forming factors* for an ensemble of many samples, with each sample represented as a vector. As mentioned in Sec. 1, MICA is designed as a *supervised learning* method for data labeled with forming factors such as *people* (subject ID), *views* and *illuminations* so the tensor in [23] is formed with four modes as *pixels × people × views × illuminations*. Thus, MICA degenerates to ICA when these factors are unknown, i.e., in unsupervised learning. In contrast, MMICA is an *unsupervised learning* method that does not require such labels. Furthermore, *hidden sources* are not defined in [23]. Hence, the MICA model can not interpret tensor data as in (6) and it cannot perform blind source separation for tensors while MMICA can do so.

4.2 MMICA vs. DTICA

MMICA and DTICA both model a number of tensors with a generative model. DTICA models tensor mixtures with N *mixing matrices and one single source tensor*, built from a *factor-analysis* point of view. In contrast, MMICA models tensor mixtures with *one single mixing tensor and N source matrices*, built from an *independent-component-analysis* point of view. Thus, MMICA can be interpreted in a similar manner as the classical ICA model [3] and perform blind source separation while DTICA cannot be similarly interpreted as mixing several sources and separate sources since its model only involves one *single source tensor*. Furthermore, the MMICA algorithm is iterative while the DTICA algorithm is noniterative, and DTICA requires resampling while MMICA does not require. In addition, DTICA is only formulated for one architecture while we have formulated both architectures for MMICA. Lastly, MMICA makes use of *regularization* to get better results than DTICA.

4.3 MMICA vs. ICA

From (4) to (8), while the classical ICA model assumes that the sources are *mutually independent*, the MMICA model assumes that the sources are *structured tensors formed from modewise matrices with independent columns* instead, which has a simpler and more compact representation when $N > 1$. For the same number of mixing parameters $P = \prod_{n=1}^{N} P_n$, the sources $\{\mathbf{S}^{(n)}\}$ to estimate in MMICA have a size of $\sum_{n=1}^{N} (I_n \times P_n)$ while those in ICA have a size of $\left(\prod_{n=1}^{N} I_n\right) \times \left(\prod_{n=1}^{N} P_n\right)$. E.g., for $N = 3$, $P = 8$, $\{P_n = 2\}$ and $\{I_n = 10\}$, MMICA sources have a size of 60, while ICA sources have a size of 8000, which is about 132 times larger. For $N = 3$, $P = 125$, $\{P_n = 5\}$ and $\{I_n = 100\}$, MMICA sources have a size of 1.5×10^3, while ICA sources have a size of 1.25×10^8, which is about 8.3×10^4 times larger.

5 Experiments

MMICA is applicable to tensors of any order, such as videos, 3-D images, and multi-way social networks [6, 7]. In particular, MMICA can be applied to domains where ICA has been applied in the past, such as biometrics [4], bioinformatics [35], and neuroimaging [36]. For easy visual illustration, this paper studies 2-D images only, which are *matrix data*, i.e., *second-order tensor data* (N = 2). We evaluate MMICA on both synthetic and real data. For synthetic data, we study its capability in estimating hidden sources given their mixtures. For real data, we test it on face recognition, which is widely-used for learning algorithm evaluation [8, 14, 25, 26] with practical importance in security-related applications such as biometric authentication and surveillance.

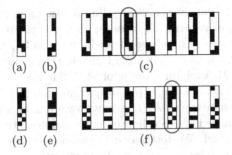

Fig. 2. Blind source separation on synthetic data: (a) true 1-mode source, (b) MMICA estimate of 1-mode source, (c) equivalent patterns of 1-mode MMICA estimate, (d) true 2-mode source, (e) MMICA estimate of 2-mode source, (f) equivalent patterns of 2-mode MMICA estimate. (The pattern matched with the true source is enclosed with an oval.)

5.1 Blind Source Separation on Synthetic Data

Data Generation: This experiment studies whether MMICA can estimate source matrices from synthetic mixture data generated according to (8). The source matrices used are as shown in Fig. 1(b), which are reproduced in Figs. 2(a) and 2(d). Each source matrix is a randomly generated simple *binary pattern* of size 10×2 ($I_n = 10, P_n = 2$). We generated 100 mixtures ($M = 100$) according to (8) by drawing the elements of mixing tensors randomly from a uniform distribution on the unit interval. Figure 1(a) shows ten such mixtures as 8-bit gray images.

Hidden Source Recovery: We applied MMICA with $Q = 100$ using Architecture I for this blind source separation task, followed by binarization to obtain binary source patterns in Figs. 2(b) and 2(e). Since ICA estimation is only unique up to sign and permutation [3], the estimated MMICA sources in Figs. 2(b) and 2(e) are equivalent to the patterns in Figs. 2(c) and 2(f), respectively. One pattern in Figs. 2(c) and 2(f) matches Figs. 2(a) and 2(d) exactly, respectively. Thus, independent modewise source patterns are estimated correctly. To the best of our knowledge, this is the first multilinear extension of ICA for tensor data with demonstrated capability of blind source separation.

Effects of Iteration and Regularization: For this binary source estimation problem, MMICA has recovered the true hidden patterns with only one iteration, showing good convergence. If there is no regularization (using (9)), the mixing tensors can be recovered exactly. Using (11) with $\eta = 10^{-3}$, the estimation has a small average error of $0.005(\pm 0.001)$.

5.2 Face Recognition Studies

Data: The Pose, Illumination, and Expression (PIE) database [37] is widely used for testing face recognition performance. It contains 68 individuals with face images captured under varying pose, illumination and expression. As using the full set leads to low recognition rates for all compared ICA algorithms, here we report the results from a subset of medium difficulty, with five frontal or near frontal poses (C05, C07, C09, C27, C29) under 14 illumination conditions (05 to 14 and 18 to 21, excluding the poorest 7 illumination conditions). Thus, there are about 70 (5×14) samples per subject and a total number of 4,754 face images (with six faces missing). All face images were manually cropped, aligned (with manually annotated eye coordinate) and normalized to 32×32 pixels, with 256 gray levels per pixel. We test face recognition performance under varying number of training samples per subject, denoted by L.

Algorithms and Settings: Since this paper focuses on examining ICA and its extensions under both Architectures I and II, we evaluate MMICA against the classical ICA, MICA in [23] and DTICA in [26]. For fair comparison, we consider *unsupervised learning* only. Hence, training data are not labeled (with image forming factors: poses, illuminations and expressions) and MICA *degenerates* to the classical ICA in this case. Effectively, we have six algorithms to compare: ICA1/MICA1, DTICA1 and MMICA1 for Architecture I, and ICA2/MICA2, DTICA2 and MMICA2 for Architecture II. For DTICA, we form the directional images with the amount of shift $l = 2$, as suggested in [26]. We fix the regularization parameter in (10) as $\eta = 10^{-3}$ for MMICA. All algorithms tested employ FastICA version 2.5[3] [27] with identical (default) settings for fair comparison. We test four values of Q (85, 90, 95, 98), the energy kept in PCA. For all six algorithms, we sort extracted features in descending class discriminability γ in (16) and take the first P features for recognition. To classify extracted features, we use the nearest neighbor classifier with Euclidean distance measure.

Gray-level face images are naturally second-order tensors (matrices), i.e., $N = 2$. Therefore, they are input directly as 32×32 tensors to DTICA and MMICA. For ICA/MICA, they are vectorized to 1024×1 vectors as input. For each subject in a face recognition experiment, $L(= 4, 6, 8, 10)$ samples were randomly selected for training and the rest were used for testing. We report the best results over Q and P, averaged from ten such random splits (repetitions).

Impact of Iterations: We first study the effect of the number of iterations K on the recognition performance of MMICA. Typical results are shown in Fig. 3 for up to 20 iterations with $P = 60$ and $Q = 98$. The figure shows that all accuracy curves are stable with respect to K, while the first a few iterations are more effective for MMICA2 than for MMICA1 in general. Based on this study, we set $K = 3$ in MMICA to reduce the computational cost.

Recognition Results: Figures 4(a) and 4(b) show the best average recognition rates for each algorithm with up to 300 features tested ($P = 1, ..., 300$) for Architectures I and II, respectively. The error bars indicate the standard deviations. Different performance variation is observed for two architectures. Using Architecture I, both DTICA1 and MMICA1 outperform ICA1/MICA1 by around 4% on average. MMICA1 outperforms ICA1/MICA1 by 5.3%, 4.9%, 4.1% and 3.5% for

---

[3] Code at http://www.cis.hut.fi/projects/ica/fastica/code/FastICA_2.5.zip

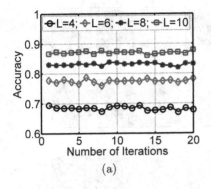
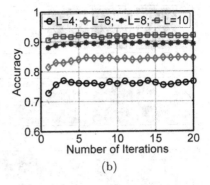

Fig. 3. The impact of iteration numbers on the face recognition accuracy of MMICA with (a) Architecture I (MMICA1) and (b) Architecture II (MMICA2), for $L = 4, 6, 8, 10$

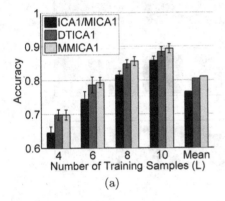
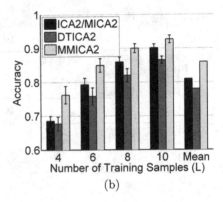

Fig. 4. The average face recognition accuracy comparison of ICA/MICA, DTICA and MMICA for $L = 4, 6, 8, 10$ from ten repetitions with (a) Architecture I and (b) Architecture II. The error bars in (a) and (b) indicate the standard deviations. The results are the best ones for each method from testing four values of Q (85, 90, 95, 98) and 300 values of P (1,...,300).

$L = 4$, 6, 8 and 10, respectively, with more advantage for a smaller L. However, the performance difference between DTICA1 and MMICA1 is small ($< 1\%$ on average). Using Architecture II, DTICA2 is inferior to ICA2/MICA2 in all cases. MMICA2 outperforms ICA2/MICA2 by 7.7%, 5.7%, 4.0% and 2.6% (mean=5%) for $L = 4$, 6, 8 and 10, respectively, again showing superior performance for a smaller L.

5.3   Feature Characteristics

Next, we examine the characteristics of learned features to gain some insight. Figure 5 depicts the eight most discriminable projection bases as 8-bit gray-level

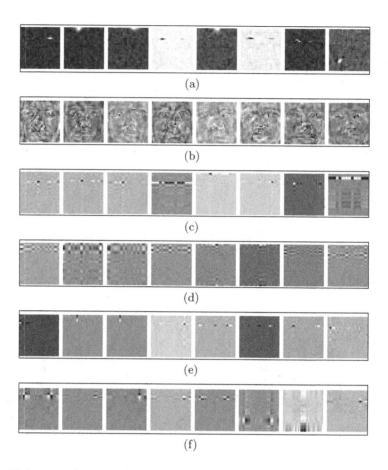

Fig. 5. Eight most discriminable bases obtained from the PIE database with $L = 10$ for (a) ICA1/MICA1, (b) ICA2/MICA2, (c) DTICA1, (d) DTICA2, (e) MMICA1, and (f) MMICA2

images for ICA/MICA, DTICA and MMICA obtained from the PIE database with $L = 10$ using Architectures I and II.

In Figs. 5(a), 5(c) and 5(e), similar to the observations in [4], each projection basis reflects the closeness of each pixel to a cluster of pixels having similar behavior across images. Therefore, these bases are sparse for all three algorithms. In particular, several DTICA1 bases share similar characteristics with MMICA1 bases, showing more structured information than ICA1/MICA1 bases. This may partly be the reason for their closer recognition performance.

With Architecture II, more global properties are encoded. Each ICA-based projection basis attempts to capture a cluster of similar images or image patches, as in Figs. 5(b), 5(d) and 5(f). While this architecture generates more face-like bases for ICA2/MICA2, we found that DTICA2 and MMICA2 bases are quite different, where each basis captures a particular local pattern of the face image instead. DTICA2 and MMICA2 bases have strong structures due to their multilinear nature, with MMICA2 sparser than DTICA2 on the whole. Although each MMICA basis has a size of $32 + 32 = 64$ while each ICA/MICA basis has a size of $32 \times 32 = 1024$, which means 15 times larger, the simpler MMICA bases have achieved much better recognition performance than the more complex ICA/MICA bases. A possible explanation is that the recognition task here has the *small sample size* problem, where the number of samples is small relative to the size of variables to be estimated. Sparser bases have more compact size, leading to *less overfitting* and *better generalization*.

6 Conclusions

We have introduced the multilinear modewise ICA for tensor data using a multilinear mixing model. MMICA extracts modewise independent sources directly from tensor representations through an iterative alternating projection method. We solved this problem by embedding ICA into the MPCA framework and examined two ICA architectures. Studies on synthetic data indicate that MMICA can recover hidden sources from their mixtures accurately. Moreover, experiments on face recognition show different behaviors under different architectures. Using Architecture I, MMICA has similar performance as DTICA while they both outperform ICA/MICA. For Architecture II, MMICA gives the best performance and it is particularly effective when there are only a small number of samples for training. We further examined the extracted features in order to understand the implications and found that MMICA features are sparser and more structured even with Architecture II.

Acknowledgments. The author would like to thank the anonymous reviewers of this work for their insightful and constructive comments.

References

1. Hyvärinen, A., Karhunen, J., Oja, E.: Independent Component Analysis. John Wiley & Sons (2001)
2. Jolliffe, I.T.: Principal Component Analysis, 2nd edn. Springer Series in Statistics (2002)
3. Hyvärinen, A., Oja, E.: Independent component analysis: Algorithms and applications. Neural Networks 13(4-5), 411–430 (2000)
4. Bartlett, M.S., Movellan, J.R., Sejnowski, T.J.: Face recognition by independent component analysis. IEEE Transactions on Neural Networks 13(6), 1450–1464 (2002)
5. Shakhnarovich, G., Moghaddam, B.: Face recognition in subspaces. In: Li, S.Z., Jain, A.K. (eds.) Handbook of Face Recognition, pp. 141–168 (2004)
6. Lu, H., Plataniotis, K.N., Venetsanopoulos, A.N.: A survey of multilinear subspace learning for tensor data. Pattern Recognition 44(7), 1540–1551 (2011)

7. Faloutsos, C., Kolda, T.G., Sun, J.: Mining large time-evolving data using matrix and tensor tools. In: International Conference on Machine Learning 2007 Tutorial (2007), http://www.cs.cmu.edu/ christos/TALKS/ICML-07-tutorial/ICMLtutorial.pdf

8. Ye, J.: Generalized low rank approximations of matrices. Machine Learning 61(1-3), 167–191 (2005)

9. Ye, J., Janardan, R., Li, Q.: GPCA: An efficient dimension reduction scheme for image compression and retrieval. In: ACM SIGKDD International Conference on Knowledge Discovery and Data Mining, pp. 354–363 (2004)

10. Lu, H., Plataniotis, K.N., Venetsanopoulos, A.N.: MPCA: Multilinear principal component analysis of tensor objects. IEEE Transactions on Neural Networks 19(1), 18–39 (2008)

11. Lu, H., Plataniotis, K.N., Venetsanopoulos, A.N.: Uncorrelated multilinear principal component analysis through successive variance maximization. In: International Conference on Machine Learning, pp. 616–623 (2008)

12. Lu, H., Plataniotis, K.N., Venetsanopoulos, A.N.: Uncorrelated multilinear principal component analysis for unsupervised multilinear subspace learning. IEEE Transactions on Neural Networks 20(11), 1820–1836 (2009)

13. Ye, J., Janardan, R., Li, Q.: Two-dimensional linear discriminant analysis. In: Advances in Neural Information Processing Systems (NIPS), pp. 1569–1576 (2004)

14. Yan, S., Xu, D., Yang, Q., Zhang, L., Tang, X., Zhang, H.J.: Discriminant analysis with tensor representation. In: IEEE Conference on Computer Vision and Pattern Recognition, pp. 526–532 (2005)

15. Tao, D., Li, X., Wu, X., Maybank, S.J.: General tensor discriminant analysis and gabor features for gait recognition. IEEE Transactions on Pattern Analysis and Machine Intelligence 29(10), 1700–1715 (2007)

16. Tao, D., Li, X., Wu, X., Maybank, S.J.: Tensor rank one discriminant analysis-a convergent method for discriminative multilinear subspace selection. Neurocomputing 71(10-12), 1866–1882 (2008)

17. Lu, H., Plataniotis, K.N., Venetsanopoulos, A.N.: Uncorrelated multilinear discriminant analysis with regularization and aggregation for tensor object recognition. IEEE Transactions on Neural Networks 20(1), 103–123 (2009)

18. Lee, S.H., Choi, S.: Two-dimensional canonical correlation analysis. IEEE Signal Processing Letters 14(10), 735–738 (2007)

19. Kim, T.K., Cipolla, R.: Canonical correlation analysis of video volume tensors for action categorization and detection. IEEE Transactions on Pattern Analysis and Machine Intelligence 31(8), 1415–1428 (2009)

20. Lu, H.: Learning canonical correlations of paired tensor sets via tensor-to-vector projection. In: The 23rd International Joint Conference on Artificial Intelligence, pp.1516–1522 (2013)

21. Lathauwer, L.D., Vandewalle, J.: Dimensionality reduction in higher-order signal processing and rank-$(R_1, R_2, ., R_N)$ reduction in multilinear algebra. Linear Algebra and its Applications 391, 31–55 (2004)

22. Beckmann, C.F., Smith, S.M.: Tensorial extensions of independent component analysis for multisubject fMRI analysis. NeuroImage 25(1), 294–311 (2005)

23. Vasilescu, M.A.O., Terzopoulos, D.: Multilinear independent components analysis. In: IEEE Conference on Computer Vision and Pattern Recognition, pp. 547–553 (2005)

24. Raj, R.G., Bovik, A.C.: MICA: A multilinear ICA decomposition for natural scene modeling. IEEE Transactions on Image Processing 17(3), 259–271 (2009)

25. Zhang, L., Gao, Q., Zhang, D.: Directional independent component analysis with tensor representation. In: IEEE Conference on Computer Vision and Pattern Recognition, pp. 1–7 (2008)
26. Gao, Q., Zhang, L., Zhang, D., Xu, H.: Independent components extraction from image matrix. Pattern Recognition Letters 31(3), 171–178 (2010)
27. Hyvärinen, A.: Fast and robust fixed-point algorithms for independent component analysis. IEEE Transactions on Neural Networks 10(3), 626–634 (1999)
28. Lathauwer, L.D., Moor, B.D., Vandewalle, J.: On the best rank-1 and rank-$(R_1, R_2, ., R_N)$ approximation of higher-order tensors. SIAM Journal of Matrix Analysis and Applications 21(4), 1324–1342 (2000)
29. Lathauwer, L.D., Moor, B.D., Vandewalle, J.: A multilinear singualr value decomposition. SIAM Journal of Matrix Analysis and Applications 21(4), 1253–1278 (2000)
30. Kolda, T.G., Bader, B.W.: Tensor decompositions and applications. SIAM Review 51(3), 455–500 (2009)
31. Tucker, L.R.: Some mathematical notes on three-mode factor analysis. Psychometrika 31, 279–311 (1966)
32. Friedman, J.H.: Regularized discriminant analysis. Journal of the American Statistical Association 84(405), 165–175 (1989)
33. Lu, H., Plataniotis, K.N., Venetsanopoulos, A.N.: Boosting discriminant learners for gait recognition using MPCA features. EURASIP Journal on Image and Video Processing 2009, Article ID 713183, 11 pages (2009), doi:10.1155/2009/713183
34. Oja, E., Yuan, Z.: The FastICA algorithm revisited: convergence analysis. IEEE Transactions on Neural Networks 17(6), 1370–1381 (2006)
35. Scholz, M., Gatzek, S., Sterling, A., Fiehn, O., Selbig, J.: Metabolite fingerprinting: detecting biological features by independent component analysis. Bioinformatics 20(15), 2447–2454 (2004)
36. Zuo, X.N., Kelly, C., Adelstein, J.S., Klein, D.F., Castellanos, F.X., Milham, M.P.: Reliable intrinsic connectivity networks: test-retest evaluation using ICA and dual regression approach. Neuroimage 49(3), 2163–2177 (2010)
37. Sim, T., Baker, S., Bsat, M.: The CMU pose, illumination, and expression database. IEEE Transactions on Pattern Analysis and Machine Intelligence 25(12), 1615–1618 (2003)

Taxonomic Prediction with Tree-Structured Covariances

Matthew B. Blaschko[1,2], Wojciech Zaremba[1,2], and Arthur Gretton[3]

[1] Center for Visual Computing, École Centrale Paris, France
[2] Équipe Galen, INRIA Saclay, Île-de-France, France
[3] Gatsby Computational Neuroscience Unit, University College London, UK
matthew.blaschko@inria.fr, {woj.zaremba,arthur.gretton}@gmail.com

Abstract. Taxonomies have been proposed numerous times in the literature in order to encode semantic relationships between classes. Such taxonomies have been used to improve classification results by increasing the statistical efficiency of learning, as similarities between classes can be used to increase the amount of relevant data during training. In this paper, we show how data-derived taxonomies may be used in a structured prediction framework, and compare the performance of learned and semantically constructed taxonomies. Structured prediction in this case is multi-class categorization with the assumption that categories are taxonomically related. We make three main contributions: (i) We prove the equivalence between tree-structured covariance matrices and taxonomies; (ii) We use this covariance representation to develop a highly computationally efficient optimization algorithm for structured prediction with taxonomies; (iii) We show that the taxonomies learned from data using the Hilbert-Schmidt Independence Criterion (HSIC) often perform better than imputed semantic taxonomies. Source code of this implementation, as well as machine readable learned taxonomies are available for download from https://github.com/blaschko/tree-structured-covariance.

1 Introduction

In many fields where large numbers of objects must be categorized, including computer vision, bioinformatics, and document classification, an underlying taxonomic structure is applied. While such taxonomies are useful visualization tools to organize data, and to talk about inter-relationships between (sub)categories, it is less clear whether taxonomies can help to perform structured learning, or whether learned taxonomies outperform those imposed by domain experts.

Several learning algorithms have been developed that make use of user-imposed taxonomies, with the main goal being to improve discriminative performance by using hierarchical structure. For example, [1] proposed a learning framework that incorporated semantic categories, and [2] implemented structured output prediction based on a fixed taxonomic structure. For the most part, these previous works have have found that taxonomic structure results in slight improvements

H. Blockeel et al. (Eds.): ECML PKDD 2013, Part II, LNAI 8189, pp. 304–319, 2013.
© Springer-Verlag Berlin Heidelberg 2013

in performance at best, while sometimes decreasing performance. The empirical results in this paper give strong evidence that this may be the result of the user-imposed taxonomy not being aligned to the feature similarities in the data.

In this paper, we make use of a non-parametric dependence measure, the Hilbert-Schmidt Independence Criterion (HSIC), to learn taxonomies. We establish the equivalence between taxonomies and tree structured covariance matrices, and show that the latter constitute a natural way to encode taxonomies in structured prediction problems (indeed, the HSIC is a regularizer for structured output SVM when taxonomies are used). Moreover, we use this tree structured covariance representation to develop a highly efficient algorithm for structured prediction with taxonomies, such that it can be used in large scale problems.

A number of approaches have been proposed for the discovery of taxonomic structure and relationships between classes. Dependency graphs and co-occurrences were modeled in [3,4]. [5] proposed to perform a top-down greedy partitioning of the data into trees. Hierarchical clustering has been employed in [6,7]. Marszałek and Schmid first made use of a semantic hierarchy [8], and later proposed to do a non-disjoint partition into a "relaxed hierarchy" which can then be used for prediction [9]. [10] assume a given taxonomy and then uses a group lasso structured sparsity regularizer with overlapping blocks conforming to the taxonomic structure. In contrast, we do not make the assumption implicit in the group lasso that individual features are exactly aligned with category concepts. [11] perform hierarchical categorization using a taxonomic feature map and loss, but perform an explicit feature map and do not gain the computational advantages arising from the use of tree structured covariance matrices. [12] consider structured prediction of hierarchically organized image labels using a latent variable method to estimate missing annotations in a weakly supervised setting. None of these methods has identified the relationship between hierarchical prediction and tree-structured covariance matrices. [2] made use of a learning framework that is perhaps the most similar to that employed here, based on structured output prediction. However, they did not learn the taxonomy using a non-parametric dependence measure as we do, but instead used a fixed taxonomic structure.

While these works all make use of some clustering objective distinct from the learning procedure, in contrast, this work employs the Hilbert-Schmidt Independence Criterion, which interestingly is coupled with the learning algorithm in its interpretation as a direct optimization of the function prior in ℓ_2 regularized risk with a taxonomic joint kernel map (cf. Equation (13) and Section 5).

Recent works addressing the machine learning aspects of taxonomic prediction include [13], which embeds a taxonomic structure into Euclidean space, while in contrast our method can efficiently learn from taxonomic structures without this approximation. [14] learn a tree structure in order to improve computational efficiency by only evaluating a logarithmic number of classifiers, while [15] relax this tree structure to a directed acyclic graph. Such greedy methods are advantageous when the number of categories is too large to evaluate exactly, while the current paper addresses the problem of efficient learning when exact evaluation is desired.

In experiments on the PASCAL VOC [16], Oxford Flowers [17], and WIPO-alpha [18] datasets, we show that learned taxonomies substantially improve over hand-designed semantic taxonomies in many cases, and never perform significantly worse. Moreover, we demonstrate that learning using taxonomies is widely applicable to large datasets, thanks to the efficiency of our algorithm.

Our paper is organized as follows: in Section 2, we review structured output SVMs, following [19]. We proceed in Section 3 to establish the equivalence of taxonomies and tree structured covariance matrices. In Section 4, we show how tree structured covariance matrices may be incorporated into a structured output learning algorithm, and in particular that this representation of taxonomic structure results in substantial computational advantages. In Section 5, we determine how to learn edge lengths of a taxonomy given a fixed topology using the Hilbert-Schmidt Independence Criterion. Finally, Section 6 contains our experimental results.

2   Taxonomic Prediction

Given a training set of data $\mathcal{S} = \{(x_1, y_1), \ldots, (x_n, y_n)\} \in (\mathcal{X} \times \mathcal{Y})^n$, a structured output SVM with slack rescaling [19,20] optimizes the following learning objective

$$\min_{w \in \mathbb{R}^d, \xi \in \mathbb{R}} \frac{1}{2} \|w\|^2 + C\xi \tag{1}$$

$$\text{s.t.} \sum_i \max_{\tilde{y}_i \in \mathcal{Y}} \left(\langle w, \phi(x_i, y_i) - \phi(x_i, \tilde{y}_i) \rangle - 1 \right) \Delta(y_i, \tilde{y}_i) \geq -\xi \tag{2}$$

$$\xi \geq 0 \tag{3}$$

where ϕ is a joint feature map, and $\Delta(y_i, \tilde{y}_i)$ measures the cost of the erroneous prediction $\tilde{y}_i$ when the correct prediction should be y_i.

Cai and Hofmann proposed a special case of this learning framework in which $\mathcal{Y}$ is taxonomically structured [21]. In that setting, $\phi(x_i, y_i)$ decomposes as $\phi_y(y_i) \otimes \phi_x(x_i)$ and $\phi_y(y_i)$ is a binary vector that encodes the hierarchical relationship between classes. In particular, a taxonomy is defined to be an arbitrary lattice (e.g. tree) whose minimal elements (e.g. leaves) correspond to the categories. $\phi_y(y_i)$ is of length equal to the number of nodes in a taxonomy (equal to the number of categories plus the number of ancestor concepts), and contains non-zero entries at the nodes corresponding to predecessors of the class node. It is straightforward to extend this concept to non-negative entries corresponding to the relative strength of the predecessor relationship. The loss function employed may depend on the length of the shortest path between two nodes [22], or it may be the length of the distance to the nearest common ancestor in the tree [21].

We show in the next two sections that structured prediction with taxonomies is intimately tied to the concept of tree-structured covariance matrices.

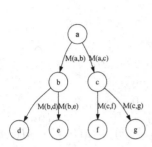

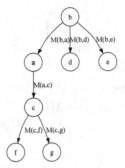

(a) A binary rooted tree. Edges are annotated by their length. The tree metric is defined by the sum of the path lengths between two leaf nodes.

(b) Rerooting the tree by setting node "b" to the root. Distances between leaf nodes are preserved regardless of the rooting.

Fig. 1. An arbitrarily rooted binary tree may be rerooted without changing the pairwise distances between leaf nodes. Furthermore, rerooting has no effect on the value of $HSIC_{\text{cov}}$ (Section 5 and Theorem 2).

3   Tree-Structured Covariance Matrices

Here we consider the structure of a covariance matrix necessary to encode taxonomic structure [23,24].

Definition 1 (Partition property). *A binary matrix V of size $k \times (2k - 1)$ has the partition property for trees of size k (i.e. having k leaves) if it satisfies the following conditions:*

1. *V contains the vector of all ones as a column*
2. *for every column w in V with more than one non-zero entry, it contains two columns u and v such that $u + v = w$.*

We now use this definition to construct a tree structured covariance matrix

Definition 2 (Tree covariance representation). *A matrix B is a tree-structured covariance matrix if and only if $B = VDV^T$ where D is a diagonal matrix with nonnegative entries and V has the partition property.*

This definition is chosen to correspond to [24, Theorem 2]. Such an encoding of tree-structured covariance matrices separates the specification of the topology of the tree, which is encoded in V, from the lengths of the tree branches, which is specified in D. As a concrete example, the tree structured covariance matrix corresponding to Figure 1(a) is

$$V = \begin{pmatrix} 1\,1\,0\,1\,0\,0\,0 \\ 1\,1\,0\,0\,1\,0\,0 \\ 1\,0\,1\,0\,0\,1\,0 \\ 1\,0\,1\,0\,0\,0\,1 \end{pmatrix},$$ (4)

$$D = \mathrm{diag}[0, M(a,b), M(a,c), M(b,d), M(b,e), M(c,f), M(c,g)]^T,$$

$$B = \begin{pmatrix} M(a,b) + M(b,d) & M(a,b) & 0 & 0 \\ M(a,b) & M(a,b) + M(b,e) & 0 & 0 \\ 0 & 0 & M(a,c) + M(c,f) & M(a,c) \\ 0 & 0 & M(a,c) & M(a,c) + M(c,g) \end{pmatrix}$$

Section 3.1 derives a mapping between tree structured covariance matrices and tree metrics, giving a one-to-one relationship and implicitly showing the NP-hardness of optimizing over tree-structured covariance matrices with arbitrary topology.

3.1   Properties of Tree-Structured Covariances and Tree Metrics

In the sequel, the following lemma will be useful

Lemma 1. B_{ij} contains the weighted path length from the root to the nearest common ancestor of nodes i and j.

Proof. Each column of V can be associated with a node in the tree. Each row of V contains a set of binary variables that are equal to 1 iff a corresponding node in the tree is on the path to the leaf associated with that row. As V is binary, $B_{ij} = V_{i:}DV_{j:}^T$ sums over those elements, m, of D for which $V_{im} = V_{jm} = 1$. These elements are exactly the lengths of the branches leading to the common ancestors of nodes i and j.   □

Definition 3 (Four point condition). *A metric M satisfies the four point condition if the following holds*

$$M(a,b) + M(c,d) \le \max(M(a,c) + M(b,d), M(a,d) + M(b,c)) \quad \forall a,b,c,d \quad (5)$$

Theorem 1 (Equivalence of the partition property and the 4 point condition). *The following statements are equivalent*

1. M *is a tree metric.*
2. M *satisfies the four point condition.*
3. $M(i,j) = B_{ii} + B_{jj} - 2B_{ij}$ *where $B = VDV^T$ is a tree-structured covariance matrix.*

Proof. $1 \iff 2$ is shown in [25].

$3 \implies 1$: Using Lemma 1, $M(i,j)$ is the length of the path from the root to node i (B_{ii}) plus the length of the path from the root to node j (B_{jj}) minus two times the length of the path to the nearest common ancestor of nodes i and j (B_{ij}). $B_{ii} - B_{ij}$ is therefore the length from node i to the nearest common ancestor of i and j, and $B_{jj} - B_{ij}$ is the length from node j to their nearest common ancestor. $M(i, j)$ is simply the sum of the two subpaths.

$1 \implies 3$ is a consequence of [24, Theorem 2].   □

We note that [25] considered unrooted trees while Definition 1 and Lemma 1 makes use of the root of a tree. This can be rectified by choosing a root arbitrarily in an unrooted tree (Figure 1). Such a choice corresponds to a degree of freedom in the construction of B that is customarily eliminated by data centering, or by working in a canonical basis as in Definition 1. This is formalized in Theorem 2.

Theorem 2 (Centering trees with different roots but identical topology). *Trees with different roots but identical topology project to the same covariance matrix when centered:*

$$H_k B_1 H_k = H_k B_2 H_k, \tag{6}$$

where B_1 and B_2 have identical topology and edge weights, but different roots, and $H_k = I - \frac{1}{k} e_k e_k^T$ is a centering matrix, e_k being the length k vector of all ones.

Proof. We first note that the linear operator defined in part 3 of Theorem 1, $B_{ii} + B_{jj} - 2B_{ij}$, projects to the same metric all tree structured covariance matrices with identical topology and edge weights, but potentially different roots. This is clear as $M(i, j)$ is simply the sum of weights along the unique path from node i to node j. Consequently, this operator applied to $B_1 - B_2$ yields the zero matrix, yielding a system of linear equations describing the null space of the operator. The null space can be summarized in compact matrix notation as follows

$$C e_k e_k^T + e_k e_k^T C \tag{7}$$

where C is an arbitrary diagonal matrix. We can consequently write any matrix with a fixed topology and edge weights as the summation of the component that lies in the null space of the operator, and the component that is orthogonal to the null space

$$B_1 = B_\perp + C_1 e_k e_k^T + e_k e_k^T C_1, \tag{8}$$

where $B_\perp$ is the component that is orthogonal to the null space, and is identical for all matrices with the same tree topology and edge weights.

We have that $H_k e_k e_k^T = e_k e_k^T H_k = \mathbf{0}$, which yields $H_k (C e_k e_k^T + e_k e_k^T C) H_k = \mathbf{0}$. This in turn implies that

$$H_k (B_1 - B_2) H_k = H_k (B_\perp + C_1 e_k e_k^T + e_k e_k^T C_1 - \tag{9}$$
$$B_\perp - C_2 e_k e_k^T - e_k e_k^T C_2) H_k = \mathbf{0}$$
$$H_k B_1 H_k = H_k B_2 H_k. \tag{10}$$

□

4   Structured Prediction with Tree-Structured Covariances

Given the concepts developed in Section 3, we find now that the specification of joint feature maps and loss functions for taxonomic prediction is much simplified. We may assume that a taxonomy is specified that encodes the loss function Δ for a given problem, which need not be the same as a taxonomy for specifying the feature map ϕ. For the minimal path distance, $\Delta(y, \tilde{y}) = M(y, \tilde{y})$ for M defined as in Theorem 1. For Δ equal to the distance to the nearest common ancestor, we may use $B_{\tilde{y}\tilde{y}} - B_{y\tilde{y}}$. We have used the minimal path distance in the experimental section whenever taxonomic loss has been employed. The standard taxonomic structured loss functions therefore only require as an input a tree-structured covariance matrix B_{loss}, which need not be the same matrix as the one used to define a feature map (0-1 loss is recovered by using the identity matrix).

We now turn to the tree-structured joint kernel map (cf. Section 2). Given a tree-structured covariance matrix B and its decomposition into $B = VDV^T$, we may compactly define $\phi_y : \mathcal{Y} \mapsto \mathbb{R}^{2k-1}$ as the function that selects the kth column of $D^{\frac{1}{2}}V^T$ when y specifies that the sample belongs to the kth class.[1] Making use of the representer theorem for structured prediction with joint kernel maps [26], we know that the solution to our structured prediction objective lies in the span of our training input data $X \subset \mathcal{X}$ crossed with the output space, $\mathcal{Y}$. Assuming a kernel matrix K_x with associated reproducing kernel Hilbert space $\mathcal{F}$ such that the i, jth entry of K_x corresponds to $\langle \phi_x(x_i), \phi_x(x_j) \rangle_{\mathcal{F}}$, we have that the solution may be written

$$\sum_{1 \leq i \leq n} \sum_{y \in \mathcal{Y}} \alpha_{iy} \phi(x_i, y) \tag{11}$$

and that the corresponding joint kernel matrix decomposes as $K_x \otimes B$. Although the size of the joint kernel matrix is $n \cdot k \times n \cdot k$, we may make use of several properties of the Kronecker product to avoid high memory storage and costly matrix operations.

Looking specifically at Tikhonov regularized risk:

$$\min_g \lambda \|g\|_{\mathcal{H}}^2 + \ell(g, \mathcal{S}) = \min_\alpha \lambda \alpha^T (K_x \otimes B) \alpha + \ell(\alpha, \mathcal{S}) \tag{12}$$

where ℓ is some loss function (we have overloaded the notation in the kernelized case). Interestingly, we may use the identity from Theorem 2.3 of [27]

$$\alpha^T (K_x \otimes B) \alpha = \text{Tr}[K_x \tilde{\alpha}^T B \tilde{\alpha}] \tag{13}$$

where $\tilde{\alpha} \in \mathbb{R}^{n \times k}$ is the matrix such that vec $\tilde{\alpha} = \alpha$.

In the case of a structured output SVM, where we have a quadratic regularizer with linear constraints, we can make use of many optimization schemes, that, e.g. require repeated efficient multiplication of a vector with the Hessian:

$$(K_x \otimes B) \alpha = \text{vec} \, B \tilde{\alpha} K_x. \tag{14}$$

---

[1] A rooted tree with k leaves can be encoded with at most $2k - 1$ nodes (Figure 1).

Using the popular SVMstruct framework [19,20] in this case generates a large number of non-sparse constraints and is very memory inefficient, requiring the storage of a number of kernel values proportional to the number of tuples in $\mathcal{X} \times \mathcal{Y} \times \mathcal{X} \times \mathcal{Y}$.[2] This indicates that the resulting memory requirements for such a scheme are $\mathcal{O}(n^2 k^2)$, while making use of optimization with Equation (14) requires only $\mathcal{O}(n^2 + k^2 + nk)$ memory, and standard large scale kernel learning methods may be applied off-the-shelf to reduce the dominating $\mathcal{O}(n^2)$ component [28]. We have used a cutting plane training to efficiently train our taxonomic predictors, giving the same convergence guarantees as SVMstruct, but with substantially less expensive computation for cutting plane inference.

Cutting plane optimization requires finding a setting of $\tilde{y}$ that minimizes the right hand side of Equation (2). In the kernelized setting, we substitute for w as in Equation (12), and search for parameters $\beta \in \mathbb{R}^{nk \times 1}$ and $\delta \in \mathbb{R}$ that give the kernel coefficients and offset of the linear constraint

$$\delta - \alpha^T (K_x \otimes B)\beta \geq \xi. \tag{15}$$

Using Equation (14) enables us to solve this cutting plane iteration efficiently, both in terms of computation and memory usage. A reference implementation of this efficient optimization scheme is available for download from https://github.com/blaschko/tree-structured-covariance.

In the next section, we discuss how to learn taxonomies from data that are suitable for learning in this structured prediction model.

5   Optimizing Tree-Structured Covariances with the Hilbert-Schmidt Independence Criterion

In this section, we show how a non-parametric dependence test may be employed to learn taxonomies that can then be employed in the construction of a joint feature map for taxonomic prediction.

The Hilbert-Schmidt Independence Criterion (HSIC) is a kernel statistical measure that may be used to measure the dependence between empirical data observations and matrices that encode the hypothesized taxonomic structure of a data set [3]. The HSIC is defined to be the Hilbert-Schmidt norm of the cross covariance operator C_{xy} between mappings from the input space $\mathcal{X}$ and from the label space $\mathcal{Y}$. For characteristic kernels [29],[3] this is zero if and only if X and Y are independent. Given a finite sample of size n from $\mathrm{Pr}_{X,Y}$, the HSIC is

$$HSIC := \mathrm{Tr}[H_n K H_n L] \tag{16}$$

where K is the Gram matrix for samples from Pr_X with (i,j)th entry $k(x_i, x_j)$, and L is the Gram matrix with kernel $l(y_i, y_j)$.

---

[2] This follows from an analogous argument to the one used in binary classification that the storage requirements of a SVM are proportional to the Bayes rate, and therefore linear in the number of i.i.d. training samples.

[3] e.g. the Gaussian Kernel on $\mathbb{R}^d$.

To define our kernel matrix on the output space, we consider a family of functions proposed several times in the literature in the context of HSIC [3,30]. In particular, we define the kernel in terms of a label matrix $\Pi \in \{0,1\}^{k \times n}$, and a covariance matrix, $B \in \mathbb{R}^{k \times k}$, that encodes the relationship between classes. Given these matrices, $L = \Pi^T B \Pi$. The HSIC with this kernel over $\mathcal{Y}$ is

$$HSIC_{\text{cov}} := \text{Tr}[H_n K H_n \Pi^T B \Pi]. \tag{17}$$

As pointed out by [31], $H_k \Pi H_n = \Pi H_n$, which in conjunction with Theorem 2 indicates that $HSIC_{\text{cov}}$ is identical regardless of how the tree is rooted (cf. Figure 1). We note that L is characteristic over $\mathcal{Y}$ whenever rank$[B] \geq k-1$ and the null space of B is empty or contains e_k.

When K_x is centered, the functional form of Equation (13) is identical to Equation (17), indicating that the regularizer is $HSIC_{\text{cov}}$ with $\tilde{\alpha}$ in place of Π. While our derivation has focused on tree-structured covariance matrices, this novel theoretical result is applicable to arbitrary covariances over $\mathcal{Y}$, indicating a tight coupling between non-parametric dependence tests and regularization in structured prediction.

With this fundamental relationship in place, we consider in turn optimizing over tree structured covariance matrices with fixed and arbitrary topology. The learned taxonomies may then be employed in structured prediction.

5.1   Optimization Over Tree-Structured Covariance Matrices

Theorem 2 gives a convenient decomposition of a tree structured covariance matrix into a binary matrix encoding the topology of the tree and a positive diagonal matrix encoding the branch lengths. One such consequence of the existence of this decomposition is

Theorem 3. *The set of trees with identical topology is a convex set.*

Proof. [24] Given two tree structured covariance matrices with the same topology, $B = VDV^T$ and $\tilde{B} = V\tilde{D}V^T$, any convex combination can be written

$$\eta B + (1 - \eta)\tilde{B} = V\left(\eta D + (1 - \eta)\tilde{D}\right)V^T \tag{18}$$

for arbitrary $0 \leq \eta \leq 1$.   □

Optimization of such covariance matrices with fixed topology is consequently significantly simplified. For D^* maximizing the HSIC subject to a norm constraint, a closed form solution is given by

$$D^* \propto \text{diag}\left[V^T \Pi^T H_n K_x H_n \Pi V\right]. \tag{19}$$

We note that this optimization is analogous to that in [3] for tree structured covariance matrices with arbitrary topology. In that work, a closed form solution for arbitrary positive definite matrices was found, which was later projected onto the space of tree-structured matrices using a *numerical taxonomy* algorithm with tight approximation bounds. We have employed the method of [3] for comparison in the experimental results section. Theorems 1 and 2 justify the equivalence of our procedures for learning tree-structured covariance matrices with both fixed and arbitrary covariance matrices.

6   Experimental Results

We perform an empirical study on two popular computer vision datasets, PASCAL VOC [16] and Oxford Flowers [17], and on the WIPO text dataset [18].

6.1   PASCAL VOC

We evaluate the performance of semantic vs. visual taxonomies on the PASCAL VOC 2007 dataset. To construct features for this data, we have employed results from the best performing submission to the 2007 classification challenge, INRIA_Genetic, which won all but one category. Our feature vector is constructed by concatenating variance normalized class prediction scores, after which a Gaussian kernel is applied, setting the σ parameter to the median of the pairwise distances in the feature space. As the parameters of the prediction functions were trained on data separate from the test images, this is a proper kernel over the test data set.[4] By construction, we are certain that the relevant visual information is contained within this feature representation, indicating that it is appropriate to use it to optimize the taxonomic structure. Furthermore, the INRIA_Genetic method did not make use of taxonomic relationships, meaning that no imputed class relationships will influence the taxonomy discovery algorithm.

The semantic taxonomy was transcribed from the one proposed by the competition organizers [16]. As they do not provide edge lengths for their taxonomy (i.e. relative similarities for each subclass), we have learned these optimally from data using Equation (19). We have also learned a taxonomy with unconstrained topology, which is presented in Figure 2. Interestingly, the semantic topology and the learned topology are very close despite the learning algorithm's not having access to any information about the topology of the semantic taxonomy.

We have performed classification on the PASCAL VOC data set using the taxonomic prediction method described in Section 2. We trained on the first 50% of the competition test set, and report results as ROC curves on the second 50%. We emphasize that the results are designed for comparison between semantic and learned visual taxonomies, and are not for comparison within the competition framework. We additionally compare to the multi-class prediction method proposed by [32]. Results are shown in Figure 3.

---

[4] The learned taxonomy is available for download from https://github.com/blaschko/tree-structured-covariance. We note that this taxonomy is not appropriate to apply to the VOC 2007 dataset as that would involve training on the test set. However, as subsequent years of the VOC challenge use a disjoint set of images but the same classes, the taxonomy is applicable in those settings.

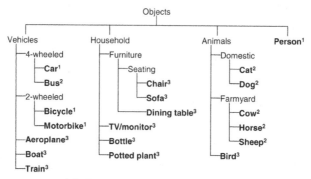

(a) Semantic taxonomy from [16].

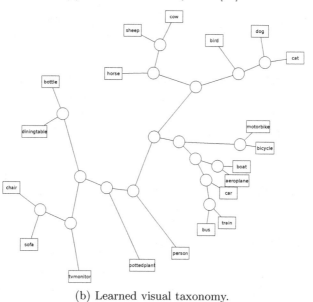

(b) Learned visual taxonomy.

Fig. 2. The semantic and learned taxonomies for the PASCAL VOC dataset. The semantic and visual taxonomies are very close, despite that the construction of the visual taxonomy made no use of the semantic relationships.

6.2   Oxford Flowers

In the second set of experiments, we have compared semantic to visual taxonomies on the Oxford Flowers data set. To construct a rich image representation, we have made use of the features designed by the authors of the dataset. The image representations consist of information encoding color, shape, (local) gradient histograms, and texture descriptors [17]. These features have resulted in high performance on this task in benchmark studies. We have constructed kernel matrices using the mean of Gaussian kernels as described in [33].

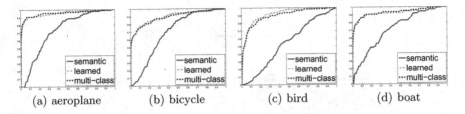

| (a) aeroplane | (b) bicycle | (c) bird | (d) boat |

Fig. 3. ROC curves for the PASCAL VOC dataset. The learned visual taxonomy performs consistently better than the semantic taxonomy. Multi-class classification was performed with a multi-label generalization of [32]. Only the first four classes are shown due to space constraints. The other classes show qualitatively the same relationship between methods.

Table 1. Classification scores for the Oxford Flowers data set. The semantic taxonomy (Figure 4(a)) gives comparatively poor performance, likely due to the strong mismatch between the biological taxonomy and visual similarity. The learned visual taxonomy (Figure 4(b)), however, maintains good performance compared with one-vs.-rest classification.

One vs. rest [33]	Semantic Taxonomy	Learned Taxonomy
84.9 ± 1.9	56.3 ± 6.3	$\mathbf{87.7 \pm 2.6}$

The topology of the semantic taxonomy was constructed using the Linnaean biological taxonomy, while edge distances were computed by optimizing D according to Equation (19). The topologies of the semantic taxonomy and the learned visual taxonomy are given in Figure 4.

We have additionally performed classification using the semantic and learned visual taxonomies. We have applied the taxonomic prediction method described in Section 2. The results are presented in Table 1. In line with previous results on taxonomic prediction, the performance of the taxonomic method with a visual taxonomy performs comparably to 1-vs.-rest classification (here we report the results from [33], which use an identical kernel matrix to our method). However, we note that the semantic taxonomy performs very poorly, while the learned taxonomy maintains good results. We hypothesize that this is due to the strong mismatch between the semantic relationships and the visual ones. In this case, it is inappropriate to make use of a semantic taxonomy, but our approach enables us to gain the benefits of taxonomic prediction without requiring an additional information source to construct the taxonomy.

6.3   Text Categorization

We present timing and accuracies on the WIPO data set [18], a hierarchically structured document corpus that is commonly used in taxonomic prediction [21]. Kernel design was performed simply using a bag of words feature representation combined with a generalized Gaussian χ^2 kernel with the bandwidth parameter

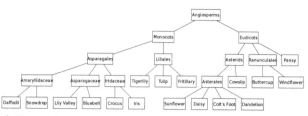

(a) Semantic taxonomy constructed using biological information.

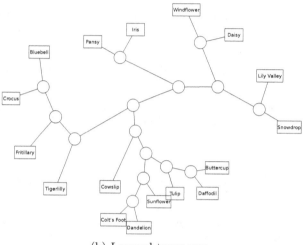

(b) Learned taxonomy.

Fig. 4. Semantic and visual taxonomies on the Oxford Flowers dataset. The topologies of the two taxonomies differ significantly, indicating a strong mismatch between the semantic hierarchy and visual similarity.

set to the median of the pairwise χ^2 distances. The topology, V, of the tree structure was constructed using the taxonomy provided by the data set organizers. The loss function, Δ, was either set to 0-1 loss, or the taxonomic distance between two concepts. The taxonomic distance between two concepts was measured as the unweighted path length between the two leaves in the taxonomy (i.e. not making use of the learned taxonomy but instead fixing edge lengths to 1).

We have computed results using a number of covariance structures, as well as a number of loss functions. Table 2 lists these settings and shows their numerical accuracies. We emphasize that the results correspond to the learning setting proposed by [21] when the covariance matrix is tree-structured. Any differences in performance for this column are due to our using a more recent version of the data set with a comparatively naïve feature representation, while Cai and Hofmann made use of an unspecified kernel function computed using a proprietary software system [21].

Table 2. Losses on the WIPO data set (lower is better). The columns correspond to varying covariance structures, while the rows correspond to different loss functions. For the covariance structures, I corresponds to a standard multi-class feature map [32], B^* is learned using the method proposed in [3] for learning taxonomies without fixed topology, and D^* is learned from Equation (19). Each system was trained with a structured output support vector machine optimizing the loss on which it is evaluated.

	I	B^*	$\left|H_k V D^* V^T H_k\right|$	$V D^* V^T$
0-1	0.281 ± 0.027	$\mathbf{0.278 \pm 0.042}$	0.284 ± 0.037	0.362 ± 0.028
taxonomic	0.950 ± 0.100	$\mathbf{0.833 \pm 0.179}$	1.125 ± 0.071	1.120 ± 0.028

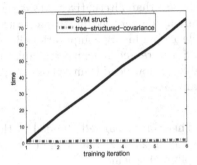

Fig. 5. Computation time for constraint generation using the proposed method of optimization vs. the popular SVMstruct optimization package [19,20]. The proposed optimization is several orders of magnitude faster than SVMstruct for this problem, and has constant computation time per iteration, while SVMstruct has computation that grows linearly with the training iteration.

We focus on the efficiency of the optimization using our strategy, and the kernelized variant of SVMstruct [19,20]. We compare the empirical time per cutting plane iteration in Figure 5. We note that timing results are presented as a fraction of the first training iteration to account for differences in vector and matrix libraries employed in our implementation vs. SVMstruct. Nevertheless, our implementation was several orders of magnitude faster than SVMstruct at all iterations of training due to the avoidance of naïve looping over cached kernel values as employed by their general purpose framework. In the SVMstruct implementation of taxonomic prediction, the joint kernel function was implemented by multiplying K_{ij} by $B_{y_i y_j}$, which were both kept in memory to optimize computation time. The computation time of our algorithm is constant per iteration, in contrast to SVMstruct, which grows approximately linearly with high slope as the number of support vectors grows. In later training iterations, a single kernelized cutting plane iteration of SVMstruct can take several minutes, while our method takes only several milliseconds. The number of cutting plane iterations required by both methods is identical.

7   Conclusions

In this work, we have compared taxonomies learned from data with semantic taxonomies provided by domain experts, where these taxonomies are used to impose structure in learning problems. While a semantic taxonomy provides a measure of prior information on class relationships, this may be unhelpful to the desired learning outcome when the features available are not in accord with this structure. Indeed, in such cases, we have shown that the imposition of prior taxonomic information may result in a significant performance penalty.

By contrast, we have observed that learned taxonomies based on feature similarity can do significantly better than hand-designed taxonomies, while never performing significantly worse than alternatives. Moreover, we have shown that the taxonomic structure may be encoded in a tree-structured covariance: as a result, we were able to develop a highly computationally efficient learning algorithm over taxonomies. Software and machine readable tree-structured covariance matrices are available for download from https://github.com/blaschko/tree-structured-covariance.

Acknowledgements. This work is partially funded by the European Research Council under FP7/ERC Grant 259112, and by the Royal Academy of Engineering through the Newton alumni scheme.

References

1. Zweig, A., Weinshall, D.: Exploiting object hierarchy: Combining models from different category levels. In: ICCV (2007)
2. Binder, A., Müller, K.R., Kawanabe, M.: On taxonomies for multi-class image categorization. IJCV (2012)
3. Blaschko, M.B., Gretton, A.: Learning taxonomies by dependence maximization. In: NIPS (2009)
4. Lampert, C.H., Blaschko, M.B.: A multiple kernel learning approach to joint multi-class object detection. In: Rigoll, G. (ed.) DAGM 2008. LNCS, vol. 5096, pp. 31–40. Springer, Heidelberg (2008)
5. Tibshirani, R., Hastie, T.: Margin trees for high-dimensional classification. JMLR 8, 637–652 (2007)
6. Fan, X.: Efficient multiclass object detection by a hierarchy of classifiers. In: CVPR (2005)
7. Griffin, G., Perona, P.: Learning and using taxonomies for fast visual categorization. In: CVPR (2008)
8. Marszałek, M., Schmid, C.: Semantic hierarchies for visual object recognition. In: CVPR (2007)
9. Marszałek, M., Schmid, C.: Constructing category hierarchies for visual recognition. In: Forsyth, D., Torr, P., Zisserman, A. (eds.) ECCV 2008, Part IV. LNCS, vol. 5305, pp. 479–491. Springer, Heidelberg (2008)
10. Zhao, B., Li, F.F.F., Xing, E.P.: Large-scale category structure aware image categorization. In: NIPS, pp. 1251–1259 (2011)

11. Mittal, A., Blaschko, M.B., Zisserman, A., Torr, P.H.S.: Taxonomic multi-class prediction and person layout using efficient structured ranking. In: Fitzgibbon, A., Lazebnik, S., Perona, P., Sato, Y., Schmid, C. (eds.) ECCV 2012, Part II. LNCS, vol. 7573, pp. 245–258. Springer, Heidelberg (2012)

12. McAuley, J., Ramisa, A., Caetano, T.: Optimization of robust loss functions for weakly-labeled image taxonomies. IJCV, 1–19 (2012)

13. Weinberger, K., Chapelle, O.: Large margin taxonomy embedding for document categorization. In: NIPS, pp. 1737–1744 (2009)

14. Bengio, S., Weston, J., Grangier, D.: Label embedding trees for large multi-class tasks. In: NIPS, pp. 163–171 (2010)

15. Gao, T., Koller, D.: Discriminative learning of relaxed hierarchy for large-scale visual recognition. In: ICCV, pp. 2072–2079 (2011)

16. Everingham, M., Van Gool, L., Williams, C.K.I., Winn, J., Zisserman, A.: The PASCAL Visual Object Classes (VOC) challenge. IJCV 88(2), 303–338 (2010)

17. Nilsback, M.E., Zisserman, A.: Delving deeper into the whorl of flower segmentation. Image and Vision Computing (2009)

18. World Intellectual Property Organization: WIPO-alpha data set (2009), http://www.wipo.int/

19. Tsochantaridis, I., Hofmann, T., Joachims, T., Altun, Y.: Support vector machine learning for interdependent and structured output spaces. In: ICML (2004)

20. Joachims, T., Finley, T., Yu, C.N.J.: Cutting-plane training of structural SVMs. Mach. Learn. 77(1), 27–59 (2009)

21. Cai, L., Hofmann, T.: Hierarchical document categorization with support vector machines. In: CIKM (2004)

22. Wang, K., Zhou, S., Liew, S.C.: Building hierarchical classifiers using class proximity. In: VLDB (1999)

23. Cavalli-Sforza, L.L., Edwards, A.W.F.: Phylogenetic analysis: Models and estimation procedures. American Journal of Human Genetics 19, 223–257 (1967)

24. Corrada Bravo, H., Wright, S., Eng, K., Keleş, S., Wahba, G.: Estimating tree-structured covariance matrices via mixed-integer programming. In: AISTATS (2009)

25. Buneman, P.: The recovery of trees from measures of dissimilarity. In: Kendall, D.G., Tautu, P. (eds.) Mathematics in the Archeological and Historical Sciences, pp. 387–395. Edinburgh University Press (1971)

26. Lafferty, J., Zhu, X., Liu, Y.: Kernel conditional random fields: representation and clique selection. In: ICML (2004)

27. Magnus, J.R., Neudecker, H.: Matrix Differential Calculus with Applications in Statistics and Econometrics. Wiley (1988)

28. Bottou, L., Chapelle, O., DeCoste, D., Weston, J.: Large-Scale Kernel Machines. MIT Press (2007)

29. Fukumizu, K., Gretton, A., Sun, X., Schölkopf, B.: Kernel measures of conditional dependence. In: NIPS, pp. 489–496 (2008)

30. Song, L., Smola, A., Gretton, A., Borgwardt, K.M.: A dependence maximization view of clustering. In: ICML (2007)

31. Blaschko, M.B., Gretton, A.: Taxonomy inference using kernel dependence measures. Technical Report 181, Max Planck Inst. for Bio. Cybernetics (2008)

32. Crammer, K., Singer, Y.: On the algorithmic implementation of multiclass kernel-based vector machines. JMLR 2, 265–292 (2002)

33. Gehler, P., Nowozin, S.: On feature combination methods for multiclass object classification. In: ICCV (2009)

Position Preserving Multi-Output Prediction

Zubin Abraham[1], Pang-Ning Tan[1], Perdinan[1], Julie Winkler[1],
Shiyuan Zhong[1], and Malgorzata Liszewska[2]

[1] Michigan State University, USA
{abraha84,perdinan,winkler,zhongs}@msu.edu, ptan@cse.msu.edu
[2] University of Warsaw, Poland
m.liszewska@icm.edu.pl

Abstract. There is a growing demand for multiple output prediction methods capable of both minimizing residual errors and capturing the joint distribution of the response variables in a realistic and consistent fashion. Unfortunately, current methods are designed to optimize one of the two criteria, but not both. This paper presents a framework for multiple output regression that preserves the relationships among the response variables (including possible non-linear associations) while minimizing the residual errors of prediction by coupling regression methods with geometric quantile mapping. We demonstrate the effectiveness of the framework in modeling daily temperature and precipitation for climate stations in the Great Lakes region. We showed that, in all climate stations evaluated, the proposed framework achieves low residual errors comparable to standard regression methods while preserving the joint distribution of the response variables.

1 Introduction

Multiple output regression (MOR) is the task of inferring the joint values of multiple response variables from a set of common predictor variables. The response variables are often related, though their true relationships are generally unknown *a priori*. An example application of multiple output regression is to simultaneously estimate the projected future values of temperature, precipitation, and other climate variables needed for climate change impact, adaptation and vulnerability (CCIAV) assessments. The projected values are used as the driving input variables for phenological and hydrological models to simulate the responses of the ecological system to future climate change scenarios. To ensure the projected values are realistic, there are certain constraints on the relationship among the response variables that must be preserved; e.g., minimum temperature must not exceed maximum temperature or liquid precipitation should be zero when temperature is below freezing. While there have been numerous multiple output regression methods developed in recent years [7,20,4,18,12], most of them are focused on fitting the conditional mean or preserving covariance structure of the outputs. Such methods do not adequately capture the full range of variability in the joint output distribution, as illustrated in Figure 1(a).

The inability of standard regression-based approaches to reproduce the shape of the true distribution of output variables, even for univariate response

H. Blockeel et al. (Eds.): ECML PKDD 2013, Part II, LNAI 8189, pp. 320–335, 2013.
© Springer-Verlag Berlin Heidelberg 2013

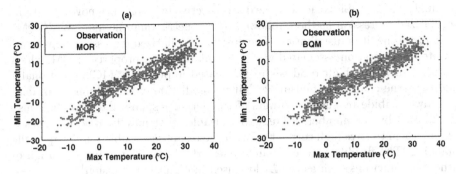

Fig. 1. Scatter plot of observed daily maximum and minimum temperature at a climate station in Michigan, USA

variables, is well-documented [2]. Univariate *distribution-driven approaches* such as quantile mapping (QM) [13] and statistical asynchronous regression (SAR) [17] have been developed to address this limitation, but the accuracy of these approaches is generally poor since they are not designed to minimize residual errors. Quantile mapping approaches map a univariate predictor variable x to its corresponding response variable y by transforming the cumulative distribution function (CDF) of x to match that of y. More recently, a bivariate quantile mapping approach (BQM) (see Figure 1(b)) has been developed to generate bivariate response values that mimic the joint distribution of the observed response data [11]. However, as will be shown in this paper, the residual error is significantly worse when compared to regression-based methods because the position and rank correlation between the predictor and response variables remain invariant under QM-based transformation, which in turn, hinders its ability to minimize residual errors. Thus, unless the predictor variable has a high rank correlation with the response variable, the residual error upon applying QM-based approaches is likely to be large.

This suggests a possible hybrid approach to improve both the residual errors and distribution fitting is by first applying a regression-based method to transform the predictor variables so that their rank correlation with respect to the response variable is high, before applying quantile mapping to adjust for the fit in distribution. However, maximizing the rank correlation of the data points is necessary but not sufficient condition for improvement in the residuals for QM, unless the response values of the data points are uniformly spaced. Hence, the need for position regularization, that would prioritize the prediction accuracy of data points whose position, when incorrectly estimated, results in high residual. The term 'position' here refers to the geometric quantile of a data point with respect to a multivariate distribution, which is analogous to the quantile of a data point in the case of univariate distribution. In this paper, we present a position-regularized, multi-output prediction framework called Multi-Output Contour Regression (MCR), that addresses the dual objective of preserving the associations among the multiple output variables as well as minimizing residuals. MCR is able to achieve the dual objective by applying a novel, position-regularized regression method, followed by geometric quantile mapping (GQM) to improve the fit in distribution. The position-regularized regression helps to alleviate the limitation associated with the rank invariant property of QM, which contributes to the high residuals of QM-based approaches. MCR additionally addresses the challenge of ensuring that its prediction of the response variables will always abide by the constraints of the actual response data. MCR is also not limited by the number of predictor variables that may be used nor does it require them to have high correlation with the response variables, unlike quantile mapping. The flexible nature of our framework allows for the incorporation of other loss functions such as the L_1 loss used in quantile regression[1].

2   Related Work

Supervised learning methods for predicting continuous-valued outputs may be categorized as either *accuracy-driven* or *distribution-driven*. Accuracy-driven approaches such as multiple linear regression (MLR), lasso regression, neural networks, and analog methods [13] are commonly used with emphasis on minimizing sum-square residual (SSR) errors. In contrast, distribution-driven approaches focus on reproducing the distribution characteristics of the output variable. These approaches include quantile mapping (QM) [13], Equidistant CDF Matching (EDCDFm), statistical asynchronous regression (SAR) [17] and the transfer functions proposed by Piani et al. [19]. These approaches are applicable even when the predictor and response variables are asynchronous and are generally susceptible to high residual errors. Given the drawbacks of accuracy-driven and distribution-driven approaches, a hybrid method known as Contour Regression (CR) [2] was developed to simultaneously minimize error and preserve the shape of the fitted distribution. CR extends the loss functions of standard regression methods (including linear and quantile regression) to regularize the area between the CDF of the response variable and the CDF of the predicted output.

In addition to the single output regression (SOR) approaches, techniques for inferring multiple response variables (MOR) simultaneously have been developed, including multi-output regression [10] and structured output regression [5]. A number of these techniques focus on penalizing the regression coefficients using low rank methods such as reduced rank regression [12]. However, these approaches do not consider the correlation among the output variables. Another common approach to multiple output prediction is to penalize the shared input space, for co-linearity, such as partial least square regression discriminant analysis (PLSDA) [18]. However these models, too, do not capture the association among response variables. Curds and Whey is an example of regression based approach that considers the output correlation [7]. However, it assumes the relationship among the response variables is linear. Multiple output SVR is another approach that takes advantage of correlation among response variables and extends Support Vector Regression (SVR) to multi-output systems by employing

---

[1] We omit the derivation for other loss function in this paper due to lack of space.

co-kriging, to account for the cross covariances between different response variables [20]. Group lasso [14], LL-MIMO [6], gaussian process MOR [3] are other examples of MOR.

However, none of the these approaches preserve the full range of variability of the joint distribution of the response variables. He et al. [11] proposed bivariate quantile mapping to extend QM to bivariate space. The method uses the intuition proposed by Buja et al. [8] regarding geometric quantiles. While this approach is capable of capturing the distribution characteristics of bivariate response variables, similar to QM, it is susceptible to high residual errors.

3   Preliminaries

Let $\mathbf{X} = [\mathbf{x}_1, .., \mathbf{x}_n]^T$ be an $(n \times d)$ data matrix and $\mathbf{Y} = [\mathbf{y}_1, .., \mathbf{y}_n]^T$ be the corresponding $(n \times q)$ response matrix, such that $\mathbf{x}_i \in \Re^d$ and $\mathbf{y}_i \in \Re^q$ are column vectors representing the respective values of predictor and response variables for the ith data point. The objective of multi-output regression (MOR) is to learn a target function $h(\mathbf{x}, \Omega)$ that best estimates the multi-output response $\mathbf{y}$, where $\Omega = (\omega_1, .., \omega_q)$ is the parameter set of the target function.

For a univariate random variable $X \in \Re$, let $F_X(x)$ be its cumulative distribution function (CDF), i.e., $F_X(x) = P(X \le x)$. The corresponding α-quantile of X is given by $\inf \{x \in \Re : F_X(x) \ge \alpha\}$. Intuitively, each quantile indicates the value in which a certain fraction of the data points are below it, and thus, provides a measure of its position in the data. For example, the median, which is equivalent to the 0.5-quantile, is the central location of the distribution. More generally, the position [16] of data point $\mathbf{z}$ relative to a set of points $\mathbf{Z} = (\mathbf{z}_1, .., \mathbf{z}_m)^T$ is given by

$$\mathbf{p}_{\mathbf{Z}}(\mathbf{z}) = \tfrac{1}{m} \sum_{i=1}^m \eta(\mathbf{z} - \mathbf{z}_i) \qquad \text{where} \qquad \eta(\mathbf{w}) = \begin{cases} \frac{\mathbf{w}}{\|\mathbf{w}\|}, & \text{if } \mathbf{w} \ne \mathbf{0} \\ 0, & \text{if } \mathbf{w} = \mathbf{0} \end{cases}$$

For univariate data, the position $p_Z(z)$ is equal to $2F_Z(z) - 1$, where $F_Z(z)$ is the cumulative distribution function of Z. The multi-dimensional equivalent of quantile function is geometric quantile [9].

Distribution correction methods such as quantile mapping is only applicable if one can match the position of a data point in one univariate distribution (say for x) to its corresponding position in another univariate distribution (say for y). This is possible using the preceding definition of position for univariate data since the values of p_Z are always fixed in the range between $[-1, +1]$ irrespective of the values in Z. Unfortunately, when extended to multivariate positions, the range of values for p_Z may vary depending on the values in Z. To overcome this problem, He et al. [11] introduce the notion of a stationary position by iteratively applying the following position transformation function until convergence:

$$\mathbf{p}_Y^k(\mathbf{z}) = \frac{1}{\kappa n} \sum_{i=1}^n \frac{\mathbf{p}_Y^{k-1}(\mathbf{z}) - \mathbf{p}_Y^{k-1}(\mathbf{y}_i)}{\| \mathbf{p}_Y^{k-1}(\mathbf{z}) - \mathbf{p}_Y^{k-1}(\mathbf{y}_i) \|}, \qquad \mathbf{p}_Y^1(\mathbf{z}) = \frac{1}{\kappa n} \sum_{i=1}^n \frac{\mathbf{z} - \mathbf{y}_i}{\| \mathbf{z} - \mathbf{y}_i \|} \quad (1)$$

Here each component in $\mathbf{y}_i$ must be converted to its marginal rank first before applying the position transformation function. Marginal rank refers to the rank of the data point divided by the largest rank and then normalized to the range $[-1, 1]$. The normalization is done to negate the effect of variables having values that correspond to different ranges. Data points with normalized marginal rank close to ± 1 correspond to extreme values for the particular variable, while those close to $\mathbf{0}$ are located near the median of the distribution. In practice, the number of iterations needed to reach a stationary distribution is quite small, typically $K > 5$ [11]. For univariate data, it can be shown that $\mathbf{P}^k$ reaches a stationary distribution at $k = 1$.

The term κ in Equation (1) is a normalization factor to ensure the distribution of the geometric positions is supported in a q-dimensional unit hypersphere. In the case of bivariate response variable $\mathbf{Y}$, the stationary geometric quantile distribution is circularly symmetric around the origin, with the radial density of $r/\sqrt{1 - r^2}$ for $r \in (0, 1)$ [11]. Therefore,

$$\kappa = \int_0^1 \frac{r}{\sqrt{1 - r^2}} dr \Rightarrow \kappa = \frac{\pi}{4}$$

In this paper, we denote the position of the multivariate data points in $\mathbf{Y}$ as $\mathbf{P}_Y = [\mathbf{p}_Y(\mathbf{y}_1), .., \mathbf{p}_Y(\mathbf{y}_n)]^T$, where $\mathbf{p}_Y(\mathbf{y}_i) \in [-1, 1]^q$. We also use the notation $\mathbf{z}_{XY} = \mathbf{p}_X^{-1}(\mathbf{p}_Y(\mathbf{y}))$ to represent a point in the domain of $\mathbf{X}$ that has the same geometric quantile position as the data point $\mathbf{y}$ in $\mathbf{Y}$, i.e., $\mathbf{p}_X(\mathbf{z}_{XY}) = \mathbf{p}_Y(\mathbf{y})$. Consequently, $\mathbf{z}_{YY}(\mathbf{y}_i) = \mathbf{y}_i$. Finally, let $\mathbf{Z}_{XY} = [\mathbf{z}_{XY}(\mathbf{y}_1)^T, .., \mathbf{z}_{XY}(\mathbf{y}_n)^T]^T$ be the geometric quantiles in X that correspond to the data points in Y.

3.1   Quantile Mapping-Based Approaches

Quantile mapping transforms a univariate predictor variable X to its corresponding response variable Y by adjusting the cumulative distribution function F_X to match that of F_Y:

$$QM : \hat{y} = F_Y^{-1}(F_X(x)) \tag{2}$$

It can be shown that QM preserves the rank correlation[2] between the variables. For instance, consider the example in Table 1 where $\mathbf{y}$ is the response variable and $\mathbf{x}_1$, $\mathbf{x}_2$ are two independent predictor variables. Let $QM(\mathbf{x}_1)$ and $QM(\mathbf{x}_2)$ be the corresponding QM outputs for $\mathbf{x}_1$ and $\mathbf{x}_2$, respectively. If we sort the vectors in ascending order, it is easy to see that the resulting rank vectors are invariant under QM transformation. As a result, the rank correlation between $\mathbf{x}_1$ (or $\mathbf{x}_2$) and $\mathbf{y}$ is identical to the rank correlation between $QM(\mathbf{x}_1)$ (or $QM(\mathbf{x}_2)$) and $\mathbf{y}$. Furthermore, the empirical CDF for $QM(\mathbf{x}_1)$ as well as $QM(\mathbf{x}_2)$ are identical to that for $\mathbf{y}$, i.e., $F_Y = F_{QM(\mathbf{x}_1)} = F_{QM(\mathbf{x}_2)}$.

---

[2] Examples of rank correlation measures include Kendall τ and Spearman's ρ coefficients.

Even though quantile mapping was able to replicate the empirical distribution of y perfectly, $QM(x_1)$ has a higher residual error than $QM(x_2)$. This can be explained by the lower rank correlation between x_1 and y compared to the rank correlation between x_2 and y. Note that the inverse relationship between rank correlation and residual error holds only if the values of the response variable are uniformly spaced. For example, if the response value y for the fourth data point changes from 0.4 to 0.7, the residual error for $QM(x_2)$ increases from 0.02 to 0.32, and is larger than the residual error for $QM(x_1)$, which remains at 0.06. In this case, a high rank correlation for x_2 does not translate to lower residual error when applying quantile mapping. A formal proof showing the relationship between rank correlation and residual error for uniformly spaced data is given in the next section.

Table 1. Quantile Mapping

x_1 x_2	y	$QM(x_1)$	$QM(x_2)$
0.6 0.7	0.2	0.1	0.2
0.8 0.6	0.1	0.3	0.1
0.7 0.9	0.3	0.2	0.4
0.9 0.8	0.4	0.4	0.3
	SSR=	0.06	0.02

Table 2. Quantile Mapping

x_3 x_4	y	$QM(x_3)$	$QM(x_4)$
0.7 0.6	0.2	0.2	0.1
0.6 0.7	0.1	0.1	0.2
0.9 0.8	0.3	0.7	0.3
0.8 0.9	0.7	0.3	0.7
	SSR=	0.32	0.02

Since most data sets are non-uniform, maximizing rank correlation is not a sufficient condition to ensure a low residual error. Nevertheless, we observe that data points associated with quantiles that are located in sparse regions (i.e., far from their next closest quantiles) will contribute to higher residual error when incorrectly ranked compared to data points associated with quantiles located in dense regions. This is demonstrated by the example shown in Table 2, where both x_3 and x_4 have the same rank correlation with respect to the response variable y, yet have different SSR. The response values for the first three data points (0.2, 0.1, and 0.3) are closer to each other than the last data point (0.7). An incorrect ranking of the fourth data point will lead to much higher residual error compared to the first three data points. Since x_3 ranked the fourth data point incorrectly, its residual error is larger than x_4 even though they both have the same rank correlation. This suggests a possible heuristic for improving both rank correlation and residual error by emphasizing on data points that contribute to high residual errors in prediction if ranked incorrectly.

3.2   Rank Correlation and Residual Errors of Quantile Mapping

This section presents several properties of the QM approach with respect to the rank correlation and residual error of its output. First, we show that quantile mapping preserves the rank correlation between the predictor and response variables.

Proposition 1. *Rank correlation is invariant under QM transformation if the values of the predictor and response variables in a data set are unique.*

Proof. Consider a data set $\mathcal{D} = \{(x_i, y_i)\}_{i=1}^n$ that contains n points. Let $\hat{y}_i$ be the quantile mapped value for the data point with predictor variable x_i. To prove that rank correlation is invariant under QM transformation, it is sufficient to show that the rank for x_i is identical to the rank of $\hat{y}_i$ after quantile mapping. Without loss of generality, assume the data points in $\mathcal{D}$ are sorted in increasing order of their x values. Thus, the rank for data point x_i is i (since the x values are unique). Equation (2) can be rewritten as follows

$$F_Y(\hat{y}_i) = F_X(x_i)$$

Since $F_X(x_i) = i/n$, therefore $F_Y(\hat{y}_i) = F_X(x_i) = i/n$. Given that the response values y_i are distinct, the rank for $\hat{y}_i$ is also i. $\diamond$

Next, we illustrate the relationship between rank correlation and residual error of QM output for data sets with uniformly spaced response values.

Proposition 2. *The SSR of QM output is negatively proportional to the rank correlation of the input and a uniformly spaced response data.*

Proof. Given n data points, let r_i and s_i be the respective ranks of the ith input data point x_i and the corresponding response output y_i. Without loss of generality, we assume that each data point has a unique rank. Since $\mathbf{y}$ is uniformly spaced, $y_i = s_i c_1 + c_0$, where c_0 and c_1 are constants. Similarly, the QM output $\hat{y}_i = r_i c_1 + c_0$. The Spearman rank correlation can be written

$$\rho = \frac{\sum_i (r_i - \bar{r})(s_i - \bar{s})}{\sqrt{\sum_i (r_i - \bar{r})^2 \sum_i (s_i - \bar{s})^2}}$$

We have $\rho = (1/c_2)(\sum_i r_i s_i + c_3)$, where, c_2 and c_3 are constant for a fixed n. Given, $SSR = \sum_i (y_i - \hat{y}_i)^2$ and the QM output $\hat{\mathbf{y}}$ is a reordered instance of $\mathbf{y}$, we have $SSR = 2(\sum_i y_i^2 - \sum_i y_i \hat{y}_i)$. $\sum_i y_i \hat{y}_i = (c_1^2) \sum_i r_i s_i + c4$. where, c_4 is a constants for a fixed n. Therefore, $SSR = 2(\sum_i y_i^2 - (c_1^2 c_2)\rho - c_3 - c_4)$. Since, c_2 is a positive constant, $(c_1^2 c_2)$ will always be positive. Hence, SSR is negatively proportional to ρ when $\mathbf{y}$ is uniformly spaced. $\diamond$

We next show that the output of QM that perfectly replicated the response variable can be improved to have lower residual errors by correcting the ranks of the predictor variable to better match the response variable.

Proposition 3. *Correcting the ranks of data points in $\mathbf{x}$ that do not match the rank of the corresponding data point in $\mathbf{y}$, maintains, if not, improves the SSR of QM output.*

Proof. Let elements of $R_{\mathbf{x1}}$, $R_{\mathbf{x2}}$ and O be the quantile positions of data points in $\mathbf{x1}$, $\mathbf{x2}$ and $\mathbf{y}$, respectively. Given, the QM output of $\mathbf{x1}$ can be improved to have lower residuals, $SSR_{\mathbf{x1}} = \sum_i \varepsilon_{\mathbf{x1}}^2(i)/n > 0$. Consequently, $\exists j$, such that

$R_{\mathbf{x1}}(j) \neq O(j)$. Let $\mathbf{x2}(i) = \mathbf{x1}(i)\ \forall i$, where $R_{\mathbf{x1}}(i) = O(i)$ and $\exists k$, such that $R_{\mathbf{x1}}(k) \neq O(k)$ and $R_{\mathbf{x2}}(k) = O(k)$ or $R_{\mathbf{x2}}(k) = R_{\mathbf{x1}}(k)$. Therefore, $\forall i, \varepsilon_{\mathbf{x1}}^2(i) \geq \varepsilon_{\mathbf{x2}}^2(i)$. And since $SSR = \sum \varepsilon_i^2$, we have $SSR_{\mathbf{x2}} \leq SSR_{\mathbf{x1}}$. Thus proving that it SSR of QM output can be improved by correcting the ranks of those data points that do not have the same rank as its corresponding response data point. $\diamond$

Improving the rank correlation of predictor variable to perfectly match the response variable would result in QM output having zeros SSR.

Proposition 4. *The residual error obtained from QM is zero when there is perfect rank correlation between predictor and response variable.*

Proof. Let the elements of R and O be the quantile positions of the data points in $\mathbf{x}$ and $\mathbf{y}$ respectively. Let $\varepsilon_i = |F_{\mathbf{y}}^{-1}(O_i) - F_{\mathbf{y}}^{-1}(R_i)|$ be the residual error of ith data point. Therefore, $SSR = \varepsilon_i^2/n$. Given a perfect rank correlation ($\Gamma = 1$) between predictor and response variable, we have $\forall i, (R_i = O_i)$. Consequently, $\varepsilon_i = |F_{\mathbf{y}}^{-1}(O_i) - F_{\mathbf{y}}^{-1}(O_i)| = 0$. Therefore, $SSR = \sum_i \varepsilon_i^2/n = 0$. $\diamond$

Hence, we propose a framework that improves on the ordering of the predictor variables to better match the response variable in order to minimize the SSR of a QM output.

4   Multi-Output Contour Regression Framework(MCR)

Since QM and regression-based approaches have their own distinct advantages which have been successfully exploited in a hybrid manner by approaches such as CR, we propose a framework that extends the intuition behind hybrid approaches that exploits the unique advantages of both QM and regression, to work in a multi-output setting. The approach uses a position regularized regression function $h(\mathbf{x}, \hat{\Omega})$ that prioritizes matching the positions of output to best match the positions of the observed response data. This step is followed by correcting the geometric quantiles of the output from the previous step to match the observed response data using the intuition of QM. This hybrid approach addresses the limitation of QM regarding the number of predictor variables that may be used as well as requirement of the predictor variables being highly correlated to the response variable. We further enhanced the hybrid approach to be flexible enough to work in a multi-output setting so as to be able to capture the multi-output associations that are often ignored.

To prioritize improving the positions of the output, the proposed multi-output contour regression (MCR) framework learns the regression function $h(\mathbf{x}, \hat{\Omega})$. The regression function $h(\mathbf{x}, \hat{\Omega})$ consists of two components. The first component is similar to conventional regression loss function where the data matrix is made to regress with respect to the observed response variable. This component emphasizes minimizing residual error of the regression function.

The second component of $h(\mathbf{x}, \hat{\Omega})$ is the position regularizer that helps improve rank correlation of $h(\mathbf{x}, \Omega)$ and $\mathbf{y}$. At a first glance, one would expect the second term to be regressing on the position of the data points. Instead of regressing on the position of the data points, we regress on the geometric quantiles of the data points obtained by inverse mapping their positions to the output response space. This is done so that the position regularizer assigns a larger penalty to those data points whose position when incorrectly estimated, results in a larger minimum residual errors. To accomplish this, the data matrix is made to regress on $\mathbf{z}_{\hat{Y}Y}$, where,

$$\hat{\mathbf{z}}_{\hat{Y}Y}(y) = \mathbf{p}_{\hat{Y}}^{-k}(\mathbf{p}_Y^k(\mathbf{y})) \tag{3}$$

is the geometric quantile value in the $h(\mathbf{x}, \hat{\Omega})$ regression output space that corresponds to the position of the observed response variable y.

The regression function of MCR is shown in Equation (4),

$$\min_{\Omega} \sum_{i=1}^{n} (\gamma \mathcal{L}(h(\mathbf{x}_i, \Omega), \mathbf{y}_i) + (1 - \gamma)\mathcal{L}(h(\mathbf{x}_i, \Omega), \mathbf{z}_{\hat{Y}Y})) \tag{4}$$

where $0 \leq \gamma \leq 1$ is a user defined parameter that may be used for either prioritizing fidelity of regression accuracy or its position correlation.

$\mathcal{L}$ can be any generic loss function such as ordinary least square (that multiple linear regression adopts), or quantile mapping (if certain quantiles are to be prioritized overs others, such as in the case of a heavy tail distribution).For instance, when the loss function $\mathcal{L}$ is ordinary least square, Equation 4 takes the form

$$\min_{\Omega} \sum_{j=1}^{q} \sum_{i=1}^{n} (\gamma(\mathbf{x}_i^T \Omega_j - \mathbf{y}_i)^2 + (1 - \gamma)(\mathbf{x}_i^T \Omega_j - \mathbf{z}_{\hat{Y}Y})^2)$$

which corresponds to the following matrix form

$$\hat{\Omega} = \arg\min_{\Omega} \ tr(\gamma(\mathbf{X}\Omega - \mathbf{Y})^T(\mathbf{X}\Omega - \mathbf{Y}) + (\mathbf{X}\Omega - \mathbf{Z}_{\hat{Y}Y})^T(\mathbf{X}\Omega - \mathbf{Z}_{\hat{Y}Y}))$$

The regression parameters $\hat{\Omega}$ is learnt in an iterative manner. At each iteration, the regression output space from the previous iteration is used to compute $\mathbf{z}_{\hat{Y}Y}$ in the second component of the regression function $h(\mathbf{x}, \hat{\Omega})$. For the very first iteration, the regression output space is that of regular multiple linear regression.

Once $h(\mathbf{x}, \hat{\Omega})$ is learnt, the MCR prediction for a given data point $\mathbf{x}$ having corresponding observed multi-output response $\mathbf{y}$ and a regression estimation of $\hat{\mathbf{y}} = h(\mathbf{x}, \hat{\Omega})$ is obtained by inverse geometrically quantile mapping $\mathbf{p}_{\hat{Y}}^k(\hat{\mathbf{y}})$ to its corresponding value in the observed response variable space, to give the MCR prediction $\hat{\mathbf{z}}_{Y\hat{Y}}$,

$$MCR : \hat{\mathbf{z}}_{Y\hat{Y}} = \mathbf{p}_Y^{-k}(\mathbf{p}_{\hat{Y}}^k(h(\mathbf{x}, \hat{\Omega}))) \tag{5}$$

where, $\mathbf{p}_Y^{-k}(\mathbf{p}_{\hat{Y}}^k(\hat{y}))$ maps the stationary geometric quantile position of $h(\mathbf{x}, \hat{\Omega})$ to its corresponding data point in $\mathbf{Y}$.

To summarize, multi-output contour regression (MCR) performs multi-output regression of the predictor variables such that the position of its output is highly correlated with respect to position of the observed response variable, thereby reducing position errors of the multi-output regression results. This multivariate regression output is then mapped to its corresponding geometric quantile counterpart in the observed multi-output response space using geometric quantiles. The rationale behind using the regularized regression results, prior to performing multi-output geometric quantile mapping in MCR, is to improve on SSR by increasing the correlation among the multivariate ranks of the predictors and response variable.

4.1   Estimating Inverse Geometric Quantile Position

The value $\hat{z}(\mathbf{p})$ that corresponds to a given geometric quantile position $\mathbf{p}$, in a multivariate distribution F_Y i.e., $\mathbf{p}_Y(\mathbf{p})$, is empirically computed by minimizing the generalized multivariate quantile loss function [9]

$$\hat{z}(\mathbf{p}) = \arg\min_{\mathbf{z} \in \Re^q} \sum_{i=1}^{n} (\|\mathbf{y}_i - \mathbf{z}\| + <\mathbf{p}, \mathbf{y}_i - \mathbf{z}>) \tag{6}$$

where, $\mathbf{p} \in \Re^q$ and $< ., . >$ denotes the Euclidean inner product. So long all the values of y_i does not fall on the same line, $\hat{z}(\mathbf{p})$ will be unique for a given $\mathbf{p}$ for $q \geq 2$ [9]. Algorithms such as Newton-Raphson's method can be used to solve the above loss function geometric quantile $\hat{z}(\mathbf{p})$ using the following update $\hat{z} \leftarrow \hat{z} - \frac{\delta}{\delta'}$ where, $\delta = \sum_{i=1}^{n}((n\kappa)\mathbf{p} - \|\mathbf{z} - \mathbf{y}_i\|^{-1}(\mathbf{z} - \mathbf{y}_i))$

$$\delta' = \sum_{i=1}^{n} \|\mathbf{z} - \mathbf{y}_i\|^{-1}(I_q - \|\mathbf{z} - \mathbf{y}_i\|^{-2} \times (\mathbf{z} - \mathbf{y}_i)(\mathbf{z} - \mathbf{y}_i)^T)$$

For a univariate distribution, F_Y, it can be easily shown that equation (6) boils down to the same loss function used to identify the αth regression quantile in a linear regression setup for quantile regression [15], where $0 < \alpha < 1$ and $p = 2\alpha - 1$. i.e, $\sum_{1}^{n}(|y_i - z| + p(y_i - z))$ is minimized for z that corresponds to the αth quantile of Y.

4.2   Alternate Approximation-Based Approach for MCR

If one can make the assumption that given the position $(\mathbf{p})$ of a test data point $(\mathbf{y}^{test})$ that belongs to the distribution F_Y, and $\exists \mathbf{y}_i \in \mathbf{Y}$ such that $\mathbf{y}^{test} \simeq \mathbf{y}_i$, then the search space for $\hat{z} = \mathbf{y}^{test}$ can be limited to data points in $\mathbf{Y}$.

Given that the search space for $\hat{z}$ is finite it will not always possible to find the exact same point in F_Y using the loss function δ, as it returns a vector. Alternatively, the following range bound approximation that is equivalent to Equation 6, can be used to find the best solution [11,9].

$$\arg\min_{\mathbf{z}} \sum_{i=1}^{n} \{\|\mathbf{y}_i - \mathbf{z}\| + \frac{1}{\kappa}(\mathbf{y}_i - \mathbf{z})^T \mathbf{p}\} \tag{7}$$

where κ in the scaling factor chosen in Equation (4).

As shown in the experiment section, there was only a marginal performance deterioration in the solution obtained from the above approximation, due to sufficient amount of training data points. Another approximation approach with even less tighter bounds than Equation 7, having $O(n)$ time complexity is to use the following Euclidean approximation.

$$\hat{\mathbf{z}} = \arg\min_{\mathbf{y}_i}((\mathbf{p} - \mathbf{p}_Y(\mathbf{y}_i))(\mathbf{p} - \mathbf{p}_Y(\mathbf{y}_i))^T) \qquad (8)$$

The R-limited approximation approach (Equations 7) as well as the Euclidean approximation approach (8) show considerable improvement in the computation time across varying training size (Figure 2.a) and test size (Figure 2.b), with minimum deterioration in terms of accuracy of the inverse geometric quantile positions estimated.

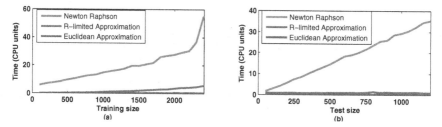

Fig. 2. Relative computation time of the various approximation-based approaches for estimating inverse geometric quantile positions

5   Experimental Results

The objective of the experiments was to evaluate the ability of MCR in replicating the associations among multiple climate response variables while minimizing sum square residuals.

All the algorithms were run using climate data obtained at fourteen weather stations in Michigan, USA. The response variables used were maximum temperature, minimum temperature, and the total precipitation for each day spanning twenty years. The predictor variables used in this study are simulated climate data obtained from Regional Climate models (RCM) that best correspond to the observed response variables at each of the fourteen weather stations. Three different RCM data sets for each of the climate stations were obtained from North American Regional Climate Change Assessment Program (NARCCAP) [1]. The three RCMs used are the Canadian Regional Climate Model (CRCM), the Weather Research and Forecasting Model (WRFG) and the Regional Climate Model Version-3 (RCM3). For the purpose of the experiments, there were a total of 126 data sets with univariate response variables, 126 data sets with bivariate responses and 42 data sets with trivariate responses.

5.1   Experimental Setup

Twenty year of predictor and response data, spanning the years 1980-1999 was split into two parts for training and testing. For the purpose of the evaluation of the relative skill in preserving associations among the multi-output responses, popular regression and quantile mapping approaches such as MLR, Ridge regression (Ridge), QM, EDCDFm, MOR, CR, BQM as well as ad-hoc approaches that sequently combine regression and quantile mapping approaches were used as baseline. An example of the ad-hoc baseline approach used is MOR in combination with BQM (RBQM) and MLR and QM (RQM). γ was set to 0.5 for all experiments. For CR and MCR based experiments, the maximum number of iterations was set to ten.

After discarding the missing values, each experiments run for each of the stations, across all the data sets, had a minimum of one thousand training and test data points. All the results provided in the following section are on test data (out-of-sample results). Kendall τ rank correlation and Spearman ρ rank correlation were the two rank correlation metrics used for evaluation univariate rank correlation. In the following experiment section, we included results of only one of the two rank correlation metrics, when their results were very similar. Root mean square error (RMSE), was used as a metric to compare the performance of the various approaches evaluated in terms of its output residual errors. Two dimensional and three dimensional scatter plots were used to visualize the relative skill of the various approaches in preserving the associations among the multi-output responses.

5.2   Results

Univariate MCR. For academic reasons, the rank correlation of the various response variables were computed in a single output MCR setting using Kendall τ rank correlation and Spearman ρ rank correlation. It was found that across all the different data sets and stations and response variables (i.e, 126 datasets), MCR consistently improved the rank correlation across both rank correlation metrics. For the purpose of comparing the intra-performance of datasets that shared similar response variables, the 126 individual data sets that corresponded to univariate response data were grouped into nine larger data sets.

Figure 3 is a box plot representing the percentage of stations in each of the nine data sets where the rank correlation regularizer used in Equation 4, improved rank correlation and reduced residuals when compared to baselines approaches.

The box plot in Figure 4 shows that in spite of MCR's reported improvement across majority of stations in terms of τ and RMSE, for both regression and quantile mapping based approaches, the improvement was not significant when compared to the regression based approaches. However, the rank correlation regularizer showed a significant improvement in terms of RMSE at each station when compared to the corresponding quantile mapping based approaches.

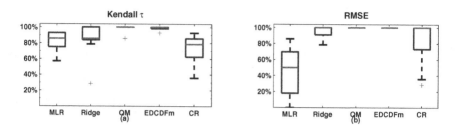

Fig. 3. Box plot of the percentage stations where MCR showed improvement over single output baselines, in terms of Kendall τ rank correlation and RMSE, across all RCM's and variables

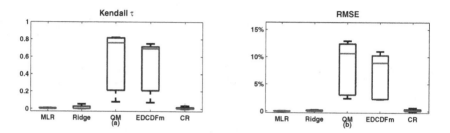

Fig. 4. Box plot of MCR's improvement over baseline approaches in terms of Kendall τ rank correlation and RMSE, across all RCM's and variables

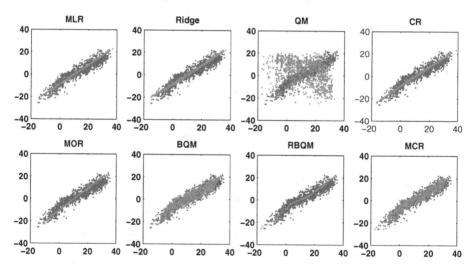

Fig. 5. Scatter plot portraying the fidelity of forecast values of various approaches replicating the observed associations among the bivariate temperature response variables

Bivariate MCR. Bivariate modeling for all the combinations of bivariate response variables were evaluated. As shown in Figure 5, MCR performed best in replicating both the bivariate associations and minimizing SSR, although BQM performed as well in terms of replicating the bivariate associations. Regression based approaches (both SOR and OMR) fared poorly in preserving associations in the 2D space, while single output quantile mapping based approaches, were very sensitive to correlation of the predictor variables with response resulting in poor bivariate associations in spite of replicating the marginal distributions of the individual responses very well.

Table 3. Performance of bivariate MCR over baseline approaches

| Data set | RMSE | | | | | | Kendall τ | | | | | |
| | % of stations outperformed baseline | | | Avg.improvement across stations over baseline | | | % of stations outperformed baseline | | | Avg.improvement across stations over baseline | | |
	MOR	QM	BQM	MOR	QM	BQM	MOR	QM	BQM	MOR	QM	BQM
$WRFG_1$	29	100	100	-0.06	0.18	0.17	64	100	100	0.03	0.40	0.41
$WRFG_2$	07	100	100	-0.08	0.16	0.16	79	100	100	0.04	0.38	0.39
$WRFG_3$	00	100	100	-0.07	0.31	0.30	0	100	100	-0.01	0.75	0.67
$CRCM_1$	93	100	100	0.06	0.25	0.25	100	100	100	0.13	0.52	0.53
$CRCM_2$	71	100	100	0.03	0.23	0.23	100	100	100	0.12	0.49	0.52
$CRCM_3$	07	100	100	-0.02	0.35	0.34	14	100	100	-0.01	0.78	0.73
$RCM3_1$	43	100	100	-0.02	0.20	0.20	79	100	100	0.06	0.46	0.46
$RCM3_2$	36	100	100	-0.03	0.19	0.18	79	100	100	0.06	0.47	0.45
$RCM3_3$	00	100	100	-0.07	0.31	0.30	0	100	100	-0.01	0.81	0.78

In terms of residuals, MCR had considerably lower residuals when compared of the various quantile mapping baseline approaches as shown in Table 3. But as expected, MCR showed marginal increase in residuals when compared to the respective SOR and MOR based approaches.

Trivariate MCR. The performance of modeling the association among three response variables was also evaluated and is shown in Figure 6. The performance is compared against single output, and multiple output models. We also use as a baseline, an trivariate extension of the bivariate BQM approach, as an additional baseline. Along with MCR, the trivariate extension of BQR fared best in replicating the observed associations among three variables when compared to the baseline approaches.

Additionally, MCR was also able to improve upon its BQM counterpart in terms of reduction of residuals. MCR produced lower RMSE for all the station across all the tri-variate datasets with an average reduction of RMSE in excess of 10%. The average improvement of the three variables in terms of rank correlation τ was found to be 0.41.

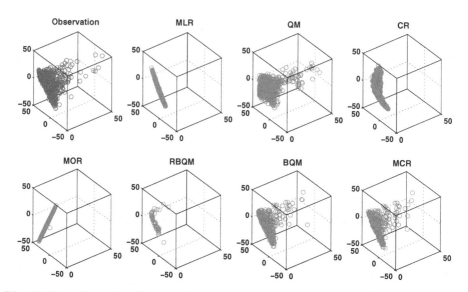

Fig. 6. Three dimensional scatter plot of the observed associations among maximum temperature, minimum temperature and precipitation as well as the respective forecasts made by the various single output and multiple output approaches

6 Conclusions

We present a multi-output regression framework that preserves the general association patterns among multiple response variables while minimizing the overall residual errors by coupling regression and geometric quantile mapping. The paper demonstrates the effectiveness of the framework in significantly reducing residuals while preserving the joint distribution of the multi-output variables, over the baseline approaches in all the climate stations evaluated.

Acknowledgments. This work is supported by NSF Award CNH 0909378. Any opinions, findings, and conclusions or recommendations expressed in this material are those of the authors and do not necessarily reflect the views of the National Science Foundation.

References

1. North American Regional Climate Change Assessment Program, http://www.narccap.ucar.edu/
2. Abraham, Z., Tan, P.-N.: Distribution regularized regression framework for climate modeling. In: SIAM International Conference on Data Mining, SDM 2013 (2013)
3. Alvarez, M., Lawrence, N.D.: Sparse convolved gaussian processes for multi-output regression. In: Advances in Neural Information Processing Systems 21, pp. 57–64 (2009)

4. Balasubramanian, K., Lebanon, G.: The landmark selection method for multiple output prediction. In: International Conference on Machine Learning, ICML (2012)
5. Blaschko, M.B., Lampert, C.H.: Learning to localize objects with structured output regression. In: Forsyth, D., Torr, P., Zisserman, A. (eds.) ECCV 2008, Part I. LNCS, vol. 5302, pp. 2–15. Springer, Heidelberg (2008)
6. Bontempi, G.: Long term time series prediction with multi-input multi-output local learning. In: Proceedings of the 2nd European Symposium on Time Series Prediction (TSP), ESTSP 2008, pp. 145–154 (2008)
7. Breiman, L., Friedman, J.H.: Predicting multivariate responses in multiple linear regression. Journal of the Royal Statistical Society: Series B (Statistical Methodology) 59(1), 3–54 (1997)
8. Buja, A., Logan, B., Reeds, J., Shepp, L.: Inequalities and positive-definite functions arising from a problem in multidimensional scaling. The Annals of Statistics, 406–438 (1994)
9. Chaudhuri, P.: On a geometric notion of quantiles for multivariate data. Journal of the American Statistical Association 91(434), 862–872 (1996)
10. Hastie, T., Tibshirani, R., Friedman, J.J.H.: The elements of statistical learning, vol. 1. Springer, New York (2001)
11. He, X., Yang, Y., Zhang, J.: Bivariate downscaling with asynchronous measurements. Journal of Agricultural, Biological, and Environmental Statistics 17(3), 476–489 (2012)
12. Izenman, A.J.: Reduced-rank regression for the multivariate linear model. Journal of Multivariate Analysis 5(2), 248–264 (1975)
13. Jakob Themeßl, M., Gobiet, A., Leuprecht, A.: Empirical-statistical downscaling and error correction of daily precipitation from regional climate models. International Journal of Climatology 31(10), 1530–1544 (2011)
14. Kim, S., Xing, E.P.: Tree-guided group lasso for multi-task regression with structured sparsity (2010)
15. Koenker, R.: Quantile Regresssion. Wiley Online Library (2005)
16. Marden, J.I.: Positions and qq plots. Statistical Science 19(4), 606–614 (2004)
17. O'Brien, T., Sornette, D., McPherron, R.: Statistical asynchronous regression: Determining the relationship between two quantities that are not measured simultaneously. Journal of Geophysical Research 106(A7), 13247–13259 (2001)
18. Pérez-Enciso, M., Tenenhaus, M.: Prediction of clinical outcome with microarray data: a partial least squares discriminant analysis (pls-da) approach. Human Genetics 112(5-6), 581–592 (2003)
19. Piani, C., Weedon, G., Best, M., Gomes, S., Viterbo, P., Hagemann, S., Haerter, J.: Statistical bias correction of global simulated daily precipitation and temperature for the application of hydrological models. Journal of Hydrology 395(3), 199–215 (2010)
20. Vazquez, E., Walter, E.: Multi-output support vector regression. In: 13th IFAC Symposium on System Identification, pp. 1820–1825 (2003)

Structured Output Learning with Candidate Labels for Local Parts*

Chengtao Li[1], Jianwen Zhang[2,**], and Zheng Chen[2]

[1] Institute for Interdisciplinary Information Sciences,
Tsinghua University, Beijing, China, 100080
lichengtao2010@gmail.com
[2] Microsoft Research Asia, Beijing, China, 100080
{jiazhan,zhengc}@microsoft.com

Abstract. This paper introduces a special setting of weakly supervised structured output learning, where the training data is a set of structured instances and supervision involves candidate labels for some local parts of the structure. We show that the learning problem with this weak supervision setting can be efficiently handled and then propose a large margin formulation. To solve the non-convex optimization problem, we propose a proper approximation of the objective to utilize the Constraint Concave Convex Procedure (CCCP). To accelerate each iteration of CCCP, a 2-slack cutting plane algorithm is proposed. Experiments on some sequence labeling tasks show the effectiveness of the proposed method.

Keywords: Structured Output Learning, Weak Supervision, Candidate Labels, Local Parts.

1 Introduction

Many applications involve predicting structured labels for a set of interdependent variables. For example, a part-of-speech tagging (POS) model needs to predict a sequence of POS tags for a sentence, one for each token. This type of problem is known as *structured output learning*. In the past decade, some effective methods have been proposed and widely used, such as Conditional Random Field (CRF) [21] and Structured SVM (SVM^{struct}) [33]. However, they are supervised methods requiring a large amount of labeled instances for training, which are expensive due to the natural complexity of the output, e.g., each token of a sentence needs labeling. Although some semi-supervised [44] and active learning methods [26,27] are proposed to reduce the number of required labels, they still require *exact* labels for the output variables. In reality, while getting exact labels as supervision is expensive, it is often cheap to get much weak/indirect supervision, e.g., some candidate labels for an instance. Thus utilizing weak/indirect

---

* Please see http://research.microsoft.com/pubs/194791/SupplementaryMaterial.pdf for all the proofs and more details of the algorithms.
** Corresponding author.

H. Blockeel et al. (Eds.): ECML PKDD 2013, Part II, LNAI 8189, pp. 336–352, 2013.
© Springer-Verlag Berlin Heidelberg 2013

supervision to train a high quality predictor is very meaningful [3,14,18,35]. In this paper, we introduce a special setting, called *Candidate Labels for Local Parts* (CLLP). In CLLP, for each instance, we only need to provide a set of candidate labels for each local part of the output variables (e.g., a chunk in a sequence), among which only one is correct.

The CLLP setting takes root in many real-world scenarios, which roughly falls into two cases: (1) There is prior knowledge from which we can provide a candidate labeling set for a local part of output variables. For example, for POS tagging, by looking up some linguistic dictionaries, we can get the candidate POS tags for a word in a sentence [23,28,25,9]. Similar scenarios exist for other sequence labeling tasks like word sense disambiguation [24], entity recognition [31], etc.. Another example is caption based image auto tagging. An image on the web is usually surrounded with tags that provide candidate labels for objects in the image [1,4,13]. (2) Noisy labels from multiple annotators. When we collect manual labels for a learning task, a labeling task is often assigned to multiple annotators, e.g., via a Crowdsourcing system. Due to labeling bias or irresponsible annotators, different annotators may give different labels for the same output variable. Thus the annotators collectively provide candidate labels for an output variable [7].

CLLP also provides a uniform viewpoint for different labeling settings of structured output learning: (1) If the candidate labeling set for each output variable is the full label space, i.e., all the possible labels, there is no useful information provided and hence it degenerates to unsupervised learning. (2) If the candidate labeling set for each output variable contains only the ground truth label, it degenerates to fully supervised learning. (3) If for some instances, we provide the candidate labels as (1) and for other instances we provide the candidate labels as (2), then it becomes semi-supervised/transductive learning [44]. (4) The general case is that each local part of the output variables is assigned with a non-trivial set of candidate labels.

In this paper, we propose a large margin approach to the CLLP setting. We maximize the margins between candidate labels and non-candidate labels, and also the margins between the predicted label and other candidate labels. Since the obtained optimization problem is non-convex, the proper approximations and Constraint Concave Convex Procedure (CCCP) are used to solve it.

The major contributions of this paper are as follows:

1. We introduce and formalize CLLP, a general type of weakly supervised setting for structured output learning and propose a large-margin approach. We find that the CLLP setting can be handled by an efficient algorithm, while some other forms of weak supervision may cause some parts of the problem to be *NP-hard*. We also show that the proposed new objective is closer to the true objective than a previous state-of-the-art method.

2. We propose a new proper approximation for the objective and propose an algorithm based on CCCP to solve the approximated problem.

3. We propose a 2-slack cutting plane algorithm to accelerate each iteration of CCCP, and give an upper bound on the number of iterations before termination.

2   Related Work

There are several related terminologies on different labeling settings of a learning task, including semi-supervised learning, multiple instance learning, and candidate label learning. Sometimes they are all generally called weakly supervised learning [18], as distinguished from traditional supervised learning requiring full and exact labels.

In semi-supervised learning (SSL) [43], a training set contains both labeled and unlabeled instances. Refs. [12] and [44] propose semi-supervised solutions for structured output learning, where some instances have exact and full labels while the remaining instances are unlabeled. Ref. [36] extends the method in [44] to incorporate domain knowledge as constraints, e.g., in POS tagging, a sentence should have at least one verb. Ref. [34] even allows a training instance itself to be partially labeled, e.g., some tokens in a sentence are labeled while the rest are unlabeled. The major difference between SSL and CLLP is: in SSL an output variable of an instance is either *exactly* labeled or unlabeled, while in CLLP the supervision is a set of candidate labels for each local part of the output variables, which dose not indicate the exact label but contains more information than when unlabeled.

Multiple instance learning (MIL) [6,40] is a classical learning problem with weak supervision. In MIL, instances are grouped into bags, and labels are given at the bag level. The original MIL only admits a binary label for a bag, and is extended to admit multiple labels later [42,16]. Some recent MIL methods consider the dependency among instances and bags, bringing the problem closer to structured output learning [41,39,5]. The difference between MIL and CLLP is: in MIL the label itself is accurate, but which instance deserves the label is ambiguous, while in CLLP the label itself is ambiguous (just a set of candidates) but which instance carries the label is clear.

Candidate label learning (CLL) [15,11] assumes a set of candidate labels is given for each instance. It is later extended to the setting of *candidate labeling set* (CLS), where instances are grouped into bags, and for each bag a set of candidate labeling vectors is given [4,14]. Each labeling vector consists of labels of all the instances in the bag. CLS looks similar with CLLP. However, CLS directly give candidate labeling vectors and has no constraints on the form of the candidate labeling set. This label setting may result in inefficiency, as shown in Theorem 1 of Section 3. We will discuss the relation between our approach and a state-of-the-art method of CLS [14] in Section 3.5, and make empirical comparisons under various tasks in Section 5.

We have noticed that in NLP literature, there are some papers on POS tagging with only POS dictionaries rather than concrete token-wise labels in sentences [23,28,8,25,10,9], which is similar to the motivation of CLLP. However, they focus on the specific domain problem and may be difficult to extend to general structured prediction or multiclass classification. In contrast, in this paper we work on providing a general formulation and an efficient algorithm for this type of problems. The proposed approach is able to solve all kinds of structured predictions or multiclass classifications with the CLLP labeling setting.

3 Learning with Candidate Labels for Local Parts

3.1 General Weak Supervision via Candidate Labeling Set

Let $\mathbf{x} \in \mathcal{X}$ denote an instance and $\mathbf{y} \subseteq \mathcal{Y}$ denote the true label that is a structured object such as a sequence, a tree, etc. $\mathcal{Y}$ is the full label space for $\mathbf{x}$ without any constraints. $Y \subseteq \mathcal{Y}$ is weak supervision for $\mathbf{x}$. Generally Y can be represented as a set of all the allowed *full* labels for $\mathbf{x}$, which is named *candidate labeling set* (CLS) [4,14]. We make the *agnostic* assumption that $\mathbf{y} \in Y$, then $\{\mathbf{y}\} \subseteq Y \subseteq \mathcal{Y}$. Given a set of weakly supervised training examples, $\{(\mathbf{x}_i, Y_i)\}_{i=1}^{N}$, the learning task is to learn a function $f : \mathbf{x} \mapsto \mathbf{y}$. Obviously, the task becomes supervised learning if $Y_i = \{\mathbf{y}_i\}, \forall i$, and degenerates to unsupervised learning when $Y_i = \mathcal{Y}_i, \forall i$.

Following the convention of structured output learning, we formulate function f by maximizing a mediate linear function $F(\mathbf{x}, \mathbf{y}; \mathbf{w})$ parameterized by $\mathbf{w}$, namely

$$f(\mathbf{x}; \mathbf{w}) = \arg\max_{\mathbf{y} \in \mathcal{Y}} F(\mathbf{x}, \mathbf{y}; \mathbf{w}) = \arg\max_{\mathbf{y} \in \mathcal{Y}} \langle \mathbf{w}, \Psi(\mathbf{x}, \mathbf{y}) \rangle, \qquad (1)$$

where Ψ is a joint feature representation of inputs and outputs, which is flexibly designed to fit various applications.

For simplicity's sake, we use $\delta\Psi_i(\mathbf{y}, \mathbf{y}')$ to denote $\Psi(\mathbf{x}_i, \mathbf{y}) - \Psi(\mathbf{x}_i, \mathbf{y}')$. The value of $\langle \mathbf{w}, \delta\Psi_i(\mathbf{y}, \mathbf{y}') \rangle$ is the cost of predicting $\mathbf{y}$ instead of $\mathbf{y}'$ given input $\mathbf{x}_i$.

Although there could be various kinds of structures for $\mathbf{y}$ with different forms of $\Psi(\mathbf{x}, \mathbf{y})$, for the simplicity of the presentation, we focus on the special case where $\mathbf{y}$ forms a sequence. It is not hard to generalize this special structured case to other general structured and non-structured cases.

3.2 Candidate Labels for Local Parts (CLLP)

CLS is a general representation for weak supervision that has been used in previous methods [14,4]. When dealing with structured output learning with the maximum margin approach, due to the huge number of constraints, the cutting plane method is usually employed to accelerate training [17]. In the cutting plane method, there should be an algorithm that is able to efficiently find the constraint that is most violated by the current solution. However, the following theorem shows that under the general CLS setting it is not possible to train efficiently:

Theorem 1. *Given a structured instance* $\mathbf{x}$ *and arbitrary candidate labeling set* Y*, there is no algorithm that can always find the most possible labels (in* Y *or not in* Y*) in* $poly(|\mathbf{x}|)$ *time unless* $P = NP$*, where* $|\mathbf{x}|$ *is the length of* $\mathbf{x}$*.*

Proof. Please refer to the supplementary material for the proofs.

But if candidate labels are given only for local parts, there exists efficient algorithms that could find the most possible labels for a sequence among its candidate/non-candidate labeling sets, as stated in the following theorem:

Theorem 2. *If the candidate labels are given marginally by local parts, namely, each Y_i in $\{\mathbf{x}_i, Y_i\}_{i=1}^{N}$ has the form $Y_i = \{Y_{i1} \otimes Y_{i2} \otimes \ldots \otimes Y_{iM_i}\} \subseteq \mathcal{Y}$, where Y_{ij} is the set of candidate labels that $\mathbf{x}_{ij}$ could possibly take, among which only one is fully correct; $\mathbf{x}_{ij}$ is the j-th local part in $\mathbf{x}_i$ whose size is upper bounded by some constant; M_i is the number of local parts in $\mathbf{x}_i$, then in the sequence structured learning there is an efficient algorithm (modified Viterbi algorithm) that can find the most possible labels among candidate and non-candidate labeling sets.*

Please note that although this theorem is for the sequence structure, by extending the Viterbi algorithm to general Belief Propagation, it is straightforward to get the same conclusion for the general graph with a limited tree width.

3.3   Loss Function

We use a loss function $\Delta : \mathcal{Y} \times \mathcal{Y} \rightarrow \mathbb{R}$ to quantify the quality of a predictor, which has the following properties:

$$\Delta(\mathbf{y}, \mathbf{y}) = 0 \tag{2}$$

$$\Delta(\mathbf{y}, \mathbf{y}') > 0, \forall \mathbf{y} \neq \mathbf{y}' \tag{3}$$

$$\Delta(\mathbf{y}_1, \mathbf{y}_2) + \Delta(\mathbf{y}_2, \mathbf{y}_3) \geq \Delta(\mathbf{y}_1, \mathbf{y}_3), \forall \mathbf{y}_1, \mathbf{y}_2, \mathbf{y}_3 \in \mathcal{Y}. \text{ (Triangle inequality)} \tag{4}$$

Among many loss functions $\Delta(\cdot, \cdot)$ satisfying the above properties, hamming loss and 0/1 loss are commonly used.

3.4   Large-Margin Formulation

The original structured SVM [32] is formulated as the following problem

$$\min_{\mathbf{w}} \sum_{i=1}^{N} C \left| \max_{\mathbf{y}_i' \in \mathcal{Y}} [\Delta(\mathbf{y}_i^*, \mathbf{y}_i') + \langle \mathbf{w}, \delta\Psi_i(\mathbf{y}_i', \mathbf{y}_i^*) \rangle] \right|_{+} + \frac{1}{2} \|\mathbf{w}\|^2. \tag{5}$$

where $| \cdot |_+$ denotes $\max\{\cdot, 0\}$ and $\mathbf{y}_i^*$ is the true label of $\mathbf{x}_i$. The formulation encourages a large margin between a true label and the runner up.

In CLLP, we are given candidate labels for each local part, which has two implications: (1) any label in the non-candidate set is not the true label; (2) some label in the candidate set is true label but we do not know which one. They imply two different types of discriminative constraints that need consideration. First, discrimination between the candidates and non-candidates. Second, discrimination between the true label and other candidates. Thus we decompose the slacks for each instance into two sets, one set for candidate labels and another for non-candidate labels. Namely, we decompose the objective as

$$\mathcal{J}_0(\mathbf{w}) = \sum_{i=1}^{N} C_1 \left| \max_{\mathbf{y}_i' \in Y_i} [\Delta(\mathbf{y}_i^*, \mathbf{y}_i') + \langle \mathbf{w}, \delta\Psi_i(\mathbf{y}_i', \mathbf{y}_i^*) \rangle] \right|_{+}$$

$$+ \sum_{i=1}^{N} C_2 \left| \max_{\mathbf{y}_i'' \in \mathcal{Y}/Y_i} [\Delta(\mathbf{y}_i^*, \mathbf{y}_i'') + \langle \mathbf{w}, \delta\Psi_i(\mathbf{y}_i'', \mathbf{y}_i^*) \rangle] \right|_{+} + \frac{1}{2} \|\mathbf{w}\|^2. \tag{6}$$

However, in contrast to the supervised case, in CLLP the true labels $\mathbf{y}_i^*$'s are unknown. We can estimate them to approximate the objective. Thus our optimization problem becomes

$$\min_{\mathbf{w},\{\mathbf{y}_i\in Y_i\}_{i=1}^N} \mathcal{J}_c(\mathbf{w},\{\mathbf{y}_i\}_{i=1}^N) = \sum_{i=1}^N C_1 \left| \max_{\mathbf{y}_i'\in Y_i} [\Delta(\mathbf{y}_i,\mathbf{y}_i') + \langle \mathbf{w}, \delta\Psi_i(\mathbf{y}_i',\mathbf{y}_i)\rangle] \right|_+$$

$$+ \sum_{i=1}^N C_2 \left| \max_{\mathbf{y}_i''\in \mathcal{Y}/Y_i} [\Delta(\mathbf{y}_i,\mathbf{y}_i'') + \langle \mathbf{w}, \delta\Psi_i(\mathbf{y}_i'',\mathbf{y}_i)\rangle] \right|_+ + \frac{1}{2}\|\mathbf{w}\|^2, \quad (7)$$

where $\mathbf{y}_i$ is the estimation of the true label $\mathbf{y}_i^*$. The intuition is that we encourage a large margin between the estimated "true" labels and the runner up in the candidate labeling set as well as another runner up in the non-candidate set. And we differentiate these two margins by C_1 and C_2.

Equation (7) looks similar to the counterparts in the Transductive Struct-SVMs [44] and the Latent Struct-SVMs [37]. However, there are three key differences. First, we do not know any true label $\mathbf{y}_i^*$ in Equation (7), while in the Transductive Struct-SVMs we know the true labels of the labeled set and in the Latent Struct-SVMs we know the true labels of the observed layer. Second, we differentiate the two types of large margin constraints. Third, in our problem, each $\mathbf{y}_i$ has its own feasible solution space Y_i.

3.5 Properties of the Objective

We compare our objective with the true objective and another objective used in the current state-of-the-art method, the Maximum Margin Set learning (MMS) [14] designed for the CLS setting.

Lemma 1. $\forall \mathbf{w}$, $\mathcal{J}_0(\mathbf{w}) \geq \min_{\{\mathbf{y}_i\in Y_i\}_{i=1}^N} \mathcal{J}_c(\mathbf{w},\{\mathbf{y}_i\}_{i=1}^N)$. Namely, the objective Equation (6) upper bounds the objective Equation (7).

Corollary 1. Let $\mathcal{J}_0^* = \min_{\mathbf{w}} \mathcal{J}_0(\mathbf{w})$, and $\mathcal{J}_c^* = \min_{\mathbf{w},\{\mathbf{y}_i\in Y_i\}_{i=1}^N} \mathcal{J}_c(\mathbf{w},\{\mathbf{y}_i\}_{i=1}^N)$, then $\mathcal{J}_0^* \geq \mathcal{J}_c^*$. Namely, the optimal value of the objective Equation (6) upper bounds that of the objective Equation (7).

On the other hand, in [14], the MMS method is proposed for tackling multiclass classification with candidate labeling sets. Actually, it can be straightforwardly extended to a structured output case by modifying $\Delta(\cdot,\cdot)$ and $\Psi(\cdot,\cdot)$. Then the problem of MMS becomes:

$$\min_{\mathbf{w}} \mathcal{J}_m(\mathbf{w}) = \frac{1}{2}\|\mathbf{w}\|^2$$

$$+ C_2 \sum_{i=1}^N \left| \max_{\mathbf{y}_i''\notin Y_i} [\Delta_{\min}(\mathbf{y}_i'', \mathcal{Y}/Y_i) + \langle \mathbf{w}, \Psi(\mathbf{x}_i,\mathbf{y}_i'')\rangle] - \max_{\mathbf{y}_i\in Y_i}\langle \mathbf{w}, \Psi(\mathbf{x}_i,\mathbf{y}_i)\rangle \right|_+ \quad (8)$$

where $\Delta_{\min}(\mathbf{y}',Y) = \min_{\mathbf{y}\in Y} \Delta(\mathbf{y}',\mathbf{y})$. Then we have the following lemma:

Lemma 2. $\forall \mathbf{w}, \min_{\{\mathbf{y}_i\in Y_i\}_{i=1}^N} \mathcal{J}_c(\mathbf{w},\{\mathbf{y}_i\}_{i=1}^N) \geq \mathcal{J}_m(\mathbf{w})$. Namely, the objective Equation (7) upper bounds the objective Equation (8).

Corollary 2. *Let* $\mathcal{J}_c^* = \min_{\mathbf{w},\{\mathbf{y}_i \in Y_i\}_{i=1}^N} \mathcal{J}_c(\mathbf{w}, \{\mathbf{y}_i\}_{i=1}^N)$, *and* $\mathcal{J}_m^* = \min_{\mathbf{w}} \mathcal{J}_m(\mathbf{w})$, *then* $\mathcal{J}_c^* \geq \mathcal{J}_m^*$. *Namely, the optimal value of the objective Equation (7) upper bounds that of the objective Equation (8).*

By combining the above lemmas and corollaries, we obtain the theorem:

Theorem 3. $\forall \mathbf{w}, \mathcal{J}_0(\mathbf{w}) \geq \min_{\{\mathbf{y}_i \in Y_i\}_{i=1}^N} \mathcal{J}_c(\mathbf{w}, \{\mathbf{y}_i\}_{i=1}^N) \geq \mathcal{J}_m(\mathbf{w})$ *and*

$$\mathcal{J}_0^* \geq \mathcal{J}_c^* \geq \mathcal{J}_m^*.$$

This theorem shows that the value of our objective (Equation (7)) lies between the value of the true objective (Equation (6)) and the value of the objective given by MMS (Equation (8)), indicating that our objective is closer to the true objective compared to MMS.

4   Optimization

4.1   Optimization with CCCP

The optimization problem defined by Equation (7) is non-convex. An effective approach to solving such a non-convex problem is the Constrained Concave-Convex Procedure (CCCP) [38,29], which requires the objective to be decomposed into a convex part and a concave part. However, the objective of Equation (7) is hard to decompose. In Equation (7), we maximize the objective with $\mathbf{y}_i'$ while minimizing it with $\mathbf{y}_i$. But the term $\Delta(\mathbf{y}_i, \mathbf{y}_i')$ correlates them together, obstructing the decomposition. The same problem exists with the term $\Delta(\mathbf{y}_i, \mathbf{y}_i'')$. Therefore, we make an approximation of the objective by decomposing each $\Delta(a, b)$ term into $(\Delta(a, c) + \Delta(c, b))$, resulting in the following objective:

$$\min_{\mathbf{w}} \sum_i \min_{\mathbf{y}_i \in Y_i} \left\{ C_1 \Big|_{\mathbf{y}_i' \in Y_i}^{\max} [\Delta(\mathbf{z}_i, \mathbf{y}_i') + \Delta(\mathbf{z}_i, \mathbf{y}_i) + \langle \mathbf{w}, \delta \Psi_i(\mathbf{y}_i', \mathbf{y}_i) \rangle] \Big|_+ + \right.$$
$$C_2 \Big|_{\mathbf{y}_i'' \in \mathcal{Y}/Y_i}^{\max} [\Delta(\mathbf{z}_i, \mathbf{y}_i'') + \Delta(\mathbf{z}_i, \mathbf{y}_i) + \langle \mathbf{w}, \delta \Psi_i(\mathbf{y}_i'', \mathbf{y}_i) \rangle] \Big|_+ \left. \right\} + \frac{1}{2}\|\mathbf{w}\|^2 \tag{9}$$

$$= \min_{\mathbf{w}} \sum_i \left\{ C_1 \Big|_{\mathbf{y}_i' \in Y_i}^{\max} [\Delta(\mathbf{z}_i, \mathbf{y}_i') + \langle \mathbf{w}, \Psi_i(\mathbf{x}_i, \mathbf{y}_i') \rangle] - \max_{\mathbf{y}_i \in Y_i} [\langle \mathbf{w}, \Psi_i(\mathbf{x}_i, \mathbf{y}_i) \rangle - \Delta(\mathbf{z}_i, \mathbf{y}_i)] \Big|_+ + \right.$$
$$C_2 \Big|_{\mathbf{y}_i'' \in \mathcal{Y}/Y_i}^{\max} [\Delta(\mathbf{z}_i, \mathbf{y}_i'') + \langle \mathbf{w}, \Psi_i(\mathbf{x}_i, \mathbf{y}_i'') \rangle] - \max_{\mathbf{y}_i \in Y_i} [\langle \mathbf{w}, \Psi_i(\mathbf{x}_i, \mathbf{y}_i) \rangle - \Delta(\mathbf{z}_i, \mathbf{y}_i)] \Big|_+ \left. \right\} + \frac{1}{2}\|\mathbf{w}\|^2 \tag{10}$$

where $\mathbf{z}_i$'s are labels initialized at the beginning of each CCCP iteration. As $\Delta(\cdot, \cdot)$ meets the triangle inequality, Equation (10) upper bounds Equation (7).

Now we can construct an upper bound for the concave term. In each CCCP iteration we substitute the concave term

$$\max_{\mathbf{y}_i \in Y_i} [\langle \mathbf{w}, \Psi(\mathbf{x}_i, \mathbf{y}_i) \rangle - \Delta(\mathbf{z}_i, \mathbf{y}_i)] \tag{11}$$

with term $\langle \mathbf{w}, \Psi(\mathbf{x}_i, \overline{\mathbf{y}_i}) \rangle - \Delta(\mathbf{z}_i, \overline{\mathbf{y}_i})$,

---

Algorithm 1. The CCCP algorithm

1: **Input:** data with weak supervision $\{(\mathbf{x}_i, Y_i)\}_{i=1}^N$
2: Initialize labels $\{\overline{\mathbf{y}_i}\}_{i=1}^N$
3: **repeat**
4:    Solve the convex optimization problem given by Equation (15)
5:    Set labels $\{\overline{\mathbf{y}_i}\}_{i=1}^N$ to be the current prediction of structured instances $\{\mathbf{x}_i\}_{i=1}^N$
      given by current model
6: **until** convergence to a local minimum

---

where

$$\overline{\mathbf{y}_i} = \arg\max_{\mathbf{y}_i \in Y_i} [\langle \mathbf{w}, \Psi(\mathbf{x}_i, \mathbf{y}_i) \rangle - \Delta(\mathbf{z}_i, \mathbf{y}_i)] \tag{12}$$

At the beginning of each CCCP iteration we initialize

$$\mathbf{z}_i = \arg\max_{\mathbf{z}_i \in Y_i} \langle \mathbf{w}, \Psi(\mathbf{x}_i, \mathbf{z}_i) \rangle \tag{13}$$

Then it follows that $\overline{\mathbf{y}_i} = \mathbf{z}_i$, indicating that we could directly initialize $\overline{\mathbf{y}_i}$'s as

$$\overline{\mathbf{y}_i} = \arg\max_{\mathbf{y}_i \in Y_i} \langle \mathbf{w}, \Psi(\mathbf{x}_i, \mathbf{y}_i) \rangle. \tag{14}$$

In this way, we are essentially setting $\overline{\mathbf{y}_i}$ to be the predicted labels of structured instances given by current model. Then the optimization problem in each iteration of CCCP becomes

$$\min_{\mathbf{w}} \sum_{i=1}^N \Bigg[-(C_1 + C_2)\langle \mathbf{w}, \Psi(\mathbf{x}_i, \overline{\mathbf{y}_i}) \rangle + C_1 \cdot \max_{\mathbf{y}_i' \in Y_i} (\Delta(\overline{\mathbf{y}_i}, \mathbf{y}_i') + \langle \mathbf{w}, \Psi(\mathbf{x}_i, \mathbf{y}_i') \rangle)$$

$$+ C_2 \cdot \max_{\mathbf{y}_i'' \in \mathcal{Y}/Y_i \cup \{\overline{\mathbf{y}_i}\}} (\Delta(\overline{\mathbf{y}_i}, \mathbf{y}_i'') + \langle \mathbf{w}, \Psi(\mathbf{x}_i \mathbf{y}_i'') \rangle) \Bigg] \tag{15}$$

where $\overline{\mathbf{y}_i}$'s are initialized as Equation (14). The CCCP procedure is shown in Algorithm 1.

4.2 Accelerating with 2-Slack Cutting Plane Algorithm

In each iteration the optimization problem of Equation (15) can be solved using standard quadratic programming. However, the huge number of constraints prevents us from solving it efficiently. We employ the Cutting Plane (CP) algorithm [19,17] to accelerate training. However, in contrast to the original CP for structured SVM [17], in this problem we have two sets of constraints (corresponding to C_1 and C_2 respectively) and we want to find the solution that satisfies them with specified precision separately. Thus we need to maintain two constraint sets Ω_1 and Ω_2, and set two precision ε_1 and ε_2 for them respectively. To find the most violated label setting in candidate labeling sets and non-candidate labeling sets, we employ a modified *Viterbi* algorithm, which will run in polynomial time of $|\mathbf{x}|$ for each instance $\mathbf{x}$ (For details of the modified Viterbi algorithm please refer to the supplementary material). The sketch of the

2-slack cutting plane algorithm is described in Algorithm 1 in the supplementary material. We also show that the algorithm will converge in at most a non-trivial fixed number of iterations. For the details please refer to Theorems 4 & 5 in Section 4 of the supplementary material.

5   Experiments

We performed experiments on three sequence labeling tasks including part-of-speech tagging (POS), chunking (CHK) and bio-entity recognition (BNE).

5.1   Tasks and Data Sets

POS: This task aims to assign each word of a sentence a unique tag indicating its linguistic category such as noun, verb, etc. We used the *Penn Treebank* [22] corpus with the parts extracted from the *Wall Street Joural (WSJ)* in 1989.

CHK: This task aims to divide a sentence into constituents that are syntactic groups such as noun groups, verb groups etc. We use the same data set in the shared task of Chunking in CoNLL 2000[1] [30].

BNE: This task aims to identify technical terms and tag them in some predefined categories. We used the same dataset in the Bio-Entity Recognition Task at BioNLP/NLPBA 2004[2] [20].

5.2   Baseline Methods

Our method, denoted by CLLP, was implemented based on the SVM[hmm] package[3] to fit with sequence labeling tasks. We compared CLLP with the following methods that are able to handle sequences with candidate labels:

Gold: We trained an SVM[hmm] predictor with ground truth full labels. The performance of Gold would be an upper bound of the performance of CLLP.

NAIVE: For each token, we randomly picked one label from its candidate labels as its true label and trained a SVM[hmm] predictor.

CL-SVM[hmm]: We treated all the candidate labels as true labels. Each sequence appeared in the training set multiple times with different labels. Then an SVM[hmm] predictor was trained on the self-contradictory labeled sequences. Similar methods have been used as baselines in [2,14].

MMS: This method was originally proposed in [14]. We made modifications as stated in Section 4 to deal with the sequence data.

All of the above methods were implemented based on the SVM[hmm] package. For all the experiments, we selected cost parameters C, C_1 and C_2 from the grids

---

[1] http://www.clips.ua.ac.be/conll2000/chunking/

[2] http://www.nactem.ac.uk/tsujii/GENIA/ERtask/report.html

[3] http://www.cs.cornell.edu/people/tj/svm_light/svm_hmm.html

[500 : 150 : 3200]. In MMS and CLLP, each CCCP iteration was a cutting plane optimization procedure whose iteration number was controlled by the parameters ε (for MMS) and ε_1 and ε_2 (for CLLP). Training too aggressively (with ε's that are too small) in the first several CCCP iterations would prevent the algorithm from recovering from the wrongly initialized labels. Thus we initialized ε (for MMS) and ε_1 and ε_2 (for CLLP) to be large at first, and then divided ε's by a discounter d in each iteration until they were less than or equal to some thresholds t, t_1 and t_2. We set the discounter to be 2 and choose thresholds from grids [0.5 : 0.5 : 3].

5.3   Experiments on Artificial Candidate Labels

Originally, these three data sets did not contain any candidate labels as supervision. We followed [14] to generate artificial candidate labels for them. In this way we were able to perform controlled experiments and study the impact of different labeling settings such as the size of the candidate set.

Candidate Label Generation
The following two methods were adopted to generate candidate labels. For both, we took each individual token as a local part, i.e., we provided candidate labels for each token, where the number of candidate labels was specified by the token's candidate labeling size. The two methods were used to control label ambiguity at the sequence level and token level respectively.

Random Generation: This method was used to control the label ambiguity at the sequence level. Each token in the sequence had an initial candidate labeling size of 1 (which is its true label). We randomly chose n tokens sequentially (not necessarily non-overlapping) and doubled their candidate labeling size. We then generated candidate labels for each token according to the label distribution in the training data, which already contained label bias.

Specified Generation: This method was used to control the label ambiguity at the token level. For all sequences, we restricted all the candidate labeling sizes to be a constant m, and randomly generated m *different* candidate labels for each token, among which only one was correct.

The NAIVE, MMS and CLLP methods require label initializations. We randomly picked one label from the candidate labels for each token as its initial label.

Results

Data Sets with Random Generation
We varied n from 1 to 16. Performances of various methods on 3 different data sets were plotted in Figure 1, from which we can make observations:

First, CLLP was more stable against different numbers of candidate labels compared to NAIVE and CL-SVM[hmm]. In addition, CL-SVM[hmm] was not scalable with a large number of candidate labels. When n exceeds 6, several days are needed for training.

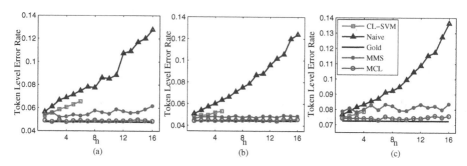

Fig. 1. The performances of various methods on data sets POS (a), CHK (b) and BNE (c). We only plotted a few points of CL-SVM^{hmm} because as the number of candidate label grows, it immediately becomes unfeasible in time.

Second, the gap between CLLP and MMS was small, especially with regards to CHK. This phenomenon resulted from the small number of candidate labels. With the random generation of candidate labels, even when n was large, there were still tokens that had only one candidate label that was exactly its true label. This fact prevented CLLP from taking advantage of its objective of better approximation then MMS, and made the gap between them negligible. However, the gap will be more visible when the number of candidate labels is large, as shall be seen in Figure 2 of Section 5.3.

Last, the CLLP method beats all the other methods and performed close to the full supervised SVM^{hmm}. This clearly shows the effectiveness and scalability of CLLP versus other methods.

Data Sets with Specified Generation

We varied m from 1 to 7 to see how the token-wise candidate labeling size affected the performance. We report the results on the CHK data set in Figure 2 (a).

The results indicate the gap between MMS and CLLP becomes more visible as m increases. This phenomenon mainly results from poor approximation of the objective of MMS.

In the objective 8, MMS considered only the margin between the most possible candidate labels and the most possible non-candidate labels. When the number of candidate labels was large, there were fewer non-candidate labels for the MMS optimizer to choose constraints from. In contrast, CLLP considers constraints from both the candidate labeling set and the non-candidate labeling set. This strategy is beneficial when the number of candidate labels is large.

We also observed that CLLP was less sensitive to the initialization error when compared to other methods, Due to the fact that initialization error increases as m increases. We conducted an auxiliary experiment at $m = 5$ to further investigate. We varied the initialization error rate from 0.5 to 0.8. Note that when the initialization error was 0.8, the initial labels were actually totally random. Under this setting, the performances of various methods are reported in Figure 2 (b). Based on the results, the initialization error rate did have a significant impact on the performances of these models. However, its influence on CLLP was limited

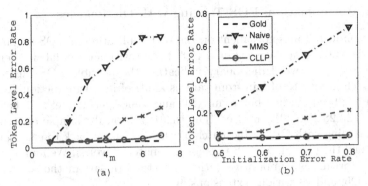

Fig. 2. Impact of model parameters on the performances of various methods. (a) Impact of m with totally random label initialization. (b) Impact of initialization error rate with $m = 5$.

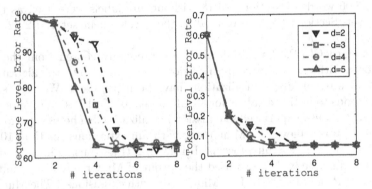

Fig. 3. The convergence curves of CLLP with different discounters d

compared to other methods, showing the stability of CLLP against the different initialization error rates.

Impact of Parameter d

We conducted this experiment to verify the impact of the discounter d on the convergence of CLLP. The experiment was done on the POS data set with random generation and n was fixed at 16. We varied the number of iterations from 1 to 10, and set the discounter d to grids $[2:5]$ to see the impact. The results are reported in Figure 3.

The results show the algorithm converged quickly, e.g., in 4 to 5 iterations. Even if we chose an inappropriate d (say, $d = 2$), the algorithm still converged in 6 iterations, showing the efficiency and robustness of the algorithm. The speed of convergence seems to be positively correlated with the value of d. However, the impact was limited. Actually it made little sense to choose a very large value for d, since the algorithm would simply do nothing in the first several CCCP iterations and would set wrong labels for the following training procedure.

5.4   Real Application: POS Tagging with Dictionaries

We also conducted an experiment on the real application of POS tagging with dictionaries. Our goal was to train a POS tagger without any labeled sentences but only require a word dictionary indicating the possible POS tags of each word, which is easy to obtain from various kinds of linguistic dictionaries. This problem has been studied before in NLP and some specific methods have been proposed [23,28,25,9]. We noticed that this is a typical example of structured output learning with CLLP: by matching the dictionary back to each sentence, we obtained the candidate POS tags for the matched words. We found that our general algorithm is competitive with the state-of-the-art methods, i.e., "Min-Greedy" [25] and its various extensions [9].

Following the settings in previous methods, we used a corpus extracted from the *Wall Street Journal (WSJ)* in 1989 in *Penn Treebank* [22]. Sections 00-07, with golden labels, were used to construct the tag dictionary. Then the first 48000/96000 words of sections 16-18 without any labels were used as the raw training set. Sections 19-21 were used as the development set and 22-24 as the test set.

In standard CLLP, the initial labels are randomly chosen from the candidate labels, without any task-specific prior incorporated into the algorithm. In [9], several ways of label initializations have been proposed. We used some of these methods to initialize labels for CLLP, noted as "CLLP + auto-supervised" and "CLLP + auto-supervised + emission initialization." The best performance shown in [9] was achieved by MinGreedy with full extensions (method 10 in the original paper), where a full-supervised HMM was trained using initialized labels output by MinGreedy. We also used the output of MinGreedy as our initialization for CLLP, noted as "CLLP + MinGreedy with extensions." The MinGreedy code is provided by the authors[4]. More details on dictionary construction and label initialization can be found in [25,9].

The results are shown in Tabel 1. CLLP outperformed all the other unitary methods without the task (POS) specific initializations. With the proper initialization of labels, CLLP is able to further improve the results. Remarkably, by using the output labels of MinGreedy with full extensions as label initialization, CLLP is able to outperform all the other methods.

5.5   Real Application: Wiki Entity Type Disambiguation

We conducted another experiment on a real problem of Wiki entity type disambiguation. This is an example of CLLP degenerating to handle multiclass classification.

Traditionally, in order to train an NER model, we need sentence-level labels. Similar to POS tagging with dictionaries, we attempted to train an NER model without any sentence labels and only requiring an entity dictionary indicating the

---

[4] https://github.com/dhgarrette/type-supervised-tagging-2012emnlp

Table 1. POS tagging accuracy of different methods

Methods	48000	96000
GOLD	94.40	94.77
NAIVE	65.33	66.79
MMS	67.68	69.51
CLLP	**76.74**	**76.56**
MinGreedy [25]	68.86	74.93
MinGreedy + auto-supervised [9]	80.78	80.90
MinGreedy + auto-supervised + emission initialization [9]	80.92	80.70
MinGreedy with extensions [9]	87.95	86.22
CLLP + auto-supervised	82.90	82.22
CLLP + auto-supervised + emission initialization	82.89	82.22
CLLP + MinGreedy with extensions	**89.87**	**88.47**

possible types of an entity, which can be easily obtained from many knowledge bases such as Freebase[5].

We conducted the experiments based on Freebase and Wikipedia articles. In some sentences of a Wikipedia article, there is an anchor link to indicate the phrase is an entity in Wikipedia and will redirect to the host article page if a user clicks it. For each entity highlighted by the anchor link, we can find the corresponding entity types in Freebase. We then obtained a set of multiclass training instances: an entity in a sentence, and the corresponding candidate labels. We selected 20 entity types plus an "Other" type. We randomly sampled 500 entities to manually label as the test set, and sampled 9991 entities as the training set. As each entity was associated with a candidate "Other" label, the "Other" class dominated other classes. Thus we sampled the "Other" class by assigning "Other" to an entity with a probability of 0.1. The the classes were therefore more balanced.

Table 2. Results on Wiki data

	F1	Precision	Recall
NAIVE	64.83	57.79	73.82
MMS	54.02	47.70	62.30
CLLP	**69.69**	**61.52**	**80.37**

The results are shown in Tabel 2. We found that MMS was even worse than NAIVE. The main reason for this phenomenon still draws from the objective that MMS aims to minimize. With the large number of "Others" labels, MMS either took it as the most possible candidate label, or simply ignored it since it was not a non-candidate label. Thus it proned to predicting many entities to "Others." Things became even worse when we added more "Others" labels to the training data. When we associated all the instances with an extra "Others"

---

[5] http://www.freebase.com

label, MMS simply predicted all the entities were "Others." In contrast, CLLP overcame this problem by using two sets of constraints and outperformed the other two methods.

6   Conclusion

In this paper, we introduced a new weakly supervised setting for structured output learning, named *candidate labels for local parts* (CLLP), where a set of candidate labels is provided for each local part of output variables. We have shown that training with this type of weak supervision can be efficiently handled. Then we proposed a large-margin formulation for the learning problem, and used proper approximations and Constraint Concave-Convex Procedure (CCCP) to deal with the non-convex optimization problem. A 2-slack cutting plane method has also been proposed to accelerate the inner loop of CCCP. Experiments on various tasks have shown the effectiveness and efficiency of the proposed method. It is interesting that the CLLP setting is rather general, and is able to degenerate to various weakly supervised setting for both structured output learning and multiclass classification. Thus the CLLP setting and the proposed large-margin learning method provide a uniform approach to formulate and solve structured output learning with different kinds of weak supervision.

References

1. Barnard, K., Duygulu, P., Forsyth, D., De Freitas, N., Blei, D., Jordan, M.: Matching words and pictures. JMLR, 1107–1135 (2003)
2. Bunescu, R., Mooney, R.: Multiple instance learning for sparse positive bags. In: ICML, pp. 105–112 (2007)
3. Chang, M., Goldwasser, D., Roth, D., Srikumar, V.: Structured output learning with indirect supervision. In: ICML (2010)
4. Cour, T., Sapp, B., Jordan, C., Taskar, B.: Learning from ambiguously labeled images. In: CVPR, pp. 919–926 (2009)
5. Deselaers, T., Ferrari, V.: A conditional random field for multiple-instance learning. In: ICML, pp. 287–294 (2010)
6. Dietterich, T., Lathrop, R., Lozano-Pérez, T.: Solving the multiple instance problem with axis-parallel rectangles. Artificial Intelligence, 31–71 (1997)
7. Dredze, M., Talukdar, P., Crammer, K.: Sequence learning from data with multiple labels. In: ECML/PKDD Workshop on Learning from Multi-Label Data (2009)
8. Ganchev, K., Gillenwater, J., Taskar, B.: Dependency grammar induction via bitext projection constraints. In: ACL, pp. 369–377 (2009)
9. Garrette, D., Baldridge, J.: Type-supervised hidden markov models for part-of-speech tagging with incomplete tag dictionaries. In: EMNLP, pp. 821–831 (2012)
10. Hall, K., McDonald, R., Katz-Brown, J., Ringgaard, M.: Training dependency parsers by jointly optimizing multiple objectives. In: EMNLP, pp. 1489–1499 (2011)
11. Hüllermeier, E., Beringer, J.: Learning from ambiguously labeled examples. In: Famili, A.F., Kok, J.N., Peña, J.M., Siebes, A., Feelders, A. (eds.) IDA 2005. LNCS, vol. 3646, pp. 168–179. Springer, Heidelberg (2005)

12. Jiao, F., Wang, S., Lee, C.H., Greiner, R., Schuurmans, D.: Semi-supervised conditional random fields for improved sequence segmentation and labeling. In: ACL, pp. 209–216 (2006)

13. Jie, L., Caputo, B., Ferrari, V.: Who's doing what: Joint modeling of names and verbs for simultaneous face and pose annotation. In: NIPS, pp. 1168–1176 (2009)

14. Jie, L., Francesco, O.: Learning from candidate labeling sets. In: NIPS, pp. 1504–1512 (2010)

15. Jin, R., Ghahramani, Z.: Learning with multiple labels. In: NIPS, pp. 897–904 (2002)

16. Jin, R., Wang, S., Zhou, Z.: Learning a distance metric from multi-instance multi-label data. In: CVPR, pp. 896–902 (2009)

17. Joachims, T., Finley, T., Yu, C.: Cutting-plane training of structural svms. Machine Learning, 27–59 (2009)

18. Joulin, A., Bach, F.: A convex relaxation for weakly supervised classifiers. In: ICML (2012)

19. Kelley Jr., J.: The cutting-plane method for solving convex programs. Journal of the Society for Industrial & Applied Mathematics, 703–712 (1960)

20. Kim, J.D., Ohta, T., Tsuruoka, Y., Tateisi, Y., Collier, N.: Introduction to the bio-entity recognition task at jnlpba. JNLPBA, 70–75 (2004)

21. Lafferty, J., McCallum, A., Pereira, F.: Conditional random fields: Probabilistic models for segmenting and labeling sequence data. In: ICML, pp. 282–289 (2001)

22. Marcus, M., Santorini, B., Marcinkiewicz, M.: Building a large annotated corpus of english: The penn treebank. Computational Linguistics, 313–330 (1993)

23. Merialdo, B.: Tagging english text with a probabilistic model. Computational Linguistics, 155–171 (1994)

24. Navigli, R.: Word sense disambiguation: A survey. ACM Comput. Surv. 41(2), 10:1–10:69 (2009)

25. Ravi, S., Vaswani, A., Knight, K., Chiang, D.: Fast, greedy model minimization for unsupervised tagging. In: COLING, pp. 940–948 (2010)

26. Roth, D., Small, K.: Margin-based active learning for structured output spaces. In: Fürnkranz, J., Scheffer, T., Spiliopoulou, M. (eds.) ECML 2006. LNCS (LNAI), vol. 4212, pp. 413–424. Springer, Heidelberg (2006)

27. Settles, B., Craven, M.: An analysis of active learning strategies for sequence labeling tasks. In: EMNLP, pp. 1070–1079 (2008)

28. Smith, N.A., Eisner, J.: Contrastive estimation: Training log-linear models on unlabeled data. In: ACL, pp. 354–362 (2005)

29. Smola, A.J., Vishwanathan, S., Hofmann, T.: Kernel methods for missing variables. In: AISTATS (2005)

30. Tjong K.S.E.F., Buchholz, S.: Introduction to the conll-2000 shared task: Chunking. In: CoNLL, pp. 127–132 (2000)

31. Tjong, K.S.E.F., De Meulder, F.: Introduction to the conll-2003 shared task: Language-independent named entity recognition. In: HLT-NAACL, pp. 142–147 (2003)

32. Tsochantaridis, I., Hofmann, T., Joachims, T., Altun, Y.: Support vector machine learning for interdependent and structured output spaces. In: ICML (2004)

33. Tsochantaridis, I., Joachims, T., Hofmann, T., Altun, Y.: Large margin methods for structured and interdependent output variables. JMLR, 1453–1484 (2005)

34. Tsuboi, Y., Kashima, H., Oda, H., Mori, S., Matsumoto, Y.: Training conditional random fields using incomplete annotations. In: COLING, pp. 897–904 (2008)

35. Vezhnevets, A., Ferrari, V., Buhmann, J.: Weakly supervised structured output learning for semantic segmentation. In: CVPR, pp. 845–852 (2012)

36. Yu, C.N.: Transductive learning of structural svms via prior knowledge constraints. In: AISTATS (2012)
37. Yu, C.N.J., Joachims, T.: Learning structural svms with latent variables. In: ICML, pp. 1169–1176 (2009)
38. Yuille, A., Rangarajan, A.: The concave-convex procedure. Neural Computation, 915–936 (2003)
39. Zhang, D., Liu, Y., Si, L., Zhang, J., Lawrence, R.: Multiple instance learning on structured data. In: NIPS, pp. 145–153 (2011)
40. Zhou, Z.: Multi-instance learning: A survey. Tech. rep., AI Lab, Department of Computer Science & Technology, Nanjing University (March 2004)
41. Zhou, Z., Sun, Y., Li, Y.: Multi-instance learning by treating instances as non-iid samples. In: ICML, pp. 1249–1256 (2009)
42. Zhou, Z., Zhang, M.: Multi-instance multi-label learning with application to scene classification. In: NIPS, pp. 1609–1616 (2006)
43. Zhu, X.: Semi-supervised learning literature survey (2005)
44. Zien, A., Brefeld, U., Scheffer, T.: Transductive support vector machines for structured variables. In: ICML (2007)

Shared Structure Learning for Multiple Tasks with Multiple Views

Xin Jin[1,2], Fuzhen Zhuang[1], Shuhui Wang[1], Qing He[1], and Zhongzhi Shi[1]

[1] Key Laboratory of Intelligent Information Processing, Institute of Computing
Technology, Chinese Academy of Sciences, Beijing 100190, China
[2] University of Chinese Academy of Sciences, Beijing 100049, China
{jinx,zhuangfz,heq,shizz}@ics.ict.ac.cn,
wangshuhui@ict.ac.cn

Abstract. Real-world problems usually exhibit dual-heterogeneity, i.e.,
every task in the problem has features from multiple views, and multiple tasks are related with each other through one or more shared views.
To solve these multi-task problems with multiple views, we propose a
shared structure learning framework, which can learn shared predictive
structures on common views from multiple related tasks, and use the consistency among different views to improve the performance. An alternating optimization algorithm is derived to solve the proposed framework.
Moreover, the computation load can be dealt with locally in each task
during the optimization, through only sharing some statistics, which significantly reduces the time complexity and space complexity. Experimental studies on four real-world data sets demonstrate that our framework
significantly outperforms the state-of-the-art baselines.

Keywords: Multi-task Learning, Multi-view Learning, Alternating
Optimization.

1 Introduction

In many practical situations, people need to solve a number of related tasks,
and multi-task learning (MTL) [5–7, 20] is a good choice for these problems.
It learns multiple related tasks together so as to improve the performance of
each task relative to learning them separately. Besides, many problems contain
different "kinds" of information, that is, they include multi-view data. Multi-view
learning (MVL) [3, 8, 19] can make better use of these different views and get
improved results. However, many real-world problems exhibit dual-heterogeneity
[14]. To be specific, a single learning task might have features in multiple views
(i.e., feature heterogeneity); different learning tasks might be related with each
other through one or more shared views (features) (i.e., task heterogeneity).
One example is the web page classification problem. If people want to identify
whether the web pages from different universities are course home pages, then
classifying each university can be seen as a task. Meanwhile, every web page
has different kinds of features, one kind is the content of the web page, and the

H. Blockeel et al. (Eds.): ECML PKDD 2013, Part II, LNAI 8189, pp. 353–368, 2013.
© Springer-Verlag Berlin Heidelberg 2013

354   X. Jin et al.

other kind is the anchor text attached to hyperlinks pointing to this page, from other pages on the web. Such problem type is Multi-task learning with Multiple Views (MTMV). However, traditional multi-task learning or multi-view learning methods are not very suitable as they cannot use all the information contained in the problem.

In supervised learning, given a labeled training set and *hypothesis space* $\mathcal{H}$, the goal of *empirical risk minimization* (ERM) is to find a predictor $f \in \mathcal{H}$ that minimizes empirical error. The error of the best predictor learned from finite sample is called the *estimation error*. The error of using a restricted $\mathcal{H}$ is often referred to as the *approximation error*, from the structure risk minimization theory [21]. One needs to select the size of $\mathcal{H}$ to balance the trade-off between approximation error and estimation error. This is typically done through model selection, which learns a set of predictors from a set of candidate hypothesis spaces $\mathcal{H}_\theta$, and then chooses the best one. When multiple related tasks are given, learning the structure parameter Θ in the predictor space becomes easier [1, 6]. Also, different tasks can share information through the structure parameter Θ, i.e., for each task t, $f_t \in \mathcal{H}_{t,\Theta}$. MTMV learning methods that can make full use of information contained in multiple tasks and multiple views are proposed [14, 25]. The transductive algorithm $IteM^2$ [14] can only deal with nonnegative feature values. $regMVMT$ algorithm [25] shares the label information among different tasks by minimizing the difference of the prediction models for different tasks. Without prior knowledge, simply restricting all the tasks are similar seems inappropriate. This paper assumes that multiple tasks share a common predictive structure, and they can do model selection collectively. Compared to other methods to learn good predictive structures, such as data-manifold methods based on graph structure, our method can learn some underlying predictive functional structures in hypothesis space, which can characterize better predictors.

To facilitate information sharing among different tasks on multi-view representation, in this paper, we propose an efficient inductive convex shared structure learning for MTMV problem (*CSL-MTMV*). Our method learns shared predictive structures on hypothesis spaces from multiple related tasks that have common views; consequently, all tasks can share information through the shared structures. In this way, the strict assumption in the previous MTMV methods that all the tasks should be similar can be discarded. We assumed that the underlying structure is a shared low-dimensional subspace, and a linear form of feature map is considered for simplicity. Furthermore, it uses the prediction consistency among different views to improve the performance. Besides, some tasks may not have all the views in many real applications. To deal with missing views, a direct extension of our algorithm is provided. Our method is more flexible than previous inductive MTMV algorithm $regMVMT$ [25]. Specifically, $regMVMT$ can be seen as a special case of *CSL-MTMV*, which means our approach is more generalized and has a potential to get better results. In addition, different from $regMVMT$, our method decouples different tasks during the model optimization, which significantly reduces the time complexity and space complexity. Therefore, our method is more scalable for problems with large number of tasks.

The rest of this paper is organized as follows. A brief review of related work is given in Section 2. The MTMV problem definition and some preliminary works are presented in Section 3. Our shared structure learning framework for MTMV problem and a convex relaxation algorithm are described in Section 4. To demonstrate the effectiveness of our algorithm, some experimental results are shown in Section 5. Conclusion is provided in Section 6.

2   Related Work

Currently, there are only a few researches on multi-task problem with multi-view data (MTMV). The traditional multi-task learning and multi-view learning methods also provide some insights for the MTMV problem. In the following, a brief description of these methods is given.

Multi-Task Learning (MTL). Multi-task learning conducts multiple related learning tasks simultaneously so that the label information in one task can be used for other tasks. The earliest MTL method [5] learns a shared hidden layer representation for different tasks. Supposing that all the tasks are similar, a regularization formulation is proposed for MTL [11]. MTL can be modeled by stochastic process methods, such as [20, 24]. Multi-task feature learning learns a low-dimensional representation which is shared across a set of multiple related tasks [2, 15]. To deal with outlier tasks, a robust multi-task learning algorithm is proposed [7]. The methods to learn predictive structures on hypothesis spaces from multiple learning tasks are also proposed [1, 6].

Multi-View Learning (MVL). The basic idea of MVL is making use of the consistency among different views to achieve better performance. One of the earliest works on multi-view learning is co-training algorithm [3], which uses one view's predictor to enlarge the training set for other views. Nigam and Ghani compared co-training, EM and co-EM methods, and showed that co-EM algorithm is the best among the three approaches [17]. Some improvements of co-training algorithm are also proposed [16, 23]. Other methods are based on co-regularization framework. Sindhwani et al. [18] proposed a learning framework for multi-view regularization. SVM-$2K$ [12] is a method which uses kernels for two views learning. Sindhwani and Rosenberg [19] constructed a single Reproducing Kernel Hilbert Spaces (RKHSs) with a data-dependent "co-regularization" norm that reduces MVL to standard supervised learning. Chen et al. [8] presented a large-margin learning framework to discover a predictive latent subspace representation shared by multiple views.

Multi-Task Learning with Multiple Views (MTMV). He and Lawrence [14] proposed a graph-based framework which takes full advantage of information among multiple tasks and multiple views, and an iterative algorithm ($IteM^2$) is proposed to optimize the model. The framework is transductive which cannot predict the unseen samples. It can only deal with problems with nonnegative feature values. $regMVMT$ [25] uses co-regularization to obtain functions that are consistent with each other on the unlabeled samples for different views.

Across different tasks, additional regularization functions are utilized to ensure the learned functions are similar. The assumption that all the tasks are similar to each other may not be appropriate. Different tasks are coupled in the computation process of *regMVMT* algorithm, making the model becoming more complex and requires more memory to store the data.

3   Preliminaries

3.1   MTMV Problem Definition

Notations. In this paper, $[m : n]$ $(n > m)$ denotes a set of integers in the range of m to n inclusively. Let $\mathbb{S}_+$ be the subset of positive semidefinite matrices. Denote $A \preceq B$ if and only if $B - A$ is positive semidefinite. Let $\text{tr}(X)$ be the trace of X, and X^{-1} be the inverse of matrix X. $\|\cdot\|$ denotes ℓ_2 norm of a vector. Unless specified otherwise, all vectors are column vectors.

In this part, a formal introduction of MTMV problem is given. The problem definition is very similar to [14, 25]. Suppose that the problem includes T tasks and V views in total. Also, N labeled and M unlabeled data samples are given. Usually, the labeled examples are insufficient while the unlabeled samples are abundant, i.e. $M \gg N$. For each task $t \in [1 : T]$, there are n_t and m_t labeled and unlabeled examples, thus we have $N = \sum_t n_t$ and $M = \sum_t m_t$. Let d_v be the number of features in the view $v \in [1 : V]$, and denote $D = \sum_v d_v$.

The feature matrix $X_t^v \in \mathbb{R}^{n_t \times d_v}$ is used to denote the labeled samples in task t for view v, the corresponding unlabeled examples is denoted $P_t^v \in \mathbb{R}^{m_t \times d_v}$. Let $y_t \in \{1, -1\}^{n_t \times 1}$ be the label vector of the labeled examples in the task t. $X_t = (X_t^1, X_t^2, \ldots, X_t^V)$, and $P_t = (P_t^1, P_t^2, \ldots, P_t^V)$ are concatenated feature matrices of the labeled and unlabeled examples for task t, respectively. It is common that in some applications not all tasks have features available from all the V views, so an indicator matrix $I_{id} \in \{1, 0\}^{T \times V}$ is used to mark which view is missing from which task, i.e. $I_{id}(t, v) = 0$ if the task t does not contain v-th view, and $= 1$ otherwise. This notation can only handle "structured" missing views [25] in the sense that if a view is present in a task, it is present in all the samples in the task; if a view is missing from a task, it is missing in all the samples in the task. Throughout the paper we use subscripts to denote tasks and superscripts to denote views. So the goal of this paper is to leverage the label information from all the tasks to help classify the unlabeled examples in each task, as well as use the consistency among different views of a single task to improve the performance.

3.2   Shared Structure Learning for MTL

Shared structure learning has been successfully used in single view multi-task learning (MTL) problems [1, 6], that is, $V = 1$ in the MTMV problem described in Section 3.1. In MTL, suppose the dimension of the feature space is d, and the objective is to learn linear predictors $f_t(x) = u_t^\top x$, for $t \in [1 : T]$, where u_t is the weight vector for the t-th task. For shared structure learning, it is assumed that the underlying structure is a shared low-dimensional subspace, and a linear form of feature map is considered for simplicity. The predictors $\{f_t\}_{t=1}^T$ can be learned simultaneously by exploiting a shared feature space. Formally, the prediction function f_t can be expressed as:

$$f_t(x) = u_t^\top x = w_t^\top x + z_t^\top \Theta x \qquad (1)$$

where the structure parameter Θ takes the form of an $h \times d$ matrix with orthonormal rows, i.e., $\Theta\Theta^\top = I$.

In [6], an improved alternating structure optimization (iASO) formulation is given:

$$\min_{\{u_t, z_t\}, \Theta^\top = I} \sum_{t=1}^{T} \left(\frac{1}{n_t} \sum_{i=1}^{n_t} L\left(f_t(x_{t,i}), y_{t,i}\right) + g_t(u_t, z_t, \Theta) \right) \qquad (2)$$

where $g_t(u_t, z_t, \Theta)$ is the regularization function defined as:

$$g_t(u_t, z_t, \Theta) = \alpha\|u_t - \Theta^\top z_t\|^2 + \beta\|u_t\|^2. \qquad (3)$$

The regularization function in Eq.(3) controls the task relatedness (via the first component) as well as the complexity of the predictor functions (via the second component) as commonly used in traditional regularized risk minimization formulation for supervised learning. α and β are pre-specified coefficients, indicating the importance of the corresponding regularization component. This formulation provides the foundation for our MTMV learning methods.

4   Shared Structure Learning for MTMV Problem

4.1   Shared Structure Learning Framework for MTMV Problem

A straightforward way to use the single view multi-task learning (MTL) methods described in Section 3.2 is as follows. First, the prediction model for each view data is learned individually, so the MTMV problem can be divided into V MTL problems. Then, a model for each view v in each task t is acquired, represented by $f_t^v(x_t^v)$ with the following formulation:

$$f_t^v(x_t^v) = u_t^{v\top} x_t^v = w_t^{v\top} x_t^v + z_t^{v\top} \Theta^v x_t^v, \qquad (4)$$

where u_t^v, w_t^v and z_t^v have similar meanings as in Eq.(1), structure parameter Θ^v represents the low-dimensional feature map for view v across different tasks. The basic assumption underlying multi-view learning for a single task is that the multiple views are conditionally independent and the predictive model of each view can be used to make predictions on data examples, then the final models are obtained according to these models. Without prior knowledge on which view contributes more to the final models than other views, it is often assumed that all views contribute equally, as described in [19]. The final model for task t in

MTMV problem is obtained by averaging the prediction results from all view functions as follows:

$$f_t(x_t) = \frac{1}{V} \sum_{v=1}^{V} f_t^v(x_t^v), \tag{5}$$

where $x_t = [{x_t^1}^\top, {x_t^2}^\top, \ldots, {x_t^V}^\top]^\top$ is the concatenated feature vector of the samples for task t.

However, in MTMV problem, it is worthwhile to make better use of the information contained in different views, not just only decompose into separate MTL problems. The models built on each single view will agree with one another as much as possible on unlabeled examples. Co-regularization is a technique to enforce such model agreement on unlabeled examples. Adding this into the model, we obtain the following formulation :

$$\min_{\{u_t^v, z_t^v, \Theta^v\}, \Theta^v({\Theta^v}^\top)=I} \sum_{t=1}^{T} \sum_{v=1}^{V} \left(\frac{1}{n_t} \sum_{i=1}^{n_t} L\left(f_t^v(x_{t,i}^v), y_{t,i}\right) + g_t^v(u_t^v, z_t^v, \Theta^v) \right.$$

$$\left. + \gamma \frac{1}{m_t} \sum_{j=1}^{m_t} \sum_{v' \neq v} \left(f_t^{v'}(p_{t,j}^{v'}) - f_t^v(p_{t,j}^v) \right)^2 \right) \tag{6}$$

where L is the empirical loss function, $x_{t,i}^v$ is the feature representation for the v-th view of i-th labeled sample in task t, $p_{t,j}^v$ ($p_{t,j}^{v'}$) is the feature representation for the v-th (v'-th) view of j-th unlabeled sample in task t, $g_t^v(u_t^v, z_t^v, \Theta^v)$ is the regularization function defined as:

$$g_t^v(u_t^v, z_t^v, \Theta^v) = \alpha \| u_t^v - {\Theta^v}^\top z_t^v \|^2 + \beta \| u_t^v \|^2, \tag{7}$$

where the structure parameter Θ^v is a $h \times d_v$ matrix. The regularization function in Eq.(7) controls the task relatedness as well as the complexity of the predictor models. So, the optimization problem described in Eq.(6) can take advantage of multiple views and multiple tasks simultaneously.

4.2   A Relaxed Convex Formulation

The problem in Eq.(6) is non-convex and difficult to solve due to its orthonormal constraints and the regularization in terms of u_t^v, z_t^v and Θ^v (suppose L is convex loss function). Converting it into a convex formulation is desirable. The optimal $\{z_t^v\}$ for the problem in Eq.(6) can be expressed as $z_t^v = \Theta^v u_t^v$. Let $U^v = [u_1^v, u_2^v, \ldots, u_T^v] \in \mathbb{R}^{d_v \times T}$ and $Z^v = [z_1^v, z_2^v, \ldots, z_T^v] \in \mathbb{R}^{h \times T}$, so $Z^v = \Theta^v U^v$. Then we denote:

$$G_0(U^v, \Theta^v) = \min_{Z^v} \sum_{t=1}^{T} g_t^v(u_t^v, z_t^v, \Theta^v) = \alpha \operatorname{tr}\left({U^v}^\top ((1+\eta)I - {\Theta^v}^\top \Theta^v) U^v \right) \tag{8}$$

where $\eta = \beta/\alpha > 0$. Eq.(8) can be reformulated into an equivalent form given by

$$G_1(U^v, \Theta^v) = \alpha\eta(1+\eta) \operatorname{tr}\left({U^v}^\top (\eta I + {\Theta^v}^\top \Theta^v)^{-1} U^v \right). \tag{9}$$

The orthonormality constraint on Θ^v is non-convex, which makes the optimization problem non-convex. One method is to relax the feasible domain of it into a convex set. Let $M^v = \Theta^{v\top}\Theta^v$, using a similar derivation as in [6], the feasible domain of the optimization problem can be relaxed into a convex set, and a convex formulation of the problem in Eq.(6) can be defined as follows:

$$\min_{\{u_t^v, M^v\}} \sum_{t=1}^{T} \sum_{v=1}^{V} \left(\frac{1}{n_t} \sum_{i=1}^{n_t} L\left(f_t^v(x_{t,i}^v), y_{t,i}\right) + \gamma \frac{1}{m_t} \sum_{j=1}^{m_t} \sum_{v' \neq v} \left(f_t^{v'}(p_{t,j}^{v'}) - f_t^v(p_{t,j}^v)\right)^2 \right)$$

$$+ \sum_{v=1}^{V} G_2(U^v, M^v), \quad \text{subject to}: \quad \text{tr}(M^v) = h, M^v \preceq I, M^v \in \mathbb{S}_+, \quad (10)$$

where $G_2(U^v, M^v)$ is defined as:

$$G_2(U^v, M^v) = \alpha\eta(1+\eta)\text{tr}\left(U^{v\top}(\eta I + M^v)^{-1}U^v\right). \quad (11)$$

Note that the problem in Eq.(10) is a convex relaxation of that in Eq.(6). The optimal Θ^v to Eq.(6) can be approximated using the top h eigenvectors (corresponding to the largest h eigenvalues) of the optimal M^v computed from Eq.(10).

4.3 Convex Shared Structure Learning Algorithm

The optimization problem in Eq.(10) is convex, so the globally optimal solution can be obtained. In this section, a convex shared structure learning algorithm for MTMV problem (*CSL-MTMV*) is presented. In *CSL-MTMV* algorithm, the two optimization variables are optimized alternately, that is, one variable is fixed, while the other one can be optimized according to the fixed one. The methods are described in the following, and the final algorithm is in Algorithm 1.

Computation of $\{U^v\}$ for Given $\{M^v\}$. In Eq.(10), if $\{M^v\}$ are given, it can be easily found that the computation of u_t^v for different tasks can be decoupled, that is, different tasks' weight vectors can be computed separately. Suppose the least square loss function is used where:

$$L\left(f_t^v(x_{t,i}^v), y_{t,i}\right) = (u_t^{v\top}x_{t,i}^v - y_{t,i})^2. \quad (12)$$

We denote the objective function in Eq.(10) as F, and the derivative regarding to each u_t^v is:

$$\frac{\partial F}{\partial u_t^v} = \frac{2}{n_t} \sum_{i=1}^{n_t} (u_t^{v\top}x_{t,i}^v - y_{t,i})x_{t,i}^v + \gamma \frac{2}{m_t} \sum_{j=1}^{m_t} \sum_{v' \neq v} \left(u_t^{v\top}p_{t,j}^v - u_t^{v'\top}p_{t,j}^{v'}\right)p_{t,j}^v$$

$$+ 2\alpha\eta(1+\eta)(\eta I + M^v)^{-1}u_t^v \quad (13)$$

For convenience, the following notations are given:

$$A_t^v = \frac{2}{n_t}X_t^{v\top}X_t^v + \gamma\frac{2}{m_t}(V-1)P_t^{v\top}P_t^v + 2\alpha\eta(1+\eta)(\eta I + M^v)^{-1}$$

$$B_t^{vv'} = -\gamma\frac{2}{m_t}P_t^{v\top}P_t^{v'}, \quad C_t^v = \frac{2}{n_t}X_t^{v\top}y_t \tag{14}$$

where X_t^v, P_t^v and y_t are described in Section 3.1. By setting Eq.(13) to zero and rearranging the terms, the following equation can be obtained:

$$A_t^v u_t^v + \sum_{v'\neq v} B_t^{vv'} u_t^{v'} = C_t^v. \tag{15}$$

From Eq.(15), u_t^v is correlated with other $u_t^{v'}$ for the same task t, i.e., the views of the same task are correlated. u_t^v and $u_{t'}^v$ from different tasks are not correlated. Therefore, the u_t^v of different tasks can be computed separately, while the different views for the same task must be solved together. Note that such an equation can be obtained for each view v in task t. By combining these equations, the following linear equation system can be obtained for each task t:

$$\mathcal{L}_t \mathcal{W}_t = \mathcal{R}_t \tag{16}$$

where $\mathcal{L}_t \in \mathbb{R}^{D\times D}$ is a symmetric block matrix with $V\times V$ blocks. The specific forms of the symbols in Eq.(16) are as follows:

$$\mathcal{L}_t = \begin{bmatrix} A_t^1 & B_t^{12} & \cdots & B_t^{1V} \\ B_t^{21} & A_t^2 & \cdots & B_t^{2V} \\ \vdots & \vdots & \ddots & \vdots \\ B_t^{V1} & B_t^{V2} & \cdots & A_t^V \end{bmatrix} \tag{17}$$

$$\mathcal{W}_t = \text{Vec}\left([u_t^1, u_t^2, \cdots, u_t^V]\right), \quad \mathcal{R}_t = \text{Vec}\left([C_t^1, C_t^2, \cdots, C_t^V]\right)$$

where Vec() denotes the function stacking the column vectors in a matrix to a single column vector. For each task t, an equation system described in Eq.(16) is constructed and solved. The optimal solution of $\{u_t^v\}$ can be easily obtained by left multiplication of the (pseudo-) inverse of matrix $\mathcal{L}_t$.

Computation of $\{M^v\}$ for Given $\{U^v\}$. For given $\{U^v\}$, in Eq.(10), different M^v are not correlated, they can be computed separately. For each view v, the following problem can be obtained:

$$\min_{M^v} \text{tr}\left(U^{v\top}(\eta I + M^v)^{-1}U^v\right), \text{subject to} : \text{tr}(M^v) = h, M^v \preceq I, M^v \in \mathbb{S}_+ \tag{18}$$

This problem is a semidefinite program (SDP), where direct optimization is computationally expensive. An efficient approach to solve it is described in the following. Let $U^v = P_1 \Sigma P_2^\top$ be its singular value decomposition (SVD), where $P_1 \in \mathbb{R}^{d_v\times d_v}$ and $P_2 \in \mathbb{R}^{T\times T}$ are column-wise orthogonal, and rank$(U^v) = q$. In general, $q \leq T \leq d_v$, we also suppose that the dimension h of the shared feature space for the T tasks satisfies $h \leq q$. Then,

$$\Sigma = \mathrm{diag}(\sigma_1, \cdots, \sigma_T) \in R^{d_v \times T}, \ \sigma_1 \geq \cdots \geq \sigma_q > 0 = \sigma_{q+1} = \cdots = \sigma_T. \quad (19)$$

Consider the following optimization problem:

$$\min_{\gamma_i} \sum_{i=1}^{q} \frac{\sigma_i^2}{\eta + \gamma_i}, \qquad \text{subject to} : \sum_{i=1}^{q} \gamma_i = h, 0 \leq \gamma_i \leq 1, \forall i, \quad (20)$$

where $\{\sigma_i\}$ are the singular values of U^v defined in Eq.(19), this optimization problem is convex [4]. The problem in Eq.(20) can be solved via many existing algorithms such as the projected gradient descent method [4].

Chen et al. [6] show that how to transform the SDP problem in Eq.(18) into the convex optimization problem in Eq.(20). Specifically, let $\{\gamma_i^*\}$ be optimal to Eq.(20) and denote $\Lambda^* = diag(\gamma_1^*, \cdots, \gamma_q^*, 0) \in \mathbb{R}^{d_v \times d_v}$. Let $P_1 \in \mathbb{R}^{d_v \times d_v}$ be orthogonal consisting of the left singular vectors of U^v. Then $M^{v*} = P_1 \Lambda^* P_1^\top$ is an optimal solution to Eq.(18). In addition, by solving the problem in Eq.(20) we obtain the same optimal solution and objective value as Eq.(18).

Algorithm 1. Convex shared structure learning algorithm for MTMV problem (*CSL-MTMV*)

Input:
$\quad \{y_t\}_{t=1}^{T}, \{X_t\}_{t=1}^{T}, \{P_t\}_{t=1}^{T}, \alpha, \beta, \gamma, h$
Output:
$\quad \{U^v\}_{v=1}^{V}, \{Z^v\}_{v=1}^{V}, \{\Theta^v\}_{v=1}^{V}$
Method:
1: Initialize $\{M^v\}_{v=1}^{V}$ that satisfy the constraints in Eq.(18);
2: **repeat**
3:    **for** $t = 1$ to T **do**
4:       Construct $A_t^v, B_t^{vv'}, C_t^v$ defined in Eq.(14);
5:       Construct $\mathcal{L}_t, \mathcal{R}_t$ defined in Eq.(17);
6:       Compute $\mathcal{W}_t = \mathcal{L}_t^{-1} \mathcal{R}_t$;
7:    **end for**
8:    **for** $v = 1$ to V **do**
9:       Compute the SVD of $U^v = P_1 \Sigma P_2^\top$;
10:      Compute the optimal values of $\{\gamma_i^*\}$ for problem in Eq.(20);
11:      Denote $\Lambda^* = diag(\gamma_1^*, \cdots, \gamma_q^*, 0)$, and compute $M^v = P_1 \Lambda^* P_1^\top$;
12:    **end for**
13: **until** convergence criterion is satisfied.
14: For each v, construct Θ^v using the top h eigenvectors of M^v;
15: Compute $Z^v = \Theta^v U^v$;
16: **return** $\{U^v\}_{v=1}^{V}, \{Z^v\}_{v=1}^{V}, \{\Theta^v\}_{v=1}^{V}$.

4.4   Dealing with Missing-View Data

In the previous sections, we only consider the ideal case that all tasks in a data set have complete data. When incomplete data is involved in the MTMV learning, the problem becomes more challenging. We aim to handle the case of "structured" missing views as described in [25]. That is, if a view is missing from a task, it is missing in all the samples in the task. Of course, partially observed views (i.e. some views are missing only in a part of samples in a task) are more difficult to deal with, which is beyond the scope of this paper.

In our MTMV learning framework, it is easy dealing with structured missing views. Let $V_t \leq V$ denote the real number of views contained in task t and $T_v \leq T$ denote the number of tasks contain view v. When computing $\{U^v\}$ for given $\{M^v\}$, if view v is missing from task t, the variables related to view v in Eq.(16) are all useless, including u_t^v in $\mathcal{W}_t$, C_t^v in $\mathcal{R}_t$, $B_t^{vv'}$ in $\mathcal{L}_t$, and the v-th block row and block column in matrix $\mathcal{L}_t$. After removing these variables, and replace V, T using V_t, T_v in the corresponding equations, a problem with smaller size can be obtained:

$$\mathcal{L}'_t \mathcal{W}'_t = \mathcal{R}'_t \tag{21}$$

When computing $\{M^v\}$ for given $\{U^v\}$, if view v is missing from task t, then in Eq.(18), the t-th column of matrix $\{U^v\}$ (i.e. u_t^v) does not exist. After removing this column and replace V, T using V_t, T_v in the corresponding equations, a similar optimization problem can be obtained.

Furthermore, if for a view v, $T_v = 1$, i.e., there is only one task that contains view v, the algorithm can still be improved. As stated above, the shared structure among multiple tasks is learned based on the relationships of these tasks, if only one task exists for a view, then there is no need to learn the shared low dimensional feature space for this view. Specifically, if only task t contains view v, then the prediction model for this view is as follows:

$$\bar{f}_t^v(x_t^v) = u_t^{v\top} x_t^v \tag{22}$$

In the optimization problem in Eq.(6), for this view, the regularization function $g_t^v(u_t^v, z_t^v, \Theta^v)$ is replaced with $\bar{g}_t^v(u_t^v) = \beta \|u_t^v\|^2$. After some direct derivation, it can be found that the A_t^v in Eq.(17) should have the new form as:

$$\bar{A}_t^v = \frac{2}{n_t} X_t^{v\top} X_t^v + 2\beta I \tag{23}$$

For this view, there is no need to compute M^v or Θ^v in every iteration.

4.5   Complexity Analysis of the Algorithm

To analyze the complexity of *CSL-MTMV* algorithm, we consider the worst case that all the tasks in the problem have features from all the views. In the algorithm, we need to construct $\mathcal{L}_t, \mathcal{W}_t, \mathcal{R}_t$ defined in Eq.(17), compute the inverse of matrix $\mathcal{L}_t$, and compute $\{M^v\}$. It can be found that the speed bottleneck is computation of T inverse of matrices $\mathcal{L}_t$, where the time complexity is $\mathbf{O}(TD^3)$.

The space requirement of the algorithm mainly depends on the size of matrix $\mathcal{L}_t$ with $O(D^2)$. The time complexity of $regMVMT$ algorithm [25] is $O(T^3D^3)$ and space complexity is $O(TD^2 + T(T-1)D)$. It can be easily found that through decoupling different tasks in the computation process, $CSL\text{-}MTMV$ can significantly reduce the time and space complexity.

4.6   Relationship with $regMVMT$ Algorithm

The $regMVMT$ algorithm [25] can be seen as a special case of our algorithm. Specifically, in Eq.(6), we set $\Theta^v = I$, $z_t^v = \frac{1}{T_v}\sum_{i=1}^{T_v} u_t^v$, and do not use the weighting factors $\frac{1}{m_t}$ and $\frac{1}{n_t}$ to compensate for the tasks with different sample numbers. With this setting, our model is transformed into the $regMVMT$ problem definition in Eq.(5) in [25]. Therefore, the problem formulation in this paper is more generalized and flexible, which is able to find good solutions with more chance. In fact, $regMVMT$ requires that the model parameters of all the tasks are similar, which is too rigorous for problems with outlier tasks. In this paper, the common structures between different tasks are learned and different tasks share information using these structures. Compared with other state-of-the-art methods, such as data-manifold methods based on graph structure, our method can learn some underlying predictive functional structures in hypothesis space, which better characterizes a set of good predictors.

5   Experiments

In this section, we conduct the experiments on four real-world data sets to validate the effectiveness of the proposed algorithm $CSL\text{-}MTMV$.

5.1   Data Sets

All the four data sets have multiple tasks with multiple views, and some statistics of them are summarized in Table 1, where N_p and N_n denote the number of positive and negative samples in each task, respectively. The first two data sets are with complete views, and the rest two are with missing views.

- The first one is the NUS-WIDE Object web image database [9] where each image is annotated by objects such as "boat","bird", and etc. We take blockwise color moments as one view and the rest features as the other one. In this data set, we remove the images associated with zero or only one object, and those tasks with too few positive or negative examples. Finally, a two-view data set with 11 tasks are obtained.
- The second one is the Leaves data set [13]. It includes leaves from one hundred plant species that are divided into 32 different genuses, and 16 samples of leaves for each plant species are presented. For each sample, a shape descriptor, fine scale margin and texture histogram are given. By selecting one species from each of the 32 different genuses, 32 tasks with three views are obtained.

- The third one is constructed from 20 Newsgroups[1], which includes 20 categories. 200 documents are randomly selected from each category. For each task, the documents from one category are regarded as positive samples, and from another different category are negative ones. We take the words appearing in all 20 tasks as the common view, and the words existing only in each task as specified view. Finally, we construct 20 tasks with totally 21 views, while each task with 19 views missing. The *tf-idf* weighting scheme is adopted, and the principal component analysis [22] is used to reduce the dimension of features to 300 for each view.
- The last one is NIST Topic Detection and Tracking (TDT2) corpus [10]. In this data set, only the largest 20 categories are selected, and for the categories containing more than 200 documents, we randomly selected 200 documents from each category. The tasks and views are similarly constructed as 20 Newsgroups. We also have 20 tasks with totally 21 views, and each task with 19 views missing.

Table 1. Description of the data sets

Data set	T	V	N_p	N_n	View Missing?
NUS-WIDE Object	11	2	$310 \sim 1220$	$2438 \sim 3348$	No
leaves	32	3	16	496	No
20 Newsgroups	20	21	200	200	Yes
TDT2	20	21	$98 \sim 200$	200	Yes

5.2 Baselines

We compare *CSL-MTMV* with the following baselines, which can handle multi-task problems with multiple views:

• *IteM²*: *IteM²* algorithm [14] is a transductive algorithm, and it can only handle nonnegative feature values. When applying $IteM^2$ algorithm to some of our data sets that have negative feature values, we add a positive constant to the feature values to guarantee its nonnegativity.

• *regMVMT*: *regMVMT* algorithm [25] is an inductive algorithm, which assumes all tasks should be similar to achieve good performance.

5.3 Experiment Setting and Evaluation Metric

Experiment Setting. In each data set, we randomly select n labeled samples and m unlabeled samples for each task as training set. The value of n is set according to the complexity of the learning problem, and m is generally $2 \sim 4$ times of n. We apply five-fold cross validation on the training set to optimize

---

[1] http://people.csail.mit.edu/jrennie/20Newsgroups/

the parameters for the algorithms CSL-$MTMV$ (including α, β and γ.) and reg-$MVMT$ (including λ, μ and γ). The parameters of $IteM^2$ are set the same as their original paper. For CSL-$MTMV$, the number of iteration is set to 20, and number of dimensionality h as $\lfloor (T-1)/5 \rfloor \times 5$ in our experiments.

Evaluation Metric. The F_1 measure is adopted to evaluate all the algorithms, since the *accuracy* measure may be vulnerable to the class unbalance, which just exists in some of our data sets. Let tp, fp and fn denote the numbers of true positive samples, false positive samples and false negative samples, respectively, then $Precision = tp/(tp+fp)$, $Recall = tp/(tp+fn)$.

$$F_1 = \frac{2 \times Precision \times Recall}{Precision + Recall}. \tag{24}$$

Each experiment is repeated 10 times, and each time we randomly select n labeled samples and m unlabeled samples for each task as training set. Finally, the average value of F_1 is recorded.

5.4   Experiment Results

Learning with Complete-View Data. The first two data sets in Table 1 are with complete views.

For NUS-WIDE Object data set, different number of labeled samples are chosen as training set to test the performance of these algorithms, i.e., the number of labeled samples are selected in the range $[100, 700]$ with interval of 100. All the results are shown in Table 2, which can be observed that, the value of F_1 increases with the increasing of the number of labeled samples, and CSL-$MTMV$ achieves the best results under various cases.

For the leaves data set, there are only 16 positive samples for each task, so the number of labeled samples is fixed as 50, among which the number of positive samples is set to $\{1,2,3,4,5,6,7\}$ separately. The experiment results are shown in Table 3, where the first line gives the numbers of positive samples. Similar results

Table 2. Experimental results on NUS-WIDE Object data set

samples #	100	200	300	400	500	600	700
$IteM^2$	0.1539	0.1529	0.1526	0.1534	0.1546	0.1522	0.1512
$regMVMT$	0.3695	0.3822	0.3875	0.3918	0.4036	0.4102	0.4159
CSL-$MTMV$	**0.3930**	**0.4075**	**0.4104**	**0.4178**	**0.4193**	**0.4211**	**0.4263**

Table 3. Experimental results on leaves data set

positive samples #	1	2	3	4	5	6	7
$IteM^2$	0.0289	0.0341	0.0397	0.0390	0.0373	0.0371	0.0392
$regMVMT$	0.0598	0.0981	0.1611	0.2637	0.3573	0.4623	0.5644
CSL-$MTMV$	**0.0802**	**0.1072**	**0.1905**	**0.3017**	**0.4045**	**0.5229**	**0.6128**

can be obtained, i.e., *CSL-MTMV* performs the best under different numbers of labeled positive samples.

Learning with Missing-View Data. In real-world problems, some tasks may not share all the views, so the problems with missing views are also considered. In the experiments, the last two data sets, 20 Newsgroups and TDT2, are with missing views.

Different number of labeled samples are also selected as training set to test the performance of these compared algorithms. The number is sampled in the range [10, 70] with interval of 10, and the results are recorded in Tables 4 and 5. We can observe the similar results as the first two data sets. Again, *CSL-MTMV* gives the best performance.

Table 4. Experimental results on 20 Newsgroups data set

samples #	10	20	30	40	50	60	70
$IteM^2$	0.4880	0.4879	0.4912	0.4776	0.4866	0.5068	0.5247
$regMVMT$	0.8570	0.9144	0.9330	0.9500	0.9566	0.9629	0.9651
CSL-MTMV	**0.8733**	**0.9256**	**0.9406**	**0.9540**	**0.9597**	**0.9652**	**0.9667**

Table 5. Experimental results on TDT2 data set

samples #	10	20	30	40	50	60	70
$IteM^2$	0.4922	0.4897	0.5142	0.5101	0.5159	0.5069	0.5160
$regMVMT$	0.9742	0.9903	0.9930	0.9941	0.9949	0.9947	0.9947
CSL-MTMV	**0.9825**	**0.9936**	**0.9946**	**0.9956**	**0.9957**	**0.9962**	**0.9958**

It is worth mentioning that, we find $IteM^2$ can not perform well on these four data sets. We conjecture there may be two reasons, 1) $IteM^2$ can only handle the data sets with non-negative values of features. 2) $IteM^2$ assumes the test set should have the same proportion of positive samples as the training set, which might also degrade classification performance.

6   Conclusions

To deal with the MTMV problems, a shared structure learning framework called *CSL-MTMV* is proposed in this paper, in which both the shared predictive structure among multiple tasks and prediction consistence among different views within a single task are considered. We also convert the optimization problem to a convex one, and develop an alternating optimization algorithm to solve it. The algorithm can decouple different tasks in the computation process, which significantly reduces the time complexity and space complexity. Moreover, *CSL-MTMV* is a general framework, since the recently proposed algorithm *regMVMT* can be regarded as a special case of ours. The experiments on four real-world data sets demonstrate the effectiveness of the proposed framework.

Acknowledgments. This work is supported by the National Natural Science Foundation of China (No. 61175052, 61203297, 60933004, 61035003), National High-tech R&D Program of China (863 Program) (No. 2013AA01A606, 2012AA011003), and National Program on Key Basic Research Project (973 Program) (No. 2013CB329502).

References

1. Ando, R.K., Zhang, T.: A framework for learning predictive structures from multiple tasks and unlabeled data. Journal of Machine Learning Research 6, 01 (2005)
2. Argyriou, A., Evgeniou, T., Pontil, M.: Multi-task feature learning. In: Advances in Neural Information Processing Systems, Vancouver, BC, Canada, pp. 41–48 (2007)
3. Blum, A., Mitchell, T.: Combining labeled and unlabeled data with co-training. In: Proceedings of the Eleventh Annual Conference on Computational Learning Theory, COLT 1998, pp. 92–100. ACM, New York (1998)
4. Boyd, S., Vandenberghe, L.: Convex optimization. Cambridge University Press (2004)
5. Caruana, R.: Multitask learning. Machine Learning 28(1), 41–75 (1997)
6. Chen, J., Tang, L., Liu, J., Ye, J.: A convex formulation for learning shared structures from multiple tasks. In: Proceedings of the 26th International Conference on Machine Learning, ICML 2009, Montreal, QC, Canada, pp. 137–144 (2009)
7. Chen, J., Zhou, J., Ye, J.: Integrating low-rank and group-sparse structures for robust multi-task learning. In: Proceedings of the ACM SIGKDD International Conference on Knowledge Discovery and Data Mining, San Diego, CA, United States, pp. 42–50 (2011)
8. Chen, N., Zhu, J., Xing, E.P.: Predictive subspace learning for multi-view data: A large margin approach. In: Annual Conference on Neural Information Processing Systems 2010, NIPS 2010, Vancouver, BC, Canada (2010)
9. Chua, T.S., Tang, J., Hong, R., Li, H., Luo, Z., Zheng, Y.: Nus-wide: A real-world web image database from national university of singapore. In: CIVR 2009 - Proceedings of the ACM International Conference on Image and Video Retrieval, Santorini Island, Greece, pp. 368–375 (2009)
10. Cieri, C., Strassel, S., Graff, D., Martey, N., Rennert, K., Liberman, M.: Corpora for topic detection and tracking. In: Allan, J. (ed.) Topic Detection and Tracking, pp. 33–66. Kluwer Academic Publishers, Norwell (2002)
11. Evgeniou, T., Pontil, M.: Regularized multi-task learning. In: Proceedings of the Tenth ACM SIGKDD International Conference on Knowledge Discovery and Data Mining, Seattle, WA, United States, pp. 109–117 (2004)
12. Farquhar, J.D., Hardoon, D.R., Meng, H., Shawe-Taylor, J., Szedmak, S.: Two view learning: Svm-2k, theory and practice. In: Advances in Neural Information Processing Systems, Vancouver, BC, Canada, pp. 355–362 (2005)
13. Frank, A., Asuncion, A.: UCI machine learning repository (2013), http://archive.ics.uci.edu/ml
14. He, J., Lawrence, R.: A graph-based framework for multi-task multi-view learning. In: Proceedings of the 28th International Conference on Machine Learning, ICML 2011, Bellevue, WA, United States, pp. 25–32 (2011)
15. Jalali, A., Ravikumar, P., Sanghavi, S., Ruan, C.: A dirty model for multi-task learning. In: 24th Annual Conference on Neural Information Processing Systems 2010, NIPS 2010, Vancouver, BC, Canada (2010)

16. Muslea, I., Minton, S., Knoblock, C.A.: Active + Semi-supervised Learning = Robust Multi-View Learning. In: International Conference on Machine Learning, pp. 435–442 (2002)
17. Nigam, K., Ghani, R.: Analyzing the effectiveness and applicability of co-training. In: International Conference on Information and Knowledge Management, pp. 86–93 (2000)
18. Sindhwani, V., Niyogi, P., Belkin, M.: A Co-Regularization Approach to Semi-supervised Learning with Multiple Views. In: Workshop on Learning with Multiple Views, International Conference on Machine Learning (2005)
19. Sindhwani, V., Rosenberg, D.S.: An rkhs for multi-view learning and manifold co-regularization. In: Proceedings of the 25th International Conference on Machine Learning, Helsinki, Finland, pp. 976–983 (2008)
20. Skolidis, G., Sanguinetti, G.: Bayesian multitask classification with gaussian process priors. IEEE Transactions on Neural Networks 22(12), 2011–2021 (2011)
21. Vapnik, V.: The nature of statistical learning theory. Springer (1999)
22. Wold, S., Esbensen, K., Geladi, P.: Principal component analysis. Chemometrics and Intelligent Laboratory Systems 2(1), 37–52 (1987)
23. Yu, S., Krishnapuram, B., Rosales, R., Bharat Rao, R.: Bayesian co-training. Journal of Machine Learning Research 12, 2649–2680 (2011)
24. Yu, S., Tresp, V., Yu, K.: Robust multi-task learning with t-processes. In: Twenty-Fourth International Conference on Machine Learning, Corvalis, OR, United States, vol. 227, pp. 1103–1110 (2007)
25. Zhang, J., Huan, J.: Inductive multi-task learning with multiple view data. In: Proceedings of the ACM SIGKDD International Conference on Knowledge Discovery and Data Mining, Beijing, China, pp. 543–551 (2012)

Using Both Latent and Supervised Shared Topics for Multitask Learning

Ayan Acharya[1], Aditya Rawal[2], Raymond J. Mooney[2], and Eduardo R. Hruschka[3]

[1] Department of ECE, University of Texas at Austin, USA
aacharya@utexas.edu
[2] Department of CS, University of Texas at Austin, USA
{aditya,mooney}@cs.utexas.edu
[3] Department of CS, University of São Paulo at São Carlos, Brazil
erh@icmc.usp.br

Abstract. This paper introduces two new frameworks, Doubly Supervised Latent Dirichlet Allocation (DSLDA) and its non-parametric variation (NP-DSLDA), that integrate two different types of supervision: topic labels and category labels. This approach is particularly useful for multitask learning, in which both latent and supervised topics are shared between multiple categories. Experimental results on both document and image classification show that both types of supervision improve the performance of both DSLDA and NP-DSLDA and that sharing both latent *and* supervised topics allows for better multitask learning.

1 Introduction

Humans can distinguish as many as 30,000 relevant object classes [7]. Training an isolated object detector for each of these different classes would require millions of training examples in aggregate. Computer vision researchers have proposed a more efficient learning mechanism in which object categories are learned via *shared* attributes, abstract descriptors of object properties such as "striped" or "has four legs" [17,25,24]. The attributes serve as an intermediate layer in a classifier cascade. The classifier in the first stage is trained to predict the attributes from the raw features and that in the second stage is trained to predict the categories from the attributes. During testing, only the raw features are observed and the attributes must be inferred. This approach is inspired by human perception and learning from high-level object descriptions. For example, from the phrase "eight-sided red traffic sign with white writing", humans can detect stop signs [25]. Similarly, from the description "large gray animals with long trunks", human can identify elephants. If the *shared* attributes transcend object class boundaries, such a classifier cascade is beneficial for *transfer learning* [28] where fewer labeled examples are available for some object categories compared to others [25].

Multitask learning (MTL) is a form of transfer learning in which simultaneously learning multiple related "tasks" allows each one to benefit from the learning of all of the others. If the tasks are related, training one task should provide helpful "inductive bias" for learning the other tasks. To enable the reuse of training information across multiple related tasks, all tasks might utilize the same latent shared intermediate representation – for example, a shared hidden layer in a multi-layer perceptron [11]. In this

H. Blockeel et al. (Eds.): ECML PKDD 2013, Part II, LNAI 8189, pp. 369–384, 2013.

case, the training examples for all tasks provide good estimates of the weights connecting the input layer to the hidden layer, and hence only a small number of examples per task is sufficient to achieve high accuracy. This approach is in contrast to "isolated" training of tasks where each task is learned independently using a separate classifier.

In this paper, our objective is to combine these two approaches to build an MTL framework that can use *both* attributes *and* class labels. The multiple tasks here correspond to different object categories (classes), and *both* observable attributes and latent properties are shared across the tasks. We want to emphasize that the proposed frameworks support general MTL; however, the datasets we use happen to be multiclass, where each class is treated as a separate "task" (as typical in multi-class learning based on binary classifiers). But, in no way are the frameworks restricted to multiclass MTL. Since attribute-based learning has been shown to support effective transfer learning in computer vision, the tasks here naturally correspond to object classes.

The basic building block of the frameworks presented in this paper is Latent Dirichlet Allocation (LDA) [9]. LDA focuses on unsupervised induction of multiple "topics" that help characterize a corpus of text documents. LDA has also been applied in computer vision where SIFT features are appropriately quantized to generate a *bag of visual words* for representing an image [35]. Since our experiments use both text and image data, we will overload the word "document" to denote either a text document or an image represented as a bag of visual words. The LDA approach has also been augmented to include two different types of supervision, document-level labels for either topics [31] or for an overall category inferred from the topics [43]. This paper introduces two new approaches, Doubly Supervised Latent Dirichlet Allocation (DSLDA) and its non-parametric variation (NP-DSLDA), that integrate both forms of supervision. At the topic level, the models assume that supervision is available for some topics during training (corresponding to the "attributes" used in computer vision), but that other topics remain latent (corresponding to the hidden layer in traditional MTL). The ability to provide supervision for *both* categories and a *subset* of topics improves the models' ability to perform accurate classification. In many applications, a variety of kinds of supervision may be naturally available from different sources at multiple levels of abstraction, such as keywords, topics, and categories for documents, or visual attribute, object, and scene labels for images. By effectively utilizing such multiple, interacting levels of supervision, DSLDA is able to learn more accurate predictors. In a supervised LDA [8,43] setting, forcing multiple tasks to share the same set of latent topics results in an LDA-based approach to MTL. By allowing supervision to also be provided for a subset of these shared topics, DSLDA and NP-DSLDA support a particularly effective form of MTL.

The rest of the paper is organized as follows. We present related literature in Section 2, followed by the descriptions of DSLDA and NP-DSLDA in Section 3 and Section 4 respectively. Experimental results on both multi-class image and document categorization are presented in Section 5, demonstrating the value of integrating both supervised and latent shared topics in diverse applications. Finally, future directions and conclusions are presented in Section 6.

Note on Notation: Vectors and matrices are denoted by bold-faced lowercase and capital letters, respectively. Scalar variables are written in italic font, and sets are denoted by calligraphic uppercase letters. Dir(), Beta() and multinomial() stand for Dirichlet, Beta and multinomial distribution respectively.

2 Related Work

2.1 Statistical Topic Models

LDA [9] treats documents as a mixture of topics, which in turn are defined by a distribution over a set of words. The words in a document are assumed to be sampled from multiple topics. In its original formulation, LDA can be viewed as a purely-unsupervised form of dimensionality reduction and clustering of documents in the topic space, although several extensions of LDA have subsequently incorporated some sort of supervision. Some approaches provide supervision by labeling each document with its set of topics [31,32]. In particular, in *Labeled LDA* (LLDA [31]), the primary objective is to build a model of the words that indicate the presence of certain topic labels. For example, when a user explores a webpage based on certain tags, LLDA can be used to highlight interesting portions of the page or build a summary of the text from multiple webpages that share the same set of tags. The words in a given training document are assumed to be sampled *only* from the supervised topics, which the document has been labeled as covering.

Some other researchers [8,43,12] assume that supervision is provided for a single *response variable* to be predicted for a given document. The response variable might be real-valued or categorical, and modeled by a normal, Poisson, Bernoulli, multinomial or other distribution (see [12] for details). Some examples of documents with response variables are essays with their grades, movie reviews with their numerical ratings, web pages with their number of hits over a certain period of time, and documents with category labels. In *Maximum Entropy Discriminative LDA* (MedLDA) [43], the objective is to infer some low-dimensional (topic-based) representation of documents which is predictive of the response variable. Essentially, MedLDA solves two problems jointly – dimensionality reduction and max-margin classification using the features in the dimensionally-reduced space. Compared to earlier versions of supervised topic models [8,12], MedLDA has simpler update equations and produces superior experimental results. Therefore, in the frameworks presented in Sections 3.2 and 4, the max-margin principle adopted in MedLDA is preferred over other supervised topic models.

2.2 Transfer and Multitask Learning

Transfer learning allows the learning of some tasks to benefit the learning of others through either simultaneous [11] or sequential [10] training. In multitask learning (MTL [11]), a single model is simultaneously trained to perform multiple related tasks. MTL has emerged as a very promising research direction for various applications including biomedical informatics [6], marketing [15], natural language processing [2], and computer vision [34].

Many different MTL approaches have been proposed over the past 15 years (*e.g.*, see [38,28,29] and references therein). These include different learning methods, such as empirical risk minimization using group-sparse regularizers [20,23,21], hierarchical Bayesian models [41,26] and hidden conditional random fields [30]. Evgeniou *et al.* [14] proposed the regularized MTL which constrained the models of all tasks to be close to each other. The task relatedness in MTL has also been modeled by constraining multiple tasks to share a common underlying structure [5,3,11]. Ando and Zhang [1] proposed a structural learning formulation, which assumed multiple predictors for different tasks shared a common structure on the underlying predictor space.

In all of the MTL formulations mentioned above, the basic assumption is that all tasks are related. In practical applications, these might not be the case and the tasks might exhibit a more sophisticated group structure. Such structure is handled using clustered multi-task learning (CMTL). In [4] CMTL is implemented by considering a mixture of Gaussians instead of single Gaussian priors. Xue *et al.* [39] introduced the Dirichlet process prior that automatically identifies subgroups of related tasks. In [19], a clustered MTL framework was proposed that simultaneously identified clusters and performed multi-task inference.

In the models presented in the next two sections, an LDA-based approach to MTL is easily obtained by maintaining a common set of topics to support the prediction of multiple response variables. This idea is analogous to implementing MTL using a common shared underlying structure [5,3,11]. We will also explain how NP-DSLDA is capable of performing CMTL.

3   Doubly Supervised LDA (DSLDA)

3.1   Task Definition

Assume we are given a training corpus consisting of N documents belonging to Y different classes (where each document belongs to exactly one class and each class corresponds to a different task). Further assume that each of these training documents is also annotated with a set of K_2 different topic "tags" (henceforth referred to as "supervised topics"). For computer vision data, the supervised topics correspond to the attributes provided by human experts. The objective is to train a model using the words in a data, as well as the associated supervised topic tags and class labels, and then use this model to classify completely unlabeled test data for which no topic tags nor class labels are provided. The human-provided supervised topics are presumed to provide abstract information that is helpful in predicting the class labels of test documents.

3.2   Generative Model

In order to include both types of supervision (class and topic labels), a combination of the approaches described in Section 2.1 is proposed. Note that LLDA uses *only* supervised topics and does not have any mechanism for generating class labels. On the other hand, MedLDA has only *latent* topics but learns a discriminative model for predicting classes from these topics. To the best of our knowledge, ours is the first LDA approach to integrate both types of supervision in a single framework. The generative process of DSLDA is described below.

For the nth document, sample a topic selection probability vector $\theta_n \sim \text{Dir}(\alpha_n)$, where $\alpha_n = \Lambda_n \alpha$ and α is the parameter of a Dirichlet distribution of dimension K, which is the total number of topics. The topics are assumed to be of two types – latent and supervised, and there are K_1 latent topics and K_2 supervised topics. Therefore, $K = K_1 + K_2$. Latent topics are never observed, while supervised topics are observed in training but not in test data. Henceforth, in each vector or matrix with K components, it is assumed that the first K_1 components correspond to the latent topics and the next K_2 components to the supervised topics. Λ_n is a diagonal binary matrix of dimension $K \times K$. The kth diagonal entry is unity if *either* $1 \leq k \leq K_1$ *or* $K_1 < k \leq K$ and the nth document is tagged with the kth topic. Also, $\alpha = (\alpha_1, \alpha_2)$ where α_1 is a parameter of a Dirichlet distribution of dimension K_1 and α_2 is a parameter of a Dirichlet distribution of dimension K_2.

For the mth word in the nth document, sample a topic $z_{nm} \sim \text{multinomial}(\theta'_n)$, where $\theta'_n = (1 - \epsilon)\{\theta_{nk}\}_{k=1}^{k_1} \epsilon\{\Lambda_{n,kk}\theta_{nk}\}_{k=1+k_1}^{K}$. This implies that the supervised topics are weighted by ϵ and the latent topics are weighted by $(1 - \epsilon)$. Sample the word $w_{nm} \sim \text{multinomial}(\beta_{z_{nm}})$, where β_k is a multinomial distribution over the vocabulary of words corresponding to the kth topic.

For the nth document, generate $Y_n = \arg\max_y r_y^T \mathbb{E}(\bar{z}_n)$ where Y_n is the class label associated with the nth document, $\bar{z}_n = \sum_{m=1}^{M_n} z_{nm}/M_n$. Here, z_{nm} is an indicator vector of dimension K. r_y is a K-dimensional real vector corresponding to the yth class, and it is assumed to have a prior distribution $\mathcal{N}(0, 1/C)$. M_n is the number of words in the nth document. The maximization problem to generate Y_n (or the classification problem) is carried out using a max-margin principle.

Note that predicting each class is effectively treated as a separate task, and that the shared topics are useful for generalizing the performance of the model across classes. In particular, when all classes have few training examples, knowledge transfer between classes can occur through the shared topics. So, the mapping from the original feature space to the topic space is effectively learned using examples from all classes, and a few examples from each class are sufficient to learn the mapping from the reduced topic space to the class labels.

3.3 Inference and Learning

Let us denote the hidden variables by $Z = \{\{z_{nm}\}, \{\theta_n\}\}$, the observed variables by $X = \{w_{nm}\}$ and the model parameters by κ_0. The joint distribution of the hidden and observed variables is:

$$p(\mathbf{X}, \mathbf{Z}|\kappa_0) = \prod_{n=1}^{N} p(\theta_n|\alpha_n) \prod_{m=1}^{M_n} p(z_{nm}|\theta'_n)p(w_{nm}|\beta_{z_{nm}}) \tag{1}$$

To avoid computational intractability, inference and estimation are performed using Variational **EM**. The factorized approximation to the posterior distribution on hidden variables $\mathbf{Z}$ is given by:

$$q(\mathbf{Z}|\{\kappa_n\}_{n=1}^{N}) = \prod_{n=1}^{N} q(\theta_n|\gamma_n) \prod_{m=1}^{M_n} q(z_{nm}|\phi_{nm}), \tag{2}$$

where $\theta_n \sim \text{Dir}(\gamma_n)$ $\forall n \in \{1, 2, \cdots, N\}$, $z_{nm} \sim$ multinomial(ϕ_{nm}) $\forall n \in \{1, 2, \cdots, N\}$ and $\forall m \in \{1, 2, \cdots, M_n\}$, and $\kappa_n = \{\gamma_n, \{\phi_{nm}\}\}$, which is the set of variational parameters corresponding to the nth instance. Further, $\gamma_n = (\gamma_{nk})_{k=1}^K$ $\forall n$, and $\phi_{nm} = (\phi_{nmk})_{k=1}^K$ $\forall n, m$. With the use of the lower bound obtained by the factorized approximation, followed by Jensen's inequality, DSLDA reduces to solving the following optimization problem[1]:

$$\min_{q, \kappa_0, \{\xi_n\}} \frac{1}{2}\|r\|^2 - \mathcal{L}(q(\boldsymbol{Z})) + C \sum_{n=1}^N \xi_n,$$

$$\text{s.t. } \forall n, y \neq Y_n : \mathbb{E}[r^T \Delta f_n(y)] \geq 1 - \xi_n; \xi_n \geq 0. \tag{3}$$

Here, $\Delta f_n(y) = f(Y_n, \bar{z}_n) - f(y, \bar{z}_n)$ and $\{\xi_n\}_{n=1}^N$ are the slack variables, and $f(y, \bar{z}_n)$ is a feature vector whose components from $(y-1)K + 1$ to yK are those of the vector $\bar{z}_n$ and all the others are 0. $\mathbb{E}[r^T \Delta f_n(y)]$ is the "expected margin" over which the true label Y_n is preferred over a prediction y. From this viewpoint, DSLDA projects the documents onto a combined topic space and then uses a max-margin approach to predict the class label. The parameter C penalizes the margin violation of the training data.

$$\phi_{nmk}^* \propto \Lambda_{n,kk} \exp\left[\psi(\gamma_{nk}) + \log(\beta_{kw_{nm}}) + \log(\epsilon')\right. \tag{4}$$

$$\left. +1/M_n \sum_{y \neq Y_n} \mu_n(y)\mathbb{E}[r_{Y_nk} - r_{yk}]\right] \quad \forall n, m, k.$$

$$\gamma_{nk}^* = \Lambda_{n,kk}\left[\alpha_k + \sum_{m=1}^{M_n} \phi_{nmk}\right] \quad \forall n, vk. \tag{5}$$

$$\beta_{kv}^* \propto \sum_{n=1}^N \sum_{m=1}^{M_n} \phi_{nmk}\mathbb{I}_{\{w_{nm}=v\}} \quad \forall k, v. \tag{6}$$

$$\mathcal{L}_{[\alpha_1/\alpha_2]} = \left[\sum_{n=1}^N \log(\Gamma(\sum_{k=1}^K \alpha_{nk})) - \sum_{n=1}^N \sum_{k=1}^K \log(\Gamma(\alpha_{nk}))\right] \tag{7}$$

$$+ \sum_{n=1}^N \sum_{k=1}^K \left[\psi(\gamma_{nk}) - \psi(\sum_{k=1}^K \gamma_{nk})\right](\alpha_{nk} - 1).$$

Let $\mathcal{Q}$ be the set of all distributions having a fully factorized form as given in (2). Let the distribution q^* from the set $\mathcal{Q}$ optimize the objective in Eq. (3). The optimal values of corresponding variational parameters are given in Eqs. (4) and (5). In Eq. (4), $\epsilon' = (1 - \epsilon)$ if $k \leq K_1$ and $\epsilon' = \epsilon$ otherwise. Since ϕ_{nm} is a multinomial distribution, the updated values of the K components should be normalized to unity. The optimal values of ϕ_{nm} depend on γ_n and vice-versa. Therefore, iterative optimization is adopted to maximize the lower bound until convergence is achieved.

---

[1] Please see [43] for further details.

During testing, one does not observe a document's supervised topics and, in principle, has to explore 2^{K_2} possible combinations of supervised tags – an expensive process. A simple approximate solution, as employed in LLDA [31], is to assume the absence of the variables $\{\Lambda_n\}$ altogether in the test phase, and just treat the problem as inference in MedLDA with K latent topics. One can then threshold over the last K_2 topics if the tags of a test document need to be inferred. Equivalently, one can also assume Λ_n to be an identity matrix of dimension $K \times K$ $\forall n$. This representation ensures that the expressions for update equations (4) and (5) do not change in the test phase.

In the M step, the objective in Eq. (3) is maximized w.r.t κ_0. The optimal value of β_{kv} is given in Eq. (6). Since β_k is a multinomial distribution, the updated values of the V components should be normalized. However, numerical methods for optimization are required to update α_1 or α_2. The part of the objective function that depends on α_1 and α_2 is given in Eq. (7). The update for the parameter r is carried out using a multi-class SVM solver [16]. With all other model and variational parameters held fixed (*i.e.* with $\mathcal{L}(q)$) held constant), the objective in Eq. (3) is optimized w.r.t. r. A reader familiar with the updates in unsupervised LDA can see the subtle (but non-trivial) changes in the update equations for DSLDA.

4   Non-parametric DSLDA

We now propose a non-parametric extension of DSLDA (NP-DSLDA) that solves the model selection problem and automatically determines the best number of latent topics for modeling the given data. A modified stick breaking construction of Hierarchical Dirichlet Process (HDP) [33], recently introduced in [36] is used here which makes variational inference feasible. The idea in such representation is to share the corpus level atoms across documents by sampling atoms with replacement for each document and modifying the weights of these samples according to some other GEM distribution [33] whose parameter does not depend on the weights of the corpus-level atoms.

The combination of an infinite number of latent topics with a finite number of supervised topics in a single framework is not trivial and ours is the first model to accomplish this. One simpler solution is to introduce one extra binary hidden variable for each word in each document which could select either the set of latent topics or the set of supervised topics. Subsequently, a word in a document can be sampled from either the supervised or the latent topics based on the value sampled by the hidden "switching" variable. However, the introduction of such extra hidden variables adversely affects model performance as explained in [13]. In NP-DSLDA, we are able to avoid such extra hidden variables by careful modeling of the HDP. This will be evident in the generative process of NP-DSLDA presented below:

- Sample $\phi_{k_1} \sim \text{Dir}(\eta_1) \forall k_1 \in \{1, 2, \cdots, \infty\}$ and $\phi_{k_2} \sim \text{Dir}(\eta_2)$ $\forall k_2 \in \{1, 2, \cdots, K_2\}$. η_1, η_2 are the parameters of Dirichlet distribution of dimension V.
- Sample $\beta'_{k_1} \sim \text{Beta}(1, \delta_0)$ $\forall k_1 \in \{1, 2, \cdots, \infty\}$.
- For the nth document, sample $\pi_n^{(2)} \sim \text{Dir}(\Lambda_n \alpha_2)$. α_2 is the parameter of Dirichlet of dimension K_2. Λ_n is a diagonal binary matrix of dimension $K_2 \times K_2$. The kth diagonal entry is unity if the nth word is tagged with the kth supervised topic.

- $\forall n, \forall t \in \{1, 2, \cdots, \infty\}$, sample $\pi'_{nt} \sim \text{Beta}(1, \alpha_0)$. Assume $\boldsymbol{\pi}_n^{(1)} = (\pi_{nt})_t$ where $\pi_{nt} = \pi'_{nt} \prod_{l<t}(1 - \pi'_{nl})$.
- $\forall n, \forall t$, sample $c_{nt} \sim \text{multinomial}(\boldsymbol{\beta})$ where $\beta_{k_1} = \beta'_{k_1} \prod_{l<k_1}(1 - \beta'_l)$. $\boldsymbol{\pi}_n^{(1)}$ represents the probability of selecting the sampled atoms in c_n. Due to sampling with replacement, c_n can contain multiple atoms of the same index from the corpus level DP.
- For the mth word in the nth document, sample $z_{nm} \sim \text{multinomial}((1 - \epsilon)\boldsymbol{\pi}_n^{(1)},$ $\epsilon \boldsymbol{\pi}_n^{(2)})$. This implies that w.p. ϵ, a topic is selected from the set of supervised topics and w.p. $(1 - \epsilon)$, a topic is chosen from the set of (infinite number of) unsupervised topics. Note that by weighting the $\boldsymbol{\pi}$'s appropriately, the need for additional hidden "switching" variable is avoided.
- Sample w_{nm} from a multinomial given by the following equation:

$$\prod_{k_1=1}^{\infty} \prod_{v=1}^{V} \phi_{k_1 v}^{\mathbb{I}_{\{w_{nm}=v\}} \mathbb{I}_{\{c_{nz_{nm}}=k_1 \in \{1, \cdots, \infty\}\}}} \prod_{k_2=1}^{K_2} \prod_{v=1}^{V} \phi_{k_2 v}^{\mathbb{I}_{\{w_{nm}=v\}} \mathbb{I}_{\{z_{nm}=k_2 \in \{1, \cdots, K_2\}\}}} (8)$$

The joint distribution of NP-DSLDA is given as follows:

$$p(\mathbf{X}, \mathbf{Z}|\boldsymbol{\kappa}_0) = \prod_{k_1=1}^{\infty} p(\phi_{k_1}|\boldsymbol{\eta}_1) p(\beta'_{k_1}|\boldsymbol{\delta}_0) \prod_{k_2=1}^{K_2} p(\phi_{k_2}|\boldsymbol{\eta}_2) \prod_{n=1}^{N} p(\boldsymbol{\pi}_n^{(2)}|\alpha_2) \quad (9)$$

$$\prod_{t=1}^{\infty} p(\pi_{nt}'^{(1)}|\alpha_0) p(c_{nt}|\beta') \prod_{m=1}^{M_n} p(z_{nm}|\boldsymbol{\pi}_n^{(1)}, \boldsymbol{\pi}_n^{(2)}, \epsilon) p(w_{nm}|\phi, c_{nz_{nm}}, z_{nm}).$$

As an approximation to the posterior distribution over the hidden variables, we use the following factorized distribution:

$$q(\mathbf{Z}|\boldsymbol{\kappa}) = \prod_{k_1=1}^{\overline{K_1}} q(\phi_{k_1}|\lambda_{k_1}) \prod_{k_2=1}^{K_2} q(\phi_{k_2}|\lambda_{k_2}) \prod_{k_1=1}^{\overline{K_1-1}} q(\beta'_{k_1}|u_{k_1}, v_{k_1}) \quad (10)$$

$$\prod_{n=1}^{N} q(\boldsymbol{\pi}_n^{(2)}|\boldsymbol{\gamma}_n) \prod_{t=1}^{T-1} q(\pi_{nt}'^{(1)}|a_{nt}, b_{nt}) \prod_{t=1}^{T} q(c_{nt}|\varphi_{nt}) \prod_{m=1}^{M_n} q(z_{nm}|\zeta_{nm}).$$

Here, $\boldsymbol{\kappa}_0$ and $\boldsymbol{\kappa}$ denote the sets of model and variational parameters, respectively. $\overline{K_1}$ is the truncation limit of the corpus-level Dirichlet Process and T is the truncation limit of the document-level Dirichlet Process. $\{\lambda_k\}$ are the parameters of Dirichlet each of dimension V. $\{u_{k_1}, v_{k_1}\}$ and $\{a_{nt}, b_{nt}\}$ are the parameters of variational Beta distribution corresponding to corpus level and document level sticks respectively. $\{\varphi_{nt}\}$ are multinomial parameters of dimension $\overline{K_1}$ and $\{\zeta_{nm}\}$ are multinomials of dimension $(T + K_2)$. $\{\boldsymbol{\gamma}_n\}_n$ are parameters of Dirichlet distribution of dimension K_2.

The underlying optimization problem takes the same form as in Eq. (3). The only difference lies in the calculation of $\Delta f_n(y) = f(Y_n, \bar{\mathbf{s}}_n) - f(y, \bar{\mathbf{s}}_n)$. The first set of dimensions of $\bar{\mathbf{s}}_n$ (corresponding to the unsupervised topics) is given by $1/M_n \sum_{m=1}^{M_n} c_{nz_{nm}}$, where c_{nt} is an indicator vector over the set of unsupervised topics. The following K_2 dimensions (corresponding to the supervised topics) are given by $1/M_n \sum_{m=1}^{M_n} z_{nm}$. After the variational approximation with $\overline{K_1}$ number of corpus level sticks, $\bar{\mathbf{s}}_n$ turns out to be of dimension $(\overline{K_1} + K_2)$ and the feature vector $f(y, \bar{s}_n)$ constitutes $Y(\overline{K_1} + K_2)$ elements. The components of $f(y, \bar{s}_n)$ from $(y-1)(\overline{K_1} + K_2) + 1$ to $y(\overline{K_1} + K_2)$ are those of the vector $\bar{s}_n$ and all the others are 0. Essentially, due to the variational approximation, NP-DSLDA projects each document on to a combined topic space of dimension $(\overline{K_1} + K_2)$ and learns the mapping from this space to the classes.

$$\zeta^*_{nmt} \propto \exp\left[[\psi(a_{nt}) - \psi(a_{nt} + b_{nt})]\mathbb{I}_{\{t<T\}} + \sum_{t'=1}^{t-1}[\psi(b_{nt'}) - \psi(a_{nt'} + b_{nt'})]\right. \quad (11)$$

$$+ \sum_{k_1=1}^{\overline{K_1}} \varphi_{ntk_1}\left[\psi(\lambda_{k_1 w_{nm}}) - \psi(\sum_{v=1}^{V}\lambda_{k_1 v})\right]$$

$$\left. + \sum_{y \neq Y_n} \mu_n(y) \sum_{k_1=1}^{\overline{K_1}} \mathbb{E}[r_{Y_n k_1} - r_{y k_1}]\varphi_{ntk_1}\right] \quad \forall n, m, t.$$

$$\zeta^*_{nm(T+k_2)} \propto \Lambda_{nk_2 k_2}\exp\left[\psi(\gamma_{nk_2}) - \psi(\sum_{k_2=1}^{K_2}\gamma_{nk_2}) + \psi(\lambda_{\overline{(K_1+k_2)}w_{nm}})\right. \quad (12)$$

$$\left. -\psi(\sum_{v=1}^{V}\lambda_{\overline{(K_1+k_2)}v}) + 1/M_n \sum_{y \neq Y_n} \mu_n(y)\mathbb{E}[r_{Y_n \overline{(K_1+k_2)}} - r_{y \overline{(K_1+k_2)}}]\right] \quad \forall n, m, k_2.$$

$$\varphi^*_{ntk_1} \propto \exp\left[[\psi(u_{k_1}) - \psi(u_{k_1} + v_{k_1})]\mathbb{I}_{\{k_1 < K_1\}}\right. \quad (13)$$

$$+ \sum_{k'=1}^{k_1-1}[\psi(v_{k'}) - \psi(u_{k'} + v_{k'})] + \sum_{m=1}^{M_n}\zeta_{nmt}\left[\psi(\lambda_{k_1 w_{nm}}) - \psi(\sum_{v=1}^{V}\lambda_{k_1 v})\right]$$

$$\left. + 1/M_n \sum_{y \neq Y_n} \mu_n(y)\mathbb{E}[r_{Y_n k_1} - r_{y k_1}]\left(\sum_{m=1}^{M_n}\zeta_{nmt}\right)\right] \quad \forall n, t, k_1.$$

Some of the update equations of NP-DSLDA are given in the above equations, where $\{\varphi_{ntk_1}\}$ are the set of variational parameters that characterize the assignment of the documents to the global set of $(\overline{K_1} + K_2)$ topics. One can see how the effect of the class labels is included in the update equation of $\{\varphi_{ntk_1}\}$ via the average value of the parameters $\{\zeta_{nmt}\}$. This follows intuitively from the generative assumption. update exists for the model parameters and hence numerical optimization has to be used. Other updates are either similar to DSLDA or the model in [36] and are omitted due to space constraints. $\{\zeta_{nm}\}$, corresponding to supervised and unsupervised topics, should be individually normalized and then scaled by ϵ and $(1 - \epsilon)$ respectively. Otherwise, the effect of the Dirichlet prior on supervised topics will get compared to that of the GEM prior on the unsupervised topics which does not follow the generative assumptions. The variational parameters $\{\lambda_k\}$ and $\{\varphi_{nt}\}$ are also normalized.

Note that NP-DSLDA offers some flexibility with respect to the latent topics that could be dominant for a specific task. One could therefore postulate that NP-DSLDA can learn the clustering of tasks from the data itself by making a subset of latent topics to be dominant for a set of tasks. Although do not have supporting experiments, NP-DSLDA is capable in principle of performing clustered multi-task learning without any prior assumption on the relatedness of the tasks.

5   Experimental Evaluation

5.1   Data Description

Our evaluation used two datasets, a text corpus and a multi-class image database, as described below.

aYahoo Data. The first set of experiments was conducted with the aYahoo image dataset from [17] which has 12 classes – carriage, centaur, bag, building, donkey, goat, jetski, monkey, mug, statue, wolf, and zebra.[2] Each image is annotated with relevant visual attributes such as "has head", "has wheel", "has torso" and 61 others, which we use as the supervised topics. Using such intermediate "attributes" to aid visual classification has become a popular approach in computer vision [25,24]. After extracting SIFT features [27] from the raw images, quantization into 250 clusters is performed, defining the vocabulary for the bag of visual words. Images with less than two attributes were discarded. The resulting dataset of size 2,275 was equally split into training and test data.

ACM Conference Data. The text corpus consists of conference paper abstracts from two groups of conferences. The first group has four conferences related to data mining – WWW, SIGIR, KDD, and ICML, and the second group consists of two VLSI conferences – ISPD and DAC. The classification task is to determine the conference at which the abstract was published. As supervised topics, we use keywords provided by the authors, which are presumably useful in determining the conference venue. Since authors usually take great care in choosing keywords so that their paper is retrieved by relevant searches, we believed that such keywords made a good choice of supervised topics. Part of the data, crawled from ACM's website, was used in [37]. A total of 2,300 abstracts were collected each of which had at least three keywords and an average of 78 (± 33.5) words. After stop-word removal, the vocabulary size for the assembled data is 13,412 words. The final number of supervised topics, after some standard pre-processing of keywords, is 55. The resulting dataset was equally split into training and test data.

5.2   Methodology

In order to demonstrate the contribution of each aspect of the overall model, DSLDA and NP-DSLDA are compared against the following simplified models:

---

[2] http://vision.cs.uiuc.edu/attributes/

- MedLDA with one-vs-all classification (MedLDA-OVA): A separate model is trained for each class using a one-vs-all approach leaving no possibility of transfer across classes.
- MedLDA with multitask learning (MedLDA-MTL): A single model is learned for all classes where the latent topics are shared across classes.
- DSLDA with only shared supervised topics (DSLDA-OSST): A model in which supervised topics are used and shared across classes but there are no latent topics.
- DSLDA with no shared latent topics (DSLDA-NSLT): A model in which only supervised topics are shared across classes and a separate set of latent topics is maintained for each class.
- Majority class method (MCM): A simple baseline which always picks the most common class in the training data.

These baselines are useful for demonstrating the utility of *both* supervised and latent shared topics for multitask learning in DSLDA. MedLDA-OVA is a non-transfer method, where a separate model is learned for each of the classes, *i.e.* one of the many classes is considered as the positive class and the union of the remaining ones is treated as the negative class. Since the models for each class are trained separately, there is no possibility of sharing inductive information across classes. MedLDA-MTL trains on examples from all classes simultaneously, and thus allows for sharing of inductive information *only* through a common set of latent topics. In DSLDA-OSST, only supervised topics are maintained and knowledge transfer can *only* take place via these supervised topics. DSLDA-NSLT uses shared supervised topics but also includes latent topics which are *not* shared across classes. This model provides for transfer *only* through shared supervised topics but provides extra modeling capacity compared to DSLDA-OSST through the use of latent topics that are not shared. DSLDA and NP-DSLDA are MTL frameworks where both supervised *and* latent topics are shared across all classes. Note that, all of the baselines can be implemented using DSLDA with a proper choice of Λ and ϵ. For example, DSLDA-OSST is just a special case of DSLDA with ϵ fixed at 1.

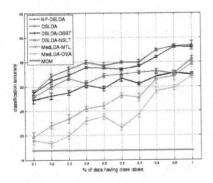

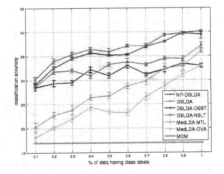

Fig. 1. $p_1 = 0.5$ (aYahoo)   **Fig. 2.** $p_1 = 0.7$ (aYahoo)

Table 1. Illustration of Latent and Supervised Topics

LT1	function, label, graph, classification, database, propagation, algorithm, accuracy, minimization, transduction
LT2	performance, design, processor, layer, technology, device, bandwidth, architecture, stack, system
CAD	design, optimization, mapping, pin, simulation, cache, programming, routing, biochip, electrode
VLSI	design, physical, lithography, optimization, interdependence, global, robust, cells, layout, growth
IR	algorithm, web, linear, query, precision, document, repair, site, search, semantics
Ranking	integration, catalog, hierarchical, dragpushing, structure, source, sequence, alignment, transfer, flattened, speedup
Learning	model, information, trajectory, bandit, mixture, autonomous, hierarchical, feedback, supervised, task

In order to explore the effect of different amounts of both types of supervision, we varied the amount of both topic-level and class-level supervision. Specifically, we provided topic supervision for a fraction, p_1, of the overall training set, and then provided class supervision for only a further fraction p_2 of this data. Therefore, only $p_1 * p_2$ of the overall training data has class supervision. By varying the number of latent topics from 20 to 200 in steps of 10, we found that $K_1 = 100$ generally worked the best for all the parametric models. Therefore, we show parametric results for 100 latent topics. For each combination of (p_1, p_2), 50 random trials were performed with $C = 10$. To maintain equal representational capacity, the total number of topics K is held the same across all parametric models (except for DSLDA-OSST where the total number of topics is K_2). For NP-DSLDA, following the suggestion of [36], we set $\overline{K}_1 = 150$ and $T = 40$, which produced uniformly good results. When required, ϵ was chosen using 5-fold internal cross-validation using the training data.

5.3 Results

Figs. 1 and 2 present representative learning curves for the image data, showing how classification accuracy improves as the amount of class supervision (p_2) is increased. Results are shown for two different amounts of topic supervision ($p_1 = 0.5$ and $p_1 = 0.7$). Figs. 3 and 4 present similar learning curves for the text data. The error bars in the curves show standard deviations across the 50 trials.

The results demonstrate that DSLDA and NP-DSLDA quite consistently outperform all of the baselines, clearly demonstrating the advantage of combining both types of

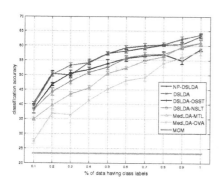

Fig. 3. $p_1 = 0.5$ (Conference)

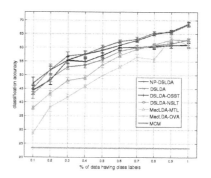

Fig. 4. $p_1 = 0.7$ (Conference)

topics. NP-DSLDA performs about as well as DSLDA, for which the optimal number of latent topics has been chosen using an expensive model-selection search. This demonstrates that NP-DSLDA is doing a good job of automatically selecting an appropriate number of latent topics.

Overall, DSLDA-OSST and MedLDA-MTL perform about the same, showing that, individually, both latent and supervised shared topics each support multitask learning about equally well when used alone. However, combining both types of topics provides a clear improvement.

MedLDA-OVA performs quite poorly when there is only a small amount of class supervision (note that this baseline uses *only* class labels). However, the performance approaches the others as the amount of class supervision increases. This is consistent with the intuition that multitask learning is most beneficial when each task has limited supervision and therefore has more to gain by sharing information with other tasks.

Shared supervised topics clearly increase classification accuracy when class supervision is limited (i.e. small values of p_2), as shown by the performance of both DSLDA-NSLT and DSLDA-OSST. When $p_2 = 1$ (equal amounts of topic and class supervision), DSLDA-OSST, MedLDA-MTL and MedLDA-OVA all perform similarly; however, by exploiting *both* types of supervision, DSLDA and NP-DSLDA still maintain a performance advantage.

5.4 Topic Illustration

In Table 1, we show the most indicative words for several topics discovered by DSLDA from the text data (with $p_1 = 0.8$ and $p_2 = 1$). LT1 and LT2 correspond to the most frequent latent topics assigned to documents in the two broad categories of conferences (data mining and VLSI, respectively). The other five topics are supervised ones. CAD and IR stand for Computer Aided Design and Information Retrieval respectively. The illustrated topics are particularly discriminative when classifying documents.

5.5 Discussion

DSLDA-NSLT only allows sharing of supervised topics and its implementation is not straightforward. Since MedLDA-OVA, MedLDA-MTL and DSLDA use K topics (latent or a combination of supervised and latent), to make the comparison fair, it is necessary to maintain the same number of topics for DSLDA-NSLT. This ensures that the models compared have the same representational capacity. Therefore, for each class in DSLDA-NSLT, k_2/Y latent topics are maintained. While training DSLDA-NSLT with examples from the yth class, only a subset of the first k_1 topics (or a subset of the supervised ones based on which of them are present in the training documents) and the next $\left(\frac{(y-1)k_2}{Y} + 1\right)^{\text{th}}$ to $\left(\frac{yk_2}{Y}\right)^{\text{th}}$ topics are considered to be "active" among the latent topics. The other latent topics are assumed to have zero contribution, implying that the parameters associated with these topics are not updated based on observations of documents belonging to class y. During testing, however, one needs to project a document onto the entire K-dimensional space, and the class label is predicted based on this representation and the parameters r.

Overall, the results support the hypothesis that DSLDA's ability to incorporate both supervised and latent topics allow it to achieve better predictive performance compared to baselines that exploit only one, the other, or neither. Furthermore, NP-DSLDA is able to automate model-selection, performing nearly as well as DSLDA with optimally chosen parameters.

6   Future Work and Conclusion

This paper has introduced Doubly Supervised LDA (DSLDA) and non-parametric DSLDA (NP-DSLDA), novel approaches that combine the following – generative and discriminative models, latent and supervised topics, and class and topic level supervision, in a principled probabilistic manner. Four ablations of this model are also evaluated in order to understand the individual effects of latent/supervised topics and multitask learning on the overall model performance. The general idea of "double supervision" could be applied to many other models, for example, in multi-layer perceptrons, latent SVMs [40] or in deep belief networks [18]. In MTL, sharing tasks blindly is not always a good approach and further extension with clustered MTL [42] is possible. Based on a very recent study [22], a sampling based algorithm could also be developed for NP-DSLDA, possibly leading to even better performance.

Acknowledgments. This research was partially supported by ONR ATL Grant N00014-11-1-0105, NSF Grants (IIS-0713142 and IIS-1016614) and by the Brazilian Research Agencies FAPESP and CNPq.

References

1. Ando, R., Zhang, T.: A framework for learning predictive structures from multiple tasks and unlabeled data. Journal of Machine Learning Research 6, 1817–1853 (2005)
2. Ando, R.K.: Applying alternating structure optimization to word sense disambiguation. In: Proceedings of Computational Natural Language Learning (2006)
3. Argyriou, A., Micchelli, C.A., Pontil, M., Ying, Y.: A spectral regularization framework for multi-task structure learning. In: Proceedings of Neural Information Processing Systsems (2007)
4. Bakker, B., Heskes, T.: Task clustering and gating for Bayesian multitask learning. Journal of Machine Learning Research 4 (2003)
5. Ben-David, S., Schuller, R.: Exploiting task relatedness for multiple task learning. In: Schölkopf, B., Warmuth, M.K. (eds.) COLT/Kernel 2003. LNCS (LNAI), vol. 2777, pp. 567–580. Springer, Heidelberg (2003)
6. Bickel, S., Bogojeska, J., Lengauer, T., Scheffer, T.: Multi-task learning for HIV therapy screening. In: Proceedings of International Conference on Machine Learning, pp. 56–63. ACM, New York (2008)
7. Biederman, I.: Recognition-by-components: A theory of human image understanding. Psychological Review 94, 115–147 (1987)
8. Blei, D.M., Mcauliffe, J.D.: Supervised topic models. In: Proceedings of Neural Information Processing Systems (2007)

9. Blei, D.M., Ng, A.Y., Jordan, M.I.: Latent Dirichlet Allocation. Journal of Machine Learning Research 3, 993–1022 (2003)
10. Bollacker, K.D., Ghosh, J.: Knowledge transfer mechanisms for characterizing image datasets. In: Soft Computing and Image Processing. Physica-Verlag, Heidelberg (2000)
11. Caruana, R.: Multitask learning. Machine Learning 28, 41–75 (1997)
12. Chang, J., Blei, D.: Relational topic models for document networks. In: Proceedings of Artificial Intelligence and Statistics (2009)
13. Eisenstein, J., Ahmed, A., Xing, E.P.: Sparse additive generative models of text. In: Proceedings of International Conference on Machine Learning, pp. 1041–1048 (2011)
14. Evgeniou, T., Micchelli, C.A., Pontil, M.: Learning multiple tasks with kernel methods. Journal of Machine Learning Research 6, 615–637 (2005)
15. Evgeniou, T., Pontil, M., Toubia, O.: A convex optimization approach to modeling consumer heterogeneity in conjoint estimation. Marketing Science 26(6), 805–818 (2007)
16. Fan, R.-E., Chang, K.-W., Hsieh, C.-J., Wang, X.-R., Lin, C.-J.: LIBLINEAR: A library for large linear classification. Journal of Machine Learning Research 9, 1871–1874 (2008)
17. Farhadi, A., Endres, I., Hoiem, D., Forsyth, D.: Describing objects by their attributes. In: Proceedings of Computer Vision and Pattern Recognition (2009)
18. Hinton, G.E., Osindero, S.: A fast learning algorithm for deep belief nets. Neural Computation 18, 2006 (2006)
19. Jacob, L., Bach, F., Vert, J.-P.: Clustered multi-task learning: A convex formulation. CoRR, abs/0809.2085 (2008)
20. Jalali, A., Ravikumar, P., Sanghavi, S., Ruan, C.: A Dirty Model for Multi-task Learning. In: Proceedings of Neural Information Processing Systems (December 2010)
21. Jenatton, R., Audibert, J., Bach, F.: Structured variable selection with sparsity-inducing norms. Journal of Machine Learning Research 12, 2777–2824 (2011)
22. Jiang, Q., Zhu, J., Sun, M., Xing, E.: Monte carlo methods for maximum margin supervised topic models. In: Proceedings of Neural Information Processing Systems, pp. 1601–1609 (2012)
23. Kim, S., Xing, E.P.: Tree-guided group lasso for multi-task regression with structured sparsity. In: Proceedings of International Conference on Machine Learning, pp. 543–550 (2010)
24. Kovashka, A., Vijayanarasimhan, S., Grauman, K.: Actively selecting annotations among objects and attributes. In: International Conference on Computer Vision, pp. 1403–1410. IEEE (2011)
25. Lampert, C.H., Nickisch, H., Harmeling, S.: Learning to detect unseen object classes by betweenclass attribute transfer. In: Proceedings of Computer Vision and Pattern Recognition (2009)
26. Low, Y., Agarwal, D., Smola, A.J.: Multiple domain user personalization. In: Proceedings of Knowledge Discovery and Data Mining, pp. 123–131 (2011)
27. Lowe, D.G.: Distinctive image features from scale-invariant keypoints. International Journal of Computer Vision 60(2), 91–110 (2004)
28. Pan, S.J., Yang, Q.: A survey on transfer learning. IEEE Transactions on Knowledge and Data Engineering 22, 1345–1359 (2010)
29. Passos, A., Rai, P., Wainer, J., Daumé III, H.: Flexible modeling of latent task structures in multitask learning. In: Proceedings of International Conference on Machine Learning (2012)
30. Quattoni, A., Wang, S., Morency, L.P., Collins, M., Darrell, T., Csail, M.: Hidden-state conditional random fields. In: IEEE Transactions on Pattern Analysis and Machine Intelligence (2007)
31. Ramage, D., Hall, D., Nallapati, R., Manning, C.D.: Labeled LDA: a supervised topic model for credit attribution in multi-labeled corpora. In: Proceedings of Empirical Methods in Natural Language Processing, pp. 248–256 (2009)

32. Rubin, T.N., Chambers, A., Smyth, P., Steyvers, M.: Statistical topic models for multi-label document classification. CoRR, abs/1107.2462 (2011)
33. Teh, Y.W., Jordan, M.I., Beal, M.J., Blei, D.M.: Hierarchical Dirichlet Processes. Journal of the American Statistical Association 101, 1566–1581 (2006)
34. Torralba, A., Murphy, K.P., Freeman, W.T.: Sharing visual features for multiclass and multi-view object detection. IEEE Trans. Pattern Anal. Mach. Intell. 29(5), 854–869 (2007)
35. Wang, C., Blei, D.M., Li, F.F.: Simultaneous image classification and annotation. In: Proceedings of Computer Vision and Pattern Recognition, pp. 1903–1910 (2009)
36. Wang, C., Paisley, J.W., Blei, D.M.: Online variational inference for the hierarchical Dirichlet process. Journal of Machine Learning Research - Proceedings Track 15, 752–760 (2011)
37. Wang, C., Thiesson, B., Meek, C., Blei, D.: Markov topic models. In: Proceedings of Artificial Intelligence and Statistics (2009)
38. Weinberger, K., Dasgupta, A., Langford, J., Smola, A., Attenberg, J.: Feature hashing for large scale multitask learning. In: Proceedings of International Conference on Machine Learning, pp. 1113–1120 (2009)
39. Xue, Y., Liao, X., Carin, L., Krishnapuram, B.: Multi-task learning for classification with Dirichlet process priors. Journal of Machine Learning Research 8, 35–63 (2007)
40. Yu, C.J., Joachims, T.: Learning structural SVMs with latent variables. In: Proceedings of International Conference on Machine Learning, pp. 1169–1176 (2009)
41. Zhang, J., Ghahramani, Z., Yang, Y.: Flexible latent variable models for multi-task learning. Machine Learning 73(3), 221–242 (2008)
42. Zhou, J., Chen, J., Ye, J.: Clustered Multi-Task Learning Via Alternating Structure Optimization. In: Proceedings of Neural Information Processing Systems (2011)
43. Zhu, J., Ahmed, A., Xing, E.P.: MedLDA: maximum margin supervised topic models for regression and classification. In: Proceedings of International Conference on Machine Learning, pp. 1257–1264 (2009)

Probabilistic Clustering for Hierarchical Multi-Label Classification of Protein Functions

Rodrigo C. Barros[1], Ricardo Cerri[1],
Alex A. Freitas[2], and André C.P.L.F. de Carvalho[1]

[1] Universidade de São Paulo, São Carlos-SP, Brazil
{rcbarros,cerri,andre}@icmc.usp.br
[2] University of Kent, Canterbury, Kent, UK
A.A.Freitas@kent.ac.uk

Abstract. Hierarchical Multi-Label Classification is a complex classification problem where the classes are hierarchically structured. This task is very common in protein function prediction, where each protein can have more than one function, which in turn can have more than one sub-function. In this paper, we propose a novel hierarchical multi-label classification algorithm for protein function prediction, namely HMC-PC. It is based on probabilistic clustering, and it makes use of cluster membership probabilities in order to generate the predicted class vector. We perform an extensive empirical analysis in which we compare our new approach to four different hierarchical multi-label classification algorithms, in protein function datasets structured both as trees and directed acyclic graphs. We show that HMC-PC achieves superior or comparable results compared to the state-of-the-art method for hierarchical multi-label classification.

Keywords: Hierarchical Multi-Label Classification, Protein Function Prediction, Probabilistic Clustering.

1 Introduction

Classification is the well-known machine learning task whose goal is to assign instances to predefined categories. Classification algorithms are given as input a set of N n-dimensional training instances $\mathbf{X} = \{\mathbf{x}_1, \mathbf{x}_2, ..., \mathbf{x}_N\}$, as well as their respective set of class labels $C = \{C_{\mathbf{x}_1}, C_{\mathbf{x}_2}, ..., C_{\mathbf{x}_N}\}$, in which $C_{\mathbf{x}_i} \in \{C_1, ...C_k\}$ in a k-class problem.

The vast majority of classification problems require the association of each instance with a single class, which means $C_{\mathbf{x}_i}$ is a single categorical value in $\{C_1, ...C_k\}$. This particular kind of problem is regarded as *flat* or *non-hierarchical* classification. Notwithstanding, there are problems in which the classes are organized in a hierarchical structure — a tree or a directed acyclic graph (DAG) — and each instance may be associated to multiple classes in multiple paths of this hierarchy. The difference between the tree and DAG hierarchies is that, in a tree, each class can have only one superclass, which implies there is just one path

H. Blockeel et al. (Eds.): ECML PKDD 2013, Part II, LNAI 8189, pp. 385–400, 2013.

between the root node and the class node. In a DAG hierarchy, each class can have more than one superclass at the same time, which means that there can be multiple paths from the root node to a class node. As an example, consider the dotted nodes in Fig. 1. While in Fig. 1(a) there is just one possible path between the root node and the dotted nodes (1/2 and 2/2/1) (tree structure), we can see that in Fig. 1(b) (DAG structure) we can reach the dotted node at the second level by two paths (1/2 and 2/2). The same can be observed at the dotted node located in the third level, which can be reached by two different paths (2/1/1 and 2/2/1).

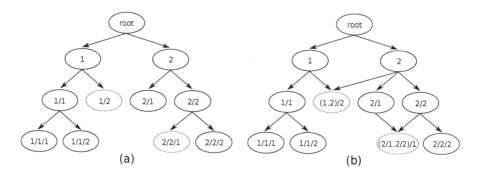

Fig. 1. Example of class hierarchy

Either structured as a DAG or tree, this particularly complex problem is known as hierarchical multi-label classification (HMC), and is the primary subject of this paper. In a HMC problem, the set of class labels can be represented as a matrix $\mathbf{V} = \{\mathbf{v}_{\mathbf{x}_1}, \mathbf{v}_{\mathbf{x}_2}, ..., \mathbf{v}_{\mathbf{x}_N}\}$, in which $\mathbf{v}_{\mathbf{x}_i}$ is the c-dimensional binary class vector associated with instance $\mathbf{x}_i$, in a hierarchy with c nodes (classes). Each position of the class vector $\mathbf{v}_{\mathbf{x}_i}$ represents a class, and it is set to 1 if $\mathbf{x}_i$ belongs to that respective class, or 0 otherwise.

Two important examples of HMC problems are the tasks of text classification [29,30,25] and protein function prediction [4,6,31]. The latter is an increasingly important research field by itself, given the recent availability of unknown proteins for analysis and determination of their respective biological functions. Protein function prediction can be seen as a predictive problem in which each protein is a dataset instance, whereas different protein features are used as predictive attributes, and ultimately the goal is to classify these proteins according to different functions they can perform. Since a protein can perform multiple functions, and these functions are usually organized in a hierarchical structure (*e.g.*, the FunCat [26] and Gene Ontology [3] protein functional-definition schemes), the protein function prediction can be regarded as a typical HMC problem.

There has been an increasing number of machine learning approaches for hierarchical multi-label classification of protein functions. Roughly speaking, these approaches can be divided into *local* and *global* approaches. In the local approach, conventional (flat) classification algorithms such as decision trees or support vector machines are trained to produce a hierarchy of classifiers, which are then executed following a top-down strategy to classify unlabeled instances [11]. In this approach, local information about the class hierarchy is used during the training process of each base classifier. Such local information can be used in different ways, depending on how the local classifiers are induced [28]. Differently from the local approach, the global approach induces a single classifier using all classes of the hierarchy at once. After the training process, the classification of a new instance occurs in just one step [31]. As global methods induce a single classifier to consider the specificities of the classification problem, they usually do not make use of conventional classification algorithms, unless these are heavily adapted to consider the hierarchy of classes.

In this paper, we propose a novel global hierarchical multi-label classification algorithm that is based on probabilistic clustering, namely Hierarchical Multi-label Classification with Probabilistic Clustering (HMC-PC). HMC-PC works according to the following assumption: instances that belong to a given cluster have similar class vectors, and hence the training instances are clustered following an expectation-maximization scheme [13], and the average class vector of the training instances from a given cluster is used to classify new unseen instances associated to the same cluster. The cluster membership probabilities are also used to tune the average class vector in each cluster. HMC-PC offers the advantages of the global methods, namely: (a) reduced time complexity when compared to the execution of multiple classifiers in the local approach; and (b) it does not suffer the problem of error propagation across levels, since the classification of a given hierarchical level is not done separately from the other levels. Finally, we show that HMC-PC presents competitive results when compared with the state of the art decision-tree-based method Clus-HMC [31].

This paper is organized as follows. Section 2 discusses related work on machine learning approaches for protein function prediction. Section 3 details HMC-PC, our novel global algorithm for hierarchical multi-label classification. Section 4 depicts the methodology employed for the experimental analysis of protein function prediction, which is in turn presented in Section 5. We present our conclusions and future work opportunities in Section 6.

2   Related Work

One of the first HMC algorithms was proposed by Clare and King [10], namely HMC4.5, which is a global method based on decision-tree induction algorithms. It is a variant of C4.5 [24] with modifications in the calculation of class entropy. The proposed modification uses the sum of the number of bits needed to describe membership or non-membership of each class, and also the information related to the size of the tree rooted by a given class. The method was used in tree-structured hierarchies.

In [18] and [19], the authors proposed a global method for the classification of Gene Ontology (GO) [3] genes based on the classification of documents from the MedLine repository that describe these genes. This method expands the sets of classes by including their ancestor classes and then applies the AdaBoost algorithm [27] in the modified dataset. Inconsistent predictions that may have occurred are corrected.

In [4], the authors proposed a local method that uses a hierarchy of SVM classifiers for the prediction of gene functions structured according to the GO. The classifiers are trained separately for each class, and the predictions are then combined using Bayesian Networks [16], aiming at finding the most probable consistent set of predictions.

In the work of Vens et al. [31], three methods based on the concept of Predictive Clustering Trees (PCT) were extensively compared. The authors make use of the global Clus-HMC method [5] that induces a single decision tree to cope with the entire hierarchical multi-label classification problem. They compared its performance with two local methods. The first one, Clus-SC, induces an independent decision tree for each class of the hierarchy, ignoring the relationships between classes. The second, Clus-HSC, explores the hierarchical relationships between the classes to induce a decision tree for each class. The authors employed the above-mentioned methods to hierarchies structured both as trees and DAGs, and discussed the modifications needed so that the algorithms could cope with both types of hierarchical structures. While in [31] the authors used the Euclidean distance to calculate the similarities and dissimilarities between instances in the decision tree, Aleksovski et al. [2] expanded this study by investigating the use of other distance measures, namely Jaccard, SimGIC, and ImageClef.

In [22], Otero et al. proposed hAnt-Miner, a global method for hierarchical single-label classification using Ant Colony Optimization (ACO) [15,14]. The authors later extended this method [23] to allow multi-label classification, considering both tree- and DAG-structured hierarchies.

Cerri et al. [8] proposed a global method that employs a Genetic Algorithm (GA) to produce HMC rules. The GA evolves the antecedents of classification rules, in order to optimize the level of coverage of each antecedent. The employed fitness (evaluation) function gives a better reward to rules with the antecedents that cover a higher number of instances. Then, the set of optimized antecedents is selected to build the corresponding consequent of the rules (set of classes to be predicted). The method was used in hierarchies structured as trees.

Both in [21] and [1], the authors propose methods that employ clustering as a substep of classification, though these approaches only deal with flat multi-label data and not with hierarchical multi-label data.

Finally, Cerri et al. [7,9] proposed a local approach that employs a sequence of connected artificial neural networks for protein function prediction. Each network is associated to a hierarchical level, and the output of the network in level l is used as the input of the network in level $l + 1$. A strategy for avoiding inconsistent predictions is employed, since a given neural network may predict a class whose superclass had not been predicted before. The method is tested over a hierarchy structured as a tree.

3   HMC-PC

In this paper, we propose Hierarchical Multi-label Classification with Probabilistic Clustering (HMC-PC), which is a novel global HMC algorithm. The general rationale behind HMC-PC is the assumption that we can discover the most fitting probability distribution for each particular group of training instances, and that those instances that were generated by the same distribution also share similar class vectors. Hence, once we have discovered the set of k distributions from the training set, a new unseen instance can be easily classified by performing two procedures: (i) discovering the distribution that most probably has generated it — *i.e.*, discovering which cluster it belongs to; and (ii) assigning to this new instance the average class vector of the training instances that were generated by the same distribution (cluster).

HMC-PC can roughly be divided into three mains steps:(i) cluster generation; (ii) class vector generation; and (iii) classification.

1. Cluster generation: the training dataset is arranged into different clusters following a probabilistic expectation-maximization (EM) scheme [13];
2. Class vector generation: for each cluster, the class vectors of the training instances that surpass a given probability threshold are averaged, and later used to classify unseen instances;
3. Classification: each test instance is assigned to the cluster it most probably belongs to. Then, the cluster's average class vector generated in the previous step is assigned to the test instance as the final prediction.

3.1   Cluster Generation

The first step of HMC-PC is to generate clusters from the training dataset $\mathbf{X}$, which is comprised of n attributes and N instances. In this step, the class vector of each instance $\mathbf{x}_i \in \mathbf{X}$, $i = 1...N$ is not used during cluster generation.

Each cluster in HMC-PC is assumed to be generated by a distinct Gaussian probability distribution. HMC-PC clustering iterates over the steps of expectation and maximization, much the same as the well-known EM algorithm [13]. We make the further naïve assumption of attribute independence, which means we are only interested in the diagonal of the covariance matrix $\mathbf{\Sigma}_i$ from the ith Gaussian. This assumption is intended to speed-up the algorithm, avoiding the cost of computing the inverse of $\mathbf{\Sigma}_i$, which is usually $O(n^3)$.

In the expectation step, the cluster membership of each instance $\mathbf{x}_i$ regarding each Gaussian distribution (cluster) C_j is computed, assuming the parameters of each of the k distributions are already known:

$$Pr[C_j|\mathbf{x}_i] = \frac{Pr[\mathbf{x}_i|C_j] \times Pr[C_j]}{Pr[\mathbf{x}_i]} \tag{1}$$

where $Pr[\mathbf{x}_i|C_j] = \mathcal{N}(\mathbf{x}_i|\mu_j, \mathbf{\Sigma}_j)$, and $Pr[C_j]$ is estimated as $\sum_{i=1}^{N} Pr[C_j|\mathbf{x}_i]/N$. Note that $Pr[\mathbf{x}_i]$ can simply be replaced by the sum of $Pr[\mathbf{x_i}|C_j] \times Pr[C_j]$ for the k distributions.

The maximization step is performed by simply updating the parameters of the k distributions, taking into account the recently computed values of $Pr[C|\mathbf{x}]$:

$$N_j = \sum_{i=1}^{N} Pr[C_j|\mathbf{x}_i] \tag{2}$$

$$Pr[C_j] = \frac{N_j}{N} \tag{3}$$

$$\mu_j = \frac{1}{N_j} \sum_{i=1}^{N} Pr[C_j|\mathbf{x}_i] \times \mathbf{x}_i \tag{4}$$

$$\mathbf{\Sigma}_j = \frac{1}{N_j} \sum_{i=1}^{N} Pr[C_j|\mathbf{x}_i] \times (\mathbf{x}_i - \mu_j)(\mathbf{x}_i - \mu_j)^T \tag{5}$$

The iteration between the expectation and maximization steps is performed until one of the two conditions occurs:

- the maximum number of 100 iterations is reached; or
- the difference between the log-likelihood of two consecutive steps is smaller than 1×10^{-6}.

The log-likelihood is computed after each expectation step:

$$\mathcal{LL} = \sum_{i=1}^{N} \ln \left(\sum_{j=1}^{k} Pr[C_j] \times Pr[\mathbf{x_i}|C_j] \right) \tag{6}$$

Since we have to assume that either the cluster memberships $Pr[C_j|\mathbf{x}_i]$ or the distribution parameters $\mu_j, \mathbf{\Sigma}_j$ are informed before the beginning of the expectation-maximization iterations, HMC-PC executes the well-known k-means algorithm [20] 10 times varying the random initialization. The partition with the smallest value of the squared error is employed to initialize the parameters of the k distributions.

The only problem that remains is the definition of the number of Gaussian distributions (clusters). In order to avoid the use of a user-defined parameter, we propose the following methodology for automatically defining the value of k:

1. Set $k = 1$;
2. Run the expectation-maximization iterations with 10-fold cross-validation over the *training set*, i.e., in each run EM is applied to 9 of the 10 training folds and the log-likelihood is assessed on the hold-out fold, averaging the results over the 10 runs.
3. If the log-likelihood has increased, increase the value of k by 1 and the procedure continues in step 2.

After the algorithm has converged and the final parameters of the k clusters are known, the N training instances are assigned to their most probable cluster, *i.e.*:

$$C_{\mathbf{x}_i} = \underset{C_j}{\mathrm{argmax}} \left(\frac{Pr[\mathbf{x}_i|C_j] \times Pr[G_j]}{Pr[\mathbf{x}_i]} \right) \tag{7}$$

3.2 Class Vector Generation

Once the training instances have been distributed throughout the clusters, HMC-PC generates one class vector per cluster. The rationale behind this step is that a future test instance will be assigned to its most probable cluster according to Eq. (7), and then classified according to the class vector generated from that cluster. HMC-PC offers two strategies to generate one class vector per cluster:

1. The class vector of cluster C_j is generated as the average class vector of the training instances that were assigned to cluster C_j, *i.e.*:

$$\bar{\mathbf{v}}_{C_j} = \frac{1}{N} \sum_{\mathbf{x}_i \in C_j} \mathbf{v}_{\mathbf{x}_i} \tag{8}$$

2. The class vector of cluster C_j is generated as the average class vector of the training instances that were assigned to cluster C_j and whose cluster membership probability surpass a given previously-defined threshold $\varDelta_j$, *i.e.*:

$$\bar{\mathbf{v}}_{C_j} = \frac{1}{N} \sum_{\substack{\mathbf{x}_i \in C_j \wedge \\ Pr[C_j|\mathbf{x}_i] \geq \varDelta_j}} \mathbf{v}_{\mathbf{x}_i} \tag{9}$$

Note that strategy 1 is a special case of strategy 2 in which $\varDelta = 0$ for all clusters. The second strategy, on the other hand, makes use of the cluster memberships to define the average class vectors. The disadvantage of the second strategy is the need of defining threshold values $\varDelta$ for each cluster. In order to overcome this problem, we propose an adaptive threshold selection strategy as follows. First, the training set is divided into two subsets: sub-training and validation. The sub-training set is used as before to generate the clusters, and its instances are distributed throughout the discovered clusters. Next, we also distribute the validation instances to their most probable cluster, also according to Eq. (7). Then, for each cluster, we evaluate the classification performance of the validation instances with the area under the precision-recall curve (AUPRC, more details in Section 4) by building the average class vector following Eq. (9). For that, we have to try different values of $\varDelta_j$, *i.e.*, $\{0, 0.1, 0.2, 0.3, 0.4, 0.5, 0.6, 0.7, 0.8, 0.9\}$. The average class vector built according to the threshold value that yielded the largest AUPRC value is then chosen to classify the test instances that are assigned to cluster C_j.

The pseudo-code of HMC-PC with the adaptive threshold selection strategy is presented in Alg. 1. The main difference between the adaptive threshold strategy

---

Algorithm 1. HMC-PC with adaptive threshold selection.

Require: Training dataset **X**
Require: Threshold set $ts = \{0, 0.1, 0.2, 0.3, 0.4, 0.5, 0.6, 0.7, 0.8, 0.9\}$
 1: Divide **X** in sub-training $\mathbf{X}^t$ and validation $\mathbf{X}^v$ sets
 2: $k \leftarrow CV(\mathbf{X}^t)$
 3: $partition \leftarrow EM(\mathbf{X}^t, k)$
 4: **for** $\mathbf{x}_i \in X^t$ **do**
 5:    $C_{\mathbf{x}_i} \leftarrow$ Eq. (7)
 6: **end for**
 7: **for** $\mathbf{x}_i \in X^v$ **do**
 8:    $C_{\mathbf{x}_i} \leftarrow$ Eq. (7)
 9: **end for**
10: **for** cluster $C_j \in partition$ **do**
11:    $bestAUPRC \leftarrow 0$
12:    **for** $\Delta_j \in ts$ **do**
13:       $\bar{\mathbf{v}}_{C_j} \leftarrow$ Eq. (9)
14:       $AUPRC \leftarrow classify(\{\mathbf{x^v}_i | \mathbf{x^v}_i \in C_j\}, \bar{\mathbf{v}}_{C_j})$
15:       **if** $AUPRC \geq bestAUPRC$ **then**
16:          $bestAUPRC \leftarrow AUPRC$
17:          **thresholds**$_j \leftarrow \Delta_j$
18:       **end if**
19:    **end for**
20: **end for**
21: $partition \leftarrow EM(\mathbf{X}^t \cup \mathbf{X}^v, k)$
22: **return thresholds**, $partition$

---

and the threshold-free strategy is the need for a validation set to automatically select the value of Δ_j for cluster C_j. Note that both training and validation data are distributed throughout the clusters (lines 5 and 8). The main loop in line 10 performs the adaptive threshold selection by evaluating the validation data that were assigned to a given cluster C_j with regard to the different threshold values. The algorithm stores in vector **thresholds** the optimized threshold value per cluster. These thresholds were the ones that maximized the AUPRC generated by Eq. (9) (line 13). Therefore, even with the threshold Δ possibly assuming different values for each cluster, the user is not required to set any ad-hoc parameter during the whole execution of HMC-PC. Finally, the method performs the expectation-maximization steps once again (line 21) with the full training set (sub-training + validation).

3.3   Classification

The last step of HMC-PC is to classify unseen instances. The classification process is straightforward: (i) assign each test instance to its most probable cluster according to Eq. (7); (ii) assuming test instance $\mathbf{x}_i$ was assigned to cluster C_j, make use of class vector $\bar{\mathbf{v}}_{C_j}$ computed from the training instances that belong to cluster C_j and have cluster membership probability greater than **thresholds**$_j$ as the class prediction for test instance $\mathbf{x}_i$.

4   Experimental Methodology

4.1   Baseline Algorithms

We employ four of the methods reviewed in Section 2 as the baseline algorithms during the experiments performed in this work. We make use of the global decision-tree based method Clus-HMC, which is considered the state-of-the-art method in the literature, since it obtained the best results so far, and we also make use of its local variants Clus-HSC and Clus-SC. The three methods are detailed in [31]. We also employ the Ant Colony Optimization-based method *hm*Ant-Miner [23], which is a global method that obtained competitive results when compared to Clus-HMC. We decided to select these algorithms because they were all applied to the same protein function prediction datasets used in the experiments (both for tree and DAG structures). In addition, they all produce the same type of output provided by HMC-PC, and their performance were analyzed with the same evaluation measure we use in this paper.

We evaluated HMC-PC's performance with the two alternative class-vector generation mechanisms presented in Section 3.2. The first version will be regarded as HMC-PC whereas the second version will be regarded as HMC-PC$_\Delta$.

4.2   Datasets

Ten freely-available numeric datasets[1] related to protein function prediction are used in the experiments, namely: cellcycle, derisi, eisen, gasch1, and gasch2 (FunCat-annotated and Gene Ontology-annotated). The option for all-numeric datasets is because the current version of HMC-PC can only cope with numeric attributes. Dealing with nominal attributes is a topic left for future work.

These datasets are related to issues like phenotype data and gene expression levels. They are organized according to two different class hierarchy structures: tree structure (FunCat-annotated data sets) and directed acyclic graph structure (Gene Ontology-annotated data sets).

Table 1 summarizes the main characteristics of the training, validation, and test datasets employed in the experiments. In the particular case of *hm*Ant-Miner and HMC-PC, the training and validation datasets are merged and used together to generate the predictive models. The PCT-based methods and HMC-PC$_\Delta$ make use of the validation datasets to optimize parameters during their executions.

A detailed description of each dataset can be found in [31]. For executing HMC-PC in these datasets, all missing values were replaced with the mean value of the respective attribute.

4.3   Evaluation Measures and Statistical Analysis

Considering that all algorithms tested in this paper output a vector of class probabilities for each instance being predicted, we make use of the area under the average PR-curve ($AU(\overline{PRC})$) as the evaluation measure to compare them.

---

[1] http://www.cs.kuleuven.be/~dtai/clus/hmcdatasets.html

Table 1. Summary of datasets: number of attributes ($|A|$), number of classes ($|C|$), total number of instances (Total) and number of multi-label instances (Multi)

Structure	Dataset	$|A|$	$|C|$	Training		Validation		Test	
				Total	Multi	Total	Multi	Total	Multi
	Cellcycle	77	499	1628	1323	848	673	1281	1059
	Derisi	61	499	1608	1309	842	671	1275	1055
Tree	Eisen	79	461	1058	900	529	441	837	719
	Gasch1	173	499	1634	1325	846	672	1284	1059
	Gasch2	52	499	1639	1328	849	674	1291	1064
	Cellcycle	77	4125	1625	1625	848	848	1278	1278
	Derisi	61	4119	1605	1605	842	842	1272	1272
DAG	Eisen	79	3573	1055	1055	528	528	835	835
	Gasch1	173	4125	1631	1631	846	846	1281	1281
	Gasch2	52	4131	1636	1636	849	849	1288	1288

To obtain a PR-curve for a given algorithm, different thresholds ranging within $[0, 1]$ are applied to the outputs of the methods, and thus different values of precision and recall are obtained, one for each threshold value. Each threshold value then represents a point within the PR space. The union of these points forms a PR-curve. In order to calculate the area below the PR-curve, the PR-points must be interpolated [12]. This interpolation guarantees that the area below the curve is not artificially increased, which would happen if the curves were constructed just connecting the points without interpolation. Given a threshold value, a precision-recall point ($\overline{Prec}, \overline{Rec}$) in the PR-space can be obtained through Eq. (10) and (11), corresponding to the micro-average of precision and recall, where i ranges from 1 to c, and TP, FP, and FN stand, respectively, for the number of true positives, false positives, and false negatives.

$$\overline{Prec} = \frac{\sum_i TP_i}{\sum_i TP_i + \sum_i FP_i} \quad (10) \qquad \overline{Rec} = \frac{\sum_i TP_i}{\sum_i TP_i + \sum_i FN_i} \quad (11)$$

In order to provide some reassurance about the validity and non-randomness of the results, we employed the Friedman and Holm statistical tests, recommended for comparisons when a control classifier is compared against other classifiers [17]. We employed a confidence level of 95% in the statistical tests.

5   Results and Discussion

Table 2 presents the comparison among the two HMC-PC versions and the baseline methods Clus-HMC, Clus-HSC, Clus-SC, and hmAnt-Miner. Given that hmAnt-Miner is a non-deterministic method, its results are averages over 15 executions. Both versions of HMC-PC and the PCT-based methods are deterministic algorithms and thus require a single execution. We highlight in bold the best absolute values for each dataset, and we provide at the end of the table the average rank of each method, following the Friedman statistical test.

We can observe that both HMC-PC versions are the best-ranked among all methods. It is interesting to see that the threshold-based version HMC-PC$_\Delta$,

Table 2. $AU\overline{(PRC)}$ values for the comparison of HMC-PC with hmAnt-Miner, Clus-HMC, Clus-HSC, Clus-SC

		HMC-PC	HMC-PC$_\Delta$	Clus-HMC	Clus-HSC	Clus-SC	hm-Ant-Miner
DAG							
	Cellcycle	0.368	0.369	0.357	**0.371**	0.252	0.332
	Derisi	0.341	0.344	**0.355**	0.349	0.218	0.334
	Eisen	0.396	**0.398**	0.380	0.365	0.270	0.376
	Gasch1	0.381	**0.382**	0.371	0.351	0.239	0.356
	Gasch2	0.369	0.367	0.365	**0.378**	0.267	0.344
Tree							
	Cellcycle	**0.200**	0.187	0.172	0.111	0.106	0.154
	Derisi	0.163	0.163	**0.175**	0.094	0.089	0.161
	Eisen	0.211	**0.214**	0.204	0.127	0.132	0.180
	Gasch1	**0.212**	0.210	0.205	0.106	0.104	0.175
	Gasch2	0.196	**0.197**	0.195	0.121	0.119	0.152
	Average Rank	2.15	1.85	2.80	4.00	5.90	4.30

which takes into consideration the cluster membership probability to build the average class vectors, is slightly better ranked than HMC-PC. This is coherent with our initial hypothesis that the cluster membership probabilities can be used to tune the cluster class vector generation process, improving the prediction of unseen instances. Indeed, by making use of the cluster memberships, HMC-PC$_\Delta$ was capable of detecting the training instances that could bring more precise information within each cluster.

Regardless of the HMC-PC version employed, we can see that it provides better results than hm-Ant-Miner for all ten datasets. The same can be said regarding Clus-SC, which is outperformed by either version of HMC-PC by a large margin.

In the comparison against Clus-HSC, we can notice that HMC-PC and HMC-PC$_\Delta$ outperform it in 7 out of 10 datasets, and they are outperformed by it in the remaining three. It should be noticed that the performance of Clus-HSC in the datasets structured as a tree is very poor, which seems to be a problem that both local PCT-based methods share.

Finally, when comparing HMC-PC and HMC-PC$_\Delta$ with Clus-HMC, we can see that both versions outperform Clus-HMC in 8 out of 10 datasets, being outperformed in only two datasets, which is quite a considerable difference considering the fact that Clus-HMC is the best-performing method in the literature.

For assessing the statistical significance of the results, we first consider the p-value provided by the Friedman test: 3.15×10^{-6}, which states that the null hypotheses in which all methods perform similarly should be rejected. Then, we take the best-ranked method as the *control* algorithm, and a set of pairwise adjusted comparisons according to Holm's procedure are performed.

Table 3 presents the p-values and adjusted α values for the Holm pos-hoc pairwise comparisons, bearing in mind that HMC-PC$_\Delta$ is the control algorithm. The statistical test rejects those hypotheses that have a p-value ≤ 0.025. Note that HMC-PC$_\Delta$ outperforms with statistical significance all methods but Clus-HMC and HMC-PC.

Table 3. Holm's procedure for $\alpha = 0.05$. HMC-PC$_\Delta$ is the control algorithm

i	Algorithm	$z = (R_0 - R_i)/SE$	p-value	Holm's adjusted α
5	Clus-SC	4.84	1.29×10^{-6}	0.0100
4	hm-Ant-Miner	2.93	3.40×10^{-3}	0.0125
3	Clus-HSC	2.57	1.02×10^{-2}	0.0167
2	Clus-HMC	1.14	2.56×10^{-1}	0.0250
1	HMC-PC	0.36	0.72×10^{-1}	0.0500

Since Clus-HMC is the best-performing baseline algorithm, we now compare it with HMC-PC$_\Delta$ in specific classes of the hierarchy in order to examine their behavior when predicting classes at different hierarchical levels. For selecting these specific classes, we used the following methodology: we selected ten classes from each dataset in which Clus-HMC presented the best per-class AUPRC values in the training set. We compare the test per-class AUPRC values between HMC-PC$_\Delta$ and Clus-HMC in the DAG-structured datasets and in the tree-structured datasets (Table 4).

By careful inspection of Table 4, we can observe that in the datasets in which HMC-PC$_\Delta$ outperformed Clus-HMC in $AU(\overline{PRC})$, it also outperformed Clus-HMC in the majority of the classes regarding the per-class AUPRC. The only exception was the Gasch2 dataset organized as a DAG, in which Clus-HMC and HMC-PC$_\Delta$ tied 5-5 in the 10 selected classes. Overall, HMC-PC$_\Delta$'s good performance is consistent across hierarchical levels.

For exemplifying this scenario, we can notice that HMC-PC outperformed Clus-HMC in several classes that lie deep in the hierarchy. Recall that these classes are associated with more specific protein functions, and the more specific the function, the more useful the information about the protein. Also, recall that the deeper the class, the fewer the number of instances assigned to it. As an example, we can cite the case of the GO term (class) GO:0006412 in datasets Cellcycle, Eisen, Gasch1, and Gasch2 (GO-annotated), in which HMC-PC$_\Delta$ consistently outperforms Clus-HMC. Figure 2 shows how deep in the DAG-structured hierarchy the GO term GO:0006412 lies.

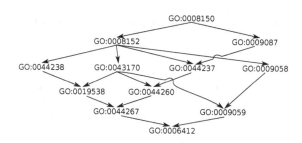

Fig. 2. Hierarchical paths that lead to term GO:0006412

Table 4. Per-class AUPRC values for 10 specific classes in each dataset. Classes that start with "GO:" belong to DAG-structured datasets, whereas the remaining belong to tree-structured datasets.

Dataset	Class	Clus-HMC	HMC-PC$_\Delta$	Class	Clus-HMC	HMC-PC$_\Delta$
Cellcycle	GO:0044464	**0.967**	0.961	GO:0044424	0.898	**0.921**
	GO:0009987	0.860	**0.876**	GO:0008152	0.726	**0.767**
	GO:0003735	0.404	**0.649**	GO:0044237	0.676	**0.699**
	GO:0044238	0.650	**0.677**	GO:0044444	0.586	**0.624**
	GO:0043170	0.580	**0.632**	GO:0006412	0.361	**0.599**
Eisen	GO:0044464	0.981	**0.983**	GO:0044424	0.926	**0.929**
	GO:0009987	**0.916**	0.909	GO:0003735	0.692	**0.713**
	GO:0008152	0.808	**0.840**	GO:0044445	0.636	**0.687**
	GO:0006412	0.607	**0.652**	GO:0044237	0.751	**0.770**
	GO:0044238	0.716	**0.726**	GO:0043170	0.664	**0.720**
Derisi	GO:0044464	0.965	**0.974**	GO:0044424	**0.888**	0.880
	GO:0009987	**0.849**	0.838	GO:0008152	**0.736**	0.731
	GO:0044237	**0.674**	0.668	GO:0044238	**0.643**	0.640
	GO:0044444	**0.582**	0.555	GO:0003735	**0.462**	0.293
	GO:0043170	**0.585**	0.556	GO:0043226	0.525	**0.540**
Gasch1	GO:0044464	0.963	**0.978**	GO:0044424	0.912	**0.927**
	GO:0009987	0.855	**0.875**	GO:0003735	**0.659**	0.614
	GO:0008152	0.733	**0.754**	GO:0044237	0.674	**0.699**
	GO:0006412	0.574	**0.583**	GO:0044238	0.660	**0.671**
	GO:0044444	**0.647**	0.639	GO:0043170	0.616	**0.659**
Gasch2	GO:0044464	**0.966**	0.961	GO:0044424	0.910	**0.926**
	GO:0009987	**0.863**	0.850	GO:0003735	**0.622**	0.609
	GO:0008152	**0.741**	0.739	GO:0044237	**0.696**	0.682
	GO:0006412	0.521	**0.536**	GO:0044238	0.667	**0.668**
	GO:0043170	0.655	**0.669**	GO:0044422	0.601	**0.619**
Cellcycle	1	0.402	**0.432**	12.01	0.330	**0.425**
	10	0.335	**0.349**	12.01.01	0.326	**0.351**
	10.01	0.195	**0.200**	14	0.303	**0.329**
	11	**0.371**	0.351	16	0.261	**0.276**
	12	0.304	**0.424**	20	0.252	**0.254**
Eisen	1	0.392	**0.462**	12.01	0.582	**0.650**
	2	**0.420**	0.301	12.01.01	0.609	**0.722**
	10	0.352	**0.364**	14	0.396	**0.370**
	11	**0.394**	0.369	14.13.01	**0.177**	0.082
	12	0.508	**0.650**	16	**0.270**	0.250
Derisi	1	0.376	**0.400**	12.01	**0.376**	0.235
	2	**0.238**	0.215	12.01.01	**0.385**	0.234
	10	0.240	**0.278**	14	**0.280**	0.278
	11	0.264	**0.337**	16	**0.237**	0.232
	12	**0.366**	0.259	20	**0.259**	0.248
Gasch1	1	0.444	**0.454**	12.01	**0.635**	0.590
	2	**0.285**	0.190	12.01.01	**0.660**	0.593
	10	0.326	**0.367**	14	0.329	**0.362**
	11	0.363	**0.439**	16	**0.279**	0.259
	12	0.566	**0.587**	20	0.300	**0.302**
Gasch2	1	0.451	**0.470**	12.01.01	**0.627**	0.478
	10	0.286	**0.290**	14	0.342	**0.352**
	11	**0.409**	0.400	16	0.248	**0.283**
	12	**0.562**	0.546	20	0.253	**0.261**
	12.01	**0.634**	0.525	42	0.236	**0.251**

We conclude by stating that HMC-PC seems to be a good alternative to Clus-HMC for the following reasons: (i) it performs better in the majority of the datasets; (ii) it is a parameter-free algorithm; (iii) it provides good AUPRC values for both shallow and deep classes in the hierarchy; and (iv) its time complexity is linear in all its input variables.

6   Conclusions

In this paper, we present a novel global hierarchical multi-label classification algorithm based on probabilistic clustering for the task of protein function prediction. The method is named Hierarchical Multi-Label Classification with Probabilistic Clustering (HMC-PC). We present two different versions of HMC-PC, namely HMC-PC and HMC-PC$_\Delta$.

HMC-PC works by clustering the protein function datasets in k clusters following an expectation-maximization scheme [13]. Then, for each of the k clusters, the average class vector is generated based on the training instances that were (hard-)assigned to each cluster. The choice of which instances will be used to define the per-cluster average class vector is based on the probabilities of cluster membership. The threshold-free version of HMC-PC assumes all instances that were assigned to a given cluster should be used for generating the cluster's average class vector, whereas HMC-PC$_\Delta$ employs an adaptive threshold selection strategy based on validation data to select the best value for Δ in each cluster.

We performed experiments using ten protein function prediction datasets (five of them structured as trees and five of them structured as DAGs). We compared HMC-PC versions with four well-known HMC algorithms: two decision tree-based local methods, namely Clus-HSC and Clus-SC; one decision tree-based global method, namely Clus-HMC; and one global method based on Ant Colony Optimization, namely hmAnt-Miner. Among all the methods previously proposed in the literature, Clus-HMC has been considered so far the state-of-the-art method for hierarchical multi-label classification [31]. We evaluated the methods using the area under the average PR-curve ($AU(\overline{PRC})$).

The comparison with the baseline methods shows that HMC-PC — particularly its threshold-based version HMC-PC$_\Delta$ — outperforms them in the majority of the datasets. We also compared HMC-PC$_\Delta$ and Clus-HMC in individual hierarchical classes, and showed that HMC-PC$_\Delta$ often obtained the best performance, including in classes that lie deep in the class hierarchy. This is particularly important, since deep class predictions tend to be more useful to biololgists than shallow class predictions.

As future work, we intend to extend HMC-PC so it can also deal with categorical attributes. We also intend to perform a deeper analysis to verify whether the thresholds in the different clusters are correlated to each other in any sense. Finally, we plan to investigate the use of the complete covariation matrix, so we do not have to make the naïve assumption of attribute independence. Nevertheless, such a modification will lead to an increased time complexity, since the complexity of finding the inverse of a $n \times n$ matrix is $O(n^3)$, whereas the current version of HMC-PC is linear in all its input variables.

Acknowledgment. The authors would like to thank Fundação de Amparo à Pesquisa do Estado de São Paulo (FAPESP), Brazil, for funding this research.

References

1. Ahmed, M.S.: Clustering guided multi-label text classification. Ph.D. thesis, University of Texas at Dallas (2012)
2. Aleksovski, D., Kocev, D., Dzeroski, S.: Evaluation of distance measures for hierarchical multilabel classification in functional genomics. In: 1st Workshop on Learning from Multi-Label Data (MLD) held in conjunction with ECML/PKDD, pp. 5–16 (2009)
3. Ashburner, M., et al.: Gene ontology: tool for the unification of biology. The Gene Ontology Consortium. Nature Genetics 25, 25–29 (2000)
4. Barutcuoglu, Z., Schapire, R.E., Troyanskaya, O.G.: Hierarchical multi-label prediction of gene function. Bioinformatics 22, 830–836 (2006)
5. Blockeel, H., Bruynooghe, M., Dzeroski, S., Ramon, J., Struyf, J.: Hierarchical multi-classification. In: Workshop on Multi-Relational Data Mining, pp. 21–35 (2002)
6. Blockeel, H., Schietgat, L., Struyf, J., Džeroski, S., Clare, A.J.: Decision trees for hierarchical multilabel classification: A case study in functional genomics. In: Fürnkranz, J., Scheffer, T., Spiliopoulou, M. (eds.) PKDD 2006. LNCS (LNAI), vol. 4213, pp. 18–29. Springer, Heidelberg (2006)
7. Cerri, R., Barros, R.C., Carvalho, A.C.P.L.F.: Hierarchical multi-label classification for protein function prediction: A local approach based on neural networks. In: Intelligent Systems Design and Applications (ISDA), pp. 337–343 (November 2011)
8. Cerri, R., Barros, R.C., Carvalho, A.C.P.L.F.: A genetic algorithm for hierarchical multi-label classification. In: Proceedings of the 27th Annual ACM Symposium on Applied Computing, SAC 2012, pp. 250–255. ACM, New York (2012)
9. Cerri, R., Barros, R.C., Carvalho, A.C.P.L.F.: Hierarchical multi-label classification using local neural networks. Journal of Computer and System Sciences (in press, 2013)
10. Clare, A., King, R.D.: Predicting gene function in saccharomyces cerevisiae. Bioinformatics 19, 42–49 (2003)
11. Costa, E.P., Lorena, A.C., Carvalho, A.C.P.L.F., Freitas, A.A., Holden, N.: Comparing several approaches for hierarchical classification of proteins with decision trees. In: Sagot, M.-F., Walter, M.E.M.T. (eds.) BSB 2007. LNCS (LNBI), vol. 4643, pp. 126–137. Springer, Heidelberg (2007)
12. Davis, J., Goadrich, M.: The relationship between precision-recall and roc curves. In: International Conference on Machine Learning, pp. 233–240 (2006)
13. Dempster, A.P., Laird, N.M., Rubin, D.B.: Maximum Likelihood from Incomplete Data via the EM Algorithm. Journal of the Royal Statistical Society 39(1), 1–38 (1977)
14. Dorigo, M.: Optimization, Learning and Natural Algorithms. Ph.D. thesis, Dipartimento di Elettronica, Politecnico di Milano, IT (1992)
15. Dorigo, M., Maniezzo, V., Colorni, A.: Positive feedback as a search strategy. Tech. rep., Dipartimento di Elettronica, Politecnico di Milano, IT (1991)
16. Friedman, N., Geiger, D., Goldszmidt, M.: Bayesian network classifiers. Machine Learning 29(2-3), 131–163 (1997)

17. García, S., Fernández, A., Luengo, J., Herrera, F.: Advanced nonparametric tests for multiple comparisons in the design of experiments in computational intelligence and data mining: Experimental analysis of power. Information Sciences 180(10), 2044–2064 (2010)

18. Kiritchenko, S., Matwin, S., Famili, A.: Functional annotation of genes using hierarchical text categorization. In: Proc. of the ACL Workshop on Linking Biological Literature, Ontologies and Databases: Mining Biological Semantics (2005)

19. Kiritchenko, S., Matwin, S., Nock, R., Famili, A.: Learning and evaluation in the presence of class hierarchies: Application to text categorization. In: Lamontagne, L., Marchand, M. (eds.) Canadian AI 2006. LNCS (LNAI), vol. 4013, pp. 395–406. Springer, Heidelberg (2006)

20. Lloyd, S.P.: Least squares quantization in pcm. IEEE Transactions on Information Theory 28(2), 129–137 (1982)

21. Nasierding, G., Tsoumakas, G., Kouzani, A.Z.: Clustering based multi-label classification for image annotation and retrieval. In: IEEE International Conference on Systems, Man and Cybernetics, pp. 4514–4519 (2009)

22. Otero, F.E.B., Freitas, A.A., Johnson, C.: A hierarchical classification ant colony algorithm for predicting gene ontology terms. In: Pizzuti, C., Ritchie, M.D., Giacobini, M. (eds.) EvoBIO 2009. LNCS, vol. 5483, pp. 68–79. Springer, Heidelberg (2009)

23. Otero, F.E.B., Freitas, A.A., Johnson, C.: A hierarchical multi-label classification ant colony algorithm for protein function prediction. Memetic Computing 2, 165–181 (2010)

24. Quinlan, J.R.: C4.5: programs for machine learning. Kaufmann Publishers Inc., San Francisco (1993)

25. Rousu, J., Saunders, C., Szedmak, S., Shawe-Taylor, J.: Kernel-based learning of hierarchical multilabel classification models. Journal of Machine Learning Research 7, 1601–1626 (2006)

26. Ruepp, A., Zollner, A., Maier, D., Albermann, K., Hani, J., Mokrejs, M., Tetko, I., Güldener, U., Mannhaupt, G., Münsterkötter, M., Mewes, H.W.: The funcat, a functional annotation scheme for systematic classification of proteins from whole genomes. Nucleic Acids Research 32(18), 5539–5545 (2004)

27. Schapire, R.E., Singer, Y.: Improved boosting algorithms using confidence-rated predictions. In: Machine Learning, vol. 37, pp. 297–336. Kluwer Academic Publishers, Hingham (1999)

28. Silla, C., Freitas, A.A.: A survey of hierarchical classification across different application domains. Data Mining and Knowledge Discovery 22, 31–72 (2011)

29. Sun, A., Lim, E.P.: Hierarchical text classification and evaluation. In: Fourth IEEE International Conference on Data Mining, pp. 521–528 (2001)

30. Sun, A., Lim, E.P., Ng, W.K., Srivastava, J.: Blocking Reduction Strategies in Hierarchical Text Classification. IEEE Transactions on Knowledge and Data Engineering 16, 1305–1308 (2004)

31. Vens, C., Struyf, J., Schietgat, L., Džeroski, S., Blockeel, H.: Decision trees for hierarchical multi-label classification. Machine Learning 73, 185–214 (2008)

Multi-core Structural SVM Training

Kai-Wei Chang, Vivek Srikumar, and Dan Roth

Dept. of Computer Science,
University of Illinois, Urbana-Champaign, IL, USA
{kchang10,vsrikum2,danr}@illinois.edu

Abstract. Many problems in natural language processing and computer vision can be framed as structured prediction problems. Structural support vector machines (SVM) is a popular approach for training structured predictors, where learning is framed as an optimization problem. Most structural SVM solvers alternate between a model update phase and an inference phase (which predicts structures for all training examples). As structures become more complex, inference becomes a bottleneck and thus slows down learning considerably. In this paper, we propose a new learning algorithm for structural SVMs called DEMI-DCD that extends the dual coordinate descent approach by decoupling the model update and inference phases into different threads. We take advantage of multi-core hardware to parallelize learning with minimal synchronization between the model update and the inference phases. We prove that our algorithm not only converges but also fully utilizes all available processors to speed up learning, and validate our approach on two real-world NLP problems: part-of-speech tagging and relation extraction. In both cases, we show that our algorithm utilizes all available processors to speed up learning and achieves competitive performance. For example, it achieves a relative duality gap of 1% on a POS tagging problem in 192 seconds using 16 threads, while a standard implementation of a multi-threaded dual coordinate descent algorithm with the same number of threads requires more than 600 seconds to reach a solution of the same quality.

1 Introduction

Many prediction problems in natural language processing, computer vision and other fields are structured prediction problems, where decision making involves assigning values to interdependent variables. The output structure can represent sequences, clusters, trees or arbitrary graphs over the decision variables. The *structural support vector machine* (structural SVM) [23] is a widely used approach for training the parameters of structured models. However, training a structural SVM is computationally expensive and this often places limits on the size of the training sets that can be used or limits the expressivity of the structures considered among the interdependent variables. Designing efficient learning algorithms for structural prediction models is therefore an important research question.

Various approaches have been proposed in the literature to learn with the structural SVM algorithm – both exact [22,12,3,4,13] and approximate [19,8]. However, these algorithms are inherently single-threaded, and extending them to a multi-core environment is not trivial. Therefore, these algorithms cannot take advantage of the multiple cores available in most modern workstations.

H. Blockeel et al. (Eds.): ECML PKDD 2013, Part II, LNAI 8189, pp. 401–416, 2013.

Existing parallel algorithms for training structural SVMs (such as [3]) use a sequence of two phases: an inference phase, where loss-augmented inference is performed using the current model, and a model update phase. Each phase is separated from the other by a barrier that prevents model update until the inference is complete and vice versa. A similar barrier exists in map-reduce implementations of the binary SVM [5] and the Perceptron [17] algorithms. Such barrier-based approaches prevent existing parallel algorithms from fully utilizing the available processors.

In this paper, we propose the DEMI-DCD algorithm (DEcoupled Model-update and Inference with Dual Coordinate Descent), a new barrier-free parallel algorithm based on the dual coordinate descent (DCD) method for the L2-loss structural SVM. DCD has been shown competitive with other optimization techniques such as the cutting plane method [23] and the Frank-Wolfe method [13]. DEMI-DCD removes the need for a barrier between the model update and the inference phases allowing us to distribute these two steps across multiple cores. We show that our approach has the following advantages:

1. DEMI-DCD requires little synchronization between threads. Therefore, it fully utilizes the computational power of multiple cores to reduce training time.
2. As in the standard dual coordinate descent approach, DEMI-DCD can make multiple updates on the structures discovered by the loss-augmented inference, thus fully utilizing the available information. Furthermore, our approach retains the convergence properties of dual coordinate descent.

We evaluate our method on two NLP applications – part-of-speech tagging and entity-relation extraction from text. In both cases, we demonstrate that not only does DEMI-DCD converge faster than existing methods to better performing solutions (according to both primal objective value and test set performance), it also fully takes advantage of all available processors unlike the other methods. For the part-of-speech tagging task, we show that with 16 threads, our approache reaches a relative duality gap of 1% in 192 seconds, while a standard multi-threaded implementation of the dual coordinate descent algorithm with 16 threads takes more than 600 seconds to reach an equivalent solution. Similarly, for the entity-relations task, our approach reaches within 1% of the optimal within 86 seconds, compared to 275 seconds for the baseline.

The rest of this paper is organized as follows. We review the structural SVM model and the DCD method in Section 2. The proposed algorithm is described in Section 3. We survey related methods in Section 4. Empirical results are demonstrated in Section 5. Section 6 provides concluding remarks and discussion.

2   Background: Structural SVM

We are given a set of training examples $\mathcal{D} = \{\mathbf{x}_i, \mathbf{y}_i\}_{i=1}^l$, where instances $\mathbf{x}_i \in \mathcal{X}$ are annotated with structures $\mathbf{y}_i \in \mathcal{Y}_i$. Here the set $\mathcal{Y}_i$ is a set of feasible structures for the ith instance. Training a structural SVMs (SSVM) [23] is framed as the problem of learning a real-valued weight vector $\mathbf{w}$ by solving the following optimization problem:

$$\min_{\boldsymbol{w},\boldsymbol{\xi}} \quad \frac{1}{2}\boldsymbol{w}^T\boldsymbol{w} + C\sum_i \ell(\xi_i) \tag{1}$$

$$s.t. \quad \boldsymbol{w}^T\Phi(\boldsymbol{x}_i,\boldsymbol{y}_i) - \boldsymbol{w}^T\Phi(\boldsymbol{x}_i,\boldsymbol{y}) \geq \Delta(\boldsymbol{y}_i,\boldsymbol{y}) - \xi_i, \quad \forall i, \boldsymbol{y} \in \mathcal{Y}_i.$$

where $\Phi(\boldsymbol{x},\boldsymbol{y})$ is the feature vector extracted from input the $\boldsymbol{x}$ and output $\boldsymbol{y}$ and $\ell(\xi)$ is the loss that needs to be minimized. The constraints in (1) indicate that for all training examples and all possible output structures, the score for the correct output structure $\boldsymbol{y}_i$ is greater than the score for other output structures $\boldsymbol{y}$ by at least $\Delta(\boldsymbol{y}_i,\boldsymbol{y})$. The slack variable $\xi_i \geq 0$ penalizes the violation. The loss ℓ is an increasing function of the slack that is minimized as part of the objective: when $\ell(\xi) = \xi$, we refer to (1) as an L1-loss structural SVM, while when $\ell(\xi) = \xi^2$, we call it an L2-loss structural SVM.[1] For mathematical simplicity, in this paper, we only consider the linear L2-loss structural SVM model, although our method can potentially be extended to other variants of the structural SVM.

Instead of directly solving (1), several optimization algorithms for SSVM consider its dual form [23,12,4] by introducing dual variables $\alpha_{i,\boldsymbol{y}}$ for each output structure $\boldsymbol{y}$ and each example $\boldsymbol{x}_i$. If $\boldsymbol{\alpha}$ is the set of all dual variables, the dual problem can be stated as

$$\min_{\boldsymbol{\alpha}>0} \quad D(\boldsymbol{\alpha}), \text{ and}$$

$$D(\boldsymbol{\alpha}) \equiv \frac{1}{2}\left\|\sum_{\alpha_{i,\boldsymbol{y}}}\alpha_{i,\boldsymbol{y}}\phi(\boldsymbol{y},\boldsymbol{y}_i,\boldsymbol{x}_i)\right\|^2 + \frac{1}{4C}\sum_i\left(\sum_{\boldsymbol{y}}\alpha_{i,\boldsymbol{y}}\right)^2 - \sum_{i,\boldsymbol{y}}\Delta(\boldsymbol{y},\boldsymbol{y}_i)\alpha_{i,\boldsymbol{y}}, \tag{2}$$

where $\phi(\boldsymbol{y},\boldsymbol{y}_i,\boldsymbol{x}_i) = \phi(\boldsymbol{y}_i,\boldsymbol{x}_i) - \phi(\boldsymbol{y},\boldsymbol{x}_i)$. The constraint $\boldsymbol{\alpha} \geq 0$ restricts all the dual variables to be non-negative (i.e., $\alpha_{i,\boldsymbol{y}} \geq 0 \forall i, \boldsymbol{y}$).

For the optimal values, the relationship between the primal optimal $\boldsymbol{w}^*$ (that is, the solution of (1)), and the dual optimal $\boldsymbol{\alpha}^*$ (that is, the solution of (2)) is

$$\boldsymbol{w}^* = \sum_{i,\boldsymbol{y}}\alpha^*_{i,\boldsymbol{y}}\phi(\boldsymbol{y},\boldsymbol{y}_i,\boldsymbol{x}_i).$$

Although this relationship only holds for the solutions, in a linear model, one can maintain a temporary vector

$$\boldsymbol{w} \equiv \sum_{i,\boldsymbol{y}}\alpha_{i,\boldsymbol{y}}\phi(\boldsymbol{y},\boldsymbol{y}_i,\boldsymbol{x}_i) \tag{3}$$

to assist the computations [10].

In practice, for most definitions of structures, the set of feasible structures for a given instance (that is, $\mathcal{Y}_i$) is exponentially large, leading to an exponentially large number of

---

[1] In L2-loss structural SVM formulation, one may replace $\Delta(\boldsymbol{y}_i,\boldsymbol{y})$ by $\sqrt{\Delta(\boldsymbol{y}_i,\boldsymbol{y})}$ to obtain an upper bound on the empirical risk [23]. However, we keep using $\Delta(\boldsymbol{y}_i,\boldsymbol{y})$ for computational and notational convenience. Thus, Eq. (1) minimizes the mean square loss with a regularization term.

dual variables. Therefore, existing dual methods [4,23] maintain an active set of dual variables $\mathcal{A}$ (also called the working set in the literature). During training, only the dual variables in $\mathcal{A}$ are considered for an update and the rest $\alpha_{i,y} \notin \mathcal{A}$ are fixed to 0. We denote the active set associated with the instance x_i as $\mathcal{A}_i$ and $\mathcal{A} = \bigcup_i \mathcal{A}_i$. The following theorem justifies the use of an active set.

Theorem 1. *Let* α^* *be the optimal solution of* (2) *and* $\mathcal{A}^* = \{\alpha^*_{i,y} \mid \alpha^*_{i,y} > 0\}$. *Then any optimal solution of*

$$\min_{\alpha \geq 0} \quad D(\alpha) \quad s.t. \ \alpha_{i,y} = 0, \quad \forall \alpha_{i,y} \notin \mathcal{A}^* \tag{4}$$

is an optimal solution of (2).

This suggests that we can reduce the size of the optimization problem by carefully identifying nonzero α's. This property of the dual is widely used for training SVMs, for example, with the cutting-plane method [23,12], with a dual coordinate descent method [3,4], and has also been used for solving binary SVM [11,10,2].

We observe that across all these methods, in a single-thread implementation, training consists of two phases:

1. Updating the values $\alpha_{i,y} \in \mathcal{A}$ (learning step), and
2. Selecting and maintaining the active set $\mathcal{A}$ (active set selection step).

The learning step usually updates each dual variable $\alpha_{i,y} \in \mathcal{A}$ several times until convergence for the current active set. The active set selection step involves solving the following loss-augmented inference problem for each example x_i:

$$\max_{y \in \mathcal{Y}_i} \quad w^T \phi(x_i, y) + \Delta(y_i, y) \tag{5}$$

Solving loss-augmented inference is usually computationally more expensive than the time for updating the model. In the traditional sequential implementations, these two steps block each other. Even if inference for each example is performed in parallel on a multi-core machine, the model update cannot be done until inference is solved for all training examples. Similarly, inference cannot start until the model update is complete. Balancing the time spent on these two parts is a crucial aspect of algorithm design. In the next section, we will show that, on a multi-core machine, we can indeed fully utilize the available computational power without the barrier between the two training steps.

3   Parallel Strategies for Structured Learning

In this section, we describe the parallel learning algorithm DEMI-DCD which decouples the model update steps from the inference steps during learning, and hence fully utilizes the computational power of multi-core machines.

Let p be the number of threads allocated for learning. DEMI-DCD first splits the training data $\mathcal{D}$ into $p-1$ disjoint parts: $\{B_j\}_{j=1}^{p-1}$ with each $B_j \subset \mathcal{D}$. Then, it generates two types of threads, a learning thread and $p-1$ active set selection threads. The jth

active set selection thread is responsible for maintaining an active set $\mathcal{A}_i$ for each example i in the part B_j. It does so by searching for candidate structures for each instance in B_j using the current model w which is maintained by the learning thread.

The learning thread loops over all the examples and updates the model w using $\alpha_{i,y} \in \mathcal{A}_i$ for the ith example. The $p - 1$ active set selection threads are independent of each other, and they share $\mathcal{A}_i$ and w with the learning thread using shared memory buffers. We will now discuss our model in detail. The algorithms executing the learning and the active set selection threads are listed as Algorithm 1 and 2 respectively.

Learning Thread. The learning thread performs a two-level iterative procedure until the stopping conditions are satisfied. It first initializes α^0 to a zero vector, then it generates a sequence of solutions $\{\alpha^0, \alpha^1, \ldots\}$. We refer to the step from α^t to α^{t+1} as the outer iteration. Within this iteration, the learning thread sequentially visits each instance x_i in the data set and updates $\alpha_{i,y} \in \mathcal{A}_i$ while all the other dual variables are kept fixed. To update $\alpha_{i,y}$, it solves the following one variable sub-problem that uses the definition of the dual objective function from Eq. (2):

$$\bar{d}_{i,y} = \arg \min_d D(\alpha + \mathbf{e}_{i,y}d) \quad \text{s.t.} \quad d\alpha_{i,y} + d \geq 0$$

$$= \arg \min_d \frac{1}{2}\|w + \phi(y, y_i, x_i)d\|^2 + \frac{1}{4C}\left(d + \sum_{y \in \mathcal{A}_i} \alpha_{i,y}\right)^2 - d\Delta(y_i, y) \quad (6)$$

$$\text{s.t.} \quad \alpha_{i,y} + d \geq 0.$$

Here, w is defined in Eq. (3) and $\mathbf{e}_{i,y}$ is a vector where only the element corresponding to (i, y) is one and the rest are zero. Eq. (6) is a quadratic optimization problem with one variable and has an analytic solution. Therefore, the update rule of $\alpha_{i,y}$ can be written as:

$$\bar{d}_{i,y} \leftarrow \frac{\Delta(y, y_i) - w^T\phi(y, y_i, x_i) - \frac{1}{(2C)}\sum_{y'} \alpha_{i,y'}}{\|\phi(y, y_i, x_i)\|^2 + \frac{1}{(2C)}}, \quad (7)$$

$$\alpha_{i,y} \leftarrow \max(\alpha_{i,y} + \bar{d}_{i,y}, 0).$$

To maintain the relation between α and w specified in Eq. (3), w is updated accordingly:

$$w \leftarrow w + d_{i,y}\phi(y, y_i, x_i). \quad (8)$$

We will now discuss two implementation issues to improve DEMI-DCD. First, during the learning, w is shared between the learning and the active set selection threads. Therefore, we would like to maintain w in a shared buffer. However, the update rules in Steps (7)-(8) can be done in $O(\bar{n})$, where $\bar{n}$ is number of average active features of $\phi(x, y)$. Thus, when the number of active features is small, the updates are quick. Hence, we require to maintain a lock to prevent the active set selection threads from accessing w when it is being updated. To reduce the cost of synchronization, we maintain a local copy of w and copy w to a shared buffer (denoted by $\bar{w}$) after every ρ updates. This reduces the overhead of synchronization.

Algorithm 1. Learning Thread

Input: Dataset $\mathcal{D}$ and the number of iterations before updating the shared buffer, ρ.
Output: The learned model w
1: $w \leftarrow 0, \alpha \leftarrow 0$, #updates $\leftarrow 0$.
2: **while** stopping conditions are not satisfied **do**
3:    **for** $i = 1 \rightarrow l$ (loop over each instance) **do**
4:       **for all** y in $\mathcal{A}_i$ **do**
5:          **if** Eq. (9) is satisfied **then**
6:             $\mathcal{A}_i \leftarrow \mathcal{A}_i \setminus \{\alpha_{i,y}\}$.
7:          **else**
8:             update corresponding $\alpha_{i,y}$ by Eq. (7)-Eq.(8),
9:             #updates $\leftarrow$ #updates + 1.
10:         **end if**
11:         **if** #updates mod $\rho = 0$ **then**
12:            Copy w to $\bar{w}$ in a shared buffer.
13:         **end if**
14:       **end for**
15:    **end for**
16: **end while**

Second, as the active set selection thread keeps adding dual variables into $\mathcal{A}$, the size of $\mathcal{A}$ grows quickly. To avoid the learning thread from wasting time on the bounded dual variables, we implement a shrinking strategy inspired by [11][2]. Specifically, if $\alpha_{i,\bar{y}}$ equals to zero and

$$-\nabla(\boldsymbol{\alpha})_{i,\bar{y}} = \Delta(\bar{\boldsymbol{y}}, \boldsymbol{y}_i) - \boldsymbol{w}^T \phi(\bar{\boldsymbol{y}}, \boldsymbol{y}_i, \boldsymbol{x}_i) - \frac{1}{2} \sum_{y \in \mathcal{A}_i} \alpha_{i,y} < \delta \qquad (9)$$

then DEMI-DCD removes $\alpha_{i,\bar{y}}$ from $\mathcal{A}_i$. Notice that the shrinking strategy is more aggressive if δ is large. For binary classification, a negative δ is usually used. This is because, in the binary classification case, the size of the data is usually large (typically millions of examples); therefore incorrectly removing an instance from the active set requires a lengthy process that iterates over all the examples to add it back. However, in our case, aggressive shrinking strategy is safe because the active set selection threads can easily add the candidate structures back. Therefore, in our implementation, we set δ to 0.01.

Active Set Selection Threads. As mentioned, the jth active set selection thread iterates over all instances $\boldsymbol{x}_i \in B_j$ and selects candidate active variables based on solving an augmented inference problem (line 4 in Algorithm 2), that is, eq. (5). Just like the learning thread, each active set selection thread maintains a local copy of w. This setting aims to prevent w from being changed while the loss augmented inference is being solved, thus avoiding a possibly suboptimal solution. The local copy of w will be updated from $\bar{w}$ in the shared buffer after each ρ iterations.

---

[2] This shrinking strategy is also related to the condition used by cutting plane methods for adding constraints into the working set.

Algorithm 2. The jth Active Set Selection Thread

Input: Part of dataset for this thread $\mathcal{B}_j$, and the number of iterations before updating the shared buffer, ρ.

1: $\boldsymbol{w} \leftarrow \boldsymbol{0}$ (a local copy), #Inference $\leftarrow$ 0,
2: **while** Learning thread is not stopped **do**
3:      **for all** $(\boldsymbol{x}_i, \boldsymbol{y}_i) \in \mathcal{B}_j$ **do**
4:          $\bar{\boldsymbol{y}} \leftarrow f_{\text{AugInf}}(\boldsymbol{w}, \boldsymbol{x}_i, \boldsymbol{y}_i)$.
5:          **if** $\bar{\boldsymbol{y}} \notin \mathcal{A}_i$ **then**
6:              $\mathcal{A}_i \leftarrow \{\mathcal{A}_i \cup \bar{\boldsymbol{y}}\}$.
7:          **end if**
8:          #Inference $\leftarrow$ #Inference + 1
9:          **if** #Inference mod $\rho = 0$ **then**
10:             Copy from the model $\bar{\boldsymbol{w}}$ in shared buffer to $\boldsymbol{w}$.
11:         **end if**
12:     **end for**
13: **end while**

Synchronization. Our algorithm requires little synchronization between threads. In fact, only the learning thread can write to $\bar{\boldsymbol{w}}$ and only the jth active set selection thread can modify $\mathcal{A}_i$ for any $i \in \mathcal{B}_j$. It is very unlikely, but possible, that the learning thread and the inference threads will be reading/writing $\bar{\boldsymbol{w}}$ or $\mathcal{A}_i$ concurrently. To avoid this, one can use a mutex lock to ensure that the copy operation is atomic. However, in practice, we found that this synchronization is unnecessary.

3.1 Analysis

In this section, we analyze the proposed multi-core algorithm. We observed that α's can be added to the $\mathcal{A}$ by the active set selection thread, but might be removed by the learning thread with the shrinking strategy. Therefore, we define $\bar{\mathcal{A}}$ to be a subset of $\mathcal{A}$ that contains $\alpha_{i,\boldsymbol{y}}$ which has been visited by the learning thread at least once and remains in the $\mathcal{A}$.

Theorem 2. *The number of variables which have been added to $\bar{\mathcal{A}}$ during the entire training process is bounded by $O(1/\delta^2)$.*

The proof follows from [4, Theorem 1]. This theorem says that the size of $\bar{\mathcal{A}}$ is bounded as a function of δ.

Theorem 3. *If $\bar{\mathcal{A}} \neq \emptyset$ is not expanded, then the proposed algorithm converges to an ϵ-optimal solution of*

$$\min_{\boldsymbol{\alpha} \geq 0} \quad D(\boldsymbol{\alpha}) \quad s.t. \ \alpha_{i,\boldsymbol{y}} = 0, \forall \boldsymbol{y} \notin \mathcal{A}_i \tag{10}$$

in $O(log(1/\epsilon))$ steps.

If $\bar{\mathcal{A}}$ is fixed, our learning thread performs standard dual coordinate descent, as in [10]. Hence, this theorem follows from the analysis of dual coordinate descent. The global convergence rate of the method can be inferred from [24] which generalizes the proof in [15].

Theorem 2 shows that the size of $\bar{A}$ will eventually stop growing. Theorem 3 shows that when $\bar{A}$ is fixed, the weight vector w converges to the optimum of (10). Hence, the local copies in the learning thread and the active set selection threads can be arbitrary close. Following the analysis in [4], the convergence of DEMI-DCD then follows.

4   Related Work

Several related works have a resemblance to the method proposed in this paper. In the following, we briefly review the literature and discuss the connections.

A Parallel Algorithm for Dual Coordinate Descent. The structural SVM package JLIS [3] implements a parallel algorithm in a Master-Slave architecture to solve Eq. (2).[3], Given p processors, it first splits the training data into p parts. Then the algorithm maintains a model w, dual variables α, and an active set A and updates these in an iterative fashion. In each iteration, the master thread sends one part of the data to a slave thread. For each slave thread, it iterates over each assigned example x_i, and picks the best structures $\bar{y}$ according to current w. Then $(x_i, \bar{y})$ is added into the active set A if the following condition is satisfied:

$$\Delta(\bar{y}, y_i) - w^T\phi(\bar{y}, y_i, x_i) - \frac{1}{2}\sum_{y \in A_i} \alpha_{i,y} > \delta. \tag{11}$$

Only after all the slave threads have finished processing all the examples, the master thread performs dual coordinate descent updates to solve the following optimization loosely:

$$\min_{\alpha \geq 0} \quad D(\alpha) \quad \text{s.t. } \alpha_{i,y} = 0, \forall y \notin A_i.$$

The algorithm stops when a stopping condition is reached. This approach is closely related to the n-slack cutting plane method for solving structural SVM [23]. However, [23] assumes that the sub-problem is solved exactly[4], while this restriction can be relaxed under a dual coordinate descent framework.

We will refer to this approach for parallelizing structural SVM training as MS-DCD (for Master-Slave dual coordinate descent) and compare it experimentally with DEMI-DCD in Section 5.

Structured Perceptron and Its Parallel Version. The Structured Perceptron [6] algorithm has been widely used in the literature. At each iteration, it picks an example x_i that is annotated with y_i and finds its best structured output $\bar{y}$ according to the current model w using an inference algorithm. Then, the model is updated as

$$w \leftarrow w + \eta(\phi(x_i, y_i) - \phi(x_i, \bar{y})),$$

---

[3] The implementation of MS-DCD can be downloaded at
http://cogcomp.cs.illinois.edu/page/software_view/JLIS
[4] The cutting-plane solver for structural SVM usually sets a tolerance parameter, and stops the sub-problem solver when the inner stopping criteria is satisfied.

where η is a learning rate. Notice that the Structural Perceptron requires an inference step before each update. This makes it different from the dual methods for structured SVMs where the inference step is used to update the active set. One important advantage of the dual methods is that they can perform multiple updates on the elements in the active set without doing inference for each update. As we will show in Section 5, this limits the efficiency of the Perceptron algorithm. Some caching strategies have been developed for structural Perceptron. For example, [7] introduces a caching technique to periodically update the model with examples on which the learner had made mistakes in previous steps. However, this approach is ad-hoc without convergence guarantees.

A parallelization strategy for structured Pereceptron (SP-IPM) has been proposed [18] using the Map-Reduce framework. It calls for first splitting the training data into several parts. In the map phase, it distributes data to each mapper and runs a separate Perceptron learner on each shard in parallel. Then, in the reduce phase, the models are mixed using a linear combination. The mixed model serves as the initial model for the next round. In this paper, we implement this map-reduce framework as a multi-thread program. SP-IPM requires barrier synchronizations between the map and reduce phases, which limits the computational performance. In addition, in the model mixing strategy, each local model is updated using exclusive data blocks. Therefore, the mappers might make updates that are inconsistent with each other. As a result, it requires many iterations to converge.

General Parallel Algorithms for Convex Optimization. Some general parallel optimization algorithms have been proposed in the literature. For example, delayed stochastic (sub-)gradient descent methods have been studied with assumptions on smoothness of the problem [1,14]. However, their applicability to structured SVM has not been explored.

5 Experiments

In this section, we show the effectiveness of the DEMI-DCD algorithm compared to other parallel structured learning algorithms.

5.1 Experimental Setup

We evaluate our algorithm on two natural language processing tasks: part-of-speech tagging (POS-WSJ) and jointly predicting entities and their relations (Entity-Relation). These tasks, which have very different output structures, are described below.

POS Tagging (POS-WSJ). POS tagging is the task of labeling each word in a sentence with its part of speech. This task is typically modeled as a sequence labeling task, where each tag is associated with emission features that capture word-tag association and transition features that capture sequential tag-tag association. For this setting, inference can be solved efficiently using the Viterbi algorithm.

We use the standard Penn Treebank Wall Street Journal corpus [16] to train and evaluate our POS tagger using the standard data split for training (sections 2-22, 39832 sentences) and testing (section 23, 2416 sentences). In our experiments, we use indicators for the conjunction between the word and tags as emission features and pairs of tags as transition features.

Entity and Relation Recognition (Entity-Relation). This is the task of assigning entity types to spans of text and identifying relations among them [20]. For example, for the sentence *John Smith resides in London*, we would predict *John Smith* to be a PERSON, LONDON to be a LOCATION and the relation LIVES-IN between them.

As in the original work, we modeled prediction as a 0-1 linear program (ILP) where binary indicator variables capture all possible entity-label decisions and entity pair label decisions (that is, relation labels). Linear constraints force exactly one label to be assigned with each entity and relation. In addition, the constraints also encode background knowledge about the types of entities allowed for each relation label. For example, a LIVES-IN relation can only connect a PERSON to a LOCATION. We refer the reader to [21] for further details. Unlike the original paper, which studied a decomposed learning setting, we jointly train the entity-relation model using an ILP for the loss-augmented inference. We used the state-of-the-art Gurobi ILP solver [9] to solve inference for this problem both during training and test. We report results using the annotated data from [21] consisting of 5925 examples with an 80-20 train-test split.

We based our implementation on the publicly available implementation of MS-DCD, which implements a dual coordinate descent method for solving structural SVM. DCD [4] has been shown competitive comparing to other L2-loss and L1-loss structural SVM solvers such as a 1-slack variable cutting-plane method [12] and a Frank-Wolfe optimization method [13] when using one CPU core.

As described in Section 3, our method is an extension of DCD and further improves its performance using multiple cores. All the algorithms are implemented in Java. We conducted our experiments on a 24-core machine with Xeon E5-2440 processors running 64-bit Scientific Linux. Unless otherwise stated, all results use 16 threads. We set the value of C to 0.1 for all experiments.

Our experiments compare the following three methods:

1. DEMI-DCD: the proposed algorithm described in Section 3. We use one thread for learning and the rest for inference (that is, active set selection).
2. MS-DCD: A master-slave style parallel implementation of dual coordinate descent method from JLIS package [3] described in Section 4.
3. SP-IPM: A parallel structural Perceptron algorithm proposed in [18]. We run 10 epochs of Perceptron updates for each shard at each outer iteration.

Note that the first two methods solve an L2-Structural SVM problem (1) and converge to the same minimum.

Our experiments answer the following research question: Can DEMI-DCD make use of the available CPU resources to converge faster to a robust structural prediction model? To answer this, for both our tasks, we first compare the convergence speed of DEMI-DCD and MS-DCD. Second, we compare the performance of the three algorithms on the test sets of the two tasks. Third, we show that DEMI-DCD maximally utilizes all available CPU cores unlike the other two algorithms. Finally, we report the results of an analysis experiment, where we show the performance of our algorithm with different number of available threads.

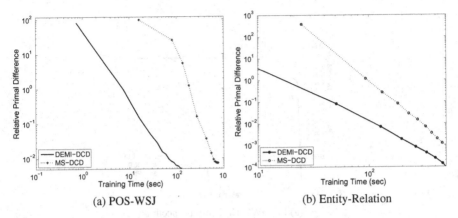

(a) POS-WSJ   (b) Entity-Relation

Fig. 1. Relative primal function value difference to the reference model versus wall clock time. See the text for more details.

5.2 Empirical Convergence Speed

First, we compare DEMI-DCD to MS-DCD in terms of their speed of convergence in terms of objective function value. We omit SP-IPM in this comparison because it does not solve the SVM optimization problem. Figure 1 shows the relative primal objective function (that is, the value of the objective in Eq. (1)) with respect to a reference model w^* as a function of wall-clock training time. In other words, if the objective function is denoted by f, we plot $(f(w) - f(w^*))/f(w^*)$ as w varies with training time for each algorithm. The reference model is the one that achieves the lowest primal value among the compared models. In the figure, both the training time and the relative difference are shown in log-scale. From the results, we see that the proposed method is faster than MS-DCD on both tasks. DEMI-DCD is especially faster than MS-DCD in the early stage of optimization. This is important because usually we can achieve a model with reasonable generative performance before solving (1) exactly. Note that the inference in Entity-Relation is much slower than inference for POS-WSJ. For example, at each iteration on POS-WSJ, MS-DCD takes 0.62 seconds to solve inference on all the training samples using 16 threads and takes 3.87 seconds to update the model using 1 thread, while it takes 10.64 seconds to solve the inference and 8.45 seconds to update the model on Entity-Relation. As a result, the difference between DEMI-DCD and MS-DCD is much higher on POS-WSJ. We will discuss the reason for this in Section 5.3.

Next, we show that the proposed method can obtain a reasonable and stable solution in a short time. Figure 2 shows the test set performance of the three algorithms as training proceeds. For POS-WSJ, we evaluate the model using token-based accuracy. For Entity-Relation, we evaluate the entity and relation labels separately and report the micro-averaged F1 over the test set. In all cases, DEMI-DCD is the fastest one to achieve a converged performance. As mentioned, DEMI-DCD is more efficient than MS-DCD in the early iterations. As a result, it takes less time to generate a reasonable model. Note that SP-IPM achieves better final performance on the POS-WSJ task. This is so because SP-IPM converges to a different model from DEMI-DCD and

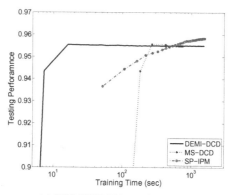

(a) POS-WSJ: Token-wise accuracy

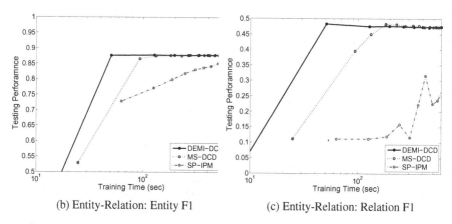

(b) Entity-Relation: Entity F1   (c) Entity-Relation: Relation F1

Fig. 2. Performance of POS-WSJ and Entity-Relation plotted as a function of wall clock training time. For Entity-Relation, we report the F1 scores of both entity and relation tasks. Note that the X-axis is in **log scale**. We see that DEMI-DCD converges faster to better performing models.

MS-DCD. For both entities and relations, SP-IPM converges slowly and, moreover, its performance is unstable for the relations. We observe that the convergence of SP-IPM is slow because the Perceptron algorithm needs to solve an inference problem before each update. When the inference is slow, the number of updates in SP-IPM is significantly smaller than DEMI-DCD and MS-DCD. For example, in the Entity-Relation task, DEMI-DCD makes around 55 million updates within 100 seconds, while SP-IPM only makes 76 thousand updates in total.[5] In other words, DEMI-DCD is faster and better than SP-IPM because it performs many inexpensive updates.

---

[5] SP-IPM has 16 threads running on different shards of data in parallel. We simply sum up all the updates on different shards.

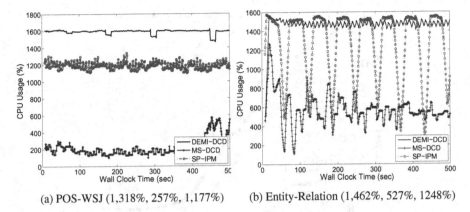

(a) POS-WSJ (1,318%, 257%, 1,177%)   (b) Entity-Relation (1,462%, 527%, 1248%)

Fig. 3. CPU usage of each method during training. The numbers listed in the caption are the average CPU usage percentage per second for DEMI-DCD, MS-DCD, SP-IPM, respectively. We show moving averages of CPU usage with a window of 2 seconds. Note that we allow all three algorithms to use 16 threads in a 24 core machine. Thus, while all the algorithms can report upto 1600% CPU usage, only DEMI-DCD consistently reports high CPU usage.

5.3   CPU Utilization

Finally, we investigate the CPU utilization of the three algorithms by plotting the *moving average* of CPU usage in Figure 3 during training, as reported by the Unix command top. Since we provide all algorithms with 16 threads, the maximum CPU utilization can be 1600%. The results show that DEMI-DCD almost fully utilizes the available CPU resources.

We see that neither baseline manages to use the available resources consistently. The average CPU usage for MS-DCD on POS-WSJ is particularly small because the inference step on this task is relatively easy (i.e. using the Viterbi algorithm). In fact, MS-DCD spends only around 25% of the time on the inference steps, which is solved by the slave threads in parallel. Both MS-DCD and SP-IPM require a barrier to ensure that the subtasks sent to the slave threads are complete. This barrier limits the CPU utilization and hence slows down the learners. Since inference is more expensive (i.e. an ILP call) for the Entity-Relation case, more time is spent on inference in the learning algorithms. Since this step is distributed across the slaves for both MS-DCD and SP-IPM, we see periods of high CPU activity followed by low activity (when only the master thread is active).

5.4   Performance with Different Number of Threads

In our final set of experiments, we study the performance of DEMI-DCD for different number of threads. Figure 4 shows the change in the primal objective function value difference as a function of training time for different number of threads. Note that the

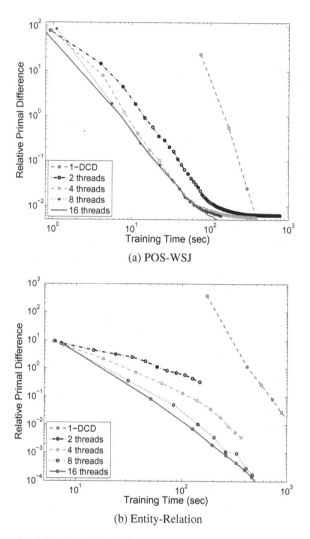

Fig. 4. Relative primal function value difference along training time using different number of threads. We also show a DCD implementation using one CPU core (1-DCD). Both x-axis and y-axis are in **log scale**.

training time and the function values are shown in *log-scale*. For comparison, we also show the performance of a DCD implementation using only one CPU core (1-DCD). As can be seen in the figure, the training time is reduced as the number of threads increases. With multiple threads, DEMI-DCD is significantly faster than 1-DCD. However, when the number of threads is more than 8, the difference is small. That is because the inference step (i.e, the active set selection step) is no longer the bottleneck.

6 Discussion and Conclusion

In this paper, we have proposed a new learning algorithm for training structural SVMs, called DEMI-DCD. This algorithm decouples the model update and inference phases of learning allowing us to execute them in parallel, thus allowing us taking advantage of multi-core machines for training. We showed that the DEMI-DCD algorithm converges to the optimum solution of the structural SVM objective. We experimentally evaluated our algorithm on the structured learning tasks of part-of-speech tagging and entity-relation extraction and showed that it outperforms existing strategies for training structured predictors in terms of both convergence time and CPU utilization.

The approach proposed in this paper opens up several directions for future research. Here, we have considered the case of a single learning thread that updates the models using the available active sets. A promising direction for future exploration is to explore the possibility of using multiple learning threads to update the models.

For some structured prediction problems, the output structure may be too complex for inference to be solved in a tractable fashion. For such cases, learning schemes with approximate inference have been proposed (e.g., [8,19]). Incorporating approximate inference into our method is an interesting topic for future study.

Acknowledgments. We thank Ming-Wei Chang for helpful discussions. This material is based on research sponsored by DARPA under agreement number FA8750-13-2-0008. The U. S. Government is authorized to reproduce and distribute reprints for Governmental purposes notwithstanding any copyright notation thereon. The views and conclusions contained herein are those of the authors and should not be interpreted as necessarily representing the official policies or endorsements, either expressed or implied, of DARPA or the U. S. Government. This work is also supported by an ONR Award on "Guiding Learning and Decision Making in the Presence of Multiple Forms of Information."

References

1. Agarwal, A., Duchi, J.: Distributed delayed stochastic optimization. In: Shawe-Taylor, J., Zemel, R., Bartlett, P., Pereira, F., Weinberger, K. (eds.) NIPS, pp. 873–881 (2011)
2. Chang, K., Roth, D.: Selective block minimization for faster convergence of limited memory large-scale linear models. In: KDD (2011)
3. Chang, M., Srikumar, V., Goldwasser, D., Roth, D.: Structured output learning with indirect supervision. In: ICML (2010)
4. Chang, M.W., Yih, W.T.: Dual coordinate descent algorithms for efficient large margin structural learning. Transactions of the Association for Computational Linguistics (2013)
5. Chu, C., Kim, S.K., Lin, Y., Yu, Y., Bradski, G., Ng, A.Y., Olukotun, K.: Map-reduce for machine learning on multicore. In: NIPS, vol. 19, p. 281 (2007)
6. Collins, M.: Discriminative training methods for hidden Markov models: Theory and experiments with perceptron algorithms. In: EMNLP (2002)
7. Collins, M., Roark, B.: Incremental parsing with the perceptron algorithm. In: ACL (2004)
8. Finley, T., Joachims, T.: Training structural SVMs when exact inference is intractable. In: ICML, pp. 304–311 (2008)

9. Gurobi Optimization, Inc.: Gurobi optimizer reference manual (2012)
10. Hsieh, C.J., Chang, K.W., Lin, C.J., Keerthi, S.S., Sundararajan, S.: A dual coordinate descent method for large-scale linear SVM. In: ICML (2008)
11. Joachims, T.: Making large-scale SVM learning practical. In: Schölkopf, B., Burges, C., Smola, A. (eds.) Advances in Kernel Methods - Support Vector Learning (1999)
12. Joachims, T., Finley, T., Yu, C.N.: Cutting-plane training of structural svms. Machine Learning (2009)
13. Lacoste-Julien, S., Jaggi, M., Mark Schmidt, P.P.: Block-coordinate Frank-Wolfe optimization for structural SVMs. In: ICML (2013)
14. Langford, J., Smola, A.J., Zinkevich, M.: Slow learners are fast. In: NIPS, pp. 2331–2339 (2009)
15. Luo, Z.Q., Tseng, P.: Error bounds and convergence analysis of feasible descent methods: a general approach. Annals of Operations Research 46, 157–178 (1993)
16. Marcus, M.P., Santorini, B., Marcinkiewicz, M.A.: Building a large annotated corpus of english: The penn treebank. Computational Linguistics
17. McDonald, R., Hall, K., Mann, G.: Distributed training strategies for the structured Perceptron. In: Human Language Technologies: The 2010 Annual Conference of the North American Chapter of the Association for Computational Linguistics, Los Angeles, California, pp. 456–464. Association for Computational Linguistics (June 2010)
18. McDonald, R.T., Hall, K., Mann, G.: Distributed training strategies for the structured perceptron. In: HLT-NAACL, pp. 456–464 (2010)
19. Meshi, O., Sontag, D., Jaakkola, T., Globerson, A.: Learning efficiently with approximate inference via dual losses. In: ICML (2010)
20. Roth, D., Yih, W.: A linear programming formulation for global inference in natural language tasks. In: Ng, H.T., Riloff, E. (eds.) CoNLL (2004)
21. Roth, D., Yih, W.: Global inference for entity and relation identification via a linear programming formulation. In: Getoor, L., Taskar, B. (eds.) Introduction to Statistical Relational Learning (2007)
22. Taskar, B., Chatalbashev, V., Koller, D., Guestrin, C.: Learning structured prediction models: a large margin approach. In: ICML (2005)
23. Tsochantaridis, I., Joachims, T., Hofmann, T., Altun, Y.: Large margin methods for structured and interdependent output variables. Journal of Machine Learning Research (2005)
24. Wang, P.W., Lin, C.J.: Iteration complexity of feasible descent methods for convex optimization. Technical Report, National Taiwan University (2013)

Multi-label Classification with Output Kernels

Yuhong Guo[1] and Dale Schuurmans[2]

[1] Department of Computer and Information Sciences
Temple University
Philadelphia, PA 19122, USA
yuhong@temple.edu
[2] Department of Computing Science
University of Alberta
Edmonton, AB, T6G 2E8, Canada
dale@cs.ualberta.ca

Abstract. Although multi-label classification has become an increasingly important problem in machine learning, current approaches remain restricted to learning in the original label space (or in a simple linear projection of the original label space). Instead, we propose to use *kernels* on output label vectors to significantly expand the forms of label dependence that can be captured. The main challenge is to reformulate standard multi-label losses to handle kernels between output vectors. We first demonstrate how a state-of-the-art large margin loss for multi-label classification can be reformulated, exactly, to handle output kernels as well as input kernels. Importantly, the pre-image problem for multi-label classification can be easily solved at test time, while the training procedure can still be simply expressed as a quadratic program in a dual parameter space. We then develop a projected gradient descent training procedure for this new formulation. Our empirical results demonstrate the efficacy of the proposed approach on complex image labeling tasks.

1 Introduction

Multi-label classification is a central problem in modern data analysis, where complex data items, such as documents, images and videos, exhibit multiple concepts of interest and thus belong to multiple non-overlapping categories. For example, in text categorization, a news article or web page is often relevant to a set of topics; similarly, in image labeling, an image can contain multiple objects and therefore be assigned multiple class labels. Although multi-label classification has been well investigated, it continues to receive significant attention. Initial work considered transforming multi-label classification to a set of independent binary classification problems [1], but this approach proved unsatisfactory as it failed to exploit label interdependence [2]. A key issue has since become capturing label dependence to improve multi-label classification accuracy. Many approaches have been developed to exploit label dependence in multi-label learning, including pairwise dependence methods [3, 4], large-margin methods [5–8],

H. Blockeel et al. (Eds.): ECML PKDD 2013, Part II, LNAI 8189, pp. 417–432, 2013.
© Springer-Verlag Berlin Heidelberg 2013

ranking based methods [9–12], and probabilistic graphical models [13–15]. Unfortunately, these methods work in the original label space, limiting their ability to capture complex dependence structure in a computationally efficient manner.

There has been recent interest in multi-label methods that work in *transformed* label spaces [16–21], primarily based on low-dimensional projections of high dimensional label vectors. For example, random projections [16], maximum eigenvalue projections [18, 17], and Gaussian random projections [21] provide techniques for mapping high dimensional label vectors to low dimensional codewords to improve the efficiency of multi-label learning. Canonical correlation analysis (CCA) has also been considered for relating inputs to label projections [20]. However, these projection approaches divide the learning problem into separate dimensionality reduction and training steps, which does not ensure that the reduced output representation is amenable to predictor training. Max margin output coding [19], on the other hand, combines output projection and prediction model learning in a joint optimization, but it must consider every label combination while ignoring the residual error from the projected representation back to the original label set. These methods primarily focus on reducing output dimension to improve efficiency, rather than attempt to explicitly capture richer label dependence. Moreover, the proposed label vector projections are limited to linear transformations, which cannot capture nonlinear label dependence.

Instead, in this paper we propose a new multi-label classification approach that uses *output kernels* to capture more complex *nonlinear* dependences between labels in a flexible yet tractable manner. Such an approach significantly expands the form of label dependences that can be captured, both at training and test time. Although kernel methods have been widely used for expanding input representations, kernels have yet to be used to explicitly capture nonlinear output structure in multi-label classification. We base our formulation on a recent large margin multi-label approach that minimizes *calibrated separation ranking loss* [8]. Such a loss achieves state-of-the-art results in multi-label classification, but it makes kernelization a challenge because it is different from any loss formulation that has been previously shown to be kernelizable. Demonstrating that a tailored multi-label loss can be equivalently re-expressed in terms of output kernels is one of the key contributions of this paper.

After reviewing related work on learning with output kernels in Section 2, we introduce the main multi-label classification formulation we use in Section 3. Our formulation is based on the calibrated separation ranking loss of [8], which we show can be equivalently re-expressed by an output kernel in Section 4. In particular, we produce a quadratic program in dual parameter space that encodes both the outputs and inputs in kernel forms. We also show that the pre-image problem for multi-label classification can be easily solved at test time. A scalable projected gradient descent optimization algorithm is then presented in Section 5. Finally, we conduct experiments on multi-label data in Section 6, and compare to standard multi-label classification. Our results demonstrate the efficacy of the proposed approach when the labels demonstrate complex dependence structure. We conclude the paper with a brief discussion of future work in Section 7.

2   Related Work: Learning with Output Kernels

Although other losses have been re-expressed in terms of output kernels, current formulations have either assumed a *least squares* loss or a simple 0-1 misclassification loss. These standard losses make the extension to output kernels straightforward, but developing a similar extension for the more complex loss we consider for multi-label classification is a greater challenge.

To re-express a problem in terms of a kernel over an output space $\mathcal{Y}$, one assumes there is a feature map $\varphi : \mathcal{Y} \to \mathcal{H}_{\mathcal{Y}}$ that maps each output label vector $\mathbf{y}$ into a new representation $\varphi(\mathbf{y})$. A kernel between output vectors can then be defined by an inner product between two output label vectors in the new representation space (formally, in a reproducing kernel Hilbert space [22, 23]). Such formulations have already been explored in machine learning, but they are often hampered by an intractable *pre-image problem* at test time [24]: for a given test instance, even though the similarity between any candidate output and training outputs can be determined easily, the search for the optimal test output can be a hard computational problem [23]. We will seek to avoid such intractability in our method.

Previous work on multi-class (not multi-label) classification learning has demonstrated that training can be equivalently expressed in terms of an output kernel when the classes are *disjoint* [25–27]. In particular, extensions to output kernels have been achieved for unsupervised and semi-supervised logistic regression training with hidden output variables [25], and convex reformulations of unsupervised and semi-supervised training of support vector machines [26, 27]. It turns out that output kernelization is trivially achieved in this special case simply by using the linear kernel between class indicator vectors. However, in these contexts, this extension is only used to achieve convex reformulations of the training process, not to expand the set of output dependence structures that can be captured. Moreover, these approaches generally involve learning an output kernel via expensive semi-definite programming.

Applying kernel methods in the output space has also been exploited in regression methods for structured output learning [28–37]. For example, in [28–30], regression models are trained by least squares to predict an output kernel matrix K_y from an input kernel matrix K_x. In [35, 36], similar methods are developed for transductive link prediction and regression to fixed output kernel values extracted from given link labels. The methods in [32–34] extend tree-based regression to kernelized output spaces for structured data, but do not exploit kernels defined over the input space. A related approach is to adopt a joint kernel over input/output pairs [38]. Unfortunately, all of these regression based approaches require the solution of a difficult pre-image problem to recover the predictions for any test instance. Furthermore, none of these methods directly address multi-label classification.

Other recent work has proposed to learn a covariance matrix between labels in a multi-label setting to capture dependence [8, 39–41]. However, these methods do not produce a kernel representation in the output space; rather, their output representations remain restricted to the original label set. Our goal in this paper is to exploit kernels to capture complex *nonlinear* dependence between labels for multi-label classification.

3   Background: Large Margin Multi-label Classification

To address multi-label classification we consider a large margin approach to classifier training. By optimizing a discriminative objective, large margin methods have proved successful in practice, achieving both good generalization performance and computational efficiency. We will therefore focus on the calibrated separation ranking loss criterion of [8], which achieves state-of-the-art multi-label classification results while retaining the simplicity and efficiency of a large margin approach. This loss expresses the sum of two large margin losses, one between the prediction value of the least positive label response and the value of a dummy threshold class, and the other between the prediction value of the least negative label response and the value of the dummy threshold class. Such an approach allows the predictions to be coordinated across different labels simply by using a shared adaptive threshold, rather than suffering the intractability of considering all label subsets [42] or even the cost of considering all label pairs (followed by a difficult labeling problem at test time) [9].

Definitions and Notation: To formulate the approach, we introduce some definitions and notation. $\mathcal{X}$ and $\mathcal{Y}$ denote the input and output spaces respectively. Below we will use capital letters to denote matrices, bold lower-case letters to denote column vectors, and regular lower-case letters to denote scalars, unless special declaration is given. Given a vector $\mathbf{x}$, $\|\mathbf{x}\|_2$ denotes its Euclidean norm. Given a matrix X, $\|X\|_F^2$ denotes its Frobenius norm. We use $X_{i:}$ to denote the ith row of a matrix X, use $X_{:j}$ to denote the jth column of X, and use X_{ij} to denote the entry at the ith row and jth column of X. For matrices, we use $\|X\|$ to refer to a generic norm on X, and use tr to denote trace. We use I_d to denote a $d \times d$ identity matrix; and use $\mathbf{1}$ to denote a column vector with all 1 entries, generally assuming its length can be inferred from context. Inequalities $\geq, \leq$ are applied entrywise. For a boolean label matrix Y we let $\bar{Y}$ denote its complement $\bar{Y} = \mathbf{11}^\top - Y$. Finally, we use $\circ$ to denote Hadamard (componentwise) product.

To introduce the underlying approach, assume one is given an input data matrix $X \in \mathbb{R}^{t \times d}$ and label indicator matrix $Y \in \{0, 1\}^{t \times L}$, where L denotes the number of classes. For convenience, we assume a feature function $\phi : \mathcal{X} \to \mathcal{H}_\mathcal{X}$ is provided that maps each input vector $\mathbf{x}$ into a new representation $\phi(\mathbf{x})$ in the Hilbert space $\mathcal{H}_\mathcal{X}$. Therefore the input data X can be putatively converted (row-wise) into a feature matrix $\Phi := \phi(X)$. Given an input instance $\mathbf{x}$, an L dimensional response vector $\mathbf{s}(\mathbf{x}) := \phi(\mathbf{x})^\top W$ can be recovered using parameter matrix W, giving a "score" for each label. These scores can then be compared to a threshold value $s_0(\mathbf{x}) := \phi(\mathbf{x})^\top \mathbf{u}$, using a parameter vector $\mathbf{u}$, to determine which labels are to be 'on' and 'off' respectively. In particular, the classification of a test example $\mathbf{x}$ is determined by

$$y_l^* = \arg \max_{y_l \in \{0,1\}} y_l(s_l(\mathbf{x}) - s_0(\mathbf{x})), \tag{1}$$

for each candidate label $l \in \{1, ..., L\}$.

To learn the parameters, W and $\mathbf{u}$, of this score based multi-label classifier, we consider the calibrated separation ranking loss of [8], given by:

$$\max_{l \in Y_{i:}}(1 + s_0(X_{i:}) - s_l(X_{i:}))_+ + \max_{\bar{l} \in \bar{Y}_{i:}}(1 + s_{\bar{l}}(X_{i:}) - s_0(X_{i:}))_+. \tag{2}$$

Intuitively, this training loss encourages the model to produce scores a minimum margin above the threshold value for 'on' labels, and a minimum margin below the threshold value for 'off' labels. In previous work, [8] demonstrates that this loss achieves state-of-the-art generalization performance across a range of multi-label data sets while retaining efficient training and test procedures.

To allow efficient optimization, training with the calibrated separation ranking loss under standard Euclidean regularization of the parameters can be formulated as a convex quadratic program (where we have rewritten the formulation given in [8] in a more compact matrix form):

$$\min_{W,\mathbf{u},\boldsymbol{\xi},\boldsymbol{\eta}} \frac{\alpha}{2}(\|W\|_F^2 + \|\mathbf{u}\|_2^2) + \mathbf{1}^\top \boldsymbol{\xi} + \mathbf{1}^\top \boldsymbol{\eta} \tag{3}$$
$$\text{s.t.} \quad \boldsymbol{\xi} \geq 0, \quad \boldsymbol{\xi}\mathbf{1}^\top \geq Y \circ (\mathbf{1}\mathbf{1}^\top + \Phi(\mathbf{u}\mathbf{1}^\top - W)),$$
$$\boldsymbol{\eta} \geq 0, \quad \boldsymbol{\eta}\mathbf{1}^\top \geq \bar{Y} \circ (\mathbf{1}\mathbf{1}^\top - \Phi(\mathbf{u}\mathbf{1}^\top - W)).$$

Training with respect to a kernel over the input space can then be easily achieved by considering the dual of this quadratic program [8], given by:

$$\max_{M,N} \mathbf{1}^\top(M + N)\mathbf{1} - \frac{1}{2\alpha}\text{tr}((M - N)^\top K(M - N)(I + \mathbf{1}\mathbf{1}^\top)) \tag{4}$$
$$\text{s.t.} \quad M \geq 0, \; M\mathbf{1} \leq 1, \; M \circ \bar{Y} = 0,$$
$$N \geq 0, \; N\mathbf{1} \leq 1, \; N \circ Y = 0,$$

where $K = \Phi\Phi^\top$, M and N are both $t \times L$ dual parameter matrices. Here the primal solution is related to the dual solution by:

$$W = \frac{1}{\alpha}X^\top(M - N), \quad \mathbf{u} = \frac{1}{\alpha}X^\top(N - M)\mathbf{1}. \tag{5}$$

Thus, one reaches the conclusion that the original training problem can be expressed in terms of a kernel on the input space. However, the target output labels appear linearly in the constraints in both the primal and dual formulations. It is not obvious how these constraints can be equivalently re-expressed in terms of a kernel between output vectors.

4   Multi-label Classification with Output Kernels

A main contribution of this paper is to derive an equivalent formulation to (4) that is expressed entirely in terms of a kernel between output vectors. Such a formulation allows one to express multi-label classification in a manner that can flexibly capture nonlinear dependence between output labels.

We start by making the assumption that $L < t$; that is, there are more training examples than labels, which is a natural assumption in many applications. First observe that, since $M \geq 0$ and $\bar{Y} \geq 0$ in (4), the constraint $M \circ \bar{Y} = 0$ can be equivalently re-expressed as $\mathrm{tr}(M^\top \bar{Y}) = 0$; similarly, the constraint $N \circ Y = 0$ can be equivalently re-expressed as $\mathrm{tr}(N^\top Y) = 0$. This allows the quadratic program (4) to be simplified somewhat to:

$$\max_{M,N} \mathbf{1}^\top (M + N)\mathbf{1} - \tfrac{1}{2\alpha}\mathrm{tr}((M - N)^\top K(M - N)(I + \mathbf{11}^\top)) \tag{6}$$

$$\text{s.t. } M \geq 0,\ M\mathbf{1} \leq \mathbf{1},\ tr(M^\top \bar{Y}) = 0,$$
$$N \geq 0,\ N\mathbf{1} \leq \mathbf{1},\ tr(N^\top Y) = 0.$$

Unfortunately it is still not obvious that (6) can be converted to a form that involves only inner products between label vectors. However, we will see now that this can be achieved in two steps.

The first key step is to consider an *expanded* set of inner products; that is, consider the set of inner products not just between label vectors in Y but also between complements of these label vectors (i.e. in $\bar{Y}$) and the canonical set of single class indicator vectors (i.e. in I_L). In particular, consider the expanded $(L + 2t) \times L$ label matrix $S = [I; Y; \bar{Y}]$ (i.e., stacked vertically) from which one can form the augmented inner product matrix

$$Q = SS^\top \ = \ \begin{bmatrix} I & Y^\top & \bar{Y}^\top \\ Y & YY^\top & Y\bar{Y}^\top \\ \bar{Y} & \bar{Y}Y^\top & \bar{Y}\bar{Y}^\top \end{bmatrix}. \tag{7}$$

This augmented kernel matrix embodies useful information for reformulating the training problem (6). For example, one important property it satisfies is:

$$Q\mathbf{1} = \begin{bmatrix} 1 + (Y + \bar{Y})^\top \mathbf{1} \\ Y\mathbf{1} + Y(Y + \bar{Y})^\top \mathbf{1} \\ \bar{Y}\mathbf{1} + \bar{Y}(Y + \bar{Y})^\top \mathbf{1} \end{bmatrix} = \begin{bmatrix} 1 + \mathbf{11}^\top \mathbf{1} \\ Y\mathbf{1} + Y\mathbf{11}^\top \mathbf{1} \\ \bar{Y}\mathbf{1} + \bar{Y}\mathbf{11}^\top \mathbf{1} \end{bmatrix} = (t+1)\begin{bmatrix} 1 \\ Y\mathbf{1} \\ \bar{Y}\mathbf{1} \end{bmatrix} = (t+1)S\mathbf{1}, \tag{8}$$

which will be helpful below.

The second key step is to apply a change of variables by the following lemma.

Lemma 1. *For any S defined as above, and for any $M \geq 0$ and $N \geq 0$, there must exist two real value matrices $\Omega \geq 0$ and $\Gamma \geq 0$ of size $t \times (L+2t)$ such that*

$$M = \Omega S \text{ and } N = \Gamma S. \tag{9}$$

Proof. First observe that $M = \Omega S$ defines a system of t linear equations where the ith equation is given by $M_{i:} = \Omega_{i:}S$. By Farkas' Lemma, given $M_{i:} \in \mathbb{R}^L$ and $S \in \mathbb{R}^{(L+2t) \times L}$, exactly one of the following two statements must be true:

1. There exists an $\boldsymbol{\omega} \in \mathbb{R}^{(L+2t)}$ such that $M_{i:} = \boldsymbol{\omega}^\top S$ and $\boldsymbol{\omega} \geq 0$.
2. There exists a $\mathbf{z} \in \mathbb{R}^L$ such that $S\mathbf{z} \geq 0$ and $M_{i:}\mathbf{z} < 0$.

Assume that there exists a $\mathbf{z} \in \mathbb{R}^L$ such that $M_{i:}\mathbf{z} < 0$. Then, since $M_{i:} \geq 0$, $\mathbf{z}$ must have at least one negative entry; i.e., $\mathbf{z}_j < 0$ for some j. However, since S has an identity submatrix, we conclude that the jth entry of $S\mathbf{z}$ must be negative. Therefore the second statement of Farkas' lemma cannot hold. According to the first statement, we therefore know that for the given S and $M_{i:} \geq 0$, there must exist an $\boldsymbol{\omega} \geq 0$ such that $M_{i:} = \boldsymbol{\omega}^\top S$. Finally, by taking all of the linear systems into consideration, we conclude that for any $M \geq 0$ there must exist an $\Omega \in \mathbb{R}^{t \times (L+2t)}$, $\Omega \geq 0$ that satisfies (9). An identical argument can be used to establish the condition between the $N \geq 0$ and $\Gamma \geq 0$ matrices. □

Next, by introducing the expanded label kernel matrix Q and by making the variable substitution suggested by Lemma 1, the main result can be established: the original training problem (6) can be re-expressed in terms of inner products between output vectors from the augmented set of label vectors S.

Proposition 1. *By applying the variable substitution justified by Lemma 1 and using (7) and (8), the quadratic program (6) can be equivalently re-expressed as:*

$$\max_{\Omega,\Gamma} \mathbf{1}^\top(\Omega+\Gamma)Q\mathbf{1} - \tfrac{1}{2\alpha}tr((\Omega-\Gamma)^\top K(\Omega-\Gamma)((t+1)Q + \tfrac{1}{t+1}Q\mathbf{1}(Q\mathbf{1})^\top)) \quad (10)$$

s.t. $\Omega \geq 0$, $\Omega Q\mathbf{1} \leq (t+1)\mathbf{1}$, $tr(\Omega QB) = 0$,
$\Gamma \geq 0$, $\Gamma Q\mathbf{1} \leq (t+1)\mathbf{1}$, $tr(\Gamma QA) = 0$;

where $A = [O_{L,t}; I_t; O_t]$, $B = [O_{L,t}; O_t; I_t]$, I_t is a $t \times t$ identity matrix, O_t is a $t \times t$ matrix with all 0 values, and $O_{L,t}$ is a $L \times t$ matrix with all 0 values.

Proof. Using the substitution (9), the objective in (6) can be rewritten as:

$$(6) = \mathbf{1}^\top(\Omega+\Gamma)S\mathbf{1} - \tfrac{1}{2\alpha}tr((\Omega-\Gamma)^\top K(\Omega-\Gamma)(SS^\top + S\mathbf{1}(S\mathbf{1})^\top)). \quad (11)$$

Next, observe that using $Q = SS^\top$ and $S\mathbf{1} = \tfrac{1}{t+1}Q\mathbf{1}$, the objective (11) can be further rewritten as:

$$(11) = \tfrac{1}{t+1}\mathbf{1}^\top(\Omega+\Gamma)Q\mathbf{1} - \tfrac{1}{2\alpha}tr((\Omega-\Gamma)^\top K(\Omega-\Gamma)(Q + \tfrac{1}{(t+1)^2}Q\mathbf{1}(Q\mathbf{1})^\top)), \quad (12)$$

which, multiplying by $t + 1$, leads to the form stated in the proposition.

Finally, we consider the constraints in (6). For the equality constraints, using the non-negativity of the matrices involved and applying the previous substitutions one obtains:

$$tr(M^\top \bar{Y}) = tr(\Omega S \bar{Y}^\top) = tr(\Omega[\bar{Y}^\top; Y\bar{Y}^\top; \bar{Y}\bar{Y}^\top]) = tr(\Omega QB) = 0, \quad (13)$$
$$tr(N^\top Y) = tr(\Gamma SY^\top) = tr(\Gamma[Y^\top; YY^\top; \bar{Y}Y^\top]) = tr(\Gamma QA) = 0. \quad (14)$$

For the middle inequality constraints, applying the same substitution (9) yields:

$$M\mathbf{1} = \Omega S\mathbf{1} = \tfrac{1}{t+1}\Omega Q\mathbf{1} \leq \mathbf{1}, \quad (15)$$
$$N\mathbf{1} = \Gamma S\mathbf{1} = \tfrac{1}{t+1}\Gamma Q\mathbf{1} \leq \mathbf{1}. \quad (16)$$

Finally, for the non-negativity constraints $M \geq 0$ and $N \geq 0$, Lemma 1 shows that these can be equivalently enforced by asserting $\Omega \geq 0$ and $\Gamma \geq 0$.
Combining the above set of derivations establishes the proposition. □

4.1   Extension to Output Kernels for Multi-label Classification

Since Proposition 1 shows that minimizing the regularized calibrated separation ranking loss can be expressed in terms of inner products between label vectors, an extension to output kernels can be achieved in the obvious way. As before, one assumes a feature map $\varphi : \mathcal{Y} \to \mathcal{H}_\mathcal{Y}$ that transforms each label vector $\mathbf{y}$ into a new representation $\varphi(\mathbf{y})$ in the Hilbert space $\mathcal{H}_\mathcal{Y}$, hence a kernel between output vectors can be defined by an inner product between two output label vectors in the new representation space (an RKHS). In practice, one chooses a positive semidefinite kernel function $\kappa_y(\cdot, \cdot)$ such that conceptually $\kappa_y(\mathbf{y}, \tilde{\mathbf{y}}) = \varphi(\mathbf{y})^\top \varphi(\tilde{\mathbf{y}})$ (where we are assuming this denotes inner product in the implied reproducing kernel Hilbert space). In this way, the matrix Q can be constructed as $Q = \kappa_y(S, S)$, where conceptually $\kappa_y(S, S) = \varphi(S)\varphi(S)^\top$.

However, there is an important catch: in this case it turns out that, unlike input kernelization (or output kernelization for least squares regression), not every valid kernel is suitable as an output representation for multi-label classification. Specifically, the optimization formulation above is only well posed for a subset of possible output kernel functions (although any input kernel can still be used).

To preserve equivalence between the output kernelized form (10) and the dual form (4) established in Proposition 1, we at least require that the kernel matrix Q be doubly non-negative; i.e., $Q \succeq 0$, and $Q \geq 0$ entrywise. Furthermore, to preserve Lemma 1, Q must also preserve orthogonality; that is, if $Y_{i:} Y_{j:}^\top = 0$ then $Q_{ij} = 0$. Therefore, overall, for any output kernel function κ_y that one would wish to use for multi-label classification training the following set of constraints must be satisfied: positive semi-definiteness, $\kappa_y(S, S) \succeq 0$ for any finite S; non-negativity, $\kappa_y(\mathbf{y}, \tilde{\mathbf{y}}) \geq 0$ for all $\mathbf{y} \in \{0, 1\}^L$ and $\tilde{\mathbf{y}} \in \{0, 1\}^L$; and orthogonality, $\mathbf{y}^\top \tilde{\mathbf{y}} = 0$ must imply $\kappa_y(\mathbf{y}, \tilde{\mathbf{y}}) = 0$.

These properties are obviously satisfied by the linear kernel used to derive Proposition 1. However, in addition to the linear kernel, other kernels common in document and language modeling are appropriate for this setting [43]. One particularly useful family of kernels that satisfy these properties are the homogeneous polynomial kernels:

$$\mathcal{K}_{poly}(\mathbf{y}, \tilde{\mathbf{y}}) = \sum_{i=1}^{k} w_i (\mathbf{y}^\top \tilde{\mathbf{y}})^i, \tag{17}$$

where $\mathbf{w} \geq 0$ is a vector of non-negative weights. Unfortunately, many standard kernels, such as the Gaussian (RBF) kernel are not suitable, since by violating the constraints it blocks all nonzero solutions to (10). Below we find that the simple weighted polynomial kernels allow sufficient flexibility in capturing nonlinear dependence to achieve positive results in some real world multi-label data sets.

4.2   Classification of Test Instances

Although Proposition 1 shows that training for multi-label classification can be formulated in terms of a kernel between label vectors, this does not imply that classifying new instances $\mathbf{x}$ at test time will be necessarily easy. In fact, for regression formulations, test prediction generally involves solving a hard pre-image problem [23, 24]. Fortunately, the pre-image problem can be efficiently solved for multi-label classification, even when there are an exponential (2^L) number of label vectors to consider.

After solving the training problem (10) one obtains the global solution (Ω, Γ), which can be used to efficiently classify a new test instance as follows. Let κ_x and κ_y denote the input and output kernels respectively. Conceptually, we can think of these as evaluating an inner product between feature representations of the inputs and outputs as $\kappa_x(\mathbf{x}, \tilde{\mathbf{x}}) = \phi(\mathbf{x})^\top \phi(\tilde{\mathbf{x}})$ and $\kappa_y(\mathbf{y}, \tilde{\mathbf{y}}) = \varphi(\mathbf{y})^\top \varphi(\tilde{\mathbf{y}})$ respectively. Then from the optimal parameters, we can conceptually recover the solution (M, N) to the original dual problem (4) via

$$M = \Omega\varphi(S) \text{ and } N = \Gamma\varphi(S). \tag{18}$$

Using (1), the optimal parameters $(W, \mathbf{u})$ for the original problem (3) are:

$$W = \tfrac{1}{\alpha}\phi(X)^\top(\Omega - \Gamma)\varphi(S), \tag{19}$$

$$\mathbf{u} = \tfrac{1}{\alpha}\phi(X)^\top(\Gamma - \Omega)\varphi(S)\mathbf{1} = \tfrac{1}{\alpha(t+1)}\phi(X)^\top(\Gamma - \Omega)Q\mathbf{1}. \tag{20}$$

Finally, recall the classification rule used for multi-label assignment (1). Given a new test instance $\mathbf{x} \in \mathbb{R}^d$, we will determine its labels by computing the score function values $\mathbf{s}(\mathbf{x}) = [\mathbf{s}_1(\mathbf{x}), \cdots, \mathbf{s}_L(\mathbf{x})]$ and $\mathbf{s}_0(\mathbf{x})$; that is, the label vector $\mathbf{y}$ for $\mathbf{x}$ is then given by a $L \times 1$ indicator vector where $\mathbf{y}_l = 1$ if $\mathbf{s}_l(\mathbf{x}) \geq \mathbf{s}_0(\mathbf{x})$, $\mathbf{y}_l = 0$ otherwise. Fortunately, these score values can be efficiently computed directly from the recovered (Ω, Γ) parameters via:

$$\mathbf{s}_0(\mathbf{x}) = \phi(\mathbf{x})\mathbf{u} = \tfrac{1}{\alpha(t+1)}\kappa_x(\mathbf{x}, X)(\Gamma - \Omega)Q\mathbf{1}, \tag{21}$$

$$\mathbf{s}_l(\mathbf{x}) = \phi(\mathbf{x})W\varphi(\mathbf{1}_l) = \tfrac{1}{\alpha}\kappa_x(\mathbf{x}, X)(\Omega - \Gamma)\kappa_y(S, \mathbf{1}_l), \forall l = 1, \cdots, L; \tag{22}$$

where $\mathbf{1}_l$ denotes a vector with 1 as its lth entry and 0 elsewhere. Thus, the multi-label assignment to test instance $\mathbf{x}$ can be efficiently computed.

5   A Scalable Training Method

One of the main challenges with this formulation is that the quadratic programming problem (10) is defined over $(L + 2t) \times t$ matrix variables, which makes the training problem challenging for standard solvers. Instead, we develop a row-wise projected gradient method to achieve a more scalable approach.

First note that the optimization problem (10) can be written in a more compact form. Replace Ω and Γ with $\Lambda = [\Omega, \Gamma]$. Let $C = [I_{L+2t}; O_{L+2t}]$ and $D = [O_{L+2t}; I_{L+2t}]$, such that $\Omega = \Lambda C$ and $\Gamma = \Lambda D$. Furthermore, let $P = \tfrac{1}{\alpha}(C - D)((t+1)Q + \tfrac{1}{t+1}Q\mathbf{1}(Q\mathbf{1})^\top))(C - D)^\top$; $E = ((C + D)Q\mathbf{1}\mathbf{1}^\top)^\top$; $\mathbf{a} = \tfrac{1}{t+1}CQ\mathbf{1}$; $\mathbf{b} = \tfrac{1}{t+1}DQ\mathbf{1}$; $G = (CQB)^\top$ and $F = (DQA)^\top$. Then (10) can be rewritten more succinctly as:

$$\min_{\Lambda} \ \tfrac{1}{2}\mathrm{tr}(\Lambda^\top K\Lambda P) - tr(\Lambda E^\top) \tag{23}$$

s.t. $\Lambda \geq 0$, $\Lambda\mathbf{a} \leq \mathbf{1}$, $\mathrm{tr}(\Lambda G^\top) = 0$, $\Lambda\mathbf{b} \leq \mathbf{1}$, $\mathrm{tr}(-\Lambda F^\top) = 0$.

The key property of this quadratic program is that the constraints decompose row-wise. This allows us to use a row-wise coordinate descent procedure to achieve scalability. Consider the ith row of Λ, assuming all other rows are fixed. An update to row $\Lambda_{i:}$ can be expressed as $\Lambda = \Lambda + \mathbf{1}_i(\mathbf{z}^\top - \Lambda_{i:})$, where $\mathbf{1}_i$ is a column vector of zeros with a single 1 in the ith position. Let $\Lambda_{\bar{i}:} := \Lambda - \mathbf{1}_i\Lambda_{i:}$. The objective function $f(\Lambda) := \tfrac{1}{2}\mathrm{tr}(\Lambda^\top K\Lambda P) - tr(\Lambda E^\top)$ of the quadratic program can be re-expressed as a function g over the single row update $\mathbf{z}$ such that:

$$\begin{aligned}
g(\mathbf{z}) &= f(\Lambda + \mathbf{1}_i(\mathbf{z}^\top - \Lambda_{i:})) = f(\Lambda_{\bar{i}:} + \mathbf{1}_i\mathbf{z}^\top) \\
&= \tfrac{1}{2}K_{ii}(\mathbf{z}^\top P\mathbf{z}) + (K_{i:}\Lambda_{\bar{i}:}P - E_{i:})\mathbf{z} + const \\
&= \tfrac{1}{2}K_{ii}(\mathbf{z}^\top P\mathbf{z}) + (K_{i:}\Lambda P - K_{ii}\Lambda_{i:}P - E_{i:})\mathbf{z} + const
\end{aligned} \tag{24}$$

which yields the row optimization problem:

$$\min_{\mathbf{z}} \ g(\mathbf{z}) \quad \text{s.t.} \quad \mathbf{z} \geq 0, \ \mathbf{z}^\top\mathbf{a} \leq 1, \ G_{i:}\mathbf{z} = 0, \ \mathbf{z}^\top\mathbf{b} \leq 1, \ F_{i:}\mathbf{z} = 0. \tag{25}$$

The update of the ith row only affects other rows through the $K_{i:}\Lambda P$ term. Therefore, we maintain a matrix $U = K\Lambda P$ that can be updated locally after an update to $\Lambda_{i:}$, by $U = U + K_{:i}(\mathbf{z}^\top - \Lambda_{i:})P$. To ensure that progress is always made, while maintaining scalability, we use a row-wise steepest descent method. For the objective function $g(\mathbf{z})$, its gradient vector is given by:

$$\mathbf{g} = \frac{dg(\mathbf{z})}{d\mathbf{z}} = K_{ii}P\mathbf{z} + (K_{i:}\Lambda P - K_{ii}\Lambda_{i:}P - E_{i:})^\top, \tag{26}$$

which can be efficiently computed. Since the constraints on $\mathbf{z}$ are simple, this gradient vector can be efficiently projected to a feasible direction $\mathbf{d}$. Because the objective f has a simple quadratic form, the optimal step size in the feasible direction d can be computed in closed form. Thus, optimal updates can be made by locally operating on each row of Λ in succession. We have found this approach to be reasonably effective in our experiments below.

6   Experiments

To evaluate the proposed approach, we conducted experiments on a multi-label classification image set, *scene*, and a set of multi-label classification tasks constructed from a real-world image data set, *MIRFlickr*. We compared the proposed output kernel approach to a number of large margin multi-label classification methods, and to an output-kernel based least square regression method.

Table 1. Properties of the multi-label data sets used in the experiments

DATA SET	#CLASSES	#INSTANCES	#FEATURES	LABEL-CARD.
SCENE-6	6	2407	294	1.1
MIRFLICKR-10	10	1484	1000	2.4
MIRFLICKR-15	15	1929	1000	2.5
MIRFLICKR-20	20	2902	1000	2.6
MIRFLICKR-25	25	3414	1000	2.7
MIRFLICKR-30	30	4057	1000	2.7

6.1 Experimental Setting

Data Sets. We focused on image data sets for these experiments, since image data usually exhibits highly nonlinear semantic dependence between labels. The *scene* [44] data set has 2407 images and only 6 classes, whereas the *MIRFlickr* [45] data set contains 25,000 images and 457 classes. Although MIRFlickr has a very large number of classes, the labels appear in a very sparse manner. One key property of multi-label data sets is their label cardinality [42]; the average number of labels assigned to each instance. If the label cardinality of a data set is close to 1, the task reduces to a standard single label classification task, and there will not be any significant label dependence to capture. The effectiveness of multi-label learning can therefore primarily be demonstrated on data sets whose label cardinality is reasonably large and complex. We thus constructed a set of multi-label classification tasks from the MIRFlickr image data set that maintained reasonable label cardinalities while ranging across a set of different numbers of classes. Specifically, we constructed five multi-label subsets, *MIRFlickr-10, MIRFlickr-15, MIRFlickr-20, MIRFlickr-25,* and *MIRFlickr-30*, by randomly selecting L classes, for $L \in \{10, 15, 20, 25, 30\}$ respectively, to achieve a reasonable level of label cardinality in each case; see Table 1 for a summary.

Approaches. Our proposed approach (*LM-K*) is based on using output kernels to capture nonlinear label dependence during training. In these experiments, we employed the homogeneous polynomial kernels as defined in (17). With $k \geq 2$, these polynomial kernels can automatically encode pairwise and higher-order label dependence structures in an expanded output space.

We compare the proposed approach to a number of state-of-the-art multi-label classification methods to investigate the consequences of using nonlinear output kernels. These competitors were: (1) the large margin method based on the calibrated separation ranking loss (*CSRL*) [8]; (2) the pairwise ranking loss SVM (*Rank*) proposed in [9], which first trains a large margin ranking model and then learns the threshold of the multi-label predications using a least-square method; and (3) the max-margin multi-label classification method (*M3L*) proposed in [7], which takes prior knowledge about the label correlations into account. None of these methods use output kernels. Therefore, we also compare the proposed method with a least squares regression method that uses output kernels (*LS-K*), thresholding its predictions for multi-label classification.

Table 2. Summary of the performance (%) for the compared methods in terms of micro-F1 (top section) and macro-F1 (bottom section)

DATA SET	RANK	M3L	CSRL	LS-K	LM-K
SCENE-6	58.8±0.5	52.7±0.7	52.8±0.7	46.0±0.7	60.6±0.4
MIRFLICKR-10	32.9±0.5	37.6±0.7	41.0±0.9	36.8±0.5	44.4±0.2
MIRFLICKR-15	26.9±0.5	26.9±0.9	33.1±0.3	28.0±0.2	34.3±0.1
MIRFLICKR-20	17.9±0.6	19.2±0.9	26.0±0.5	22.1±0.2	27.9±0.1
MIRFLICKR-25	16.0±0.4	17.2±0.7	22.4±0.5	18.6±0.2	24.4±0.1
MIRFLICKR-30	13.8±0.3	14.6±0.6	18.7±0.6	15.8±0.2	21.8±0.1
SCENE-6	60.0±0.5	52.0±0.9	51.8±0.9	45.3±0.7	60.3±0.6
MIRFLICKR-10	30.9±0.4	32.2±0.6	38.1±1.0	34.7±0.4	42.6±0.2
MIRFLICKR-15	20.7±0.4	22.0±0.4	28.8±0.4	24.7±0.2	32.5±0.2
MIRFLICKR-20	14.0±0.4	15.2±0.7	22.8±0.5	19.5±0.2	26.8±0.1
MIRFLICKR-25	12.8±0.3	12.5±0.4	19.1±0.5	16.0±0.2	23.2±0.1
MIRFLICKR-30	10.7±0.4	10.1±0.4	15.6±0.5	13.3±0.2	20.9±0.1

6.2   Experimental Results

Classification Results. We first conducted a set of experiments on the six multi-label data sets by randomly sampling 300 labeled images as training data and holding out the remaining as test data. The intent is to investigate how well each approach can exploit label dependence when there are limited training instances available. In the experiments, we set the trade-off parameter $\alpha = 0.1$ for the proposed approach and *CSRL*, and set the trade-off parameters for *Rank* and *M3L* correspondingly with $C = 10$. We used the linear input kernel for all methods. For *LS-K* and *LM-K*, we used the polynomial output kernel given in (17) with maximum degree $k = 2$, with weights $w_1 = w_2 = 1$. This polynomial kernel automatically encodes all pairwise label dependency structures within the induced high dimensional output space. We repeated each experiment 10 times and report the average multi-label classification performance in terms of micro-F1 and macro-F1 in Table 2.

From Table 2, one can observe that the difficulty of the learning problem increases with label set size, causing degradation in the performance of all methods. However, with a nonlinear output kernel, the proposed approach *LM-K* consistently outperforms the three state-of-the-art large margin multi-label classification methods, *Rank*, *M3L* and *CSRL*, across all the data sets with different numbers of classes. It also significantly outperforms least-squares regression method with the same output kernel, *LS-K*. These results suggest that a nonlinear output kernel is indeed useful for improving multi-label classification models in a setting with interesting label dependencies. Here the proposed approach appears to provide an effective method for exploiting nonlinear dependence structure through the use of a polynomial output kernel.

Polynomial Kernels. Based on the definition of homogeneous polynomial kernels given in (17), one can produce many different kernels with different weights

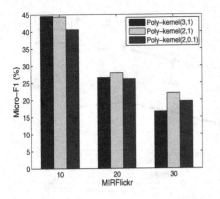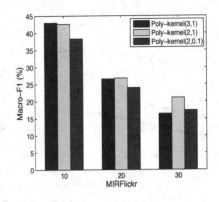

Fig. 1. Comparison of different polynomial output kernels on the MIRFlickr data sets

$\{w_i\}$ and different maximum degree k. We next investigated the influence of alternative output kernels on multi-label classification. We considered three different polynomial kernels by varying the degree number and the weights: (1) Poly-kernel(3,1) uses degree up to $k = 3$ and weights $w_1 = w_2 = w_3 = 1$; (2) Poly-kernel(2,1) uses degree up to $k = 2$ and weights $w_1 = w_2 = 1$; and (3) Poly-kernel(2,0.1) uses degree up to $k = 2$ but with weights $w_1 = 1$ and $w_2 = 0.1$. Evidently the first polynomial kernel with maximum degree 3 considers triplet-wise label dependencies, whereas the other two kernels only consider pairwise label dependence. Moreover, the last kernel put relatively less weight on the higher order dependence features.

We conducted experiments on three MIRFlickr data sets, *MIRFlickr-10, MIRFlickr-20, and MIRFlickr-30*, using the same setting as above. The results are reported in Figure 1, in terms of micro-F1 measure and macro-F1 measure. From these results one can see that even though it embodies more complex label features, Poly-kernel(3,1) demonstrates inferior performance when the class number increases, compared to the less complex Poly-kernel(2,1). This suggests that Poly-kernel(3,1) can over-fit when the classification problem gets more complex given limited training data. On the other hand, Poly-kernel(2,0.1) further suppresses the influence of the higher order label features, and demonstrates inferior performance compared to Poly-kernel(3,1) when there are fewer labels. The intermediate Poly-kernel(2,1) demonstrates good performance on all three data sets. These results suggest selecting output kernels with the right complexity is important, and pairwise label features are very useful for encoding label dependence structure, somewhat vindicating an original intuition about multi-label classification [9]. A proper output kernel should give proper consideration over the pairwise feature expansions, and the complexity of the problem.

Performance vs. Training Size. With a modest training size, we have demonstrated that the proposed approach can effectively improve multi-label classification performance by exploiting the label dependence information and structure through the nonlinear output kernel. There remains a question of how the

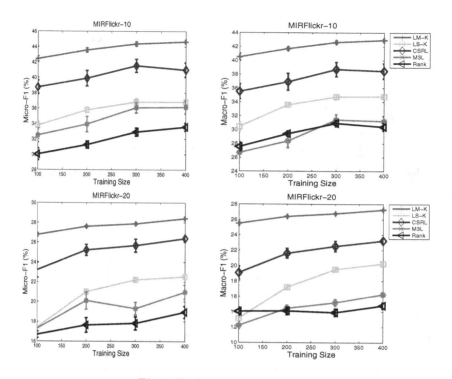

Fig. 2. Performance vs training size

behavior of the various methods would change with increasing sample size. To answer this question, we conducted experiments with a number of different training sizes, $t \in [100, 200, 300, 400]$ on two of the data sets, *MIRFlickr-10* and *MIRFlickr-20*. We otherwise used the same experimental setting as above. The average results and standard deviations in terms of micro-F1 and macro-F1 measure on these two data sets are plotted in Figure 2. Here, one can see that with increasing training size, the performance of all methods generally improves. However, the proposed approach with polynomial output kernel consistently outperforms the other methods across all training sizes, evaluation measures, and data sets. These results again demonstrate the efficacy of the proposed approach for using nonlinear output kernel to capture label dependency of multi-label learning.

7 Conclusion

We have introduced a new form of multi-label classification learning that uses an output kernel between multi-label output vectors to capture a rich set of non-linear dependences between output labels, while retaining a tractable equivalent formulation as a quadratic program. Although the resulting quadratic programs are expanded, a scalable training algorithm can be based on example-wise projected gradient descent. The resulting method demonstrates advantages in multi-label image classification experiments over standard linear-output approaches.

In addition to investigating the benefits of alternative output kernels and alternative scaling strategies, an important direction for future research is to investigate other important loss formulations in machine learning, to determine whether they too might be amenable to an equivalent kernelized approach.

References

1. Joachims, T.: Text categorization with support vector machines: learn with many relevant features. In: Proc. of ECML (1998)
2. McCallum, A.: Multi-label text classification with a mixture model trained by EM. In: AAAI Workshop on Text Learning (1999)
3. Zhu, S., Ji, X., Xu, W., Gong., Y.: Multi-labelled classification using maximum entropy method. In: SIGIR 2005 (2005)
4. Petterson, J., Caetano, T.: Submodular multi-label learning. In: Advances in Neural Information Processing Systems, NIPS (2011)
5. Kazawa, H., Izumitani, T., Taira, H., Maeda, E.: Maximal margin labeling for multi-topic text categorization. In: NIPS 17 (2004)
6. Godbole, S., Sarawagi, S.: Discriminative methods for multi-labeled classification. In: Dai, H., Srikant, R., Zhang, C. (eds.) PAKDD 2004. LNCS (LNAI), vol. 3056, pp. 22–30. Springer, Heidelberg (2004)
7. Hariharan, B., Vishwanathan, S., Varma, M.: Efficient max-margin multi-label classification with applications to zero-shot learning. Machine Learning 88 (2012)
8. Guo, Y., Schuurmans, D.: Adaptive large margin training for multilabel classification. In: Proc. of AAAI (2011)
9. Elisseeff, A., Weston, J.: A kernel method for multi-labelled classification. In: NIPS (2001)
10. Schapire, R., Singer, Y.: Boostexter: A boosting-based system for text categorization. Machine Learning Journal, 135–168 (2000)
11. Shalev-Shwartz, S., Singer, Y.: Efficient learning of label ranking by soft projections onto polyhedra. JMLR 7, 1567–1599 (2006)
12. Fuernkranz, J., Huellermeier, E., Mencia, E., Brinker, K.: Multilabel classification via calibrated label ranking. Machine Learning 73(2)
13. Ghamrawi, N., McCallum, A.: Collective multi-label classification. In: Proc. of CIKM (2005)
14. Zaragoza, J., Sucar, L., Morales, E., Bielza, C., Larranaga, P.: Bayesian chain classifiers for multidimensional classification. In: Proc. of IJCAI (2011)
15. Guo, Y., Gu, S.: Multi-label classification using conditional dependency networks. In: Proc. of IJCAI (2011)
16. Hsu, D., Kakade, S., Langford, J., Zhang, T.: Multi-label prediction via compressed sensing. In: Proceedings NIPS (2009)
17. Chen, Y., Lin, H.: Feature-aware label space dimension reduction for multi-label classification. In: Proceedings NIPS (2012)
18. Tai, F., Lin, H.: Multi-label classification with principal label space transformation. In: Proc. 2nd International Workshop on Learning from Multi-Label Data (2010)
19. Zhang, Y., Schneider, J.: Max margin output coding. In: Proc. ICML (2012)
20. Zhang, Y., Schneider, J.: Multi-label output codes using canonical correlation analysis. In: Proceedings AISTATS (2011)
21. Zhou, T., Tao, D., Wu, X.: Compressed labeling on distilled labelsets for multi-label learning. Machine Learning 88, 69–126 (2012)

22. Kimeldorf, G., Wahba, G.: Some results on tchebycheffian spline functions. Journal of Mathematical Analysis and Applications 33, 82–95 (1971)
23. Schoelkopf, B., Smola, A.: Learning with Kernels: Support Vector Machines, Regularization, Optimization, and Beyond. MIT Press (2002)
24. Huang, D., Tian, Y., De la Torre, F.: Local isomorphism to solve the pre-image problem in kernel methods. In: Proceedings CVPR (2011)
25. Guo, Y., Schuurmans, D.: Convex relaxations of latent variable training. In: Proceedings of Advances in Neural Information Processing Systems, NIPS (2007)
26. Xu, L., Schuurmans, D.: Unsupervised and semi-supervised multi-class support vector machines. In: Proceedings AAAI (2005)
27. Xu, L., Wilkinson, D., Southey, F., Schuurmans, D.: Discriminative unsupervised learning of structured predictors. In: Proceedings ICML (2006)
28. Cortes, C., Mohri, M., Weston, J.: A general regression technique for learning transductions. In: Proceedings ICML (2005)
29. Weston, J., Chapelle, O., Elisseeff, A., Schoelkopf, B., Vapnik, V.: Kernel dependency estimation. In: Proceedings NIPS (2002)
30. Wang, Z., Shawe-Taylor, J.: A kernel regression framework for SMT. Machine Translation 24(2), 87–102 (2010)
31. Micchelli, C., Pontil, M.: On learning vector-valued functions. Neural Computation 17(1), 177–204 (2005)
32. Geurts, P., Wehenkel, L., d'Alché Buc, F.: Kernelizing the output of tree-based methods. In: Proceedings ICML (2006)
33. Geurts, P., Wehenkel, L., d'Alché Buc, F.: Gradient boosting for kernelized output spaces. In: Proceedings ICML (2007)
34. Geurts, P., Touleimat, N., Dutreix, M., d'Alché Buc, F.: Inferring biological networks with output kernel trees. BMC Bioinformatics 8(S-2) (2007)
35. Brouard, C., d'Alché Buc, F., Szafranski, M.: Semi-supervised penalized output kernel regression for link prediction. In: Proceedings ICML (2011)
36. Brouard, C., Szafranski, M.: Regularized output kernel regression applied to protein-protein interaction network inference. In: NIPS MLCB Workshop (2010)
37. Kadri, H., Duflos, E., Preux, P., Canu, S., Davy, M.: Nonlinear functional regression: a functional RKHS approach. In: Proceedings AISTATS (2010)
38. Weston, J., Schölkopf, B., Bousquet, O.: Joint kernel maps. In: Cabestany, J., Prieto, A.G., Sandoval, F. (eds.) IWANN 2005. LNCS, vol. 3512, pp. 176–191. Springer, Heidelberg (2005)
39. Zhang, Y., Yeung, D.: A convex formulation for learning task relationships in multi-task learning. In: Proceedings UAI (2010)
40. Dinuzzo, F., Fukumizu, K.: Learning low-rank output kernels. In: Proceedings ACML (2011)
41. Dinuzzo, F., Ong, C., Gehler, P., Pillonetto, G.: Learning output kernels with block coordinate descent. In: Proceedings ICML (2011)
42. Tsoumakas, G., Katakis, I.: Multi-label classification: An overview. International Journal of Data Warehousing and Mining (2007)
43. Shawe-Taylor, J., Cristianini, N.: Kernel Methods for Pattern Analysis. Cambridge University Press (2004)
44. Boutell, M., Luo, J., Shen, X., Brown, C.: Learning multi-label scene classiffication. Pattern Recognition 37(9), 1757–1771 (2004)
45. Huiskes, M., Lew, M.: The MIR flickr retrieval evaluation. In: Proc. of ACM International Conference on Multimedia Information Retrieval (2008)

Boosting for Unsupervised Domain Adaptation

Amaury Habrard, Jean-Philippe Peyrache, and Marc Sebban

Université Jean Monnet de Saint-Etienne
Laboratoire Hubert Curien, UMR CNRS 5516
18 rue du Professeur Benoit Lauras - 42000 Saint-Etienne Cedex 2 - France
{amaury.habrard,jean-philippe.peyrache,marc.sebban}@univ-st-etienne.fr

Abstract. To cope with machine learning problems where the learner receives data from different source and target distributions, a new learning framework named *domain adaptation* (DA) has emerged, opening the door for designing theoretically well-founded algorithms. In this paper, we present SLDAB, a self-labeling DA algorithm, which takes its origin from both the theory of boosting and the theory of DA. SLDAB works in the difficult unsupervised DA setting where source and target training data are available, but only the former are labeled. To deal with the absence of labeled target information, SLDAB jointly minimizes the classification error over the source domain and the proportion of margin violations over the target domain. To prevent the algorithm from inducing degenerate models, we introduce a measure of divergence whose goal is to penalize hypotheses that are not able to decrease the discrepancy between the two domains. We present a theoretical analysis of our algorithm and show practical evidences of its efficiency compared to two widely used DA approaches.

1 Introduction

In many learning algorithms, it is usually required to assume that the training and test data are drawn from the same distribution. However, this assumption does not hold in many real applications challenging common learning theories such as the PAC model [20]. To cope with such situations, a new machine learning framework has been recently studied leading to the emergence of the theory of *domain adaptation* (DA) [1,14]. A standard DA problem can be defined as a situation where the learner receives labeled data drawn from a *source* domain (or even from several sources [13]) and very few or no labeled points from the *target* distribution. DA arises in a large spectrum of applications, such as in computer vision [16], speech processing [11,18], natural language processing [3,5], etc. During the past few years, new fundamental results opened the door for the design of theoretically well-founded DA-algorithms. In this paper, we focus on the scenario where the training set is made of labeled source data and *unlabeled* target instances. To deal with this more complex situation, several solutions have been presented in the literature (see, e.g., surveys [15,17]). Among them, *instance weighting-based methods* are used to deal with covariate shift where the labeling functions are supposed to remain unchanged between the two domains. On the

H. Blockeel et al. (Eds.): ECML PKDD 2013, Part II, LNAI 8189, pp. 433–448, 2013.

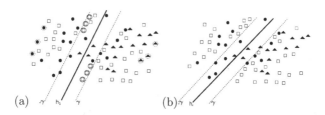

Fig. 1. Underlying principle of DASVM. (a): black examples are labeled source data (circle or triangle). Squares are unlabeled target data. A first SVM classifier h_1 is learned from the labeled source data. Then, DASVM iteratively changes some source data by semi-labeled target examples selected in a margin band (black source instances in a dashed circle and target squares in a circle). (b): new hypothesis h_2 learned using the newly semi-labeled data. h_2 works well on the source and satisfies some margin constraints on the target.

other hand, *feature representation approaches* aim at seeking a domain invariant feature space by either generating latent variables or selecting a relevant subset of the original features. In this paper, we focus on a third class of approaches, called *iterative self-labeling methods*. For example, in DASVM [4], a SVM classifier is learned from the labeled source examples. Then, some of them are replaced by target data selected within a margin band (to allow slight modifications of the current classifier) but at a reasonable enough distance from the hyperplane (to have a sufficient confidence in those unlabeled examples). A new classifier is then learned using these newly *semi-labeled* target data (see Figures 1(a) and 1(b)). The process is repeated until having only semi-labeled data in the training set.

In the context of self-labeling DA, DASVM has become during the past few years a reference method. However, beyond algorithmic constraints due to the resolution of many non trivial optimization problems, it faces an important limitation: it is based on the strong assumption that, if a classifier h works well on the source data, the higher the distance from h, the higher the probability for an unlabeled sample to be correctly classified. It is worth noting that such an assumption holds only if the underlying DA problem does not require to substantially move closer the source and target distributions. As suggested by the theoretical frameworks presented in [1,14], a DA algorithm may have not only to induce a classifier that works well on the source but also to reduce the divergence between the two distributions. This latter condition essentially enables us to have confidence in the ability of the hypothesis learned from the source to correctly classify target data. It is important to note that DASVM has not been designed for such a discrepancy reduction. In this paper, our objective is to fill the gap between the iterative self-labeling strategy and these theoretical recommendations. We present a novel DA algorithm which takes its origin from both the theory of boosting [7] and the theory of DA. Let us remind that boosting (via its well known ADABOOST algorithm) iteratively builds a combination of weak classifiers. At each step, ADABOOST makes use of an update rule which increases (resp. decreases) the weight of those instances misclassified (resp. correctly classified) by previous classifiers. It is worth noting that boosting has already been exploited in DA methods but mainly in supervised situations where the learner receives some labeled target instances. In [6], TRADABOOST uses the standard weighting scheme of ADABOOST on the target data, while the weights of the source instances are monotonically decreased according to their margin. A generalization of TRADABOOST to multiple sources is presented in [21]. On the other hand, some boosting-based approaches relax the constraint of having labeled target examples. However, they are proposed in the context of semi-supervised ensemble methods, *i.e.* assuming that the source and the target domains are (sufficiently) similar [2,12].

In this paper, we present SLDAB, a boosting-like DA algorithm which both optimizes the *source classification error* and *margin constraints* over the unlabeled target instances. However, unlike state of the art self-labeling DA methods, SLDAB aims at also reducing the divergence between the two distributions in the space of the learned hypotheses. In this context, we introduce the notion of weak DA assumption which takes into account a measure of divergence. This classifier-induced measure is exploited in the update rule so as to penalize hypotheses inducing a large discrepancy. This strategy tends to prevent the algorithm from building degenerate models which would, e.g., perfectly classify the source data while moving the target examples far away from the learned hyperplane (and thus satisfying any margin constraint). We present a theoretical analysis of SLDAB and derive several theoretical results that, in addition to good experimental results, support our claims.

The rest of this paper is organized as follows: notations and definitions are given in Section 2; SLDAB is presented in Section 3 and theoretically analyzed in Section 4; We discuss the way to compute the divergence between the source and target domains in Section 5; Finally, we conduct two series of experiments and show practical evidences of the efficiency of SLDAB in Section 6.

2   Definitions and Notations

Let S be a set of labeled data (x', y') drawn from a source distribution $\mathcal{S}$ over $X \times \{-1, +1\}$, where X is the instance space and $\{-1, +1\}$ is the set of labels. Let T be a set of unlabeled examples x drawn from a target distribution $\mathcal{T}$ over X. Let $\mathcal{H}$ be a class of hypotheses and $h_n \in \mathcal{H} : X \rightarrow [-1, +1]$ a hypothesis learned from S and T and their associated empirical distribution D_n^S and D_n^T. We denote by $g_n \in [0, 1]$ a measure of divergence induced by h_n between S and T. Our objective is to take into account g_n in our new boosting scheme so as to penalize hypotheses that do not allow the reduction of the divergence between S and T. To do so, we consider the function $f_{DA} : [-1, +1] \rightarrow [-1, +1]$ such that $f_{DA}(h_n(x)) = |h_n(x)| - \lambda g_n$, where $\lambda \in [0, 1]$. $f_{DA}(h_n(x))$ expresses the ability of h_n to not only induce large margins (a large value for $|h_n(x)|$), but also to reduce the divergence between S and T (a small value for g_n). λ plays the role of a trade-off parameter tuning the importance of the margin and the divergence.

Let $T_n^- = \{x \in T | f_{DA}(h_n(x)) \leq \gamma\}$. If $x \in T_n^- \Leftrightarrow |h_n(x)| \leq \gamma + \lambda g_n$. Therefore, T_n^- corresponds to the set of target points that either violate the margin condition (indeed, if $|h_n(x)| \leq \gamma \Rightarrow |h_n(x)| \leq \gamma + \lambda g_n$) or do not satisfy sufficiently that margin to compensate a large divergence between S and T (i.e. $|h_n(x)| > \gamma$ but $|h_n(x)| \leq \gamma + \lambda g_n$). In the same way, we define $T_n^+ = \{x \in T | f_{DA}(h_n(x)) > \gamma\}$ such that $T = T_n^- \cup T_n^+$. Finally, from T_n^- and T_n^+, we define $W_n^+ = \sum_{x \in T_n^+} D_n^T(x)$ and $W_n^- = \sum_{x \in T_n^-} D_n^T(x)$ such that $W_n^+ + W_n^- = 1$.

Let us remind that the weak assumption presented in [7] states that a classifier h_n is a weak hypothesis over S if it performs at least a little bit better than random guessing, that is $\hat{\epsilon}_n < \frac{1}{2}$, where $\hat{\epsilon}_n$ is the empirical error of h_n over S w.r.t. D_n^S. In this paper, we extend this weak assumption to the DA setting.

Definition 1 (Weak DA Learner). *A classifier h_n learned at iteration n from a labeled source set S drawn from $\mathcal{S}$ and an unlabeled target set T drawn from $\mathcal{T}$ is a weak DA learner for T if $\forall \gamma \leq 1$:*

1. h_n is a weak learner for S, i.e. $\hat{\epsilon}_n < \frac{1}{2}$.

2. $\hat{L}_n = \mathbb{E}_{x \sim D_n^T}[|f_{DA}(h_n(x))| \leq \gamma] = W_n^- < \frac{\gamma}{\gamma + max(\gamma, \lambda g_n)}$.

Condition 1 means that to adapt from $\mathcal{S}$ to $\mathcal{T}$ using a boosting scheme, h_n must learn something new at each iteration about the source labeling function. Condition 2 takes into account not only the ability of h_n to satisfy the margin γ but also its capacity to reduce the divergence between S and T. From Def.(1), it turns out that:

1. if $max(\gamma, \lambda g_n) = \gamma$, then $\frac{\gamma}{\gamma + max(\gamma, \lambda g_n)} = \frac{1}{2}$ and Condition 2 looks like the weak assumption over the source, except the fact that $\hat{L}_n < \frac{1}{2}$ expresses a margin condition while $\hat{\epsilon}_n < \frac{1}{2}$ considers a classification constraint. Note that if this is true for any hypothesis h_n, it means that the divergence between the source and target distributions is rather small, and thus the underlying task looks more like a semi-supervised problem.
2. if $max(\gamma, \lambda g_n) = \lambda g_n$, then the constraint imposed by Condition 2 is stronger (that is $\hat{L}_n < \frac{\gamma}{\gamma + max(\gamma, \lambda g_n)} < \frac{1}{2}$) in order to compensate a large divergence between S and T. In this case, the underlying task requires a domain adaptation process in the weighting scheme.

In the following, we make use of this weak DA assumption to design a new boosting-based DA algorithm, called SLDAB.

3 SLDAB Algorithm

The pseudo-code of SLDAB is presented in Algorithm 1. Starting from uniform distributions over S and T, it iteratively learns a new hypothesis h_n that verifies the weak DA assumption of Def.(1). This task is not trivial. Indeed, while learning a stump (i.e. a one-level decision tree) is sufficient to satisfy the weak

---

Algorithm 1. SLDAB

Input: a set S of labeled data and T of unlabeled data, a number of iterations N, a margin $\gamma \in [0,1]$, a trade-off parameter $\lambda \in [0,1]$, $l = |S|$, $m = |T|$.
Output: two source and target classifiers H_N^S and H_N^T.
Initialization: $\forall (x',y') \in S, D_1^S(x') = \frac{1}{l}, \forall x \in T, D_1^T(x) = \frac{1}{m}$.
for $n = 1$ to N **do**
  Learn a weak DA hypothesis h_n by solving Problem (1).
  Compute the divergence value g_n (see Section 5 for details).
  $\alpha_n = \frac{1}{2} \ln \frac{1 - \hat{e}_n}{\hat{e}_n}$ and $\beta_n = \frac{1}{\gamma + \max(\gamma, \lambda g_n)} \ln \frac{\gamma W_n^+}{\max(\gamma, \lambda g_n) W_n^-}$
  $\forall (x',y') \in S, D_{n+1}^S(x') = D_n^S(x') . \frac{e^{-\alpha_n sgn(h_n(x')).y'}}{Z_n'}$.
  $\forall x \in T, D_{n+1}^T(x) = D_n^T(x) . \frac{e^{-\beta_n f_{DA}(h_n(x)).y^n}}{Z_n}$,
  where $y^n = sgn(f_{DA}(h_n(x)))$ if $|f_{DA}(h_n(x))| > \gamma$,
  $y^n = -sgn(f_{DA}(h_n(x)))$ otherwise,
  and Z_n' and Z_n are normalization coefficients.
end for
$\forall (x',y') \in S, F_N^S(x') = \sum_{n=1}^{N} \alpha_n sgn(h_n(x'))$,

$\forall x \in T, F_N^T(x) = \sum_{n=1}^{N} \beta_n sgn(h_n(x))$.
Final source and target classifiers: $H_N^S(x') = sgn(F_N^S(x'))$ and $H_N^T(x) = sgn(F_N^T(x))$.

---

assumption of ADABOOST, finding an hypothesis fulfilling Condition 1 on the source and Condition 2 on the target is more complicated. To overcome this problem, we present in the following a simple strategy which tends to induce hypotheses that satisfy the weak DA assumption.

First, we generate $\frac{k}{2}$ stumps that satisfy Condition 1 over the source and $\frac{k}{2}$ that fulfill Condition 2 over the target. Then, we seek a convex combination $h_n = \sum_k \kappa_k h_n^k$ of the k stumps that satisfies simultaneously the two conditions of Def.(1). To do so, we propose to solve the following convex optimization problem:

$$\underset{\kappa}{\operatorname{argmin}} \sum_{(x',y') \in S} D_n^S(x') \left[-y' \sum_k \kappa_k sgn(h_n^k(x')) \right]_+ + \sum_{x \in T} D_n^T(x) \left[1 - \left(\sum_k \kappa_k marg(f_{DA}(h_n^k(x))) \right) \right]_+ \quad (1)$$

where $[1-x]_+ = max(0, 1-x)$ is the hinge loss, and $marg(f_{DA}(h_n^k(x)))$ returns -1 if $f_{DA}(h_n^k(x))$ is lower than γ (i.e. h_n does not achieve a sufficient margin w.r.t. g_n) and $+1$ otherwise. Solving this optimization problem tends to fulfill Def.(1). Indeed, minimizing the first term of Eq.(1) tends to reduce the empirical risk over the source data, while minimizing the second term tends to decrease the number of margin violations over the target data.

Note that in order to generate a simple weak DA learner, we start the process with $k = 2$. If the weak DA assumption is not satisfied, we increase the dimension of the induced hypothesis h_n. Moreover, if the optimized combination does not satisfy the weak DA assumption, we draw a new set of k stumps.

Once h_n has been learned, the weights of the labeled and unlabeled data are modified according to two different update rules. Those of source examples are updated using the same strategy as that of ADABOOST. Regarding the target examples, their weights are changed according to their location in the space. If a target example x does not satisfy the condition $f_{DA}(h_n(x)) > \gamma$, a pseudo-class $y^n = -sgn(f_{DA}(h_n(x)))$ is assigned to x that simulates a misclassification. Note that such a decision has a geometrical interpretation: it means that we exponentially increase the weights of the points located in an extended margin band of width $\gamma + \lambda g_n$. If x is outside this band, a pseudo-class $y^n = sgn(f_{DA}(h_n(x)))$ is assigned leading to an exponential decrease of $D_n^T(x)$ at the next iteration.

4   Theoretical Analysis

In this section, we present a theoretical analysis of SLDAB. Recall that the goodness of a hypothesis h_n is measured by its ability to not only correctly classify the source examples but also to classify the unlabeled target data with a large margin w.r.t. the classifier-induced divergence g_n. Provided that the weak DA constraints of Def.(1) are satisfied, the standard results of ADABOOST directly hold on $\mathcal{S}$. In the following, we show that the loss $\hat{L}_{H_N^T}$, which represents after N iterations the proportion of margin violations over T (w.r.t. the successive divergences g_n), also decreases with N.

4.1   Upper Bound on the Empirical Loss

Theorem 1. *Let $\hat{L}_{H_N^T}$ be the proportion of target examples of T with a margin smaller than γ w.r.t. the divergences g_n ($n = 1 \ldots N$) after N iterations of SLDAB:*

$$\hat{L}_{H_N^T} = \mathbb{E}_{x \sim T}[\mathbf{y}\mathbf{F_N^T}(x) < 0] \leq \frac{1}{|T|} \sum_{x \sim T} e^{-\mathbf{y}\mathbf{F_N^T}(x)} = \prod_{n=1}^{N} Z_n, \qquad (2)$$

where $\mathbf{y} = (y^1, \ldots, y^n, \ldots, y^N)$ is the vector of pseudo-classes and $\mathbf{F_N^T}(x) = (\beta_1 f_{DA}(h_1(x)), \ldots, \beta_n f_{DA}(h_n(x)), \ldots, \beta_N f_{DA}(h_N(x)))$.

Proof. The proof is the same as that of [7] except that $\mathbf{y}$ is the vector of pseudo-classes (which depend on λg_n and γ) rather than the vector of true labels.   □

4.2   Optimal Confidence Values

Theorem 1 suggests the minimization of each Z_n to reduce the empirical loss $\hat{L}_{H_N^T}$ over T. To do this, let us rewrite Z_n as follows:

$$Z_n = \sum_{x \in T_n^-} D_n^T(x) e^{-\beta_n f_{DA}(h_n(x))y^n} + \sum_{x \in T_n^+} D_n^T(x) e^{-\beta_n f_{DA}(h_n(x))y^n}. \qquad (3)$$

The two terms of the right-hand side of Eq.(3) can be upper bounded as follows:

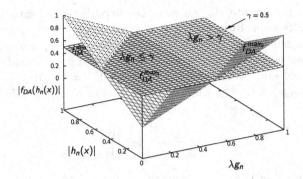

Fig. 2. Upper bounds of the components of Z_n for an arbitrary value $\gamma = 0.5$. When $x \in T_n^+$, the upper bound is obtained with $|f_{DA}| = \gamma$ (see the plateau f_{DA}^{min}). When $x \in T_n^-$, we get the upper bound with $\max(\gamma, \lambda g_n)$, that is either γ when $\lambda g_n \le \gamma$ (see $f_{DA}^{max_1}$) or λg_n otherwise (see $f_{DA}^{max_2}$).

- $\forall x \in T_n^+$, $D_n^T(x)e^{-\beta_n f_{DA}(h_n(x))y^n} \le D_n^T(x)e^{-\beta_n\gamma}$.
- $\forall x \in T_n^-$, $D_n^T(x)e^{-\beta_n f_{DA}(h_n(x))y^n} \le D_n^T(x)e^{\beta_n \max(\gamma, \lambda g_n)}$.

Figure 2 gives a geometrical explanation of these upper bounds. When $x \in T_n^+$, the weights are decreased. We get an upper bound by taking the smallest drop, that is $f_{DA}(h_n(x))y^n = |f_{DA}| = \gamma$ (see f_{DA}^{min} in Figure 2). On the other hand, if $x \in T_n^-$, we get an upper bound by taking the maximum value of f_{DA} (i.e. the largest increase). We differentiate two cases: (i) when $\lambda g_n \le \gamma$, the maximum is γ (see $f_{DA}^{max_1}$), (ii) when $\lambda g_n > \gamma$, Figure 2 shows that one can always find a configuration where $\gamma < f_{DA} \le \lambda g_n$. In this case, $f_{DA}^{max_2} = \lambda g_n$, and we get the upper bound with $|f_{DA}| = \max(\gamma, \lambda g_n)$.

Plugging the previous upper bounds in Eq.(3), we get:

$$Z_n \le W_n^+ e^{-\beta_n\gamma} + W_n^- e^{\beta_n \max(\gamma, \lambda g_n)} = \tilde{Z}_n. \tag{4}$$

Deriving the previous convex combination w.r.t. β_n and equating to zero, we get the optimal values for β_n in Eq.(3)[1]:

$$\frac{\partial \tilde{Z}_n}{\beta_n} = 0 \Rightarrow \max(\gamma, \lambda g_n)W_n^- e^{\beta_n \max(\gamma, \lambda g_n)} = \gamma W_n^+ e^{-\beta_n\gamma}$$

$$\Rightarrow \beta_n = \frac{1}{\gamma + \max(\gamma, \lambda g_n)} \ln \frac{\gamma W_n^+}{\max(\gamma, \lambda g_n)W_n^-}. \tag{5}$$

It is important to note that β_n is computable if

$$\frac{\gamma W_n^+}{\max(\gamma, \lambda g_n)W_n^-} \ge 1 \Leftrightarrow \gamma(1-W_n^-) \ge \max(\gamma, \lambda g_n)W_n^- \Leftrightarrow W_n^- < \frac{\gamma}{\gamma + max(\gamma, \lambda g_n)},$$

---

[1] Note that the approximation $\tilde{Z}_n$ used in Eq.(4) is essentially a linear upper bound of Eq.(3) on the range $[-1; +1]$. Clearly, other upper bounds which give a tighter approximation could be used instead (see [19] for more details).

that is always true because h_n is a weak DA hypothesis and satisfies Condition 2 of Def.(1). Moreover, from Eq.(5), it is worth noting that β_n gets smaller as the divergence gets larger. In other words, a hypothesis h_n of weights W_n^+ and W_n^- (which depend on the divergence g_n) will have a greater confidence than a hypothesis $h_{n'}$ of same weights $W_{n'}^+ = W_n^+$ and $W_{n'}^- = W_n^-$ if $g_n < g_{n'}$.

Let $\max(\gamma, \lambda g_n) = c_n \times \gamma$, where $c_n \geq 1$. We can rewrite Eq.(5) as follows:

$$\beta_n = \frac{1}{\gamma(1 + c_n)} \ln \frac{W_n^+}{c_n W_n^-}, \tag{6}$$

and Condition 2 of Def.(1) becomes $W_n^- < \frac{1}{1+c_n}$.

4.3   Convergence of the Empirical Loss

The following theorem shows that, provided the weak DA constraint on T is fulfilled (that is, $W_n^- < \frac{1}{1+c_n}$), Z_n is always smaller than 1 that leads (from Theorem 1) to a decrease of the empirical loss $\hat{L}_{H_N^T}$ with the number of iterations.

Theorem 2. *If H_N^T is the linear combination produced by* SLDAB *from N weak DA hypotheses, then* $\lim_{N \to \infty} \hat{L}_{H_N^T} = 0$.

Proof. Plugging Eq.(6) into Eq.(4) we get:

$$Z_n \leq W_n^+ \left(\frac{c_n W_n^-}{W_n^+} \right)^{\frac{1}{(1+c_n)}} + W_n^- \left(\frac{W_n^+}{c_n W_n^-} \right)^{\frac{c_n}{(1+c_n)}} \tag{7}$$

$$= \left(W_n^+ \right)^{\frac{c_n}{(1+c_n)}} \left(W_n^- \right)^{\frac{1}{(1+c_n)}} \left(c_n^{\frac{1}{(1+c_n)}} + c_n^{-\frac{c_n}{(1+c_n)}} \right)$$

$$= \left(W_n^+ \right)^{\frac{c_n}{(1+c_n)}} \left(W_n^- \right)^{\frac{1}{(1+c_n)}} \left(\frac{c_n + 1}{c_n^{\frac{c_n}{(1+c_n)}}} \right) = u_n \times v_n \times w_n, \tag{8}$$

where $u_n = (W_n^+)^{\frac{c_n}{(1+c_n)}}$, $v_n = (W_n^-)^{\frac{1}{(1+c_n)}}$ and $w_n = \left(\frac{c_n+1}{c_n^{\frac{c_n}{(1+c_n)}}} \right)$. Computing the derivative of u_n, v_n and w_n w.r.t. c_n, we get

$$\frac{\partial u_n}{\partial c_n} = \frac{\ln W_n^+}{(c_n + 1)^2} \left(W_n^+ \right)^{\frac{c_n}{(1+c_n)}}, \quad \frac{\partial v_n}{\partial c_n} = -\frac{\ln W_n^-}{(c_n + 1)^2} \left(W_n^- \right)^{\frac{1}{(1+c_n)}},$$

$$\frac{\partial w_n}{\partial c_n} = -\frac{\ln c_n}{(c_n + 1)^2} \frac{c_n + 1}{c_n^{\frac{c_n}{(1+c_n)}}}.$$

We deduce that

$$\frac{\partial Z_n}{\partial c_n} = \left(\frac{\partial u_n}{\partial c_n} \times v_n + \frac{\partial v_n}{\partial c_n} \times u_n \right) \times w_n + \frac{\partial w_n}{\partial c_n} \times u_n \times v_n$$

$$= \left(W_n^+\right)^{\frac{c_n}{(1+c_n)}} \times \left(W_n^-\right)^{\frac{1}{(1+c_n)}} \times \left(\frac{c_n+1}{c_n^{\frac{c_n}{(1+c_n)}}}\right) \times \frac{1}{(c_n+1)^2} \times \left(\ln W_n^+ - \ln W_n^- - \ln c_n\right)$$

$$= \left(W_n^+\right)^{\frac{c_n}{(1+c_n)}} \times \left(W_n^-\right)^{\frac{1}{(1+c_n)}} \times \frac{c_n^{\frac{-c_n}{(1+c_n)}}}{c_n+1} \times \left(\ln W_n^+ - \ln W_n^- - \ln c_n\right).$$

The first three terms of the previous equation are positive. Therefore,

$$\frac{\partial Z_n}{\partial c_n} > 0 \Leftrightarrow \ln W_n^+ - \ln W_n^- - \ln c_n > 0 \Leftrightarrow W_n^- < \frac{1}{c_n+1},$$

that is always true because of the weak DA assumption. Therefore, $Z_n(c_n)$ is a monotonic increasing function over $[1, \frac{W_n^+}{W_n^-}[$, with:

- $Z_n < 2\sqrt{W_n^+ W_n^-}$ (standard result of ADABOOST) when $c_n = 1$,
- and $\lim\limits_{c_n \to \frac{W_n^+}{W_n^-}} Z_n = 1$.

Therefore, $\forall n, Z_n < 1 \Leftrightarrow \lim\limits_{N \to \infty} \hat{L}_{H_N^T} < \lim\limits_{N \to \infty} \prod\limits_{n=1}^{N} Z_n = 0.$ □

Let us now give some insight about the nature of the convergence of $\hat{L}_{H_N^T}$. A hypothesis h_n is DA weak if $W_n^- < \frac{1}{1+c_n} \Leftrightarrow c_n < \frac{W_n^+}{W_n^-} \Leftrightarrow c_n = \tau_n \frac{W_n^+}{W_n^-}$ with $\tau_n \in]\frac{W_n^-}{W_n^+}; 1[$. τ_n measures how close is h_n to the weak assumption requirement. Note that β_n gets larger as τ_n gets smaller. From Eq.(8) and $c_n = \tau_n \frac{W_n^+}{W_n^-}$ (that is $W_n^- = \frac{\tau_n}{\tau_n + c_n}$), we get (see Appendix 1 for more details):

$$Z_n \leq \left(W_n^+\right)^{\frac{c_n}{(1+c_n)}} \left(W_n^-\right)^{\frac{1}{(1+c_n)}} \left(\frac{c_n+1}{c_n^{\frac{c_n}{(1+c_n)}}}\right) = \left(\frac{\tau_n^{\frac{1}{1+c_n}}}{\tau_n + c_n}\right)(c_n+1).$$

We deduce that

$$\prod_{n=1}^{N} Z_n = exp \sum_{n=1}^{N} \ln Z_n \leq exp \sum_{n=1}^{N} \left(\ln\left(\left(\frac{\tau_n^{\frac{1}{1+c_n}}}{\tau_n + c_n}\right)(c_n+1)\right)\right)$$

$$= exp \sum_{n=1}^{N} \left(\frac{1}{1+c_n} \ln \tau_n + \ln(\frac{c_n+1}{\tau_n + c_n})\right).$$

Theorem 2 tells us that the term between brackets is negative (that is $\ln Z_n < 0, \forall Z_n$). Therefore, the empirical loss decreases exponentially fast towards 0 with the number of iterations N. Moreover, let us study the behaviour of $\ln Z_n$

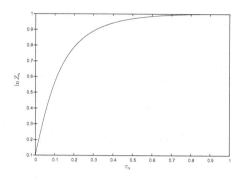

Fig. 3. Evolution of $\ln Z_n$ w.r.t. τ_n

w.r.t. τ_n. Since Z_n is a monotonic increasing function of c_n over $[1, \frac{W_n^+}{W_n^-}[$, it is also a monotonic increasing function of τ_n over $[\frac{W_n^-}{W_n^+}; 1[$. In other words, the smaller τ_n the faster the convergence of the empirical loss $\hat{L}_{H_N^T}$. Figure 3 illustrates this claim for an arbitrarily selected configuration of W_n^+ and W_n^-. It shows that $\ln Z_n$, and thus $\hat{L}_{H_N^T}$, decreases exponentially fast with τ_n.

5 Measure of Divergence

From DA frameworks [1,14], a good adaptation is possible when the mismatch between the two distributions is small while maintaining a good accuracy on the source. In our algorithm, the latter condition is satisfied via the use of a standard boosting scheme. Concerning the mismatch, we inserted in our framework a measure of divergence g_n, induced by h_n. An important issue of SLDAB is the definition of this measure. A solution is to compute a divergence with respect to the considered *class of hypotheses*, like the well-known $\mathcal{H}$-divergence[2] [1]. We claim that such a divergence is not suited to our framework because SLDAB rather aims at evaluating the discrepancy induced by a *specific classifier h_n*. We propose to consider a divergence g_n able to both evaluate the mismatch between the source and target data and avoid degenerate hypotheses.

For the first objective, we use the recent *Perturbed Variation* measure [8] that evaluates the discrepancy between two distributions while allowing small permitted variations assessed by a parameter $\epsilon > 0$ and a distance d:

Definition 2 ([8]). *Let P and Q two marginal distributions over X, let $M(P,Q)$ be the set of all joint distributions over $X \times X$ with P and Q. The perturbed variation w.r.t. a distance $d : \mathcal{X} \times \mathcal{X} \to \mathbb{R}$ and $\epsilon > 0$ is defined by*

$$PV(P,Q) = \inf_{\mu \in M(P,Q)} Proba_\mu[d(P',Q') > \epsilon]$$

---

[2] The $\mathcal{H}$-divergence is defined with respect to the hypothesis class $\mathcal{H}$ by: $\sup_{h,h' \in \mathcal{H}} |\mathbb{E}_{x \sim \mathcal{T}}[h(x) \neq h'(x)] - \mathbb{E}_{x' \sim \mathcal{S}}[h(x') \neq h'(x')]|$, it can be empirically estimated by learning a classifier able to discriminate source and target instances [1].

Algorithm 2. Computation of $\hat{PV}(S,T)$ [8].

Input: $S = \{x'_1, \ldots, x'_n\}$, $T = \{x_1, \ldots, x_m\}$, $\epsilon > 0$ and a distance d

1. Define the graph $\hat{G} = (\hat{V} = (\hat{A}, \hat{B}), \hat{E})$ where $\hat{A} = \{x'_i \in S\}$ and $\hat{B} = \{x_j \in T\}$, Connect an edge $e_{ij} \in \hat{E}$ if $d(x'_i, x_j) \leq \epsilon$
2. Compute the maximum matching on $\hat{G}$
3. S_u and T_u are the number of unmatched vertices in S and T respectively
4. Output $\hat{PV}(S,T) = \frac{1}{2}\left(\frac{S_u}{n} + \frac{T_u}{m}\right) \in [0,1]$

---

over all pairs $(P', Q') \sim \mu$ s.t. the marginal of P' (resp. Q') is P (resp. Q).

Intuitively two samples are similar if every target instance is close to a source one w.r.t. d. This measure is consistent and the empirical estimate $\hat{PV}(S,T)$ from two samples $S \sim P$ and $T \sim Q$ can be efficiently computed by a maximum graph matching procedure summarized in Algorithm 2. In our context, we apply this empirical measure on the classifier outputs: $S_{h_n} = \{h_n(x'_1), \ldots, h_n(x'_{|S|})\}$, $T_{h_n} = \{h_n(x_1), \ldots, h_n(x_{|T|})\}$ with the L_1 distance as d and use $1 - \hat{PV}(S_{h_n}, T_{h_n})$.

For the second point, we take the following entropy-based measure:

$$ENT(h_n) = 4 \times p_n \times (1 - p_n)$$

where $p_n{}^3$ is the proportion of target instances classified as positive by h_n:
$p_n = \frac{\sum_{i=1}^{|T|}[h_n(x_i) \geq 0]}{|T|}$. For the degenerate cases where all the target instances have the same class, the value of $ENT(h_n)$ is 0, otherwise if the labels are equally distributed this measure is close to 1.

Finally, g_n is defined by 1 minus the product of the two previous similarity measures allowing us to have a divergence of 1 if one of the similarities is null.

$$g_n = 1 - (1 - \hat{PV}(S_{h_n}, T_{h_n})) \times ENT(h_n).$$

6   Experiments

To assess the practical efficiency of SLDAB and support our claim of Section 2, we perform two kinds of experiments, respectively in the DA and semi-supervised settings. We use two different databases. The first one, MOONS [4], corresponds to two inter-twinning moons in a 2-dimensional space where the data follow a uniform distribution in each moon representing one class. The second one is the UCI SPAM database[4], containing 4601 e-mails (2788 considered as "non-spams" and 1813 as "spams") in a 57-dimensional space.

6.1   Domain Adaptation

Moons Database. In this series of experiments, the target domain is obtained by rotating anticlockwise the source domain, corresponding to the original data.

---

[3] True labels are assumed well balanced, if not p_n has to be reweighted accordingly.
[4] http://archive.ics.uci.edu/ml/datasets/Spambase

Table 1. On the left: error rates (in%) on MOONS, the Average column reports the rate averages along with average standard deviations. On the right: error rates on SPAMS.

Angle	$20°$	$30°$	$40°$	$50°$	$60°$	$70°$	$80°$	$90°$	Average	Algorithm	Error rate (in%)
SVM	10.3	24	32.2	40	43.3	55.2	67.7	80.7	44.2 ± 0.9	SVM	38
AdaBoost	20.9	32.1	44.3	53.7	61.2	69.7	77.9	83.4	55.4 ± 0.4	AdaBoost	59.4
DASVM	0.0	21.6	28.4	33.4	38.4	74.7	78.9	81.9	44.6 ± 3.2	DASVM	37.5
SVM-W	6.8	12.9	9.5	26.9	48.2	59.7	66.6	67.8	37.3 ± 5.3	SVM-W	37.9
SLDAB-$\mathcal{H}$	6.9	11.3	18.1	32.8	37.5	45.1	55.2	59.7	33.3 ± 2.1	SLDAB-$\mathcal{H}$	37.1
SLDAB-g_n	1.2	3.6	7.9	10.8	17.2	39.7	47.1	45.5	21.6 ± 1.2	SLDAB-g_n	35.8

We consider 8 increasingly difficult problems according to 8 rotation angles from 20 degrees to 90 degrees. For each domain, we generate 300 instances (150 of each class). To estimate the generalization error, we make use of an independent test set of 1000 points drawn from the target domain. Each adaptation problem is repeated 10 times and we report the average results obtained on the test sample without the best and the worst draws.

We compare our approach with two non DA baselines: the standard AD-ABOOST, using decision stumps, and a SVM classifier (with a Gaussian kernel) learned only from the source. We also compare SLDAB with DASVM (based on a LibSVM implementation) and with a reweighting approach for the co-variate shift problem presented in [9]. This unsupervised method (referred to as SVM-W) reweights the source examples by matching source and target distributions by a kernel mean matching process, then a SVM classifier is inferred from the reweighted source sample. Note that all the hyperparameters are tuned by cross-validation. Finally, to confirm the relevance of our divergence measure g_n, we run SLDAB with two different divergences: SLDAB-g_n uses our novel measure g_n introduced in the previous section and SLDAB-$\mathcal{H}$ is based on the $\mathcal{H}$-divergence. We tune the parameters of SLDAB by selecting, threw a grid search, those able to fulfill Def.(1) and leading to the smallest divergence over the final combination F_N^T. As expected, the optimal λ grows with the difficulty of the problem.

Results obtained on the different adaptation problems are reported in Table 1. We can see that, except for 20 degrees (for which DASVM is slightly better), SLDAB-g_n achieves a significantly better performance, especially on important rotation angles. DASVM that is not able to work with large distribution shifts diverges completely. This behaviour shows that our approach is more robust to difficult DA problems. Finally, despite good results compared to other algorithms, SLDAB-$\mathcal{H}$ does not perform as well as the version using our divergence g_n, showing that g_n is indeed more adapted to our approach.

Figure 4(a) illustrates the behaviour of our algorithm on a 20 degrees rotation problem. First, as expected by Theorem 2, the empirical target loss converges very quickly towards 0. Because of the constraints imposed on the target data, the source error $\hat{\epsilon}_{H_N^S}$ requires more iterations to converge than a classical AD-ABOOST procedure. Moreover, the target error $\epsilon_{H_N^T}$ decreases with N and keeps dropping even when the two empirical losses have converged to zero. This confirms the benefit of having a low source error with large target margins.

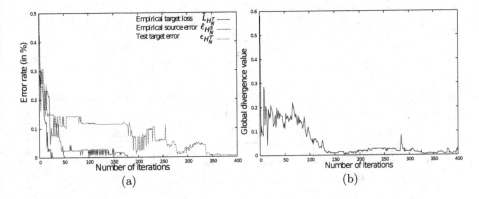

Fig. 4. (a): loss functions on a 20° task. (b): evolution of the global divergence.

Figure 4(b) shows the evolution throughout the iterations of the divergence g_n of the combination $H_n^T = \sum_{k=1}^{n} \beta_k h_k(x)$. We can see that our boosting scheme allows us to reduce the divergence between the source and the target data.

Spams Database. To design a DA problem from this UCI database, we first split the original data in three different sets of equivalent size. We use the first one as the learning set, representing the source distribution. In the two other samples, we add a gaussian noise to simulate a different distribution. As all the features are normalized in the [0,1] interval, we use, for each feature n, a random real value in [-0.15,0.15] as expected value μ_n and a random real value in [0,0.5] as standard deviation σ_n. We then generate noise according to a normal distribution $\mathcal{N}(\mu_n, \sigma_n)$. After having modified these two samples jointly with the same procedure, we keep one as the target learning set, the other as the test set.

This operation is repeated 5 times. The average results of the different algorithms are reported in Table 1. As for the moons problem, we compare our approach with the standard ADABOOST and a SVM classifier learned only from the source. We also compare it with DASVM and SVM-W. We see that SLDAB is able to obtain better results than all the other algorithms on this real database. Moreover, SLDAB used with our divergence g_n leads again to the best result.

6.2 Semi-supervised Setting

Our divergence criterion allows us to quantify the distance between the two domains. If its value is low all along the process, this means that we are facing a problem that looks more like a semi-supervised task. In a semi-supervised setting, the learner receives few labeled and many unlabeled data generated from the same distribution. In this series of experiments, we study our algorithm on two semi-supervised variants of the MOONS and SPAMS databases.

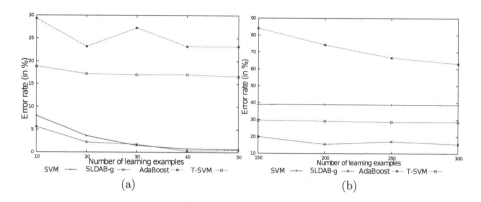

Fig. 5. (a): error rate of different algorithms on the moons semi-supervised problem according to the size of the training set. (b): error rate of different algorithms on the spam recognition semi-supervised problem according to the size of the training set.

Moons Database. We generate randomly a learning set of 300 examples and an independent test set of 1000 examples from the same distribution. We then draw n labeled examples from the learning set, from $n = 10$ to 50 such that exactly half of the examples are positives, and keep the remaining data for the unlabeled sample. The methods are evaluated by computing the error rate on the test set. For this experiment, we compare SLDAB-g_n with ADABOOST, SVM and the transductive SVM T-SVM introduced in [10] which is a semi-supervised method using the information given by unlabeled data to train a SVM classifier. We repeat each experiment 5 times and show the average results in Figure 5(a).

Our algorithm performs better than the other methods on small training sets and is competitive to SVM for larger sizes. We can also remark that ADABOOST using only the source examples is not able to perform well. This can be explained by an overfitting phenomenon on the small labeled sample leading to poor generalization performances. Surprisingly, T-SVM performs quite poorly too. This is probably due to the fact that the unlabeled data are incorrectly exploited, with respect to the small labeled sample, producing wrong hypotheses.

Spams Database. We use here the same set up as in the semi-supervised setting for MOONS. We take the 4601 original instances issued from the same distribution and split them into two sets: one third for the training sample and the remaining for the test set used to compute the error rate. From the training set, n labeled instances are drawn as labeled data, n varying from 150 to 300, the remaining part is used as unlabeled data as in the previous experiment. This procedure is repeated 5 times for each n and the average results are provided in Figure 5(b).

All the approaches are able to decrease their error rate according to the size of the labeled data (even if it is not significant for SVM and T-SVM), which is an expected behaviour. SVM and even more ADABOOST (that do not use unlabeled data), achieve a large error rate after 300 learning examples. T-SVM is able to take advantage of the unlabeled examples, with a significant gain compared to SVM. Finally, SLDAB outperforms the other algorithms by at least 10 percentage points. This confirms that SLDAB is also able to perform well in a semi-supervised learning setting. This feature makes our approach very general and relevant for a large class of problems.

7   Conclusion

In this paper, we have presented a new boosting-based DA algorithm called SLDAB. This algorithm, working in the difficult unsupervised DA setting, iteratively builds a combination of weak DA learners able to minimize both the source classification error and margin violations on the unlabeled target instances. The originality of this approach is to introduce the use of a new distribution divergence during the iterative process for avoiding bad adaptation due to the production of degenerate hypotheses. This divergence gives more importance to classifiers able to move closer source and target distributions with respect to the outputs of the classifiers. In this context, we have theoretically proved that our approach converges exponentially fast with the number of iterations. Our experiments have shown that SLDAB performs well in a DA setting both on synthetic and real data. Moreover, SLDAB is also general enough to work well in a semi-supervised case, making our approach widely applicable.

Even if our experiments have shown good results, we did not prove yet that the generalization error decreases. Such a result deserves further investigation but we conjecture that this is true for SLDAB. Indeed, the minimization of the margin violations on the target instances implies a minimization of our divergence in the space induced by the classifiers $\beta_n h_n$. Classical DA frameworks indicate that good generalization capabilities arise when a DA algorithm is able both to ensure a good performance on the source domain and to decrease the distribution mismatch, which is what SLDAB does. A perspective is then to show that the specific divergence we propose is able to ensure good generalization guarantees up to the ϵ used in the perturbed variation measure. Another one is to extend our approach to allow the use of a small labeled target sample.

References

1. Ben-David, S., Blitzer, J., Crammer, K., Kulesza, A., Pereira, F., Vaughan, J.: A theory of learning from different domains. Mach. Learn. 79(1-2), 151–175 (2010)
2. Bennett, K., Demiriz, A., Maclin, R.: Exploiting unlabeled data in ensemble methods. In: KDD, pp. 289–296 (2002)
3. Blitzer, J., Dredze, M., Pereira, F.: Biographies, bollywood, boom-boxes and blenders: Domain adaptation for sentiment classification. In: ACL (2007)
4. Bruzzone, L., Marconcini, M.: Domain adaptation problems: a DASVM classification technique and a circular validation strategy. T. PAMI 32(5), 770–787 (2010)
5. Chelba, C., Acero, A.: Adaptation of maximum entropy capitalizer: Little data can help a lot. Computer Speech & Language 20(4), 382–399 (2006)

6. Dai, W., Yang, Q., Xue, G., Yu, Y.: Boosting for transfer learning. In: ICML, pp. 193–200 (2007)
7. Freund, Y., Schapire, R.: Experiments with a new boosting algorithm. In: ICML, pp. 148–156 (1996)
8. Harel, M., Mannor, S.: The perturbed variation. In: Proceedings of NIPS, pp. 1943–1951 (2012)
9. Huang, J., Smola, A., Gretton, A., Borgwardt, K., Schölkopf, B.: Correcting sample selection bias by unlabeled data. In: NIPS, pp. 601–608 (2006)
10. Joachims, T.: Transductive inference for text classification using support vector machines. In: Proceedings of ICML, ICML 1999, pp. 200–209 (1999)
11. Leggetter, C., Woodland, P.: Maximum likelihood linear regression for speaker adaptation of continuous density hidden markov models. Computer Speech & Language, 171–185 (1995)
12. Mallapragada, P., Jin, R., Jain, A., Liu, Y.: Semiboost: Boosting for semi-supervised learning. IEEE T. PAMI 31(11), 2000–2014 (2009)
13. Mansour, Y., Mohri, M., Rostamizadeh, A.: Domain adaptation with multiple sources. In: NIPS, pp. 1041–1048 (2008)
14. Mansour, Y., Mohri, M., Rostamizadeh, A.: Domain adaptation: Learning bounds and algorithms. In: COLT (2009)
15. Margolis, A.: A literature review of domain adaptation with unlabeled data. Tech. rep., Univ. Washington (2011)
16. Martínez, A.: Recognizing imprecisely localized, partially occluded, and expression variant faces from a single sample per class. IEEE T. PAMI 24(6), 748–763 (2002)
17. Quionero-Candela, J., Sugiyama, M., Schwaighofer, A., Lawrence, N.: Dataset Shift in Machine Learning. MIT Press (2009)
18. Roark, B., Bacchiani, M.: Supervised and unsupervised pcfg adaptation to novel domains. In: HLT-NAACL (2003)
19. Schapire, R.E., Singer, Y.: Improved boosting algorithms using confidence-rated predictions. Machine Learning 37(3), 297–336 (1999)
20. Valiant, L.: A theory of the learnable. Commun. ACM 27(11), 1134–1142 (1984)
21. Yao, Y., Doretto, G.: Boosting for transfer learning with multiple sources. In: CVPR, pp. 1855–1862 (2010)

Appendix 1

If the weak DA assumption is satisfied, $c_n = \tau_n \frac{W_n^+}{W_n^-}$ with $\tau_n \in]\frac{W_n^-}{W_n^+}; 1[$. We deduce that $W_n^- = \frac{\tau_n}{\tau_n + c_n}$.

$$Z_n < \left(W_n^+\right)^{\frac{c_n}{(1+c_n)}} \left(W_n^-\right)^{\frac{1}{(1+c_n)}} \left(\frac{c_n + 1}{c_n^{\frac{c_n}{(1+c_n)}}}\right)$$

$$= \left(1 - \frac{\tau_n}{\tau_n + c_n}\right)^{\frac{c_n}{(1+c_n)}} \left(\frac{\tau_n}{\tau_n + c_n}\right)^{\frac{1}{(1+c_n)}} \left(\frac{c_n + 1}{c_n^{\frac{c_n}{(1+c_n)}}}\right)$$

$$= \frac{c_n^{\frac{c_n}{(1+c_n)}}}{(\tau_n + c_n)^{\frac{c_n}{(1+c_n)}}} \cdot \frac{\tau_n^{\frac{1}{(1+c_n)}}}{(\tau_n + c_n)^{\frac{1}{(1+c_n)}}} \cdot \frac{c_n + 1}{c_n^{\frac{c_n}{(1+c_n)}}}$$

$$= \left(\frac{\tau_n^{\frac{1}{1+c_n}}}{\tau_n + c_n}\right)(c_n + 1).$$

Automatically Mapped Transfer between Reinforcement Learning Tasks via Three-Way Restricted Boltzmann Machines

Haitham Bou Ammar[1], Decebal Constantin Mocanu[1], Matthew E. Taylor[2], Kurt Driessens[1], Karl Tuyls[1], and Gerhard Weiss[1]

[1] Maastricht University, Department of Knowledge Engineering, Netherlands
{haitham.bouammar,kurt.driessens,k.tuyls,
gerhard.weiss}@maastrichtuniversity.nl,
d.mocanu@student.maastrichtuniversity.nl
[2] School of EECS, Washington State University, USA
taylorm@eecs.wsu.edu

Abstract. Existing reinforcement learning approaches are often hampered by learning *tabula rasa*. Transfer for reinforcement learning tackles this problem by enabling the reuse of previously learned results, but may require an inter-task mapping to encode how the previously learned task and the new task are related. This paper presents an autonomous framework for learning inter-task mappings based on an adaptation of restricted Boltzmann machines. Both a full model and a computationally efficient factored model are introduced and shown to be effective in multiple transfer learning scenarios.

Keywords: Transfer Learning, Reinforcement Learning, Inter-task mapping, Boltzmann Machines, Least Squares Policy Iteration.

1 Introduction

Reinforcement learning (RL) has become a popular framework for autonomous behavior generation from limited feedback [4,13], but RL methods typically learn *tabula rasa*. Transfer learning [17] (TL) aims to significantly improve learning by providing informative knowledge from a previous (source) task to a learning agent in a novel (target) task. If the agent is to be fully autonomous, it must: (1) automatically select a source task, (2) learn how the source task and target tasks are related, and (3) effectively use transferred knowledge when in the target task. While fully autonomous transfer is not yet possible, this paper advances the state of the art by focusing on (2) above. In particular, this work proposes methods to automatically learn the relationships between pairs of tasks and then use this learned relationship to transfer effective knowledge.

In TL for RL, the source task and target task may differ in their formulations. In particular, when the source task and target task have different state and/or action spaces, an *inter-task mapping* [18] that describes the relationship between

H. Blockeel et al. (Eds.): ECML PKDD 2013, Part II, LNAI 8189, pp. 449–464, 2013.

the two tasks is needed. While there have been attempts to discover this mapping automatically, finding an optimal way to construct this mapping is still an open question. Existing technique either rely on restrictive assumptions made about the relationship between the source and target tasks, or adopt heuristics that work only in specific cases.

This paper introduces an autonomous framework for learning inter-task mappings based on *restricted Boltzmann machines* [1] (RBMs). RBMs provide a powerful but general framework that can be used to describe an abstract common space for different tasks. This common space is then used to represent the inter-task mapping between two tasks and can successfully transfer knowledge about transition dynamics between the two tasks.

The contributions of this paper are summarized as follows. First, a novel RBM is proposed that uses a three-way weight tensor (i.e., TrRBM). Since this machine has a computational complexity of $\mathcal{O}(N^3)$, a factored version (i.e., FTr-RBM) is then derived that reduces the complexity to $\mathcal{O}(N^2)$. Experiments then transfer samples between pairs of tasks, showing that the proposed method is capable of successfully learning a useful inter-task mapping. Specifically, the results demonstrate that FTrRBM is capable of:

1. Automatically learning an inter-task mapping between different MDPs.
2. Transferring informative samples that reduce the computational complexity of a sample-based RL algorithm.
3. Transferring informative instances which reduce the time needed for a sample-based RL algorithm to converge to a near-optimal behavior.

2   Preliminaries

This section provides a brief summary of background knowledge needed to understand the remaining of the paper.

2.1   Reinforcement Learning

In an RL problem, an agent must decide how to sequentially select actions to maximize its expected return. Such problems are typically formalized as a Markov decision process (MDP), defined by $\langle S, A, P, R, \gamma \rangle$. S is the (potentially infinite) set of states, A is the set possible actions that the agent may execute, $P : S \times A \times S \to [0, 1]$ is a state transition probability function, describing the task dynamics, $R : S \times A \times S \to \mathbb{R}$ is the reward function measuring the performance of the agent, and $\gamma \in [0, 1)$ is the discount factor. A policy $\pi : S \times A \to [0, 1]$ is defined as a probability distribution over state action pairs, where $\pi(s, a)$ represents the probability of selecting action a in state s. The goal of an RL agent is to find a policy $\pi^\star$ which maximizes the cumulative future rewards. It can be attained by taking greedy actions according to the optimal Q-function $Q^\star(s, a) = \max_\pi \mathbb{E}_\pi \left[\sum_{t=0}^{\infty} \gamma^t R(s_t, a_t) | s = s_0, a = a_0 \right]$. In tasks with continuous state and/or action spaces, Q and π cannot be represented in a table format, typically requiring sampling and function approximation techniques. This paper uses one such common technique, Least Squares Policy Iteration [4] (LSPI).

2.2 Transfer Learning for Reinforcement Learning

Transfer learning (TL) aims to improve learning times and/or behavior of an agent by re-using knowledge from a solved task. Most often, TL uses a single source task, $\mathcal{T}_1$, and a single target task, $\mathcal{T}_2$. Each task is described by MDPs, which may differ significantly. Specifically, task $\mathcal{T}_1$ is described by $\langle S_1, A_1, P_1, R_1, \gamma_1 \rangle$ and $\mathcal{T}_2$ by $\langle S_2, A_2, P_2, R_2, \gamma_2 \rangle$. To enable transfer between tasks with different state and/or action spaces, an inter-task mapping χ is required, so that information from a source task is applicable to the target task. Typically, χ is divided into two mappings: (1) an inter-state mapping χ_S, and (2) an inter-action mapping χ_A. The first relates states from the source task to the target task, while the second maps actions from the source task to the target task. This paper learns such mappings for samples in a pair of tasks. We define the inter-task mapping to relate source and target transitions (i.e., $\chi : S_1 \times A_1 \times S_1 \to S_2 \times A_2 \times S_2$). This allows the algorithm to discover dynamical similarities between tasks and construct the an inter-task mapping accordingly, enabling the transfer of near-optimal[1] transitions.

2.3 Restricted Boltzmann Machines

Restricted Boltzmann machines (RBMs) are energy-based models for unsupervised learning. They use a generative model of the distribution of the data for prediction. These models are stochastic with stochastic nodes and layers, making them less vulnerable to local minima [15]. Further, due to multiple layers and the neural configurations, RBMs possess excellent generalization abilities. For example, they have successfully discovered informative hidden features in unlabeled data [2]. Formally, an RBM consists of two binary layers: one visible and one hidden. The visible layer represents the data and the hidden layer increases the learning capacity by enlarging the class of distributions that can be represented to an arbitrary complexity [15]. This paper follows standard notation where i represents the indices of the visible layer, j those of the hidden layer, and $w_{i,j}$ denotes the weight connection between the ith visible and j^{th} hidden unit. We further use v_i and h_j to denote the state of the ith visible and j^{th} hidden unit, respectively. According to the above definitions, the energy function is given by:

$$E(v,h) = -\sum_{i,j} v_i h_j w_{ij} - \sum_i v_i b_i - \sum_j h_j b_j \tag{1}$$

where b_i and b_j represent the biases. The first term, $\sum_{i,j} v_i h_j w_{ij}$, represents the energy between the hidden and visible units with their associated weights. The second, $\sum_i v_i b_i$, represents the energy in the visible layer, while the third term represents the energy in the hidden layers. The joint probability of a state of the hidden and visible layers is defined as:

$$P(v,h) \propto \exp\left(-E(v,h)\right)$$

---

[1] When using function approximation techniques, RL algorithms typically learn near-optimal behaviors.

To determine the probability of a data point represented by a state v, the marginal probability is used, summing out the state of the hidden layer:

$$p(v) = \sum_h P(v, h) \propto \sum_h \left(\exp \left(-\sum_{i,j} v_i h_j w_{ij} - \sum_i v_i b_i - \sum_j h_j b_j \right) \right) \quad (2)$$

The above equations can be used for any given input to calculate the probability of either the visible or the hidden configuration to be activated. The calculated probabilities can then be used to perform inference to determine the conditional probabilities in the model.

To maximize the likelihood of the model, the gradient of the log-likelihood with respect to the weights from the previous equation must be calculated. The gradient of the first term, after some algebraic manipulations, can be written as:

$$\frac{\partial \log \left(\sum_h \exp \left(-E(v, h) \right) \right)}{\partial w_{ij}} = v_i \cdot P(h_j = 1 | v)$$

However, computing the gradient of the second term (i.e., $\frac{\partial \log(\sum_{x,y} \exp(-E(x,y)))}{\partial w_{ij}} = P(v_i = 1, h_j = 1)$) is intractable.

2.4 Contrastive Divergence Learning

Because of the difficulty of computing the derivative of the log-likelihood gradients, Hinton proposed an approximation method called *contrastive divergence* (CD) [6]. In maximum likelihood, the learning phase actually minimizes the Kullback-Leiber (KL) measure between the input data distribution and the approximate model. In CD, learning follows the gradient of $CD_n = D_{KL}(p_0(\mathbf{x}) \| p_\infty(\mathbf{x})) - D_{KL}(p_n(\mathbf{x}) \| p_\infty(\mathbf{x}))$ where $p_n(.)$ is the distribution of a Markov chain starting from $n = 0$ and running for a small number of n steps. To derive the update rules of w_{ij} for the RBM, the energy function is re-written in a matrix form as $E(\mathbf{v}, \mathbf{h}; \mathbf{W}) = -\mathbf{h}^T \mathbf{W} \mathbf{v}$. $\mathbf{v} = [v_1, \ldots, v_{n_v}]$, where v_i is the value of the ith visible neuron and n_v is the index of the last visible neuron. $\mathbf{h} = [h_1, \ldots, h_{n_h}]$, where h_j is the value of the jth hidden neuron and n_h is the index of the last hidden neuron. $\mathbf{W} \in \mathbb{R}^{n_h \times n_v}$ is the matrix of all weights. Since the visible units are conditionally independent given the hidden units and vice versa, learning in such an RBM is easy. One step of Gibbs sampling can be carried in two half-steps: (1) update all the hidden units, and (2) update all the visible units. Thus, in CD_n the weight updates are done as follows:

$$w_{ij}^{\tau+1} = w_{ij}^\tau + \alpha \left(\langle \langle h_j v_i \rangle_{p(\mathbf{h}|\mathbf{v};\mathbf{W})} \rangle_0 - \langle h_j v_i \rangle_n \right)$$

where τ is the iteration, α is the learning rate, $\langle \langle h_j v_i \rangle_{p(\mathbf{h}|\mathbf{v};\mathbf{W})} \rangle_0 = \frac{1}{N} \sum_{n=1}^N v_i^{(n)} P(h_i^{(n)} = 1 | \mathbf{h}; \mathbf{W})$, and $\langle h_j v_i \rangle_n = \frac{1}{N} \sum_{n=1}^N v_i^{(n)G_l} P(h_j^{(n)G_l} | \mathbf{h}^{(n)G_l}; \mathbf{W})$ with N the total number of input instances and G_l indicating that the states are obtained after l iterations of Gibbs sampling from the Markov chain starting at $p_0(.)$. In this work, a variant of the CD algorithm is used to better learn the neural configuration of the proposed FTrRBM model and is explained in Section 3.2.

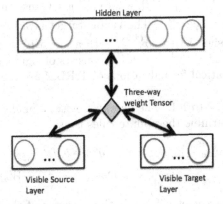

Fig. 1. This picture diagrams the overall structure of the proposed full model. The three-way weight tensor is shown in the middle with its connections to the visible source layer, visible target layer, and hidden layer.

3   RBMs for Transfer Learning

The core hypothesis of this paper is that RBMs can automatically build an inter-task mapping using source task and target task samples, because an RBM can discover the latent similarities between the tasks, implicitly encoding an inter-task mapping. To construct an algorithm to test this hypothesis, the TrRBM framework consists of three layers as shown in Figure 1. The first is the source task visible layer (describing source samples), the second is the target task visible layer (describing target samples), and the third is the hidden layer that encodes similarities between the two tasks. This hidden layer therefore encodes a type of inter-task mapping, which will be used to transfer samples from the source to the target.

The next section presents a derivation of the full model. However, this model is computationally expensive, and a factored version of the model is developed in Section 3.2.

3.1   Transfer Restricted Boltzmann Machine

TrRBM is used to automatically learn the inter-task mapping between source and target tasks. TrRBM consists of three layers: (1) a visible source task task layer, (2) a visible target task layer, and (3) a hidden layer. The number of units in the visible layers is equal to dimensionality of each of the source and target task transitions (*i.e.*, $\langle s_t^i, a_t^i, s_t^{i'} \rangle$) for $i \in \{1,2\}$, respectively. Since the inputs might be of continuous nature, the units in the visible units are set to be Gaussians with means, which are learned as described later in this section. These three layers are connected with a three-way weight tensor. Formally, the visible source layer is $\mathcal{V}_S = [v_S^{(1)}, \ldots, v_S^{(V_S)}]$, where $v_S^{(i)}$ is a Gaussian $\mathcal{N}(\mu_S^{(i)}, \sigma_S^{(i)2})$, with a mean $\mu_S^{(i)}$ and a variance $\sigma_S^{(i)2}$. The visible target layer $\mathcal{V}_T = [v_T^{(1)}, \ldots, v_T^{(V_T)}]$, where $v_T^{(k)}$ is a Gaussian $\mathcal{N}(\mu_T^{(j)}, \sigma_T^{(j)2})$, with a mean $\mu_T^{(j)}$ and a variance $\sigma_T^{(j)2}$, and the hidden layer $H = [h^{(1)}, \ldots, h^{(H)}]$ consists of sigmoidal activations.

Next, the mathematical formalizations of TrRBM are represented.[2]

Energy of the Full Model. The energy function, analogous to Equation 1, that is needed to determine the update rules is as:

$$E_1(\mathcal{V}_S, \mathcal{V}_T, \mathcal{H}) = -(v_{S,\alpha} - a_\alpha)\Sigma_S^{\alpha\alpha}(v_S^\alpha - a^\alpha) - b_\beta h^\beta$$
$$- (v_{T,\gamma} - c_\gamma)\Sigma_T^{\gamma\gamma}(v_T^\gamma - c^\gamma) - W_{\alpha\beta\gamma}\Sigma_T^{\gamma\gamma}v_T^\gamma h^\beta \Sigma_S^{\alpha\alpha}v_S^\alpha$$

where $v_{.,\alpha}$ is the covector of v_s^α, a_α is the covector of the visible layer biases a^α, b_β is the covector of the hidden layer biases b^β, and c_γ is the covector of the visible layer biases c^γ. $\Sigma^{\alpha\alpha}$ is a second order diagonal tensor representing the variances of the different layers, and $W_{\alpha\beta\gamma}$ is the third order tensor representing the weight connections.

Inference in the Full Model. Because there are no connection between the units in each of the three layers, inference is conducted in parallel. Formally,

$$\mu_S^{\alpha::} = W_{\beta\gamma}^{\alpha::}v_T^\gamma h^\beta + a^{\alpha::}$$
$$\mu_T^{::\gamma} = W_{\alpha\beta}^{::\gamma}h^\beta v_S^\alpha + b^{::\gamma}$$
$$s^{:\beta:} = W_{\alpha\gamma}^{:\beta:}v_T^\alpha v_S^\alpha + c^{:\beta:}$$

where α :: are slices in the tensor field in the α direction, : β : are slices in the tensor field in the β direction, and :: γ are these in the γ direction. Therefore,

$$p(\mathcal{V}_S|\mathcal{V}_T, \mathcal{H}) = \times_\alpha \mathcal{N}(\mu_S^{\alpha::}, \Sigma^{:\alpha\alpha})$$
$$p(\mathcal{V}_T|\mathcal{V}_S, \mathcal{H}) = \times_\gamma \mathcal{N}(\mu_T^{::\gamma}, \Sigma^{:\gamma\gamma})$$
$$p(\mathcal{H}|\mathcal{V}_T, \mathcal{V}_S) = \times_\beta sigmoid(s^{:\beta:})$$

where $\times$. are tensor products in the corresponding $\cdot$ fields.

Update Rules of the Full Model. In this section, the update rules to learn the weights and the biases of TrRBM are described. These rules are attained by deriving the energy function of Equation 1 with respect to weight tensor yielding the following:

---

[2] **A word on notation:** Because of space concerns, we resorted to Einstein's 1916 tensor index notation and conventions [5] for the mathematical details. As we realize this is not standard in ML literature, a more expansive derivation using more standard notation can be found in the expanded version of this paper.

$$\nabla_{W_{\alpha\beta\gamma}} = \left\langle v_T^\alpha h_\beta v_{S,\alpha} \right\rangle_0 - \left\langle v_T^\alpha h_\beta v_{S,\alpha} \right\rangle_\lambda$$

$$\Delta a^\alpha \propto \langle v_S^\alpha \rangle_0 - \langle v_S^\alpha \rangle_\lambda$$

$$\Delta b^\beta \propto \langle h^\beta \rangle_0 - \langle h^\beta \rangle_\lambda$$

$$\Delta c^\gamma \propto \langle v_T^\gamma \rangle_0 - \langle v_T^\gamma \rangle_\lambda$$

where $\langle . \rangle_0$ is the expectation over the original data distribution and $\langle . \rangle_\lambda$ is the expected reconstruction determined by a Markov Chain of length λ, attained through Gibbs sampling that started at the original data distribution.

3.2   Factored Transfer Restricted Boltzmann Machine

TrRBM, as proposed in the previous section, is computationally expensive. Because TL is a speedup technique, any TL method must be efficient or it will not be useful in practice. This section presents a factored version of the algorithm, FTrRBM. In particular, the three-way weight tensor is factored into sums of products through a factoring function, thus reducing the computational complexity from $\mathcal{O}(N^3)$ for TrRBM to $\mathcal{O}(N^2)$ for FTrRBM.

Energy of the Factored Model. As mentioned previously, the three-way weight tensor among the different layers is now factored. Therefore, the energy function is now defined as:

$$E(\mathcal{V}_S, \mathcal{V}_T, \mathcal{H}) = -(v_{S,\alpha} - a_\alpha)\Sigma_S^{\alpha\alpha}(v_S^\alpha - a^\alpha) - b_\beta h^\beta$$
$$- (v_{T,\gamma} - c_\gamma)\Sigma_T^{\gamma\gamma}(v_T^\gamma - c^\gamma) - w_{f\alpha}^{(V_S)}\Sigma_S^{\alpha\alpha}v_S^\alpha w_{f\beta}^{(h)} w_{f\gamma}^{(V_T)}\Sigma_T^{\gamma\gamma}v_T^\gamma$$

where f is the number of factors used to decompose the weight tensor.

Inference in the Factored Model. Inference in the factored version is done in a similar manner to that of the full model with different inputs for the nodes. In particular, because there are no connections between the units in the same layer, inference is done in parallel for each of the nodes. Mathematically these are derived as:

$$\mu_S^{:\alpha} = w_f^{(V_S):\alpha} w_{f\beta}^{(h)} h^\beta w_{f\gamma}^{(V_T)} v_T^\gamma + a^{:\alpha}$$

$$\mu_T^{:\gamma} = w_f^{(V_T):\gamma} w_{f\beta}^{(h)} h^\beta w_{f\alpha}^{(V_S)} v_S^\alpha + b^{:\gamma}$$

$$s^{:\beta} = w_f^{(h):\beta} w_{f\alpha}^{(V_S)} v_S^\alpha w_{f\gamma}^{(V_T)} v_T^\gamma + c^{:\beta}$$

Update Rules for the Factored Model. Learning in the factored model is done using a modified version of Contrastive Divergence. The derivatives of the energy function are computed again, this time yielding:

$$\Delta w_{:f\alpha}^{(V_S)} \propto \left\langle \Sigma_S^{\alpha:} v_S^{\alpha:} w_{:f\beta}^{(h)} h^\beta w_{:f\gamma}^{(V_T)} \Sigma_T^{\gamma\gamma} v_T^\gamma \right\rangle_0$$

$$- \left\langle \Sigma_S^{\alpha:} v_S^{\alpha:} w_{:f\beta}^{(h)} h^\beta w_{:f\gamma}^{(V_T)} \Sigma_T^{\gamma\gamma} v_T^\gamma \right\rangle_\lambda$$

$$\Delta w_{:f\gamma}^{(V_T)} \propto \left\langle \Sigma_S^{\gamma:} v_T^{\gamma:} w_{:f\beta}^{(h)} h^\beta w_{:f\alpha}^{(V_S)} \Sigma_S^{\alpha\alpha} v_S^\alpha \right\rangle_0$$

$$- \left\langle \Sigma_S^{\gamma:} v_T^{\gamma:} w_{:f\beta}^{(h)} h^\beta w_{:f\alpha}^{(V_S)} \Sigma_S^{\alpha\alpha} v_S^\alpha \right\rangle_\lambda$$

$$\Delta w_{:f\beta}^{(h)} \propto \left\langle h^{:\beta} w_{:f\gamma}^{(V_T)} \Sigma_T^{\gamma\gamma} v_T^\gamma w_{:f\alpha}^{(V_S)} \Sigma_S^{\alpha\alpha} v_S^\alpha \right\rangle_0$$

$$- \left\langle h^{:\beta} w_{:f\gamma}^{(V_T)} \Sigma_T^{\gamma\gamma} v_T^\gamma w_{:f\alpha}^{(V_S)} \Sigma_S^{\alpha\alpha} v_S^\alpha \right\rangle_\lambda$$

where f is the index running over the number of factors, $\langle . \rangle_0$ is the expectation from the initial probability distribution (i.e., data), and $\langle . \rangle_\lambda$ is the expectation of the Markov chain, starting at the initial probability distribution, and sampled λ steps using a Gibbs sampler. The update rules for the biases are the same as for the full model.

Unfortunately, learning in this model cannot be done with normal CD. The main reason is that if CD divergence was used as is, FTrRBM will learn to correlate *random* samples from the source task to *random* samples in the target. To tackle this problem, as well as ensure computational efficiency, a modified version of CD is proposed. In *Parallel Contrastive Divergence* (PCD), the data sets are first split into batches of samples. Parallel Markov chains run to a certain number of steps on each batch. At each step of the chain, the values of the derivatives are calculated and averaged to perform a learning step. This runs for a certain number of epochs. At the second iteration the same procedure is followed but with randomized samples in each of the batches. Please note that randomizing the batches is important to avoid fallacious matchings between source and target triplets.

4   Using the Inter-task Mapping

Using FTrRBMs for transfer in RL is done using two phases. First, the inter-task mapping is learned through source and target task samples. Second, samples are transferred from the source to the target, to be used as starting samples for a sample-based RL algorithm (which proceeds normally from this point onward).

Algorithm 1. Overall Transfer Framework

1: **Input:** Random source task samples $\mathcal{D}_S = \{\langle s_S^{(i)}, a_S^{(i)}, s_s'^{(i)}\rangle\}_{i=1}^m$, random target task samples $\mathcal{D}_T = \{\langle s_T^{(j)}, a_T^{(j)}, s_T'^{(j)}\rangle\}_{j=1}^n$, optimal source task policy $\pi_S^\star$
2: Use $\mathcal{D}_S$ and $\mathcal{D}_T$ to learn the intertask mapping using FTrRBM.
3: Sample source task according to $\pi_S^\star$ to attain $\mathcal{D}_S^\star$.
4: Use the learned RBM to transfer $D_S^\star$ and thus attain $\mathcal{D}_T^0$.
5: Use $\mathcal{D}_T^0$ to learn using a sample-based RL algorithm.
6: **Return:** Optimal target task policy $\pi_T^\star$.

4.1 Learning Phase

When FTrRBM learns, weights and biases are tuned to ensure a low reconstruction error between the original samples and the predicted ones from the model. The RBM is initially provided with random samples from both the source and the target tasks. Triplets from the source task (i.e., $\{\langle s_1^{(i)}, a_1^{(i)}, s_1'^{(i)}\rangle\}_{i=1}^m$) and target task (i.e., $\{\langle s_2^{(j)}, a_2^{(j)}, s_2'^{(j)}\rangle\}_{j=1}^n$) are inputs to the two visible layers of the RBM. These are then used to learn good hidden and visible layer feature representations. Note that these triplets should come from random sampling—the RBM is attempting to learn an inter-task mapping that covers large ranges in both the source and target tasks' state and actions spaces. If only "good" samples were used the mapping will be relevant in only certain narrow areas of both source and target spaces.

4.2 Transfer Phase

After learning, the FTrRBM encodes an inter-task mapping from the source to the target task. This encoding is then used to transfer (near-)optimal sample transitions from the source task, forming sample transitions in the target task. Given a (near-)optimal source task policy, $\pi_S^\star$, the source task is sampled greedily according to $\pi_S^\star$ to acquire optimal state transitions. The triplets are passed through the visible source layer of FTrRBM and are used to reconstruct initial target task samples at the visible target layer, effectively transferring samples from one task to another. If the source and target task are close enough[3], then the transferred transitions are expected to aid the target agent in learning an (near-)optimal behavior. They are then used in a sample based RL algorithm, such as LSPI to learn an optimal behavior in the target task (i.e., $\pi_T^\star$). The overall process of the two phases is summarized in Algorithm 1.

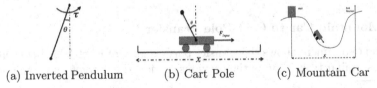

(a) Inverted Pendulum   (b) Cart Pole   (c) Mountain Car

Fig. 2. Experimental domains

---

[3] Note that defining a similarity metric between tasks is currently an open problem and beyond the scope of this paper.

5 Experiments and Results

To asses the efficiency of the proposed framework, experiments on different RL benchmarks were performed. Two different transfer experiments were conducted using the tasks shown in Figure 2[4].

5.1 Inverted Pendulum to Cart-Pole Transfer

To test the transfer capabilities of FTrRBMs, transfer was conducted from the Inverted Pendulum (IP) (i.e., Figure 2(a)) to the (CP) (i.e., Figure 2(b)).

Source Task. The state variables of the pendulum are $\langle \theta, \dot{\theta} \rangle$. The action space is a set of two torques $\{-10, 10\}$ in Newton meters. The goal of the agent is again to balance the pole in an upright position with $\langle \theta = 0, \dot{\theta} = 0 \rangle$. A reward of $+1$ is given to the agent when the pole's angle is in $-\frac{\pi}{12} < \theta < \frac{\pi}{12}$ and -1 otherwise.

Target Task. The target task was a CP with, $l = 0.5$. The target action space, transition probability and reward functions are different from the source task. The action space of the target agent was changed to $A_T = \{-10, 0, 10\}$ and the reward function was changed to $cos(\theta)$, giving the agent a maximum value of $+1$ when the pole is the upright position.

Experiment. 3000 random source task samples and 2000 target task samples were used to learn the inter-task mapping. The RBM contained 80 hidden units and 25 factors. Learning was performed as before, with FTrRBM converging in about 3.5 minutes. Transfer was accomplished and tested similarly to the previous experiment. The results are reported in Figure 3. It is again clear that transfer helps the target agent in his learning task. LSPI again converged with fewer iterations when using transfer. LSPI convergence time also decreased on different transferred samples. For example, LSPI converged with only 9 iterations on 5000 transferred samples compared to 12 using random ones and with 17 compared to 19 on 8000 transferred and random samples, respectively. The time needed to reach near optimal behavior was reduced from 22 to 17 minutes by using a transferred policy to initialize LSPI. Therefore,

ConclusionI: FTrRBM is capable of learning a relevant inter-task mapping between a pair of dissimilar tasks.

5.2 Mountain Car to Cart-Pole Transfer

A second experiment shows the transfer performance of FTrRBMs between pairs even less similar tasks than in the previous section. The target task remained the

---

[4] The samples required for learning the inter-task mapping were not measured as extra samples for the random learner in the target. Please note, that even if these were included the results as seen from the graphs will still be in favor of the proposed methods.

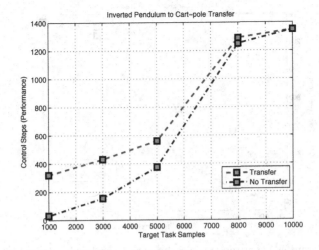

Fig. 3. Transfer versus no transfer comparison on different tasks

same cart-pole as before, while the source task was chosen to be the Mountain-Car (MC) problem. Although very different, successful transfer results between these tasks had previously been shown in our previous work [3].

Source Task. The system is described with two state variable $\langle x, \dot{x} \rangle$. The agent can choose between two actions $\{-1, 1\}$. The goal of the agent is to drive the car up the hill to the end position. The car's motor is not sufficient to drive the car directly to its goal state — the car has to oscillate in order to acquire enough momentum to drive to the goal state. The reward of the agent is -1 for each step the car did not reach the end position. If the car reaches the goal state, the agent receives a positive reward of $+1$ and the episode terminates. Learning in the source task was performed using SARSA.

Experiment. 4000 random source task samples and 2000 target task samples were used to lean the inter-task mapping as before. The RBM contained 120 hidden units and 33 factors and converged in about 3.5 minutes to the lowest reconstruction error. The results of transfer (performed as before) are reported in Figure 4. It is clear that transfer helps even when tasks are highly dissimilar. As before, LSPI converged with fewer iterations when using transfer than without using transfer. For example, LSPI converged with only 10 iterations on 5000 transferred samples compared to 12 using random ones. LSPI converged to an optimal behavior in about 18 minutes compared to 22 minutes for the non-transfer case. Therefore,

ConclusionII: FTrRBM is capable of learning a relevant inter-task mapping between a pair of highly dissimilar tasks.

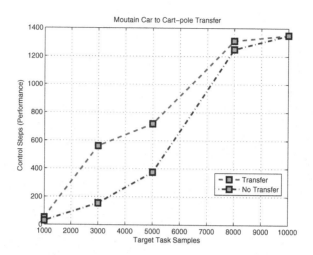

Fig. 4. Transfer versus no transfer comparison on highly dissimilar tasks

5.3 FTrRBM Robustness

To provide a comprehensive comparison between FTrRBM and the work in [3], two additional experiments were conducted. The source task was either the IP or the MC, while the target task was the CP system. 1000 and 500 target samples were used to learn an intertask mapping using either FTrRBM or a method from our previous [3]. Having these intertask mappings, (near-)optimal source task samples[5] were then transferred to provide an initial batch for the target RL agent to learn on. Performance, measured by the number of successful control steps in the target, was then reported in Figures 5 and 6.

Figure 5 shows two comparison graphs. The left graph reports the performance when using 1000 target samples to learn the intertask mapping. These clearly demonstrate that FTrRBM performs better than sparse coded intertask mappings, where for example, FTrRBM attains about 570 control steps compared to 400 in the sparse coded case at 5000 transferred samples. As the number of control steps increases, the performance of both methods also increases, to reach around 1300 control steps for FTrRBM compared to 1080 in the sparse coded case at 10000 transferred samples. The right graph shows the results of the same experiments, however, when using only 500 target samples to learn the intertask mapping. Again these results show that apart from the first two points, FTrRBM outperforms the Sparse coded intertask mapping.

In Figure 6 the results of the same experiments on highly dissimilar tasks are shown. In the left graph, 1000 target samples were used to learn an intertask mapping using either FTrRBM or the approach of [3]. The results clearly manifest the superiority of FTrRBM compared to the sparse coded approach, where at 5000 transferred samples FTrRBM attains 600 control steps, with 410 steps for the sparse coded intertask mapping. This performance increases to reach

---

[5] The optimal policy in the source was again attained using SARSA.

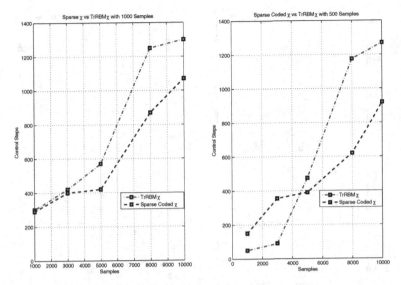

Fig. 5. These graphs compare the performance of transfer from IP to CP using FTr-RBM or Sparse Coded inter-task mappings [3]. The left graph shows the results of transfer when using 1000 target samples to learn the intertask mapping, while the right presents the results when using 500 samples to learn such a mapping.

about 1300 control steps for FTrRBM with 1050 for the sparse coded approach on 10000 transferred samples. In the right graph the same experiments were repeated by using 500 samples to learn the intertask mapping. It is again clear that FTrRBM outperforms the approach of [3].

6 Related Work

Learning an inter-task mapping has been of major interest in the transfer learning community [17] because of its promise of a fully autonomous speed-up method for lifelong learning. However, the majority of existing work assumes that 1) the source task and target task are similar enough that no mapping is needed, or 2) an inter-task mapping is provided to the agent.

For example, many authors have considered transfer between two agents which are similar enough that learned knowledge in the source task can be directly used in the target task. For instance, the source and target task could have different reward functions (e.g., *compositional learning* [12]) or have different transition functions (e.g., changing the length of a pole over time in the cart pole task [11]). More difficult are cases in which the source task and target task agents have different state descriptions or actions. Some researchers have attempted to allow transfer between such agents without using an inter-task mapping. For example, a shared *agent space* [7] may allow transfer between such pairs of agents, but requires the agents to share the same set of actions and an agent-centric mapping. The primary contrast between these methods and

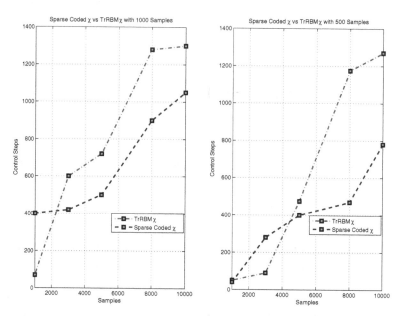

Fig. 6. These graphs compare the performance of transfer from MC to CP using FTr-RBM or Sparse Coded inter-task mappings [3]. The left graph shows the results of transfer when using 1000 target samples to learn the intertask mapping, while the right presents the results when using 500 samples to learn such a mapping. It is clear that FTrRBM transfer outperforms that described in [3].

the current work is that this paper is interested in *learning* a mapping between states and actions in pairs of tasks, rather than assuming that it is provided, or rendered unnecessary because of similarities between source task and target task agents, a requirement for fully autonomous transfer. There has been some recent work on learning such mappings. For example, semantic knowledge about state features between two tasks may be used [8,10], background knowledge about the range or type of state variables can be used [14,19], or transition models for each possible mapping could be generated and tested [16].

There are currently no general methods to learn an inter-task mapping without requiring 1) background knowledge that is not typically present in RL settings, or 2) an expensive analysis of an exponential number (in the size of the action and state variable sets) of inter-task mappings. This paper overcomes these problems by automatically discovering high-level features and using them to transfer knowledge between agents without suffering from an exponential explosion. The closest work to the proposed method is that of our previous work [3]. This method is based on sparse coding, sparse projection, and sparse Gaussian processes to learn an inter-task mapping between MDPs with arbitrary variations. However, the we relied on a Euclidean distance correlation between source task and target task triplets, which may fail in highly dissimilar tasks. The work in the current paper overcomes these problems by adopting a more robust feature extraction technique and by avoiding the need for such a distance correlation as shown in the experiments of Section 5.3.

Others have focused on transferring samples between tasks. For instance, Lazaric et al. [9] transfers samples in batch reinforcement learning using a compliance measure. The main difference to this work is that we neither assume any similarities between the transition probabilities, nor restrict the framework to similar state and/or action feature representations. In contrast to all existing methods (to the best of our knowledge), this paper allows for differences between all variables describing Markov Decision Processes for the source and target tasks and robustly learns an *inter-task mapping*, rather than a mapping based on state features.

7 Discussion and Conclusions

This paper introduces a theoretically grounded method for learning an inter-task mapping based on RBMs. The approach was validated through experimental evidence. The proposed technique successfully learned a useful inter-task mapping between highly dissimilar pairs of tasks. Furthermore, a comparison between the proposed technique and our earlier work [3] showed that FTrRBM outperforms sparse coded inter-task mappings when fewer samples are available.

Although successful, the approach is not guaranteed to provide useful transfer. To clarify, the reward was not included in the definition of the inter-task mapping, but when transferring near-optimal behaviors sampled according to near-optimal policies such rewards are implicitly taken into account and thus, attaining successful transfer results.

Despite that these experiments showed transfer between tasks with different reward functions, negative transfer may occur if the rewards of the source task and target tasks were highly dissimilar. Such a mismatch may lead to an incorrect mapping because the reward is not considered in the presented method. A solution to this potential problem is left for future work, but will likely require incorporating the sampled reward into the current approach.

A second potential problem may occur during the learning phase of FTrRBM, which could be traced back to quality of the random samples. If the number of provided samples is low and very sparse[6], the learned mapping may be uninformative. This problem is also left for future work, but could possibly be tackled by using a deep belief network to increase the level of abstraction.

Acknowledgments. This work was supported in part by NSF IIS-1149917.

---

[6] Sparse in this context means that the samples arrive from very different locations in the state-space and areas of the state space are not sufficiently visited.

References

1. Ackley, H., Hinton, E., Sejnowski, J.: A learning algorithm for boltzmann machines. Cognitive Science, 147–169 (1985)
2. Bengio, Y.: Learning deep architectures for AI. Foundations and Trends in Machine Learning 2(1), 1–127 (2009); Also published as a book. Now Publishers (2009)
3. Bou-Ammar, H., Taylor, M., Tuyls, K., Driessens, K., Weiss, G.: Reinforcement learning transfer via sparse coding. In: Proceedings of AAMAS (2012)
4. Buşoniu, L., Babuška, R., De Schutter, B., Ernst, D.: Reinforcement Learning and Dynamic Programming Using Function Approximators. CRC Press, Boca Raton (2010)
5. Einstein, A.: The Foundation of the General Theory of Relativity. Annalen Phys. 49, 769–822 (1916)
6. Hinton, G.E.: Training Products of Experts by Minimizing Contrastive Divergence. Neural Computation 14(8), 1771–1800 (2002)
7. Konidaris, G., Barto, A.: Autonomous shaping: knowledge transfer in reinforcement learning. In: Proceedings of ICML (2006)
8. Kuhlmann, G., Stone, P.: Graph-based domain mapping for transfer learning in general games. In: Kok, J.N., Koronacki, J., Lopez de Mantaras, R., Matwin, S., Mladenič, D., Skowron, A. (eds.) ECML 2007. LNCS (LNAI), vol. 4701, pp. 188–200. Springer, Heidelberg (2007)
9. Lazaric, A., Restelli, M., Bonarini, A.: Transfer of samples in batch reinforcement learning. In: Proceedings of ICML (2008)
10. Liu, Y., Stone, P.: Value-function-based transfer for reinforcement learning using structure mapping. In: Proceedings of AAAI (2006)
11. Selfridge, O.G., Sutton, R.S., Barto, A.G.: Training and tracking in robotics. In: IJCAI (1985)
12. Singh, S.: Transfer of learning by composing solutions of elemental sequential tasks. Machine Learning, 323–339 (1992)
13. Sutton, R.S., Barto, A.G.: Reinforcement learning: An introduction (1998)
14. Talvitie, E., Singh, S.: An experts algorithm for transfer learning. In: Proceedings of IJCAI (2007)
15. Taylor, G.W., Hinton, G.E., Roweis, S.T.: Two distributed-state models for generating high-dimensional time series. Journal of Machine Learning Research 12, 1025–1068 (2011)
16. Taylor, M.E., Kuhlmann, G., Stone, P.: Autonomous transfer for reinforcement learning. In: Proceedings of AAMAS (2008)
17. Taylor, M.E., Stone, P.: Transfer learning for reinforcement learning domains: A survey. Journal of Machine Learning Research 10, 1633–1685 (2009)
18. Taylor, M.E., Stone, P., Liu, Y.: Transfer learning via inter-task mappings for temporal difference learning. Journal of Machine Learning Research 8(1), 2125–2167 (2007)
19. Taylor, M.E., Whiteson, S., Stone, P.: Transfer via inter-task mappings in policy search reinforcement learning. In: Proceedings of AAMAS (2007)

A Layered Dirichlet Process for Hierarchical Segmentation of Sequential Grouped Data

Adway Mitra[1], Ranganath B.N.[1], and Indrajit Bhattacharya[2]

[1] CSA Department, Indian Institute of Science, Bangalore, India
{adway,ranganath}@csa.iisc.ernet.in
[2] IBM India Research Lab, Bangalore, India
indrajitb@gmail.com

Abstract. We address the problem of hierarchical segmentation of sequential grouped data, such as a collection of textual documents, and propose a Bayesian nonparametric approach for this problem. Existing Bayesian nonparametric models such as the sticky HDP-HMM are suitable only for single-layer segmentation. We propose the Layered Dirichlet Process (LaDP), where each layer has a countable set of Dirichlet Processes, draws from which define a distribution over the countable set of Dirichlet Processes at the next layer. Each data item gets assigned to a distribution (index) from each layer of the hierarchy, leading to hierarchical segmentation of the sequence. The complexity of inference depends upon the exchangeability assumptions for the measures at different layers. We propose a new notion of exchangeability called Block Exchangeability, which lies between Markov Exchangeability (used in HDP-HMM) and Complete Group Exchangeability (used in HDP), and allows for faster inference than Markov Exchangeability. Using experiments on a news transcript dataset and a product review dataset, we show that LaDP generalizes better than existing non-parametric models for sequential data, and by simultaneously segmenting at multiple levels, outperforms existing models in terms of single-layer segmentation. We also show empirically that using Block Exchangeability greatly speeds up inference and allows trading off accuracy for execution time.

1 Introduction

We address the problem of hierarchical segmentation of sequential grouped data. For example, consider transcripts of news broadcast on television or radio. Here, each transcript represents a group of data points, which are the words in this case. The words in each transcript or group form a sequence that needs to be segmented. The segmentation needs to be at two layers — news categories, such as politics, sports, etc, and individual stories within a category. There are benefits to segmenting transcripts simultaneously, instead of individually. Stories are typically shared across transcripts, and transition patterns between stories (more important stories often come earlier) and categories (e.g. sports rarely comes before other categories) may also be shared across transcripts. Also, there are benefits to simultaneous segmentation into stories and categories. Inferring

H. Blockeel et al. (Eds.): ECML PKDD 2013, Part II, LNAI 8189, pp. 465–482, 2013.

a story strongly suggests a category, while inferring a category increases the posterior probability for certain stories. Finally, while the number of categories may often be known or guessed, this is not true for the number of stories. In this paper, we propose a Bayesian approach for this problem, which is both hierarchical and non-parametric. For the news example, each story can be modeled as a distribution over words, and each category as a distribution over stories. The same stories and categories are shared between all news transcripts. Being non-parametric, the model does not require the number of stories to be specified.

The Dirichlet Process[1] is a measure over measures and is useful as a prior in Bayesian nonparametric mixture models, where the number of mixture components is not specified a-priori, and is allowed to grow with the number of data points. The Hierarchical Dirichlet Process(HDP)[5] hierarchically extends DP for grouped data, such as words partitioned into documents, so that mixture components are shared between groups. The DP is a completely exchangeable model (probability of the data is invariant to permutations in the sequence), while the HDP is completely exchangeable within each group (Group Exchangeable). As a result, these are not suitable as statistical models for segmentation. HDP variants such as the HDP-HMM [5] and sticky HDP-HMM [6], which satisfy Markov exchangeability, are more suitable for segmentation. However, these perform segmentation at a single layer and not at several layers simultaneously.

We propose the Layered Dirichlet Process, where each layer has a countable set of DP-distributed measures over integers. The integers at each layer serve as indices for the measures at the next layer. Each data item filters down this layered structure, where a measure at each layer assigns it to a measure at the next layer. Such assignments for each data item in the sequence results in a hierarchical segmentation of the sequence. The assignment of a measure to each data item at each layer depends on the exchangeability assumption at that layer. For Complete Group Exchangeability, it depends only on its assignment at the previous layer. For other partial Group Exchangeabilities, it additionally depends on the assignments of other data items at that layer. We perform inference for LaDP using collapsed Gibbs sampling. Since the assignments are coupled across layers, inference is naturally complex. We propose a new notion of exchangeability called Block Exchangeability. We show that this relaxes Complete Exchangeability to capture sequential patterns, but is stricter than Markov Exchangeability with significantly lower inference complexity.

Using experiments on multiple real datasets, we show that by modeling grouping at multiple layers simultaneously, LaDP is able to generalize better that state-of-the-art non-parametric models. We also show that simultaneous segmentation at multiple layers improves segmentation accuracy over single layer segmentation. Additionally, using Block Exchangeability leads to significantly faster inference compared to Markov Exchangeability, while incurring negligible increase in segmentation error and perplexity in some cases, and actually improving performance in some others. Interestingly, Block Exchangeability has the novel ability of trading off efficiency for accuracy.

2   Background and Related Work

In this section we briefly review some existing nonparametric models used as priors for infinite mixture models, and existing notions of exchangeability.

DP Mixture Model and Complete Exchangeability: A random measure G on Θ is said to be distributed according to a Dirichlet Process (DP) [1] ($G \sim DP(\alpha, H)$) with base distribution H and concentration parameter α if, for every finite partition $\{\Theta_1, \Theta_2, \ldots, \Theta_k\}$ of Θ, $(G(\Theta_1), G(\Theta_2), \ldots, G(\Theta_k)) \sim Dir(\alpha H(\Theta_1), \alpha H(\Theta_2), \ldots, \alpha H(\Theta_k))$. The stick-breaking representation shows the discreteness of draws G from a DP:

$$\theta_k \sim H; \; \beta_k = \hat{\beta}_k \prod_{i=1}^{k-1}(1 - \hat{\beta}_i); \; \hat{\beta}_i \sim Beta(1, \alpha); \; G \triangleq \sum_k \beta_k \delta_{\theta_k}$$

We write $\beta_k \sim GEM(\alpha)$. Given n independent draws $\{\theta_i\}_{i=1}^n$ from G as above, the predictive distribution of the next draw, on integrating out G, is given by $p(\theta_{n+1}|\theta_1 \ldots \theta_n) \propto \sum_{k=1}^K n_k \delta_{\phi_k} + \alpha H$, where $\{\phi_k\}_{k=1}^K$ be the K unique values taken by $\{\theta_i\}_{i=1}^n$ with corresponding counts $\{n_k\}_{k=1}^K$. This shows the clustering nature of the DP. Using the DP as a prior results in an 'infinite mixture model' for data $\{w_i\}_{i=1}^n$ with the following generative process:

$$G \sim DP(\alpha, H); \; \theta_i \stackrel{iid}{\sim} G; \; w_i \stackrel{iid}{\sim} F(\theta_i), \; i = 1 \ldots n$$

where F is a measure defined over Θ. This is called the DP mixture model [1]. This can alternatively be represented using the stick-breaking construction and integer latent variables z_i as follows:

$$\beta \sim GEM(\alpha); \; \theta_k \sim H, \; k = 1 \ldots \infty \; ; \; z_i \sim \beta; \; w_i \sim F(\theta_{z_i}), \; i = 1 \ldots n$$

An important notion for hierarchical Bayesian modeling is that of exchangeability [11,2]. Given any assignment $\{\bar{z}_1, \bar{z}_2, \ldots, \bar{z}_n\}$ to a sequence of random variables $\{z_n\} \in \mathcal{S}$, where $\mathcal{S}$ is a space of sequences, exchangeability (under joint distribution P on $\mathcal{S}$) defines which permutations $\{\bar{z}_{\pi(1)}, \bar{z}_{\pi(2)}, \ldots, \bar{z}_{\pi(N)}\}$ of the assignment have the same probability (under P). In general, any notion of exchangeability E is defined using a statistic, which we call Exchangeability Statistic $S_E(z)$. A model, defining a joint distribution P, is said to satisfy exchangeability E if $S_E(z_1) = S_E(z_2)$ implies $P(z_1) = P(z_2)$, for all $z_1, z_2 \in \mathcal{S}$.

Given a sequence $z \in \mathcal{S}$, define $S_C(z) = \{n_i\}_{i=1}^K$ as the vector of counts of the K unique values occurring in it, where n_i is the count of the ith unique value. Using $S_C(z)$ as the exchangeability statistic leads to the definition of Complete Exchangeability (CE), under which all permutations are equiprobable.

De Finetti's Theorem [3] shows that if an infinite sequence of random variables z is infinitely exchangeable (meaning that every finite subset is completely exchangeable) under a joint distribution $P(z)$, then the joint distribution can be equivalently represented as a Bayesian hierarchy: $P(z) = \int_\theta P(\theta) \prod_i P(z_i|\theta) d\theta$. It can be shown that a sequence drawn from a DP mixture model, using a similar hierarchical generation process, satisfies Complete Exchangeability.

HDP Mixture Model and Group Exchangeability: Now consider grouped data of the form $\{w_i, g_i\}_{i=1}^n$, where $g_i \in \{1, G\}$ indicates the group to which w_i belongs. The Hierarchical Dirichlet Process (HDP) [5] allows sharing of mixture components $\{\phi_k\}$ across groups using two levels of DPs:

$$\phi_k \sim H, \ k = 1 \ldots \infty; \ \beta \sim GEM(\gamma), \ \pi_j \sim DP(\alpha, \beta), \ j = 1 \ldots G$$
$$z_i \sim \pi_{g_i}; \ w_i \sim F(\phi_{z_i}), \ i = 1 \ldots n \qquad (1)$$

This generative procedure for the data is called the HDP mixture model. We have modified the representation to make the group variable explicit, which we can build upon for our work. Note that the HDP can also be represented directly using measures instead of indices. The HDP mixture model can be shown to satisfy a notion of partial exchangeability called Group Exchangeability. For grouped data of the form $\{z_i, g_i\}_{i=1}^n$, where the z_i and g_i variables take K and G unique values respectively, define $S_G(z, g) = \{\{n_{j,k}\}_{k=1}^K\}_{j=1}^G$, where $n_{j,k} = \sum_{i=1}^n \delta(z_i, k)\delta(g_i, j)$. Group Exchangeability (GE) is characterized by the exchangeability statistic $S_G(z, g)$. For GE models, all intra-group permutations are equiprobable, but probability changes with exchange of values across groups.

Other Group Exchangeable Nonparametric Models: For grouped data $\{w_i, g_i\}_{i=1}^n$, the Nested Dirichlet Process (NDP) [7] proposes the following generative model with two layers of latent variables (z^2, z^1) for each data item:

$$\phi_{k,l} \sim H, \ k, l = 1 \ldots \infty; \beta_k^1 \sim GEM(\beta), \ k = 1 \ldots \infty; \beta^2 \sim GEM(\alpha);$$
$$z_g^2 \sim \beta^2, \ g = 1 \ldots G; z_i^1 \sim \beta_{z_{g_i}^2}^1 ; w_i \sim \phi_{z_{g_i}^2, z_i^1}, \ i = 1 \ldots n$$

Unlike the HDP, only some groups share mixture components. Additionally, unlike the HDP they also share distributions over these components.

The MLC-HDP [9] models data of the form $\{w_i, g_i^1, g_i^2, g_i^3\}_{i=1}^n$, which is grouped at 3 different levels, and proposes the following generative process:

$$\phi_k \sim H, \ k = 1 \ldots \infty; \beta^3 \sim GEM(\gamma^3), \ \beta^2 \sim GEM(\gamma^2), \ \beta^1 \sim GEM(\gamma^1);$$
$$\pi^3 \sim DP(\alpha^3, \beta^3), \ \pi_k^2 \sim DP(\alpha^2, \beta^2), \ \pi_l^1 \sim DP(\alpha^1, \beta^1), \ k, l = 1 \ldots \infty;$$
$$z_a^3 \sim \pi^3 \ \forall a; \ z_{ab}^2 \sim \pi_{z_a^3}^2 \ \forall a \forall b; \ z_{abc}^1 \sim \pi_{z_{ab}^2}^1 \ \forall a \forall b \forall c; \ w_i \sim \phi_{z_{g_i^3, g_i^2, g_i^1}^1}, \ i = 1 \ldots n$$

Here the mixture components can be shared by all groups, and two groups can have identical distributions over these components with non-zero probability.

Segmentation, HDP-HMM and Markov Exchangeability: Now we come to the segmentation problem for a sequence $\{w_i, z_i\}$ where the the variables w_i are observed while $z_i \in \{1, 2 \ldots\}$ are latent, with distribution $P(w, z) = P(z)P(w|z)$. Given any assignment to the $\{z_i\}$ variables, segments are defined as maximal sub-sequences (s, e) such that $z_e = z_s = z_i$ for $s \leq i \leq e$. Since $\{z_i\}$ variables are random, a natural definition for the segmentation problem is to first perform inference to find the optimal assignment to $\{z_i\}$ according to the posterior distribution $P(z|w)$, and then identifying segments for this assignment. Instances of this problem include segmentation according to topics for textual documents, and according to speaker in conversational audio. Naturally, distinguishing between different permutations is critical for segmentation of grouped (un-grouped) data, and GE (CE) assumptions for $P(z)$ are not appropriate, since all permutations are equiprobable. Therefore, HDP (DP) mixture models are not suitable for such segmentation tasks. These call for more discerning models that satisfy other notions of exchangeability that distinguish between different segmentations of $\{w_i, z_i\}$ represented by different assignments to $\{z_i\}$.

To model (ungrouped) data $\{w_i\}$ with such properties, the HDP-HMM [6] considers the mixture components z_i as states of an HMM with infinite state-space. This is done by identifying the groups as well as the mixture components in the HDP with the HMM states. Now $\pi_j \sim DP(\alpha, \beta)$ is considered as transition distribution for the jth state, and is used to generate the next state:

$$\pi_j \sim DP(\alpha, \beta), \ j = 1 \ldots \infty; \ z_i \sim \pi_{z_{i-1}}; \ w_i \sim F(\theta_{z_i}), \ i = 1 \ldots n \qquad (2)$$

A special case of this is the Sticky HDP-HMM (sHDP-HMM) [6], which increases the probability of self-transition as $\pi_j \sim DP(\alpha + \kappa, \frac{\alpha\beta + \kappa\delta_j}{\alpha + \kappa})$, to enforce sequential continuity of mixture components which occur naturally in speech (where a mixture component represents a speaker) and text (where a mixture component represents a topic). Though originally developed for single sequences, the HDP-HMM and sHDP-HMM models can also be extended for grouped data.

Consider the following statistic: $S_M(z) = (\{n_{ij}\}_{i=1,j=1}^{K,K}, s)$, where n_{ij} is the number of transitions from the ith unique value to the jth unique value in the sequence z, and $z_1 = s$. Using S_M as the exchangeability statistic leads to the definition of Markov Exchangeability (ME) [2]. Intuitively, this means that two different sequences are equiprobable under the joint distribution, if they begin with the same value and preserve the transition counts between unique values. Representation theorems, similar to De Finetti's theorem, exist for Markov Exchangeability as well [2]. It can be shown that the HDP-HMM and sticky HDP-HMM mixture models satisfy Markov Exchangeability.

3 Hierarchical Segmentation and LaDP

We now discuss hierarchical segmentation of grouped data and propose Bayesian nonparametric models for it, using existing notions of partial exchangeability.

Hierarchical Segmentation: Consider grouped data of the form $\{w_i, g_i\}$, where $g_i \in \{1 \ldots G\}$ indicates the group to which each data point w_i belongs. The data $\{w_i : g_i = g\}$ in each of the G groups forms a sequence. In the news transcript example, each group corresponds to one transcript, and the words in each transcript form a sequence. We call such data sequential grouped data.

Our task is to segment the sequential data in each of the groups at multiple layers. We define an L-layer segmentation of the data as follows. Instead of a single latent variable z_i as before, we associate L latent variables $\{z_i^l\}_{l=1}^L$, each taking integer values, with the ith data point. We call z_i^l the group for the ith data point at layer l. We will assume that the grouped structure of the input data provides the grouping at the highest layer, i.e. $z_i^{L+1} = g_i$. Given any assignment to these n sets of group variables, the hierarchical segmentation at any layer l ($1 \leq l \leq L$) is defined using $\{\{z_i^{l'}\}_{l' \geq l}\}_{i=1}^n$, which are all the group variables at layer l or higher. Two data points at positions i and j ($i < j$) belong to the same hierarchical segment at layer l if the group variables of intermediate data points k are identical at layers l and above: $z_i^{l'} = z_k^{l'} = z_j^{l'}$, $\forall k, i \leq k \leq j$, $l \leq l' \leq L+1$. This may also be defined recursively. Two data points belong to the same segment at layer l if they belong to the same segment at layer $l + 1$, and all intermediate group variables z_k^l have the same value at layer l. In the case of news transcripts, the group variables z_i^{L+1} at the highest layer indicate which transcript the ith data point (word) belongs to, which is provided as input. Imagine the next layers to correspond to categories (layer 2) and stories (layer 1). Then, two words would belong to the same category segment (layer 2), if they are in the same transcript and share the same category label with all intermediate words. Similarly, two words belong to the same story segment (layer 1) if they belong to the same category segment and have the same story labels as all intermediate words. The problem is to find the L-layer hierarchical segmentation at layer 1.

Completely Exchangeable Layered Dirichlet Process: We define a joint probability distribution over the n sets of group variables $\{\{z_i^l\}_{l=1}^L\}_{i=1}^n$ and the n data points $\{w_i\}_{i=1}^n$ using a hierarchical Bayesian approach. For each layer l, $1 \leq l \leq L$, we have a countable set of measures $\{\pi_g^l\}_{g=1}^\infty$ defined over positive integers. The group variables $\{z_i^l\}_{i=1}^n$ at layer l serve as indexes for these measures. Using this countable property, the atoms of all of these measures at layer l, which are integers, correspond one-to-one with the measures at the next layer $l - 1$. This gives us a hierarchy of measures, in the sense that each π_g^l forms a measure over the measures $\{\pi_{g'}^{l-1}\}_{g'=1}^\infty$ at the next layer. Finally, at the lowest layer, each $F(\phi_k)$ is a measure over the space $\mathcal{W}$ of the observations $\{w_i\}$. For discrete text data, these are multinomial distributions over the vocabulary.

Next we need to define the measures $\{\pi_g^l\}_{g=1}^\infty$ and the exchangeability properties at each layer l. In LaDP, we define each of these distributions to be DP-distributed. We begin with the simplest case, which assumes complete exchangeability at every layer. The generative process looks as follows:

$$\phi_k \sim H, \ k = 1 \ldots \infty$$
$$\beta_g^l \sim GEM(\gamma^l); \ \pi_g^l \sim DP(\alpha^l, \beta_g^l), \ g = 1 \ldots \infty, \ l = L \ldots 1$$
$$z_i^l \sim \pi_{z_i^{l+1}}^l, \ l = L \ldots 1, \quad w_i \sim F(\phi_{z_i^1}), \ i = 1 \ldots n \quad (3)$$

In each layer, a countable set of measures is first constructed by drawing from a DP with a distribution over integers as a base distribution. These measures as a result also have support over integers, which serve as indexes to the measures at the next lower layer, which also form a countable set. Once we have this hierarchy of measures, the group variable z_i^l for each data point at each layer l is sampled from the measure indexed by the group z_i^{l+1} assigned at the previous (higher) layer. The measures at the lowest layer (layer 1) are sampled from a suitable base distribution H. H could be Dirichlet when each ϕ_g is a multinomial parameter. It is easy to verify that the above process satisfies Complete Exchangeability (CE). As such, we call this model the CE-LaDP.

Layered Dirichlet Process for Segmentation: Since CE models are not useful for segmentation, we next incorporate Markov Exchangeability within LaDP. The key to incorporating ME is to relax the *iid* assumption for the group variables, within a layer as in the HDP-HMM, and additionally across layers, and generate z_i^l conditioned on some of the previously sampled groups $\{z_j^{l'} : j < i,\ l' > l\}$. The HDP-HMM identifies groups with states and makes the Markovian independence assumption that $P(z_i|z_{<i}) = P(z_i|z_{i-1})$. Accordingly, it defines transition distribution π_g over next states for each group (state) g. In our case, we make the following independence assumption: $P(z_i^l|z^{>l}, z_{<i}^l) = P(z_i^l|z_i^{l+1}, z_{p(i,l)}^l)$, where $z^{>l} \equiv \{z_i^{l'} : l' > l\}$, $z_{<i}^l \equiv \{z_{i'}^l : i' < i\}$, and $p(i,l) \equiv \{j : z_j^{l+1} = z_i^{l+1}, j < i,\ z_k^{l+1} \neq z_i^{l+1}, j < k < i\}$ is the previous datapoint having the same group as i at layer $l+1$. This means that the group assignment to data point i at layer l depends on its group at the layer $l+1$ (like in CE-LaDP), and also on the group assignment at layer l of its parent datapoint $p(i,l)$. (We later overload the notation $p(i,l)$ for brevity to refer to the group value $z_{p(i,l)}^l$ as well. We accordingly define transition distribution $\pi_{g,g'}^l$ over next groups from each parent group g' at layer l, in each assigned group g in layer $(l+1)$. The generative process for layer l ($L \geq l \geq 1$) is defined as:

$$\beta_g^l \sim GEM(\gamma^l),\ \pi_{g,g'}^l \sim DP(\alpha^l, \beta_g^l),\ g, g' = 1 \ldots \infty,$$
$$z_i^l \sim \pi_{z_i^{l+1}, p(i,l)}^l,\ i = 1 \ldots n$$

Part of the graphical model is shown in Fig. 1.

For the first data point in any group in layer $(l+1)$, $p(i,l)$ is undefined, and z_i^l is sampled from $\beta_{z_i^{l+1}}^l$. It can be shown that this generative process satisfies ME within each group at layer l. When this process is used at all layers, we call the model ME-LaDP. As in sticky HDP-HMM, we may add more probability κ^l for self-transitions: $\pi_{g,g'}^l \sim DP(\alpha^l + \kappa^l, \frac{\alpha^l \beta_g^l + \kappa^l \delta_{g'}}{\alpha^l + \kappa^l})$, where κ^l is a continuity parameter. This is done to encourage the same mixture component for adjacent data points. This captures the temporally smooth nature of most real-world data, and also encourages segmentations (based on group index assignments).

Layer-Specific Exchangeability: We have defined CE-LaDP as using CE at all layers, and ME-LaDP as using ME at all layers. However, each of the processes can be defined specific to a single layer, and it is possible to use layer-specific exchangeability assumptions, as demanded by particular applications. Indeed, we use such *mixed exchangeability models* in our experiments. As example, the generative process of such a model is described in the Appendix [13].

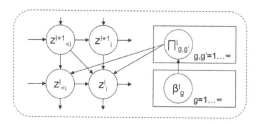

Fig. 1. Graphical model of LaDP focused on the ith data point in two adjacent layers

Incorporation of Domain Knowledge: The $\{\beta_g^l\}$ variables at each layer l in Eqns. 3 and 4 are group-specific distributions over indexes (and measures) at the next layer $l-1$. These are useful for incorporating domain knowledge such as distribution over categories for specific news transcripts. For example, we can indicate that the jth transcript is dominated by category index k by setting distribution β_j^L over category indexes at the appropriate layer $(L = 2)$ to $\sum_{c=1}^{\infty} \delta_k(c)$.

In some cases, one may also wish to bias the ϕ-distributions using domain knowledge. One option is to directly specify these ϕ_k. For weaker supervision, we may introduce an additional layer $l = 0$:

$$\beta_g^0 \sim GEM(\gamma^0); \ \pi_g^0 \sim DP(\alpha^0, \beta_g^0), g = 1 \ldots \infty$$
$$z_i^0 \sim \pi_{z_i^1}^0; \ w_i = \mathcal{W}_{z_i^0}, \ i = 1 \ldots n \tag{4}$$

Now, on specifying some of the $\{\beta_g^0\}$ distributions, the corresponding distributions $\{\pi_g^0\}$ will be similar to these, depending on the concentration parameter α^0, and the words will be drawn from these $\{\pi_g^0\}$ distributions. Observe that we use complete (group) exchangeability at layer $l = 0$.

Relation with Other Models: Observe the relation between the CE-LaDP (Eqn. 3) and the HDP mixture model (Eqn. 1). Recall that the group at the highest layer z_i^{L+1} is the input group label g_i. For $L = 1$, this is exactly the HDP mixture model. However, by separating the group index in the HDP generative model, and identifying the z_i variable as the random group variable leading to the next layer, the CE-LaDP naturally extends the HDP generative process to generate layered grouping. A similar relation holds between the ME-LaDP with $L = 1$ and the HDP-HMM. The MLC-HDP [9] extends HDP to 3 layers, with each data point w_i having input group indices g_i^3, g_i^2, g_i^3. When the group indices z^3 are observed (rather than sampled from π^3, as in [9]) and are identified with the indices g_i for LaDP, and additionally the input group indices $g_i^3 = i$, $g_i^2 = 1$ and $g_i^1 = 1$ are shared by data points w_i, we get back the CE-LaDP with $L = 2$. Thus the LaDP framework can be used to generalize existing models to any number of layers. Secondly, the LaDP enables incorporation of domain knowledge in all layers. Among existing models, only the recently-proposed DP-MRM [10] is equipped to incorporate such domain knowledge, though only for a single layer. Finally, while all existing methods only use a single exchangeability property

(CE or ME), LaDP has the attractive property that different layers can have different exchangeability properties. In the next section, we define a new notion of exchangeability, and show how it can be incorporated in any layer of LaDP.

4   Block-Exchangeability and BE-LaDP

The models that we have introduced for segmentation satisfy ME. However, as we analyze later, there is a significant price to be paid in terms of complexity of inference as we move from CE and GE to ME. This is particularly severe for us, since we need to segment simultaneously at multiple layers. In this section, we explore an alternative notion of exchangeability, called Block Exchangeability (BE), that allows segmentation, but is less expensive than ME for inference.

Block Exchangeability. Consider the following statistic for a sequence z (with k unique values in it): $S_B(z) = (\{n_{i,i}, n_{i,-i}\}_{i=1}^{k}, e)$, where $n_{i,i}$ is the number of transitions from the i-th unique value of z to itself, $n_{i,-i}$ is the number of transitions from the i-th unique value to all other values and e is the value at the last position. Using $S_B(z)$ as the exchangeability statistic defines Block Exchangeability (BE) for a sequence z with distribution $P(z)$, or for a model that defines $P(z)$. First we observe some properties of a block exchangeable model.

Theorem 1. *If a model defining a joint distribution P is Completely Exchangeable then it is necessarily Block Exchangeable, but not the converse.*

Theorem 2. *If a model defining a joint distribution P is Block Exchangeable then it is necessarily Markov Exchangeable, but not the converse.*

BE-DP Mixture Model: Consider grouped data of the form $\{w_i, z_i, g_i\}_{i=1}^{n}$, where $g_i \in \{1 \ldots G\}$ indicates the pre-assigned group corresponding to the ith data point, and $z_i \in \{1 \ldots \infty\}$ is the (latent) index of the mixture component corresponding to w_i. We now define a DP-based non-parametric mixture model for sequential grouped data that satisfies Block Exchangeability, as follows:

$$\phi_k \sim H, \ k = 1 \ldots \infty; \ \pi_g \sim DP(\alpha, \beta_g), \ g = 1 \ldots G$$
$$q_{gk} \sim Beta(1, \kappa); \ \hat{\pi}_{gk} = q_{gk}\pi_g + (1 - q_{gk})\delta_k, \ g = 1 \ldots G, \ k = 1 \ldots \infty$$
$$z_i \sim \hat{\pi}_{g_i, p(i)}; \ w_i \sim F(\phi_{z_i}), \ i = 1 \ldots n \quad (5)$$

The first two lines describe the BE-DP prior, and the last line shows data generation using the mixture indices z_i. $p(i)$ is the group assignment to the data point just before i in group g_i. For the first data point in any group, $z_i \sim \pi_{g_i}$.

Theorem 3. *The BE-DP prior distribution and the corresponding mixture model satisfy Block Exchangeability.*

The proofs of the theorems are available in our supplementary material [13].

We now provide an alternative representation of the BE-DP equivalent to that in Eqn. 5, but which provides a justification for its nomenclature by capturing the structure of equi-probable permutations of any sequence. Consider the sequence $\{w_i, z_i, c_i, g_i\}_{i=1}^n$, where the variables w_i, z_i and g_i are as before, and we have added a binary variable $c_i \in \{0, 1\}$ for each data point. We change the generative process in Eqn. 5 to include the c_i variables as follows.

$$c_i \sim Ber(q_{g_i, p(i)}); \quad z_i = p(i), \text{ if } c_i = 0; \; z_i \sim \pi_{g_i}, \text{ else, } i = 1 \ldots n \qquad (6)$$

Clearly, this version is equivalent to the generative process in Eqn. 5, in the sense that the marginal $P(\mathbf{w}, \mathbf{z}, \mathbf{g})$ obtained by summing out $\mathbf{c}$ from $P(\mathbf{w}, \mathbf{z}, \mathbf{c}, \mathbf{g})$ is identical to that obtained from Eqn. 5. But this version has the advantage that the introduction of the auxiliary variables makes inference more tractable, as we will see in Section 5.

Separately, introduction of the the c variables provides some new insights into Block Exchangeability. Observe that as long as c_i remains 0, z_i retains the value $p(i)$ of the previous mixture index in its group g_i. When c_i takes value 1, z_i takes a new random value based on a group-specific distribution over mixtures π_{g_i}, that does not depend on the previous mixture index $p(i)$. Thus c_i acts as a change-point indicator variable. The distribution of c_i, and therefore the continuity of the the current mixture component, depends on the group and also the mixture component. Thus mixture components are characterized by how long they persist, but not by what follows them in the sequence. We use the term *block* to refer to a sub-sequence $\{s, s + 1, \ldots, s + m\}$ such that $p(i + 1) = p(i), \forall i \in [s + 1, s + m]$, $c_s = 1$ and $c_{s+k} = 0 \; \forall k \in [1, m]$. Note that this implies $z_s = z_{s+1} = \cdots = z_{s+m}$. For a block exchangeable sequence, permutations of blocks as a whole does not change probability of the sequence. Consider two different assignments x and y to $\{c_i, z_i\}_{i=1}^n$. We say that a block of x corresponds to another block of y if they are of same length, and have the same value of $\{z_i\}$ for the data points within them. Then, if there exists a bijection between the blocks of x and those of y, then they should have same probability for a BE model.

BE-LaDP: We now show that BE can be incorporated into any layer l of the LaDP instead of CE or ME. Consider the generative process in Eqn. 3 or in Eqn. 4. We modify the random variables for layer l as follows:

$$\beta_g^l \sim GEM(\gamma^l), \quad \pi_g^l \sim DP(\alpha^l, \beta_g^l), \quad g = 1 \ldots \infty$$

$$q_{g,g'}^l \sim Beta(1, \kappa); \quad \pi_{g,g'}^l = q_{g,g'}^l \pi_g^l + (1 - q_{g,g'}^l)\delta_{g'}, g, g' = 1 \ldots \infty$$

$$z_i^l \sim \pi_{z_i^{l+1}, p(i,l)}^l \quad i = 1 \ldots n \qquad (7)$$

where $p(i, l)$ is now the layer specific parent group. Note that κ again plays the role of a continuity parameter as for the ME-LaDP. When BE is used in every layer of LaDP as in Eqn. 7 we call the model BE-LaDP.

5   Inference Using LaDP

The inference problem in LaDP, given observations $\{w_i\}$, is to find posterior distributions over the group variables $\{z_i^l\}$ at all layers l for each data point.

As for models such as HDP, HDP-HMM and sHDP-HMM, exactly computing this posterior distribution is not tractable, and we resort to Gibbs Sampling for approximate inference as for the other models. One possibility is to perform collapsed Gibbs Sampling using only the group variables after integrating out all the parameter variables such as π_g^l and β_g^l. When the β_g^l variable takes the same value across groups in any layer l, the distribution of the variables at that layer is identical to the HDP. The predictive distribution of the z_i^l in that case is given by the CRF equations as for the HDP [5]. However, in cases where some of the β_g^l distributions are specified through domain knowledge, we integrate out only the π_g^l distributions.

Predictive Distributions: For the different LaDP models, we first derive the predictive distributions for z_i^l, the ith group variable in the lth layer, given the assignments to all group variables in the layers above (denoted $z^{>l}$), and the first $i-1$ group variables in layer l (denoted $z_{<i}^l$), after integrating out the $\pi_{g,g'}^l$ distributions from which they are drawn.

If the lth layer uses CE (Eq. 3), the predictive distribution is given by

$$p(z_i^l = a | z_{<i}^l, z^{l+1}) \propto n_{z_i^{l+1},i,a}^l + \alpha^l \beta_{z_i^{l+1}}^l(a)$$

where $n_{j,i,a}^l = |\{t : z_t^l = a, z_t^{l+1} = j, t \in [1, i-1]\}|$. This is the number of data points before datapoint i in group j of layer $l+1$ were assigned to group a in layer l. If the lth layer uses ME (Eq. 4), the predictive distribution becomes

$$p(z_i^l = a | z_{<i}^l, z^{l+1}) \propto n_{z_i^{l+1},i,p(i,l),a}^l + \kappa\delta(p(i,l),a) + \alpha^l \beta_{z_i^{l+1}}^l(a)$$

where $n_{j,i,b,a}^l = |\{t : z_t^l = a, p(t,l) = b, z_t^{l+1} = j, t \in [1, i-1]\}|$ is the number of times successive data points before datapoint i in group j of layer $l+1$ assigned to groups b and a respectively in layer l. For BE at layer l, we consider joint predictive distribution of z_i^l and the change-point indicator c_i^l at layer l. The conditional probabilities are as follows:

$$P(c_i^l = 0, z_i^l = p(i,l) | c_{<i}^l, z_{<i}^l, z^{l+1}) \propto a_{z_i^{l+1},i,0,p(i,l)}^l + \kappa$$

$$P(c_i^l = 1, z_i^l = k | c_{<i}^l, z_{<i}^l, z^{l+1}) \propto (a_{z_i^{l+1},i,1,p(i,l)}^l + 1)(v_{z_i^{l+1},i,1,k}^l + \kappa\beta_{z_i^{l+1}}^l(k))$$

where $a_{j,i,c,p}^l \equiv |\{t : c_t^l = c, p(t,l) = p, z_t^{l+1} = j, t \in [1, i-1]\}|$ is the number of times data points before datapoint i in group j of layer $l+1$ were assigned to group p in layer l, and the adjacent change-point value in group j is c; and $v_{j,i,c,a}^l = |\{t : c_t^l = c, z_t^l = a, z_t^{l+1} = j, t \in [1, i-1]\}|$ is the number of times data points before datapoint i in group j of layer $l+1$ were assigned to group a in layer l, and the change-point value at the same position is c.

Inference Using Gibbs Sampling: We sample each of the z_i^l variables conditioned on all the others sequentially in each iteration until convergence. In each iteration we traverse all group variables for one data point before moving to the next data point, and for a specific data point we traverse layers top down.

The conditional distribution is given by $p(z_i^l | z_{-i}^l, z^{l-1}, z^{l+1}) \propto p(z_i^l | z_{-i}^l, z^{l+1})$ $p(z^{l-1}|z^l)$. The second term can be computed using the chain rule and the predictive distributions described above: $p(z^{l-1}|z^l) = p(z_1^{l-1}|z_1^l) \prod_{i=2}^{n} p(z_i^{l-1}|z_{<i}^{l-1}, z_i^l)$. At layer $l = 1$ this is the likelihood of the data, conditioned on the table assignments of layer 1. The form of the first term depends on the exchangeability assumption.

If layer l uses CE the ith variable can be swapped with the last to get

$$p(z_i^l = a | z_{-i}^l, z^{l+1}) \propto n_{-i, z_i^{l+1}, a}^l + \alpha^l \beta_{z_i^{l+1}}^l(a)$$

where $n_{-i,j,a}^l = |\{t \neq i : z_t^l = a, z_t^{l+1} = j\}|$. Swapping is possible by CE property.

If layer l uses ME with sticky transitions, we make use of the conditional distribution for the sHDP-HMM [6] to get:

$$p(z_i^l = a | z_{-i}^l, z^{l+1}) = (\alpha^l \beta_j^l(a) + s(p(i,l),a) + \kappa\delta(p(i,l),a)) \times$$

$$\frac{\alpha^l \beta_j^l(c(i,l)) + s(a,c(i,l)) + \kappa\delta(c(i,l),a) + \delta(c(i,l),a)\delta(p(i,l),a)}{\alpha^l + s(a,.) + \kappa + \delta(p(i,l),a)}$$

where $j = d_i^{l+1}$, $s(a,b) = |\{t : z_t^l = a, c(t,l) = b\}|$. $p(i,l)$ is as defined before Eqn. 4, and $c(i,l)$ is defined analogously with $i + 1 \leq j \leq n$ instead of $1 \leq j \leq i - 1$.

Recall the $p(i,l)$ and $c(i,l)$ notations defined for the z_i^l variables. To define the conditionals for BE, we need to extend these to have equivalent notations for the c_i^l variables. So, we will use $p^z(i,l)$ and $c^z(i,l)$ for the earlier definitions, and $p^c(i,l)$ and $c^c(i,l)$ for equivalent definitions using c_i^l instead of z_i^l. Then the joint conditional $p(c_i^l, z_i^l | c_{-i}^l, z_{-i}^l, z^{l+1})$ has the following cases (omitting conditioning variables for notational brevity): For $c^c(i,l) = 0$

$$p(c_i^l = 1, z_i^l = c^z(i,l)|.) = \qquad\qquad 1 \qquad\qquad \text{if } p^z(i,l) \neq c^z(i,l)$$

$$p(c_i^l = 0, z_i^l = p^z(i,l)|.) \propto \quad (\kappa + n_2^{z,c}(p^z(i,l),0)) \quad \text{if } p^z(i,l) = c^z(i,l)$$

$$p(c_i^l = 1, z_i^l = p^z(i,l)|.) \propto \frac{(n_1^{z,c}(p^z(i,l),1) + \alpha_j^l \beta_j^l(p^z(i,l)))}{(n_1^c(1) + \delta)}$$
$$\times (1 + n_2^{z,c}(p^z(i,l),1))$$

For $(c^c(i,l) = 1)$

$$p(c_i^l = 0, z_i^l = p^z(i,l)|.) \propto \frac{(\kappa + n_2^{z,c}(p^z(i,l),0))}{(1 + \kappa + n_2^{z,c}(p^z(i,l),.))}$$

$$p(c_i^l = 1, z_i^l = b|.) \propto \frac{(n_1^{z,c}(b,1) + \alpha_j^l \beta_j^l(b))}{(n_1^c(1) + \alpha_j^l)} \frac{(1 + n_2^{z,c}(b,1))}{(1 + \kappa + n_2^{z,c}(b,.))}$$

where $n_1^c(1)$ is the count for $c_t^l = 1$ where $t \neq i$, $n_1^{z,c}(u,k)$ for $z_t^l = u$ and $c_t^l = k$, for all $t(\neq i)$ satisfying $z_t^{l+1} = z_i^{l+1}$, $n_2^{z,c}(u,k)$ for $p^z(t,l) = u$ and $c_t^l = k$, for all $t(\neq i)$ satisfying $z_t^{l+1} = z_i^{l+1}$. The equations are modified appropriately for the first and last data points of each group.

Complexity of Inference: ME vs. BE: Consider the conditional distributions at any layer l. For ME, for each data point, we need to sample from a k dimensional multinomial, where k is the current number of unique group values at layer l. This leads to a complexity of $\mathcal{O}(nk)$. In case of BE, the variable c_i^l determines whether the ith data point continues with the value of $p(i, l)$. When $c_i^l = 1$, we need to sample a new value of z_i^l from a k-dimensional multinomial. Hence, the complexity of each iteration of inference in BE is $\mathcal{O}(n + bk)$, where b is the number of data-points with $c_i^l = 1$. This can be significantly less than $\mathcal{O}(nk)$, particularly for high values of k. The value of b depends on κ in Eqn. 7.

Discussion: The number of unobserved variables in the model grows linearly with the number of layers, and dependencies also become more complex, leading to slower mixing of the Gibbs sampler. Our hypothesis is that the order in which the variables are sampled in each iteration of Gibbs sampling influences the rate of convergence. Specifically, updating directly dependent variables immediately after updating any variable may lead to faster convergence. Note that there are two kinds of the dependencies in the model. There are 'vertical dependencies' between group variables for the same data point across layers, and 'horizontal dependencies' between group variables of data points and their parents within a layer. Currently, our inference algorithm orders variable updates only based on vertical dependencies — we sample all the $\{z_i^l\}_{l=1}^L$ variables corresponding to the data point i, before moving to the next data point. Future work would include investigating other possibilities.

In hierarchical Bayesian non-parametric models, the conditional distributions of latent variables, given assignments to earlier ones are typically associated with restaurant analogies. For the LaDP, we may consider a hypothetical restaurant that has layers consisting of infinite number of tables, each layer possibly corresponding to one course in the menu. Each customer, unlike in a formal dinner, has to move from one layer to the next after each course. The restaurant has multiple entrances, corresponding to each input group, and in the first layer, each customer randomly chooses a table based on table assignments of previous customers who came in through the same entrance. After completing the ith course, each customer randomly chooses a table for the next $(i + 1)^{th}$ course based on tables assigned to previous customers who shared his table in the ith course. Clearly, the dependencies are more complex for ME and BE.

6   Experiments

In this section, we empirically evaluate our proposed models on two datasets for the tasks of document modeling and segmentation. We first check if learning multiple layers of grouping leads to better fit on held-out data, and also if the resultant simultaneous segmentation at multiple layers is better than single layer segmentation performed by models such as the sticky HDP-HMM. We also compare the performance using the proposed notion of BE and that using ME in terms of generalization ability, segmentation quality and execution time.

Datasets: Our first dataset is a set of semi-synthetic **News** transcripts. We crawled archived pages from 5 news websites (Yahoo! News, The Hindu, The Times of India, Deccan Herald, The Telegraph) for a 30 day period (April 1-30, 2012), where news articles for each day were arranged in sequence like news transcripts. We selected stories from 5 categories — politics, national affairs, international affairs, business and sports, to create one transcript for each day for each news source. This produced a dataset of $150(30 \times 5)$ virtual news transcripts, consisting of 2600 individual news articles, spread over the 5 categories. From these, 60 transcripts were used for training and the rest for testing. After eliminating stop-words and rare words, we had a vocabulary of size 7204, with a total of 0.4 million tokens in the complete dataset. Our second dataset is on customer-generated laptop **Reviews** from Amazon.com. Here each document is a single review, consisting of parts discussing different product facets, like appearance, weight, screen size, image clarity, connectivity etc. The vocabulary size was 7147 and there were 1.5 million tokens in the entire dataset. We used 11510 documents for training, and 1000 for testing. In 100 of the test documents, we annotated the facet segments manually for use as gold standard segmentation for evaluation.[1]

Weak Supervision: Our models can accept weak supervision through the group-specific β_g^l base distributions at any layer l. In the news dataset we have gold-standard on the category labels. In the topmost layer L of any LaDP model, the groups correspond to news documents, each belonging to one of the 5 categories. For some of the models (as discussed later) the training documents j were provided supervision by setting β_j^L to a δ-distribution peaked at the label of the category. We do not have such unique labels for stories. Separately, we ran HDP in advance on the entire set of news articles (considering each article as a document) and manually selected 136 meaningful topics, which we used as β_g^0 (for $g = 1 \ldots 136$), which serve as base distributions for the stories. (Eqn. 4).

Evaluated Models: We evaluate models with different number of layers, different exchangeability properties at each layer, and with and without supervision at specific layers. We choose a naming convention that clearly identifies these choices. For example, the name $ME_s^r\text{-}BE\text{-}CE_s\text{-}LaDP$ indicates that the model has 2 layers, with ME used at layer 2, BE at layer 1 and CE at layer 0 for words. The s subscripts indicate that supervision is used at layers 2 and 0. The r superscript indicates that the number of mixture components is restricted at layer 2, instead of an infinite mixture. All of our models use complete exchangeability at the layer of words, but we still include it in the name of the model, since we have the option of using supervision at that layer. In our experiments, we use 2 and 1 layer models (i.e. with $L = 2$ and $L = 1$). Note that CE-CE-LaDP is the same as HDP [5], ME-CE-LaDP as shDP-HMM [6], and CE-CE-CE-LaDP as MLC-HDP [9].

---

[1] The data is available at
http://clweb.csa.iisc.ernet.in/adway/ladp/data.tar.gz

Performance Measures: We aim to evaluate generalization ability and segmentation of the models. To evaluate generalization ability, we measure *perplexity*(PP) [11] on test data: $exp(-\frac{\sum_d \log p(W_d)}{\sum_d N_d})$, where W_d are the words, and N_d the number of words in the dth test document. A lower value of perplexity indicates better performance. For evaluating *segmentation*, we use the P_k measure [12], which is the probability that two tokens, k positions apart in the same document, are reported to be in different segments despite being in same gold-standard segment, or the other way round. Since different models perform well for different ranges of k, we report the average over three different values of k (short, medium and long), which we denote as $S2$ for layer 2 and $S1$ for layer 1. The performance of the proposed models involving ME or BE depends critically on the parameter κ (Eqns. 4 and 7). We tune these parameters for all models using a validation set of 5 transcripts to optimize performance.

Experiments on News: For the news dataset, we have gold standard segmentation at the level of categories as well as at the level of stories. We evaluate five versions of 2-layered LaDP ($L = 2$), considering layer 2 as categories and layer 1 as stories. The first four use combinations of BE and ME for layers 2 and 1, with the number of components restricted to 5 at $l = 2$. We test a version that uses CE at both layers, (MLC-HDP [9] model restricted at layer 2). For models with $L = 1$, we consider $l = 1$ to correspond to stories, leading to the models ME-CE-LaDP (sHDP-HMM) and BE-CE-LaDP, which can be evaluated for story segmentation ($S1$) and perplexity (PP). These models do not have a layer corresponding to categories. Alternatively, we could consider $l = 1$ to correspond to categories, with no layer for stories, leading to ME^r-CE-$LaDP$ and BE^r-CE-$LaDP$, and evaluate for category segmentation ($S2$). For $S2$ we use k values of 700(long), 200(medium), and 50(short), while for $S1$ we use 160(long), 50(medium) and 20(short), based on the typical lengths of category and story segments in the gold-standard. The results are shown in Table 1. We separately evaluate all these models with weak supervision at layers $l = 2$ and $l = 0$ as discussed. The results are shown in Table 2.

Table 1. Perplexity and Segmentation Error for News without supervision

Model	PP	S2	S1
CE-CE	5245	-	0.60
$CE^r - CE$	5969	0.69	-
ME-CE	3751	-	0.59
$ME^r - CE$	7204	**0.33**	-
BE-CE	3371	-	0.61
$BE^r - CE$	3975	0.69	-
CE^r-CE-CE	3656	0.68	0.61
ME^r-ME-CE	**3326**	0.45	0.53
ME^r-BE-CE	3856	0.68	0.61
BE^r-ME-CE	4475	0.49	0.42
BE^r-BE-CE	3713	0.45	**0.37**

Table 2. Perplexity and Segmentation Error for News with supervision

Model	PP	S2	S1
CE-CE_s	5309	-	0.60
$CE^r - CE_s$	6248	0.69	-
ME-CE_s	2763	-	0.59
$ME^r_s - CE_s$	7204	0.45	-
BE-CE_s	**2173**	-	0.59
$BE^r_s - CE_s$	2830	0.44	-
CE^r_s-CE-CE_s	3632	0.68	0.61
ME^r_s-ME-CE_s	2546	0.33	**0.42**
ME^r_s-BE-CE_s	2830	0.46	0.49
BE^r_s-ME-CE_s	3000	0.49	**0.42**
BE^r_s-BE-CE_s	3184	**0.28**	0.44

From the Tables 1 and 2, we first observe that supervision significantly improves performance of the models in terms of PP, and also often in terms of $S2$ or $S1$. Also, low perplexity and high segmentation errors for $CE^r\text{-}CE\text{-}CE$ and $CE_s^r\text{-}CE\text{-}CE_s$ confirm that capturing the sequential nature of the data is essential. More importantly, we can see that, in general, joint segmentation at two layers improves performance over independent segmentation at each layer. Only $ME^r\text{-}CE$ in the unsupervised case for category segmentation performs better than two-layer models. Two-layer models are also better in general in terms of perplexity. Only $BE\text{-}CE_s$ in the supervised case achieves better perplexity than two-layer models. Secondly, comparing BE and ME models, we see that the best perplexity is achieved by $BE\text{-}CE_s$, while $BE^r\text{-}BE\text{-}CE$ has the best $S1$ among unsupervised models. Among supervised models, $BE_s^r\text{-}BE\text{-}CE_s$ has the best $S2$, while $BE_s^r\text{-}ME\text{-}CE_s$ (jointly) has the best $S1$. This improved performance using BE can be attributed to the fact that its Exchangeability Statistic S_B is simpler than that of ME (S_M), and so it has to learn fewer parameters.

Table 3. Perplexity and Segmentation Error for Reviews

Model	PP	$S2$	$S1$
$CE\text{-}CE$	703	-	0.49
$ME\text{-}CE$	399	-	0.46
$BE\text{-}CE$	258	-	0.4
$CE\text{-}CE\text{-}CE$	1786	0.50	0.50
$ME\text{-}ME\text{-}CE$	1549	0.49	**0.38**
$ME\text{-}BE\text{-}CE$	1136	0.41	0.44
$BE\text{-}ME\text{-}CE$	1058	0.46	0.43
$BE\text{-}BE\text{-}CE$	477	0.43	0.46
$ME\text{-}CE\text{-}CE$	742	**0.37**	0.50
$BE\text{-}CE\text{-}CE$	**184**	0.41	0.50
$CE\text{-}ME\text{-}CE$	1913	0.48	0.48
$CE\text{-}BE\text{-}CE$	1787	0.49	0.42

Table 4. BE-ME comparison results

Model	Review			News		
	IT	$S1$	PP	IT	$S1$	PP
ME	53	0.46	399	4	0.59	2763
BE4	5.8	0.43	245	1.0	0.59	2173
BE3	5.2	0.40	258	0.7	0.56	2996
BE2	3.1	0.39	431	0.2	0.32	8650
BE1	0.6	0.48	614	0.1	0.39	13839

)

Experiments on Reviews: Recall that documents in this dataset only require a single layer of segmentation. However, it is still meaningful to use 2-layer models, and then use either the first or the second layer segmentation. The corresponding measures of segmentation error are $S1$ and $S2$. We also evaluate single-layer models, with $S1$ being the segmentation error. For both $S1$ and $S2$, we consider k values 4 (short), 8 (medium) and 16 (long), again based on the typical lengths of segments (by facet) in the gold-standard. Since we did not have product facet labels for this dataset, we did not provide any supervision. As before, we used CE at layer 0 (words). Since segmentation is needed at only one layer in this case, we considered all combinations of ME, BE and CE at layers 2 and 1, leading to 3 1-layer models, and 9 2-layer models.

    The results are shown in Table 3. We note that the baselines models HDP (CE-CE), shDP-HMM (ME-CE) and MLC-HDP (CE-CE-CE) are outperformed on all measures. More interestingly, though the data has segment information only at one layer, the best performance in terms of both PP and $S1$ is obtained by 2-layer models. Finally, BE performs well in terms of PP. Though BE does not outperform ME in terms of segmentation accuracy for this experiment, its usefulness becomes apparent in our next experiment.

Execution Time: Finally, we evaluate the effect of the continuity parameter κ (Eqn. 7) on the execution time and accuracy of Block Exchangeability. We compare the single-layer models BE-CE-LaDP and ME-CE-LaDP (sHMM-HDP) on News and Reviews in terms of inference time (per-iteration) (IT) (measured in seconds) during training, segmentation error $S1$ and perplexity PP using different values of κ. We consider 4 representative parameter settings for BE, denoted by BE1, BE2, BE3 and BE4 with κ values of 100000, 5000, 1000, 10 for news and 4000, 1500, 500, 300 for reviews. The results are shown in Table 4. To begin with, BE4 matches ME in segmentation and does better on perplexity while taking significantly less time (53 secs vs 5.8 secs for Reviews, 4 secs vs 1 sec for News). On decreasing κ, inference time reduces further with gradual degradation of perplexity, while average segmentation error decreases much below that of ME (for BE2) and then increases again. This happens because segmentation error for short k decreases monotonically with increase in block length i.e, decrease in κ, while that for long k increases monotonically. This demonstrates that using block exchangeability it is possible to trade off inference time for segmentation and modeling accuracy, unlike any existing exchangeability notion.

7   Conclusion

In this paper, we have addressed the problem of hierarchical segmentation of a collection of sequences, and proposed a Bayesian nonparametric model named the Layered Dirichlet Process, where data points filter down a layered structure of Dirichlet Processes, and get assigned to a group at every layer, depending on the exchangeability properties at that layer, leading to a hierarchical segmentation. We propose a new notion of exchangeability, that allows for more efficient inference compared to Markov exchangeability while enabling segmentation unlike complete exchangeability. We have demonstrated experimentally that using the proposed models joint segmentation at multiple layers is better than independent single-layer segmentation, and we are additionally able to trade off execution time for modeling and segmentation accuracy unlike any existing model.

References

1. Ferguson, T.: Bayesian analysis of some nonparametric problems. Annals of Statistics 1(2), 209–230 (1973)
2. Diaconis, P., Freedman, D.: De Finetti's generalizations of exchangeability. Studies in Inductive Logic and Probability 2, 233–249 (1980)
3. de Finetti, B.: Theory of probability, vol. 1-2 (1975)

4. Sethuraman, J.: A constructive definition of Dirichlet priors. Stat. Sinica 4, 639–650 (1994)
5. Teh, Y.W., Jordan, M.I., Beal, M.J., Blei, D.: Hierarchical Dirichlet Processes. Journal of American Statistics Association 101(476) (2006)
6. Fox, E.B., Sudderth, E.B., Jordan, M.I., Willsky, A.S.: An HDP-HMM for Systems with State Persistence. In: Intl. Conf. on Machine Learning, pp. 312–319 (2008)
7. Rodriguez, A., Dunson, D.B., Gelfand, A.E.: The nested Dirichlet process. Journal of the American Statistical Association 103(483), 1131–1154 (2008)
8. Blei, D.M., Griffiths, T.L., Jordan, M., Tanenbaum, J.B.: The Nested Chinese Restaurant Process and Bayesian Nonparametric Inference of Topic Hierarchies. Journal of the ACM 57(2) (2010)
9. Wulsin, D., Jensen, S., Litt, B.: A Hierarchical Dirichlet Process Model with Multiple Levels of Clustering for Human EEG Seizure Modeling. In: Intl. Conf. on Machine Learning (2012)
10. Kim, D., Kim, S., Oh, A.: Dirichlet Process with Mixed Random Measures: A Nonparametric Topic Model for Labeled Data. In: Intl. Conf. on Machine Learning (2012)
11. Blei, D.M., Ng, A.Y., Jordan, M.: Latent Dirichlet Allocation. Journal of Machine Learning Research 3, 993–1022 (2003)
12. Beeferman, D., Berger, A., Lafferty, J.: Statistical models for text segmentation. Machine Learning 34(1-3) (1999)
13. http://clweb.csa.iisc.ernet.in/adway/ladp/appendix.pdf

A Bayesian Classifier for Learning from Tensorial Data

Wei Liu[1,2,*], Jeffrey Chan[1], James Bailey[1,2], Christopher Leckie[1,2],
Fang Chen[2], and Kotagiri Ramamohanarao[1,2]

[1] Department of Computing and Information Systems, The University of Melbourne
{jeffrey.chan,baileyj,caleckie,kotagiri}@unimelb.edu.au
[2] ATP and Victoria Research Laboratory, National ICT Australia
{wei.liu,fang.chen}@nicta.com.au

Abstract. Traditional machine learning methods characterize data observations by feature vectors, where an entry of a vector denotes a scalar feature value of a data instance. While this data representation facilitates the application of conventional machine learning algorithms, in many cases it is not the best way of extracting all useful information from the data observations. In this paper we relax the (often unstated) assumption of *vectorizing* features of data instances, and allow a more natural representation of the data in a tensor format. Tensors are multimode (aka multi-way) arrays, of whom vectors (i.e., one-mode tensors) and matrices (i.e., two-mode tensors) are special cases. We show that the tensor representation captures useful information that is difficult to provide in the conventional vector format. More importantly, to effectively utilize the rich information contained in tensors, we propose a novel semi-naive Bayesian tensor classification method (which we call Bat) that builds predictive models directly on data in tensor form (instead of on their vectorizations). We apply Bat to supervised learning problems, and perform comprehensive experiments on classifying text documents and graphs, which demonstrate *(1)* the advantage of the tensor representation over conventional feature-vectorization approaches, and *(2)* the superiority of the proposed Bat tensor classifier over other existing learners.

1 Introduction

A major challenge in machine learning is finding an appropriate representation to characterize observed data. Given a set of data observations (instances), traditional feature extraction methods seek to enumerate a list of features that are associated with an instance, and thus interpret an instance by a vector of feature values. This feature formulation strategy is a common, though often unstated, presumption of many supervised and unsupervised learning algorithms that build machine learning models on feature vectors [1].

In this paper, we relax the assumption that features of data observations are linearized into vectors. Instead, we allow a more natural representation of a data

---

* Corresponding author.

H. Blockeel et al. (Eds.): ECML PKDD 2013, Part II, LNAI 8189, pp. 483–498, 2013.

instance in the form of a tensor. Tensors are multi-mode (also known as multi-way) arrays, whose one-mode special cases are vectors and two-mode cases are matrices.

Motivation: A major motivation for using tensors to represent instances is that they capture the "interactions" among features in different modes of the tensorial format, which are normally hard to capture using feature vectors. Although there is a rich literature in computer vision and other related fields on the study of data observations that are originally represented by tensors, such as two-mode images and three-mode videos, the more general scenarios where data observations are usually represented by vectors (such as text documents by frequent words, and graphs by frequent subgraphs) have been less well studied. To clearly motivate the use of using tensor representations, in the following we give examples to illustrate how we use *tensors of features* to describe data instances, and their advantages compared to using *vectors of features*.

In the following three motivating examples, we first explain that the tensor formulation can be applied to any conventional data set, and then introduce some examples on document and graph retrieval problems. In Examples 2 and 3, the tensor representation is capturing new information that is not explicit in the original representation, whereas in Example 1 it is capturing information that is explicit in the original representation. The different nature of the three examples is a reflection on the ability of tensor representations in addressing a diverse range of domain problems.

Example 1 *(Data representation of a data set from an arbitrary domain).* For an arbitrary standard data set, one can first discover a set of closed frequent patterns from the data's categorical features (or discretized continuous features) as shown in Fig. 1(b). Then for each instance one can construct a tensor that uses the frequent patterns as its dimensions in each mode (shown in Fig. 1(c)). In this way, an entry in this tensor indicates whether two closed frequent patterns (if the tensor is of two modes) co-occur in the same instance. This formulation can also be generalized to construct tensors of n modes, where an entry tells whether n distinct patterns co-occur in the same instances.

A special case of the formulation in Example 1 is that one can ignore the step of discovering frequent patterns, but use each unique value of each feature as a dimension in a mode. However, in this case the number of dimensions in each mode could be extremely large even for data sets of small sizes (i.e., when the total number of all features' unique values is very large), which is impractical for use in real domains. Hence we make use of the frequent pattern discovery step, which reduces the number of dimensions in each mode and also keeps the main variance of the original data.

Example 2 *(Data representation in document classification problems).* While a document is conventionally represented by a vector of frequent words (denoted by "Fw"), a tensor representation can capture more information than a vector. As shown in Fig. 2, a two-mode tensor contains in its diagonal entries the frequency of each Fw in the document, while its off-diagonal entries record the numbers of

f1	f2	f3	f4
1	1	0	1

(a) An arbitrary standard data instance characterized by a list of four binary features (f).

Fp1 = {f1=1,f3=0,f4=1}, Fp2 = {f1=0,f2=0}, Fp3 = {f2=1,f3=0}
Fp4 = {f2=1,f4=1}, Fp5 = {f3=1,f4=0}

(b) Five frequent patterns (Fp) discovered from the overall training data.

	Fp1	Fp2	Fp3	Fp4	Fp5
Fp1	1		1	1	
Fp2					
Fp3	1		1	1	
Fp4	1		1	1	
Fp5					

(c) The data instance in (a) characterized by a two-mode tensor of five frequent patterns.

Fig. 1. Vector and (two-mode) tensor representations of a data instance from an arbitrary domain. We assume binary feature values in this example.

"pairs" of features that co-occur in the same paragraph. For example, Fw1 and Fw2 co-occur twice (in paragraph 1 and 3) so the entries at locations (Fw1, Fw2) and (Fw2, Fw1) are 2. In the document representation, we use paragraphs to capture the relations among words, since paragraphs are natural segmentations of the original documents. We note that both local and global weighting methods, including *tf-idf* (term frequency-inverse document frequency), can be applied to the entries of tensors in the same manner to entries of Fw vectors. We also note that it is straightforward to generalize the representation of the two-mode tensor to a n-mode tensor that describes co-occurrences of n frequent words in paragraphs.

Example 3 *(Data representation in graph classification problems).* Given a graph database, it is common for a data miner to first discover closed frequent subgraphs, and then use these to describe each graph instance [2, 3]. The example shown in Fig. 3 is a graph that contains six closed frequent subgraphs (denoted by "Sg"). Based on the graph's layout (Fig. 3(a)), conventional graph-based machine learning methods (e.g., [2–6]) construct a graph instance by defining the six subgraphs as features and the frequency of each subgraph found in the original graph as feature values (Fig. 3(b)). While this feature vector can provide information on the components of a graph, it does not capture relations among the components that could potentially be very useful for machine learning tasks. In this regard, we use the representation of tensors to capture both the occurrences of subgraphs and their hidden internal relations. As shown in Fig. 3(c), for the same graph instance, we can use a two-mode tensor to first capture in its diagonal entries all information stored in the instance's feature vector, and then record in the tensor's off-diagonal entries the relations/interactions among

(Paragraph 1) ···, Fw1, ···, Fw2, ···, Fw3, ···.

(Paragraph 2) ···,Fw4, ···, Fw1, ···, Fw5, ···, Fw6, ···.

(Paragraph 3) ···, Fw2, ···, Fw6, ···, Fw1, ···.

(a) The original layout of a text document, showing only frequent words (Fw).

Fw1	Fw2	Fw3	Fw4	Fw5	Fw6
3	2	1	1	1	2

(b) The text document in (a) characterized by a vector of frequent words.

	Fw1	Fw2	Fw3	Fw4	Fw5	Fw6
Fw1	3	2	1	1	1	1
Fw2	2	2	1			1
Fw3	1	1	1			
Fw4	1			1	1	1
Fw5	1			1	1	1
Fw6	1	1		1	1	2

(c) The text document In (a) characterized by a two-mode tensor (a matrix).

Fig. 2. Vector and (two-mode) tensor representations of a text document

different features. Note that to enclose more information from the original graph, we give an *asymmetric* design to the tensor: the upper-diagonal entries indicate how many "column-wise features" are connected to "row-wise features" and the lower-diagonal entries tell how the number of "row-wise features" that are connected to "column-wise features". For example, there are two "Sg2" connected to one "Sg1", so the upper-diagonal entry in location (Sg1,Sg2) is 2, and the lower-diagonal entry in location (Sg2,Sg1) is 1. Besides the two-mode tensor representation shown in Fig. 3(c), we note that by using the same formulation one can have a generalized n-mode tensor to describe a graph instance (e.g., a three-mode tensor can capture the triad connection among Sg1, Sg2, and Sg3 in Fig. 3(a)).

Having obtained the tensor representation of data observations, a challenge one faces is how to build machine learning models that can discover knowledge from data in the tensor format, so that one can fully utilize the rich information contained in tensors. This is a non-trivial challenge since most standard learning algorithms assume data instances are feature vectors, and it is not straightforward to apply these algorithms on tensorial data. A simple solution is to linearize the tensors into new vectors, and use the new vectors in a conventional learning algorithm. However, such tensor linearizations will *break* the relations among

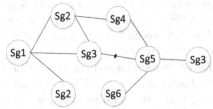

(a) The original structure of a graph instance, represented by closed frequent subgraphs (Sg).

Sg1	Sg2	Sg3	Sg4	Sg5	Sg6
1	2	2	1	1	1

(b) The graph instance in (a) characterized by a vector of subgraphs.

	Sg1	Sg2	Sg3	Sg4	Sg5	Sg6
Sg1	1	2	1			
Sg2	1	2	1	1		
Sg3	1	1	2		1	
Sg4		1		1	1	
Sg5			2	1	1	1
Sg6					1	1

(c) The graph instance in (a) characterized by a two-mode tensor (a matrix).

Fig. 3. Vector and (two-mode) tensor representations of a graph instance

features in different modes, which is equivalent to assuming independence among entries of tensors[1].

To tackle this tensor learning challenge, in this paper we proposed a generative semi-naive Bayesian classifier, which can be trained directly on data in tensor formats with respect to relations among features in different modes. The semi-naive property of our method enables the learning of inter-mode relations in an effective manner. We analyse why the assumption made in simple naive Bayes classifiers is not ideal for learning tensorial representations, and also why other existing semi-naive Bayesian classifiers are not suitable for capturing the precise information represented in tensors. These shortcomings of existing Bayesian classifiers motivate us to design a novel type of learning algorithm that can make the best use of tensorial representations.

In brief, the contributions we make in this research are as follows:

1. We propose to characterize data instances by using tensors of features, which contain much richer information than using feature vectors.

---

[1] This is because the ordering of entries in the linearization is not used by a classifier, but the ordering of entries in the original tensor can potentially play an important role in disclosing the relations among features in different modes.

2. We introduce a semi-naive Bayesian tensor (Bat) learning method, which builds classifiers by making use of the relations among features in different modes of a tensor. The Bat method can be applied directly to data in tensor format without tensor vectorization.

3. We apply the tensor representations and the Bat method to graph and document classification problems, and comprehensively evaluate our method with comparison to existing naive and semi-naive Bayesian methods.

The rest of the paper is organized as follows. We review related work in Section 2. Section 3 introduces our Bat method and discusses the differences between Bat and other existing Bayesian learners. Experimental results are presented in Section 4. We conclude in Section 5.

2   Related Work

Existing methods on tensor analysis mostly focus on decomposing tensors into factor matrices [7–9], whereas the problem of how to build classifiers directly on tensor data has not been well studied.

Tao *et al.* [10] proposed supervised tensor learning (STL) as a generalization of support vector machines, where the authors learn weight vectors separately from each mode of a data tensor. However, there is no theoretical guarantee that the weight vector learned on separate modes will provide global optima of training bias minimization. In the recent Bayesian learning literature, averaged one-dependence estimators (AODE) [11] have been introduced as a special form of the one dependence estimator (ODE), which relaxes the naive Bayes' independence assumption by making each feature a parent of other features. This method is improved by weightily averaged one-dependence estimators (WAODE) [12], which give different levels of importance to parent features by examining the mutual information between those features and the class variable. Since making each feature a parent of other features incurs very high computational costs, subsumption resolution of AODE (AODEsr) has been proposed to speed up AODE's learning process by eliminating features that are generalizations or specializations of another feature. Another way of building dependencies among features is to look at the hidden relationship between pairs of features, which gives the proposal of hidden naive Bayes (HNB) [13]. Bayesian networks are also popular ways to discover the dependencies among features, among which K2 [14] and TAN [15] have been two of the most popular methods.

3   Bayesian Tensor Classification

The strong feature-independence assumption used in NB ignores possible correlations among features. Hence when the data has multiple modes, the potentially useful interactions among features will not be taken into the classification rule of NB, which could degrade the classification performance. To address this problem, in the following we introduce a method that is specifically designed to tackle data with multiple modes. Our proposed method belongs to the taxonomy of semi-naive Bayesian learning models [16], since it enhances the conditional probability estimation of naive Bayes by relaxing its attribute independence assumption.

In contrast to using a vector of features (e.g., $\mathbf{x}$) to represent a data observation, we describe an instance by using a n-mode tensor of features, denoted as $\mathcal{X} \in \mathbb{R}^{m \times m \ldots \times m}$. Vectors ($\mathbb{R}^m$) and matrices ($\mathbb{R}^{m \times m}$) are specifications of (1-mode and 2-mode) tensors. Without loss of generality, we present our learning method by using tensors of two modes (i.e., $\mathcal{X} \in \mathbb{R}^{m \times m}$). Then each entry of $\mathcal{X}$ can be viewed as a new feature value (denoted by $\mathcal{X}_{i,j}$ which represents the relations between features i and j), and the tensor $\mathcal{X}$ represents a set of m^2 features. To relax the assumption of conditional independence made in NB, we assign a "parent" feature to other features that share the same dimension with the parent feature. The scenario when entries share the "same dimension" of a tensor is analogical to when entries are in the "same row" or "same column" of a matrix. All entries are in the same dimension in a one-mode tensor (a vector).

The notion of the "parent" feature is the same as the concept of a parent node/vertex used in graphical models such as Bayesian networks, where features that are not connected represent variables that are conditionally independent of each other. In other words, the features that are independent in NB conditioned on the class variable will only be independent in our model given both the class and the parent. Therefore, instead of computing $P(y, \mathbf{x})$ by $P(y)P(\mathbf{x}|y)$ as in NB, we estimate $P(y, \mathbf{x})$ by

$$P_{\mathcal{X}_{p_1,p_2}}(y, \mathbf{x}) = P(y, \mathcal{X}_{p_1,p_2})P(\mathbf{x}|y, \mathcal{X}_{p_1,p_2}) = P(y, \mathcal{X}_{p_1,p_2}) \prod_{i,j=1}^{m} P(\mathcal{X}_{i,j}|y, \mathcal{X}_{p_1,p_2}) \quad (1)$$

where the first step assigns a parent feature $\mathcal{X}_{p_1,p_2} \in \mathcal{X}$, while the second step utilizes conditional feature-independence (given the class and the parent feature), where p_1 and p_2 are respectively the row and column indices. For ease of interpretation, in Eq. 1 we use $\prod_{i,j=1}^{m}$ to represent the operation $\prod_{i=1}^{m} \prod_{j=1}^{m}$. Since we assume the relation among features exists only if they share the same dimension in the tensor, for a given parent feature $\mathcal{X}_{p_1,p_2}$, the actual conditional probabilities we use are:

$$P_{\mathcal{X}_{p_1,p_2}}(y, \mathbf{x})$$
$$= P(y, \mathcal{X}_{p_1,p_2}) \prod_{i=1}^{m} P(\mathcal{X}_{i,p_2}|y, \mathcal{X}_{p_1,p_2}) \prod_{j=1}^{m} P(\mathcal{X}_{p_1,j}|y, \mathcal{X}_{p_1,p_2})P(y) \prod_{i \neq p_1, j \neq p_2} P(\mathcal{X}_{i,j}|y)$$
$$(2)$$

where the first two products ensure $\mathcal{X}_{p_1,p_2}$ is a parent feature of $\mathcal{X}_{i,p_2}$ and $\mathcal{X}_{p_1,j}$ only when they are from the same row or column of a two-mode tensor. And the last product means when they are not from the same row or column, we use standard NB (the naive feature-independence assumption) to compute the posterior probabilities.

To make an unbiased selection of parent features, we make each element in tensor $\mathcal{X}$ a parent at a time, and use the average of the probabilities conditioned on each parent as the final classification rule. It is reported in [12] that the performance of semi-naive Bayesian models can be improved by taking into account the mutual information between the parent feature and the class variable. By using the definition from information theory, the mutual information $I_{i,j}$ (i.e., assume a two-mode tensor) between a feature $\mathcal{X}_{i,j}$ and the class variable y is

$$I_{i,j} = \sum_{\mathcal{X}_{i,j}} P(\mathcal{X}_{i,j}, y) \log \frac{P(\mathcal{X}_{i,j}, y)}{P(\mathcal{X}_{i,j})P(y)} \tag{3}$$

where the summation is on all unique values in feature $\mathcal{X}_{i,j}$. After applying this mutual information, the label of an instance $\mathcal{X}$ is determined by our Bat method using:

$$label = \arg\max_{y} P(y|\mathbf{x}) \propto \arg\max_{y} \frac{\sum_{p_1,p_2=1}^{m} I_{p_1,p_2} P_{\mathcal{X}_{p_1,p_2}}(y, \mathbf{x})}{\sum_{p_1,p_2=1}^{m} I_{p_1,p_2}} \tag{4}$$

where $P_{\mathcal{X}_{p_1,p_2}}(y, \mathbf{x})$ is defined in Eq. 2. Eq. 4 is the final classification rule of our Bayesian tensor classifier Bat.

For tensorial training data with two modes, the time complexity of estimating Eq. 4 is $O(2mt)$, where t is the number of training instances, and its space complexity is $O(k(mv)^2)$, where k is the number of classes and v is the average number of values in each feature.

3.1   Advantages of Bat

Similar to NB, Bat only needs to update the conditional probabilities when a new training instance becomes available, hence one advantage of Bat is that it is capable of incremental learning.

AODE [11] is a special case of Bat when each training instance is of one mode (i.e., a feature vector). However, when training instances are of more than one mode, AODE does not provide a way to handle the data. If we linearize the data tensor into a vector (like what we do to make tensor data learnable to other classifiers), AODE will have to enumerate each entry in the vector (linearized tensor) to be a parent of other entries that are in the same dimension of a tensor. This will lead to a training time complexity of $O(tm^2)$. However in Bat, tensor linearization is not needed, and an entry will be a parent of another entry if only they share the same row index or column index. Therefore, besides the specific focus on inter-relations of features, another advantage of Bat is that its training time complexity (i.e., $O(2mt)$) is an order of magnitude lower than that of AODE (i.e., $O(tm^2)$).

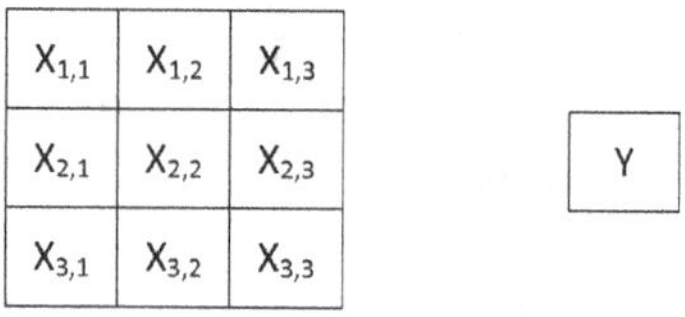

$X_{1,1}$	$X_{1,2}$	$X_{1,3}$	
$X_{2,1}$	$X_{2,2}$	$X_{2,3}$	Y
$X_{3,1}$	$X_{3,2}$	$X_{3,3}$	

Fig. 4. An illustrative example to explain the differences among NB, AODE, and Bat (see Example 4 at Sec 3.1).

Example 4 *(Differences among NB, AODE, and Bat)*. Suppose we have a data set associated with a class variable y and a 3×3 two-mode tensor (9 features). As shown in Fig. 4, the 9 features are indexed as $\mathcal{X}_{1,1}$, $\mathcal{X}_{1,2}$, ..., $\mathcal{X}_{3,3}$. Then taking feature $\mathcal{X}_{2,2}$ as an example, NB estimates the joint likelihood (from the Bayesian rule) of this feature and class variable simply by $P(y)P(\mathcal{X}_{2,2}|y)$, which means $\mathcal{X}_{2,2}$ is (conditionally) independent of other features. AODE estimates this joint likelihood by $\prod_{i,j=1}^{3} P(y, \mathcal{X}_{i,j})P(\mathcal{X}_{2,2}|y, \mathcal{X}_{i,j})$, which means AODE has to enumerate other features to be parent features of $\mathcal{X}_{2,2}$. This is a reflection of the fact that AODE cannot distinguish data of vector features from tensor features. In contrast, Bat estimates the joint likelihood by $P(y, \mathcal{X}_{1,2})P(\mathcal{X}_{2,2}|y, \mathcal{X}_{1,2})P(y, \mathcal{X}_{2,1})P(\mathcal{X}_{2,2}|y, \mathcal{X}_{2,1})P(y, \mathcal{X}_{2,3})P(\mathcal{X}_{2,2}|y, \mathcal{X}_{2,3})$ $P(y, \mathcal{X}_{3,2})P(\mathcal{X}_{2,2}|y, \mathcal{X}_{3,2})$, which means Bat only uses the features that are at the same column or row of the target feature (i.e., $\mathcal{X}_{2,2}$) to be parent features (i.e., $\mathcal{X}_{1,2}$, $\mathcal{X}_{2,1}$, $\mathcal{X}_{2,3}$, and $\mathcal{X}_{3,2}$). This design of Bat gives it the advantage of learning the specific structures of tensors.

Bat has a lower risk of overfitting the training data compared to AODE, since Bat still assumes conditional independence between features that are not in the same rows or columns (just like NB in this case), while AODE would have to assume none of the features are conditionally independent from each other given only the class variable.

The AODEsr method [17] is also closely related to Bat. However, different to our method, it infers the interdependence relationship by inspecting generalization/specialization or duplications among features. For example, given two features *Gender* and *Pregnant*, the feature value *"Gender=female"* is a generalization of *"Pregnant=yes"*. Such types of so-called subsumption resolutions are used by AODEsr to discover highly correlated features.

Relation to Bayesian Networks: Bat can be viewed as a special case of Bayesian networks, where the network structure is pre-defined in order to learn the underlying knowledge hidden behind the correlation among features. Such a pre-defined structure specialised by Bat captures the dependence among variables in a common dimension of a tensor, which is generally not captured by a vectorized format. It is possible that Bayesian networks can also learn some network structures to approximate the feature dependencies, but this would involve manual tuning on the selection of structure search algorithms. The final structure of a Bayesian network could also be too complicated to capture all useful tensorial information, which makes it infeasible in practice.

While classification error is commonly used to estimate the performance of classifiers, two other factors, namely *bias* and *variance*, can be decomposed from classification results that contribute to the error. The bias of a classifier is the difference between *the expected value (i.e., the central tendency) of the class variable returned by the classifier* and *the true values of the labels*. The variance of a classifier is the portion of the total error that is due to deviations from the expected value (i.e., the central tendency) of the classifier [18]. An ideal classifier is the one that has both low bias and low variance.

Why Is Bat Conjectured to Outperform Other Methods? One reason that Bat is conjectured to outperform alternative methods is that Bat makes weaker conditional independence assumptions than NB, hence the bias of Bat is expected to be lower[2] than that of NB. Since Bat considers dependencies among features from different modes, the bias of AODE is conjectured to be higher than that of Bat. Furthermore, because AODE assumes parent-child relationships among *all* features, its variance is conjectured to be higher than that of Bat. These comparisons of bias, variance, and errors are what we use to evaluate Bat against other classifiers in the experiment section.

Another reason that puts Bat at an advantage in learning tensorial data is that Bat uses a representation that we believe is more likely to capture dependencies between features and allows this dependency information to be used explicitly as part of the training process. Bayesian networks also could potentially learn this feature dependence, but it would require complicated structure search which could be infeasible in practice.

3.2   Limitations of Bat

Bat can only be trained on tensorial data with a *given* number of modes. The number of modes needed in a tensor to best describe the original data can be domain specific, so the proposed Bat is not designed with a mechanism that can automatically find the best number of modes. Another limitation is that Bat can only learn from categorical features values. However we note that this is also the case for all NB type classifiers, and is not a problem specific to Bat. Numerical feature values can be applied to Bat after they are discretized.

4   Evaluation

The objectives of our experiments are to evaluate *(1)* the effectiveness of the tensor formulations compared to traditional feature vectors, and *(2)* the classification performance of Bat compared to naive and other semi-naive Bayesian learners. We implement Bat in Weka [19], and evaluate our method with comparisons to NB, AODE [11], AODEsr [17], HNB [13], K2 [14], TAN [15], STL [10], logistic regression, SVMs with RBF/Sigmoid/Soft margin kernels, and decision trees. All the methods in our comparison are from Weka (version 3.6.7) and we use their default parameter settings (e.g., 10 single trees for a random forest, $\gamma = 0.01$ for SVM with RBF kernels etc.). The results are obtained from 5-fold cross validation with 10 repeated runs. We use the same bias and variance estimation method as was used in [18].

4.1   Data Sets

We use two types of data sets in our experiments: graphs and text documents. Instances in all of the data sets are transformed into tensors of two modes. We

---

[2] However, because the classification rule of Bat is derived from higher dimensions than that of NB, its variance might be larger than NB.

Table 1. Statistics of graph (chemical compound) data sets. "#Inst" represents the number of instances in each data set.

Name	#Inst	#Classes	Sources
AID1481	217968	2	ATPase Inhibition
AID83	27784	2	Breast Cancer
AID81	40700	2	Colon Cancer
AID1446	217968	2	Janus Kinase
AID123	40152	2	Leukemia
AID1	40460	2	Lung Cancer
AID1531	289475	2	Mek Inhibitors
AID33	40209	2	Melanoma
AID47	40447	2	Nerve Cancer
AID109	40691	2	Ovarian Cancer
PTC-M	336	5	Mice Toxic
PTC-R	349	5	Rats Toxic

Table 2. Statistics of document data sets. "#Fw" represents the number of frequent words extracted from each data set.

Name	#Fw	#Classes	Sources
oh0	3183	10	OHSUMED collection
oh5	3013	10	OHSUMED collection
oh10	3239	10	OHSUMED collection
oh15	3101	10	OHSUMED collection
re0	2887	13	Reuters-21578
re1	3759	25	Reuters-21578
tr11	6430	9	TREC
tr12	5805	8	TREC
tr23	5833	6	TREC

note that the ways to model the relationship between subgraphs or between words are the same as the ones we demonstrated in Examples 2 and 3.

We use chemical compounds as graphs where atoms in compounds are treated as nodes of graphs, and bonds that connect atoms are treated as edges of graphs. The labels of the chemical compounds are obtained from two sources: *(1)* Bioassays of anti-cancer activity and kinase inhibition (AID)[3]: the task is to predict whether a compound is positive or negative in anti-cancer activities or in kinase inhibition activities. The original data sets contain a large number of compounds (shown in Table 1). We randomly sample 1000 compounds from each data set for evaluation. *(2)* Toxicology prediction (PTC)[4]: the task is to predict the carcinogenicity of compounds on mice and rats. Each chemical compound is associated with a carcinogenicity class from {CE, SE, P, E, EE, IS, NE, N}. Following the settings of [2], we use {CE, SE, P} as positive classes, {NE, N} as negative ones, and discard other neutral classes.

The text document data sets are obtained from various sources, including the OHSUMED collection [20], the Reuters-21578 text collection[5], and the TREC

---

[3] http://pubchem.ncbi.nlm.nih.gov
[4] http://www.predictive-toxicology.org/ptc/
[5] http://www.daviddlewis.com/resources/testcollections/reuters21578/

Table 3. Performance of each classifier on data sets in vector and tensor formats

Data sets	NB Vector	NB Tensor*	AODE Vector	AODE Tensor*	AODEsr Vector	AODEsr Tensor*	HNB Vector	HNB Tensor*	WAODE Vector	WAODE Tensor*	K2 Vector	K2 Tensor*
ATPase	.9347	.9708	.9572	.9696	.9383	.9758	.9614	.9789	.8998	.9758	.9533	.9752
Breast	.9292	.9480	.8599	.9554	.9061	.9622	.9304	.9715	.9304	.9715	.8956	.9597
Colon	.8525	.8964	.8853	.9007	.9003	.9236	.9081	.9168	.9182	.9266	.8315	.9081
Jak2	.8981	.9296	.9278	.9309	.9171	.9446	.8875	.9440	.9212	.9465	.9066	.9390
Leuk.	.8694	.9608	.9134	.9677	.9196	.9801	.9316	.9776	.9681	.9832	.8939	.9695
Lung	.8459	.9317	.8540	.9366	.9339	.9594	.8874	.9538	.9324	.9594	.8768	.9403
Mek	.9674	.9725	.9063	.9713	.8941	.9744	.9080	.9763	.9333	.9763	.9731	.9738
Mela.	.8679	.9370	.9222	.9401	.8639	.9580	.8974	.9586	.9081	.9586	.8876	.9444
Nerve	.8602	.8840	.8543	.8870	.8423	.9086	.8964	.8994	.8665	.9080	.8628	.8975
Ovar.	.8924	.9318	.8931	.9362	.9225	.9554	.9371	.9436	.8752	.9591	.8600	.9467
Mice	.8897	.9413	.8545	.9475	.9102	.9666	.9262	.9567	.9189	.9691	.9566	.9567
Rats	.8499	.9383	.9255	.9402	.9465	.9568	.9037	.9544	.8677	.9581	.9038	.9476
oh0	.3860	.4028	.6213	.6795	.7158	.7871	.6727	.7198	.7150	.7637	.1246	.1301
oh5	.3182	.3529	.5534	.5915	.6642	.7282	.6223	.6487	.6230	.6890	.1242	.1302
oh10	.4106	.4233	.6507	.6762	.6734	.7333	.6596	.7186	.7101	.7276	.1248	.1352
oh15	.3402	.3658	.6054	.6172	.6891	.7076	.6079	.6550	.6480	.6950	.1235	.1276
re0	.4763	.5103	.6447	.6735	.6616	.7035	.6092	.6745	.6815	.7018	.2071	.2247
re1	.3785	.4001	.5414	.5688	.6392	.6406	.5726	.6017	.5911	.6337	.1616	.1696
tr11	.3404	.3659	.7350	.7440	.7505	.7838	.7069	.7307	.7293	.7838	.1697	.1703
tr12	.3280	.3514	.5743	.6102	.7211	.7444	.6433	.6502	.6759	.6805	.0832	.0847
tr23	.3105	.3162	.6252	.6397	.6848	.6961	.6374	.6789	.6783	.6961	.2411	.2598
Win	16		19		17		19		18		20	
Tie	5		2		4		2		3		1	
Loss	0		0		0		0		0		0	
t-test	2×10^{-6}		5×10^{-6}		2×10^{-7}		3×10^{-8}		1×10^{-7}		1×10^{-4}	

*: The "tensor" data are *linearized*, since existing classifiers can only handle feature vectors.

Table 4. Bias of each classifier. "*F. test*" represents the Friedman significance test, which compares classifiers by their rankings.

Data sets	Bat	NB	AODE	AODEsr	HNB	WAODE	K2	TAN	STL
Win	124	27	40	104	95	112	25	35	37
Tie	40	13	18	50	39	43	39	35	37
Loss	4	128	110	14	34	13	104	98	94
F. test	Base	4×10^{-6}	4×10^{-6}	0.016	2×10^{-4}	2×10^{-4}	3×10^{-5}	4×10^{-6}	4×10^{-6}

repository[6]. We use the frequent words that are originally extracted by Han *et al.* [21] to construct tensors. Details of the document data sets are listed in Table 2.

4.2   Effectiveness of Tensor Formulation

We first test the effectiveness of using tensors to represent data instances, with comparisons to the traditional way of using feature vectors. To inspect the influence of tensors, we compare the classification accuracy of different representations using the same classifier on the same data set. Since most of the existing classifiers can only handle data instances by using feature vectors, the data in tensor formats are linearized into one mode before building classifiers on them.

---

[6] http://trec.nist.gov

Table 5. Variance of each classifier

Data sets	Variance of each learner's classification performance								
	Bat	NB	AODE	AODEsr	HNB	WAODE	K2	TAN	STL
Win	83	61	80	108	71	112	43	10	29
Tie	42	30	46	33	39	35	43	12	38
Loss	43	77	42	27	58	21	82	146	101
F. test	Base	0.002	1	0.016	0.827	0.049	0.016	5×10^{-5}	2×10^{-4}

Table 6. Average AUC-PR of each classifier

Data sets	AUC-PR of each learner's classification performance								
	Bat	NB	AODE	AODEsr	HNB	WAODE	K2	TAN	STL
Win	148	27	49	125	101	129	39	38	46
Tie	9	8	10	20	13	14	11	11	12
Loss	11	133	109	23	54	25	118	119	110
F. test	Base	4×10^{-6}	4×10^{-6}	0.016	2×10^{-4}	0.016	4×10^{-6}	4×10^{-6}	2.1×10^{-4}

Table 7. Comparisons between **Bat** and other classifiers that are either not Bayesian based or not generative learners. "LogReg" is short for logistic regression.

Data sets	AUC-PR of each learner's classification performance						
	Bat	LogReg	RBF	Sigmoid	Soft	C4.5	Forest
Win	124	53	5	30	68	24	85
Tie	1	25	6	19	23	12	16
Loss	1	48	115	77	35	90	25
F. test	Base	8×10^{-4}	1×10^{-8}	5×10^{-5}	6×10^{-4}	1×10^{-6}	0.016

As some of the data sets (i.e., the document data) have multiple labels, we use average precision (AvgPrec) as the evaluation metric for a class variable. AvgPrec evaluates the ranking performance of queried objects, which is also geometrically referred to as the area under the precision-recall curve (AUC-PR) [22]. Since AvgPrec only evaluates the performance of rankings of one class label, we use the mean of the AvgPrec of all class labels to examine the performance of a classifier, which is equivalent to the mean of AUC-PR of all class labels. For clarity, in the remainder of the paper we use AUC-PR to denote the mean of the AvgPrec of all class labels in the classification tasks.

The classifiers' performance on data in vector formats and tensor formats are shown in Table 3. For each classifier, we compare the list of AUC-PR values on all data sets between their feature vector representations and their tensor representations. Demšar et al. [23] have reported that t-tests are appropriate to compare pairs of classifiers. Hence we perform t-tests between vector and tensor data formats on each classifier, under the null hypothesis that the AUC-PR on vector and tensor formats are not significantly different. As shown in the bottom of Table 3, the p-values are all extremely small for each classifier. This suggests that the rich information contained in the tensorial format has significantly improved the performance of all classifiers.

4.3   Effectiveness of Bat Tensor Learning

In the evaluation of our Bat method, besides AUC-PR values of each classifier
we also look at the biases and variances[7] that contribute to the errors of classification on each data set. Comparisons between Bat and other classifiers, in
terms of bias, variance and accuracy, are respectively shown in Tables 4, 5 and
6. Classifiers that have to take data instances by their feature vectors are trained
on linearized tensors. Due to page limits, we only present statistics concluded
from the comparisons in Tables 4 to 7.

In contrast to previous subsections where pairs of classifiers (i.e., trained on
data of vector and tensor formats respectively) are compared, in this experiment we compare multiple classifiers all together with multiple data sets. In
such multiple classifier comparisons, Demšar et al. [23] have reported that the
most appropriate measure is to perform Friedman tests on the rankings of the
classifiers. So we rank the classifiers on each data set by their bias, variance
and AUC-PR values, and conduct Friedman tests under the null hypothesis that
their rankings are not significantly different. p-values from these tests that are
lower than 0.05 reject the null hypothesis with 95% confidence. In addition, we
also perform t-tests in each data set separately to summarize the number of
wins, ties, and losses of each classifier, under the 95% significance level.

As we can observe from Tables 4 and 5, the bias of Bat is almost always the
lowest among all classifiers, while its variance is slightly higher than those of
AODEsr and WAODE. This phenomenon confirms our preceding analysis that
Bat reduces the bias by taking into account the interactions among features of
different dimensions. It also shows that the introduction of feature interactions
increases the dimensionality of the learner, which usually comes at the cost of
increased variance. However as shown in the bottom of Table 5, the increased
variance of Bat is *not significantly* different to that of AODE and HNB. Note
that in these tables, the comparisons are performed for multiple classifiers on
21 datasets. So when we have 8 alternative classifiers in the evaluation, each
classifier will need to be compared 21 × 8 times (and hence the total count of
win/tie/loss is 21 × 8). In other words, the win/tie/loss states how often the
classifier in that column scores better/neutral/worse than classifiers in any other
columns.

The figures in Table 6 show that the overall errors of Bat, which are affected
by both its bias and variance, are significantly better than all other methods
in the comparison. We can also see that NB generates the worst results, which
indicates that the *naive* conditional independence assumption made in NB is
detrimental to the learning process when the data contains richer information
in tensor formats than vector formats. It is also noticeable that AODEsr and
HNB both perform better than AODE, K2 and TAN, which suggests that the
"feature elimination" strategy used in AODEsr and the "hidden variable" used
in HNB are more beneficial to tensorial data than using averaged dependence
estimators or using Bayesian networks. However, the constraint that AODEsr

---

[7] We use the same bias and variance estimation method as in [18].

and HNB require feature vectors as learning inputs ignores the relations among different modes of features, and hence limits their performance in tensorial data.

In addition, to validate the advantages of Bat in learning from tensorial data, we also conduct comparisons with other classifiers that are not dependent on the Bayesian setting, or are not generative classifiers. These comparisons include logistic regression, SVMs with the radial basis function kernel (RBF), Sigmoid kernel, as well as the linear soft margin kernel. We also include decision tree based methods, such as C4.5 [24] and random forests. It is important to note that, like other existing classifiers, these non-generative or non-Bayesian models can only handle vector features (instead of tensors) and hence the experiments on these models are done on linearized tensors. The comparison results are shown in Fig. 7. It is easy to see that Bat is able to statistically outperform all of the other methods in the comparisons.

5   Conclusion and Future Work

In this research we propose to formulate data observations by using tensorial formats, which capture more information than traditional feature vector representations. To effectively learn from the tensorial data, we designed a novel semi-naive Bayesian tensor learner Bat, which builds classifiers directly on data of tensors without linearizing them into vectors. Bat uses feature dependence by learning the interactions of features among different modes of the training data. This gives it the advantage that it can fully utilize the rich information contained in tensorial data, which leads to much higher classification accuracy compared to existing Bayesian methods. We evaluate Bat using data of text documents and chemical compound graphs, whose classification results confirm the advantage of using tensor formats to represent observations and the superiority of Bat in learning tensorial data.

In the future, we plan to apply Bat to other domains such as image classification and video semantic analysis. It is also interesting to examine the performance of Bat when data are represented by tensors in three or even higher modes.

References

1. Han, J., Kamber, M., Pei, J.: Data Mining: Concepts and Techniques, 3rd edn. Morgan Kaufmann Publishers Inc. (2011)
2. Kong, X., Yu, P.: Semi-supervised feature selection for graph classification. In: Proceedings of the 16th ACM SIGKDD Conference on Knowledge Discovery and Data Mining (KDD), pp. 793–802 (2010)
3. Kong, X., Fan, W., Yu, P.: Dual active feature and sample selection for graph classification. In: Proceedings of the 17th ACM SIGKDD Conference on Knowledge Discovery and Data Mining (KDD), pp. 654–662 (2011)
4. Yan, X., Han, J.: Closegraph: mining closed frequent graph patterns. In: Proceedings of the 9th ACM SIGKDD Conference on Knowledge Discovery and Data Mining (KDD), pp. 286–295 (2003)

5. Fei, H., Huan, J.: Structure feature selection for graph classification. In: Proceedings of the 17th ACM Conference on Information and Knowledge Management (CIKM), pp. 991–1000 (2008)

6. Saigo, H., Nowozin, S., Kadowaki, T., Kudo, T., Tsuda, K.: gBoost: a mathematical programming approach to graph classification and regression. Machine Learning 75(1), 69–89 (2009)

7. Kolda, T., Bader, B.: Tensor decompositions and applications. SIAM Review 51(3), 455 (2009)

8. Liu, W., Kan, A., Chan, J., Bailey, J., Leckie, C., Pei, J., Kotagiri, R.: On compressing weighted time-evolving graphs. In: Proceedings of CIKM 2012, pp. 2319–2322 (2012)

9. Liu, W., Chan, J., Bailey, J., Leckie, C., Ramamohanarao, K.: Mining labelled tensors by discovering both their common and discriminative subspaces. In: Proceedings of SDM 2013 (2013)

10. Tao, D., Li, X., Wu, X., Hu, W., Maybank, S.: Supervised tensor learning. Knowledge and Information Systems 13(1), 1–42 (2007)

11. Webb, G., Boughton, J., Wang, Z.: Not so naive bayes: Aggregating one-dependence estimators. Machine Learning 58(1), 5–24 (2005)

12. Jiang, L., Zhang, H.: Weightily averaged one-dependence estimators. In: Yang, Q., Webb, G. (eds.) PRICAI 2006. LNCS (LNAI), vol. 4099, pp. 970–974. Springer, Heidelberg (2006)

13. Jiang, L., Zhang, H., Cai, Z.: A novel Bayes model: Hidden naive Bayes. IEEE Transaction on Knowledge and Data Engineering 21(10), 1361–1371 (2009)

14. Cooper, G.F., Herskovits, E.: A Bayesian method for the induction of probabilistic networks from data. Machine Learning 9(4), 309–347 (1992)

15. Friedman, N., Geiger, D., Goldszmidt, M.: Bayesian network classifiers. Machine Learning 29(2-3), 131–163 (1997)

16. Kononenko, I.: Semi-naive bayesian classifier. In: Kodratoff, Y. (ed.) EWSL 1991. LNCS, vol. 482, pp. 206–219. Springer, Heidelberg (1991)

17. Zheng, F., Webb, G., Suraweera, P., Zhu, L.: Subsumption resolution: an efficient and effective technique for semi-naive bayesian learning. Machine Learning 87(1), 93–125 (2012)

18. Webb, G.: Multiboosting: A technique for combining boosting and wagging. Machine Learning 40(2), 159–196 (2000)

19. Hall, M., Frank, E., Holmes, G., Pfahringer, B., Reutemann, P., Witten, I.: The WEKA data mining software: an update. SIGKDD Explorations Newsletter 11(1), 10–18 (2009)

20. Hersh, W., Buckley, C., Leone, T., Hickam, D.: Ohsumed: an interactive retrieval evaluation and new large test collection for research. In: Proceedings of SIGIR, pp. 192–201 (1994)

21. Han, E.-H.(S.), Karypis, G.: Centroid-based document classification: Analysis and experimental results. In: Zighed, D.A., Komorowski, J., Żytkow, J.M. (eds.) PKDD 2000. LNCS (LNAI), vol. 1910, pp. 424–431. Springer, Heidelberg (2000)

22. Davis, J., Goadrich, M.: The relationship between precision-recall and ROC curves. In: Proceedings of the 23rd International Conference on Machine Learning (ICML), pp. 233–240 (2006)

23. Demšar, J.: Statistical comparisons of classifiers over multiple data sets. Journal of Machine Learning Research 7, 1–30 (2006)

24. Quinlan, J.R.: C4.5: programs for machine learning, vol. 1. Morgan kaufmann (1993)

Prediction with Model-Based Neutrality

Kazuto Fukuchi[1], Jun Sakuma[1], and Toshihiro Kamishima[2]

[1] University of Tsukuba, 1-1-1 Tennodai, Tsukuba, Ibaraki, 305-8577 Japan
kazuto@mdl.cs.tsukuba.ac.jp, jun@cs.tsukuba.ac.jp
[2] National Institute of Advanced Industrial Science and Technology (AIST),
AIST Tsukuba Central 2, Umezono 1-1-1, Tsukuba, Ibaraki, 305-8568 Japan
mail@kamishima.net

Abstract. With recent developments in machine learning technology, the resulting predictions can now have a significant impact on the lives and activities of individuals. In some cases, predictions made by machine learning can result unexpectedly in unfair treatments to individuals. For example, if the results are highly dependent on personal attributes, such as gender or ethnicity, hiring decisions might be deemed discriminatory. This paper investigates the neutralization of a probabilistic model with respect to another probabilistic model, referred to as a viewpoint. We present a novel definition of neutrality for probabilistic models, η-neutrality, and introduce a systematic method that uses the maximum likelihood estimation to enforce the neutrality of a prediction model. Our method can be applied to various machine learning algorithms, as demonstrated by η-neutral logistic regression and η-neutral linear regression.

Keywords: neutrality, fairness, discrimination, logistic regression, linear regression, classification, regression, social responsibility.

1 Introduction

With recent developments in machine learning technology, the resulting predictions can now have a significant impact on the lives and activities of individuals. In some cases, there are safeguards in place so that the predictions do not cause unfair treatment, discrimination, or biased views of individuals [1]. The following two examples describe situations in which predictions made by machine learning can cause unfair treatments.

Example 1 (hiring decision). A company collects personal information from employees and job applicants; this information includes age, gender, race or ethnicity, place of residence, and work experience. The company uses machine learning to predict the work performance of the applicants, using information collected from employees. The hiring decision is then based on this prediction.

Example 2 (personalized advertisement and recommendation). A company that provides web services records user behavior, including usage history

H. Blockeel et al. (Eds.): ECML PKDD 2013, Part II, LNAI 8189, pp. 499–514, 2013.
© Springer-Verlag Berlin Heidelberg 2013

Table 1. Summary of learning algorithms with neutrality guarantee

method	neutrality guarantee	domain of target	domain of viewpoint	model of viewpoint
elimination of viewpoint variable	no guarantee	any	any	×
CV2NB [2]	CV Score	multiple	multiple	×
PR [8]	mutual information	any	multiple	×
Lipschitz property [4]	statistical parity	multiple	multiple	×
η-neutral logistic regression (proposal)	η-neutrality	multiple	multiple	√
η-neutral linear regression (proposal)	η-neutrality	continuous	continuous	√

and search logs, and uses machine learning to predict user attributes and preferences. The advertisements or recommendations displayed on web pages are thus personalized so that they match the predicted user attributes and preferences.

In the hiring-decision example, if the results are highly dependent on personal attributes, such as gender or ethnicity, hiring decisions might be deemed discriminatory. In the second example, when recommendations are accurately pinpointed to sensitive issues, such as political or religious affiliation, the result may be increasingly biased views. This is known as the problem of the filter bubble [10]. For example, suppose supporters of the Democratic Party wish to read news articles related to politics. If the recommended articles are all related to their party and are absent of criticism, they may develop a biased view of the political situation. In the web-service example, showing advertisements that suit the user's attributes, such as gender or age, would improve the service for some users. Other users, however, may object to advertisements that are apparently based on their race, ethnicity, or gender. Thus, it is difficult to clearly distinguish personalization from discrimination.

We now introduce some terms that will be useful in the following discussion. The input and output of a prediction model are referred to as *input* variables (e.g., race, ethnicity, or web-usage history) and *target* variables (hiring decisions or website recommendations). Factors that might result in discrimination or bias are referred to as *viewpoint* variables (e.g., race, ethnicity, or political affiliation).

The objective of machine learning is to learn prediction functions that predict target variables from given examples. In the example above, if the viewpoint variables (e.g., race or ethnicity) are dependent on the predicted target variables (e.g., hiring decisions), the prediction function cause unfair treatment. In this paper, we introduce a systematic way to remove this dependency from prediction models and neutralize them with respect to a given viewpoint.

1.1 Related Works

Several techniques that take account of fairness or discrimination have recently received attention [4][6][12]. One of the easiest ways to suppress unfair treatment is to remove the values of the viewpoint from the input values before the learning process with the prediction model. If there is no correlation between the input and viewpoint variables, no discrimination or bias will appear after elimination. However, if another input variable is dependent on the viewpoint variable, then even after the viewpoint values are eliminated, the target variable will retain dependency on the viewpoint variable (Table 1, line 1). For example, assume that the race or ethnicity attribute is eliminated in Example 1. Even so, hiring decisions may be dependent on race or ethnicity if there is a correlation between individuals' addresses and their race or ethnicity; this is known as the redlining effect [2][11].

Calders et al. presented the *Calders–Verwer 2 Naive Bayes method* (CV2NB), which proactively removes the redlining effect [2]. Let $y \in \{y_+, y_-\}$ be the binary target variable, and let $v \in \{v_+, v_-\}$ be the binary viewpoint variable. Then, the Calders–Verwer (CV) score is defined by $CV(\mathcal{D}) = p(y_+|v_+) - p(y_+|v_-)$. The CV2NB modifies the naive Bayes classifier in such a way that the CV score becomes zero with respect to the given examples $\mathcal{D}$. The CV2NB guarantees the elimination of discrimination in terms of the CV score. The limitation of the CV2NB is that it cannot be used when the target or viewpoint variables are continuous. Related to the CV2NB, it has been shown [14] that positive CV scores do not necessarily cause discrimination in some situations. There is also a method [9] that uses the kth-nearest neighbor to test for the existence of discrimination. Both these methods are based on the CV2NB, so they share its limitations.

Kamishima et al. have introduced the *prejudice remover regularizer* (PR) for fairness-aware classification [8]. The PR regularizer penalizes the loss function if there is a high correlation between the target variable and the viewpoint variable. The penalty is evaluated based on the information that is shared by the target variable y and the viewpoint variable v. This penalty function can work with a continuous target variable if it is approximated by a histogram, as demonstrated by Kamishima et al. [7]. Continuous viewpoint variables, however, cannot be treated by the PR method.

Dwork et al. have presented a classification method that uses a fairness-aware framework, in which statistical parity is used as the measure of fairness [4]. Intuitively, statistical parity occurs when the demographics of those receiving positive (or negative) classifications are identical to the demographics of the population as a whole. In their fairness-aware framework, the classification is made to be fair by minimizing the empirical risk while satisfying certain constraints that are called the Lipschitz property. As is the case with the CV2NB and PR methods, this framework assumes that the viewpoint variables are binary or multiple; continuous viewpoint variables are not considered.

1.2   Our Contribution

Modeling Viewpoint Variables. In this manuscript, we provide a method to neutralize the target prediction model with respect to a probabilistic model of a given viewpoint. Existing methods assume the viewpoint is observed and is explicitly provided in the input, but this is not always the case. For instance, consider the recommendation of articles neutralized with respect to political affiliation, as in Example 2. Political affiliation is not explicitly included in the input, but given as input the logs of keyword searches or subscribed news articles, modern machine learning techniques can easily predict party affiliation. In such a case, our method neutralizes the target prediction model with respect to the probability model of such a "hidden viewpoint".

In order to neutralize a model with respect to a viewpoint, we represent the viewpoint as a probabilistic model and define η-neutrality (Section 2), which is a measure of the dependency of the target prediction model on the viewpoint prediction model. With η-neutrality, we can check the neutrality of a target prediction model with respect to any hidden viewpoint, as long as we have a probabilistic model of the viewpoint variable (Table 1, the rightmost column). Furthermore, since η-neutrality is measured with respect to probabilistic models, the neutrality of the prediction model with respect to unseen examples is expected to be effectively guaranteed, and this is demonstrated by experiments (Section 5).

Maximum Likelihood Estimation with η-Neutrality. Following the definition of η-neutrality, we introduce a systematic method that removes this dependency from the prediction model obtained by the maximum likelihood estimation (Section 2). Our methods can treat target and viewpoint variables that are either discrete (Table 1, line 5) or continuous (Table 1, line 6), as demonstrated by η-neutrality with logistic regression (Section 3) and linear regression (Section 4). The effectiveness of our methods is examined by both artificial and real datasets in Section 5.

2 η-Neutrality

We propose a novel definition of neutrality, η-neutrality. We then present a general maximum likelihood estimation method that has a guarantee of neutrality.

Let $\mathcal{D} = \{(x_i, y_i) \in \mathcal{X} \times \mathcal{Y}\}_{i=1}^{N}$ be a set of training examples that are assumed to be i.i.d. samples drawn from a probability distribution $\Pr(X, Y)$. The random variables X and Y are referred to as the input and target, respectively. In the following discussion, the prediction function of the target variable is represented as a probabilistic model $f(Y|X; \theta) = \Pr(Y|X)$, parametrized by θ. The target prediction model can be obtained by minimization of the negative log-likelihood with respect to the parameter θ:

$$\theta^* = \operatorname{argmin}_{\theta \in \Theta} L(\theta),$$

where

$$L(\theta) = - \sum_{(x_i, y_i) \in \mathcal{D}} \ln f(y_i | x_i; \theta). \tag{1}$$

2.1   Definition of η-Neutrality

In addition to the input random variable X and the target random variable Y, we now introduce the viewpoint random variable V. Let $\mathcal{V}$ be the domain of V. The realized values of the variables are denoted by the corresponding lowercase letters. Thus, the random variable X can take the value x. In the following discussion, we assume the input random variable X is continuous. We can treat a discrete X by replacing the integral with a sum. For Y and V, the discussion below is valid for both discrete and continuous variables.

As we did for the target random variable, we assume that the prediction model of the viewpoint variable is represented as a conditional probability $\Pr(V|X)$. Noting that the values of the target and the viewpoint variables are predicted independently, the joint probability is

$$\Pr(X, Y, V) = \Pr(X)\Pr(Y|X)\Pr(V|X).$$

With this assumption, we consider the dependency of the target random variable Y and the viewpoint random variable V. When V and Y are statistically independent, for any $y \in \mathcal{Y}$ and $v \in \mathcal{V}$, $\Pr(v, y)/\Pr(v)\Pr(y) = 1$. When $\Pr(v, y)/\Pr(v)\Pr(y) > 1$, v and y are more dependent than independent. Hence, our neutrality definition is defined as the ratio of the marginal probabilities, as follows.

Definition 1 (η-neutrality). *Let X and Y be the input and target random variables, respectively. Let V denote the viewpoint random variable. Given $\eta \geq 0$, the probability distribution $\Pr(X, Y, V)$ is η-neutral if*

$$\forall v \in \mathcal{V}, y \in \mathcal{Y}, \quad \frac{\Pr(v, y)}{\Pr(v)\Pr(y)} \leq 1 + \eta. \tag{2}$$

Noting $\sum_{y \in \mathcal{Y}, v \in \mathcal{V}} \Pr(y, v) = 1$ holds, the dependency represented by $\Pr(v, y)/\Pr(v)\Pr(y) < 1$ is no need to consider.

Next, given the probabilistic models of $\Pr(Y|X)$ and $\Pr(V|X)$, we derive conditions that the model of the joint probability distribution satisfies η-neutrality. The target and the viewpoint prediction models are described by the probability distributions $f(Y|X; \theta) = \Pr(Y|X)$ and $g(V|X; \phi) = \Pr(V|X)$, respectively, where θ and ϕ are the model parameters.

Thus, given the target prediction model $f(Y|X; \theta)$ and the viewpoint prediction model $g(V|X; \phi)$, the probabilistic model of $\Pr(X, Y, V)$ becomes

$$M(X, Y, V; \theta, \phi) = \Pr(X)f(Y|X; \theta)g(V|X; \phi). \tag{3}$$

In what follows, we assume the viewpoint prediction model is fixed, and so the model parameter ϕ is omitted and g is described by $g(V|X)$. The following theorem shows the condition that the model of Eq. 3 is empirically η-neutral.

Theorem 1. *Suppose the joint probability distribution of input X, target Y, and viewpoint V follows the model $M(X, Y, V; \theta) = \Pr(X)f(Y|X; \theta)g(V|X)$. Then M is η-neutral if $\forall v \in \mathcal{V}, y \in \mathcal{Y}$,*

$$\int_x \Pr(x)f(y|x;\theta)\left[g(v|x) - (1+\eta)\bar{g}(v)\right]dx \leq 0, \tag{4}$$

where $\bar{g}(v) = \int_x \Pr(x)g(v|x)dx$.

Proof. By the marginalization of $\Pr(y, v)$, $\Pr(y)$, and $\Pr(v)$, we have

$$\Pr(y, v) = \int_x \Pr(x, y, v)dx = \int_x \Pr(x)f(y|x;\theta)g(v|x)dx,$$

$$\Pr(y) = \int_x \int_v \Pr(x, y, v)dvdx = \int_x \Pr(x)f(y|x;\theta)dx,$$

$$\Pr(v) = \int_x \int_y \Pr(x, y, v)dydx = \int_x \Pr(x)g(v|x)dx = \bar{g}(v).$$

By substituting the above equations into Eq. 2, we have

$$\forall v, y, \int_x \Pr(x)f(y|x;\theta)g(v|x)dx - (1+\eta)\bar{g}(v)\int_x \Pr(x)f(y|x;\theta)dx \leq 0,$$

$$\forall v, y, \int_x \Pr(x)f(y|x;\theta)\left[g(v|x) - (1+\eta)\bar{g}(v)\right]dx \leq 0.$$

Thus, M is η-neutral if Eq. 4 holds.

2.2   Approximation of η-Neutrality

When $\Pr(x)$ cannot be obtained, η-neutrality can be empirically evaluated with respect to the frequency distribution $\tilde{\Pr}(x)$ of the examples $\mathcal{D}$. The neutrality condition with respect to this frequency distribution is derived in a similar manner, as follows. Given examples $\mathcal{D}$, we approximate η-neutrality with respect to the frequency distribution

$$\tilde{\Pr}(X = x) = \frac{1}{N}\sum_{i=1}^N I(x_i = x),$$

where $I(\cdot)$ denotes the indicator function. From this, we have

$$\tilde{\Pr}(X, Y, V) = \tilde{\Pr}(X)\Pr(Y|X)\Pr(V|X),$$

and an approximation of η-neutrality is defined by this $\tilde{\Pr}(X, Y, V)$.

Definition 2 (Empirical η-neutrality). *Let X and Y be the input and target random variables, respectively. Let V denote the viewpoint random variable. Let $\tilde{\Pr}(X)$ be the frequency distribution of X obtained from $\mathcal{D}$. Given $\eta \geq 0$, if $\tilde{\Pr}(X, Y, V)$ is η-neutral, $\Pr(X, Y, V)$ is said to be empirically η-neutral with respect to the dataset $\mathcal{D}$.*

The following theorem shows the condition that the model of Eq. 3 is η-neutral with respect to the given examples.

Theorem 2. *Suppose the joint probability distribution of the input X, target Y, and viewpoint V follows the model $M(X, Y, V; \theta) = \Pr(X)f(Y|X; \theta)g(V|X)$. Then, given $\mathcal{D} = \{(x_i, y_i)\}_{i=1}^{N}$, M is empirically η-neutral if*

$$\forall y, v, \sum_{i=1}^{N} f(y|x_i; \theta) \left[g(v|x_i) - (1 + \eta)\tilde{g}(v) \right] \leq 0, \tag{5}$$

where $\tilde{g}(v) = \frac{1}{N} \sum_{i=1}^{N} g(v|x_i)$.

Proof. Theorem 1 states that $\Pr(X, Y, V)$ is η-neutral if Eq. 4 holds. By substituting $\tilde{\Pr}(X)$ into Eq. 4, the neutrality condition is rewritten as

$$\forall y, v, \frac{1}{N} \sum_{i=1}^{N} f(y|x_i) \left[g(v|x_i) - (1 + \eta)\tilde{g}(v) \right] \leq 0.$$

Thus, M is empirically η-neutral if Eq. 5 holds.

For convenience in the following discussion, the neutrality condition is notated as

$$N(y, v) = \sum_{i=1}^{N} f(y|x_i) \left[g(v|x_i) - (1 + \eta)\tilde{g}(v) \right] \leq 0. \tag{6}$$

2.3   Maximum Likelihood Estimation with η-Neutrality

Given examples and a viewpoint prediction model, we performed maximum likelihood estimations with the guarantee of η-neutrality. We wanted a target prediction model that would achieve the maximum log-likelihood with respect to the given data. At the same time, we wanted a target prediction function that would make $\Pr(X, Y, V)$ empirically η-neutral with respect to the given data and viewpoint prediction model. This problem is the following constrained optimization problem:

$$\text{minimize } L(\theta) \quad \text{subject to } N(y, v; \theta) \leq 0, \quad \forall y, v.$$

Existing neutrality indexes measure neutrality with certain statistics, such as differences in the conditional probabilities [2] or mutual information [8]. If such measures are used to guarantee neutrality, the neutrality of the model is statistically guaranteed for the set of given examples. In principle, it is desirable to guarantee neutrality with respect to each individual contained in the given examples. However, such prediction functions tend to overfit to the given examples and do not provide neutrality of unseen examples.

Assuming the model of the viewpoint correctly represents the true distribution, a model that satisfies our η-neutrality condition guarantees statistical independency between every combination of target value y and viewpoint value v. Note that η-neutrality can be realized even when the viewpoint values are not contained in the given examples because the neutrality is evaluated with respect to each combination of possible target value y and viewpoint value v.

2.4   Prediction Model for Viewpoints

In principle, we assume $g(V|X)$ accurately represents the true probabilistic distribution $\Pr(V|X)$, but in reality, this does not always hold. In this subsection, we consider three types of possible viewpoint models.

The first case assumes an extreme example; model $g(V|X)$ is the probabilistic model that outputs random or constant values independent of input x. If we have no knowledge of the viewpoint, we have no choice other than this. Since $g(V|X)$ takes a constant value independent of X, η-neutrality is guaranteed for any $f(Y|X;\theta)$ in this model; however, such neutralization is meaningless.

The second case assumes that model $g(V|X)$ is taken as the empirical distribution of the training examples. Existing methods, including CV2NB, statistical parity, and PR, achieve neutralization with respect to this empirical distribution. This model realizes neutralization with respect to the given training examples, but neutralization with respect to unseen examples is not guaranteed.

The third case considers the situation that is our focus; model $g(V|X)$ is given as a parametrized probabilistic model. In this case, if $g(V|X)$ accurately represent the true distribution without overfitting, the output of the target prediction model is expected to be neutralized with respect not only to the training examples, but also to the unseen examples; this is demonstrated in the following sections by experiments.

The definition of η-neutrality contains all of the above cases, but we specifically consider only the third case, the parametric model.

2.5   Equivalence of η-Neutrality and Statistical Parity

In this subsection, in order to discuss the equivalence of η-neutrality and the statistical parity[4], we assume examples $\mathcal{D}$ contains the viewpoint values. The statistical parity defines the neutrality considering the difference of two probabilistic distribution of target y, $P(y)$ and $Q(y)$,

$$D_{tv}(P,Q) = \frac{1}{2}\sum_{y \in \mathcal{Y}} |P(y) - Q(y)|.$$

Given $\epsilon \geq 0$ as a neutrality parameter, the statistical parity is defined by

$$D_{tv}(\Pr(Y|v_+), \Pr(Y|v_-)) \leq \epsilon,$$

where $\Pr(Y|V)$ is empirically evaluated with the given example set $\mathcal{D}$.

If the empirical distribution of the example set $\mathcal{D}$ is used as the model of the viewpoint in the empirical η-neutrality, and letting the distance function of the statistical parity

$$D_\eta(P,Q) = \max_{y \in \mathcal{Y}} \frac{\max\{P(y), Q(y)\}}{\tilde{\Pr}(v_+)P(y) + \tilde{\Pr}(v_-)Q(y)},$$

the statistical parity with parameter η is equivalent to the η-neutrality. The proof of the equivalence will be presented in the journal version of this manuscript.

In the following two sections, we demonstrate two applications of maximum likelihood estimation with a guarantee of empirical η-neutrality: η-neutral logistic regression and η-neutral linear regression.

3 η-Neutral Logistic Regression

In this section, we incorporate our neutrality definition into logistic regression. In logistic regression, the domain of the input variable is $\mathcal{X} = \mathbb{R}^d$, and the domain of the target variable is binary, $\mathcal{Y} = \{0, 1\}$. Letting $\boldsymbol{\theta} \in \mathbb{R}^d$ be the model parameter, the target prediction model for logistic regression is

$$f(y|\boldsymbol{x}; \boldsymbol{\theta}) = \sigma(\boldsymbol{\theta}^T \boldsymbol{x})^y (1 - \sigma(\boldsymbol{\theta}^T \boldsymbol{x}))^{1-y}, \tag{7}$$

where $\sigma(a)$ is the logistic sigmoid function.

Letting Eq. 7 be the target prediction model, the log-likelihood is given by Eq. 1, and then the problem of η-neutral logistic regression is

$$\text{minimize } L(\boldsymbol{\theta}) \quad \text{subject to } N(y, v; \boldsymbol{\theta}) \leq 0, \quad \forall v, y.$$

Note that the viewpoint prediction model $g(v|\boldsymbol{x})$ can be any probabilistic model.

We consider the optimization of η-neutral logistic regression. The gradient and Hessian matrix of $L(\boldsymbol{\theta})$ with respect to $\boldsymbol{\theta}$ are, respectively,

$$\nabla L(\boldsymbol{\theta}) = \sum_{i=1}^{N} \left(\sigma(\boldsymbol{\theta}^T \boldsymbol{x}_i) - y_i \right) \boldsymbol{x}_i,$$

$$\nabla^2 L(\boldsymbol{\theta}) = \sum_{i=1}^{N} \sigma(\boldsymbol{\theta}^T \boldsymbol{x}_i)(1 - \sigma(\boldsymbol{\theta}^T \boldsymbol{x}_i)) \boldsymbol{x}_i \boldsymbol{x}_i^T.$$

Due to the nature of the logistic sigmoid function, the Hessian matrix is positive semidefinite. Hence, the log-likelihood function is convex.

Next, we examine the convexity of the constraints associated with the η-neutrality condition. Since $N(y, v; \boldsymbol{\theta})$ is a linear combination of f, the convexity of f is investigated. The gradient of f with respect to the parameter $\boldsymbol{\theta}$ is

$$\nabla f(y, \boldsymbol{x}; \boldsymbol{\theta}) = \nabla \exp\left(\ln f(y|\boldsymbol{x}; \boldsymbol{\theta}) \right) = \left(y - \sigma(\boldsymbol{\theta}^T \boldsymbol{x}) \right) f(y|\boldsymbol{x}; \boldsymbol{\theta}) \boldsymbol{x}.$$

The Hessian is similarly obtained as

$$\nabla^2 f(y|\boldsymbol{x}; \boldsymbol{\theta}) = \alpha(\boldsymbol{x}, y, \boldsymbol{\theta}) f(y|\boldsymbol{x}; \boldsymbol{\theta}) \boldsymbol{x} \boldsymbol{x}^T,$$

where $\alpha(\boldsymbol{x}, y, \boldsymbol{\theta}) = 2\sigma(\boldsymbol{\theta}^T \boldsymbol{x})^2 + y^2 - (2y + 1)\sigma(\boldsymbol{\theta}^T \boldsymbol{x})$.

Since $\alpha(\boldsymbol{x}, y, \boldsymbol{\theta}) \in \mathbb{R}$ can be negative, the Hessian is not positive definite, and f is nonconvex with respect to $\boldsymbol{\theta}$. Thus, unfortunately, the neutrality condition in logistic regression is nonconvex, regardless of the choice of $g(v|\boldsymbol{x})$.

In our experiments with η-neutral logistic regression, we used Shor's r-algorithm based on adaptive space dilation [13]. Shor's r-algorithm can be initialized with any solution. We set the initial solution to the result of the logistic regression without the neutrality constraint. Although the constraint is nonconvex, in Section 5 we show by experiment η-neutrality can be achieved without sacrificing too much of the accuracy of the prediction. This nonconvexity arises in part from the nonconvexity of the probability distribution. Further research on convexifying the neutrality constraint is left as an area of future work.

4 η-Neutral Linear Regression

We now consider η-neutral linear regression and demonstrate that maximum likelihood estimation with η-neutrality can work with continuous viewpoint variables. In linear regression, the domain of the target variable is $\mathcal{Y} = \mathbb{R}$, and the input domain is $\mathcal{X} = \mathbb{R}^d$. The target prediction function is given by

$$f(y|\boldsymbol{x};\boldsymbol{w},\beta) = \sqrt{\tfrac{\beta}{2\pi}} \exp\left[-\tfrac{\beta(\boldsymbol{w}^T\boldsymbol{x}-y)^2}{2}\right].$$

The linear regression problem is solved by the minimization of the negative log-likelihood, as given by Eq. 1.

The domain of the viewpoint is $\mathcal{V} = \mathbb{R}$. Similarly, we assume the viewpoint prediction model is

$$g(v|\boldsymbol{x};\boldsymbol{w}_v,\beta_v) = \sqrt{\tfrac{\beta_v}{2\pi}} \exp\left[-\tfrac{\beta(\boldsymbol{w}_v^T\boldsymbol{x}-v)^2}{2}\right].$$

Predictions of the target random variable Y and the viewpoint random variable V are obtained, respectively, by

$$\hat{y} = \operatorname*{argmax}_{y} f(y|\boldsymbol{x};\boldsymbol{w},\beta), \quad \hat{v} = \operatorname*{argmax}_{v} g(v|\boldsymbol{x};\boldsymbol{w}_v,\beta_v).$$

Then, η-neutral linear regression is formulated as an optimization problem with the same constraints as in Eq. 6:

$$\text{minimize } \tfrac{1}{2}\boldsymbol{w}^T\boldsymbol{X}^T\boldsymbol{X}\boldsymbol{w} - \boldsymbol{y}^T\boldsymbol{X}\boldsymbol{w} \text{ subject to } \max_{\boldsymbol{x}\in\mathcal{D}}\{N(\boldsymbol{w}^T\boldsymbol{x},\boldsymbol{w}_v^T\boldsymbol{x};\boldsymbol{w},\beta)\} \leq 0,$$

where $\boldsymbol{X} = (\boldsymbol{x}_1^T, \boldsymbol{x}_2^T, ..., \boldsymbol{x}_N^T)^T$ is the matrix of input vectors and $\boldsymbol{y} = (y_1, y_2, ..., y_N)^T$ is the vector of target values.

As in the case with η-neutral logistic regression, we investigate the convexity of the neutrality constraint given models f and g by investigating the convexity of f. The gradient and Hessian matrix of f are, respectively,

$$\nabla_{\boldsymbol{w}} f(y|\boldsymbol{x};\boldsymbol{w},\beta) = \nabla_{\boldsymbol{w}} \exp(-\ln f(y|\boldsymbol{x};\boldsymbol{w},\beta)) = -\beta(\boldsymbol{w}^T\boldsymbol{x}-y)f(y|\boldsymbol{x};\boldsymbol{w},\beta)\boldsymbol{x},$$

$$\nabla_{\boldsymbol{w}}^2 f(y|\boldsymbol{x};\boldsymbol{w},\beta) = \alpha(\boldsymbol{x},y,\boldsymbol{w},\beta)\beta f(y|\boldsymbol{x};\boldsymbol{w},\beta)\boldsymbol{x}\boldsymbol{x}^T,$$

where $\alpha(\boldsymbol{x},y,\boldsymbol{w},\beta) = \beta(\boldsymbol{w}^T\boldsymbol{x}-y)^2 - 1$.

Since, depending on $\boldsymbol{w}$, $f(y|\boldsymbol{x};\boldsymbol{w},\beta) \geq 0$ and $\alpha(\boldsymbol{x},y,\boldsymbol{w},\beta) \in \mathbb{R}$ can take negative values, the Hessian is not positive definite. Hence, unfortunately, f is not convex with respect to $\boldsymbol{w}$. For this nonlinear constraint optimization, we again use Shor's r-algorithms for the experiments.

5   Experiments

5.1   Classification

Settings. In order to examine and compare the classification performance and the neutralization effect of the η-neutral logistic regression with other methods, we performed experiments on two real data sets, Adult [5] and the Dutch Census [3]. Table 2 summarizes the specifications of each dataset. In both datasets, the target and viewpoint variables are set to "income (large/small)" and "gender (male/female)", respectively. Our method does not necessarily require that the viewpoint value (gender, in this case) be explicitly provided in the given dataset, but for comparison with other methods, it was chosen from the input variable of the dataset.

We compared the following methods: logistic regression (LR, no neutrality guarantee), logistic regression that learns without using the values of viewpoint (LRns), the Naive Bayes classifier (NB, no neutrality guarantee), the Naive Bayes classifier that learns without the values of viewpoint (NBns), CV2NB [2], logistic regression that uses the PR [7], and η-neutral logistic regression with viewpoint neutrality (VN, proposal).

In the PR method, the regularizer parameter λ, which balances the loss minimization and neutralization, was varied as $\lambda \in \{0, 5, 10, 15, 20, 30\}$. The neutrality parameter η, which determines the degree of neutrality, was varied as $\eta \in \{0.00, 0.01, ..., 0.40\}$

As neutrality indices of prediction models, the normalized prejudice index (NPI) and $\hat{\eta}$ are introduced. NPI is defined as the normalized mutual information of the target random variable Y and the viewpoint random variable V, normalized by the entropy of Y and V [8]:

$$\text{NPI} = \frac{I(X;Y)}{\sqrt{H(Y)H(V)}},$$

where $I(X;Y)$ is the mutual information of target Y and viewpoint V, $I(X;Y)/H(Y)$ is the ratio of information of V used for predicting Y, and $I(X;Y)/H(V)$ is the ratio of information that is exposed if a value of Y is known. Thus NPI can be interpreted as the geometrical mean of these two ratios. The range of this NPI is $[0, 1]$.

The neutrality measure $\hat{\eta}$ is defined as

$$\hat{\eta} = \max_{y \in \mathcal{Y}, v \in \mathcal{V}} \frac{\tilde{\Pr}(v, y)}{\tilde{\Pr}(v)\tilde{\Pr}(y)} - 1,$$

where $\hat{\eta}$ can be interpreted as the degree of the dependency of y and v with which the largest dependency occurs. If Y and V are mutually independent, $\hat{\eta} = 0$. If the neutrality measure with respect to a target prediction model is $\hat{\eta}$, it means the model of Eq. 3 is empirically $\hat{\eta}$-neutral with respect to the given examples.

We compared the three measures: the accuracy, the normalized prejudice index (NPI), and the $\hat{\eta}$ of η-neutrality. These indices was evaluated with five-fold cross validation.

Table 2. Specification of datasets. $\#y_+$ and $\#v_+$ represent the number of positive target and viewpoint values, respectively. The prediction accuracy (logistic regression) of the target $(\mathrm{Acc}(y)$ w.r.t. income) and viewpoint $(\mathrm{Acc}(v)$ w.r.t. gender) variables are also shown.

dataset	Adult	Dutch Census
#Instances	16281	60420
#Attributes	13	10
$\#y_+$	3846(23.6%)	31657(52.4%)
$\#v_+$	10860(66.7%)	30273(50.1%)
$\mathrm{Acc}(y)$	0.851	0.835
$\mathrm{Acc}(v)$	0.842	0.665

The values used for the learning of $f(y|x)$, the guarantee of neutrality, and the measurement of neutrality are summarized in Table 3. For the guarantee of neutrality, we consider the following two cases.

Case 1 assumes that the values of the viewpoint random variable are provided in examples. In this case, our method performs neutralization with respect to the model of the viewpoint learned from the examples, whereas other methods perform neutralization with respect to the actual viewpoint values provided.

Case 2 assumes that the values of the viewpoint are not provided. Instead, the model of the viewpoint variable, $g(v|x)$, is provided. In this case, our method again learns the model of the target without using values of the viewpoint and performs neutralization with respect to the given model g. Other methods need the values of the viewpoint, so these are estimated as $\hat{v} = \mathrm{argmax}_v g(v|x)$. Other methods then learn the model of the target with $(x, \hat{v})$, and neutralization is performed with respect to $\hat{v}$.

As a measurement of neutrality, all methods used the true viewpoint value v in both cases.

Results. Figure 1 shows the experimental results. In the graphs, the best result is at the left top. Comparing the results of NB and NBns in Case 1, we can see that the improvement of neutrality by elimination of the viewpoint variable is limited. The same applies to LR and LRns.

In Case 1, CV2NB achieves a neutrality of nearly 0 in terms of both NPI and $\hat{\eta}$ in both datasets. In addition, the decrease in the accuracy of the prediction is less than 1% in the Adult dataset and 5% in the Dutch Census. Thus, neutralization by CV2NB works successfully in Case 1. On the other hand, neutralization by CV2NB does not work well in Case 2; the neutralization level is almost the same as it is for NBns. CV2NB modifies the target prediction model so that the CV score with respect to the given examples becomes zero. This can cause the prediction model to overfit the given examples. Hence, the NPI and $\hat{\eta}$ of CV2NB with respect to the unseen values of the viewpoint are large, as seen in the results of Case 2.

In Case 1, PR successfully balances the NPI and the accuracy for the Adult dataset, but it fails to balance the accuracy and $\hat{\eta}$. This is because NPI evaluates

Table 3. Summary of the treatment of the viewpoint random variables in two settings

case	method	learning of $f(y\|x)$	neutrality guarantee	neutrality measure
Case 1	others	x, v	v	$\hat{y}, v$
	ours	x, v	$g(v\|x)$	$\hat{y}, v$
Case 2	others	$x, \hat{v}$	$\hat{v}$	$\hat{y}, v$
	ours	x	$g(v\|x)$	$\hat{y}, v$

neutrality with respect to the average of the given examples, while $\hat{\eta}$ evaluates the lowest neutrality for all values of y and v as the worst case. This result indicates that the dependency of the predictions of target Y to the predictions of viewpoint V can be strong for some y and v, even when the NPI is kept small. In Case 2, neutralization with PR did not work well in either dataset. This was again due to overfitting; this can be confirmed by the fact that the NPI of v and y is large in Case 2 even when the NPI of $\hat{v}$ and y is kept small (these results are omitted due to space limitations).

In both cases, our proposal, VN, successfully balances neutralization and accuracy of the predication by changing η. Furthermore, the decrease in the accuracy of the prediction was at most 5%, even after strong neutralization with small η. In some cases, the accuracy of VN becomes unstable with small η. The reason is thought to be the nonconvexity of the neutrality constraint. VN always guarantees neutrality of the prediction model, but the accuracy of the prediction can suddenly drop if the solution is captured by a local optimum.

5.2 Regression

Settings. In order to investigate the behaviors of neutralization in linear regression, we performed experiments of η-neutral linear regression with the Housing dataset [5]. This dataset contains 506 examples with 14 attributes; the MEDV (median value of owner-occupied homes, in \$1000s) and the LSTAT (% lower status of the population) were used as the target and viewpoint values, respectively. Letting the regression parameters of the target f and viewpoint g be w and w_v, respectively, the predicted values were $\hat{y} = w^T x$ and $\hat{v} = w_v^T x$. The accuracy of the prediction was measured by the root-mean-square error (RMSE), and $\hat{\eta}$ was used as the measure of neutrality.

Results. Figure 2 shows the scatter plots of $(\hat{y}, y)$ (the top row) and $(\hat{y}, \hat{v})$ (the bottom row). From left to right, the neutrality parameter η was varied as $\eta \in \{1.0, 3.0, 10.0\}$. The $(\hat{y}, \hat{v})$ plot represents the prediction accuracy of the regression model. When the model achieves a better RMSE, the points in the $(\hat{y}, y)$ plot concentrate more along the diagonal line. At the same time, the $(\hat{y}, \hat{v})$ plot represents neutrality. If the neutrality is low, any correlation between $\hat{y}$ and $\hat{v}$ appears in the $(\hat{y}, \hat{v})$ plot.

In Figure 2 (h), a strong negative correlation between $\hat{y}$ and $\hat{v}$ can be found. Thus, this regression model has a low neutrality if no neutralization is performed.

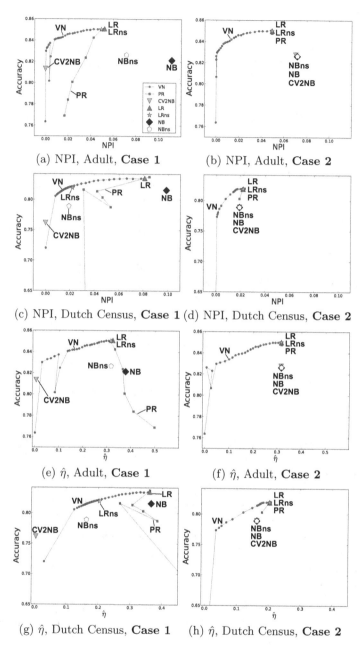

(a) NPI, Adult, **Case 1**　　(b) NPI, Adult, **Case 2**

(c) NPI, Dutch Census, **Case 1** (d) NPI, Dutch Census, **Case 2**

(e) $\hat{\eta}$, Adult, **Case 1**　　(f) $\hat{\eta}$, Adult, **Case 2**

(g) $\hat{\eta}$, Dutch Census, **Case 1**　(h) $\hat{\eta}$, Dutch Census, **Case 2**

Fig. 1. Accuracy vs. neutrality measure

In Figure 2, the level of neutralization increases from right to left. The plots show that the dependency of $\hat{y}$ on $\hat{v}$ becomes weaker as η decreases. This result indicates that our method can use η to successfully control the neutralization level of the regression model. The RMSE increases as η is decreases. In Figure 2 (e), we can

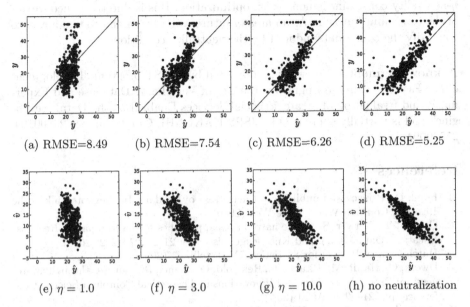

(a) RMSE=8.49   (b) RMSE=7.54   (c) RMSE=6.26   (d) RMSE=5.25

(e) $\eta = 1.0$ (f) $\eta = 3.0$ (g) $\eta = 10.0$ (h) no neutralization

Fig. 2. Top row: scatter plots of target prediction value $\hat{y}$ and true target value y. Bottom row: scatter plots of target prediction value $\hat{y}$ and viewpoint prediction value $\hat{v}$. Correlation in the $\hat{y} - \hat{v}$ plots means that the neutralization level of the regression model is low.

see that the regression model of the target value that has high neutrality outputs almost constant values; such regression is useless even if the model is well neutralized. Thus, tuning of η is important to obtain a neutralized regression model with high accuracy.

6   Conclusion

In this paper, we proposed a framework in which to use a maximum likelihood estimation for learning probabilistic models with neutralization. There are two key points in which our proposal is different from existing methods.

First, our method guarantees neutrality of the target prediction model with respect to a given viewpoint prediction model. Due to this model-based neutralization, our method allows neutralization of target prediction models with respect to viewpoints arbitrarily defined by users, as long as the viewpoint prediction model is provided in the form of a probabilistic distribution.

Second, our neutrality measure, η-neutrality, is based on the principle that the model should guarantee neutrality with respect to every combination of target and viewpoint value that appears in the dataset.

Experimental results show that our method with model-based neutralization achieves neutralization even when only a model of the viewpoint is provided. In addition, it balances the accuracy of the target prediction with the neutrality. As discussed in Section 3 and Section 4, likelihood maximization with the

η-neutrality constraint is nonconvex optimization; this is due the nonconvexity of the constraint function. As an area of future work, we intend to find a way to convexify the constraints induced by the neutrality condition.

Acknowledgments. The work is supported by FIRST program "Development of the Fastest Database Engine for the Era of Very Large Database and Experiment and Evaluation of Strategic Social Services Enabled by the Database Engine" and is partially supported by JSPS KAKENHI, Grant Number 24500194, 25540094.

References

1. Boyd, D.: Privacy and publicity in the context of big data. In: Keynote Talk of the 19th Int. Conf. on World Wide Web (2010)
2. Calders, T., Verwer, S.: Three naive bayes approaches for discrimination-free classification. Data Mining and Knowledge Discovery 21(2), 277–292 (2010)
3. Dutch Central Bureau for Statistics: "Volkstelling" (2001)
4. Dwork, C., Hardt, M., Pitassi, T., Reingold, O., Zemel, R.: Fairness through awareness. In: Proceedings of the 3rd Innovations in Theoretical Computer Science Conference, pp. 214–226. ACM (2012)
5. Frank, A., Asuncion, A.: UCI machine learning repository (2010), http://archive.ics.uci.edu/ml
6. Kamiran, F., Calders, T., Pechenizkiy, M.: Discrimination aware decision tree learning. In: 2010 IEEE 10th International Conference on Data Mining (ICDM), pp. 869–874. IEEE (2010)
7. Kamishima, T., Akaho, S., Asoh, H., Sakuma, J.: Enhancement of the neutrality in recommendation. In: Proceedings of the 2nd Workshop on Human Decision Making in Recommender Systems (Decisions@RecSys), pp. 8–14 (2012)
8. Kamishima, T., Akaho, S., Asoh, H., Sakuma, J.: Fairness-aware classifier with prejudice remover regularizer. In: Flach, P.A., De Bie, T., Cristianini, N. (eds.) ECML PKDD 2012, Part II. LNCS, vol. 7524, pp. 35–50. Springer, Heidelberg (2012)
9. Luong, B.T., Ruggieri, S., Turini, F.: k-nn as an implementation of situation testing for discrimination discovery and prevention. In: Proceedings of the 17th ACM SIGKDD International Conference on Knowledge Discovery and Data Mining, KDD 2011, pp. 502–510. ACM, New York (2011)
10. Pariser, E.: The Filter Bubble: What The Internet Is Hiding From You. Viking, London (2011)
11. Pedreshi, D., Ruggieri, S., Turini, F.: Discrimination-aware data mining. In: Proceeding of the 14th ACM SIGKDD International Conference on Knowledge Discovery and Data Mining, pp. 560–568. ACM (2008)
12. Ruggieri, S., Pedreschi, D., Turini, F.: Dcube: Discrimination discovery in databases. In: Proceedings of the 2010 International Conference on Management of Data, pp. 1127–1130. ACM (2010)
13. Shor, N.Z., Kiwiel, K.C., Ruszcayński, A.: Minimization methods for non-differentiable functions. Springer-Verlag New York, Inc., New York (1985)
14. Žliobaitė, I., Kamiran, F., Calders, T.: Handling conditional discrimination. In: Proceedings of the 2011 IEEE 11th International Conference on Data Mining, ICDM 2011, pp. 992–1001. IEEE Computer Society, Washington, DC (2011)

Decision-Theoretic Sparsification for Gaussian Process Preference Learning

M. Ehsan Abbasnejad[1], Edwin V. Bonilla[2], and Scott Sanner[2]

[1] Australian National University (ANU) and National ICT Australia (NICTA)*
eabbasnejad@anu.edu.au
[2] National ICT Australia (NICTA) and Australian National University (ANU)
ebonilla@nicta.com.au, ssanner@nicta.com.au

Abstract. We propose a decision-theoretic sparsification method for Gaussian process preference learning. This method overcomes the loss-insensitive nature of popular sparsification approaches such as the Informative Vector Machine (IVM). Instead of selecting a subset of users and items as inducing points based on uncertainty-reduction principles, our sparsification approach is underpinned by decision theory and directly incorporates the loss function inherent to the underlying preference learning problem. We show that by selecting different specifications of the loss function, the IVM's differential entropy criterion, a value of information criterion, and an upper confidence bound (UCB) criterion used in the bandit setting can all be recovered from our decision-theoretic framework. We refer to our method as the *Valuable Vector Machine* (VVM) as it selects the most useful items during sparsification to minimize the corresponding loss. We evaluate our approach on one synthetic and two real-world preference datasets, including one generated via Amazon Mechanical Turk and another collected from Facebook. Experiments show that variants of the VVM significantly outperform the IVM on all datasets under similar computational constraints.

1 Introduction

Preference learning has become an important subfield in machine learning transcending multiple disciplines such as economics, operations research and social sciences. A wide range of applications in areas such as recommender systems, autonomous agents, human-computer interaction and e-commerce has motivated machine learning researchers to investigate flexible and effective ways to construct predictive preference models from preference observations. This is a challenging problem since complex relations between users and their preferred products (items) must be uncovered. Furthermore, flexible and principled ways to handle uncertainty over the users' preferences are required in order to balance what the system knows. To address these challenges, non-parametric Bayesian approaches based on Gaussian processes (GPs) have shown to be effective in

* NICTA is funded by the Australian Government as represented by the Department of Broadband, Communications and the Digital Economy and the Australian Research Council through the ICT Centre of Excellence program.

H. Blockeel et al. (Eds.): ECML PKDD 2013, Part II, LNAI 8189, pp. 515–530, 2013.

real applications [1,2,3,4]. However, one of the major limitations of preference learning approaches based on GPs is their poor scalability when dealing with a large number of users and items.

Scalability issues in GPs are not exclusive to preference learning and they are common in other settings such as regression and classification. It is customary in these settings to adopt *sparsification* approaches, where a subset of training examples is selected as inducing points, considerably reducing the time complexity of posterior approximation and prediction [5,6,7]. A popular approach to GP sparsification is the Informative Vector Machine [8], where inducing points are selected according to an information-theoretic criterion. A key characteristic of the IVM is that it can be embedded naturally in sequential algorithms such as Assumed Density Filtering [9] or Expectation Propagation [10]. These algorithms provide efficient computation of the quantities of interest (i.e., posterior variances) to be used by the IVM's sparsification criterion.

Nevertheless, the IVM's purely entropic sparsification criterion fails at addressing the varying loss functions that may be of interest to the final decision-theoretic task — especially those tasks that naturally arise in preference learning. For example, we might be interested in (a) optimizing the utility of the best recommendation, (b) giving a ranking of all items (or a subset), or (c) correctly classifying all pairwise preferences. In each case we seek to optimize a loss for a different decision-theoretic task and when we need to approximate in a Bayesian setting, it is important that our approximation is loss-calibrated [11]. We note that the uncertainty reduction principle inherent to the IVM approximation is not loss-calibrated for all tasks (a)–(c).

In this paper, we continue to bridge the gap between decision theory and approximate Bayesian inference [11] in a direction that leverages the efficiency of the IVM approach for GP sparsification, while overcoming its loss-insensitive approximation. We show that the IVM's differential entropy criterion, a value of information criterion, and an upper confidence bound [12] criterion can *all* be recovered in our framework by specifying the appropriate loss.

An additional important aspect of the preference learning problem that distinguishes it from standard machine learning settings is that the complexity of making predictions does not directly depend upon the number of observations (i.e. preference relations), but rather the number of users and items. Our method takes this into consideration and adopts an item-driven sparsification strategy that retains the items that best encode the users' preferences. Our experiments show that this is an effective way of reducing the complexity of inference in preference learning with GPs while addressing the objective function of interest directly. We refer to our generic method as the *Valuable Vector Machine* (VVM) since it incorporates the loss function directly into its sparsification mechanism.

The rest of this paper is organized as follows: we outline the use of GPs for multi-user preference learning and prediction in Section 2 followed by our our proposed VVM sparsification framework in Section 3 and empirical evaluation in Section 4. We differentiate our approach from related work in Section 5 and conclude in Section 6.

2 GPs for Preference Learning

In this section, we define a general approximation framework for Bayesian preference learning via Gaussian Processes that we will adapt to the loss-calibrated setting in the next section.

Let $U = \{u_1, u_2, \ldots, u_n\}$ be a set of n users and let $X = \{x_1, x_2, \ldots, x_m\}$ be a set of m items and denote the set of observed preferences of each user $u \in U$ with $\mathcal{D}^u = \{x_i \succ x_j\}$ where $1 \le i \le m$ and $1 \le j \le m$. Given the preferences $\mathcal{D}^u$ for u, satisfaction of the von Neumann-Morgenstern axioms [13] justify the existence of utilities $f_i^u \in \mathbb{R}$ for each item $x_i \in X$ s.t. $x_i \succ x_j \in D^u$ iff $f_i^u > f_j^u$. In order to model the distribution over these utilities, we build upon the model proposed by [2]. We denote a latent utility vector f for *all* users and items with $f = [f_1^{u_1}, f_2^{u_1}, \ldots, f_m^{u_n}]^T$. Then, we can define the likelihood over all the preferences given the latent functions as:

$$p(\mathcal{D}|f) = \prod_{u \in U} \prod_{x_i \succ x_j \in \mathcal{D}^u} p(x_i \succ x_j | f_i^u, f_j^u) \tag{1}$$

$$\text{with} \qquad p(x_i \succ x_j | f_i^u, f_j^u) = \Phi\left(\frac{f_i^u - f_j^u}{\alpha}\right), \tag{2}$$

where $\Phi(x) = \int_{-\infty}^x \mathcal{N}(y)dy$ and $\mathcal{N}(y)$ is a zero-mean Gaussian distribution with unit variance. In this model, $p(f)$ is the prior over the latent utilities f and is defined via a GP with zero-mean function and a covariance function that factorizes over users and items [2]. Therefore:

$$p(f) = \mathcal{N}(f; 0, K), \quad K = K_u \otimes K_x, \tag{3}$$

where K is the kernel matrix composed of the Kronecker product of the kernel matrix over the users K_u and the kernel matrix over the items K_x. One interesting feature of this model is the inherent transfer of preferences across users through the correlated prior, which will subsequently help the prediction on those users for which there are not many preferences recorded. Additionally, as we shall see later, having a fully factorized likelihood across users and items will facilitate the application of sequential approximate posterior inference methods such as *Expectation Propagation* (EP) [10].

The posterior of the latent functions f given all the preferences is:

$$p(f|\mathcal{D}) = \frac{1}{Z}p(f)p(\mathcal{D}|f), \tag{4}$$

with Z being the normalizer. This posterior is analytically intractable due to the non-Gaussian nature of the likelihood. Therefore, we need to resort to approximations. Here we use EP which approximates the posterior $p(f|\mathcal{D})$ by a tractable distribution $q(f)$. EP assumes that each likelihood term $p(x_i \succ x_j | f_i^u, f_j^u)$ can be approximated by a distribution $q(f_i^u, f_j^u)$ such that the approximated posterior $q(f)$ factorizes over $q(f_i^u, f_j^u)$. Then EP iteratively approximates each $q(f_i^u, f_j^u)$ in turn by dividing it out from the approximated posterior $q(f)$ (obtaining the cavity distribution), multiplying in the true likelihood $p(x_i \succ x_j | f_i^u, f_j^u)$, and projecting the result back to its factorized form by matching its moments to an updated $q(f_i^u, f_j^u)$. This overall procedure is motivated by the aim to minimize the KL−divergence between the true posterior $p(\mathbf{f}|\mathcal{D})$ and its approximation $q(\mathbf{f})$.

In the preference learning case we detailed earlier, we can approximate the posterior with a Gaussian:

$$q(\mathbf{f}) = \frac{1}{Z} p(\mathbf{f}) \prod_{u \in U} \prod_{\{\{i,j\}|\mathbf{x}_i \succ \mathbf{x}_j \in \mathcal{D}^u\}} q(f_i^u, f_j^u) = \mathcal{N}(\mathbf{f}; \boldsymbol{\mu}, \boldsymbol{\Sigma}). \tag{5}$$

We are interested in locally approximating each likelihood term in Equation 1 as:

$$p(\mathbf{x}_i \succ \mathbf{x}_j | f_i^u, f_j^u) \approx q(f_i^u, f_j^u) \tag{6}$$
$$= \tilde{Z}_{i,j}^u \mathcal{N}(f_i^u, f_j^u; \tilde{\boldsymbol{\mu}}_{u,[i,j]}, \tilde{\boldsymbol{\Sigma}}_{u,[i,j]}),$$

where $\mathcal{N}(f_i^u, f_j^u; \tilde{\boldsymbol{\mu}}_{u,[i,j]}, \tilde{\boldsymbol{\Sigma}}_{u,[i,j]})$ denotes the local two-dimensional Gaussian over $[f_i^u, f_j^u]^T$ with mean $\tilde{\boldsymbol{\mu}}_{u,[i,j]}$ and covariance $\tilde{\boldsymbol{\Sigma}}_{u,[i,j]}$ corresponding to items i and j.

Hence we can approximate the posterior as:

$$q(\mathbf{f}) = \frac{1}{Z} p(\mathbf{f}) \prod_{u \in U} \prod_{\{i,j\} \in \mathcal{D}} q(f_i^u, f_j^u) = \mathcal{N}(\mathbf{f}; \boldsymbol{\mu}, \boldsymbol{\Sigma}), \tag{7}$$

where
$$\boldsymbol{\mu}_{u,[i,j]} = \boldsymbol{\Sigma}_{u,[i,j]} \tilde{\boldsymbol{\Sigma}}_{u,[i,j]}^{-1} \tilde{\boldsymbol{\mu}}_{u,[i,j]} \tag{8}$$

$$\boldsymbol{\Sigma}_{u,[i,j]}^{-1} = (\mathbf{K}_{u,[i,j]}^{-1} + \tilde{\boldsymbol{\Sigma}}_{u,[i,j]}^{-1}). \tag{9}$$

This means that in order to determine the parameters of our approximate posterior, we need to compute estimates of the local parameters $\tilde{\boldsymbol{\mu}}$ and $\tilde{\boldsymbol{\Sigma}}$. To show these updates, we need to define additional distributions: (a) the *cavity* distribution which we will denote with the backslash symbol "\" and (b) the *unnormalized marginal posterior*, which we will denote with the hat symbol "ˆ".

Here we only show how to compute the parameters necessary to estimate the posterior[1]. We iterate through the following steps:

1. Update the Cavity Distribution: The cavity distribution $q_{\backslash}(f_i^u, f_j^u)$ results from multiplying the prior by all the local approximate likelihood terms except $q(f_i^u, f_j^u)$ and marginalizing all latent dimensions except f_i^u and f_j^u. This is done in practice simply by removing the current approximate likelihood term from the approximate posterior. Hence we obtain:

$$q_{\backslash}(f_i^u, f_j^u) = \mathcal{N}(f_i^u, f_j^u; \boldsymbol{\mu}_{\backslash u,[i,j]}, \boldsymbol{\Sigma}_{\backslash u,[i,j]}) \tag{10}$$

$$\boldsymbol{\mu}_{\backslash u,[i,j]} = \boldsymbol{\Sigma}_{\backslash u,[i,j]} (\boldsymbol{\Sigma}_{u,[i,j]}^{-1} \boldsymbol{\mu}_{u,[i,j]} - \tilde{\boldsymbol{\Sigma}}_{u,[i,j]}^{-1} \tilde{\boldsymbol{\mu}}_{u,[i,j]}) \tag{11}$$

$$\boldsymbol{\Sigma}_{\backslash u,[i,j]} = (\boldsymbol{\Sigma}_{u,[i,j]}^{-1} - \tilde{\boldsymbol{\Sigma}}_{u,[i,j]}^{-1})^{-1}. \tag{12}$$

---

[1] Similar updates for the single user case are given in [1].

2. Update the Unnormalized Marginal Posterior : This results from finding the unnormalized Gaussian that best approximates the product of the cavity distribution and the exact likelihood:

$$\hat{q}(f_i^{\mathbf{u}}, f_j^{\mathbf{u}}) \approx p(\mathbf{x}_i \succ \mathbf{x}_j | f_i^{\mathbf{u}}, f_j^{\mathbf{u}}) q_{\backslash}(f_i^{\mathbf{u}}, f_j^{\mathbf{u}}) \tag{13}$$

$$\hat{q}(f_i^{\mathbf{u}}, f_j^{\mathbf{u}}) = \hat{Z}^{-1} \mathcal{N}(f_i^{\mathbf{u}}, f_j^{\mathbf{u}}; \hat{\boldsymbol{\mu}}_{\mathbf{u},[i,j]}, \hat{\boldsymbol{\Sigma}}_{\mathbf{u},[i,j]}) \quad \text{with} \tag{14}$$

$$\hat{Z} = \Phi(r_{i,j}) \tag{15}$$

$$\hat{\boldsymbol{\mu}}_{\mathbf{u},[i,j]} = \boldsymbol{\mu}_{\backslash \mathbf{u},[i,j]} + \boldsymbol{\Sigma}_{\backslash \mathbf{u},[i,j]} \mathbf{w}_{\mathbf{u},[i,j]} \tag{16}$$

$$\hat{\boldsymbol{\Sigma}}_{\mathbf{u},[i,j]} = \boldsymbol{\Sigma}_{\backslash \mathbf{u},[i,j]} \tag{17}$$

$$- \boldsymbol{\Sigma}_{\backslash \mathbf{u},[i,j]} (\mathbf{w}_{\mathbf{u},[i,j]} \mathbf{w}_{\mathbf{u},[i,j]}^{\mathsf{T}} \hat{r}_{i,j} \mathbf{w}_{\mathbf{u},[i,j]} \mathbf{1}_1^{\mathsf{T}}) \boldsymbol{\Sigma}_{\backslash \mathbf{u},[i,j]},$$

where

$$\mathbf{w}_{\mathbf{u},[i,j]} = \frac{\mathcal{N}(r_{i,j})}{\Phi(r_{i,j})(\alpha^2 + \mathrm{tr}(\boldsymbol{\Sigma}_{\backslash \mathbf{u},[i,j]} \mathbf{1}_2))} \mathbf{1}_1,$$

$$r_{i,j} = \frac{\boldsymbol{\mu}_{\backslash \mathbf{u},[i,j]} \mathbf{1}_1}{\alpha^2 + \mathrm{tr}(\boldsymbol{\Sigma}_{\backslash \mathbf{u},[i,j]} \mathbf{1}_2)}, \quad \hat{r}_{i,j} = \frac{r_{i,j}}{\alpha^2 + \mathrm{tr}(\boldsymbol{\Sigma}_{\backslash \mathbf{u},[i,j]} \mathbf{1}_2)}$$

$$\text{and} \quad \mathbf{1}_1 = \begin{bmatrix} 1 \\ -1 \end{bmatrix}, \quad \mathbf{1}_2 = \begin{bmatrix} 1 & -1 \\ -1 & 1 \end{bmatrix}. \tag{18}$$

3. Update the Local Factor Approximation: by performing moment matching, we can calculate the corresponding parameters in $q(f_i^{\mathbf{u}}, f_j^{\mathbf{u}})$ as:

$$\tilde{\boldsymbol{\mu}}_{\mathbf{u},[i,j]} = \tilde{\boldsymbol{\Sigma}}_{\mathbf{u},[i,j]} (\hat{\boldsymbol{\Sigma}}_{\mathbf{u},[i,j]}^{-1} \hat{\boldsymbol{\mu}}_{\mathbf{u},[i,j]} - \boldsymbol{\Sigma}_{\backslash \mathbf{u},[i,j]}^{-1} \boldsymbol{\mu}_{\backslash \mathbf{u},[i,j]})$$

$$\tilde{\boldsymbol{\Sigma}}_{\mathbf{u},[i,j]} = (\hat{\boldsymbol{\Sigma}}_{\mathbf{u},[i,j]}^{-1} - \boldsymbol{\Sigma}_{\backslash \mathbf{u},[i,j]}^{-1})^{-1}. \tag{19}$$

At each iteration once we have local factor parameters $\tilde{\boldsymbol{\mu}}$ and $\tilde{\boldsymbol{\Sigma}}$, we can compute the parameters of the full posterior approximation using 7. We iterate through all the factors and update the local approximations sequentially.

2.1 Prediction

Given a pair of items $\mathbf{x}_1^*, \mathbf{x}_2^*$ for a particular user, we will be able to determine the predictive distribution over the latent utility functions as:

$$p(f_1^*, f_2^* | \mathcal{D}) = \int_{-\infty}^{\infty} p(f_1^*, f_2^* | \mathbf{f}) p(\mathbf{f} | \mathcal{D}) d\mathbf{f} = \mathcal{N}(\boldsymbol{\mu}^*, \mathbf{C}^*) \tag{20}$$

$$\text{with} \quad \boldsymbol{\mu}^* = \mathbf{K}^* (\mathbf{K} + \tilde{\boldsymbol{\Sigma}})^{-1} \boldsymbol{\mu} \tag{21}$$

$$\mathbf{C}^* = \boldsymbol{\Sigma}^* - \mathbf{K}^{*\mathsf{T}} (\mathbf{K} + \tilde{\boldsymbol{\Sigma}})^{-1} \mathbf{K}^*, \tag{22}$$

where $\boldsymbol{\Sigma}^*$ is the 2×2 kernel matrix built from the item pair $\mathbf{x}_1^*$ and $\mathbf{x}_2^*$; $\mathbf{K}^* = \mathbf{K}_u^* \otimes \mathbf{K}_x^*$ that represents the kernel matrix of the test user and items with all the users and items in the training set; $\mathbf{K}_u^*$ is the $1 \times n$ kernel matrix of the queried user with other users; and $\mathbf{K}_x^*$ is the $2 \times m$ kernel matrix of the queried pair of items with other items. Subsequently, their preference for a user is determined by integrating out the latent utility functions:

$$p(\mathbf{x}_1^* \succ \mathbf{x}_2^* | \mathcal{D}) = \int \int p(\mathbf{x}_1^* \succ \mathbf{x}_2^* | f_1^*, f_2^*, \mathcal{D}) p(f_1^*, f_2^* | \mathcal{D}) df_1^* df_2^*$$

$$= \Phi \left(\frac{\mu_1^* - \mu_2^*}{\alpha^2 + C_{1,1}^* + C_{2,2}^* - 2C_{1,2}^*} \right). \tag{23}$$

We see that the mean and covariance of the predictive distribution require the inversion of a (possibly) very large matrix. This matrix is, in general, of dimensions $nm \times nm$. Even though the inverse matrix can be reused for multiple query points, this is intractable for any real application. Hence, we will focus on how to sparsify this matrix by selecting a subset of *inducing items*. Note the main difference with other machine learning settings where there is a one-to-one correspondence between the number of observations and the dimensionality of the corresponding matrix. In our case, the observations (preference relations) affect the dimensionality of this matrix only indirectly and we are more concerned with the number of users and items. More importantly, as we shall see in the following section, we will make use of decision theory for sparsification. Our method, which we will refer to as the Valuable Vector Machine (VVM), selects the most useful items during sparsification so as to minimize the loss inherent to the preference learning problem. Therefore, the prediction time which is cubic in the number of items is improved in VVM over the case where all items are used.

Since our focus is on improving prediction time and this scales cubically with the number of items we need to obtain a risk-sensitive posterior approximation. EP is well-suited for this case because it considers all data efficiently and locally.

2.2   Optimizing the Kernel Hyper-parameters

One of the inherent advantages of GPs over other non-Bayesian kernel methods is its capability of optimizing the hyper-parameters. This can be easily done by maximizing the marginal likelihood in a gradient descent algorithm. The marginal likelihood can be obtained from the normalizer $\tilde{Z}$ in Equation 5 as:

$$\tilde{Z} = \int p(\mathbf{f}) \prod_{u \in U} \prod_{\{i,j\} \in \mathcal{D}} q(f_i^u, f_j^u) d\mathbf{f} \tag{24}$$

where both $p(\mathbf{f})$ and $q(f_i^u, f_j^u)$ are Gaussian distributions and their product produces an unnormalized Gaussian distribution. Therefore, the log likelihood is:

$$\log(\tilde{Z}) = -\frac{1}{2}\tilde{\mu}^\top (\mathbf{K} + \tilde{\Sigma})^{-1}\tilde{\mu} - \frac{1}{2}\log\det(\mathbf{K} + \tilde{\Sigma}) - \frac{n}{2}\log 2\pi \tag{25}$$

The derivative of the marginal likelihood with respect to the kernel hyper-parameters can be used in a gradient descent algorithm to optimize the kernel.

Table 1. Loss and corresponding risk to minimize for VVM variants. Let $q(\mathbf{f}) := q_{S''}(\mathbf{f})$ and $a \in \{\mathbf{x}_i^u\}$.

Algorithm	IVM (item)	VVM-VOI	VVM-UCB
Loss Type	Log loss	Regret	Risk-seeking
$\mathcal{L}(\mathbf{f}, a, \mathbf{u})$	$-\log(q(f_i^u))$	$-\mathbb{I}[f_i^u > f^{u,*}](f_i^u - f^{u,*})$	$-\exp(\beta f_i^u), \beta > 0$
$Risk_{\mathcal{L}}(\mathbf{S}', \mathbf{x}_i, \mathbf{u})$	$H(q(f_i^u))$	$\sigma(\mathbf{u}, \mathbf{x}_i)[c\Phi(c) + \mathcal{N}(c)]$	$1 + \beta\mu(\mathbf{u}, \mathbf{x}_i) + \frac{\beta^2}{2}\bar{\sigma}^2(\mathbf{u}, \mathbf{x}_i)$
Selection Time	$\mathcal{O}(1)$	$\mathcal{O}(n \log n)$	$\mathcal{O}(1)$

3   Decision-Theoretic Sparsification

To recap, our multi-user preference learning objective is to approximate a posterior $q(\mathbf{f}) = \mathcal{N}(\mathbf{f}; \boldsymbol{\mu}, \boldsymbol{\Sigma})$ over latent utilities $\mathbf{f} = [f_1^{u_1}, f_2^{u_1}, \ldots, f_m^{u_n}]^T$ for users $\mathbf{u} \in U$ and items $\mathbf{x}_i \in X$. In the previous section, we showed how to learn $q(\mathbf{f})$ from preference data by EP; in this section due to computational considerations, we wish to sparsify this Gaussian posterior in a loss-calibrated manner. We note that in the special case of GP-based preference learning, there are at least two different ways one might approach sparsification: observation-driven sparsification and item-driven sparsification.

3.1   Sparsification

Observation-Driven Sparsification. In this approach, we incrementally select a subset of observations (in this case preferences) in order to approximate the posterior $q(\mathbf{f})$. More formally, recall that $\mathcal{D}^{\mathbf{u}} = \{\mathbf{x}_i \succ \mathbf{x}_j\}$ and let $\mathcal{D}' \subseteq \mathcal{D}$ be a subset of selected preferences. Observation-driven sparsification simply chooses the data subset $\mathcal{D}'$ according to some criterion to obtain a posterior approximation $q_{\mathcal{D}'}(\mathbf{f}) \approx p(\mathbf{f}|\mathcal{D}')$ (e.g., via EP as outlined in the last section).

As a concrete example, the original *Informative Vector Machine* [8] initializes $\mathcal{D}'$ to a small random subset and then incrementally builds $\mathcal{D}' := \mathcal{D}' \cup \{d^*\}$ for the d^* that maximizes information gain

$$d^* = \underset{d \in \mathcal{D} \setminus \mathcal{D}'}{\arg\max} H(q_{\mathcal{D}' \cup \{d\}}(\mathbf{f})) - H(q_{\mathcal{D}'}(\mathbf{f})), \qquad (26)$$

where $q_{\mathcal{D}'}(\mathbf{f}) \approx p(\mathbf{f}|\mathcal{D}')$ and $q_{\mathcal{D}' \cup \{d\}}(\mathbf{f}) \approx p(\mathbf{f}|\mathcal{D}' \cup \{d\})$. This repeats until the desired level of observation sparsity has been reached. Since $\mathcal{D}'$ is fixed at each iteration and thus $H(q_{\mathcal{D}'}(\mathbf{f}))$ is a constant, each incremental selection in the IVM is equivalent to choosing the d^* that maximizes entropy, i.e., $d^* = \arg\max_{d \in \mathcal{D} \setminus \mathcal{D}'} H(q_{\mathcal{D}' \cup \{d\}}(\mathbf{f}))$.

Item-Driven Sparsification: Valuable Vector Machine. Inclusion of a preference observation entails a 2-dimensional update to our GP posterior; however, since preferences may overlap, there is not a direct relationship between the number of included preferences and the dimensionality of the posterior and hence the computational complexity of prediction described in section 2.1. A more direct way to control the sparsity level of our Gaussian posterior is to simply retain the *items* for users (equivalently dimensions f_i^u of our Gaussian posterior) that minimizes some criterion.

This item-driven approach underlies the *Valuable Vector Machine* (VVM) that we propose in this paper for different decision-theoretic settings. First we introduce some notation. Let $S' \subseteq S = \{(\mathbf{x}_i, \mathbf{u})\}$ be a selected subset of user-item pairs corresponding to latent utility dimensions f_i^u of $\mathbf{f}$ with cardinality $|S'|$. Let $q(\mathbf{f}_{[S']}) = \mathcal{N}(\mathbf{f}_{[S']}; \boldsymbol{\mu}_{[S']}, \boldsymbol{\Sigma}_{[S',S']})$ where $\mathbf{f}_{[S']}$ and $\boldsymbol{\mu}_{[S']}$ respectively represent the subvectors of $\mathbf{f}$ and $\boldsymbol{\mu}$ for selected dimensions S' (i.e. selected users and items in set S') and $\boldsymbol{\Sigma}_{[S',S']}$ the corresponding submatrix of $\boldsymbol{\Sigma}$. Motivated by the observation-driven IVM, after running EP, let us incrementally select dimensions $s^* \in S$ of our Gaussian posterior to retain so that initializing $S' = \emptyset$, at each iteration we update $S' := S' \cup \{s^*\}$ to obtain an improved posterior $q_{S'}(\mathbf{f})$ until some dimensionality limit has been reached.

In decision theory, our objective is to select an action $a^* \in A$ from a possible space of actions A so as to minimize the expectation of some loss $\mathcal{L}(a)$ w.r.t. uncertainty (in this case utility uncertainty over $\mathbf{f}$), i.e. $a^* = \arg\min_a \mathbb{E}_{\mathbf{f}} \mathcal{L}(\mathbf{f}, a)$. Our specific task at each iteration of the VVM is to propose an item-user dimension $\mathbf{x}_i^u$ for inclusion in the posterior — hence the action space $A = \{\mathbf{x}_i\}$ — and to select the item s^* that minimizes expected loss (risk)

$$s^* = \arg\min_{(\mathbf{x}_i, \mathbf{u}) \in S \setminus S'} Risk_\mathcal{L}(S', \mathbf{x}_i, \mathbf{u});$$

$$\text{where } Risk_\mathcal{L}(S', \mathbf{x}_i, \mathbf{u}) := \mathbb{E}_{\mathbf{f} \sim q_{S''}} [\mathcal{L}(\mathbf{f}, \mathbf{x}_i, \mathbf{u})], \tag{27}$$

and $S'' = S' \cup \{\mathbf{x}_i^u\}$. In the following, we will detail choices of loss functions and their respective $Risk_\mathcal{L}(S', \mathbf{x}_i, \mathbf{u})$ yielding the VVM variants as summarized in Table 1 and its corresponding method in Algorithm 1.

In each iteration, VVM selects the action (i.e. item) that minimizes the expected loss for each user until desired predefined dimensionality is reached. Our experiments with a variable number of items per user led to worse performance since it often overemphasizes item selection for the noisiest users. Hence, we found that a constant number of items enforces fairness of GP modeling effort per user.

It should also be noted that the greedy selection here is fairly general and in special cases such as submodular losses, one can prove further convergence guarantees [14].

3.2   Loss Functions and Risk

Log Loss and IVM. Log-loss is appropriate when we want to maximize the log posterior over all preferences. Here we see that we can actually recover an

---

Algorithm 1. Valuable Vector Machine

---

input: $X, U, \mathcal{D}, r$ // r is the percentage of items selected for each user
while not converged **do**
  for $\mathbf{x}_i \succ \mathbf{x}_j \in \mathcal{D}$ **do**
    1. Update the cavity distribution $\mu_{\backslash \mathbf{u},[i,j]}$, $\Sigma_{\backslash \mathbf{u},[i,j]}$ from Equation 11 and 12.
    2. Update the unnormalized marginal posterior $\hat{\mu}, \hat{\Sigma}$ from Equation 16 and 17.
    3. Update the local factor approximation $\tilde{\mu}, \tilde{\Sigma}$ from Equation 19.
    5. Update μ and Σ from Equation 7.
  end for
end while
for each $\mathbf{u} \in U$ // Selection of best items for each user **do**
  $S'^{\mathbf{u}} = \{\}$ //The user $\mathbf{u}$'s subset
  while $|S'^{\mathbf{u}}| < r$ **do**
    $\mathbf{x}_i^* = \arg\min_{\mathbf{x}_i \in X, \mathbf{x}_i \notin S'^{\mathbf{u}}} Risk_{\mathcal{L}}(S', \mathbf{x}_i, \mathbf{u})$// Table 1
    $S'^{\mathbf{u}} = S'^{\mathbf{u}} \cup \{\mathbf{x}_i^*\}$
  end while
  $S = S \cup S'^{\mathbf{u}}$
end for
return $\mu_{[S]}, \Sigma_{[S,S]}$ // Subset of posterior parameters

---

item-based variant of the IVM when using log-loss. Specifically, letting $q(f_i^{\mathbf{u}})$ refer to the marginal of $q(\mathbf{f})$ over $f_i^{\mathbf{u}}$ then

$$Risk_{\mathcal{L}}(S', \mathbf{x}_i, \mathbf{u}) = \int_{-\infty}^{\infty} -q_{S''}(f_i^{\mathbf{u}})[\log q_{S''}(f_i^{\mathbf{u}})]df_i^{\mathbf{u}}$$
$$= H(q_{S''}(f_i^{\mathbf{u}})), \tag{28}$$

which corresponds to the entropic criterion used by the IVM. Recall that the second entropy term in the standard IVM information gain calculation is constant and can be omitted as noted for (26).

Valuable Vector Machine – Value Of Information (VVM-VOI). In the case that our end objective is to predict or recommend the best item $\mathbf{x}_i$ for user $\mathbf{u}$, the loss we might consider minimizing is the *regret*, $\mathbb{I}[f_i^{\mathbf{u}} - f^{\mathbf{u},*} > 0](f_i^{\mathbf{u}} - f^{\mathbf{u},*})$ where we could define $f^{\mathbf{u},*} = \arg\max_i f_i^{\mathbf{u}}$; in words, we want to minimize how much utility we lose for recommending a suboptimal item. In expectation, we might simply fix $f^{\mathbf{u},*} = \max_i \mathbb{E}_{q_{S''}}[f_i^{\mathbf{u}}]$ (the best current item in expectation) where expected loss minimization leads us to the following risk:

$$Risk_{\mathcal{L}}(S', \mathbf{x}_i, \mathbf{u}) = \int_{-\infty}^{\infty} \mathbb{I}[f_i^{\mathbf{u}} > f^{\mathbf{u},*}](f_i^{\mathbf{u}} - f^{\mathbf{u},*})q_{S''}(f_i^{\mathbf{u}})df_i^{\mathbf{u}}$$
$$= \underbrace{\sigma(\mathbf{u}, \mathbf{x})[c\Phi(c) + \mathcal{N}(c)]}_{VOI} \tag{29}$$

where $q_{S''}(f_i^{\mathbf{u}}) = \mathcal{N}(f_i^{\mathbf{u}}; \mu(\mathbf{u}, \mathbf{x}), \sigma^2(\mathbf{u}, \mathbf{x}))$ and $c = \frac{\mu(\mathbf{u}, \mathbf{x}) - \hat{f}^{\mathbf{u}}}{\sigma(\mathbf{u}, \mathbf{x})}$. This is precisely the statement of *Value of Information* (VOI) [15] under a Gaussian assumption

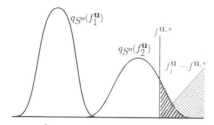

Fig. 1. Illustration of Value of Information: it is the product of the shaded area under the normal curve when the utility is higher than the optimal value and the linear function of their difference. This value corresponds to the expectation of the difference of the utility of the item and the optimal under the shaded mass. As it is observed, $f_1^{\mathbf{u}}$ has negligible mass above $f^{\mathbf{u},*}$ point, therefore the item corresponding to $f_2^{\mathbf{u}}$ is selected.

of uncertainty — quite simply, the more probability mass an item utility has in its tail above the best item in expectation, the higher its chance of being the best item — hence the higher VOI associated with selecting this item in S''.

Valuable Vector Machine – Upper Confidence Bound (VVM-UCB).
It is well-known that a concave or convex valuation of underlying utility respectively encourages risk-averse or risk-seeking behavior w.r.t. utility uncertainty. Risk-seeking behavior from a convex utility function will encourage including "potentially optimal" items according to how uncertain we are regarding their utility function value. A natural convex utility transformation is $\exp(\beta f_{\mathbf{x}}^{\mathbf{u}})$, which leads to the following risk

$$Risk_{\mathcal{L}}(S', \mathbf{x}_i, \mathbf{u}) = -\int_{-\infty}^{\infty} q_{S''}(f_i^{\mathbf{u}}) \exp(\beta f_i^{\mathbf{u}}) df_i^{\mathbf{u}}$$

$$= -\int_{-\infty}^{\infty} q_{S''}(f_i^{\mathbf{u}}) \left(1 + \beta f_i^{\mathbf{u}} + \frac{\beta^2}{2} f_i^{\mathbf{u}2} + \ldots\right) df_i^{\mathbf{u}}$$

$$\approx 1 + \beta \cdot UCB(\mathbf{u}, \mathbf{x}_i)$$

$$\text{where } UCB(\mathbf{u}, \mathbf{x}_i) = \mu(\mathbf{u}, \mathbf{x}_i) + \frac{\beta}{2} \bar{\sigma}^2(\mathbf{u}, \mathbf{x}_i). \tag{30}$$

where $\bar{\sigma}^2(\mathbf{u}, \mathbf{x}_i) = \mathbb{E}_{q_{S''}}[f_i^{\mathbf{u}2}]$. Here, we first replaced $\exp(\beta f_i^{\mathbf{u}})$ with its Taylor expansion and approximated it by truncating third-order terms and above. When doing this, we see that the dimension $\mathbf{x}_i^u$ selected by the VVM will be the one with the greatest *Upper Confidence Bound* (UCB) [16] used in bandit problems, where larger $\beta > 0$ encourages more risk-seeking behavior.

4   Empirical Evaluation

In this section we evaluate the performance of our algorithms (VVM-VOI and VVM-UCB) compared to the IVM and the full GP, i.e. a GP-preference model that does not use sparsification, in terms of two losses: the 0/1 loss and recommendation loss. The *0/1 loss* is the percentage of incorrectly predicted preferences and the *recommendation loss* is the proportion of items that are incorrectly predicted as the best for recommendation. In other words, if the set of items that a user considers to be the best (as induced by her preferences) is denoted by T and the predicted set of best items is T^*, the recommendation loss is $\frac{|\bar{T}|}{m} \times 100$, where m is the number of items and $\bar{T} = \{\mathbf{x} \in T^* | \mathbf{x} \notin T\}$. We report the results as a function of the proportion of items selected for sparsification.

Our experimental rationale is to exhibit how different risk-sensitive sparsifications perform across two different important losses related to preference learning compared to IVM. As such, we use IVM in its original form that works with ADF since it was argued by [17] that it performs better than running full EP, which we observed as well. Hence we chose the IVM variant that offered best performance and compared it against VVM.

We consider three datasets: a synthetic dataset and two real-world datasets. The synthetic experiment assesses the effectiveness of our approach in a controlled setting and the real-world datasets include users' preferences over cars that we have collected using Amazon Mechanical Turk and a Facebook dataset that we have obtained via an in-house application that collects user preferences over web links.

In all these datasets we are given a set of users and items and their corresponding features along with each user's preferences over item pairs. For each user and item we augment their feature vectors with their ID index (this is a common practice in collaborative filtering) and transform their categorical features into binary variables. We split each user's set of preferences into 60% for training and 40% for testing.

We use the squared exponential covariance kernel with automatic relevance determination (ARD) (see [18], Page 106) for both users and items and optimize the hyper-parameters by maximizing the marginal likelihood under the EP approximation as detailed in section 2.2. Finally, we have set $\alpha = 3$ (see Equation (2)) and $\beta = 1$ (see Equation (30)) for all experiments on all datasets.

4.1 Datasets

Synthetic Dataset: In this experiment we created a synthetic dataset where the utility function value for each item is known beforehand and is subsequently used to generate users' preferences. A set of hypothetical users and items are created and identified by their IDs. For each user, items are randomly split into two sets to indicate the ones that are liked (with a constant utility value of 10) and disliked (with a constant utility value of 5). From these utility functions we generate full sets of preferences for 10 items and 50 users. Consequently, for each user, 5 items have higher utility value and are naturally preferred to the other half.

Facebook Data: This dataset has been created using a Facebook App that recommends web links to users every day. The users may give their feedback on the links indicating whether they liked/disliked them. At its peak usage, 111 users had elected to install the Facebook app developed for this project. We also collected user information consisting of ID, age and gender and the link features including count of link's total "likes", count of link's "shares" and count of total link comments. The Facebook App recommended three links per day to avoid position bias and information overload. The preference set is built such that the links that are liked are considered *preferred* to the ones disliked in the batch of three recommended each day. We used 20% of users with the highest number of preferences over 50 links commonly recommended to all users.

Car Preference Dataset Using Amazon Mechanical Turk: We set up an experiment using Amazon Mechanical Turk[2] (AMT) to collect real pair-wise preferences over users. In this experiment users are presented with a choice of a car over another based on their attributes. The car attributes used are (1) body type: sedan, SUV, hatchback, (2) engine capacity: 2.5L, 3.5L, 4.5L, etc, (3) transmission: manual, automatic, (4) fuel consumption: hybrid, non-hybrid, and (5) engine/transmission layout: all-wheel-drive (AWD), forward-wheel-drive (FWD). The dataset has been collected so that 20 unique cars (items) are considered but users are required to answer only 20% of all 190 possible pair-wise preferences. We targeted US users mainly to have a localized preference dataset. For all 60 unique users that participated in this experiment, a set of attributes in terms of general questions (age range, education level, residential region and gender) has been collected as user features.

4.2   Results

We evaluate our algorithms in a cross-validation setting using 60% of preferences for training and 40% for testing and repeated each experiment 40 times. Results are averaged over the number of test users. We analyze the performance of the algorithms as a function of the level of sparsification, as given by the percentage of items selected for inference. The larger the percentage of items selected, the smaller level of sparsification and the closer the algorithms are to the Full-GP method. The performances of the different algorithms on all datasets using the recommendation loss and the 0/1 loss are shown in Figure 2.

Figures 2(a) and 2(b) show the results on the synthetic dataset. Because of the clear distinction between the items that are preferred for each user, all algorithms perform very well when using at least 40% of the items. While IVM and VVM-VOI have very similar performance, VVM-UCB's performance is outstanding, requiring only a very small number of items to achieve perfect prediction.

As seen in Figures 2(c) and 2(d), on the Facebook dataset both VVM-VOI and VVM-UCB outperform (or have equal performance to) IVM when using at least 30% of the items.

It is interesting to note here that the risk-seeking behavior of VVM-UCB leads to a better approximation of the Full-GP which is particularly visible in the Facebook dataset where the number of items are larger. We conjecture that the excellent performance of VVM-UVB with this larger number of items is because it manages to quickly find and refine the set of highest value items, more effectively than even VVM-VOI. This simultaneously lowers recommendation

---

[2] http://mturk.com

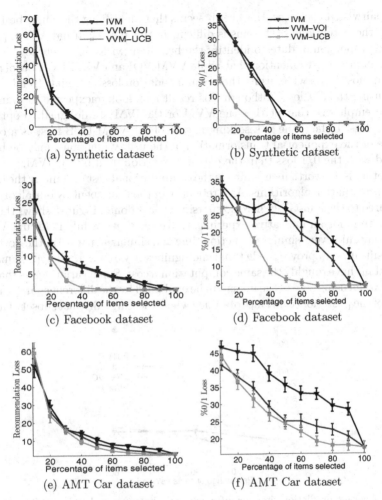

(a) Synthetic dataset

(b) Synthetic dataset

(c) Facebook dataset

(d) Facebook dataset

(e) AMT Car dataset

(f) AMT Car dataset

Fig. 2. Performance of the sparsification methods in terms of the *recommendation loss* (the proportion of items that are incorrectly predicted as the best item for recommendation) in the first column and the 0/1 *loss* (percentage of wrongly predicted preferences) in the second, as a function of the proportion of items selected for sparsification. The larger the number of items the lower the level of sparsification and the closer the algorithms are to the Full-GP method.

loss by finding a near-optimal item and 0/1 loss since the best items can then be identified with certainty in most pairwise comparisons.

Figures 2(e) and 2(f) show the results on the AMT Car dataset. In this dataset, where true preferences have been collected, VVM-VOI and VVM-UCB consistently outperform IVM. Similar to the Facebook results, we conjecture that VVM-VOI's and VVM-UCB's better performance than the IVM (most notably on 0/1 loss where all preferences matter) stems from the fact that they both select the potentially best items first and this helps identify the dominant item in all pairwise preferences. However it seems that identifying the single best item among the potentially best items is difficult in this particular dataset, requiring a large proportion of data to identify the best item with high accuracy.

It is interesting to mention that while VVM-VOI and VVM-UCB outperform IVM in most cases when using the recommendation loss, a similar trend is seen when using the 0/1 loss. Although this result may look unexpected, it is important to emphasize that neither the VVM or the IVM are designed to optimize the 0/1 loss. In fact, the risk-seeking nature of the VVM-UCB loss, as a consequence of the exponential transformation of the utility functions, may be better aligned with the 0/1 loss than the entropic criterion used by the IVM.

Another issue worth mentioning is the computational cost of running the different approximation algorithms. As a reference of the time spent by our algorithms compared to the Full-GP (where no sparsification is done), Figure 3 shows the prediction time for an indicative experiment. We see that – while IVM and VVM-UCB may enjoy very similar prediction time and similar structure in the posterior – sparsification improves prediction time significantly and that all approximation algorithms have roughly the same computational cost. Small variations as that observed when using 80% of the items can be explained by the different sparsity properties of the posterior covariance obtained when selecting a distinct subset of items.

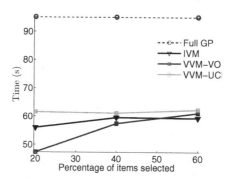

Fig. 3. Average prediction time for inference with 200 users and 10 items. The number calculated as the time consumed to make a series of predictions on the preferences of the test set.

5   Related Work

Probabilistic models for utility functions in preference learning and elicitation have previously been proposed in the machine learning community (e.g. [19,20]). Extensions to non-parametric models have also been developed. In particular, [21] proposed a preference learning framework based on Gaussian processes and [22] used this framework for active learning with discrete choice data. Multi-user GP-based preference models have been given by [23] and [2]. Our method builds upon the model proposed by [2], where the Laplace method was used to approximate the posterior.

In standard machine learning settings, low-rank approximations to the Gram matrix are commonly used by practitioners and researchers (see e.g. Chapter 8 of [18]) to deal with large datasets. A unifying framework in which most of these approximations can be formulated has been given by [24]. This framework includes the fully independent training conditional (FITC) approximation, which makes better use of all the data and can be combined with our approach to approximate the covariance at a higher cost. The work proposed in [7] considers sparsification approaches where the inducing points are latent variables and their values are optimized within a consistent probabilistic latent variable model. However, none of these algorithms addresses the sparsification problem from a decision-theoretic perspective.

Our approach is analogous to the IVM in that we borrow ideas from active learning in order to carry out sparsification during approximate inference. For example, upper confidence bounds (UCB) are used in [12] for GP optimization within the bandit setting. We note that, unlike this latter experimental design scenario, in our sparsification framework we see the data beforehand and decide to include it in our approximation afterwards.

An information theoretic active learning algorithm for classification and preference learning is proposed in [25]. In the preference learning case, this method exploits the reduction of the preference learning problem to a classification setting [26]. This work is complementary to ours in that we can use it along with the FITC approximation in order to devise more effective decision-theoretic sparsification methods for multi-user preference learning. We leave the study of such an approach for future work.

The most relevant work to ours has been proposed in [11] where the use of loss functions in Bayesian methods is considered by formulating an EM algorithm that alternates between variational inference and risk minimization. We take the idea of bridging the gap between decision theory and approximate Bayesian inference [11] in a direction that leverages the efficiency of the IVM approach for GP sparsification, while overcoming its loss-insensitive approximation.

6   Conclusion

We proposed a decision-theoretic sparsification method for Gaussian process preference learning. We referred to our method as the valuable vector machine (VVM) to emphasize the importance of considering a loss-sensitive sparsification approach. We show that the IVM's differential entropy criterion, a value of information criterion, and an upper confidence bound (UCB) criterion can *all* be recovered in a generalized decision-theoretic framework by specifying the appropriate loss. Overall, our approach contributes to the goal of bridging the gap between decision theory and approximate Bayesian inference in the context of loss-sensitive sparsification approaches for efficient Gaussian Process preference learning.

References

1. Chu, W., Ghahramani, Z.: Extensions of gaussian processes for ranking: semi-supervised and active learning. In: Workshop on Learning to Rank at NIPS (2005)
2. Bonilla, E.V., Guo, S., Sanner, S.: Gaussian process preference elicitation. In: NIPS, pp. 153–160 (2010)

3. Platt, J.C., Burges, C.J.C., Swenson, S., Weare, C., Zheng, A.: Learning a Gaussian Process Prior for Automatically Generating Music Playlists. In: NIPS (2002)
4. Xu, Z., Kersting, K., Joachims, T.: Fast active exploration for link-based preference learning using Gaussian processes. In: Balcázar, J.L., Bonchi, F., Gionis, A., Sebag, M. (eds.) ECML PKDD 2010, Part III. LNCS, vol. 6323, pp. 499–514. Springer, Heidelberg (2010)
5. Smola, A.J., Bartlett, P.: Sparse greedy Gaussian process regression. In: NIPS (2001)
6. Keerthi, S.S., Chu, W.: A Matching Pursuit Approach to Sparse Gaussian Process Regression. In: NIPS (2005)
7. Snelson, E., Ghahramani, Z.: Sparse Gaussian processes using pseudo-inputs. In: NIPS (2006)
8. Lawrence, N., Seeger, M., Herbrich, R.: Fast sparse Gaussian process methods: The informative vector machine. In: NIPS (2003)
9. Minka, T.P.: A family of algorithms for approximate bayesian inference. PhD thesis, Massachusetts Institute of Technology (2001)
10. Minka, T.P.: Expectation propagation for approximate Bayesian inference. In: UAI, vol. 17, pp. 362–369 (2001)
11. Lacoste-Julien, S., Huszar, F., Ghahramani, Z.: Approximate inference for the loss-calibrated Bayesian. In: AISTATS, pp. 416–424 (2011)
12. Srinivas, N., Krause, A., Kakade, S.M., Seeger, M.: Gaussian process optimization in the bandit setting: No regret and experimental design. In: NIPS (2009)
13. Neumann, J.V., Morgenstern, O.: Theory of Games and Economic Behavior. Princeton University Press (1944)
14. Krause, A., Golovin, D.: Submodular Function Maximization. In: Tractability: Practical Approaches to Hard Problems. Cambridge University Press (2012)
15. Howard, R.: Information value theory. IEEE Transactions on Systems Science and Cybernetics 2(1), 22–26 (1966)
16. Auer, P., Cesa-Bianchi, N., Fischer, P.: Finite-time analysis of the multiarmed bandit problem. Mach. Learn. 47(2-3), 235–256 (2002)
17. Seeger, M.W.: Bayesian Gaussian Process Models: PAC-Bayesian Generalisation Error Bounds and Sparse Approximations. PhD thesis, University of Edinburgh (2003)
18. Rasmussen, C.E., Williams, C.: Gaussian Processes for Machine Learning (2006)
19. Chajewska, U., Koller, D.: Utilities as random variables: Density estimation and structure discovery. In: UAI, pp. 63–71. Morgan Kaufmann Publishers Inc. (2000)
20. Guo, S., Sanner, S.: Real-time multiattribute Bayesian preference elicitation with pairwise comparison queries. In: AISTATS (2010)
21. Chu, W., Ghahramani, Z.: Preference learning with Gaussian processes. In: ICML, pp. 137–144. ACM, New York (2005)
22. Eric, B., Freitas, N.D., Ghosh, A.: Active preference learning with discrete choice data. In: NIPS, pp. 409–416 (2008)
23. Birlutiu, A., Groot, P., Heskes, T.: Multi-task preference learning with an application to hearing aid personalization. Neurocomputing 73 (2010)
24. Quiñonero-Candela, J., Rasmussen, C.: A unifying view of sparse approximate Gaussian process regression. JMLR, 1939–1959 (2005)
25. Houlsby, N., Huszár, F., Ghahramani, Z., Lengyel, M.: Bayesian Active Learning for Classification and Preference Learning. arXiv:1112.5745 (2011)
26. Houlsby, N., Hernández-Lobato, J.M., Huszár, F., Ghahramani, Z.: Collaborative Gaussian Processes for Preference Learning. In: NIPS (2012)

Variational Hidden Conditional Random Fields with Coupled Dirichlet Process Mixtures

Konstantinos Bousmalis[1], Stefanos Zafeiriou[1], Louis–Philippe Morency[2],
Maja Pantic[1], and Zoubin Ghahramani[3]

[1] Imperial College London, SW7 2AZ, UK
{k.bousmalis,s.zafeiriou,m.pantic}@imperial.ac.uk
[2] University of Southern California, Playa Vista, CA 90094 USA
morency@ict.usc.edu
[3] University of Cambridge, CB2 1PZ, UK
zoubin@eng.cam.ac.uk

Abstract. Hidden Conditional Random Fields (HCRFs) are discriminative latent variable models which have been shown to successfully learn the hidden structure of a given classification problem. An infinite HCRF is an HCRF with a countably infinite number of hidden states, which rids us not only of the necessity to specify a priori a fixed number of hidden states available but also of the problem of overfitting. Markov chain Monte Carlo (MCMC) sampling algorithms are often employed for inference in such models. However, convergence of such algorithms is rather difficult to verify, and as the complexity of the task at hand increases, the computational cost of such algorithms often becomes prohibitive. These limitations can be overcome by variational techniques. In this paper, we present a generalized framework for infinite HCRF models, and a novel variational inference approach on a model based on coupled Dirichlet Process Mixtures, the HCRF–DPM. We show that the variational HCRF–DPM is able to converge to a correct number of represented hidden states, and performs as well as the best parametric HCRFs —chosen via cross–validation— for the difficult tasks of recognizing instances of agreement, disagreement, and pain in audiovisual sequences.

Keywords: nonparametric models, discriminative models, hidden conditional random fields, dirichlet processes, variational inference.

1 Introduction

Over the past decade, nonparametric methods have been successfully applied to many existing graphical models, allowing them to grow the number of latent states as necessary to fit the data [1–6]. Infinite HCRFs were first presented in [7] and since exact inference for such models with an infinite number of parameters is intractable, inference was based on a Markov chain Monte Carlo (MCMC) sampling algorithm. Although MCMC algorithms have been successfully applied on numerous applications, they have some significant drawbacks: they are notoriously slow to converge, it is hard to verify their convergence,

H. Blockeel et al. (Eds.): ECML PKDD 2013, Part II, LNAI 8189, pp. 531–547, 2013.

and they often don't scale well to larger datasets and higher model complexity. Moreover, the model presented in [7] is not readily able to handle continuous input features.

In this work, we consider a deterministic alternative to MCMC sampling for infinite HCRFs with a variational inference [8] approach. Variational inference will allow us to converge faster, verify convergence and scale without a prohibitive computational cost. The model we present in this paper allows a countably infinite number of shared, among labels, hidden states via the use of multiple Dirichlet Process Mixtures (DPMs). Specifically, we present a novel mean field variational approach that uses DPM constructions in the model potentials to allow for the representation of a potentially infinite number of hidden states. Furthermore, we show that our model, the HCRF–DPM, is a generalization of the model presented in [7] and is able to handle continuous features naturally.

In the following section, we consicely present the theoretical background necessary to understand this paper. We present in Section 3 our variational HCRF–DPM model. Finally, we evaluate our model in Section 4.2, and conclude in Section 5.

2   Theoretical Background

The HCRF–DPM, like many other infinite models, relies on DPMs. We present in this section a brief introduction to Dirichlet Processes and Hidden Conditional Random Fields.

2.1   Dirichlet Processes

A Dirichlet Process (DP) is a distribution of distributions, parameterized by a scaling parameter α and a probability measure Ξ. The latter is the basis around which the distributions $G \sim \mathrm{DP}(\alpha, \Xi)$ are drawn, with variability governed by the α parameter. [9] presented the so–called "stick–breaking" construction for DPs, which is based on random variables $(\beta'_k)_{k=1}^{\infty}$ and $(h_k)_{k=1}^{\infty}$, where $\beta'_k | \alpha, \Xi \sim Beta(1, \alpha)$ and $h_k | \alpha, \Xi \sim \Xi$:

$$\beta_k = \beta'_k \prod_{l=1}^{k-1}(1 - \beta'_l) \qquad G = \sum_{k=1}^{\infty} \beta_k \delta_{h_k} , \qquad (1)$$

where δ is the Dirac delta function. By letting $\boldsymbol{\beta} = (\beta_k)_{k=1}^{\infty}$ we abbreviate this construction as $\boldsymbol{\beta} | \alpha \sim GEM(\alpha)$. A Dirichlet Process Mixture (DPM) model is a hierarchical Bayesian model that uses a DP as a nonparametric prior:

$$G | \alpha, \Xi \sim \mathrm{DP}(\alpha, \Xi), \quad c_t | G \sim G, \quad s_t \sim p(s_t | c_t) , \qquad (2)$$

where $(s_t)_{t=1}^{T}$ is a dataset of size T, governed by a distribution conditioned on $(c_t)_{t=1}^{T}$, auxiliary index variables that get assigned each to one of the clusters $(h_k)_{k=1}^{\infty}$. As new datapoints are drawn, the number of components in this mixture model grows. In the model we present in this paper, as we explain later, we employ a number of DP priors coupled together at the data generation level, i.e. s_t above is a function of auxiliary index variables drawn from all different DPs.

2.2 Finite Hidden Conditional Random Fields

HCRFs —discriminative undirected models that contain hidden states— were first presented in [10] and used to capture temporal dependencies across frames and recognize different gesture classes. They did so successfully by learning a state distribution among the different gesture classes in a discriminative manner, allowing them to not only uncover the distinctive configurations that uniquely identify each class, but also to learn a shared common structure among the classes. Conditional Random Fields and HCRFs can be defined in arbitrary graph structures but in our paper, driven by our application field, we assume data to be sequences that correspond to undirected chains. Our work, however, can be readily applied to tree–structured models.

We represent T observations as $\mathbf{X} = [\mathbf{x}_1, \mathbf{x}_2, \ldots, \mathbf{x}_T]$. Each observation at time $t \in \{1, \ldots, T\}$ is represented by a feature vector $\mathbf{f}_t \in \Re^d$, where d is the number of features, that can include any features of the observation sequence. We wish to learn a mapping between observation sequence $\mathbf{X}$ and class label $y \in \mathcal{Y}$, where $\mathcal{Y}$ is the set of available labels. The HCRF does so by estimating the conditional joint distribution over a sequence of latent variables $\mathbf{s} = [s_1, s_2, \ldots, s_T]$, each of which is assigned to a hidden state $h_k \in \mathcal{H}$, and a label y, given $\mathbf{X}$. One of the main representational power of HCRFs is that the latent variables can depend on arbitrary features of the observation sequence. This allows us to model long–range contextual dependencies: s_t, the latent variable at time t, can depend on observations that happened earlier or later than t. An HCRF models the conditional probability of a class label given an observation sequence by:

$$p(y \mid \mathbf{X}, \boldsymbol{\theta}) = \sum_{\mathbf{s}} p(y, \mathbf{s} \mid \mathbf{X}, \boldsymbol{\theta}) = \frac{\sum_{\mathbf{s}} \mathcal{F}(y, \mathbf{s}, \mathbf{X}, \boldsymbol{\theta})}{\sum_{y' \in \mathcal{Y}, \mathbf{s}} \mathcal{F}(y', \mathbf{s}, \mathbf{X}, \boldsymbol{\theta})}. \tag{3}$$

The potential function $\mathcal{F}(y, \mathbf{s}, \mathbf{X}, \boldsymbol{\theta}) \in \Re$ is parameterized by $\boldsymbol{\theta}$, which measures the compatibility between a label y, a sequence of observations $\mathbf{X}$ and a configuration of the latent variables $\mathbf{s}$. The model is discriminative because it doesn't model a joint distribution that includes input $\mathbf{X}$, but it only models the distribution of a label y conditioned on $\mathbf{X}$. The graph of a linear–chain HCRF is a chain where each node corresponds to a latent variable s_t at time t. For such a model, the potential function is usually defined as:

$$\mathcal{F}(y, \mathbf{s}, \mathbf{X}, \boldsymbol{\theta}) = \exp\left\{ \sum_{t=1}^{T} \sum_{i=1}^{d} \theta_x(s_t, i) f_t(i) + \theta_y(s_t, y) + \sum_{t=2}^{T} \theta_e(s_t, s_{t-1}, y) \right\} \tag{4}$$

In this paper, we use the notation $\theta_x(h_k, i)$ to refer to the weight that measures the compatibility between the feature indexed by i and state $h_k \in \mathcal{H}$. Similarly, $\theta_y(h_k, y)$ stand for weights that correspond to class y and state h_k, whereas $\theta_e(h_k, h', y)$ measure the compatibility of y with a transition from h' to h_k.

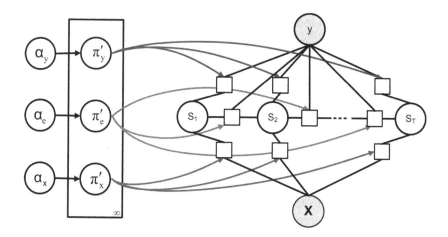

Fig. 1. Factor graph representation of our HCRF–DPM

3   Hidden Conditional Random Fields with Coupled Dirichlet Process Mixtures

For an infinite HCRF we allow an unbounded number of potential hidden states in $\mathcal{H}$. This becomes possible, by introducing random variables $\{\pi_x(h_k|i)\}_{k=1}^{\infty}$, $\{\pi_y(h_k|y)\}_{k=1}^{\infty}$, $\{\pi_e(h_k, y|h_a)\}_{k=1,y=1}^{\infty,|\mathcal{Y}|}$ for an observation feature indexed by i, label y, and an assignment $s_{t-1} = h_a$. These new random variables are drawn by distinct processes that are able to model such quantities and are subsequently incorporated in the node and edge potentials of our HCRF. We present in this paper the HCRF–DPM, a model that uses DPMs to define these random quantities (see its factor graph representation in Fig. 1). These variables, even though drawn by distinct processes, are coupled together by a common latent variable assignment in our graphical model. We redefine our potential function $\mathcal{F}$ from (4) as follows:

$$\mathcal{F}(y, \mathbf{s}, \mathbf{X}, \boldsymbol{\theta}) = \exp\left\{ \sum_{t=1}^{T}\sum_{i=1}^{d} \theta_x(s_t, i) f_t(i) \log \pi_x(s_t|i) + \theta_y(s_t, y) \log \pi_y(s_t|y) + \right.$$
$$\left. \sum_{t=2}^{T} \theta_e(s_t, s_{t-1}, y) \log \pi_e(s_t, y|s_{t-1}) \right\} . \quad (5)$$

We assume that random variables $\{\pi_x(h_k|i)\}_{k=1}^{\infty}$, $\{\pi_y(h_k|y)\}_{k=1}^{\infty}$, $\{\pi_e(h_k, y|h_a)\}_{k=1,y=1}^{\infty,|\mathcal{Y}|}$ are between 0 and 1. These are in effect the quantities that will allow the model to 'select' an appropriate number of useful hidden states for a given classification task. $\mathbf{f}_t$ are nonnegative features extracted from the observation sequence $\mathbf{X}$ and, as before, they can include arbitrary features of the input. We assume that $\boldsymbol{\theta}$ are nonnegative parameters and, as in (4), they model the relationships between hidden states and features ($\boldsymbol{\theta}_x$), labels $(\boldsymbol{\theta}_y)$ and transitions $(\boldsymbol{\theta}_e)$. These nonnegativity constraints for $\boldsymbol{\theta}$ and $\mathbf{f}$ are essential in this model, since the π-quantities are random variables and influence the probabilities of the hidden states: a negative parameter or feature would make an otherwise improbable state very likely to be chosen. Moreover, these constraints ensure compliance with the positivity constraints of our variational parameter updates (25)-(30), as we shall see later in this section. Finally, it is important to note that the positivity of $\boldsymbol{\theta}$ is not theoretically restrictive for our model due to the HCRF normalization factor $\frac{1}{Z(\mathbf{X})}$ in (3) where

$$Z(\mathbf{X}) = \sum_{y' \in \mathcal{Y}, \mathbf{s}} \mathcal{F}(y', \mathbf{s}, \mathbf{X}, \boldsymbol{\theta}) . \tag{6}$$

The HCRF–DPM model is an infinite HCRF where the quantities $\{\pi_x(h_k|i)\}_{k=1}^{\infty}$, $\{\pi_y(h_k|y)\}_{k=1}^{\infty}$, $\{\pi_e(h_k, y|h_a)\}_{k=1,y=1}^{\infty,|\mathcal{Y}|}$ in (5) are driven by coupled DPMs. It is important to understand that for the DPMs driving the π_e quantities in the edge features, h_k and y are treated as a single random variable –their product– $\omega_\mu = \{h_k, y\}$ that effectively has a state–space of size $|\mathcal{Y}| \times |\mathcal{H}|$, still an infinite number. According to the stick–breaking properties of DPs, we construct $\boldsymbol{\pi} = \{\pi_x, \pi_y, \pi_e\}$ conditioned on a new set of random variables $\boldsymbol{\pi}' = \{\pi'_x, \pi'_y, \pi'_e\}$ that follow $Beta$ distributions:

$$\pi'_x(h_k|i) \sim Beta(1, \alpha_x), \qquad \pi_x(h_k|i) = \pi'_x(h_k|i) \prod_{j=1}^{k-1} (1 - \pi'_x(h_j|i)) \tag{7}$$

$$\pi'_y(h_k|y) \sim Beta(1, \alpha_y), \qquad \pi_y(h_k|y) = \pi'_y(h_k|y) \prod_{j=1}^{k-1} (1 - \pi'_y(h_j|y)) \tag{8}$$

$$\pi'_e(\omega_\mu|h_a) \sim Beta(1, \alpha_e), \qquad \pi_e(\omega_\mu|h_a) = \pi'_e(\omega_\mu|h_a) \prod_{j=1}^{\mu-1} (1 - \pi'_e(\omega_\mu|h_a)) \tag{9}$$

This process can be made clearer by examining Fig. 2, where we visualize the stick breaking construction of an HCRF–DPM model with 2 observation features, 3 labels, and 10 'important' hidden states. The π_e-sticks have an important —for the implementation of our model— difference to the π_x and π_y-sticks in that the hidden states are intertwined with the labels, with each stick piece representing an ω–state. This means there are $|\mathcal{Y}|$ such states corresponding to one h–state. This becomes particularly important later on when we calculate our variational updates.

By using (5) the sequence of latent variables $\mathbf{s} = \{s_1, ...s_T\}$ can then be generated by the following process:

1. Draw $\pi'_x|\alpha_x \sim Beta(1, \alpha_x)$, $\pi'_y|\alpha_y \sim Beta(1, \alpha_y)$, $\pi'_e|\alpha_e \sim Beta(1, \alpha_e)$
2. Calculate $\boldsymbol{\pi}$ from (7)-(9). Note that this will only need to be calculated for a finite number of hidden states, due to our variational approximation.
3. For the tth latent variable, using (5) we draw

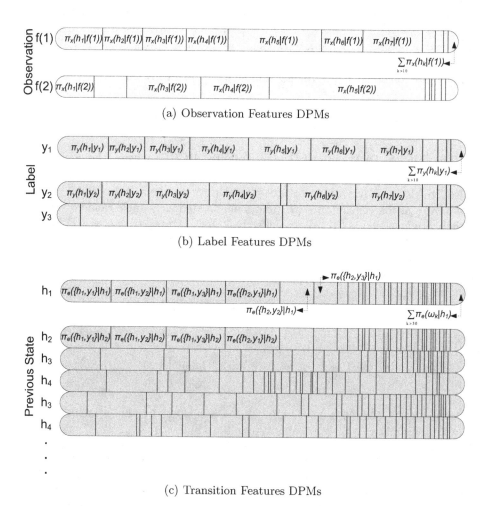

(a) Observation Features DPMs

(b) Label Features DPMs

(c) Transition Features DPMs

Fig. 2. Visualization of the π-'sticks' used to construct the infinite states in our HCRF–DPM. The fictitious model presented here has 2 observation features $f(1), f(2)$, 3 labels y_1, y_2, y_3 and fewer than 10 important hidden states $h_1, h_2, h_3 \ldots$. Each 'stick' sums up to 1, and the last piece always represents the sum of the lengths that correspond to all hidden states after the 10th state. Notice that for the π_e-'sticks' this corresponds to 30 ω–states. For example $\pi_e(h_1, y_3|h_2)$ controls the probability of transitioning from h_2 to h_1 in a sequence with label y_3. See text for more details.

$$s_t | \{\boldsymbol{\pi}_x', \boldsymbol{\pi}_y', \boldsymbol{\pi}_e', s_{t-1}, y, \mathbf{X}\} \sim Mult\Big(\exp\Big\{ \sum_{i=1}^{d} \theta_x(s_t, i) f_t(i) \log \pi_x(s_t | i) +$$

$$\theta_y(s_t, y) \log \pi_y(s_t | y) +$$

$$\theta_e(s_t, s_{t-1}, y) \log \pi_e(\{s_t, y\} | s_{t-1}) \Big\} \Big) \tag{10}$$

Rather than expressing the model in terms of $\boldsymbol{\pi}$, we use $\boldsymbol{\pi}' = \{\boldsymbol{\pi}_x', \boldsymbol{\pi}_y', \boldsymbol{\pi}_e'\}$ resulting in the folowing joint distribution that describes the HCRF–DPM:

$$p(y, \mathbf{s}, \boldsymbol{\pi}', \mathbf{X}, \theta) = p(y, \mathbf{s} \mid \boldsymbol{\pi}', \mathbf{X}, \theta) p(\boldsymbol{\pi}_x') p(\boldsymbol{\pi}_y') p(\boldsymbol{\pi}_e') \tag{11}$$

with

$$p(y, \mathbf{s} \mid \boldsymbol{\pi}', \mathbf{X}, \theta) = \frac{1}{Z(\mathbf{X})} \mathcal{F}(y, \mathbf{s}, \boldsymbol{\pi}', \mathbf{X}, \boldsymbol{\theta}) \tag{12}$$

where $Z(\mathbf{X}) = \sum_{y' \in \mathcal{Y}, \mathbf{s}} \mathcal{F}(y', \mathbf{s}, \boldsymbol{\pi}', \mathbf{X}, \boldsymbol{\theta})$. We assume independence of all $\boldsymbol{\pi}'$ variables above, so for example $p(\boldsymbol{\pi}_x') = \prod_{k=1}^{\infty} \prod_{i=1}^{d} \pi_x'(h_k | i)$.

Comparison with Previous Work. It is important at this stage to compare our model described by (5) with the MCMC model (IHCRF–MCMC) presented in [7]. The latter work defined potentials for each of the relationships between hidden states and features, labels and transitions and the potential function $\mathcal{F}$ as their product along the model chain:

$$\mathcal{F}(y, \mathbf{s}, \mathbf{X}) = \mathcal{F}_x(\mathbf{s}, \mathbf{X}) \mathcal{F}_y(y, \mathbf{s}) \mathcal{F}_e(y, \mathbf{s}) \tag{13}$$

$$\mathcal{F}_x(\mathbf{s}, \mathbf{X}) = \prod_{t=1}^{T} \prod_{i=1}^{d} \pi_x(s_t | i)^{f_t(i)} \tag{14}$$

$$\mathcal{F}_y(y, \mathbf{s}) = \prod_{t=1}^{T} \pi_y(s_t | y) \tag{15}$$

$$\mathcal{F}_e(y, \mathbf{s}) = \prod_{t=2}^{T} \pi_e(y, s_t | s_{t-1}) \tag{16}$$

The quantities $\boldsymbol{\pi}_x, \boldsymbol{\pi}_y, \boldsymbol{\pi}_e$ above are conceptually the same as in our model, except for the fact that in [7] they have Hierarchical Dirichlet Process (HDP) priors instead of DP priors, as we do in this paper.[1]

---

[1] Using HDP priors allows separate DPMs to be linked together via an identical base probabilistic measure, which is itself a DP. It would be interesting to use such priors for our model, but we were able to obtain satisfactory results without introducing higher complexity and additional hyperparameters into the variational model we experimented with. Notice that our model allows for such flexibility: using HDP priors would simply change the updates for our variational coordinate descent algorithm.

The potential function (13) above can be rewritten as follows:

$$\mathcal{F}(y, \mathbf{s}, \mathbf{X}) = \exp\left\{\sum_{t=1}^{T}\sum_{i=1}^{d} f_t(i)\log \pi_x(s_t|i) + \log \pi_y(s_t|y) + \sum_{t=2}^{T}\log \pi_e(s_t, y|s_{t-1})\right\}$$

(17)

A comparison between (17) and (5) makes it clear that our model is a generalization of the model presented in [7], which assumes, according to our framework, that $\boldsymbol{\theta}$-parameters are set to 1. The introduction of these parameters is not redundant, but allows for a more powerful and flexible models. Also, when dealing with classification problems involving continuous observation features using (5) for the potential function of an infinite HCRF is more suitable than (17), as we show in the experimental section. In those cases it is known that θ-parameters are of particular importance as they are able to capture the scaling of each input feature. The former model is not guaranteed to perform well unless some non-trivial normalization is applied on the observation features.

3.1   Variational Inference for the HCRF–DPM

Since inference on our model (11) is intractable, we need to approximate the marginal probabilities along the chain of our graphical model, and the π-quantities in (5). We shall do so with a mean–field variational inference approach. We use the following approximation for the joint distribution of our model:

$$q(y, \mathbf{s}, \boldsymbol{\pi}', \mathbf{X}) = q(y, \mathbf{s}|\mathbf{X})q(\boldsymbol{\pi}'_x)q(\boldsymbol{\pi}'_y)q(\boldsymbol{\pi}'_e)$$

(18)

where,

$$q(y, \mathbf{s}|\mathbf{X}) = q(y, s_1|\mathbf{X})\prod_{t=2}^{T} q(y, s_t|s_{t-1}, \mathbf{X})$$

$$= \prod_{i=1}^{d} q(s_1|i)q(s_1|y)\prod_{t=2}^{T}\prod_{i=1}^{d} q(s_t|i)q(s_t|y)q(s_t, y|s_{t-1}) .$$

(19)

Each individual approximate $q(\pi'_x), q(\pi'_y), q(\pi'_e)$ follows a *Beta* distribution with variational parameters $\boldsymbol{\tau}_x, \boldsymbol{\tau}_y, \boldsymbol{\tau}_e$ respectively. Explicitly, for features indexed by i, labels indexed by y, and hidden states indexed by k, k':

$$q(\pi'_x(h_k|i)) = Beta\left(\tau_{x,1}(k, i), \tau_{x,2}(k, i)\right)$$

(20)

$$q(\pi'_y(h_k|y)) = Beta\left(\tau_{y,1}(k, y), \tau_{y,2}(k, y)\right)$$

(21)

$$q(\pi'_e(y, h_k|h_{k'})) = Beta\left(\tau_{e,1}(y, k, k'), \tau_{e,2}(y, k, k')\right)$$

(22)

We approximate all $\boldsymbol{\pi}$ variables by employing a truncated stick–breaking representation which approximates the infinite number of hidden states with a finite number L [11]. This is the crux of our variational approach, and it effectively means that we set a truncation threshold L, above which the above quantities are set to 0: $\forall k > L$, $q(\pi'_x(h_k|i)) = 0$, $q(\pi'_y(h_k|y)) = 0$, $q(\pi'_e(y, h_k|h_{k'})) = 0$. Note that using this approximation is statistically rather different from using a finite model: an HCRF–DPM simply approximates the infinite number of states and will still reduce the number of useful hidden states to something smaller than L. It is finally important to stress that by constraining our θ–parameters and observation features to be positive, we effectively make the number of the θ–parameters that matter finite: changing a θ–parameter associated with a hidden state $k > L$ will not change our model.

3.2   Model Training

A trained variational HCRF–DPM model is defined as the set of optimal parameters θ^* and optimal variational parameters τ^*. In this work we obtain these with a training algorithm that can be divided in two distinct phases: (i) the optimization of our variational paramaters through a coordinate descent algorithm using the updates derived below and (ii) the optimization of parameters θ through a gradient descent method. Although it would be possible to have a fully Bayesian model with θ being random variables in our model, inference would become more difficult. Moreover, having a single value for our θ parameters is good for model interpretability and makes the application of a trained model to test data much easier.

Phase 1: Optimization of Variational Parameters τ. Now that we have defined an approximate model distribution in (18), we can approximate the necessary quantities for our inference. These approximations, as one can see later in this section, depend solely on our variational parameters τ. We calculate those by minimizing the Kullback-Liebler divergence (KL) between approximate and actual joint distributions of our model, (11) and (18), using a coordinate descent algorithm:

$$KL[q||p] = \log Z(\mathbf{X}) - \langle \log \mathcal{F}(y, \mathbf{x}, \pi', \mathbf{X})p(\pi') \rangle_{q(y,\mathbf{s},\pi'|\mathbf{X})} + \langle \log q(y, \mathbf{s}|\mathbf{X})q(\pi') \rangle_{q(y,\mathbf{s},\pi'|\mathbf{X})} \qquad (23)$$

where $\langle \cdot \rangle_q$ is the expectation of $\cdot$ with respect to q. Thus, the energy of the configuration of our random variables y, $\mathbf{s}$, and π' is $\log \mathcal{F}(y, \mathbf{x}, \pi', \mathbf{X})p(\pi')$ and the free energy of the variational distribution:

$$\mathcal{L}(q) = - \langle \log \mathcal{F}(y, \mathbf{x}, \pi', \mathbf{X})p(\pi') \rangle_{q(y,\mathbf{s},\pi'|\mathbf{X})} + \langle \log q(y, \mathbf{s}|\mathbf{X})q(\pi') \rangle_{q(y,\mathbf{s},\pi'|\mathbf{X})} \qquad (24)$$

Since $\log Z(\mathbf{X})$ is constant for a given observation sequence, minimizing the free energy $\mathcal{L}(q)$ minimizes the KL divergence.

We will obtain the variational updates for the two groups of latent variables $q(y, \mathbf{s}|\mathbf{X})$ and $q(\pi')$ by setting the partial derivative with respect to each group of $\mathcal{L}(q)$ to 0 and solving for the approximate distribution of each group of latent variables. The updates for the *Beta* parameters of $q(\pi')$ from (20)-(22) are:

$$\tau_{x,1}(k,i) = \sum_t f_t[i]\theta_x(k,i)q(s_t = h_k|i) + 1 \tag{25}$$

$$\tau_{x,2}(k,i) = \sum_t f_t[i]\theta_x(k,i)q(s_t > h_k|i) + \alpha_x \tag{26}$$

$$\tau_{y,1}(y,i) = \sum_t \theta_y(k,y)q(s_t = h_k|y) + 1 \tag{27}$$

$$\tau_{y,2}(y,i) = \sum_t \theta_y(k,i)q(s_t > h_k|y) + \alpha_y \tag{28}$$

$$\tau_{e,1}(y,k,k') = \sum_t \theta_e(k,k',y)q(s_t = h_k, y, s_{t-1} = h_{k'}) + 1 \tag{29}$$

$$\tau_{e,2}(y,k,k') = \sum_t \theta_e(k,k',y)q(s_t > h_k, y, s_{t-1} = h_{k'}) + \alpha_e \tag{30}$$

Quantities $q(s_t = h_k|i)$, $q(s_t = h_k|y)$, and $q(s_t = h_k, y, s_{t-1} = h_{k'})$ can be obtained by the forward–backward algorithm. The latter requires only conditional approximate likelihoods $q(s_t = h_k|i,y,h_{k'})$, which can be also be calculated by setting the derivative of $\mathcal{L}(q)$ to zero:

$$
\begin{aligned}
q(s_t = h_k|i,y,h_{k'}) \propto \exp\Bigg\{ \\
f_t(i)\theta_x(k,i)\left(\langle\log\pi'_x(s_t = h_k|i)\rangle_{q(\pi')} + \sum_{j=k+1}^{L}\langle\log(1-\pi'_x(s_t = h_j|i))\rangle_{q(\pi')}\right) \\
\theta_y(k,y)\left(\langle\log\pi'_y(s_t = h_k|y)\rangle_{q(\pi')} + \sum_{j=k+1}^{L}\langle\log(1-\pi'_y(s_t = h_j|y))\rangle_{q(\pi')}\right) \\
\theta_e(k,k',y)\bigg(\langle\log\pi'_e(s_t = h_k, y|s_{t-1} = h_{k'})\rangle_{q(\pi')} + \\
\sum_{j=k+1}^{L}\langle\log(1-\pi'_e(s_t = h_j, y|s_{t-1} = h_{k'}))\rangle_{q(\pi')}\bigg)\Bigg\}
\end{aligned}
\tag{31}
$$

Since all $\boldsymbol{\pi}'$ follow a Beta distribution, the expectations above are known.

Phase 2: Optimization of Parameters $\boldsymbol{\theta}$. We find our optimal parameters $\boldsymbol{\theta}^* = \arg\max\log p(y|\mathbf{X},\boldsymbol{\theta})$ based on a training set by using a common HCRF quasi–Newton gradient descent method (LBFGS), which requires the gradient of the log–likelihood with respect to each parameter. These gradients for our model are:

$$\frac{\partial \log p(y|\mathbf{X}, \boldsymbol{\theta})}{\partial \theta_x(k, i)} = \sum_t p(s_t = h_k|y, \mathbf{X}, \boldsymbol{\theta}) f_t(i) \log \pi_x(h_k|i) -$$

$$\sum_{y' \in \mathcal{Y}, t} p(s_t = h_k, y'|\mathbf{X}, \boldsymbol{\theta}) f_t(i) \log \pi_x(h_k|i) \qquad (32)$$

$$\frac{\partial \log p(y|\mathbf{X}, \boldsymbol{\theta})}{\partial \theta_y(k, y)} = \sum_t p(s_t = h_k|y, \mathbf{X}, \boldsymbol{\theta}) \log \pi_y(h_k|y) -$$

$$\sum_{y' \in \mathcal{Y}, t} p(s_t = h_k, y'|\mathbf{X}, \boldsymbol{\theta}) \log \pi_y(h_k|y) \qquad (33)$$

$$\frac{\partial \log p(y|\mathbf{X}, \boldsymbol{\theta})}{\partial \theta_e(k, k', y)} = \sum_t p(s_t = h_k, s_{t-1} = h_{k'}|y, \mathbf{X}, \boldsymbol{\theta}) \log \pi_e(h_k, y|h_{k'})$$

$$- \sum_{y' \in \mathcal{Y}, t} p(s_t = h_k, s_{t-1} = h_{k'}, y'|\mathbf{X}, \boldsymbol{\theta}) \log \pi_e(h_k, y|h_{k'}) \qquad (34)$$

We make this gradient descent tractable by using the variational approximations for the intractable quantities in the above equations. However, there is a significant difference with other CRF and HCRF models that use such techniques to find optimal parameters: we are constrained to only positive θ-parameters Since we are using a quasi–Newton method with Armijo backtracking line search, we can use the gradient projection method of [12, 13] to enforce this constrain. Finally, it is important to stress here that, although our model includes parameters that are not treated probabilistically, we have not seen signs of overfitting in our experiments (see Fig. 4).

4   Experimental Results

4.1   Performance on a Synthetic Dataset with Continuous Features

In an effort to demonstrate the ability of our HCRF–DPM to model sequences with continuous features correctly, we created a synthetic dataset, on which we compared its performance to that of the IHCRF–MCMC model [7]. The simple dataset was generated by two HMMs, with 4 Gaussian hidden states each. Two of the states were shared between the two HMMs, resulting in a total of 6 unique hidden states, out of a total of 8 for the two labels.

We trained 10 randomly initialized models of the finite HCRF, IHCRF–MCMC and HCRF–DPM on 100 training sequences and chose in each case the best one based on their performance on an evaluation set of 100 different sequences. The performance of the models was finally evaluated by comparing the F1 measure achieved on a test set of 100 other sequences. All sets had an equal number of samples from each label. The IHCRF–MCMC model was unable to solve this simple two–label sequence classfication problem with continuous-only input features: it consistently selected Label 1. On the other hand, the finite HCRF and the new HCRF–DPM model were successful in achieving a perfect F1 score of 100% on the test set (see Table 1).

4.2   Application to the Audiovisual Analysis of Human Behavior

The problem of automatically classifying episodes of high–level emotional states, such as pain, agreement and disagreement, based on nonverbal cues in audiovisual sequences of spontaneous human behavior is rather complex [14]. Although humans are particularly good at interpreting such states, automated systems perform rather poorly. Infinite models are particularly attractive for modeling human behavior as we usually cannot have a solid intuition regarding the number of hidden states in such applications. Furthermore, it opens up the way of analyzing the hidden states these models converge to, which might provide social scientists with valuable information regarding the temporal interaction of groups of behavioral cues that are different or shared in these behaviors. We therefore decided to evaluate our novel approach on behavior analysis and specifically the recognition of agreement, disagreement and pain in recordings of spontaneous human behavior. We expected that our HCRF–DPM models would find a good number of shared hidden states and perform at least as well as the best cross-validated finite HCRF and IHCRF–MCMC models.

In this work we used an audiovisual dataset of spontaneous agreement and disagreement and a visual dataset of pain to evaluate the performance of the proposed model on four classification problems: *(1)* ADA2, agreement and disagreement recognition with two labels (agreement vs. disagreement); *(2)* ADA3, agreement and disagreement recognition with three labels (agreement vs. disagreement vs. neutral); *(3)* PAIN2, pain recognition with two labels (strong pain vs. no pain); and *(4)* PAIN3, pain recognition with three labels (strong pain vs. moderate pain vs. no pain). We show that *(1)* our model is capable of finding a good number of useful states; and *(2)* HCRF–DPMs perform better than the best performing finite HCRF and IHCRF–MCMC models in all of these problems with the exception of ADA3, where the performance of the HCRF–DPM is similar to that of the finite model.

The dataset of agreement and disagreement comprises 53 episodes of agreement, 94 episodes of disagreement, and 130 neutral episodes of neither agreement or disagreement. These feature 28 participants and they occur over 11 political debates. We used automatically extracted prosodic features (continuous), and manually annotated hand and head gestures (binary). We compared the finite HCRF and the IHCRF–MCMC to our HCRF–DPM based on the F1 measure they achieved. In each case, we evaluated their performance on a test set consisting of sequences from 3 debates. We ran all models with 60 random initializations, selecting the best trained model each time by examining the F1 achieved on a validation set consisting of sequences from 3 debates. It is important to stress that each sequence belonged uniquely to either the training, the validation, or the testing set.

The database of pain we used includes 25 subjects expressing various levels of pain in 200 video sequences. Our features were based on the presence (binary) of each of the 45 observable facial muscle movements–Action Units (AUs) [15]. For our experiments, we compared the finite HCRF and the IHCRF–MCMC to our HCRF–DPM based on the F1 measure they achieved. We evaluated the performance of the models on 25 different folds (leave–7–subjects–out for testing). In each case we concatenated the predictions for every test sequence of each fold and calculated the F1 measure for each label. The measure we used was the average F1 over all labels. We ran all experiments with 10 random initializations, selecting the best model each time by examining the F1 achieved on a validation set consisting of the sequences from 7 subjects. In every fold our training, validation and testing sets comprised not only of unique sequences but also of unique subjects.

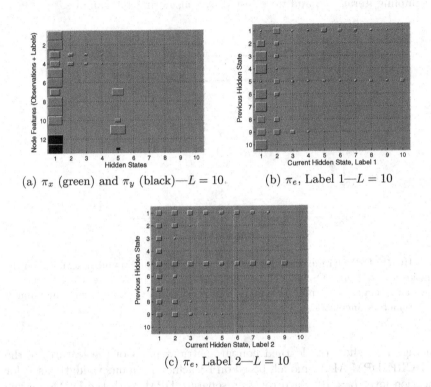

(a) π_x (green) and π_y (black)—$L = 10$ (b) π_e, Label 1—$L = 10$

(c) π_e, Label 2—$L = 10$

Fig. 3. Hinton Diagrams of π-quantities in node and edge features of variational HCRF-DPM models with truncation level $L = 10$ for ADA2. The first column presents the π-quantities for node features: π_x for observation features in green, π_y for labels in black. The second and third columns present the π_e-quantities for labels 1 and 2 respectively. See text for additional details.

For all four tasks, in addition to the random initializations the best HCRF model was also selected by experimenting with different number of hidden states and different values for the HCRF L2 regularization coefficient. Specifically, for each random initialization we considered models with 2, 3, 4, and 5 hidden states and an L2 coefficient of 1, 10, and 100. This set of values for the hidden states was selected after preliminary results deemed a larger number of hidden states only resulted in severe overfitting for all problems. We did not use regularization for our HCRF-DPM models and all of them had their truncation level set to $L = 10$ and their hyperparameters to $s_1 = 1000$ and $s_2 = 10$. Finally, our finite HCRF models were trained with a maximum of 300 iterations for the gradient ascent method used [10], whereas our HCRF-DPM models were trained with a maximum of 1200 variational coordinate descent iterations and a maximum of 600 iterations of gradient descent. All IHCRF–MCMC models were trained according to the experimental protocol of [7]. They had their initial number of represented hidden states set to $K = 10$, they were trained with 100 sampling iterations, and were tested by considering 100 samples.

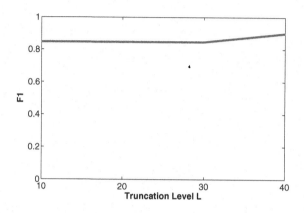

Fig. 4. HCRF–DPM F1 measure (higher F1 means higher perfomance) achieved on the validation set of ADA2. Our model does not show signs of overfitting: the F1 achieved on the validation set does not decrease as the truncation level L, and thus the number of $\boldsymbol{\theta}$–parameters, increases.

In Fig. 3 we show the learned nonparametric π parts of the features of the best HCRF–DPM ADA2 model, based on F1 achieved on our validation set, for truncation level $L = 10$. Each row is a separate DPM; with the DPMs for the edge potentials spanning across labels. Recall from Fig. 2 that these quantities have to sum to 1 across each row. As one can see in these figures, setting the truncation level $L = 10$ was a reasonable choice. Paying particular attention to the first column (node features), it seems that HCRF–DPMs converge to a small number of utilized hidden states —the equivalent table for a finite HCRF would be dense with each state being used. One can see unique and shared states, a feature of HCRFs that makes them particularly appealing for classification tasks. Fig. 3a clearly shows that the model uses only two states, one of them (state 1) being shared among both labels –features 12 and 13 in this Hinton diagram– and another (state 5) being used only by label 2.

Since we have introduced parameters $\boldsymbol{\theta}$ it is sensible to test our methodology for signs of overfitting. The only value linked with the number of our parameters is our truncation level L: their number increases as we increase L. In Fig. 4 we show the F1 measure achieved on the validation set of ADA2 for HCRF–DPMs with L=10, 20, 30, 40. This graph is a strong indication that HCRF–DPMs do not show signs of overfitting. We would see such signs if by increasing L the performance (F1 measure) for our validation set would decrease. However, as we see here, performance on the validation sets remains roughly the same as we increase L.

Table 1. F1 measure achieved by our HCRF-DPM vs. the best, in each fold of each problem, finite HCRF and IHCRF-MCMC. **Synthetic:** Two–label classification for an HMM–generated dataset with continuous–only features **ADA2:** Two–label classification for the Canal9 Dataset of agreement and disagreement; **ADA3:** Three–label classification for the Canal9 Dataset; **PAIN2:** Two–label classification for the UNBC dataset of shoulder pain; **PAIN3:** Three–label classification for the UNBC dataset

Dataset	Finite HCRF	IHCRF–MCMC	Our HCRF–DPMs
Synthetic	100.0%	33.3%	**100.0%**
ADA2	58.4%	61.2%	**76.1%**
ADA3	50.7%	**60.3%**	49.8%
PAIN2	83.9%	88.4%	**89.2%**
PAIN3	53.9%	57.7%	**59.0%**

Table 1 shows the average over all labels of the F1 measure on the test sets for all our problems. Since the nonparametric model structure is not specified a priori but is instead determined from our data, the HCRF–DPM model is more flexible than the finite HCRF and is able to achieve better performance in all cases, with the exception of 3-label classification problem of agreement/disagreement (ADA3), where the HCRF–DPM seems to perform almost equally well with the finite model. The HCRF–DPM perfomed better than the IHCRF–MCMC in all problems with the exception of ADA3. An analysis of a IHCRF–MCMC model trained for ADA3 shows that the model ignored the two continuous dimensions and used only the binary features to model the dataset, which evidently resulted in slightly better performance.

5   Conclusion

In this paper we have presented a novel variational approach to learning an infinite Hidden Conditional Random Field, the HCRF–DPM, a discriminative nonparametric sequential model with latent variables. This deterministic approach overcomes the limitations of sampling techniques, like the one presented in [7]. We have also shown that our model is in fact a generalization of the IHCRF–MCMC presented in [7] and is able to handle sequence classification problems with continuous features naturally. In support of the latter claim, we conducted an experiment with a Gaussian HMM–generated synthetic dataset of continuous–only features which showed that HCRF–DPMs are able to perform well on classification problems where the IHCRF–MCMC fails. Furthermore, we conducted experiments with four challenging tasks of classification of naturalistic human behavior. HCRF–DPMs were able to find a good number of shared hidden states, and to perform well in all problems, without showing signs of overfitting.

Acknowledgements. This work has been funded in part by the European Community's 7th Framework Programme [FP7/20072013] under the grant agreement no 231287 (SSPNet). K. Bousmalis is a recipient of the Google Europe Fellowship in Social Signal Processing, and this research is supported in part by this Google Fellowship. The work of Maja Pantic is funded in part by the European Research Council under the ERC Starting Grant agreement no. ERC-2007-StG-203143 (MAHNOB). This material is based upon work supported by the National Science Foundation under Grant No. 1118018 and the U.S. Army Research, Development, and Engineering Command. Any opinions, findings, and conclusions or recommendations expressed in this material are those of the author(s) and do not necessarily reflect the views of the National Science Foundation or the Government.

References

1. Rasmussen, C.: The Infinite Gaussian Mixture Model. In: Proc. Advances in Neural Information Processing Systems, pp. 554–560 (2000)
2. Beal, M., Ghahramani, Z., Rasmussen, C.: The Infinite Hidden Markov Model. In: Proc. Advances in Neural Information Processing Systems, pp. 577–584 (2002)
3. Fox, E., Sudderth, E., Jordan, M., Willsky, A.S.: An HDP–HMM for systems with state persistence. In: Proc. Int'l Conf. on Machine Learning (2008)
4. Orbanz, P., Buhmann, J.: Nonparametric Bayes Image Segmentation. Int'l Journal in Computer Vision 77, 25–45 (2008)
5. Van Gael, J., Teh, Y., Ghahramani, Z.: The Infinite Factorial Hidden Markov Model. Advances in Neural Information Processing Systems 21, 1697–1704 (2009)
6. Chatzis, S., Tsechpenakis, G.: The infinite hidden Markov random field model. IEEE Trans. Neural Networks 21(6), 1004–1014 (2010)
7. Bousmalis, K., Zafeiriou, S., Morency, L.P., Pantic, M.: Infinite hidden conditional random fields for human behavior analysis. IEEE Trans. Neural Networks and Learning Systems 24(1), 170–177 (2013)
8. Ghahramani, Z., Beal, M.: Propagation Algorithms for Variational Bayesian Learning. Advances in Neural Information Processing Systems 13, 507–513 (2001)
9. Sethuraman, J.: A Constructive Definition of Dirichlet Priors. Statistica Sinica 4, 639–650 (1994)
10. Quattoni, A., Wang, S., Morency, L., Collins, M., Darrell, T.: Hidden Conditional Random Fields. IEEE Trans. Pattern Analysis and Machine Intelligence, 1848–1852 (2007)
11. Blei, D., Jordan, M.: Variational Inference for Dirichlet Process Mixtures. Bayesian Analysis 1(1), 121–144 (2006)

12. Bertsekas, D.: On the Goldstein-Levitin-Polyak Gradient Projection Method. IEEE Trans. on Automatic Control 21, 174–184 (1976)
13. Bertsekas, D.: Nonlinear Programming. Athena Scientific (1999)
14. Vinciarelli, A., Pantic, M., Bourlard, H.: Social Signal Processing: Survey of an Emerging Domain. Image and Vision Computing 27(12), 1743–1759 (2009)
15. Ekman, P., Friesen, W.V., Hager, J.C.: Facial Action Coding System. Research Nexus, Salt Lake City (2002)

Sparsity in Bayesian Blind Source Separation and Deconvolution

Václav Šmídl and Ondřej Tichý

Institute of Information Theory and Automation, Prague, Czech Republic
{smidl,otichy}@utia.cas.cz

Abstract. Blind source separation algorithms are based on various separation criteria. Differences in convolution kernels of the sources are common assumptions in audio and image processing. Since it is still an ill posed problem, any additional information is beneficial. In this contribution, we investigate the use of sparsity criteria for both the source signal and the convolution kernels. A probabilistic model of the problem is introduced and its Variational Bayesian solution derived. The sparsity of the solution is achieved by introduction of unknown variance of the prior on all elements of the convolution kernels and the mixing matrix. Properties of the model are analyzed on simulated data and compared with state of the art methods. Performance of the algorithm is demonstrated on the problem of decomposition of a sequence of medical data. Specifically, the assumption of sparseness is shown to suppress artifacts of unconstrained separation method.

1 Introduction

The aim of blind source separation is to recover the original form of signals that can be observed only via their superposition. A classical example of such a situation is the cocktail party problem [9], where multiple speakers are recorded by multiple microphones. The aim is to separate audio signal of the individual speakers. This requires specification of the separability criteria. One such criteria is the assumption of temporal properties of the source, expressed via different convolution kernels [15]. Since the convolution kernels are also unknown, the problem is that of blind deconvolution within blind separation. Algorithms for this problem include optimization of information theoretic measures [4] and the EM algorithm [2]. In the image processing literature, the problem is closely related to the multi-channel blind deconvolution [18].

The presented algorithm was primarily motivated by application in medical image analysis which has the following specific issues: (i) the sources are physiological organs which will be further analyzed by medical experts for final diagnosis, and (ii) poor signal to noise conditions, where a weak signal is hard to separate from the noise. The medical experts expect that the results will respect physiological nature which is very hard to formalize mathematically. The model of source activity by convolution of common input function is one of a few mathematical models that is generally accepted. The assumption of sparsity is also

H. Blockeel et al. (Eds.): ECML PKDD 2013, Part II, LNAI 8189, pp. 548–563, 2013.
© Springer-Verlag Berlin Heidelberg 2013

natural in this application. However, such assumptions are not unique to medical imaging and the resulting algorithm may be used in any other application domain.

The poor signal-to-noise conditions of the domain motivated our choice of the Bayesian approach. It has been successfully applied in situations when the number of the sources is lower than the number of the channels, [13]. The ability to marginalize provides an automatic Occam's razor that suppresses the spurious sources and thus provides automatic denoising. Since exact marginalization may not be always possible, approximate methods of Bayesian calculus has been developed. One such formalism is the Variational Bayes method [5,16]. Its use for selection of the number of principal components has been demonstrated in [5], via the use of priors with unknown variance. In connection with the Variational Bayes approximation it favors sparse solutions. Since its introduction, this mechanism has been used in image deconvolution [19], or sparse blind source separation [17]. We introduce this modeling assumption on the convolution kernel and the mixing matrix.

The resulting algorithm is applied to the problem of image sequence decomposition. This problem has been studied independently for many years [3,6], however, it has been recognized as a special case of the blind source separation problem [13,16]. The specific nature of this problem is in interpretation of the resulting components which are further used for medical diagnosis. Even a small improvement in the estimation may have significant impact on the diagnostic quality of the results.

2   Sparsity in Bayesian Analysis

The Bayesian inference is concerned with evaluation of the full posterior density of the parameters θ from the observed data D. It requires a parametric probabilistic model of the data in the form of a probability distribution, $p(D|\theta)$, conditioned by knowledge of the parameters, θ. The prior state of knowledge of θ is quantified by the *prior* distribution, $p(\theta)$. Our state of knowledge of θ after observing D is quantified by the *posterior distribution*, $p(\theta|D)$. These functions are related via Bayes' rule:

$$p(\theta|D) = \frac{p(\theta, D)}{p(D)} = \frac{p(D|\theta)\,p(\theta)}{\int p(D|\theta)\,p(\theta)\,d\theta},\tag{1}$$

where integration in the denominator of (1) is over the whole support of the involved distributions. We will refer to $p(\theta, D)$ as the joint distribution of parameters and data, or, more concisely, as the *joint distribution*.

2.1   The Variational Bayes Approximation

The Variational Bayes (VB) approximation is a deterministic technique for approximation of the Bayes rule (1), in the sense of the following theorem [16].

Theorem 1. *Let $p(\theta|D)$ be the posterior distribution of multivariate parameter, $\theta = [\theta_1', \theta_2']'$, and $p^*(\theta|D)$ be an approximate distribution restricted to the set of conditionally independent distributions:*

$$p^*(\theta|D) = p^*(\theta_1, \theta_2|D) = p^*(\theta_1|D)\, p^*(\theta_2|D). \tag{2}$$

Any minimum of the Kullback-Leibler divergence from $p^*(\cdot)$ to $p(\cdot)$

$$KL(p^*(\theta|D)\,||\,p(\theta|D)) = \int p^*(\theta|D) \ln \frac{p(\theta|D)}{p^*(\theta|D)} d\theta, \tag{3}$$

is achieved when $p^*(\cdot) = \tilde{p}$ where

$$\tilde{p}(\theta_i) \propto \exp\left(\mathsf{E}_{\tilde{p}(\theta_{/i})}\left(\ln\left(p\left(\theta, D\right)\right)\right)\right), \ i = 1, 2. \tag{4}$$

Here, $\theta_{/i}$ denotes the complement of θ_i in θ and $\mathsf{E}_{p(\theta)}(g(\theta))$ denotes expected value of function $g(\theta)$ with respect to distribution $p(\theta)$.

Theorem 1 is also known as mean-field approximation [14] and provides a powerful tool for approximation of joint pdfs in *separable form* [16]:

$$\ln p(\theta_1, \theta_2, D) = g(\theta_1, D)'\, h(\theta_2, D). \tag{5}$$

Here, $g(\theta_1, D)$ and $h(\theta_2, D)$ are finite-dimensional vectors. Using (5) in (4),

$$\tilde{p} \propto \exp\left(g(\theta_1, D)'\, \widehat{h(\theta_2, D)}\right), \tag{6}$$

where $\widehat{h(\cdot)} = \mathsf{E}_{\tilde{p}(\theta_2|D)}(h(\cdot))$ are the moments of θ_2, and similarly for θ_1. In cases, where (6) are from exponential family, $h(\cdot)$ form its sufficient statistics [8]. An iterative moment-swapping algorithm is implied [16].

2.2 Automatic Relevance Determination

The mechanism of automatic relevance determination (ARD) is based on joint estimation of the parameters of the prior (hyper-parameters) with the data [5]. Specifically, the prior of an unknown vector parameter θ that is assumed to have elements redundant for the observed data is chosen as

$$p(\boldsymbol{\theta}|\boldsymbol{\omega}) = \mathcal{N}(0, \text{diag}(\boldsymbol{\omega})), \qquad p(\omega_i) = G(\alpha_0, \beta_0), \ \forall i, \tag{7}$$

where $\boldsymbol{\omega}$ is the vector of unknown precisions (inverse variances) of the prior on the parameter $\boldsymbol{\theta}$ and it is assumed to have conjugate Gamma prior with scalar parameters α_0, β_0. The Bayes rule is then used to estimate both $\boldsymbol{\theta}$ and $\boldsymbol{\omega}$. When the parameter is redundant, the expected value of the prior variance ψ approaches zero. This effect is known as the ARD principle and it is demonstrated on the following example.

Example 1 (Multiplicative scalar decomposition). Consider the following model of scalar measurement d being explained as a product of two unknown parameter, a and x:

$$d = ax + e, \ e \sim \mathcal{N}(0, r_e). \tag{8}$$

where variance r_e is assumed to be known. The likelihood function of the model parameters is

$$p(d|a, x) = \mathcal{N}(ax, r_e), \tag{9}$$

and has maximum anywhere on the manifold defined by the signal estimate:

$$\widehat{ax} = d. \tag{10}$$

Separation of the signal from the noise is possible only with additional assumptions. One such assumption is the choice of prior on the x variable as $p(x) = \mathcal{N}(0, r_x)$, with a chosen variance r_x. Maximum of the marginal $p(a|d) = \int p(d|a, x, r_e)p(x)dx$ is then

$$\hat{a}_{marg} = \begin{cases} \frac{\sqrt{d^2 - r_e}}{\sqrt{r_x}} & \text{if } d > \sqrt{r_e}, \\ 0 & \text{otherwise.} \end{cases} \tag{11}$$

Note the inference bound on the signal, $d > \sqrt{r_e}$, i.e. the signal should be higher than the standard deviation of the noise. This bound enforces sparsity of the solution since estimates of the parameters for a weak signal are zeros.

The ARD is based on introduction of the hyper-parameters (7) on any variable, a or x, or both. For example, a fixed prior on a, $p(a) = \mathcal{N}(0, \sigma_a)$, and the ARD prior on x, i.e. $p(x|\omega_x) = \mathcal{N}(0, \omega_x^{-1})$, with unknown precision ω_x with prior $p(\omega_x) = G(\alpha_0, \beta_0)$, yields the variational posteriors of the form

$$\tilde{p}(x|d) = \mathcal{N}(\hat{x}, \sigma_x), \qquad \tilde{p}(\omega_x|d) = G(\frac{3}{2}, \gamma_x), \qquad \tilde{p}(a|d) = \mathcal{N}(\hat{a}, \sigma_a) \tag{12}$$

with shaping parameters satisfying the following set of implicit equations:

$$\hat{x} = \frac{\sigma_x}{r_e}d\hat{a}, \qquad \sigma_x = ((\hat{a}^2 + \sigma_a)r_e^{-1} + \frac{3}{2\gamma_x})^{-1},$$

$$\hat{a} = \frac{\sigma_a}{r_e}d\hat{x}, \qquad \gamma_x = \sigma_x + \hat{x}^2. \tag{13}$$

Numerical solution of this set is achieved by the iterative algorithm [1]. We choose an initial value of $\sigma_x^{(0)}$ and $\hat{a}^{(0)}$ and then iteratively evaluate equations (13) in the order: $\hat{x}$, γ_x, σ_x, $\hat{a}$.

We note the following:

- The choice of σ_a fixes the value of $\hat{a}$ at a constant for all significant values of d, and the free parameter that grows with d is the $\hat{x}$.

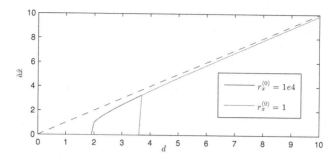

Fig. 1. Product of expected values $\hat{a}$ and $\hat{x}$ of variational posteriors from (12) for two initial conditions of $\sigma_x^{(0)}$. The dashed line denotes maximum likelihood solution (10).

- The product of the $\hat{a}\hat{x}$ is displayed in Fig. 1 for a range of values d. Note the presence of the inference bound similar to (11). In this case, the estimates are zeros for $d < 2\sqrt{r_e}$, i.e. the ARD property of the VB inference enforces sparse estimates more aggressively than the marginalization. We conjecture that this is a consequence of the variance underestimation of the VB approximation [12].
- The converged results are insensitive to the choice of the $\hat{a}^{(0)}$ parameter, and to some extent even to the $\sigma_x^{(0)}$ parameter. The hyper-parameters α, β of both variables were set to zero to yield the Jeffreys' prior. For $\sigma_x^{(0)} > 1$ the results correspond to those of $\sigma_x^{(0)} = 1e4$ in Fig. 1. However, the converged values differ for $\sigma_x^{(0)} \leq 1$, Fig. 1, which illustrates the existence of local minima in the VB procedure [16].
- The Variational PCA [5] is a multivariate extension of this model with ARD applied on the columns of the mixing matrix.

Remark 1 (Symmetric ARD). It is possible to introduce ARD on both variables a and x. However, reliable estimation is achieved only with enforced positivity of a and x via truncated Normal prior. In this case, the estimation results are closely similar to those in Fig. 1.

3   Blind Source Separation and Deconvolution

The task of blind source separation arise when the observed signal is assumed to be a superposition of the source signal. In this Section we assume that the source signals are generated via convolution of the common input function with source-specific kernels.

3.1   Signal Superposition and Convolution

The basic formulation of the blind source separation assumes that the vector of observations at time t, $\mathbf{d}_t$ is a linear superposition of all source signals at the same time, $\overline{\mathbf{x}}_t$:

$$\mathbf{d}_t = A\overline{\mathbf{x}}_t, \tag{14}$$

where A is the mixing matrix, $\overline{\mathbf{x}}_t = [x_{t,1}, \ldots, x_{t,r}]$, and r is the number of sources. The number of observations is p and the length of the source signal is n. The full observed sequence can be written in matrix notation as:

$$D = AX', \tag{15}$$

where $D = [\mathbf{d}_1, \ldots, \mathbf{d}_n]$, and the columns of matrix X are the source signals $X = [\mathbf{x}_1, \ldots, \mathbf{x}_r]$ ($\overline{\mathbf{x}}_t$ are the rows of matrix X). In this case, we assume the number of sources r to be unknown with conditions $r < n$, $r < p$.

The kth source is assumed to be the result of convolution of the common input function $\mathbf{b}$, and source-specific convolution kernels $\mathbf{u}_k$:

$$\mathbf{x}_k = \mathbf{b} * \mathbf{u}_k = B\mathbf{u}_k, \tag{16}$$

where matrix B is defined as follows:

$$B = \begin{pmatrix} b_1 & 0 & 0 & 0 \\ b_2 & b_1 & 0 & 0 \\ \ldots & b_2 & b_1 & 0 \\ b_n & \ldots & b_2 & b_1 \end{pmatrix}. \tag{17}$$

The full model of the data is thus

$$D = AX' = AU'B', \tag{18}$$

where $U = [\mathbf{u}_1, \ldots, \mathbf{u}_k]$ and all parameters A, U, B are unknown. In the sequel, we will assume that all these parameters are positive. This is motivated by the application area in image analysis.

3.2 Probabilistic Model

The deterministic assumptions in the previous Section are valid only approximately. For example, the data vectors $\mathbf{d}_t$ are subject to the observation noise. The observed data $\mathbf{d}_t$ are thus modeled as random realizations from the probability density function. For Gaussian distributed noise, the matrix of observations is assumed to be distributed as

$$p(D|A, B, U, \omega) = \mathcal{N}(AU'B', \omega^{-1}I_p \otimes I_n) = \prod_{t=1}^{n} \mathcal{N}(A\overline{\mathbf{x}}_t, \omega^{-1}I_p), \tag{19}$$

where I_p denotes identity matrix of size $p \times p$, $\mathcal{N}(.,.)$ of matrix argument denotes the matrix normal distribution [16] and symbol $\otimes$ denotes the Kronecker product. Prior distributions for all unknown parameters A, B, U, ω need to be specified.

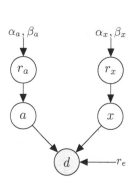
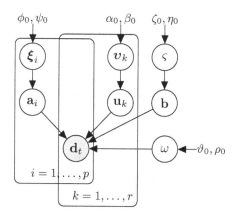

Fig. 2. Graphical model of the scalar multiplicative decomposition from Remark 1 (left) and the proposed sparse blind source separation and deconvolution (right)

The parameter ω is a precision parameter of a Gaussian density and thus it has a conjugate prior in the form of Gamma density

$$p(\omega) = G(\vartheta_0, \rho_0),$$

with chosen constants ϑ_0, ρ_0. These may be chosen to approach $\vartheta_0 \to 0, \rho_0 \to 0$ yielding an uninformative Jeffrey's prior on the scale parameter [10]. The input function $\mathbf{b}$ is assumed to have all positive elements. This assumption is modeled by Normal distributed prior with its support truncated to positive values (Appendix B.1), the truncated normal distribution is denoted $t\mathcal{N}()$ with the same arguments the Normal distribution since the truncation interval is always $\langle 0, \infty \rangle$. The precision of the prior is also unknown:

$$p(\mathbf{b}|\varsigma) = t\mathcal{N}(0, \varsigma^{-1} I_n), \qquad p(\varsigma) = G(\zeta_0, \eta_0). \qquad (20)$$

The only assumption on the mixing matrix and the convolution kernels is sparsity. In both cases it will be achieved by the ARD property (Section 2.2). Specifically, the ARD prior (7) is used for all elements of matrices A and U. In vector notation, the ARD corresponds to a variance with unknown diagonal:

$$p(\mathbf{u}_k|\boldsymbol{v}_k) = t\mathcal{N}(\mathbf{0}_{n,1}, \text{diag}(\boldsymbol{v}_k)^{-1}), \qquad (21)$$
$$p(v_{j,k}) = G(\alpha_{jk,0}, \beta_{jk,0}), \quad j = 1, \ldots, n. \qquad (22)$$

Here, $\text{diag}(.)$ denotes a matrix with the argument vector in its diagonal and zeros otherwise, and $v_{j,k}$ are elements of $\boldsymbol{v}_k$. For notational convenience, we define prior on the rows of matrix A, $\overline{\mathbf{a}}_i, i = 1, \ldots p$.

$$p(\overline{\mathbf{a}}_i|\boldsymbol{\xi}_i) = t\mathcal{N}(\mathbf{0}_{1,r}, \text{diag}(\boldsymbol{\xi}_i)^{-1}), \qquad p(\xi_i) = \prod_{k=1}^{r} G(\phi_{ik,0}, \psi_{ik,0}). \qquad (23)$$

The joint distribution of the data is then

$$p(D, A, \mathbf{b}, U, \omega) = p(D|A, \mathbf{b}, U, \omega) \prod_{i=1}^{p} [p(\overline{\mathbf{a}}_i|\boldsymbol{\xi}_i)p(\boldsymbol{\xi}_i)]$$

$$\prod_{k=1}^{r} [p(\mathbf{u}_k|\boldsymbol{v}_k)p(\boldsymbol{v}_k)] \, p(\mathbf{b}|\varsigma)p(\varsigma)p(\omega). \quad (24)$$

Graphical model of (24) is displayed in Fig. 2. The model differs from the Variational PCA [5] and its positive version [13] in the form where the ARD is applied. While one ARD parameter is common to the whole column of matrix A in the former, every element of the matrices A and U has its own relevance determination parameter in our model. The model has thus much more parameters to estimate from the data.

3.3   The Variational Bayes Posterior

We seek the variational solution in the same form as in (24). The variational posterior distributions (4) for model (24) are found to have functional form:

$$\tilde{p}(\mathbf{u}_k|D, r) = t\mathcal{N}(\mu_{\mathbf{u}_k}, \Sigma_{\mathbf{u}_k}), \qquad \tilde{p}(\boldsymbol{v}_k|D, r) = \prod_{j=1}^{n} G(\alpha_{jk}, \beta_{jk}), \qquad (25)$$

$$\tilde{p}(\mathbf{b}|D, r) = t\mathcal{N}(\mu_{\mathbf{b}}, \Sigma_{\mathbf{b}}), \qquad \tilde{p}(\varsigma|D, r) = G(\zeta, \eta), \qquad (26)$$

$$\tilde{p}(\overline{\mathbf{a}}_i|D, r) = t\mathcal{N}(\mu_{\mathbf{a}_i}, \Sigma_{\mathbf{a}_i}), \qquad \tilde{p}(\boldsymbol{\xi}_i|D, r) = \prod_{k=1}^{r} G(\phi_{ik}, \psi_{ik}), \qquad (27)$$

$$\tilde{p}(\omega|D, r) = G(\vartheta, \rho). \qquad (28)$$

The shaping parameters of the posterior distributions are given in Appendix Appendix A:. Together with moments of distributions (25)–(28) they form a set of implicit equations that needs to be solved.

3.4   Iterative Solution

Solution of the implicit set of equations in Appendix A is found using the variation iterative algorithm [1,16]. The algorithm is based on sequential evaluation of the shaping parameters in Appendix A in the following order: 1) image sources $\mathbf{a}_i, \boldsymbol{\xi}_i$, 2) convolution kernels $\mathbf{u}_k, \boldsymbol{v}_k$, 3) input function $\mathbf{b}, \varsigma$, 4) noise precision ω. This order was found to yield the fastest convergence.

Since the Variational Bayes approximation contains local minima, initialization of the iterative algorithm is critical. Similarly to the scalar decomposition example, we set values of all Gamma hyper-parameters, $\phi_0, \psi_0, \alpha_0, \beta_0, \zeta_0, \eta_0, \vartheta_0, \rho_0$, to 10^{-10} to yield uninformative prior. The most sensitive parameter to initialization is the input function $\mathbf{b}$. We propose to initialize the iterative algorithm at $\mathbf{b}^{(0)} = [1, 0, \ldots, 0]$, for which the matrix B is the identity matrix and the convolution kernels have the role of the sources. This point is known to be an important local extrema in the image deconvolution problems. The convolution kernels $\mathbf{u}_k$ were initialized randomly.

Care is needed with numerical implementation of the iterative algorithm. Specifically, when eigenvalues of the inverted matrices in (29) and (31) are almost equal, the resulting estimates of the convolution kernels contain artifacts (jagged curves). This have been prevented by the use of pseudo-inverse with removal of the smallest eigenvalues. The jagging effect is also suppressed by using (41) to estimate the second moment of $p(U)$.

The resulting algorithm will be denoted as Sparse Blind Source Separation and DeConvolution (S-BSS-DC). It is implemented in matlab and can be downloaded from: `http://www.utia.cas.cz/AS/softwaretools/image_sequences`

4 Results

In this Section, we study properties of the proposed algorithm on a simulated data and demonstrate its practical use on the data from dynamic medical imaging. In both cases, we use a non-standard interpretation of blind source separation which is now briefly introduced.

4.1 Image Sequence Decomposition

The blind source separation model (14) has been used in image sequence analysis for a long time, usually as a model of principal components [3]. The interpretation of the model parameters is slightly different from the cocktail party problem. The observation $\mathbf{d}_t$ is a vector of pixels of the image observed at time t, where the pixels are stored column-wise. The columns of matrix A are images of activity (e.g. measured by PET, SPECT or fMRI) of the underlying biological organs stored in the same form as pixels of $\mathbf{d}_t$. The elements of $\overline{\mathbf{x}}_t$ are activities of the underlying images at time t. The columns $\mathbf{a}_k$ of the matrix A and the source vectors $\mathbf{x}_k$ are thus considered to belong to each other, where the $\mathbf{a}_k$ is the image of the biological organ and $\mathbf{x}_k$ its activity in time. These will be denoted as *source images* and *source curves*, respectively.

This problem has been addressed by the Variational Bayes approach e.g. in [13,16]. The Variational Bayes method of image decomposition with positivity constraints and ARD on image sources was proposed in [13] and will be used for comparison under label BSS+. Sparsity of the image has been modeled by mixture priors, where the parameters of the mixture had to be selected [13], or discrete hidden variable [17].

In medical applications, the sources $\mathbf{x}$ correspond to the flow of biological fluids in the organism. This flow can be modeled by a compartment model, which yields model of the source as convolution of the activity of the blood stream and the tissues specific kernels [11,7]. However, the parametric convolution kernels are typically used [21,7]. Parameters of the convolution kernels are very important for estimation of diagnostic coefficients [11]. We will study the use of general convolution kernels with the ARD prior.

4.2   Phantom Study

A synthetic phantom study for the sparse blind source separation and decon-volution was proposed in [7]. The data are generated using three sources with parametric convolution kernel in the parametric form assumed in [7], so that their CAM-CM algorithm can be used to estimate them. The original phantom data are displayed in Fig. 3, top right. Each source curve generated as a con-volution between a common input function $\mathbf{b} = \exp(-\frac{t}{3})$ and source-specific convolution kernels, $\mathbf{u}_1 = \exp(-\frac{t}{10})$, $\mathbf{u}_2 = 100\exp(-4t)$, and $\mathbf{u}_3 = \frac{1}{2}\exp(\frac{t}{100})$. Each source image has resolution 50×50 pixels, i.e. $p = 2500$, and the sequence contains 50 images, i.e. $n = 50$.

The generated data intentionally contain many overlapping regions. The com-mon assumption in many image decomposition techniques is that there is at least one pixel in each image that do not overlap with others [7]. Many decomposition techniques thus separate the unique areas well, but struggle with assignment of the overlaps.

The results of the proposed algorithm are displayed in Fig. 3, bottom, via the estimated variances of each pixel, $\hat{\xi}_{i,k}$ displayed in the same order of pixels as in the estimated image, image estimates $\hat{a}_k$, source estimates $\hat{x}_k = \hat{B}\hat{u}_k$, and estimated convolution kernels $\hat{u}_k$, respectively. Note that the first convolution kernel is a pulse, hence the corresponding source curve is the estimated input function. Both the CAM-CM solution and the estimated pixel variances $\hat{\xi}_{i,k}$ (ARD on elements of A), Fig 3. bottom left, tend to select the areas where the images are unique. However, the resulting estimates of the images $\hat{a}_k$ of the S-BSS-DC have correctly assigned the overlapping parts.

The default starting points of the iterative algorithm (Section 3.4) were used in analysis of the sequence with the following observations of sensitivity:

- Initialization of the input function by the impulse is not a local minima due to the sparsity prior on the convolution kernels. The ARD prior favors sparse kernels and thus the impulse function is typically recovered in one of the convolution kernels. However, initialization of the input function by random values is unreliable and often converges to a local minima.
- Initialization of the convolution kernels by random starts is rather reliable and no local minima were observed.
- The initial estimate of the precision of the observation noise was selected using the mean of the eigenvalues of matrix $D'D$, see [16] for justification. The same results were obtained even with minimum and maximum of the eigenvalues. Local minima were observed only with extreme values of $\hat{\omega}^{(0)}$.
- The results are sensitive to the selected maximum number of sources r. When the number of sources is greater than the simulated, the strongest source is split into two factors with complementary convolution kernel.

4.3   Real Data Experiment

Validity of the model assumptions is now tested on real clinical data from renal scintigraphy. The tested dataset is a selection of dataset 28 from [20] where

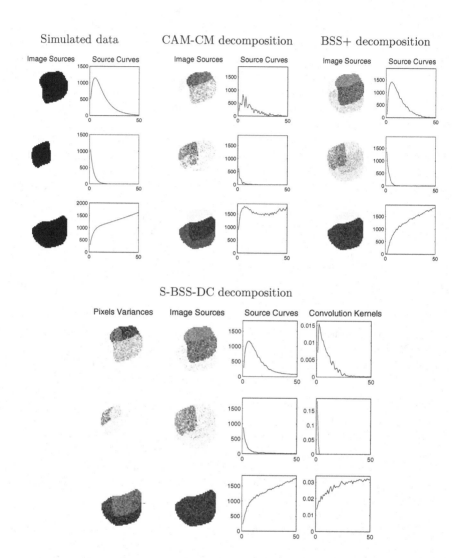

Fig. 3. Generated synthetic dataset using model [7]. **Top left**: simulated source images and curves. **Top middle**: decomposition of the data using the CAM-CM algorithm [7]. **Top right**: decomposition of the data using the BSS+ algorithm [13] **Bottom**: decomposition using the proposed S-BSS-DC algorithm.

a rectangular region of left part of the body and 99 time steps were selected. The images are obtained by counting radioactive particles, hence the observation noise is assumed to be Poisson-distributed. Therefore, we use the correspondence analysis which was found to be optimal conversion of this kind of noise to the homogeneous Gaussian noise [16]. In this application, we have good knowledge of the typical shapes of the input function and the convolution kernels. Thus, we initialize the iterative algorithm by the expected convolution kernels of a typical healthy patient. The convergence of the algorithm is thus significantly faster.

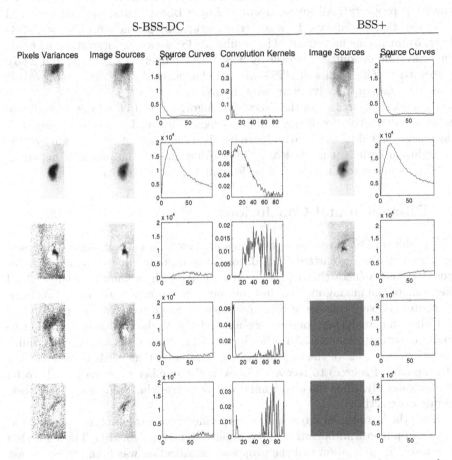

Fig. 4. Results of the proposed algorithm on a real dataset from renal scintigraphy. The columns of the S-BSS-DC algorithm are: the estimate of the variance of the image source prior, the estimate of the source image, the source curve, and its convolution kernel, respectively. The columns of the common BSS+ method are the estimates of the source images and the source curves respectively.

The results of the proposed S-BSS-DC algorithm on this data set are displayed in Fig. 4, via the estimated source pixel variance $\hat{\xi}_{i,k}$, image estimates $\hat{a}_k$, source estimates $\hat{x}_k = \hat{B}\hat{u}_k$, and estimated convolution kernels $\hat{u}_k$, respectively. Once again, the first convolution kernel was estimated to be a pulse, hence the first estimated source curve is equal to the estimated input function. For comparison, the results of blind source separation with positivity constraints (BSS+)are displayed in Fig. 4, right. Comparison with CAM-CM is omitted since its parametric form of the convolution kernels does not correspond to the data.

Note that S-BSS-DC decomposed the observed sequence to 5 sources, the BSS+ method found only 3 meaningful sources (the remaining were removed by the ARD property). All sources recovered by S-BSS-DC has very good medical interpretation as follows: 1) vascular structure, 2) parenchyma, 3) pelvis, 4) liver, 5) unspecific movement. The results of BSS+ can be interpreted as being superposition of sources discover by S-BSS-DC, namely: 1+4), 2+3) and 3+5), respectively. The results of BSS+ are not diagnostically relevant, due to their inability to separate pelvis and parenchyma.

An undesired artifact of the S-BSS-DC algorithm is its tendency to estimate non-smooth convolution kernel, see Fig. 4 right. This tendency is increasing with decreasing signal-to-noise ratio. More detailed modeling of the structure of the convolution kernels (e.g. via two unknown diagonals of the precision matrix in (21)) is required to allow reliable performance in these conditions.

5   Discussion and Conclusion

The problem of blind source separation and deconvolution is in general ill-posed and needs to be regularized by additional assumptions. In this paper, we proposed to use a hierarchical probabilistic model with unknown variance of all elements of the mixing matrix, and the convolution kernels. In effect, this prior promotes sparse estimates of these parameters. Since the proposed model does not allow for analytical solution, we applied the Variational Bayes method to find approximate solution. All other hyper-parameters are chosen to yield uninformative prior, hence the only additional parameter that needs to be chosen is the number of sources to recover. However, the number of sources needs to be chosen carefully, since the algorithm does not posses the ability to recover their number correctly.

Since the Variational Bayes is known to suffer from local minima, we proposed a general-purpose initialization of the implied iterative algorithm. The algorithm was tested in simulation and the proposed initialization was found to be robust and reliable. In specific applications, more appropriate choices can be made to speed up convergence of the algorithm.

The algorithm was also applied to the problem of decomposition of sequence of medical images. The proposed algorithm was able to identify diagnostically relevant sources better than conventional blind source separation methods.

Acknowledgment. This work was supported by the Czech Science Foundation, grant No. 13-29225S. The authors would like to thank Prof. Šámal from the 1st Faculty of Medicine, Charles University Prague for medical data and support with their evaluation.

References

1. Attias, H.: A Variational Bayesian framework for graphical models. In: Leen, T. (ed.) Advances in Neural Information Processing Systems, vol. 12. MIT Press (2000)
2. Attias, H.: New em algorithms for source separation and deconvolution with a microphone array. In: Proceedings of the 2003 IEEE International Conference on Acoustics, Speech, and Signal Processing, ICASSP 2003, vol. 5, pp. 297–300. IEEE (2003)
3. Barber, D.C.: The use of principal components in the quantitative analysis of gamma camera dynamic studies. Physics in Medicine and Biology 25, 283–292 (1980)
4. Bell, A.J., Sejnowski, T.J.: An information-maximization approach to blind separation and blind deconvolution. Neural Computation 7(6), 1129–1159 (1995)
5. Bishop, C.M.: Variational principal components. In: Proc. of the Ninth Int. Conference on Artificial Neural Networks, ICANN (1999)
6. Buvat, I., Benali, H., Paola, R.: Statistical distribution of factors and factor images in factor analysis of medical image sequences. Physics in Medicine and Biology 43, 1695 (1998)
7. Chen, L., Choyke, P., Chan, T., Chi, C., Wang, G., Wang, Y.: Tissue-specific compartmental analysis for dynamic contrast-enhanced mr imaging of complex tumors. IEEE Transactions on Medical Imaging 30(12), 2044–2058 (2011)
8. Ghahramani, Z., Beal, M.: Graphical models and variational methods. In: Opper, M., Saad, D. (eds.) Advanced Mean Field Methods. MIT Press (2001)
9. Haykin, S., Chen, Z.: The cocktail party problem. Neural Computation 17(9), 1875–1902 (2005)
10. Jeffreys, H.: Theory of Probability, 3rd edn. Oxford University Press (1961)
11. Lawson, R.: Application of mathematical methods in dynamic nuclear medicine studies. Physics in Medicine and Biology 44, R57–R98 (1999)
12. MacKay, D.: Information theory, inference, and learning algorithms. Cambridge University Press (2003)
13. Miskin, J.W.: Ensemble Learning for Independent Component Analysis. Ph.D. thesis, University of Cambridge (2000)
14. Opper, M., Saad, D.: Advanced Mean Field Methods: Theory and Practice. The MIT Press, Cambridge (2001)
15. Pedersen, M.S., Larsen, J., Kjems, U., Parra, L.C.: A survey of convolutive blind source separation methods. In: Multichannel Speech Processing Handbook, pp. 1065–1084 (2007)
16. Šmídl, V., Quinn, A.: The Variational Bayes Method in Signal Processing. Springer (2005)
17. Šmídl, V., Tichý, O.: Automatic Regions of Interest in Factor Analysis for Dynamic Medical Imaging. In: 2012 IEEE International Symposium on Biomedical Imaging (ISBI). IEEE (2012)

18. Sroubek, F., Flusser, J.: Multichannel blind deconvolution of spatially misaligned images. IEEE Transactions on Image Processing 14(7), 874–883 (2005)
19. Tzikas, D.G., Likas, A.C., Galatsanos, N.P.: Variational Bayesian sparse kernel-based blind image deconvolution with student's-t priors. IEEE Transactions on Image Processing 18(4), 753–764 (2009)
20. Praha, V.F.N.: Database of dynamic renal scintigraphy (June 2013), http://www.dynamicrenalstudy.org
21. Šmídl, V., Tichý, O., Šámal, M.: Factor analysis of scintigraphic image sequences with integrated convolution model of factor curves. In: Proceedings of the Second International Conference on Computational Bioscience. IASTED (2011)

Appendix A: Shaping Parameters of Posterior Distributions

The shaping parameters of posterior distributions (25) - (28) are as follows:

$$\Sigma_{\mathbf{u}_k} = \left(\left(\widehat{\omega} \widehat{B'B} (\widehat{\mathbf{a}_k' \mathbf{a}_k}) \right) + \mathrm{diag}(\widehat{\upsilon_k}) \right)^{-1}, \tag{29}$$

$$\mu_{\mathbf{u}_k} = \Sigma_{\mathbf{u}_k} \left(\left(-\widehat{\omega} \sum_{l=1, l \neq k}^{r} \widehat{B'B} \widehat{\mathbf{u}}_l (\widehat{\mathbf{a}_k' \mathbf{a}_l}) \right) + \widehat{\omega} \widehat{B'} D' \widehat{\mathbf{a}_k} \right), \tag{30}$$

$$\Sigma_{\mathbf{b}} = \left(\widehat{\varsigma} I_n + \widehat{\omega} \sum_{i,j=1}^{r} (\widehat{\mathbf{a}_i' \mathbf{a}_j}) \left(\sum_{k,l=0}^{n-1} \Delta_k' \Delta_l (\mathbf{u}_{k+1,j} \widehat{\mathbf{u}_{l+1,i}}) \right) \right)^{-1}, \tag{31}$$

$$\mu_{\mathbf{b}} = \Sigma_{\mathbf{b}} \widehat{\omega} \sum_{k=1}^{r} \left(\sum_{j=0}^{n-1} \Delta_j \widehat{\mathbf{u}_{j+1,k}} \right)' D' \widehat{\mathbf{a}_k}, \tag{32}$$

$$\alpha_k = \alpha_{k,0} + \frac{1}{2} \mathbf{1}_{n,1}, \quad \beta_k = \beta_{k,0} + \frac{1}{2} \mathrm{diag} \left(\widehat{\mathbf{u}_k \mathbf{u}_k'} \right), \tag{33}$$

$$\zeta = \zeta_0 + \frac{n}{2}, \qquad \eta = \eta_0 + \frac{1}{2} \mathrm{tr} \left(\widehat{\mathbf{b}' \mathbf{b}} \right), \tag{34}$$

$$\vartheta = \vartheta_0 + \frac{pn}{2}, \qquad \rho = \rho_0 + \frac{1}{2} \mathrm{tr} \left(DD' - 2\widehat{A}\widehat{X}'D' \right) + \frac{1}{2} \mathrm{tr} \left(\widehat{A'A}\widehat{X'X} \right), \tag{35}$$

$$\Sigma_{\mathbf{a}_i} = \left(\widehat{\omega} \sum_{j=1}^{n} (\widehat{\mathbf{x}_j' \mathbf{x}_j}) + \mathrm{diag}(\widehat{\xi_i}) \right)^{-1}, \quad \mu_{\mathbf{a}_i} = \left(\Sigma_{\mathbf{a}_i} \left(\widehat{\omega} \sum_{j=1}^{n} (\widehat{\mathbf{x}_j} d_{i,j})' \right) \right)', \tag{36}$$

$$\phi_i = \phi_{i,0} + \frac{1}{2} \cdot \mathbf{1}_{r,1}, \qquad \psi_i = \psi_{i,0} + \frac{1}{2} \mathrm{diag} \left(\widehat{\mathbf{a}_i' \mathbf{a}_i} \right). \tag{37}$$

The auxiliary matrix $\Delta_k \in \mathbf{R}^{n \times n}$ is defined as

$$(\Delta_k)_{i,j} = \begin{cases} 1, & \text{if } i - j = k, \\ 0, & \text{otherwise.} \end{cases}$$

The moments of variables are computed using expectations of their probability density function, Appendix B.1.

Appendix B: Required Probability Distributions

B.1   Truncated Normal Distribution

Truncated normal distribution is defined for scalar random variable $x = N_x(\mu, \sigma)$ on interval $a < x \leq b$ as follows:

$$x \sim t\mathcal{N}(\mu, \sigma, a, b) = \frac{\sqrt{2}\exp((x-\mu)^2)}{\sqrt{\pi}\sigma(erf(\beta) - erf(\alpha))}\chi_{(a;b]}(x), \tag{38}$$

where $\alpha = \frac{a-\mu}{\sqrt{2}\sigma}$, $\beta = \frac{b-\mu}{\sqrt{2}\sigma}$, $\chi_{(a,b]}(x)$ is a characteristic function of interval $(a, b]$ defined as $\chi_{(a,b]}(x) = \begin{cases} 1 & x \in (a, b] \\ 0 & x \notin (a, b] \end{cases}$, and $erf(t) = \frac{2}{\sqrt{\pi}}\int_0^t e^{-u^2}du$.

Moments of the truncated normal distribution are given as

$$\widehat{x} = \mu - \sqrt{\sigma}\frac{\sqrt{2}[\exp(-\beta^2) - \exp(-\alpha^2)]}{\sqrt{\pi}(erf(\beta) - erf(\alpha))}, \tag{39}$$

$$\widehat{x^2} = \sigma + \mu\widehat{x} - \sqrt{\sigma}\frac{\sqrt{2}[b\exp(-\beta^2) - a\exp(-\alpha^2)]}{\sqrt{\pi}(erf(\beta) - erf(\alpha))}. \tag{40}$$

B.2   Multivariate Truncated Normal Distribution

Truncation of the multivariate Normal distribution $\mathbf{x} \sim \mathcal{N}(\mu, \Sigma)$ is formally simple, however, its moments can not be expressed analytically. Therefore, we approximate the moments of $\mathbf{x}$ of the truncated Normal distribution by the moments of

$$\tilde{\mathbf{x}} \sim t\mathcal{N}(\mu, \text{diag}(\boldsymbol{\sigma})),$$

where $\boldsymbol{\sigma}$ is a vector of diagonal elements of Σ. This corresponds to approximation of the posterior by a product of marginals (38) with mean value $\widehat{\mathbf{x}}$ with elements given by (39) and $\widehat{\mathbf{xx}^T} = \widehat{\mathbf{x}}\widehat{\mathbf{x}}^T + \text{diag}(\widehat{\boldsymbol{\sigma}})$, where $\widehat{\sigma}_i = \widehat{x_i^2} - \widehat{x}_i\widehat{x}_i$. However, it may be too coarse approximation since it ignores covariance of the elements. An alternative is to approximate

$$\widehat{\mathbf{xx}^T} = \widehat{\mathbf{x}}\widehat{\mathbf{x}}^T + \text{diag}(\mathbf{o})\Sigma\text{diag}(\mathbf{o}), \tag{41}$$

where $\mathbf{o}$ is a vector of elements $o_i = \widehat{\sigma}_i^{1/2}\sigma_i^{-1/2}$. Heuristics (41) is motivated by the observation that for a Normal distribution with the main mass far from the truncation lines, $o_i \to 1$ and (41) becomes equivalent to the moment of the non-truncated Normal distribution.

Nested Hierarchical Dirichlet Process for Nonparametric Entity-Topic Analysis

Priyanka Agrawal[1,*], Lavanya Sita Tekumalla[2,*], and Indrajit Bhattacharya[1]

[1] IBM India Research Lab
{priyanka.svnit,indrajitb}@gmail.com
[2] Indian Institute of Science
lavanya@csa.iisc.ernet.in

Abstract. The Hierarchical Dirichlet Process (HDP) is a Bayesian nonparametric prior for grouped data, such as collections of documents, where each group is a mixture of a set of shared mixture densities, or topics, where the number of topics is not fixed, but grows with data size. The Nested Dirichlet Process (NDP) builds on the HDP to cluster the documents, but allowing them to choose only from a set of specific topic mixtures. In many applications, such a set of topic mixtures may be identified with the set of entities for the collection. However, in many applications, multiple entities are associated with documents, and often the set of entities may also not be known completely in advance. In this paper, we address this problem using a nested HDP (nHDP), where the base distribution of an outer HDP is itself an HDP. The inner HDP creates a countably infinite set of topic mixtures and associates them with entities, while the outer HDP associates documents with these entities or topic mixtures. Making use of a nested Chinese Restaurant Franchise (nCRF) representation for the nested HDP, we propose a collapsed Gibbs sampling based inference algorithm for the model. Because of couplings between two HDP levels, scaling up is naturally a challenge for the inference algorithm. We propose an inference algorithm by extending the direct sampling scheme of the HDP to two levels. In our experiments on two real world research corpora, we show that, even when large fractions of author entities are hidden, the nHDP is able to generalize significantly better than existing models. More importantly, we are able to detect missing authors at a reasonable level of accuracy.

1 Introduction

Dirichlet Process mixture models [1] allow for nonparametric or infinite mixture modeling, where the number of densities or mixture components is not fixed ahead of time, but grows (slowly) with the number of data items. They do so by using as a prior the Dirichlet Process (DP), which is a distribution over distributions, and has the additional property that draws from it are discrete (w.p. 1) with infinite support [1,6]. However, many applications require joint

---

* The first two authors have contributed equally to the paper.

H. Blockeel et al. (Eds.): ECML PKDD 2013, Part II, LNAI 8189, pp. 564–579, 2013.
© Springer-Verlag Berlin Heidelberg 2013

analysis of groups of data, such as a collection of text documents, where the mixture components, or topics (as they are called for text data), are shared across the documents. This calls for a coupling of multiple DPs, one for each document, where the base distribution is discrete, and shared. The hierarchical Dirichlet Process (HDP) [16] does so by placing a DP prior on a shared base distribution, so that the model now has two levels of DPs. The HDP has since been used extensively as a prior for non-parametric modeling of text collections. The popular LDA model [3] may be considered as a parametric restriction of the HDP mixture model.

The HDP mixture model (and LDA) belongs to the family of admixture models [5], where each composite data item or group gets assigned to a mixture over the mixture components or topics. While this adds more flexibility to the groups of data items, the ability to cluster groups is lost, since each group now has a distinct mixture of topics associated with it. This additional capability is desired in many applications, such as analysis of patient profiles in hospitals [13], where the hospitals need to be clustered in addition to shared grouping of patients in individual hospitals. Alternatively, imagine a corpus containing descriptions related to entities, such as a shared set of researchers who have authored a large body of scientific literature, or a shared set of personalities discussed across news articles, such that each entity can be represented as a mixture of topics. Here, topic mixtures, corresponding to entities, are required to be shared across data groups or documents, in addition to the topics themselves. This can be captured using the nested DP (nDP) [13], which has a DP corresponding to each group, which are coupled through the same base distribution, which is a DP itself, unlike being DP distributed, as in the HDP. This results in a distribution over distributions over distributions, unlike the HDP and the DP, which are distributions over distributions. The nDP can be imagined as first creating a discrete set of mixtures over topics, each mixture representing an entity, and then choosing exactly one of these entities for each document. In this sense, the nDP is a mixture of admixtures.

One major shortcoming of the nDP for entity analysis is the restrictive assumption of a single entity being associated with a document. In research papers, multiple authors are associated with any document, and any news article typically discusses multiple people, organizations etc. This requires each document to have a distribution over entities. In other words, we need a model that is an admixture of admixtures. The Author-Topic Model (ATM) [14], which models authors associated with documents, belongs to this class, but is restrictive in that it requires the authors to be observed for documents, and also assumes the number of topics to be known.

In this paper, we address the problem of nonparametric modeling of entities and topics, where the number of topics is not known in advance, and additionally the set of entities for each document is either partly or completely unknown. For this, we propose the nested HDP (nHDP), where the base distribution of an HDP is itself an HDP. This belongs to the same class as the nDP, since it specifies a distribution over distributions (entities) over distributions (topics). However, unlike the nDP, it first creates a discrete set of entities, and models each group as a mixture over these entities. To the best of our knowledge, ours is the first entity-topic model that is nonparametric in both entities and topics. The Author Topic Model falls out as a parametric version of this model, when the entity set is observed for each document, and the number of topics is fixed.

For inference using the nHDP, we propose the nested CRF (nCRF), which extends the Chinese Restaurant Franchise (CRF) analogy of the HDP to two levels by integrating out the two levels of HDPs. However, due to strong coupling between the CRF layers, inference using the nCRF poses computational challenges. We use a direct sampling scheme, based on that for the HDP, where the entity and topic indexes are directly sampled, based on the counts of table assignments and stick-breaking weights at the two levels. Using experiments over publication datasets involving author entities from NIPS and DBLP, we show that the nHDP generalizes better under different levels of available author information. More interestingly, the model is able to detect authors completely hidden in the entire corpus with reasonable accuracy.

2 Related Work

In this section, we review existing literature on Bayesian nonparametric modeling and entity-topic analysis.

Bayesian Nonparametric Models: We review the Dirichlet Process (DP) [6], the Hierarchical Dirichlet Process (HDP) [16] and the nested Dirichlet Process (nDP) [13] in detail in the Sec. 3.

The MLC-HDP[17] is a 3-layer model proposed for human brain seizures data. The 2-level truncation of the model is closely related to the HDP and the nDP. Like the HDP, it shares mixture components across groups (documents) and assigns individual data points to the same set of mixtures, and like the nDP it clusters each of the groups or documents using a higher level mixture. In other words, this is a nonparametric mixture of admixtures, while our proposed nested HDP is a nonparametric admixture of admixtures.

The nested Chinese Restaurant Process (nCRP) [2] extends the Chinese Restaurant Process analogy of the Dirichlet Process to an infinitely-branched tree structure over restaurants to define a distribution over finite length paths of trees. This can be used as a prior to learn hierarchical topics from documents, where each topic corresponds to a node of this tree, and each document is generated by a random path over these topics. An extension to this model, also called the nested HDP, has recently been proposed on Arvix [11]. In the spirit of the HDP, which has a top level DP and providing base distributions for document specific DPs, this model has a top level nCRP, which becomes the base distribution for document specific nCRPs. In contrast, our model has nested HDPs, in the sense that one HDP directly serves as the base distribution for another HDP, like in the nested DP [13], where one DP serves as the base distribution for another DP. This parallel with the nested DP motivates the nomenclature of our model as the nested HDP.

Entity-Topic Models: Next, we briefly review prior work on simultaneously modeling of entities and topics in documents. The literature mostly contains parametric models, where the number of topics and entities are known ahead of time. The LDA model [3] is the most popular parametric topic model, and has a distribution θ_d over T topics for each document, and the topic label z_{di} for each word in the document is sampled iid from θ_d. The author-topic model (ATM)[14] extends the LDA to capture *known* authors A_d of each document. Each author now has his own distribution π_a over topics K, and the words are generated by first sampling one of the known authors uniformly, followed by sampling a topic from the topic distribution of that author:

$$\phi_k \sim Dir(\beta), \ \ k = 1 \ldots T; \ \ \pi_a \sim Dir(\alpha), \ \ a = 1 \ldots A$$
$$a_{di} \sim \pi_d \equiv U(A_d); \ \ z_{di} \sim \theta_{a_{di}}; \ \ w_{di} \sim Mult(\phi_{z_{di}}) \tag{1}$$

The Author Recipient Topic model[9] distinguishes between sender and recipient entities and learns the topics and topic distributions of sender-recipient pairs. Newman et. al[10] analyze entity-topic relationships from textual data containing entity words and topic words, which are pre-annotated. The Entity Topic Model[8] proposes a generative model which is parametric in both entities and topics and assumes observed entities for each document.

There has been very little work on nonparametric entity-topic modeling, which would enable discovery of entities in settings where entities are partially or completely unobserved in documents. The Author Disambiguation Model[4] is a nonparametric model for the author entities along with topics. Primarily focusing on author disambiguation from noisy mentions of author names in documents, this model treats entities and topics symmetrically, generating entity-topic pairs from a DP prior. Contrary to this approach, our model treats the entity as a distribution over topics, thus explicitly modeling the fact that authors of documents have preferences over specific topics.

3   Background

Consider a setting where observations are organized in groups. Let x_{ji} denote the ith observation in j^{th} group. For a corpus of documents, x_{ji} is the ith word occurrence in the jth document. In the context of this paper, we will use group synonymously with document, data item with word in a document. We assume that each x_{ji} is independently drawn from a mixture model and has a mixture component parameterized by a *factor*, say θ_{ji}, representing a topic, associated with it. For each group j, let the associated factors $\boldsymbol{\theta}_j = (\theta_{j1}, \theta_{j2}, \ldots)$ have a prior distribution G_j. Finally, let $F(\theta_{ji})$ denote the distribution of x_{ji} given factor θ_{ji}. Therefore, the generative model is given by

$$\theta_{ji}|G_j \sim G_j; \ \ x_{ji}|\theta_{ji} \sim F(\theta_{ji}), \ \forall j, i \tag{2}$$

The central question in analyzing a corpus of documents is the parametrization of the G_j distributions — what parameters to share and what priors to place on them. We start with the Dirichlet Process that considers each of the G_j distributions in isolation, then the Hierarchical Dirichlet Process that ensures sharing of atoms among the different G_js, and finally the nested Dirichlet Process that additionally clusters the groups by ensuring that all the G_js are not distinct.

Dirichlet Process: Let $(\Theta, \mathcal{B})$ be a measurable space. A Dirichlet Process (DP) [6,1] is a measure over measures G on that space. Let G_0 be a finite measure on the space. Let α_0 be a positive real number. We say that G is DP distributed with concentration parameter α and base distribution G_0, written $G \sim DP(\alpha_0, G_0)$, if for any finite measurable partition $(A_1, \ldots, A_r)$ of Θ, we have

$$(G(A_1), \ldots G(A_r)) \sim Dir(\alpha_0 G_0(A_1), \ldots, \alpha_0 G_0(A_r)).$$

The *stick-breaking representation* provides a constructive definition for samples drawn from a DP. It can be shown [15] that draw G from $DP(\alpha_0, G_0)$ can be written as

$$\phi_k \overset{iid}{\sim} G_0, \ k = 1 \ldots \infty; \quad w_i \sim Beta(1, \alpha_0); \quad \beta_i = w_i \prod_{j=1}^{i-1}(1 - w_j)$$

$$G = \sum_{k=1}^{\infty} \beta_k \delta_{\phi_k}, \tag{3}$$

where the atoms ϕ_k are drawn independently from G_0 and the corresponding weights $\{\beta_k\}$ follow a stick breaking construction. This is also called the GEM distribution: $(\beta_k)_{k=1}^{\infty} \sim GEM(\alpha_0)$. The stick breaking construction shows that draws from the DP are necessarily discrete, with infinite support, and the DP therefore is suitable as a prior distribution on mixture components for 'infinite' mixture models. Subsequently, $\{\theta_{ji}\}$ are drawn from each G_j. When drawing of $\{\theta_{ji}\}$ is followed by draws $\{x_{ji}\}$ according to Eqn. 2, the model is known as the Dirichlet Process mixture model [6].

Another commonly used perspective of the DP is the *Chinese Restaurant Process* (CRP) [12], which shows that DP tends to clusters the observations. Let $\{\theta_i\}$ denote the sequence of draws from G, and let $\{\phi_k\}$ be the atoms of G. The CRP considers the predictive distribution of the ith draw θ_i given the first $i - 1$ draws $\theta_1 \ldots \theta_{i-1}$ after integrating out G:

$$\theta_i | \theta_1, \ldots, \theta_{i-1}, \alpha_0, G_0 \sim \sum_{k=1}^{K} \frac{m_k}{i - 1 + \alpha_0} \delta_{\phi_k} + \frac{\alpha_0}{i - 1 + \alpha_0} G_0$$

where $m_k = \sum_{i'=1}^{i-1} \delta(\theta_{i'}, \phi_k)$. The above conditional may be understood in terms of the following restaurant analogy. Consider an initially empty 'restaurant' that can accommodate an infinite number of 'tables'. The ith 'customer' entering the restaurant chooses a table θ_i for himself, conditioned on the seating arrangement of all previous customers. He chooses the k-th table with probability proportional to m_k, the number of people already seated at the table, and with probability proportional to α_0, he chooses a new (currently unoccupied) table. Whenever a new table is chosen, a new 'dish' ϕ_k is drawn ($\phi_k \sim G_0$) and associated with the table. The CRP readily lends itself to sampling-based inference strategies for the DP.

Hierarchical Dirichlet Process: Now reconsider our grouped data setting. If each G_j is drawn independently from a DP, then w.p. 1 the atoms $\{\phi_{jk}\}_{k=1}^{\infty}$ for each G_j are distinct. This would mean that there is no shared topic across documents, which is undesirable. The Hierarchical Dirichlet Process (HDP) [16] addresses this problem by modeling the base distribution G_0 of the DP prior in turn as DP distributed. Since draws from a DP are discrete, this ensures that the same atoms $\{\phi_k\}$ are shared across all the G_js. Specifically, given a distribution H on the space $(\Theta, \mathcal{B})$ and positive real numbers $(\alpha_j)_{j=1}^{M}$ and γ, we denote as $\text{HDP}(\alpha, \gamma, H)$ the following generative process:

$$G_0|\gamma, H \sim DP(\gamma, H)$$
$$G_j|\alpha_j, G_0 \sim DP(\alpha_j, G_0) \quad \forall j. \tag{4}$$

When this is followed by generation of $\{\theta_{ji}\}$ and $\{x_{ji}\}$ as in Eqn. 2, we get the *HDP mixture model*.

Using the stick-breaking construction, the global measure G_0 distributed as Dirichlet process can be expressed as $G_0 = \sum_{k=1}^{\infty} \beta_k \delta_{\phi_k}$, where the topics ϕ_k are drawn from H independently ($\phi_k \sim H$) and $\{\beta_k\} \sim \text{GEM}(\gamma)$ represent 'global' popularities of these topics. Since G_0 has as its support the topics $\{\phi_k\}$, each group-specific distribution G_j necessarily has support at these topics, and can be written as follows:

$$G_j = \sum_{k=1}^{\infty} \pi_{jk} \delta_{\phi_k}; \quad (\pi_{jk})_{k=1}^{\infty} \sim DP(\alpha_j, \beta) \tag{5}$$

where $\pi_j = (\pi_{jk})_{k=1}^{\infty}$ denotes the topic popularities for the jth group.

Analogous to the CRP for the DP, the Chinese Restaurant Franchise provides an interpretation of predictive distribution for the next draw from an HDP after integrating out the G_js and G_0. Let $\{\theta_{ji}\}$ denote the sequence of draws from each G_j, $\{\psi_{jt}\}$ the sequence of draws from G_0, and $\{\phi_k\}$ the sequence of K draws from H. Then the conditional distribution of θ_{ji} given $\theta_{j1}, \ldots, \theta_{j,i-1}$ and G_0, after integrating out G_j is as follows:

$$\theta_{ji}|\theta_{j1}, \ldots, \theta_{j,i-1}, \alpha_0, G_0 \sim \sum_{t=1}^{m_{j.}} \frac{n_{jt.}}{i-1+\alpha_0} \delta_{\psi_{jt}} + \frac{\alpha_0}{i-1+\alpha_0} G_0 \tag{6}$$

where $n_{jtk} = \sum_{i'=1}^{i-1} \delta(\theta_{ji'}, \psi_{jt}) \delta(\psi_{jt}, \phi_k)$, $m_{jk} = \sum_t \delta(\psi_{jt}, \phi_k)$ and dots indicate marginal counts. As G_0 is also distributed according to a Dirichlet Process, we can integrate it out similarly to get the conditional distribution of ψ_{jt}:

$$\psi_{jt}|\psi_{11}, \psi_{12}, \ldots, \psi_{21}, \ldots, \psi_{jt-1}, \gamma, H \sim \sum_{k=1}^{K} \frac{m_{.k}}{m_{..}+\gamma} \delta_{\phi_k} + \frac{\gamma}{m_{..}+\gamma} H \tag{7}$$

These equations may be interpreted using a two-level restaurant analogy. Consider a set of restaurants, one corresponding to each group. Customers entering each of the restaurants select a table θ_{ji} according a group specific CRP (Eqn

6). The restaurants share a common menu of dishes $\{\phi_k\}$. Dishes are assigned to the tables of each restaurant according to another CRP (Eqn 7). Let t_{ji} be the (table) index of the element of $\{\psi_{jt}\}$ associated with θ_{ji}, and let k_{jt} be the (dish) index of the element of $\{\phi_k\}$ associated with ψ_{jt}. Then the two conditional distributions above can also be written in terms of the indexes $\{t_{ji}\}$ and $\{k_{jt}\}$ instead of referring to the distributions directly. If we draw ψ_{jt} by choosing a summation term, we set $\psi_{jt} = \phi_k$ and let $k_{jt} = k$ for the chosen k. If the second term is chosen, we increment K by 1 and draw $\phi_K \sim H$ and set $\psi_{jt} = \phi_K$ and $k_{jt} = K$. This CRF analogy leads to efficient Gibbs sampling-based inference strategies for the HDP mixture model [16].

Nested Dirichlet Process: In other applications of grouped data, we may additionally be interested in clustering the data groups themselves. For example, when analyzing patient records in hospitals, we may want to cluster the hospitals as well in terms of the profiles of patients coming there. The HDP cannot do this, since each group specific mixture G_j is distinct. This problem is addressed by the nested Dirichlet Process [13], which first defines a set $\{G'_k\}_{k=1}^\infty$ of distributions over an infinite support:

$$G'_k = \sum_{l=1}^\infty w_{lk}\delta_{\theta'_{lk}}, \ \theta'_{lk} \sim H, \ \{w_{lk}\}_{l=1}^\infty \sim GEM(\gamma) \tag{8}$$

and then draws the group specific distributions G_j from a mixture over these:

$$G_j \sim G_0 \equiv \sum_{k=1}^\infty \pi_k\delta_{G'_k}, \ \{\pi_k\} \sim GEM(\alpha)$$

We denote the generation process as $\{G_j\} \sim nDP(\alpha, \beta, H)$. The process ensures non-zero probability of different groups selecting the same G'_k, leading to clustering of the groups. Using Eqn. 3, the draws $\{G_j\}$ can be characterized as:

$$G_j \sim G_0, \ G_0 \sim DP(\alpha, DP(\gamma, H)) \tag{9}$$

where the base distribution of the outer DP is in turn another DP, unlike the HDP where it is DP distributed. Thus the nDP can be viewed as a distribution on the space of distributions on distributions.

Given this characterization of the nDP, it appears to be useful for restricted entity-topic analysis, where we additionally want to label each document with a single entity from a countable set, with each entity associated with a distribution over topics. However, note that the support $\{\theta'_{lk}\}$ of each G'_k in Eqn 8 is distinct, which implies that different entities do not share any topics. Further, we would like to associate a distribution over entities for each document. This makes the nDP unsuitable even for such restricted entity-topic analysis.

4   Nonparametric Entity-Topic Analysis

We now present our nested Hierarchical Dirichlet Process (nHDP) model for nonparametric entity-topic analysis. We first present the model where each group or document is associated with a single entity, then extend it for multiple entities.

Single-Entity Documents: Recall that the nDP is unsuitable for entity-topic analysis, since the entity distributions do not share topic atoms. This can be modified by first creating a set of entity distributions $\{G_{k'}\}_{k'=1}^{\infty}$ such that they share atoms. One way to do this is to follow the HDP construction:

$$G_{k'} \sim HDP(\{\alpha_{k'}\}, \gamma, H) \tag{10}$$

This can be followed by drawing each group specific distribution from a mixture over the $G_{k'}$s:

$$G'_j \sim G'_0 \equiv \sum_{k'=1}^{\infty} \beta'_{k'} \delta_{G_{k'}}, \ \beta' \sim GEM(\gamma') \tag{11}$$

Using Eqn. 3, we observe that $G'_0 \sim DP(\gamma', HDP(\{\alpha_{k'}\}, \gamma, H))$. Observe the relationship with the nDP (Eqn. 9). Like the nDP, this also defines a distribution over the space of distributions on distributions. But, instead of a DP base distribution for the outer DP, we have achieved sharing of atoms using a HDP base distribution. We will write $G'_j \sim DP\text{-}HDP(\gamma', \{\alpha_{k'}\}, \gamma, H)$.

Sampling G'_j may be imagined as choosing the entity for the jth document. As before, G'_j can now be used as prior for sampling topics $\{\theta_{ji}\}$ for individual words in document j, followed by sampling of the words themselves, using Eqn 2. We will call this the DP-HDP mixture model.

Nested HDP for Multi-Entity Documents: The DP-HDP model above associates a single distribution over topics G'_j with the jth document, and the topic θ_{ji} for each word x_{ji} in the document is drawn from G'_j. In the context of entity-topic analysis, this means that a single entity is associated with a document, and words are drawn from the preferred topics of this entity. However, many applications, such as analyzing entities in news articles and authors from scientific literature, require associating multiple entities with each document, and each word in the document is drawn from a preferred topic of one of these entities. In this section, we extend the earlier model for multi-entity documents.

As before, we first create a set of distributions $\{G_{k'}\}_{k'=1}^{\infty}$ over the same set of (topic) distributions $\{\phi_k\}_{k=1}^{\infty}$ ($\phi_k \sim H$) by drawing independently from an HDP, and creating a global mixture over them:

$$G_{k'} \sim HDP(\{\alpha_{k'}\}, \gamma, H); \quad \beta' \sim GEM(\gamma'); G'_0 \equiv \sum_{k'=1}^{\infty} \beta'_{k'} \delta_{G_{k'}}$$

This may be interpreted as creating a countable set of entities by defining topic preferences (distributions over topics) for each of them, and then defining a 'global popularity' of the entities. Earlier, for single entity documents, the only entity was sampled from this global popularity. Now, we define a document-specific local popularity for entities, derived from this global popularity:

$$G'_j \equiv \sum_{k'=1}^{\infty} \pi'_{jk'} \delta_{G_{k'}}, \ \{\pi'_{jk'}\} \sim DP(\alpha'_j, \beta') \tag{12}$$

Now, sampling each factor θ_{ji} in document j is preceded by choosing an entity $\theta'_{ji} \sim G'_j$ by sampling according to local entity popularity. Note that $P(\theta'_{ji} = G_{k'}) = \pi'_{jk'}$.

Using the stick breaking definition of the HDP in Eqn. 5, we can see that G'_j is drawn from a HDP. The base distribution of that HDP has to be the distribution from which the atoms $\{G_{k'}\}$ are drawn, which is again an HDP. Therefore, we can write:

$$\theta'_{ji} \sim G'_j \sim HDP(\{\alpha'_j\}, \gamma', HDP(\{\alpha_{k'}\}, \gamma, H)) \tag{13}$$

We refer to the two relevant HDPs as the inner and outer HDPs. Observing the parallel with the nDP definition in Eqn. 9, we call this the nested HDP (nHDP), and write $\theta'_{ji} \sim nHDP(\{\alpha'_j\}, \gamma', \{\alpha_{k'}\}, \gamma, H)$. Similar to the nDP, and the DP-HDP (Eqn. 11), this again defines a distribution over the space of distributions over distributions. The complete nHDP mixture model is defined by subsequently sampling $\theta_{ji} \sim \theta'_{ji}$, followed by $x_{ji} \sim F(\theta_{ji})$.

An alternative characterization of the nHDP mixture model is using the topic index z_{ji} and entity index z'_{ji} corresponding to x_{ji}:

$$\beta \sim GEM(\gamma); \ \pi_{k'} \sim DP(\alpha, \beta); \ \phi_k \sim H, \ k, k' = 1 \ldots \infty$$
$$\beta' \sim GEM(\gamma') \ ; \pi'_j \sim DP(\alpha, \beta'), \ j = 1 \ldots M$$
$$z'_{ji} \sim \pi'_j \ ; \ z_{ji} \sim \pi_{z'_{ji}}; \ x_{ji} \sim F(\phi_{z_{ji}}), \ i = 1 \ldots n_j \tag{14}$$

This may be understood as first creating entity-specific distributions $\pi_{k'}$ over topics using global topic popularities β, followed by creation of document-specific distributions π'_j over entities using global entity popularities β'. Using these parameters, the content of the jth document is generated by sampling repeatedly in iid fashion an entity index z'_{ji} using π'_j, a topic index z_{ji} using $\pi_{z'_{ji}}$ and finally a word using $F(\phi_{z_{ji}})$.

Observe the connection with the ATM in Eqn. 1. The main difference is the the set of entities and topics is infinite. Additionally, each document now has a distinct non-uniform distribution π'_j over entities.

Also, observe that we have preserved the HDP notation to the extent possible, to facilitate understanding. To distinguish between variables corresponding to the two HDPs in the model, we use dashes ($'$) as superscripts on symbols corresponding to the outer HDP. Going forward, we follow the same convention for naming variables in the nested CRF.

Nested Chinese Restaurant Franchise: In this section, we derive the predictive distribution for the next draw θ'_{ji} from the nHDP given previous draws, after integrating out $\{G'_j\}$ and G'_0, and then the predictive distribution for the draw θ_{ji} after integrating out $\{\theta'_{ji}\}$ and G_0. We also provide an interpretation for these using two nested CRFs, corresponding to the inner and outer HDPs. These will be useful for the inference algorithm that we describe in Section 5.

Let $\{\theta'_{ji}\}$ denote the sequence of draws from G'_j, and $\{\psi'_{jt'}\}_{t'=1}$ denote the sequence of draws from G'_0. Then the conditional distribution of θ'_{ji} given all previous draws after integrating out G'_j looks as follows:

$$\theta'_{ji}|\theta'_{j1:i-1},\alpha'_j,G'_0 \sim \sum_{t'=1}^{m'_{j.}} \frac{n'_{jt'\cdot}}{i-1+\alpha'_j}\delta_{\psi'_{jt'}} + \frac{\alpha'_j}{i-1+\alpha'_j}G'_0 \qquad (15)$$

where $n'_{jt'k'} = \sum_{i'} \delta(\theta'_{ji'},\psi'_{jt'})\delta(\psi'_{jt'},G_{k'})$, $m'_{jk'} = \sum_{t'} \delta(\psi'_{jt'},G_{k'})$. Next, we integrate out G'_0, which is also distributed according to Dirichlet process:

$$\psi'_{jt'}|\psi'_{11},\psi'_{12},\ldots,\psi'_{21},\ldots,\psi'_{j,t'-1},\gamma',\mathrm{HDP}(\alpha,\gamma,H) \sim$$
$$\sum_{k'=1}^{K'} \frac{m'_{\cdot j}}{m'_{..} + \gamma'}\delta_{G_{k'}} + \frac{\gamma'}{m'_{..} + \gamma'}\mathrm{HDP}(\alpha,\gamma,H) \qquad (16)$$

Observe that each θ'_{ji} variable gets assigned to one of the $G_{k'}$ variables. Let $\{\theta_{ji}\}$ denote the sequence of draws from respective $\{\theta'_{ji}\}$ (i.e. from the corresponding $G_{k'}$), $\{\psi_{k't}\}$ the sequence of draws from G_0, and $\{\phi_k\}_{k=1}^\infty$ the sequence of draws from H. Let $\theta_{k':ji}$ denote the set of θ variables already drawn from $G_{k'}$ before sampling θ_{ji}, i.e. $\theta_{k':ji} \equiv \{\theta_{j'i'} : \theta'_{j'i'} = G'_k, \forall i', j' \le j, \text{ and } i' < i, j' = j\}$. Then, the conditional distribution of θ_{ji} given $\theta_{k':ji}$ and G_0, after integrating out $G_{k'}$ (corresponding to θ'_{ji}) is as follows:

$$\theta_{ji}|\theta_{k':ji},\alpha_0,G_0 \sim \sum_{t=1}^{m_{k'.}} \frac{n_{k't\cdot}}{i-1+\alpha_0}\delta_{\psi_{k't}} + \frac{\alpha_0}{n_{k'..} + \alpha_0}G_0 \qquad (17)$$

where $n_{k'tk} = \sum_i \delta(\theta_{k'i},\psi_{k't})\delta(\psi_{k't},\phi_k)$, $m_{k'k} = \sum_t \delta(\psi_{k't},\phi_k)$ and dots indicate marginal counts. As G_0 is also distributed according to a Dirichlet Process, we can integrate it out similarly and write the conditional distribution of $\psi_{k't}$ as follows:

$$\psi_{k't}|\psi_{11},\psi_{12},\ldots,\psi_{21},\ldots,\psi_{k't-1},\gamma,H \sim \sum_{k=1}^{K} \frac{m_{\cdot k}}{m_{..} + \gamma}\delta_{\phi_k} + \frac{\gamma}{m_{..} + \gamma}H \qquad (18)$$

Note that both conditional distributions for θ'_{ji} and θ_{ji} are similar to that for CRF (Eqns. 6 and 7). We interpret these two distributions as a *nested Chinese Restaurant Franchise*, involving one inner CRF and one outer CRF.

Consider a set of outer restaurants, one corresponding to each group. Customers entering each of these restaurants select a table θ'_{ji} according a group specific CRP (Eqn 15). The restaurants share a common set of inner restaurants $\{G_{k'}\}$. Inner restaurants are assigned to the tables of each outer restaurant according to another CRP (Eqn 16). Next, the customers go to the inner restaurant assigned to them (by some outer restaurant) and select a table θ_{ji} according to the inner restaurant specific CRP (Eqn 17). These inner restaurants share a common menu of dishes $\{\phi_k\}$. Dishes are assigned to the tables of each inner restaurant according to another CRP (Eqn 18).

Let t'_{ji} be the (outer table) index of the $\psi'_{jt'}$ associated with θ'_{ji}, and let k'_{jt} be the (inner restaurant) index of the $G_{k'}$ associated with $\psi'_{jt'}$. Let t_{ji} be the (inner table) index of the $\psi_{k't}$ associated with θ_{ji}, and let $k_{k't}$ be the (dish) index of the

ϕ_k associated with $\psi_{k't}$. Then the two conditional distributions above can also be written in terms of the indexes $\{t'_{ji}\}, \{k'_{jt'}\}, \{t_{ji}\}$ and $\{k_{k't}\}$ instead of referring to the distributions directly. For the jth outer restaurant and its ith customer, we draw θ'_{ji} using Eqn. 15. If the first summation is chosen, we set $\theta'_{ji} = \psi'_{jt'}$ and let $t'_{ji} = t'$ for the chosen t'. If the second term is chosen, then we increment $m'_{j\cdot}$ by one, draw $\psi'_{jm'_{j\cdot}} \sim G'_0$ using (Eqn 16) and set $\theta'_{ji} = \psi'_{jm'_{j\cdot}}$ and $t'_{ji} = m'_{j\cdot}$. If we draw $\psi'_{jt'}$ via choosing a summation term, we set $\psi'_{jt'} = G_{k'}$ and let $k'_{jt'} = k'$ for the chosen k'. If the second term is chosen, then we increment the current distinct entity count M by one, draw $G_M \sim \text{HDP}(\boldsymbol{\alpha}, \gamma, H)$ and set $\psi'_{jt'} = G_M$ and $k'_{jt'} = M$. Next, we similarly draw samples of θ_{ji} for each j and i using Eqn. 17. If new sample from G_0 is needed, we use Eqn. 18 to obtain a new sample $\psi_{k't}$.

5   Inference for Nested HDP

We use Gibbs sampling for approximate inference, as exact inference is intractable for this problem. The conditional distributions for the nCRF scheme lend themselves to an inference algorithm where we sample $t'_{ji}, t_{ji}, k'_{jt'}$ and $k_{k't}$. The conditionals for these variables are similar to those in equations 15, 17, 18 and 16 respectively. However, in such an approach, there exists a tight coupling between the variables t'_{ji}, t_{ji} and $k'_{jt'}$, which would call for computationally expensive joint sampling of variables.

Instead, we adopt a technique similar to the direct sampling scheme in HDP[16], where variables G_0 and G'_0 are explicitly sampled instead of being integrated out, by sampling the stick breaking weights β and β' respectively. Further, we directly sample z_{ji} (the topic) and z'_{ji} (the author) for each word in the jth document avoiding explicit table assignments to the t'_{ji} and t_{ji} variables. However, in order to sample β and β', the table information is maintained in the form of the number of tables in each outer and inner restaurant, $m'_{jk'}$ and $m_{k'k}$ respectively. Thus the latent variables that need to be sampled in our Gibbs sampling scheme are $z_{ji}, z'_{ji}, \beta, \beta', m'_{jk'}$ and $m_{k'k}$.

We introduce the following notation for the rest of this section. Let $\mathbf{x} = \{x_{ji} :$ all $j, i\}$, $\mathbf{x}_{-ji} = \{x_{j'i'} : j' \neq j, i' \neq i\}$, $\mathbf{m} = \{m_{k'k} : \text{all } k', k\}$, $\mathbf{m}' = \{m'_{jk'} :$ all $j, k'\}$, $\mathbf{z} = \{z_{ji} : \text{all } j, i\}$, $\mathbf{z}' = \{z'_{ji} : \text{all } j, i\}$, $\mathbf{z}_{-ji} = \{z_{j'i'} : j' \neq j, i' \neq i\}$, $\mathbf{z}'_{-ji} = \{z'_{j'i'} : j' \neq j, i' \neq i\}$, $\beta_{new} = \sum_{k=K+1}^{\infty} \beta_k$ and $\beta'_{new} = \sum_{k'=K'+1}^{\infty} \beta'_{k'}$

Sampling z_{ji}: The conditional distribution for topic index z_{ji} is

$$p(z_{ji} = z | \mathbf{z}_{-ji}, \mathbf{z}'_{-ji}, z'_{ji} = z', \mathbf{m}, \mathbf{m}', \beta, \beta', \mathbf{x}) \propto p(z_{ji} = z | \mathbf{z}_{-ji}, z'_{ji} = z') p(x_{ji} | z_{ji} = z, \mathbf{x}_{-ji})$$

For existing topics, $p(z_{ji} = z | \mathbf{z}_{-ji}, z'_{ji} = z')$ can be split into two terms, one from picking any of the existing tables from entity (inner restaurant) z' with topic z and the other from creating a new table for entity z' and assigning the topic z to it. For a new topic, a new table is always created for entity z'. Hence,

$$p(z_{ji} = z | \mathbf{z}_{-ji}, z'_{ji} = z') \propto \begin{cases} \frac{n_{k'\cdot k} + \alpha\beta_k}{n_{k'\cdot\cdot} + \alpha} & \text{Existing } z \\ \frac{\alpha\beta_{new}}{n_{k'\cdot\cdot} + \alpha} & \text{New topic } z \end{cases} \quad (19)$$

The other term $p(x_{ji}|z_{ji} = z, \mathbf{x}_{-ji})$ is the conditional density of x_{ji} under topic z given all data items except x_{ji}. Assuming each topic is sampled from a V dimensional symmetric Dirichlet prior over the vocabulary with parameter η, i.e $\phi_k \sim Dir(\eta)$, the above probability can be simplified to the following expression, by integrating out ϕ:

$$p(x_{ji} = w|z_{ji} = z, \mathbf{x}_{-ji}) \propto \frac{n_{zw} + \eta}{n_{z.} + V\eta}$$

where n_{zw} is the number of occurrences of topic z with word w in the vocabulary.

Sampling z'_{ji}: The conditional distribution for the entity index z'_{ji} is

$$p(z'_{ji} = z'|\mathbf{z}'_{-ji}, \mathbf{z}_{-ji}, z_{ji} = z, \mathbf{m}, \mathbf{m}', \beta, \beta', \mathbf{x}) \propto p(z'_{ji} = z|\mathbf{z}'_{-ji})p(z_{ji} = z|\mathbf{z}_{-ji}, z'_{ji} = z')$$

Again, $p(z_{ji} = z|\mathbf{z}_{-ji}, z'_{ji} = z')$ can be split into two terms, one from picking an existing outer table with entity z' and the other from creating a new outer table and assigning the entity z' to it. Further, creation of a new entity always involves the creation of a new outer table. Hence,

$$p(z'_{ji} = z'|\mathbf{z}'_{-ji}) \propto \begin{cases} \frac{n'_{j.k'} + \alpha' \beta'_{k'}}{n'_{j..} + \alpha'} \text{Existing } z' \\ \frac{\alpha' \beta'_{new}}{n'_{j..} + \alpha'} \quad\quad \text{New } z' \end{cases}$$

$p(z_{ji} = z|\mathbf{z}_{-ji}, z'_{ji} = z')$ follows from Eqn. 19.

Sampling β and β': The posterior of G_0, conditioned on samples $\psi_{k't}$ from it, is also distributed as a DP due to Dirichlet-Multinomial conjugacy, and the stick breaking weights of G_0 can be sampled as follows: $(\beta_1, \beta_2 \ldots \beta_K, \beta_{new}) \sim Dir(m_{.1}, m_{.2} \ldots m_{.K}, \gamma)$ Similarly, the stick breaking weights β' can be sampled from the posterior distribution of G'_0 conditioned on samples from G'_0 in the form of $m'_{jk'}$ as follows: $(\beta'_1, \beta'_2 \ldots \beta'_K, \beta'_{new}) \sim Dir(m'_{.1}, m'_{.2} \ldots m'_{.K'}, \gamma')$

Sampling m and m': $m_{k'k}$ is the number of inner tables generated as $n_{k'.k}$ samples are drawn from G'_k corresponding to a particular topic k. This is the number of partitions generated as samples are drawn from a Dirichlet Process with concentration parameter $\alpha\beta_k$ and is distributed according to a closed form expression [1]. We adopt a different method [7] for sampling $m_{k'k}$ by drawing a total of $n_{k'.k}$ samples with topic k, and incrementing the count $m_{k'k}$ whenever a new inner table is created with topic assignment k. Similarly, $m'_{jk'}$ is sampled by drawing a total of $n'_{j.k'}$ samples with entity k' and incrementing the count $m'_{jk'}$ whenever a new outer table is created with entity assignment k'.

Sampling Concentration Parameters: We place a vague gamma prior on the concentration parameters α, γ, α', γ' with hyper parameters (α_a, α_b), (γ_a, γ_b), (α'_a, α'_b) and (γ'_a, γ'_b) respectively. We use Gibbs sampling scheme for sampling the concentration parameters using the technique outlined in HDP[16].

We use the conditional distributions above to perform inference under three different settings. In the **"no observed entities"** setting, the conditional distributions above are repeatedly sampled from until convergence. For initialization, we first initialize the topic variables z_{ji} using an online scheme, and then initialize the entities z'_{ji} using the topics. In the **"completely observed entities"** setting, the set of entities A_j is given for every document j. Since no other entities are deemed possible for the jth document, $p(z'_{ji} = z'|\mathbf{z}'_{-ji}, \mathbf{z}_{-ji}, z_{ji} = z, \mathbf{m}, \mathbf{m}', \beta, \beta', \mathbf{x})$ is set to 0 for new entities and for all $z' \notin A_j$. In the **"partially observed entities"** setting, a partial list of known entities A_j is available for document j, but other entities are also considered possible. We perform an initialization step, similar to that in the completely observed setting, using the known entities A_j. No new authors are added in this initial step. During later iterations, we allow all assignments $z'_{ji} = z$ — one of the known entities from A_j, entities of other documents j', and new entities. However, we introduce a bias towards the known authors $z' \in A_j$ using an additional small positive term to their probability mass.

6   Experiments

In this section, we experimentally evaluate the proposed nHDP model for the task of modeling author entities who have collaboratively written research papers, and compare its performance against available baselines. Specifically, we evaluate two different aspects: (1) how well the model is able to learn from the training samples and fit held-out data, first (1a) when all the authors are observed in training and test documents, and secondly (1b) when some or all of the authors are unobserved in training and test documents, (2) how accurately the model discovers hidden authors, who are not mentioned at all in the corpus.

We consider the following models for the experiments: (i) The author-topic model (**ATM**) (Eqn. 1) where the number of topics is pre-specified, and all authors are observed for all documents. This is used as a baseline for (1a) above. (ii) The Hierarchical Dirichlet Process (**HDP**) (Eqn. 4) using the direct assignment inference scheme for fair comparison. We use our own implementation for this. Recall that the HDP infers the number of topics, and does not use author information. (iii) nHDP with completely observed entities (**nHDP-co**), which assumes complete entity information to be available for all documents, but learns topics in a nonparametric fashion. This can be imagined as an improvement over ATM where the number of topics does not need to be specified. (iv) nHDP with partially observed entities (**nHDP-po**), which makes use of available entity information, but admits the possibility of entities being hidden globally from the corpus, or locally from individual documents. (v) nHDP with no observed entities (**nHDP-no**), which does not make use of any entity information and assumes all entities to be globally hidden in the corpus. For task (1a) above, the applicable models are the ATM, HDP (which ignores the entity information) and nHDP-co. For task (1b), the ATM does not apply. We evaluate HDP, nHDP-po and nHDP-no. It is important to point out that there are no available baselines for task (2) above.

Table 1. Perplexity of ATM, HDP and nHDP-co for NIPS

Model	ATM	HDP	nHDP-co
Perplexity	2783	1775	1247

We use the following publicly available publication datasets for our experimental analysis. The **NIPS** dataset[1] is a collection of papers from Neural Information Processing Systems (NIPS) conference proceedings (volume 0-12). This collection contains 1,740 documents contributed by a total of 2,037 authors, with total 2,301,375 word tokens resulting in a vocabulary of 13,649 words. A subset of the **DBLP Abstracts** dataset[2] containing 12,000 documents by 15,252 authors collected from 20 conferences records on the Digital Bibliography and Library Project (DBLP). Each document is represented as a bag of words present in abstract and title of the corresponding paper, resulting in a vocabulary size of 11,771 words.

1. Generalization Ability: We now come to our first experiment, where we evaluate the ability of the models, whose parameters are learnt from a training set, to predict words in new unseen documents in a held-out test set. We evaluate performance of a model M on a test collection D using the standard notion of perplexity [3]: $exp(-\sum_{d \in D} p(w_d)|M)$.

In experiment (1a), all authors are observed in training and test documents. To favor the ATM, which cannot handle new authors in test document, we create test-train splits ensuring that each author in the test collection occurs in at least one training document.

Perplexity results are shown in Table 1. Recall that HDP and nHDP find the best number of topics, while for ATM we have recorded its best performance across different value of K. The results show that while knowledge of authors is useful, the ability of non-parametric topic models to infer the number of topics clearly leads to better generalization.

Next, in experiment (1b), we first create training-test distributions with reasonable author overlap by letting each author vote with probability 0.7 whether to send a document to train or test, and majority decision is taken for each document. Next, authors are partially hidden from both the test and the train documents as follows. We iterate over the global list of authors and remove each author from all training and test documents with probability p_g. We then iterate over each training and test document, and remove each remaining author of that document with probability p_l. We experiment with different values of p_g and p_l to simulate different extents of missing information on authors.

The results are shown in Table 2. We can see that when more information is available about the authors, the ability to fit held-out data improves. More interestingly, even when no / very little author information is available, just the assumption about the existence of authors, or a discrete set of topic mixtures,

---

[1] http://www.arbylon.net/resources.html
[2] http://www.cs.uiuc.edu/~hbdeng/data/kdd2011.htm

Table 2. Perplexity for HDP and nHDP with varying percentage of hidden authors

Model	HDP	nHDP-no	nHDP-po	nHDP-po	nHDP-po	nHDP-co
p_g,p_l	1,1	1,1	0.6,0.6	0.4,0.4	0.2,0.2	0,0
Perplexity NIPS	2572	1882	1434	1266	1109	987
Perplexity DBLP	1027	997	935	869	676	394

leads to better generalization ability, as can be seen from the relative performance of HDP and nHDP-no.

2. Discovering Missing Authors: Beyond data fitting, the most significant ability of the nHDP mixture model is to discover entities which are relevant for documents in the corpus, but are never mentioned. We perform a case study with the top 6 most prolific authors in NIPS, by removing them completely from the corpus, and then checking the ability of the model to discover them in a completely unsupervised fashion. While it is possible to define as a classification problem the task of detecting of *locally missing* authors in individual documents when the author is observed in other documents, we reiterate that there is no existing baseline when an author is *globally hidden*.

We evaluate the accuracy of discovering hidden author as follows. For each hidden author $h \in \{1 \ldots H\}$, we create a m-dimensional vector c_h, where m is the corpus size, with $c_h[j]$ indicating his authorship in the jth document. We explored two possibilities for this 'true' indicator vector: (a) binary indicators using the gold-standard author names for documents, and (b) the number of words written by that author in the document according to nHDP with completely observed authors (nHDP-co). Similarly, we create an m-dimensional vector for each new author $n \in \{1 \ldots N\}$ discovered by the nHDP-po, with $c_n[j]$ indicating his contribution (no. of authored words) in the jth document. We now check how well the vectors $\{c_n\}$ correspond to the 'true' vectors $\{c_h\}$. This is done by defining two variables C_n and C_h, taking values $1 \ldots H$ and $1 \ldots N$ respectively, and defining a joint distribution over them as $P(h,n) = \frac{1}{Z}\text{sim}(c_h, c_n)$, where Z is a normalization constant. For $\text{sim}(c_h, c_n)$, we use cosine similarity between normalized versions of c_h and c_n. Mutual information $I(C_h, C_n) = \sum_{h,n} p(h,n) \log \frac{p(h,n)}{p(h)p(n)}$ measures the information that C_h and C_y share. We used its normalized variant $NMI(C_h, C_n) = \frac{I(C_h,C_n)}{[H(C_h)+H(C_n)]/2}$ ($H(X)$ indicating entropy of X) which takes values between 0 and 1, higher values indicating more shared information.

First, we note that the best NMI achievable for this task, by replacing the true vectors $\{c_h\}$ for the discovered vectors $\{c_n\}$, is 0.86 for case (a) and 0.98 for case (b) above. In comparison, using nHDP-po, we achieve NMI scores of 0.59 for case (a) and 0.72 for case (b). This indicates that the actual author distributions that the model discovers not only help in fitting the data, but also have reasonable correspondence with the true hidden authors. We believe that this is a promising initial step in addressing this difficult problem.

7   Conclusions

In this paper, we have addressed the problem of entity-topic analysis from document corpora, where the set of document entities are either completely or partially hidden. For such problems, we have proposed as a prior distribution the nested Hierarchical Dirichlet Process, which consists of two levels of Hierarchical Dirichlet Processes, where one is the base distribution of the other. Using a direct sampling scheme for inference, we have shown that the nHDP is able to generalize better than existing models under varying available knowledge about authors in research publications, and is additionally able to discover completely hidden authors in the corpus.

References

1. Antoniak, C.: Mixtures of Dirichlet Processes with applications to Bayesian non-parametric problems. Ann. Statist. 2(6), 1152–1174 (1974)
2. Blei, D., Griffiths, T., Jordan, M., Tanenbaum, J.: The nested chinese restaurant process and bayesian nonparametric inference of topic hierarchies. J. ACM (2010)
3. Blei, D., Ng, A., Jordan, M.: Latent dirichlet allocation. JMLR (2003)
4. Dai, A., Storkey, A.: Author disambiguation: A nonparametric topic and co-authorship model. In: NIPS Workshop on Applications for Topic Models Text and Beyond, pp. 1–4 (2009)
5. Erosheva, E., Fienberg, S., Lafferty, J.: Mixed-membership models of scientific publications. PNAS 101(suppl. 1) (2004)
6. Ferguson, T.: A Bayesian analysis of some nonparametric problems. Ann. Statist. 1(2), 209–230 (1973)
7. Fox, E., Sudderth, E., Jordan, M., Willsky, A.: A Sticky HDP-HMM with Application to Speaker Diarization. Annals of Applied Stats. 5(2A), 1020–1056 (2011)
8. Kim, H., Sun, Y., Hockenmaier, J., Han, J.: Etm: Entity topic models for mining documents associated with entities. In: ICDM, pp. 349–358 (2012)
9. McCallum, A., Corrada-Emmanuel, A., Wang, X.: The author recepient topic model for topic and role discovery in social networks (2004)
10. Newman, D., Chemudugunta, C., Smyth, P.: Statistical entity-topic models. In: ACM SIGKDD, KDD 2006, pp. 680–686. ACM, New York (2006)
11. Paisley, J., Wang, C., Blei, D., Jordan, M.: Nested hierarchical dirichlet processes. Arxiv (2012)
12. Pitman, J.: Gibbs sampling methods for stick-breaking priors. Lecture Notes for St. Flour Summer School (2002)
13. Rodriguez, A., Dunson, D., Gelfand, A.: The nested dirichlet process. Journal of the American Statistical Association 103(483), 1131–1154 (2008)
14. Rosen-Zvi, M., Griffiths, T., Steyvers, M., Smyth, P.: The author-topic model for authors and documents. In: UAI (2004)
15. Sethuraman, J.: A constructive definition of Dirichlet Priors. Statistica Sinica 4, 639–650 (1994)
16. Teh, Y., Jordan, M., Beal, M., Blei, D.: Hierarchical Dirichlet processes. Journal of the American Statistical Association (2006)
17. Wulsin, D., Jensen, S., Litt, B.: A hierarchical dirichlet process model with multiple levels of clustering for human eeg seizure modeling. In: ICML (2012)

Knowledge Intensive Learning: Combining Qualitative Constraints with Causal Independence for Parameter Learning in Probabilistic Models

Shuo Yang and Sriraam Natarajan

School of Informatics and Computing, Indiana University, USA
{shuoyang,natarasr}@indiana.edu

Abstract. In Bayesian networks, prior knowledge has been used in the form of causal independencies between random variables or as qualitative constraints such as monotonicities. In this work, we extend and combine the two different ways of providing domain knowledge. We derive an algorithm based on gradient descent for estimating the parameters of a Bayesian network in the presence of causal independencies in the form of Noisy-Or and qualitative constraints such as monotonicities and synergies. Noisy-Or structure can decrease the data requirements by separating the influence of each parent thereby reducing greatly the number of parameters. Qualitative constraints on the other hand, allow for imposing constraints on the parameter space making it possible to learn more accurate parameters from a very small number of data points. Our exhaustive empirical validation conclusively proves that the synergy constrained Noisy-OR leads to more accurate models in the presence of smaller amount of data.

1 Introduction

Human advice or input is generally provided in learning Bayesian networks using the structure of the Bayesian network [1]. Given this network structure, most methods use some form of optimization to learn the parameters of the models. Initially, advice giving methods simply served to constrain the structure of the network. While the use of prior structure does reduce the number of examples required to learn a reasonable network, learning parameters can still require certain amount of examples to converge to a reasonable estimate. However many domains, such as medicine, can be data poor (for example, number of positive examples of a disease can possibly be quite low) but knowledge rich due to several decades of research. This domain knowledge is mostly about the influential relationships between the random variables of interest in the domain.

One of the most prominent methods of providing domain knowledge to a probabilistic learner is to provide the set of causal independencies that exist in the domain [2]. Also called as Independence of Causal Influences (ICI) [3–6], this form of knowledge identifies *sets of parents* that are independent of each other

H. Blockeel et al. (Eds.): ECML PKDD 2013, Part II, LNAI 8189, pp. 580–595, 2013.

when affecting the target random variable. The effects of these sets of random variables can typically then be combined using a function such as Noisy-Or. The key advantage of such knowledge is that these lead to a drastic reduction in the number of parameters associated with the conditional distributions (from exponential in the total number of parents to exponential in the size of these sets). This reduction can greatly affect the number of examples required for training an accurate model. While this is very attractive, ICI can be very restrictive and easily violated in many domains.

An equally alternative and more recent method of providing advice to learners is based on qualitative influences [7–11]. Qualitative influence (QI) statements essentially outline how the change of one variable affects the change of another variable. A classic example of such QI statements is monotonicity [7, 8, 12] where an increase in value of one random variable (say cholesterol) increases the probability that another variable (say risk of heart attack) takes a higher value. Another direction has been in combining context-specific independencies [13] with QI statements [9] and showing that learning with such constraints is a special case of isotonic regression [14].

In this work, we extend and combine these different methods of specifying domain knowledge. More precisely, we extend the research in two directions – First, current methods for QI can handle monotonicity statements while we extend these directions by allowing for synergistic interactions [7] between random variables. While monotonicities model the qualitative dependency between two random variables, synergistic interactions allow for richer influence relationships. For instance, with synergies, it is possible to specify statements such as "Increase in blood sugar level increases the risk of heart attack in high cholesterol level patients more than it does in low cholesterol level patients". This statement explains how sugar level and cholesterol level interact when influencing heart attack. Second, we use such synergistic and monotonicity statements and combine them with the concept of ICI i.e., we treat each "set" of monotonicity and synergistic interaction as independent of each other and their influences are combined with a combining rule. In this work we employ Noisy-Or [5] for combining the independent influences. While previous work has used context-specific independences, we generalize them to using ICI.

Following prior work [8], we convert the monotonicity and synergy statements to constraints on the parameter space of the conditional distributions. We then combine the different conditional distributions using Noisy-Or and derive the overall objective function. We adopt a gradient descent algorithm with exterior penalty method to optimize the objective function and outline the algorithm for learning in the presence of qualitative and conditional influences.

To summarize, we make the following contributions: first, we extend the qualitative influences to include synergies. Second, we combine these qualitative influences with the independence of causal influence (ICI) such as Noisy-Or and derive a new objective function that includes these influences as constraints on the parameter space. Third, we derive an algorithm for parameter learning in the presence of sparse data by exploiting these influences. Finally, we perform an exhaustive evaluation in 11 different standard domains to understand the impacts and influences of the different types of influence statements. Our results show clearly that the use of such influences helps the learning algorithm improve its performance in the presence of sparse data.

2   Background

We provide a brief background on qualitative and conditional influences. First, we introduce some basic notations used throughout the paper. In a Bayesian network with n discrete valued nodes, we denote the parameters by θ_{ijk} ($i \in \{1, 2, ..., n\}, j \in \{1, 2, ..., v_i\}, k \in \{1, 2, ..., r_i\}$) which means the conditional probability of X_i to be its k-th value given the j-th configuration of its parents (i.e. $P(X_i^k|pa_i^j)$). r_i denotes the number of states of the discrete variable X_i; Pa_i represents the parent set of node X_i; the number of configurations of Pa_i is $v_i = \prod_{X_t \in Pa_i} r_t$; j is the index of a particular configuration of node X_i's parents pa_i^j.

2.1   Qualitative Influences – Monotonic Constraints

Qualitative influence, specifically monotonicity has been explored extensively in previous work [7–9]. Specifically, Altendorf et al. [8] used monotonicities in the context of learning Bayesian networks. Monotonic influence means that stochastically, higher values of a random variable, say X result in higher (or lower) values of another variable Y, and is denoted as $X_\succ^{M+}Y$ (or $X_\succ^{M-}Y$). The interpretation is that increasing values of X shifts the cumulative distribution function of Y to the right (i.e., higher values of Y are more likely). This means that $P(Y \leq y|X = x_1) \leq P(Y \leq y|X = x_2)$ (where $x_1 \geq x_2$). Note that the same definition can be extended in the presence of multiple parents by fixing the values of the other parents. If one of X_i's parents (denoted by X_c) has monotonic constraint on X_i, this relationship still stands given the same configuration of other parents, the general form of monotonic constraints of X_c on X_i is

$$P(X_i \leqslant k_c|X_c^m, C_i^n) \geqslant P(X_i \leqslant k_c|X_c^{m+1}, C_i^n) \tag{1}$$

where $k_c \in \{1, 2, ..., r_i - 1\}, m \in \{1, 2, ..., r_c - 1\}, X_c^m \leqslant X_c^{m+1}$, C_i^n represents all possible configurations of X_i's parents other than X_c, n is the index.

Altendorf et al. [8] used these qualitative constraints to learn the parameters of a Bayes net by introducing a penalty to the objective function when the constraints are violated. Assume there is a monotonic constraint: $P(X_i \leq k_c|pa_i^{j_2}) \leq P(X_i \leq k_c|pa_i^{j_1})$. The constraint function δ with margin ϵ is defined as:

$$\delta = P(X_i \leq k_c|pa_i^{j_2}) - P(X_i \leq k_c|pa_i^{j_1}) + \epsilon \tag{2}$$

The corresponding penalty function is $P_{j_1,j_2}^{i,k_c} = I_{(\delta>0)}\delta^2$ (where I=1 when $\delta > 0$ and I=0 when $\delta \leqslant 0$). In order to rule out the need for the simplex constraints ($\sum_{k=1}^{r_i} \theta_{ijk} = 1$), Altendorf et al. defined μ_{ijk} such that

$$\theta_{ijk} = \frac{exp(\mu_{ijk})}{\sum_{k'=1}^{r_i} exp(\mu_{ijk'})} \tag{3}$$

They then derived the gradient of the objective function wrt μ and used exterior penalty method to learn from data. We refer to their work for more details. Tong et al. [11] and de Campos et al. [10], considered the problem of facial recognition from images and applied qualitative constraints to learning for recognizing these faces. They took an EM approach for learning the parameters of these qualitative constraints (that possibly could include synergy). We on the other hand, take a gradient descent approach that allows for including conditional influences such as Noisy-Or.

2.2 Noisy-Or

The term *independence of causal influence* was first used by Heckerman and Breese [3] to model the situation where there are several variables that influence a random variable independently but their collective influence can be computed using some deterministic or stochastic combination function. Typical examples of ICI include Noisy-Or, Noisy-And, Noisy-Min, Noisy-Max, Noisy-Add etc. Representing and learning with such ICI relationships have long been explored in the context of Bayes nets [3–6]. In this work, we consider a particular type of ICI, the most popular one – Noisy-Or. Simply put, if there are n independent causes $\{X_1, ..., X_n\}$ for a random variable Y and assuming for simplicity that Y is binary, then the target distribution $P(Y = 1|X_1 = x_1, ..., X_n = x_n)$ is given by

$$P(Y = 1|X_1 = x_1, ..., X_n = x_n) = 1 - \prod_i P(Y = 0|X_i = x_i) \qquad (4)$$

The interpretation is that if any parent, say X_i takes value x_i, Y will take the value 1, unless there is an effect of inhibition. This inhibition has a probability of $P(Y = 0|X_i = x_i)$ [6] and these inhibitory effects are assumed to be independent $(1 - q_i$ for $i^{th} parent)$.

3   Qualitative Constraints – Synergies

In this section, we extend the previous work on montonicities [8] by allowing for synergistic interactions. After presenting the definition of synergies, we derive the gradient for learning with such knowledge from data.

In the presence of a small amount of training data, the parameters in conditional probability tables (CPT) estimated only based on the data are most likely inaccurate, and in some cases can result in even uniform distributions due to the lack of data about certain configurations of the parents. Fortunately, in many of the real world problems, domain experts can provide sufficient prior knowledge about the influences that exist in the domain. We consider a particular type of the domain knowledge namely, qualitative influence statements that allow us to apply some constraints on the CPTs. These constraints aid in obtaining more accurate estimates of the parameters of the CPTs. More specifically, we propose to exploit the monotonicity and synergy constraints and combine them with a rule such as Noisy-Or when learning the parameters.

When multiple parents influence the resultant independently, we can simply employ monotonicities as presented in the previous section. Synergies on the other hand, allow for richer interactions between the parents where the set of parents can influence the resultant variable dependently. We use Wellman's definition on qualitative synergy [7]. Assume that two variables X_1 and X_2 affect a third variable Y synergistically (where each of them has the $X_1 \overset{M+}{\succ} Y$ and $X_2 \overset{M+}{\succ} Y$ relationship with the target). This is denoted as $X_1, X_2 \overset{S+}{\succ} Y$ (sub-synergy is denoted as S-)[1]. In simple terms, this means that increasing X_1 has a greater (lesser for sub-synergy) effect on Y for high values of X_2 than low values of X_2; likewise for increasing X_2 with fixed X_1. Note that two causes having the same monotonic influence is the premise of their synergistic interaction, which means by our definition, there cannot be a synergy or sub-synergy relation between X_1 and X_2 if $X_1 \overset{M}{\succ} Y$ while $X_2 \overset{M-}{\succ} Y$ i.e., the parents in the synergy relationship cannot have different types of monotonic influences to the target.

Consider for example a medical diagnosis problem and assume that the target of interest is heart attack. An example of a synergistic statement in the domain is, *cholesterol and blood pressure interact synergistically when influencing heart attack.* In simpler terms, the above statement simply means that hypertension increases the risk of heart attack in high cholesterol level patients more than it does in low cholesterol level patients. This defines how the two risk factors (cholesterol and blood pressure) interact with heart attack. Note that each of the cholesterol level and blood pressure has a monotonic relationship with heart attack when considered individually (i.e. $Chl \overset{M+}{\succ} HA$ and $BP \overset{M+}{\succ} HA$). A classic example of a sub-synergy in medical research is that coronary heart disease (CHD) is markedly more common in men than in women; CHD risk increases with age in both sexes, but the increase is sharper in women [15]. Hence gender and age interact sub-synergistically when influencing CHD.

Formally, based on the definition above, assume $x_i^j \leqslant x_i^{j+1}$ where x_i^j is the jth value of variable X_i. Since $P(Y \leq k_c | X_1^i, X_2^j) \geqslant P(Y \leq k_c | X_1^{i+1}, X_2^j)$ (implied by $X_1 \overset{M+}{\succ} Y$), X_1's effect on Y at low level of X_2 is

$$P(Y \leq k_c | X_1^i, X_2^j) - P(Y \leq k_c | X_1^{i+1}, X_2^j)$$

and similarly, at high level of X_2 is

$$P(Y \leq k_c | X_1^i, X_2^{j+1}) - P(Y \leq k_c | X_1^{i+1}, X_2^{j+1})$$

The synergistic constraint on conditional probability distribution can be mathematically represented as:

$$P(Y \leq k_c | X_1^i, X_2^j) - P(Y \leq k_c | X_1^{i+1}, X_2^j) \leq$$
$$P(Y \leq k_c | X_1^i, X_2^{j+1}) - P(Y \leq k_c | X_1^{i+1}, X_2^{j+1})$$

---

[1] Note that what we use the terminology of sub-synergy due to Wellman. This same concept is also called as anti-synergy in the literature.

where $i \in \{1, 2, ..., r_1 - 1\}$ and $j \in \{1, 2, ..., r_2 - 1\}$. Note the above inequation is essentially X_1's effect on Y with fixed X_2. Similarly the synergistic constraint of X_2 on Y with fixed X_1 is

$$P(Y \leq k_c | X_1^i, X_2^j) - P(Y \leq k_c | X_1^i, X_2^{j+1}) \leq$$
$$P(Y \leq k_c | X_1^{i+1}, X_2^j) - P(Y \leq k_c | X_1^{i+1}, X_2^{j+1})$$

Note that by definition, both of the above inequations need to be satisfied to make X_1 and X_2 a synergistic pair. We can generalize the above two inequalities into one inequality constraint by simply moving the subtractors to the other side of the inequality, which is the general form of synergy that we consider.

$$P(Y \leq k_c | X_1^i, X_2^j) + P(Y \leq k_c | X_1^{i+1}, X_2^{j+1}) \leq$$
$$P(Y \leq k_c | X_1^{i+1}, X_2^j) + P(Y \leq k_c | X_1^i, X_2^{j+1}) \tag{5}$$

Assume Y is binary, X_1 and X_2 are both ternary. Now, the synergy constraints between X_1 and X_2 that affect Y are as presented in Table 1.

Table 1. Synergy constraints

Synergy Constraints of X_1 and X_2 on CPT of Y
$P(Y = 0 \| x_1^1, x_2^1) + P(Y = 0 \| x_1^2, x_2^2) \leqslant P(Y = 0 \| x_1^1, x_2^2) + P(Y = 0 \| x_1^2, x_2^1)$
$P(Y = 0 \| x_1^1, x_2^2) + P(Y = 0 \| x_1^2, x_2^3) \leqslant P(Y = 0 \| x_1^1, x_2^3) + P(Y = 0 \| x_1^2, x_2^2)$
$P(Y = 0 \| x_1^2, x_2^1) + P(Y = 0 \| x_1^3, x_2^2) \leqslant P(Y = 0 \| x_1^2, x_2^2) + P(Y = 0 \| x_1^3, x_2^1)$
$P(Y = 0 \| x_1^2, x_2^2) + P(Y = 0 \| x_1^3, x_2^3) \leqslant P(Y = 0 \| x_1^2, x_2^3) + P(Y = 0 \| x_1^3, x_2^2)$

The key difference to monotonicity is that the constraints are on a set of parents (two in our example) rather than a single parent.

3.1 Derivation of the Gradient for the Synergy Qualitative Influence

We now derive the gradients by extending the prior work [8]. Let us redefine the parameters of the conditional distributions as shown in Equation 3. This allows us to define a constraint function δ for the synergistic constraints:

$$P(X_i \leq k_c | pa_i^{j1}) + P(X_i \leq k_c | pa_i^{j4}) \leq P(X_i \leq k_c | pa_i^{j2}) + P(X_i \leq k_c | pa_i^{j3})$$

The constraint function for the above definition is:

$$\delta = P(X_i \leq k_c | pa_i^{j1}) + P(X_i \leq k_c | pa_i^{j4}) - P(X_i \leq k_c | pa_i^{j2}) - P(X_i \leq k_c | pa_i^{j3}) + \epsilon \tag{6}$$

The above definition is similar to the monotonicity case. Then the gradient of the penalty function can be computed as:

$$\frac{\partial}{\partial \mu_{ijk}} P^{i,k_c}_{j_1,j_2,j_3,j_4} = \frac{\partial}{\partial \mu_{ijk}} I_{(\delta \geq 0)} \delta^2$$

$$= 2I_{(\delta \geq 0)} \delta \Big(\frac{\partial}{\partial \mu_{ijk}} \frac{Z^i_{j_1 k_c}}{Z^i_{j_1}} + \frac{\partial}{\partial \mu_{ijk}} \frac{Z^i_{j_4 k_c}}{Z^i_{j_4}} - \frac{\partial}{\partial \mu_{ijk}} \frac{Z^i_{j_2 k_c}}{Z^i_{j_2}} - \frac{\partial}{\partial \mu_{ijk}} \frac{Z^i_{j_3 k_c}}{Z^i_{j_3}} + \frac{\partial}{\partial \mu_{ijk}} \epsilon \Big)$$

$$= 2I_{(\delta \geq 0)} \delta \Big[\frac{Z^i_{j_1} I_{(j=j_1 \wedge k \leq k_c)} exp(\mu_{ijk}) - Z^i_{j_1 k_c} I_{(j=j_1)} exp(\mu_{ijk})}{(Z^i_{j_1})^2}$$

$$+ \frac{Z^i_{j_4} I_{(j=j_4 \wedge k \leq k_c)} exp(\mu_{ijk}) - Z^i_{j_4 k_c} I_{(j=j_4)} exp(\mu_{ijk})}{(Z^i_{j_4})^2}$$

$$- \frac{Z^i_{j_2} I_{(j=j_2 \wedge k \leq k_c)} exp(\mu_{ijk}) - Z^i_{j_2 k_c} I_{(j=j_2)} exp(\mu_{ijk})}{(Z^i_{j_2})^2}$$

$$- \frac{Z^i_{j_3} I_{(j=j_3 \wedge k \leq k_c)} exp(\mu_{ijk}) - Z^i_{j_3 k_c} I_{(j=j_3)} exp(\mu_{ijk})}{(Z^i_{j_3})^2} \Big]$$

$$= 2I_{(\delta \geq 0)} \delta \, exp(\mu_{ijk})(I_{(j=j_1)} + I_{(j=j_4)} - I_{(j=j_2)} - I_{(j=j_3)}) \frac{I_{(k \leq k_c)} Z^i_j - Z^i_{jk_c}}{(Z^i_j)^2} \quad (7)$$

where I is the indicator function, $Z^i_{jk_c} = \sum_{k=1}^{k_c} exp(\mu_{ijk})$, and $Z^i_j = Z^i_{jr_i}$. This gradient is very similar to the one obtained by Altendorf et al. [8]. The key difference is that in their formalism, every constraint inequality applied to two parameters, but our constraint inequality is applied to four parameters (assuming two parents of a random variable and all the variables are binary valued). This is due to the fact that monotonicities are associated with a single parent but synergies exist in a set of parents where each of the parents has a monotonic relationship with the target. Note that while we define these gradients with only two parents for brevity, they can be easily extended to sets of variables.

It should be mentioned that the definition of synergy we focus in this paper is different from the definition of Xiang and Jia [16]. It can be easily proved that the reinforcement in their work is equivalent with the positive monotonicity we defined in our paper and as defined by Altendorf et al. [8]. We clearly show this relationship in the appendix A. While their work focuses on the representation of monotonicities using ICI, we go further and combine synergies with Noisy-Or. In addition, we also derive the gradient for this combination and develop a learning algorithm in the next section.

4   Learning Parameters in Presence of Qualitative and Independence Knowledge

In the previous section, we presented the idea of using monotonicities and synergies as domain knowledge that makes it possible to learn from sparse data.

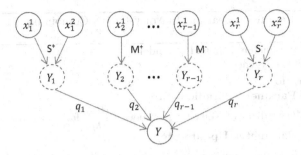

Fig. 1. Noisy-Or Bayesian network with qualitative constraints

However, it is a tedious work to list all constraints inequalities when there are a large number of parents and unnecessary when the qualitative constraint sets are independent with each other. If there are totally n parents and one of them (say X_c) has monotonic constraints on the resultant CPT, then the number of the constraints inequalities is proportional to $\prod_{X_j \in Pa_i \backslash X_c} r_j$, which is exponential in the number of total parents. The Noisy-Or structure, however, can make the number of constraints linear in the number of parent sets. In this work, we propose to use Noisy-Or to separate the influence of the different *sets* of qualitative constraints. So the inherent assumption is that the different sets of influences are independent of each other and the final structure is simply a Noisy-Or of the resulting distributions. It can be shown that introducing an extra layer of hidden nodes can still preserve the qualitative constraints of the features on the output, which exist between them in the original BN (see appendix B).

An example scenario is presented in Figure 1 where x_i^j represents the jth nodes in i^{th} constraint set. As can be seen the sets of parents can have a synergistic effect $(S+)$, sub-synergistic effect $(S\text{-})$, monotonic $(M+)$ or anti-monotonic $(M\text{-})$ effect. Each of these parent sets yields a distribution over the target (which is essentially a hidden node Y_i that is not observed in the data). These different distributions are then combined using the Noisy-Or combining rule where each of these can have an inhibition probability $(1 - q_i)$. In this work, we learn the parameters of the conditional distributions and the inhibition parameters.

Algorithm 1 presents the process of learning the parameters of conditional distributions and inhibitions given these qualitative statements and conditional influences (where α and β indicate the descent step size of CPT parameter and Noisy-Or parameter). The qualitative constraints are only applied on the CPTs of hidden node. So, the objective function of Noisy-Or parameters q_i is the log-likelihood function while the objective function J of CPT parameters is log-likelihood function minus the sum of all involved penalty functions times a penalty weight ω. It is an iterative procedure where we first learn the inhibition probabilities of the different combinations. Then using these *estimated* probabilities, we estimate the parameters of the conditional distributions subjected to the appropriate qualitative influences. This procedure is continued till convergence. It is possible that the algorithm sometimes may not converge to a feasible solution that satisfies all the constraints. In such cases, we increase the weight of

---

Algorithm 1. Parameter Learning in Noisy-Or BN Combining Qualitative Constraints

---

1. Initialize the parameters μ_{ijk} and q_i randomly
2. Repeat untill convergence:

   for $i = 1 \rightarrow r_l$ **do**

     Noisy-Or Parameter Gradient Step:

     Compute the gradient of Noisy-Or parameters $\dfrac{\partial LL}{\partial q_i}$ for all the q_i.

     Noisy-Or Parameter Update Step:

     Update each q_i by $q_i = q_i + \beta \dfrac{\partial LL}{\partial q_i}$

     for $j = 1 \rightarrow v_i$ **do**

       CPT Parameter Gradient Step:

       Compute the gradient of CPT parameters

   $$\frac{\partial J}{\partial \mu_{ijk}} = \frac{\partial LL}{\partial \mu_{ijk}} - \omega \sum \frac{\partial P_j^{i,k_c}}{\partial \mu_{ijk}}$$

       for each hidden node Y_i given every possible configuration of its parents

       CPT Parameter Update Step:

       Update each μ_{ijk} by $\mu_{ijk} = \mu_{ijk} + \alpha \dfrac{\partial J}{\partial \mu_{ijk}}$

     end for

   end for

3. If outside the feasible region, increase the penalty weight ω and repeat step 2

---

the penalty so that the solution does not go outside the feasible region. It must be mentioned that we are *not learning* the qualitative relationships but assume that these are given.

We use e_l to indicate the lth training example, r_l to denote the number of qualitative constraints sets the lth instance have, $X_{l,i}$ to represent the input vector of ith constraints set in l^{th} training example, q_i as the conditional probability $P(Y = 1 | Y_i = 1)$. The loglikelihood function in Noisy-Or BN combining multiple constraints sets is given by $LL = \sum_l log(P(y_l | e_l))$ where $P(y = 1 | e_l)$ is

$$P(y = 1 | e_l) = 1 - \prod_{i=1}^{r_l} [P_i(y = 0 | \mathbf{X}_{l,i}) + (1 - q_i) P_i(y = 1 | \mathbf{X}_{l,i})]$$

$$= 1 - \prod_{i=1}^{r_l} [P_i(y = 0 | \mathbf{X}_{l,i}) + (1 - q_i)(1 - P_i(y = 0 | \mathbf{X}_{l,i}))]$$

$$= 1 - \prod_{i=1}^{r_l} [1 - q_i + q_i P_i(y = 0 | \mathbf{X}_{l,i})] \tag{8}$$

Substitute Equation 3, we get:

$$P(y = 1 | e_l) = 1 - \prod_{i=1}^{r_l} [1 - q_i + q_i \frac{exp(\mu_{ij0})}{exp(\mu_{ij0}) + exp(\mu_{ij1})}] \tag{9}$$

Taking the derivative of the loglikelihood function with respect to μ_{ijk}, we get

$$\frac{\partial LL}{\partial \mu_{ijk}} = \sum_l \frac{1}{P(y_l|e_l)} \frac{\partial P(y_l|e_l)}{\partial \mu_{ijk}} = \sum_l [\frac{1}{P(y_l|e_l)}(-1)^y(-1)^k \varphi(e_l)] \qquad (10)$$

Where,

$$\varphi(e_l) = q_i [\frac{\partial}{\partial \mu_{ij0}}(\frac{exp(\mu_{ij0})}{exp(\mu_{ij0}) + exp(\mu_{ij1})})] \prod_{i' \neq i}[1 - q_{i'} + q_{i'} P_{i'}(y = 0|\mathbf{X}_{l,i'})]$$

$$= q_i [\frac{exp(\mu_{ij0})exp(\mu_{ij1})}{(exp(\mu_{ij0}) + exp(\mu_{ij1}))^2}] \prod_{i' \neq i}[1 - q_{i'} + q_{i'} P_{i'}(y = 0|\mathbf{X}_{l,i'})] \qquad (11)$$

The gradient of loglikelihood function with respect to q_i is given by:

$$\frac{\partial LL}{\partial q_i} = \sum_l \frac{1}{P(y_l|e_l)} \frac{\partial P(y_l|e_l)}{\partial q_i} = \sum_l [\frac{1}{P(y_l|e_l)}(-1)^y \phi(e_l)] \qquad (12)$$

$$\phi(e_l) = (P_i(y = 0|\mathbf{X}_{l,i}) - 1) \prod_{i' \neq i}[1 - q_{i'} + q_{i'} P_{i'}(y = 0|\mathbf{X}_{l,i'})] \qquad (13)$$

Once this gradient is obtained, we perform the iterative update of the Noisy-Or parameters and the CPT parameters as shown in Algorithm 1.

The natural question to ask is, where does the knowledge come from? We believe that, in many domains such as medicine, obtaining this knowledge is natural – for instance, there exists published research in understanding interactions of risk factors when predicting a disease, say heart attack. From this perspective, our proposed work here can be considered as enabling domain experts to provide more information to guide the algorithms in their search through the space of parameters. In addition, our algorithms can provide a method to evaluate the extent to which the domain knowledge is correct – it can determine the violations of the constraints in the training data. So we provide a method by which the domain experts can include some knowledge that is fully satisfied by the data and their best guesses at other relationships. Our algorithms can naturally fit the true knowledge and determine how much of the guesses are true. As we show in our experiments, there are some cases, where the independence between the sets of relationships may not be always true and in some cases, the monotonicities are as valuable as synergies. We aim to understand the interplay between the qualitative constraints and Noisy-Or and aim to determine if the combination is indeed a powerful method to exploit prior knowledge.

5 Experimental Evaluation

In this section, we present the results of evaluating our algorithm on 11 different standard machine learning domains from the UCI repository. The key questions that we seek to ask in our experiments are:

Q1: How does the use of qualitative constraints compared to not using any influence statements?

Q2: How valid is the independence assumption (i.e., how good is only using Noisy-Or)?

Q3: How does synergy compare to monotonicity?

Q4: How does the addition of the conditional influences with qualitative constraints help?

For each dataset, we learn the parameters by implementing six algorithms: (i) learning merely from data, (ii) with monotonic constraints, (iii) with synergy constraints, (iv) learning with Noisy-Or structure, (v) monotonic constraints plus Noisy-Or and (vi) synergy constraints plus Noisy-Or. All used features are discretized into two states under the following rules: i) nominal variables such as sex, race are assigned a class based on their qualitative relationships with their children nodes; ii) ordinal variables (e.g. {small, med, big}) are divided into two classes based on their distributions; iii) continuous values such as blood pressure, thyroxine are discretized according to practical thresholds in corresponding domains. The AUC-ROC and P values are calculated to compare their performances. We perform 10-fold cross-validation on all the domains for parameter selection and present the results on test set.

Table 2. Details of the experimental domains

Domain	Target Attribute	Num of Parents	Num of Samples
Heart Disease	Diagnosis of HD	5	297
Breast Cancer	BC recurrence	5	286
Credit Approval	Card Approval	5	300
Car Evaluation	Acceptable or not	6	300
Pima Indian Diabetes	Diabetes status	4	300
Census Income	$> 50K$ or $\leqslant 50K$	7	300
Iris	Versicolour or Virginica	4	100
Glass Identification	float or non_float(Building_windows)	5	146
Ecoli	Protein localization	5	284
Thyroid Disease	TD status	5	185
Hepatitis	Death of hepatitis	5	144

Table 2 presents the target attribute of interest in each domain in the second column. The third column lists the number of parents and the final column presents the number of all instances (sum of training set and test set whose proportion is about 10:1 in every domain). For the different domains, we provide prior knowledge– synergies and monotonicities whenever applicable. An example of such a network is presented in Figure 2. As can be seen, this is in the breast cancer domain where the goal is to predict recurrence of breast cancer based on 5 different attributes {age, menopause, tumor-size, deg-malig, irradiation}.

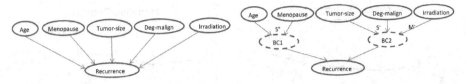

Fig. 2. An example domain without (left) and with (right) qualitative influences and Noisy-Or

In its Noisy-Or model, irradiation has a negative monotonic effect on the probability of recurrence and the others have positive monotonic effects. Parent set {age, menopause} has a synergistic interaction while {tumor-size, deg-malig} is sub-synergistic.

The results of using the different algorithms are presented in Figure 3 where the consolidated AUC-ROC over all the domains is presented. The first bar graph of every domain is a simple inverted naive Bayesian network where every feature is considered to be a parent of the target and the parameters are learned. The subsequent ones are (in that order) – *Noisy-Or, monotonicity* constraints [8], *synergies, monotonicity with Noisy-Or* and *synergy with Noisy-Or*. Hence, the last three bar graphs are the algorithms presented in this work. As can be seen, very clearly, in all the domains, the use of qualitative constraints and qualitative constraints with Noisy-Or outperform simply learning the conditional distributions from data. Hence **Q1** can be answered affirmatively.

The interesting observation is that Noisy-Or assumption seems to be a strong one in several domains. In many domains, the use of Noisy-Or is better than assuming no knowledge in almost all the domains. But the use of only qualitative statements such as monotonicity and synergy yield significantly better results in 9 domains when compared to only using Noisy-Or. Hence, it is clear that the answer to **Q2** is that only using Noisy-Or is not sufficient for a majority

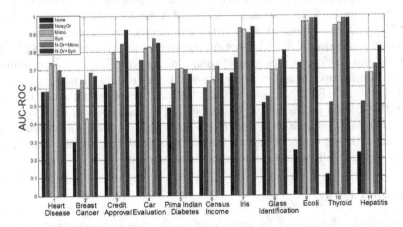

Fig. 3. Performance of the different algorithmic settings in several domains. Best viewed in color.

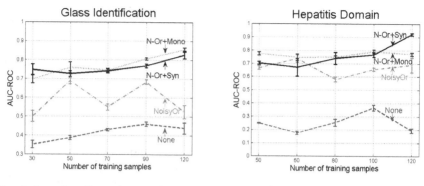

Fig. 4. Learning Curve in two domains with no knowledge, Noisy-Or and qualitative influences + Noisy-Or

of domains. Comparing monotonicity and synergies, it is clear that there is not much difference in several domains – except for breast cancer domain where synergy seems to be significantly worse than monotonicity and Noisy-Or. Hence, in answer to **Q3**, there is no significant difference between using monotonicity and synergy constraints. It remains interesting to understand the situations in which the use of synergy is more useful than the monotonicities.

Finally, the combination of qualitative and conditional influences seems to perform the best in most of the domains. The results are comparable to or better than simple qualitative constraints in all the domains. In 7 domains, the use of conditional influences seems to improve upon the use of only qualitative constraints. While in 3 others, there is no significant change in performance by adding conditional influences. Only in one domain (Breast Cancer), there is a very small dip (that is **not** statistically significant) in the AUC-ROC values. Hence, to answer **Q4**, we can affirmatively state that the use of conditional influences improves the performance of qualitative influence relationships in a majority of domains. Interestingly, the use of Noisy-Or with synergies improves upon Noisy-Or with monotonicity in three domains while in other domains the results are comparable. This is very similar to the observation about **Q3** where synergies and monotonicities exhibit comparable performance in most domains. All the significance results reported here are the results of using t-test with p-values < 0.05. Figure 4 presents the learning curves comparing the use of qualitative influence plus Noisy-Or against simple Noisy-Or and using no knowledge in two sample domains. The performance using prior knowledge has a jump start and faster convergence in both the domains, justifying the use of qualitative and conditional influence statements in these domains.

6   Conclusion

We presented a framework for combining qualitative and conditional influence statements when biasing probabilistic learners. We formalized the notion of synergistic interactions and derived the gradients for learning in the presence of such statements. We then extended our model to include conditional influences such as Noisy-Or and derived an algorithm for learning in presence of these two types of constraints. We evaluated our algorithms on 11 different domains and the results conclusively proved that the use of qualitative influences when combined with conditional influences yields a better performance in a majority of the domains. Our goal is to next understand the different types of conditional influences and their interactions with qualitative constraints. Exploring the use of such constraints in learning the structure of a full Bayesian network remains a very interesting direction for the future research.

Acknowledgments. Sriraam Natarajan gratefully acknowledges support of the DARPA DEFT Program under the Air Force Research Laboratory (AFRL) prime contract no. FA8750-13-2-0039. Any opinions, findings, and conclusion or recommendations expressed in this material are those of the authors and do not necessarily reflect the view of the DARPA, AFRL, or the US government.

References

1. Heckerman, D., Geiger, D., Chickering, D.: Learning Bayesian networks: The combination of knowledge and statistical data. Machine Learning 20, 197–243 (1995)
2. Vomlel, J.: Noisy-or classifier. International Journal of Intelligent Systems 21(3), 381–398 (2006)
3. Heckerman, D., Breese, J.: A new look at causal independence. In: UAI (1994)
4. Srinivas, S.: A generalization of the Noisy-Or model. In: UAI, pp. 208–215. Morgan Kaufmann, San Francisco (1993)
5. Pearl, J.: Probabilistic reasoning in intelligent systems: Networks of plausible inference. Morgan Kaufmann Publishers Inc. (1988)
6. Vomlel, J.: Exploiting functional dependence in Bayesian network inference. In: UAI, Edmonton, Canada, pp. 528–535 (2002)
7. Wellman, M.: Fundamental concepts of qualitative probabilistic networks. Artificial Intelligence 44, 257–303 (1990)
8. Altendorf, E., Restificar, A., Dietterich, T.: Learning from sparse data by exploiting monotonicity constraints. In: UAI, pp. 18–26 (2005)
9. Feelders, A., van der Gaag, L.: Learning Bayesian network parameters with prior knowledge about context-specific qualitative influences. In: UAI, pp. 193–200 (2005)
10. de Campos, C.P., Tong, Y., Ji, Q.: Constrained maximum likelihood learning of Bayesian networks for facial action recognition. In: Forsyth, D., Torr, P., Zisserman, A. (eds.) ECCV 2008, Part III. LNCS, vol. 5304, pp. 168–181. Springer, Heidelberg (2008)
11. Tong, Y., Ji, Q.: Learning Bayesian networks with qualitative constraints. In: CVPR (2008)
12. Brunk, H.: Maximum likelihood estimates of monotone parameters. The Annals of Mathematical Statistics 26(4), 607–616 (1955)
13. Boutilier, C., Friedman, N., Goldszmidt, M., Koller, D.: Context-specific independence in Bayesian networks. In: UAI, pp. 115–123 (1996)
14. Robertson, T., Wright, F., Dykstra, R.: Order Restricted Statistical Inference. Wiley, Chichester (1988)
15. Pekka, J., Erkki, V., Jaakko, T., Pekka, P.: Sex, age, cardiovascular risk factors, and coronary heart disease: a prospective follow-up study of 14786 middle-aged men and women in finland. Circulation 99, 1165–1172 (1999)

16. Xiang, Y., Jia, N.: Modeling causal reinforcement and undermining for efficient cpt elicitation. IEEE Transactions on Knowledge and Data Engineering 19(12), 1708–1718 (2007)

Appendix A: Equivalence of Reinforcement to Positive Monotonicity

If variable Y is resulted from a set of causes $\mathbf{X}$, the causes in $\mathbf{X}$ are said to *reinforce* each other if for any subset $\mathbf{X'} \subset \mathbf{X}$ the following holds [16]: $P(Y \ is \ true | \mathbf{X'} \ is \ true) \leqslant P(Y \ is \ true | \mathbf{X} \ is \ true)$

Proof. Assume variable Y has three parents $\{x_1, x_2, x_3\}$ all of which have positive monotonic influence on Y ($x_{1\succ}^{M+}Y$, $x_{2\succ}^{M+}Y$, $x_{3\succ}^{M+}Y$)and all variables are binary. The monotonic constraints of $x_{1\succ}^{M+}Y$ at the context of C=($\{x_2 = 1, x_3 = 1\}, \{x_2 = 1, x_3 = 0\}, \{x_2 = 0, x_3 = 1\}$) is

$$P(Y = 1|x_1^1, x_2^1, x_3^1) \geqslant P(Y = 1|x_1^0, x_2^1, x_3^1) \tag{14}$$

$$P(Y = 1|x_1^1, x_2^1, x_3^0) \geqslant P(Y = 1|x_1^0, x_2^1, x_3^0) \tag{15}$$

$$P(Y = 1|x_1^1, x_2^0, x_3^1) \geqslant P(Y = 1|x_1^0, x_2^0, x_3^1) \tag{16}$$

$x_{2\succ}^{M+}Y$ at the context of C=($\{x_1 = 1, x_3 = 1\}, \{x_1 = 1, x_3 = 0\}$) is

$$P(Y = 1|x_1^1, x_2^1, x_3^1) \geqslant P(Y = 1|x_1^1, x_2^0, x_3^1) \tag{17}$$

$$P(Y = 1|x_1^1, x_2^1, x_3^0) \geqslant P(Y = 1|x_1^1, x_2^0, x_3^0) \tag{18}$$

$x_{3\succ}^{M+}Y$ at the context of $C = (\{x_1 = 1, x_2 = 1\})$ is

$$P(Y = 1|x_1^1, x_2^1, x_3^1) \geqslant P(Y = 1|x_1^1, x_2^1, x_3^0) \tag{19}$$

Inequ.18 and **Inequ.19** $\Rightarrow P(Y = 1|x_1^1, x_2^1, x_3^1) \geqslant P(Y = 1|x_1^1, x_2^0, x_3^0)$
Inequ.15 and **Inequ.19** $\Rightarrow P(Y = 1|x_1^1, x_2^1, x_3^1) \geqslant P(Y = 1|x_1^0, x_2^1, x_3^0)$
Inequ.16 and **Inequ.17** $\Rightarrow P(Y = 1|x_1^1, x_2^1, x_3^1) \geqslant P(Y = 1|x_1^0, x_2^0, x_3^1)$

The inequalities above can be presented as the probability of Y is true given all the causes $\{x_1, x_2, x_3\}$ are activated is no less than that of only part of the causes ($\{x_1\}, \{x_2\}, \{x_3\}, \{x_2, x_3\}, \{x_1, x_3\}, \{x_1, x_2\}$) is activated, which is the definition of reinforce.

Appendix B: Sub-synergy and Synergy Constraints Can Be Preserved in Noisy-Or Structure.

Assume variable Y has four parents $\{x_1, x_2, x_3, x_4\}$, all the variables are binary and $x_{1\succ}^{M+}Y$, $x_{2\succ}^{M+}Y$ sub-synergistically (as shown in Figure 5). Sub-synergy constraints of x_1 and x_2 on variable Y given the context $\{x_3^i, x_4^j\}$ is given by:

$$P(Y = 0|x_1^1, x_2^1, x_3^i, x_4^j) + P(Y = 0|x_1^0, x_2^0, x_3^i, x_4^j) \geqslant$$
$$P(Y = 0|x_1^1, x_2^0, x_3^i, x_4^j) + P(Y = 0|x_1^0, x_2^1, x_3^i, x_4^j) \tag{20}$$

In the Noisy-Or structure, we can introduce two hidden nodes Y_1 and Y_2 the sub-synergy constraint of x_1 and x_2 on hidden node Y_1 is given by:

$$P(Y_1 = 0|x_1^1, x_2^1) + P(Y_1 = 0|x_1^0, x_2^0) \geqslant P(Y_1 = 0|x_1^1, x_2^0) + P(Y_1 = 0|x_1^0, x_2^1) \quad (21)$$

Based on Equation 8, we have

$$P(Y = 0|x_1^1, x_2^1, x_3^i, x_4^j) = [1 - q_1 + q_1 P(Y_1 = 0|x_1^1, x_2^1)] \times [1 - q_2 + q_2 P(Y_2 = 0|x_3^i, x_4^j)]$$
$$P(Y = 0|x_1^0, x_2^0, x_3^i, x_4^j) = [1 - q_1 + q_1 P(Y_1 = 0|x_1^0, x_2^0)] \times [1 - q_2 + q_2 P(Y_2 = 0|x_3^i, x_4^j)]$$
$$P(Y = 0|x_1^1, x_2^0, x_3^i, x_4^j) = [1 - q_1 + q_1 P(Y_1 = 0|x_1^1, x_2^0)] \times [1 - q_2 + q_2 P(Y_2 = 0|x_3^i, x_4^j)]$$
$$P(Y = 0|x_1^0, x_2^1, x_3^i, x_4^j) = [1 - q_1 + q_1 P(Y_1 = 0|x_1^0, x_2^1)] \times [1 - q_2 + q_2 P(Y_2 = 0|x_3^i, x_4^j)]$$

Since q_1 is a probability which is no less than zero, multiply q_1 to Inequality 21 we get

$$q_1 P(Y_1 = 0|x_1^1, x_2^1) + q_1 P(Y_1 = 0|x_1^0, x_2^0) \geqslant q_1 P(Y_1 = 0|x_1^1, x_2^0) + q_1 P(Y_1 = 0|x_1^0, x_2^1)$$

Add $(1 - q_1)$ to every item,

$$[1 - q_1 + q_1 P(Y_1 = 0|x_1^1, x_2^1)] + [1 - q_1 + q_1 P(Y_1 = 0|x_1^0, x_2^0)] \geqslant$$
$$[1 - q_1 + q_1 P(Y_1 = 0|x_1^1, x_2^0)] + [1 - q_1 + q_1 P(Y_1 = 0|x_1^0, x_2^1)]$$

Since $1 - q_2 + q_2 P(Y_2 = 0|x_3^i, x_4^j) = 1 - q_2 P(Y_2 = 1|x_3^i, x_4^j)$, which is no less than zero, multiply it with the above inequality,

$$[1 - q_1 + q_1 P(Y_1 = 0|x_1^1, x_2^1)][1 - q_2 + q_2 P(Y_2 = 0|x_3^i, x_4^j)]$$
$$+ [1 - q_1 + q_1 P(Y_1 = 0|x_1^0, x_2^0)][1 - q_2 + q_2 P(Y_2 = 0|x_3^i, x_4^j)] \geqslant$$
$$[1 - q_1 + q_1 P(Y_1 = 0|x_1^1, x_2^0)][1 - q_2 + q_2 P(Y_2 = 0|x_3^i, x_4^j)]$$
$$+ [1 - q_1 + q_1 P(Y_1 = 0|x_1^0, x_2^1)][1 - q_2 + q_2 P(Y_2 = 0|x_3^i, x_4^j)]$$

which is equivalent with Inequality 20.

It is easy to prove the transitivity of monotonic constraints in the proposed model. The process is similar as this one, which will not be shown here.

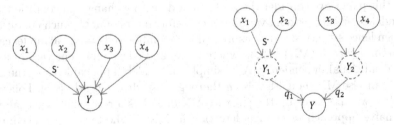

Fig. 5. Sub-synergy in one layer BN (left) and Noisy-Or BN (right)

Direct Learning of Sparse Changes in Markov Networks by Density Ratio Estimation

Song Liu[1], John A. Quinn[2], Michael U. Gutmann[3], and Masashi Sugiyama[1]

[1] Tokyo Institute of Technology, 2-12-1 O-okayama, Meguro, Tokyo 152-8552, Japan
song@sg.cs.titech.ac.jp, sugi@cs.titech.ac.jp
[2] Makerere University, P.O. Box 7062, Kampala, Uganda
jquinn@cit.ac.ug
[3] University of Helsinki and HIIT, Finland, P.O. Box 68, FI-00014, Finland
michael.gutmann@helsinki.fi

Abstract. We propose a new method for detecting changes in Markov network structure between two sets of samples. Instead of naively fitting two Markov network models separately to the two data sets and figuring out their difference, we *directly* learn the network structure change by estimating the ratio of Markov network models. This density-ratio formulation naturally allows us to introduce sparsity in the network structure change, which highly contributes to enhancing interpretability. Furthermore, computation of the normalization term, which is a critical computational bottleneck of the naive approach, can be remarkably mitigated. Through experiments on gene expression and Twitter data analysis, we demonstrate the usefulness of our method.

1 Introduction

Changes in the structure of interactions between random variables are interesting in many real-world phenomena. For example, genes may interact with each other in different ways when external stimuli change, co-occurrence between words may disappear/appear when the domains of text corpora shift, and correlation among pixels may change when a surveillance camera captures anomalous activities. Discovering such changes in interactions is a task of great interest in machine learning and data mining, because it provides useful insights into underlying mechanisms in many real-world applications.

In this paper, we consider the problem of detecting changes in conditional independence among random variables between two sets of data. Such conditional independence structure can be expressed as an undirected graphical model called a *Markov network* (MN) [1,2,3], where nodes and edges represent variables and their conditional dependency. As a simple and widely applicable case, the 2nd-order pairwise MN model has been thoroughly studied recently [4,5]. Following this line, we also focus on the pairwise MN model as a representative example.

A naive approach to change detection in MNs is the two-step procedure of first estimating two MNs separately from two sets of data by *maximum likelihood estimation* (MLE), and then comparing the structure of learned MNs. However,

H. Blockeel et al. (Eds.): ECML PKDD 2013, Part II, LNAI 8189, pp. 596–611, 2013.

Fig. 1. The rationale of direct structural change learning

MLE is often computationally expensive due to the normalization factor included in the density model. There are estimation methods which do not rely on knowing the normalization factor [6], but Gaussianity is often assumed for computing the normalization factor analytically [7]. However, this Gaussian assumption is highly restrictive in practice.

Another conceptual weakness of the above two-step procedure is that structure change is not directly learned. This indirect nature causes a problem, for example, if we want to learn a sparse structure change. For learning sparse changes, we may utilize ℓ_1-regularized MLE [8,9,5], which produces sparse MNs and thus the change between MNs also becomes sparse. However, this approach does not work if MNs are rather dense but change is sparse.

To mitigate this indirect nature, the *fused lasso* [10] is useful, where two MNs are simultaneously learned with a sparsity-inducing penalty on the difference between two MN parameters [11]. Although this fused-lasso approach allows us to learn sparse structure change naturally, the restrictive Gaussian assumption is still necessary to obtain the solution in a computationally efficient way.

A *nonparanormal* assumption [12,13] is a useful generalization of the Gaussian assumption. A nonparanormal distribution is a *semi-parametric Gaussian copula* where each Gaussian variable is transformed by a non-linear function. Nonparanormal distributions are much more flexible than Gaussian distributions thanks to the feature-wise non-linear transformation, while the normalization factors can still be computed analytically.

Thus, the fused-lasso method combined with nonparanormal models would be the state-of-the-art approach to change detection in MNs. However, the fused-lasso method is still based on separate modeling of two MNs, and its computation for more general non-Gaussian distributions is challenging.

In this paper, we propose a more direct approach to structural change learning in MNs based on *density ratio estimation* (DRE) [14]. Our method does not separately model two MNs, but directly models the *change* in two MNs. This idea follows Vapnik's principle [15]:

> If you possess a restricted amount of information for solving some problem, try to solve the problem directly and never solve a more general problem as an intermediate step. It is possible that the available information is sufficient for a direct solution but is insufficient for solving a more general intermediate problem.

This principle was used in the development of *support vector machines* (SVMs): Rather than modeling two classes of samples, SVM directly learns a decision boundary that is sufficient for performing pattern recognition. In the current context, estimating two MNs is more general than detecting changes in MNs (Figure 1). This direct approach means that we halve the number of parameters, from two MNs to one MN-difference.

Furthermore, the normalization factor in our DRE-based method can be approximated efficiently, because the normalization term in a density ratio function takes the form of an expectation and thus it can be simply approximated by sample averages without sampling.

The remainder of this paper is structured as follows. In Section 2, we formulate the problem of detecting structural changes and review currently available approaches. We then propose our DRE-based structural change detection method in Section 3. Results of illustrative and real-world experiments are reported in Section 4 and Section 5, respectively. Finally, we conclude our work and show future directions in Section 6.

2   Problem Formulation and Related Methods

In this section, we formulate the problem of change detection in Markov network structure and review existing approaches.

2.1   Problem Formulation

Consider two sets of samples drawn separately from two probability distributions P and Q on $\mathbb{R}^d$:

$$\{\boldsymbol{x}_i^P\}_{i=1}^{n_P} \overset{\text{iid}}{\sim} p(\boldsymbol{x}) \text{ and } \{\boldsymbol{x}_i^Q\}_{i=1}^{n_Q} \overset{\text{iid}}{\sim} q(\boldsymbol{x}).$$

We assume that p and q belong to the family of *Markov networks* (MNs) consisting of univariate and bivariate factors:

$$p(\boldsymbol{x}; \boldsymbol{\alpha}) = \frac{1}{Z(\boldsymbol{\alpha})} \exp\left(\sum_{i=1}^{d} \boldsymbol{\alpha}_i^\top \boldsymbol{g}_i(x_i) + \sum_{i,j=1,i>j}^{d} \boldsymbol{\alpha}_{i,j}^\top \boldsymbol{g}_{i,j}(x_i, x_j)\right), \qquad (1)$$

where $\boldsymbol{x} = (x_1, \ldots, x_d)^\top$, $\boldsymbol{\alpha}_i, \boldsymbol{\alpha}_{i,j}$ are parameters, $\boldsymbol{g}_i, \boldsymbol{g}_{i,j}$ are univariate and bivariate vector-valued basis functions, and $Z(\boldsymbol{\alpha})$ is the normalization factor. $q(\boldsymbol{x}; \boldsymbol{\alpha})$ is defined in the same way.

For notational simplicity, we unify both univariate and bivariate factors as

$$p(\boldsymbol{x}; \boldsymbol{\theta}) = \frac{1}{Z(\boldsymbol{\theta})} \exp\left(\sum_t \boldsymbol{\theta}_t^\top \boldsymbol{f}_t(\boldsymbol{x})\right), \quad \text{where } Z(\boldsymbol{\theta}) = \int \exp\left(\sum_t \boldsymbol{\theta}_t^\top \boldsymbol{f}_t(\boldsymbol{x})\right) \mathrm{d}\boldsymbol{x}.$$

$q(\boldsymbol{x}; \boldsymbol{\theta})$ is also simplified in the same way.

Our goal is to detect the change in conditional independence between random variables between P to Q.

2.2 Sparse MLE and Graphical Lasso

Maximum likelihood estimation (MLE) with group ℓ_1-regularization has been widely used for estimating the sparse structure of MNs [16,4,5]:

$$\max_{\theta} \sum_{i=1}^{n} \log p(x_i^P; \theta) - \lambda \sum_{t} \|\theta_t\|, \tag{2}$$

where $\|\cdot\|$ denotes the ℓ_2-norm. As λ increases, θ_t for pairwise factors may drop to 0. Thus, this method favors an MN that encodes more conditional independencies among variables. For computing the normalization term $Z(\theta)$ in Eq.(1), sampling techniques such as Markov-chain Monte-Carlo (MCMC) and importance sampling are usually employed. However, obtaining a reasonable value by these methods becomes computationally more expensive as the dimension d grows.

To avoid this computational problem, the Gaussian assumption is often imposed [9,17]. If we consider a zero-mean Gaussian distribution, the following $p(x; \Theta)$ can be used to replace the density model in Eq.(2):

$$p(x; \Theta) = \frac{\det(\Theta)^{1/2}}{(2\pi)^{d/2}} \exp\left(-\frac{1}{2} x^\top \Theta x\right),$$

where Θ is the inverse covariance matrix (a.k.a. the precision matrix) and $\det(\cdot)$ denotes the determinant. Then Θ is learned by

$$\max_{\Theta} \log \det(\Theta) - \operatorname{tr}(\Theta S^P) - \lambda \|\Theta\|_1,$$

where S^P is the sample covariance matrix of $\{x_i^P\}_{i=1}^{n}$. $\|\Theta\|_1$ is the ℓ_1-norm of Θ, i.e., the absolute sum of all elements. This formulation has been studied intensively in [8], and a computationally efficient solution called the *graphical lasso* [9] has been proposed.

Sparse changes in conditional independence structure between P and Q can be detected by comparing two MNs separately estimated using sparse MLE. However, this approach implicitly assumes that two MNs are sparse, which is not necessarily true even if the change is sparse.

2.3 Fused-Lasso Method

To more naturally handle sparse changes in conditional independence structure between P and Q, a method based on *fused lasso* [10] has been developed [11]. This method jointly maximizes the conditional likelihood in a feature-wise manner for P and Q with a sparsity penalty on the *difference* between parameters. More specifically, for each element x_s ($s = 1, \ldots, d$) of x,

$$\max_{\theta_s^P, \theta_s^Q} \ell_s^P(\theta_s^P) + \ell_s^Q(\theta_s^Q) - \lambda_1(\|\theta_s^P\|_1 + \|\theta_s^Q\|_1) - \lambda_2 \|\theta_s^P - \theta_s^Q\|_1,$$

where $\ell_s^P(\boldsymbol{\theta})$ is the log conditional likelihood for the s-th element $x_s \in \mathbb{R}$ given the rest $\boldsymbol{x}_{-s} \in \mathbb{R}^{d-1}$:

$$\ell_s^P(\boldsymbol{\theta}) = \sum_{i=1}^{n_P} \log p(x_{i,s}^P | \boldsymbol{x}_{i,-s}^P; \boldsymbol{\theta}).$$

$\ell_s^Q(\boldsymbol{\theta})$ is defined in the same way as $\ell_s^P(\boldsymbol{\theta})$. In this fused-lasso method, Gaussianity is usually assumed to cope with the normalization issue described in Section 2.2.

2.4   Nonparanormal Extensions

In the above methods, Gaussianity is required in practice to compute the normalization factor efficiently, which is a highly restrictive assumption.

To overcome this restriction, it has become popular to perform structure learning under the *nonparanormal* settings [12,13], where the Gaussian distribution is replaced by a *semi-parametric Gaussian copula*. $\boldsymbol{x} = (x_1, \ldots, x_d)^\top$ is said to follow a *nonparanormal* distribution, if there exists a set of monotone and differentiable functions, $\{h_i(x)\}_{i=1}^d$, such that $\boldsymbol{h}(\boldsymbol{x}) = (h_1(x^{(1)}), \ldots, h_d(x^{(d)}))^\top$ follows the Gaussian distribution. Nonparanormal distributions are much more flexible than Gaussian distributions thanks to the non-linear transformation $\{h_i(x)\}_{i=1}^d$, while the normalization factors can still be computed in an analytical way.

3   Direct Learning of Structural Changes via Density Ratio Estimation

The fused-lasso method can more naturally handle sparse changes in MNs than separate sparse MLE, and its nonparanormal extension is more flexible than the Gaussian counterpart. However, the fused-lasso method is still based on separate modeling of two MNs, and its computation for more general non-Gaussian distributions is challenging.

In this section, we propose to directly learn structural changes based on *density ratio estimation* [14], which does not involve separate modeling of each MN and which allows us to approximate the normalization term efficiently.

3.1   Density Ratio Formulation for Structural Change Detection

Our key idea is to consider the ratio of p and q:

$$\frac{p(\boldsymbol{x}; \boldsymbol{\theta}^P)}{q(\boldsymbol{x}; \boldsymbol{\theta}^Q)} \propto \exp\left(\sum_t (\boldsymbol{\theta}_t^P - \boldsymbol{\theta}_t^Q)^\top \boldsymbol{f}_t(\boldsymbol{x})\right).$$

Here $\boldsymbol{\theta}_t^P - \boldsymbol{\theta}_t^Q$ encodes the difference between P and Q for factor $\boldsymbol{f}_t$, i.e., $\boldsymbol{\theta}_t^P - \boldsymbol{\theta}_t^Q$ is zero if there is no change in the t-th factor.

Once we consider the ratio of p and q, we actually do not have to estimate $\boldsymbol{\theta}_t^P$ and $\boldsymbol{\theta}_t^Q$; instead an estimate of their difference $\boldsymbol{\theta}_t = \boldsymbol{\theta}_t^P - \boldsymbol{\theta}_t^Q$ is sufficient for change detection:

$$r(\boldsymbol{x};\boldsymbol{\theta}) = \frac{1}{N(\boldsymbol{\theta})} \exp\left(\sum_t \boldsymbol{\theta}_t^\top f_t(\boldsymbol{x})\right), \quad \text{where} \quad N(\boldsymbol{\theta}) = \int q(\boldsymbol{x}) \exp\left(\sum_t \boldsymbol{\theta}_t^\top f_t(\boldsymbol{x})\right) \, d\boldsymbol{x}. \tag{3}$$

The normalization term $N(\boldsymbol{\theta})$ guarantees[1] $\int q(\boldsymbol{x}) r(\boldsymbol{x};\boldsymbol{\theta}) \, d\boldsymbol{x} = 1$. Thus, in this density ratio formulation, we are no longer modeling each p and q separately, but we model the change from p to q *directly*. This direct nature would be more suitable for change detection purposes according to Vapnik's principle that encourages avoidance of solving more general problems as an intermediate step [15]. This direct formulation also allows us to halve the number of parameters from both $\boldsymbol{\theta}^P$ and $\boldsymbol{\theta}^Q$ to only $\boldsymbol{\theta}$.

Furthermore, the normalization factor $N(\boldsymbol{\theta})$ in the density ratio formulation can be easily approximated by sample average over $\{\boldsymbol{x}_i^Q\}_{i=1}^{n_Q} \overset{\text{iid}}{\sim} q(\boldsymbol{x})$, because $N(\boldsymbol{\theta})$ is the expectation over $q(\boldsymbol{x})$:

$$N(\boldsymbol{\theta}) \approx \frac{1}{n_Q} \sum_{i=1}^{n_Q} \exp\left(\sum_t \boldsymbol{\theta}_t^\top f_t(\boldsymbol{x}_i^Q)\right).$$

3.2 Direct Density-Ratio Estimation

Density ratio estimation (DRE) methods have been recently introduced to the machine learning community [14] and are proven to be useful in a wide range of applications. Here, we concentrate on a DRE method called the *Kullback-Leibler importance estimation procedure* (KLIEP) for a log-linear model [18,19].

For a density ratio model $r(\boldsymbol{x};\boldsymbol{\theta})$, the KLIEP method minimizes the Kullback-Leibler divergence from $p(\boldsymbol{x})$ to $\widehat{p}(\boldsymbol{x}) = q(\boldsymbol{x}) r(\boldsymbol{x};\boldsymbol{\theta})$:

$$\mathrm{KL}[p\|\widehat{p}] = \int p(\boldsymbol{x}) \log \frac{p(\boldsymbol{x})}{q(\boldsymbol{x}) r(\boldsymbol{x};\boldsymbol{\theta})} \, d\boldsymbol{x} = \text{Const.} - \int p(\boldsymbol{x}) \log r(\boldsymbol{x};\boldsymbol{\theta}) \, d\boldsymbol{x}. \tag{4}$$

Note that our density-ratio model (3) automatically satisfies the non-negativity and normalization constraints:

$$r(\boldsymbol{x};\boldsymbol{\theta}) \geq 0 \quad \text{and} \quad \int q(\boldsymbol{x}) r(\boldsymbol{x};\boldsymbol{\theta}) \, d\boldsymbol{x} = 1.$$

---

[1] An alternative normalization term $N'(\boldsymbol{\theta},\boldsymbol{\theta}^Q) = \int q(\boldsymbol{x};\boldsymbol{\theta}^Q) r(\boldsymbol{x};\boldsymbol{\theta}) d\boldsymbol{x}$ may also be considered. However, the expectation with respect to a model distribution can be computationally expensive as in the case of MLE, and this alternative form requires an extra parameter $\boldsymbol{\theta}^Q$ which is not our main interest. It is noteworthy that the use of $N(\boldsymbol{\theta})$ as a normalization factor guarantees the consistency of density ratio estimation [18].

In practice, we maximize the empirical approximation of the second term in the right-hand side of Eq.(4):

$$\ell_{\text{KLIEP}}(\boldsymbol{\theta}) = \frac{1}{n_P} \sum_{i=1}^{n_P} \log r(\boldsymbol{x}_i^P; \boldsymbol{\theta})$$

$$= \frac{1}{n_P} \sum_{i=1}^{n_P} \sum_t \boldsymbol{\theta}_t^\top \boldsymbol{f}_t(\boldsymbol{x}_i^P) - \log \frac{1}{n_Q} \sum_{i=1}^{n_Q} \exp\left(\sum_t \boldsymbol{\theta}_t^\top \boldsymbol{f}_t(\boldsymbol{x}_i^Q) \right).$$

Because $\ell_{\text{KLIEP}}(\boldsymbol{\theta})$ is concave with respect to $\boldsymbol{\theta}$, its global maximizer can be numerically found by standard optimization techniques such as gradient ascent or quasi-Newton methods: The gradient of ℓ_{KLIEP} with respect to $\boldsymbol{\theta}_t$ is given by

$$\nabla_{\boldsymbol{\theta}_t} \ell_{\text{KLIEP}}(\boldsymbol{\theta}) = \frac{1}{n_P} \sum_{i=1}^{n_P} \boldsymbol{f}_t(\boldsymbol{x}_i^P) - \frac{\frac{1}{n_Q} \sum_{i=1}^{n_Q} \exp\left(\sum_t \boldsymbol{\theta}_t^\top \boldsymbol{f}_t(\boldsymbol{x}_i^Q) \right) \boldsymbol{f}_t(\boldsymbol{x}_i^Q)}{\frac{1}{n_Q} \sum_{j=1}^{n_Q} \exp\left(\sum_t \boldsymbol{\theta}_t^\top \boldsymbol{f}_t(\boldsymbol{x}_j^Q) \right)}.$$

3.3   Sparsity-Inducing Norm

To find a sparse change in P and Q, we may regularize our KLIEP solution with a sparsity-inducing norm $\sum_t \|\boldsymbol{\theta}_t\|$. Note that the motivation for introducing sparsity in KLIEP is different from MLE. In the case of MLE, both $\boldsymbol{\theta}^P$ and $\boldsymbol{\theta}^Q$ are sparsified and then consequently the difference $\boldsymbol{\theta}^P - \boldsymbol{\theta}^Q$ is also sparsified. On the other hand, in our case, only the difference $\boldsymbol{\theta}^P - \boldsymbol{\theta}^Q$ is sparsified; thus our method can still work well even if $\boldsymbol{\theta}^P$ and $\boldsymbol{\theta}^Q$ are dense.

In practice, we may use the following *elastic-net* penalty [20] to better control overfitting to noisy data:

$$\max_{\boldsymbol{\theta}} \left[\ell_{\text{KLIEP}}(\boldsymbol{\theta}) - \lambda_1 \|\boldsymbol{\theta}\|^2 - \lambda_2 \sum_t \|\boldsymbol{\theta}_t\| \right],$$

where $\|\boldsymbol{\theta}\|^2$ penalizes the magnitude of the entire parameter vector.

4   Numerical Experiments

In this section, we compare the proposed KLIEP-based method with the Fused-lasso (Flasso) method [11] and the Graphical-lasso (Glasso) method [9]. Results are reported on datasets with three different underlying distributions: multivariate Gaussian, nonparanormal, and a non-Gaussian "diamond" distribution.

4.1   Setup

Performance Metrics: by taking the advantage of knowing the ground truth of structural changes in artificial experiments, we measure the performance of change detection methods using the *precision-recall (P-R) curve*. For KLIEP and Flasso, a precision and recall curve can be plotted by varying the group-sparsity control parameter λ_2; we fix $\lambda_1 = 0$ because the artificial datasets are noise-free. For Glasso, we vary the sparsity control parameters as $\lambda = \lambda^P = \lambda^Q$.

Model Selection: for KLIEP, we use the log-likelihood of an estimated density ratio on a hold-out dataset, which we refer to as *hold-out log-likelihood* (HOLL). More precisely, given two sets of hold-out data $\{\widetilde{\boldsymbol{x}}_i^P\}_{i=1}^{\widetilde{n}_P} \overset{\text{iid}}{\sim} P$, $\{\widetilde{\boldsymbol{x}}_i^Q\}_{i=1}^{\widetilde{n}_Q} \overset{\text{iid}}{\sim} Q$ for $\widetilde{n}_P = \widetilde{n}_Q = 3000$, we use the following quantity:

$$\ell_{\text{HOLL}} = \frac{1}{\widetilde{n}_P} \sum_{i=1}^{\widetilde{n}_P} \log \frac{\exp\left(\sum_t \widehat{\boldsymbol{\theta}}_t^\top f_t(\widetilde{\boldsymbol{x}}_i^P)\right)}{\frac{1}{\widetilde{n}_Q} \sum_{j=1}^{\widetilde{n}_Q} \exp\left(\sum_t \widehat{\boldsymbol{\theta}}_t^\top f_t(\widetilde{\boldsymbol{x}}_j^Q)\right)}.$$

In case such a hold-out dataset is not available, the *cross-validated log-likelihood* (CVLL) may be used instead.

For the Glasso and Flasso methods, we perform model selection by adding the hold-out/cross-validated likelihoods on $p(\boldsymbol{x};\boldsymbol{\theta})$ and $q(\boldsymbol{x};\boldsymbol{\theta})$ together:

$$\frac{1}{\widetilde{n}_P} \sum_{i=1}^{\widetilde{n}_P} \log p(\widetilde{\boldsymbol{x}}_i^P;\widehat{\boldsymbol{\theta}}^P) + \frac{1}{\widetilde{n}_Q} \sum_{i=1}^{\widetilde{n}_Q} \log q(\widetilde{\boldsymbol{x}}_i^Q;\widehat{\boldsymbol{\theta}}^Q).$$

Basis Function: we consider two types of f_t: a power nonparanormal f_{npn} and a polynomial transform f_{poly}.

The pairwise nonparanormal transform with power k is defined as

$$f_{\text{npn}}(x_i, x_j) := [\text{sign}(x_i)x_i^k \text{sign}(x_j)x_j^k, 1].$$

This transforms the original data by the power of k, so that the transformed data are jointly Gaussian (see Section 4.3). The univariate nonparanormal transform is defined as $f_{\text{npn}}(x_i) := f_{\text{npn}}(x_i, x_i)$.

The polynomial transform up to degree of k is defined as:

$$f_{\text{poly}}(x_i, x_j) := [x_i^k, x_j^k, x_i x_j^{k-1}, \ldots, x_i^{k-1} x_j, x_i^{k-1}, x_j^{k-1}, \ldots, x_i, x_j, 1].$$

The univariate polynomial transform is defined as $f_{\text{poly}}(x_i) := f_{\text{poly}}(x_i, 0)$.

4.2 Multivariate Gaussian

First, we investigate the performance of each learning method under Gaussianity.

Consider a 40-node sparse Gaussian MN, where its graphical structure is characterized by precision matrix $\boldsymbol{\Theta}^P$ with diagonal elements equal to 2. The off-diagonal elements are randomly chosen[2] and set to 0.2, so that the overall sparsity of $\boldsymbol{\Theta}^P$ is 25%. We then introduce changes by randomly picking 15 edges and reducing the corresponding elements in the precision matrix by 0.1. The resulting precision matrices $\boldsymbol{\Theta}^P$ and $\boldsymbol{\Theta}^Q$ are used for drawing samples as

$$\{\boldsymbol{x}_i^P\}_{i=1}^n \overset{\text{iid}}{\sim} \mathcal{N}(\boldsymbol{0}, (\boldsymbol{\Theta}^P)^{-1}) \quad \text{and} \quad \{\boldsymbol{x}_i^Q\}_{i=1}^n \overset{\text{iid}}{\sim} \mathcal{N}(\boldsymbol{0}, (\boldsymbol{\Theta}^Q)^{-1}).$$

Datasets of size $n = 50$ and $n = 100$ are tested.

---

[2] We set $\Theta_{i,j} = \Theta_{j,i}$ for not breaking the symmetry of the precision matrix.

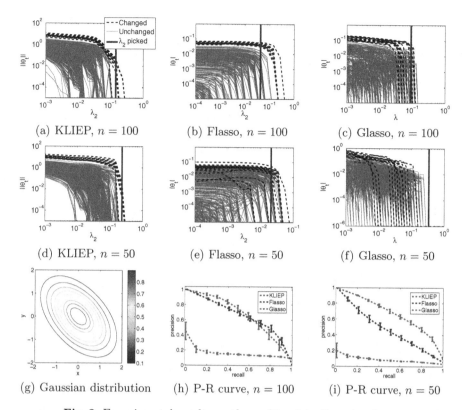

Fig. 2. Experimental results on the multivariate Gaussian dataset

We repeat the experiments 20 times with randomly generated datasets and report the results in Figure 2. The top 6 graphs are examples of regularization paths (black and red color represents the ground truth) and the bottom 3 graphs are the data generating distribution and averaged P-R curves with standard error. The top row is for $n = 100$ while the middle row is for $n = 50$. The regularization parameters picked by the model selection procedures described in Section 4.1 are marked with blue vertical lines. In this experiment, the Gaussian model (the nonparanormal basis function with power $k = 1$) is used for KLIEP. Because the Gaussian model is also used in Flasso and Glasso, the difference in performance is caused only by the difference of estimation methods.

When $n = 100$, KLIEP and Flasso clearly distinguish changed (black) and unchanged (red) edges in terms of parameter magnitude. However, when the sample size is halved, the separation is visually rather unclear in the case of Flasso. In contrast, the paths of changed and unchanged edges are still almost disjoint in the case of KLIEP. The Glasso method performs rather poorly in both cases. A similar tendency can be observed also in the averaged P-R curve. When the sample size is 100, KLIEP and Flasso work equally well, but KLIEP gains its lead when the sample size is reduced. Glasso does not perform well in both cases.

4.3 Nonparanormal

We post-process the dataset used in Section 4.2 to construct nonparanormal samples: simply, we apply the power function

$$h_i^{-1}(x) = \text{sign}(x)|x|^{\frac{1}{2}}$$

to each dimension of $\boldsymbol{x}^P$ and $\boldsymbol{x}^Q$, so that $h(\boldsymbol{x}^P) \sim \mathcal{N}(\mathbf{0}, (\boldsymbol{\Theta}^P)^{-1})$ and $h(\boldsymbol{x}^Q) \sim \mathcal{N}(\mathbf{0}, (\boldsymbol{\Theta}^Q)^{-1})$.

In order to cope with the non-linearity, we apply the nonparanormal basis function with power 2, 3 and 4 in KLIEP and choose the one that maximizes the peak HOLL value. For Flasso and Glasso, we apply the nonparanormal transform described in [12] before the structural change is learned.

The experiments are conducted on 20 randomly generated datasets with $n = 50$ and 100, respectively. The regularization paths, data generating distribution, and averaged P-R curves are plotted in Figure 3. The results show that Flasso clearly suffers from the performance degradation compared with the Gaussian case, perhaps because the number of samples is too small for the complicated nonparanormal distribution. Due to the two-step estimation scheme, the performance of Glasso is poor. In contrast, KLIEP separates changed and unchanged edges still clearly for both $n = 50$ and $n = 100$. The P-R curves also show the same tendency.

4.4 "Diamond" Distribution with No Pearson Correlation

In the previous experiment, though samples are non-Gaussian, the *Pearson correlation* is not zero. Therefore, methods assuming Gaussianity can still capture the linear correlation between random variables. In this experiment, we consider a more challenging case with a diamond-shaped distribution within the exponential family that has zero Pearson correlation coefficient between dependent variables. Thus, the methods assuming Gaussianity (i.e., Glasso and Flasso) can not extract any information in principle from this dataset.

The probability density function of the diamond distribution is defined as follows (Figure 4(a)):

$$p(\boldsymbol{x}) \propto \exp\left(-\sum_i 2x_i^2 - \sum_{(i,j):A_{i,j}\neq 0} 20x_i^2 x_j^2 \right), \tag{5}$$

where the adjacency matrix $\boldsymbol{A}$ describes an MN structure. Note that this distribution can not be transformed into a Gaussian distribution by any nonparanormal transformations. Samples from the above distribution are drawn by using a *slice sampling* method [21]. However, since generating samples from a high-dimensional distribution is non-trivial and time-consuming, we focus on a relatively low-dimensional case to avoid sampling errors which may mislead the experimental evaluation.

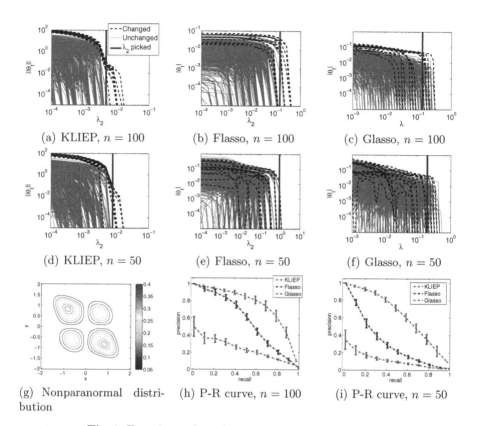

(a) KLIEP, $n = 100$ (b) Flasso, $n = 100$ (c) Glasso, $n = 100$

(d) KLIEP, $n = 50$ (e) Flasso, $n = 50$ (f) Glasso, $n = 50$

(g) Nonparanormal distri-   (h) P-R curve, $n = 100$ (i) P-R curve, $n = 50$
bution

Fig. 3. Experimental results on the nonparanormal dataset

We set $d = 9$ and $n_P = n_Q = 5000$. $\boldsymbol{A}^P$ is randomly generated with 35% sparsity, while $\boldsymbol{A}^Q$ is created by randomly removing edges in $\boldsymbol{A}^P$ so that the sparsity level is dropped to 15%.

In this experiment, we compare the performance of all three methods with their available transforms: KLIEP ($f_{\mathrm{poly}}, k = 2, 3, 4$), KLIEP ($f_{\mathrm{npn}}, k = 2, 3, 4$), KLIEP ($f_{\mathrm{npn}}, k = 1$; same as the Gaussian model), Flasso (nonparanormal), Flasso (Gaussian), Glasso (nonparanormal) and Glasso (Gaussian). The averaged P-R curves are shown in Figure 4(c). As expected, except KLIEP (f_{poly}), all other methods do not work properly. This means that the polynomial kernel is indeed very helpful in handling completely non-Gaussian data. However, as discussed in Section 2.2, it is difficult to use such a kernel in the MLE-based approaches (Glasso and Flasso) because computationally demanding sampling is involved in evaluating the normalization term. The regularization path of KLIEP (f_{poly}) illustrated in Figure 4(b) shows the usefulness of the proposed method in change detection under non-Gaussianity.

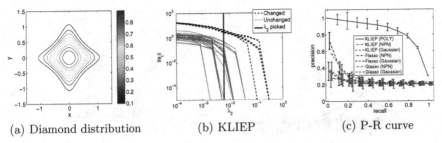

(a) Diamond distribution   (b) KLIEP   (c) P-R curve

Fig. 4. Experimental results on the diamond dataset. "NPN" and "POLY" denote the nonparanormal and polynomial models, respectively. Note that the precision rate of 100% recall for a random guess is approximately 20%.

5 Applications

In this section, experiments are conducted on a synthetic gene expression dataset and on a Twitter dataset, respectively. We consider only the KLIEP and Flasso methods here. For KLIEP, the polynomial transform function with $k \in \{2, 3, 4\}$ is used. The parameter λ_1 in KLIEP and Flasso is tested with choices $\lambda_1 \in \{0.1, 1, 10\}$. The performance reported for the experiments in Section 5.1 and 5.2 are obtained using the models selected by HOLL and 5-fold CVLL (see Section 4.1), respectively.

5.1 Synthetic Gene Expression Dataset

A gene regulatory network encodes interactions between DNA segments. However, the way genes interact may change due to environmental or biological stimuli. In this experiment, we focus on detecting such changes. We use *SynTReN*, which is a generator of gene regulatory networks used as the benchmark validation of bioinformatics algorithms [22].

To test the applicability of the proposed method, we first choose a sub-network containing 13 nodes from an existing signalling network in *Saccharomyces cerevisiae* (shown in Figure 5(a)). Three types of interactions are modelled: activation (ac), deactivation (re), and dual (du). 50 samples are generated in the first stage, after which we change the types of interactions in 6 edges, and generate 50 samples again. Four types of changes are considered in such case: ac → re, re → ac, du → ac, and du → re.

The regularization paths for KLIEP and Flasso are plotted in Figures 5(b) and 5(d). Averaged precision-recall curves over 20 simulation runs are shown in Figure 5(c). Clearly from the example of KLIEP regularization paths shown in Figure 5(d), the magnitude of estimated parameters on the changed pairwise interactions is much higher than that of the unchanged ones, and hits zero only at the final stage. On the other hand, Flasso gives many false alarms by assigning non-zero parameters to the unchanged interactions, even after some changed ones hit zeros. Reflecting a similar pattern, the P-R curves plot in Figure 5(c)

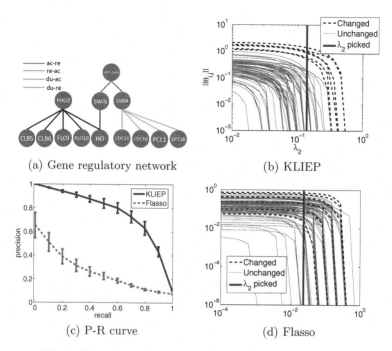

(a) Gene regulatory network (b) KLIEP

(c) P-R curve (d) Flasso

Fig. 5. Experiments on synthetic gene expression datasets

show that the proposed KLIEP method achieves significant improvement over the Flasso method.

5.2 Twitter Story Telling

In this experiment, we use KLIEP and Flasso as event detectors from Twitter. More specifically, we choose the *Deepwater Horizon oil spill*[3] as the target event, and we hope that our method can recover some story lines from Twitter as the news event develops. Counting the frequencies of 10 keywords (BP, oil, spill, Mexico, gulf, coast, Hayward, Halliburton, Transocean, and Obama), we obtain a dataset by sampling 1061 times (4 per day), from February 1st, 2010 to October 15th, 2010.

To conduct our experiments, we segment the data into two parts. The first 300 samples collected before the day of oil spill (April 20th, 2010) are regarded as conforming to a 10-dimensional joint distribution Q, while the second set of samples that are drawn in an arbitrary 50-day window approximately after the event happened is regarded as following distribution P.

The MN of Q encodes the original conditional independence of frequencies between 10 keywords, and the underlying MN of P has changed since an event occurred. Thus, unveiling a change in MNs between P and Q may recover popular topic trends on Twitter in terms of the dependency among keywords.

[3] http://en.wikipedia.org/wiki/Deepwater_Horizon_oil_spill

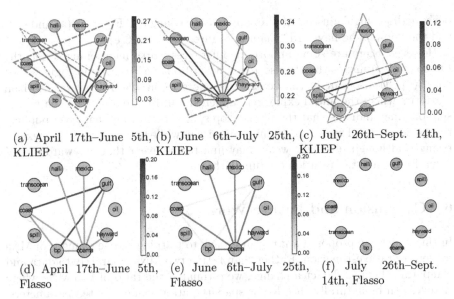

(a) April 17th–June 5th, (b) June 6th–July 25th, (c) July 26th–Sept. 14th, KLIEP   KLIEP   KLIEP

(d) April 17th–June 5th, (e) June 6th–July 25th, (f) July 26th–Sept. Flasso   Flasso   14th, Flasso

Fig. 6. Change graphs captured by the proposed KLIEP method (top) and the Flasso method (bottom). The date range beneath each figure indicates when P was sampled, while Q is fixed to dates from February 1st to April 17th. Notable structures shared by the graph of both methods are surrounded by the green dashed lines. Unique structures that only appear in the graph of the proposed KLIEP method are surrounded by the red dashed lines.

The detected change graphs (i.e. the graphs with only detected changing edges) on 10 keywords are illustrated in Figure 6. The edges are selected at a certain value of λ_2 indicated by the maximal CVLL. Since the edge set that is picked by CVLL may not be sparse in general, we sparsify the graph based on the permutation test as follows: we randomly shuffle the samples between P and Q and repeatedly run change detection algorithms for 100 times; then we observe detected edges by CVLL. Finally, we select the edges that are detected using the original non-shuffled dataset and remove those that were detected in the shuffled datasets for more than 5 times. In Figure 6, we plot detected change graphs which are generated using samples of P starting from April 17th, July 6th, and July 26th.

The initial explosion happened on April 20th, 2010. Both methods discover dependency changes between keywords. Generally speaking, KLIEP captures more conditional independence changes between keywords than the Flasso method, especially when comparing Figure 6(c) and Figure 6(f). At the first two stages (Figures 6(a), 6(b), 6(d) and 6(e)), the keyword "Obama" is very well connected with other keywords in the results given by both methods. Indeed, at the early development of this event, he lies in the center of the news stories, and his media exposure peaks after his visit to the Louisiana coast (May 2nd, May 28nd, and June 5th) and his meeting with BP CEO Tony Hayward on June 16th. Notably, both methods highlight the "gulf-obama-coast" triangle in Figures 6(a) and 6(d) and the "bp-obama-hayward" chain in Figures 6(b) and 6(e).

However, there are some important differences worth mentioning. First, the Flasso method misses the "transocean-hayward-obama" triangle in Figures 6(d) and 6(e). Transocean is the contracted operator in the Deepwater Horizon platform, where the initial explosion happened. On Figure 6(c), The chain "bp-spill-oil" may indicate that the phrase "bp spill" or "oil spill" has been publicly recognized by the Twitter community since then, while the "hayward-bp-mexico" triangle, although relatively weak, may link to the event that Hayward steped down from the CEO position on July 27th.

6   Conclusion and Future Work

In this paper, we proposed a *direct* approach to learning sparse changes in MNs by density ratio estimation. Rather than fitting two MNs separately to data and comparing them to detect a change, we estimated the ratio of two MNs where changes can be naturally encoded as sparsity patterns in estimated parameters. Through experiments on artificial and real-world datasets, we demonstrated the usefulness of the proposed method.

Compared with the conventional two-stage MLE approach, a notable advantage of our method is that the normalization term in the density ratio model can be approximated by a sample average without sampling. This considerably loosens the restriction on applicable distributions. Moreover, thanks to its direct modeling nature with density ratios, the number of parameters is halved.

We only considered MNs with pairwise factors in this paper. However, such a model may be misspecified when higher order interactions exist. For example, combination with the idea *hierarchical log-linear model* presented in [16] may lead to a promising solution to this problem, which will be investigated in our future work.

Acknowledgement. SL is supported by the JST PRESTO program and the JSPS fellowship, JQ is supported by the JST PRESTO program, and MS is supported by the JST CREST program. MUG is supported by the Finnish Centre-of-Excellence in Computational Inference Research COIN (251170).

References

1. Bishop, C.M.: Pattern Recognition and Machine Learning. Springer, New York (2006)
2. Wainwright, M.J., Jordan, M.I.: Graphical models, exponential families, and variational inference. Foundations and Trends® in Machine Learning 1(1-2), 1–305 (2008)
3. Koller, D., Friedman, N.: Probabilistic graphical models: principles and techniques. MIT Press (2009)

4. Ravikumar, P., Wainwright, M.J., Lafferty, J.D.: High-dimensional ising model selection using ℓ_1-regularized logistic regression. The Annals of Statistics 38(3), 1287–1319 (2010)

5. Lee, S.I., Ganapathi, V., Koller, D.: Efficient structure learning of Markov networks using ℓ_1-regularization. In: Schölkopf, B., Platt, J., Hoffman, T. (eds.) Advances in Neural Information, vol. 19, pp. 817–824. MIT Press, Cambridge (2007)

6. Gutmann, M., Hyvärinen, A.: Noise-contrastive estimation of unnormalized statistical models, with applications to natural image statistics. Journal of Machine Learning Research 13, 307–361 (2012)

7. Hastie, T., Tibshirani, R., Friedman, J.: The Elements of Statistical Learning: Data Mining, Inference, and Prediction. Springer, New York (2001)

8. Banerjee, O., El Ghaoui, L., d'Aspremont, A.: Model selection through sparse maximum likelihood estimation for multivariate Gaussian or binary data. Journal of Machine Learning Research 9, 485–516 (2008)

9. Friedman, J., Hastie, T., Tibshirani, R.: Sparse inverse covariance estimation with the graphical lasso. Biostatistics 9(3), 432–441 (2008)

10. Tibshirani, R., Saunders, M., Rosset, S., Zhu, J., Knight, K.: Sparsity and smoothness via the fused lasso. Journal of the Royal Statistical Society: Series B (Statistical Methodology) 67(1), 91–108 (2005)

11. Zhang, B., Wang, Y.: Learning structural changes of Gaussian graphical models in controlled experiments. In: Proceedings of the Twenty-Sixth Conference on Uncertainty in Artificial Intelligence (UAI 2010), pp. 701–708 (2010)

12. Liu, H., Lafferty, J., Wasserman, L.: The nonparanormal: Semiparametric estimation of high dimensional undirected graphs. The Journal of Machine Learning Research 10, 2295–2328 (2009)

13. Liu, H., Han, F., Yuan, M., Lafferty, J., Wasserman, L.: The nonparanormal skeptic. In: Proceedings of the 29th International Conference on Machine Learning, ICML 2012 (2012)

14. Sugiyama, M., Suzuki, T., Kanamori, T.: Density Ratio Estimation in Machine Learning. Cambridge University Press, Cambridge (2012)

15. Vapnik, V.N.: Statistical Learning Theory. Wiley, New York (1998)

16. Schmidt, M.W., Murphy, K.P.: Convex structure learning in log-linear models: Beyond pairwise potentials. Journal of Machine Learning Research - Proceedings Track 9, 709–716 (2010)

17. Meinshausen, N., Bühlmann, P.: High-dimensional graphs and variable selection with the lasso. The Annals of Statistics 34(3), 1436–1462 (2006)

18. Sugiyama, M., Suzuki, T., Nakajima, S., Kashima, H., von Bünau, P., Kawanabe, M.: Direct importance estimation for covariate shift adaptation. Annals of the Institute of Statistical Mathematics 60(4), 699–746 (2008)

19. Tsuboi, Y., Kashima, H., Hido, S., Bickel, S., Sugiyama, M.: Direct density ratio estimation for large-scale covariate shift adaptation. Journal of Information Processing 17, 138–155 (2009)

20. Zou, H., Hastie, T.: Regularization and variable selection via the elastic net. Journal of the Royal Statistical Society, Series B 67(2), 301–320 (2005)

21. Neal, R.M.: Slice sampling. The Annals of Statistics 31(3), 705–741 (2003)

22. Van den Bulcke, T., Van Leemput, K., Naudts, B., van Remortel, P., Ma, H., Verschoren, A., De Moor, B., Marchal, K.: SynTReN: A generator of synthetic gene expression data for design and analysis of structure learning algorithms. BMC Bioinformatics 7(1), 43 (2006)

Greedy Part-Wise Learning of Sum-Product Networks

Robert Peharz, Bernhard C. Geiger, and Franz Pernkopf

Signal Processing and Speech Communication Laboratory
Graz, University of Technology
{robert.peharz,geiger,pernkopf}@tugraz.at

Abstract. Sum-product networks allow to model complex variable interactions while still granting efficient inference. However, most learning algorithms proposed so far are explicitly or implicitly restricted to the image domain, either by assuming variable neighborhood or by assuming that dependent variables are related by their magnitudes over the training set. In this paper, we introduce a novel algorithm, learning the structure and parameters of sum-product networks in a greedy bottom-up manner. Our algorithm iteratively merges probabilistic models of small variable scope to larger and more complex models. These merges are guided by statistical dependence test, and parameters are learned using a maximum mutual information principle. In experiments our method competes well with the existing learning algorithms for sum-product networks on the task of reconstructing covered image regions, and outperforms these when neither neighborhood nor correlations by magnitude can be assumed.

1 Introduction

Recently, a new type of probabilistic graphical models called sum-product network (SPN) was proposed [1]. Motivated by arithmetic circuits [2, 3] and aiming at expressive models, still allowing efficient inference, they represent the network polynomial of a Bayesian network [2] with a deep network architecture containing sum and product nodes. In that way, SPNs combine the domains of deep learning and probabilistic graphical models. On the one hand, SPNs can be interpreted as deep neural networks with sum and product nodes as neurons, where the sum nodes compute a *weighted* sum (with non-negative weights) of its inputs. Besides the network structure, the weights determine the network input-output function, i.e. they represent the parameters of the network. In order to allow efficient inference, the SPN should fulfill certain constraints on the network structure, namely completeness and consistency or decomposability [1]. On the other hand, SPNs represent Bayesian networks (BNs) with rich latent structure – since sum nodes can be interpreted as hidden variables being summed out – with a high degree of context-specific independence among the hidden variables. The observable variables are placed as leaves of the BN, interacting with each other only via their latent parents. The BN interpretation opens the door for learning techniques from probabilistic graphical models, such as EM.

H. Blockeel et al. (Eds.): ECML PKDD 2013, Part II, LNAI 8189, pp. 612–627, 2013.
© Springer-Verlag Berlin Heidelberg 2013

In [1], a learning algorithm tailored for image processing was proposed. This algorithm recursively divides an image into pairs of smaller rectangles, and learns the weights of the allotted sum nodes using a kind of hard EM, penalizing the evocation of non-zero weights[1]. This algorithm relies on locality of image regions to define the basic SPN structure, and cannot be easily applied to domains without notions of locality. In [5], a hard gradient descent method optimizing the conditional likelihood was proposed, showing convincing results on image classification tasks. The used structure is a 4-layered network on top of a image-feature generation process proposed in [6]. Therefore, also this algorithm is restricted to the image domain. Dennis and Ventura [7] use the same algorithm as in [1] for learning the SPN parameters, but propose an algorithm for finding the basic structure automatically. Their algorithm recursively splits so-called *regions*, i.e. sets of observable random variables, into pairs of smaller regions, using a heuristic application of k-means. This approach clusters variables together which have similar magnitude trends over the dataset. Therefore, also this approach is primarily useful for the image domain, and the prior knowledge about locality is implicitly given by the fact that neighboring pixels typically have similar color values. Furthermore, as the authors note, the application of k-means in this manner is rather unusual and lacks justification. Recently, after we had submitted the first version of this paper, Gens and Domingos [8] proposed a structure learning framework which is applicable to general domains: they recursively apply splits on training instances (clustering) – leading to sum nodes, and splits on variables – leading to product nodes. Further related work, which proposes to learn a tractable Markov network by using a representation as arithmetic circuit, can be found in [9].

In this paper, we propose a novel algorithm for learning SPNs, where our structure learning mechanism is well justified and can be applied to discrete random variables, continuous random variables, and mixtures thereof. Our method does not rely on explicit or implicit locality assumptions, but learns the SPN structure guided by independence tests and a maximum mutual information principle. It constructs SPNs starting from simple models over small variable scopes, and grows models over larger and larger variable scopes, building successively more expressive models (bottom-up approach). This gives an alternative to the top-down approaches proposed in [1, 7, 8], which determine the SPN structure by recursive splits of variable scopes. Therefore, our method is closer in spirit to training of deep belief networks [10–12], which also aim to extract successively more abstract features in a bottom-up manner.

The paper is organized as follows: In section 2, we introduce our notation and formally review SPNs. In section 3, we introduce our approach for learning SPNs in a bottom-up manner. In section 4, we experimentally show that our method competes well with the existing approaches in the task of image completion, and outperforms them when their underlying assumptions are not met. Section 5 concludes the paper, and gives possible directions for future work.

---

[1] The claimed ℓ^0-norm penalization in [1] is not truly implemented in the provided software [4], since already evoked non-zero weights are *not* penalized any more.

2   Background and Notation

Assume a set of random variables (RVs) $\mathbf{X} = \{X_1, \ldots, X_D\}$, where each X_d can take values out of the set $\mathbf{val}(X_d) \subseteq \mathbb{R}$. When $\mathbf{val}(X_d)$ is finite, we say that X_d is *discrete*. In this case, inputs concerning X_d are represented using $|\mathbf{val}(X_d)|$ binary *indicator* nodes. When $\mathbf{val}(X_d)$ is infinite, inputs can be represented by *distribution* nodes (e.g. Gaussian). For the moment, let us assume that all RVs are discrete. Let $x_d^j \in \mathbf{val}(X_d)$ be the jth state of X_d, and I_d^j be the corresponding indicator node, which can assume values out of $\{0, 1\}$.

An SPN structure [1] is a connected acyclic directed graph, whose leaves are the indicator nodes for RVs $\mathbf{X}$, and all non-leaves are either *sum* or *product* nodes. A product node calculates the product of its child nodes. A sum node calculates a *weighted* sum of its child nodes, where the weights are non-negative parameters. We assume SPNs organized in layers, where sum and product layers alternate when proceeding to higher layers. The first layer is an input layer, the second is a product layer and the last (output) layer is a sum layer. All nodes are allowed to receive input only from a strictly lower layer. We call these SPNs (organized in layers and feed-forward) *layered* SPNs. In a layered SPN, we have L sum and L product layers, such that in total the SPN contains $2L + 1$ layers, where the first layer, the input layer, contains the indicator (or distribution) nodes. Let P^l be the lth product layer (i.e. the $(2\,l)^{\text{th}}$ layer in the SPN), and S^l the lth sum layer (i.e. the $(2\,l + 1)^{\text{th}}$ layer in the SPN). Let S_k^l be the kth sum node in the lth sum layer, and like-wise P_k^l for product layers. In graphical representations of SPN structures, we assume that nodes within one layer are numerated from left to right. The *parents* of some node N are denoted as pa(N), and the children are denoted as ch(N). Let the *scope* sc(N) of a node be a sub-set of the index set $\{1, \ldots, D\}$ of the RVs $\mathbf{X}$. For an indicator node I_d^j, the scope is defined as sc(I_d^j) = $\{d\}$. For sum and product nodes, the scope is recursively defined as sc(N) = $\bigcup_{C \in \text{ch}(N)}$ sc(C). Let $\mathbf{X}_{\text{sc}(N)}$ be the sub-set of $\mathbf{X}$ which is indexed by sc(N). A *root* is a node R with pa(R) = $\emptyset$. In [1, 7], only SPNs with a single root R were considered, and where sc(R) = $\{1, \ldots, D\}$. In this paper, we also strive for SPNs with a single root, representing the full variable scope; however, as intermediate step, we also consider SPNs with multiple roots, and roots whose scope is a strict sub-set of $\{1, \ldots, D\}$ (see section 3). For now, let us assume SPNs with a single root R. A *sub-SPN* induced by some node N is the SPN defined by the sub-graph induced by N and all its descendants, including the corresponding parameters. N is the (single) root of its induced sub-SPN.

Let $\mathbf{e} = \left(e_1^1, \ldots, e_1^{|\mathbf{val}(X_1)|}, \ldots, e_D^1, \ldots e_D^{|\mathbf{val}(X_D)|} \right)$ denote some input to the SPN, i.e. a binary pattern for the indicator nodes. Let $N(\mathbf{e})$ denote the value of node N for input $\mathbf{e}$. For indicator nodes, $I_d^j(\mathbf{e}) = e_d^j$. To input *complete* evidence, i.e. a variable assignment $\mathbf{x} = (x_1, \ldots, x_D)$, the value indicator node are set $e_d^j = 1$ if $x_d = x_d^j$, and $e_d^j = 0$ otherwise. When $\mathbf{e}$ encodes some complete evidence $\mathbf{x}$, we write $\mathbf{e} \sim \mathbf{x}$, and also use $N(\mathbf{x})$ for $N(\mathbf{e})$. The values for sum and product nodes are determined by an upward pass in the network. The root-value $R(\mathbf{x})$ is the *output* of an SPN for assignment $\mathbf{x}$. An SPN defines the probability distribution

$$P(\mathbf{x}) := \frac{R(\mathbf{x})}{\sum_{\mathbf{x}' \in \mathbf{val}(\mathbf{X})} R(\mathbf{x}')}. \tag{1}$$

While (1) can also be defined for standard neural networks with non-negative outputs, SPNs become truly powerful when they are *valid* [1], which means that for all state collections $\boldsymbol{\xi}_d \subseteq \mathbf{val}(X_d)$, $d \in \{1, \ldots, D\}$, it holds that

$$\sum_{x_1 \in \boldsymbol{\xi}_1} \cdots \sum_{x_D \in \boldsymbol{\xi}_D} P(x_1, \ldots, x_D) = \frac{R(\mathbf{e})}{\sum_{\mathbf{x}' \in \mathbf{val}(\mathbf{X})} R(\mathbf{x}')}, \tag{2}$$

where here $e_d^j = 1$ if $x_d^j \in \boldsymbol{\xi}_d$, and otherwise $e_d^j = 0$. In words, a valid SPN allows to efficiently marginalize over *partial* evidence by a single upward pass. The efficient marginalization in SPNs stems from a compact representation of the network polynomial of an underlying Bayesian network [1, 2]. Poon and Domingos [1] give sufficient conditions for the validity of an SPN, namely *completeness* and *consistency*:

Definition 1. *An SPN is* complete, *if for each sum node S all children of S have the same scope.*

Definition 2. *An SPN is* consistent, *if for each product node P and each two of its children $C, C' \in \mathrm{ch}(P), C \neq C'$, it holds that when an indicator I_d^j is a descendant of C, no indicator $I_d^{j'}$, $j \neq j'$, is a descendant of C'.*

Completeness and consistency are sufficient, but not necessary for validity; however, these conditions are necessary when also every sub-SPN rooted at some node N should be valid [1]. Definition 2 is somewhat cumbersome, and it is also questionable how consistency should be interpreted in the case of continuous RVs. Therefore, Poon and Domingos provide a simpler and more restrictive condition, which implies consistency, namely decomposability:

Definition 3. *An SPN is* decomposable, *if for each product node P and each two of its children $C, C' \in \mathrm{ch}(P), C \neq C'$, it holds that $\mathrm{sc}(C) \cap \mathrm{sc}(C') = \emptyset$.*

To end this section, we illustrate how continuous data can be modeled using SPNs. Following [1], one can simply use *distribution nodes* (instead of indicator nodes) for continuous RVs, e.g. with nodes returning the value of a Gaussian PDF (Gaussian nodes) as output. A simple example, showing a 4-component GMM with diagonal covariance matrix, is shown in Fig. 1. The parameters of the Gaussians, mean and variance, are considered as parameters of the Gaussian nodes in the input layer and are not shown in the figure.

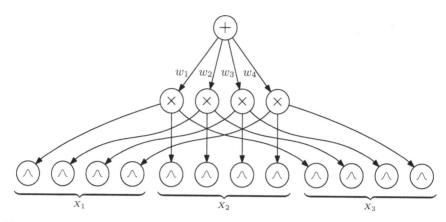

Fig. 1. SPN representing a Gaussian mixture model over three variables X_1, X_2, X_3 with 4 components and diagonal covariance matrix. The component priors (sum weights) are w_1, w_2, w_3, w_4, satisfying $w_1 + w_2 + w_3 + w_4 = 1$, $w_1, w_2, w_3, w_4 \geq 0$.

3   Greedy Part-Wise Learning of SPNs

In this section, we present our approach for part-wise learning of SPNs, where we restrict ourselves to complete and decomposable SPNs. We start with some observations serving as guidelines for our approach. First of all, an SPN with a single root R defines a probability distribution over $\mathbf{X}_{\mathrm{sc}(R)}$ according to (1). Consequently, an SPN with multiple roots $R_1, \ldots, R_r$ defines *multiple* probability distributions over $\mathbf{X}_{\mathrm{sc}(R_1)}, \ldots, \mathbf{X}_{\mathrm{sc}(R_r)}$, respectively, where in general scopes $\mathrm{sc}(R_i)$ and $\mathrm{sc}(R_j)$, $i \neq j$, can differ from each other. The representations of these distributions potentially share computational results and parameters. For example, the SPN in Fig. 2 has 4 roots, where the roots S_1^2 and S_2^2 represent two (in general distinct) distributions over the whole scope $\{X_1, X_2, X_3\}$, and roots S_3^1 and S_4^1 represent distributions over scope $\{X_2, X_3\}$. Furthermore, we see that each sub-SPN is again an SPN over the scope $\mathrm{sc}(N)$, where in the simplest case an SPN consists of a single node in the input layer. We call these single-node SPNs *atomic* SPNs. In Fig. 2, all atomic SPNs are indicator nodes[2]. However, as already noted in section 2, atomic SPNs are not restricted to be indicator nodes, but can also be distribution nodes. Even further, atomic SPNs can be probability models with arbitrarily large scopes, not only modeling single variables – they are merely not represented as SPNs in this framework, but represent some external "input"-probabilistic models. Product nodes represent distributions which assume *independence* between the variable sets indexed by their child nodes. Sum nodes represent *mixtures of distributions* represented by product nodes. We recognize that larger SPNs are simply *composite smaller* SPNs, where the basis of this inductive principle are atomic SPNs. This recursive view of SPNs is also followed in the recent paper of Gens and Domingos [8].

---

[2] An indicator node I_d^j is an SPN, which represents the distribution assigning all probability mass to the event $X_d = x_d^j$.

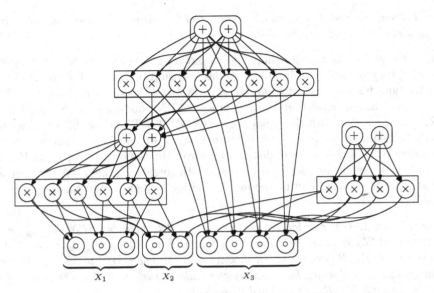

Fig. 2. Example of a multi-root SPN over variables X_1, X_2, X_3 with 3, 2 and 4 states, respectively. Nodes with ∘ denote indicator nodes. Weights of sum nodes are omitted.

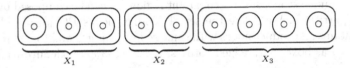

Fig. 3. Trivial multi-root SPN variables X_1, X_2, X_3 with 3, 2 and 4 states, respectively

Using this interpretation of multi-root SPNs, we can define the *trivial* multi-root SPN which merely contains atomic nodes. Fig. 3 shows the trivial SPN for the same RVs as in Fig. 2. This SPN consists merely of the indicator nodes of X_1, X_2, X_3, which are at the same time roots *and* atomic distributions. The key idea of our approach is to start from the trivial SPN containing only atomic distributions, and generate larger and larger SPNs with successively increasing scope, until we eventually obtain an SPN whose root has a scope over all RVs **X** we aim to model. In this paper, the final model will have a single root, where as intermediate step a series of multi-root SPNs is generated.

To make our approach precise, we adopt the notion of *regions, partitions*, and *region graphs* [7], which represents SPNs on a larger scale. The notion of a region is inspired by image modeling, i.e. when RVs **X** represent pixels of an image. However, the approach developed here is not necessarily restricted to the image domain.

Definition 4. *Given a layered, complete and decomposable SPN, the region $\mathcal{R}$ with scope $sc(\mathcal{R}) \subseteq \{1, \ldots, D\}$ is the set of atomic or sum nodes, which have all the same scope $sc(\mathcal{R})$. Regions containing only atomic nodes (e.g. indicator*

or distribution nodes) are called atomic *regions. Regions containing only sum nodes are called* composite *regions*.

In this paper, we assume for simplicity that regions are either atomic or composite, i.e. they do not contain atomic nodes and sum nodes simultaneously. This restriction, however, is not essential, since we could model a region both with "external" atomic models and with composite SPNs. For some scope s, we define $\mathcal{R}(s)$ as the region with this scope, i.e. $sc(\mathcal{R}(s)) = s$. While two SPN nodes can have the same scope, regions per definition have a *unique* scope. Regions can be interpreted as *dictionaries* of distributions over the same scope $sc(\mathcal{R})$. In Fig. 2 and Fig. 3, regions are depicted as boxes with rounded corners. We now define partitions [7], which describe the decomposition of regions into smaller disjoint regions.

Definition 5. *Given a layered, complete and decomposable SPN, let $\mathcal{R}_p$ be a region and $\mathbf{R}_c$ be a set of disjoint regions, with $sc(\mathcal{R}_p) = \bigcup_{\mathcal{R} \in \mathbf{R}_c} sc(\mathcal{R})$. The partition $\mathcal{P}(\mathcal{R}_p, \mathbf{R}_c)$ is the set of product nodes whose parent nodes are all contained in $\mathcal{R}_p$, and which have exactly one child in each $\mathcal{R} \in \mathbf{R}_c$. The scope of a partition $sc(\mathcal{P}(\mathcal{R}_p, \mathbf{R}_c))$ is defined as $sc(\mathcal{R}_p)$.*

Note that since we only consider layered, complete and decomposable SPNs, each product node has to be contained in *exactly one partition*. Partitions do *not* have a unique scope, since each parent region $\mathcal{R}_p$ can be composed by several *different* partitions. We define the set of product nodes $\mathcal{P}(\mathcal{R}_p) := \bigcup \mathcal{P}(\mathcal{R}_p, \cdot)$, which contains all product nodes with same scope. In Fig. 2 and Fig. 3, partitions are depicted as boxes with edged corners. A *region graph* is defined as follows.

Definition 6. *Given a layered, complete and decomposable SPN, the* region graph $\mathcal{G}$ *of this SPN is a bipartite directed acyclic graph, with two distinct set of nodes $\mathbf{R}$ and $\mathbf{P}$, where $\mathbf{R}$ are all* non-empty *regions and $\mathbf{P}$ are all* non-empty *partitions of the SPN. Region nodes are connected only with partition nodes, and vice versa. $\mathcal{R} \in \mathbf{R}$ is a parent region of $\mathcal{P}(\mathcal{R}_p, \mathbf{R}_c)$ if and only if $\mathcal{R} = \mathcal{R}_p$. $\mathcal{R} \in \mathbf{R}$ is a child region of $\mathcal{P}(\mathcal{R}_p, \mathbf{R}_c)$ if and only if $\mathcal{R} \in \mathbf{R}_c$.*

Using the notion of a region graph, we can define the *parts* of a region.

Definition 7. *Let $\mathcal{G}$ be the region graph of a layered, complete and decomposable SPN. The* parts *of a region $\mathcal{R} \in \mathcal{G}$ is the set of regions*

$$\text{parts}(\mathcal{R}) := \{\mathcal{R}' | \exists \mathcal{P}(\mathcal{R}, \mathbf{R}_c) \in \mathcal{G} : \mathcal{R}' \in \mathbf{R}_c\}. \tag{3}$$

We are now ready to sketch our general approach, which is shown in Algorithm 1. We start with the trivial multi-root SPN containing only atomic regions. In each iteration, some disjoint regions $\mathbf{R}_c$ are selected and *merged* into a *parent region* $\mathcal{R}_p$, generating a new partition $\mathcal{P}(\mathcal{R}_p, \mathbf{R}_c)$. Note that while in each iteration a partition $\mathcal{P}(\mathcal{R}_p, \mathbf{R}_c)$ is newly generated, i.e. it was not in the region graph before, the region $\mathcal{R}_p$ might already have been generated by an earlier merge. A collection of sum nodes, one from each child region, is combined by product nodes. Here, a particular selection of child region nodes is called a *feature*

Algorithm 1. General Merge Learning

1: **Initialization:** Make trivial SPN and corresponding region graph $\mathcal{G}$.
2: **while** stopping criterion not met **do**
3:     **Select Merge Candidates:** Select set of $A \geq 2$ disjoint regions $\mathcal{R}_c = \{\mathcal{R}_{c,1}, \ldots, \mathcal{R}_{c,A}\}$, with $\forall a \ \mathcal{R}_{c,a} \in \mathcal{G}$ and $\mathcal{P}(\mathcal{R}_p, \mathcal{R}_c) \notin \mathcal{G}$, where $\mathcal{R}_p = \mathcal{R}(\bigcup_{a=1}^{A} \mathrm{sc}(\mathcal{R}_{c,a}))$.
4:     **if** $\mathcal{R}_p = \emptyset$ **then**
5:         Insert K sum nodes in $\mathcal{R}_p$.
6:     **end if**
7:     **Select Set of Features** $\mathcal{F} \subseteq \mathcal{R}_{c,1} \times \cdots \times \mathcal{R}_{c,A}$.
8:     **for** $f = (N_1, \ldots, N_A) \in \mathcal{F}$ **do**
9:         Generate product node P_f.
10:        Connect P_f as parent of N_a, $a = 1, \ldots, A$.
11:        Connect P_f as child of all sum nodes in $\mathcal{R}_p$.
12:    **end for**
13:    **Learn parameters of sum nodes in** $\mathcal{R}_p$ **and its ancestor regions.**
14:    Update region graph $\mathcal{G}$.
15: **end while**

(cf. step 4), where each feature corresponds to a product node. The number of generated sum nodes K will be typically $K \ll |\mathcal{F}|$, i.e. the sum nodes *represent a compression* of the generated features.

Algorithm 1 describes a general scheme for greedy part-wise learning of SPNs. Depending on the strategy of selecting the merge candidates, of selecting features, and of learning parameters, we obtain different learning algorithms. Further questions are to select K and the stopping criterion. We treat these questions in the following sub-sections, where our approach is guided by the concept of *winner* variables.

3.1   Winner Variables

For each region $\mathcal{R}$ in the region graph of some intermediate multi-root SPN, we define a winner variable

$$W_\mathcal{R} := \arg\max_{i:N_i \in \mathcal{R}} N_i(\mathbf{X}), \tag{4}$$

where we assume some arbitrary ordering of the nodes in $\mathcal{R}$. As already noted, a region can be interpreted as dictionary of distributions N_i over the same scope. $W_\mathcal{R}(\mathbf{x})$ is the indicator of the distribution in $\mathcal{R}$ which describes $\mathbf{X}_{\mathrm{sc}(\mathcal{R})}$ best for sample $\mathbf{x}$, since the corresponding node represents the model with *highest-likelihood*. With respect to some multi-root SPN, each variable $W_\mathcal{R}$ represents some *abstract* information of variables $\mathbf{X}_{\mathrm{sc}(\mathcal{R})}$. The goal in our approach is to preserve and to abstract this information, when proceeding to higher SPN levels.

3.2 Selecting Merge Candidates

Similar as in [1, 7] we set $A = 2$, i.e. we consider decompositions of a parent-region into two sub-regions. The target for selecting merge candidates is twofold: (i) we aim to find merge candidates which are "advantageous", and (ii) we want to pursue a merging strategy which yields quickly an SPN with complete scope $\{1, \ldots, D\}$, i.e. which models all variables in $\mathbf{X}$. When we neglect the latter point, our algorithm will proceed slowly, exhaust memory and overfit the SPN. To decide when a merge is advantageous, we use independence tests, motivated by BN structure learning [13]. In BNs, an edge between two variables should be present when they are statistically dependent. The major criticism about this method is the unreliability of statistical (in)dependence tests, which either causes unreliable models, or models with high inference cost. In SPNs, the variables to be modeled are not directly connected by edges, but their interaction happens over latent parents. Here the unreliability of statistical dependence tests does not harm as much as in BNs, since introducing a new, possibly spurious partition, does increase the inference cost only marginally.

In this paper, we use the Bayesian-Dirichlet independence test proposed in [14], for two winner variables $W_{\mathcal{R}'}$ and $W_{\mathcal{R}''}$:

$$
\mathrm{BD}(W_{\mathcal{R}'}, W_{\mathcal{R}''}) = \left[1 + \frac{\frac{\Gamma(\gamma)}{\Gamma(\gamma+M)} \prod_{k=1}^{|\mathcal{R}|} \prod_{l=1}^{|\mathcal{R}'|} \frac{\Gamma(\gamma_{k,l}+c_{k,l})}{\Gamma(\gamma_{k,l})}}{\left(\frac{\Gamma(\alpha)}{\Gamma(\alpha+M)} \prod_{k=1}^{|\mathcal{R}|} \frac{\Gamma(\alpha_k+a_k)}{\Gamma(\alpha_k)} \right) \left(\frac{\Gamma(\beta)}{\Gamma(\beta+M)} \prod_{l=1}^{|\mathcal{R}'|} \frac{\Gamma(\beta_l+b_l)}{\Gamma(\beta_l)} \right)} \right]^{-1}
$$

$$(5)$$

Here a_k, b_l are the number of times, counted over all training samples, where $W_{\mathcal{R}'}$ and $W_{\mathcal{R}''}$ are in their k^{th} and l^{th} states, respectively. $c_{k,l}$ is the number of times where $W_{\mathcal{R}'}$ and $W_{\mathcal{R}''}$ are jointly in their k^{th} and l^{th} states. α_k, β_l, and $\gamma_{k,l}$ are Dirichlet priors, set uniformly to 1, and $\alpha = \sum_k \alpha_k$, $\beta = \sum_l \beta_l$, $\gamma = \sum_{k,l} \gamma_{k,l}$. M is the number of samples in the training set. The lower $\mathrm{BD}(W_{\mathcal{R}'}, W_{\mathcal{R}''})$, the more the winner variables $W_{\mathcal{R}'}$, $W_{\mathcal{R}''}$ are dependent, and the more $\mathcal{R}'$ and $\mathcal{R}''$ "prefer" to merge.

To encourage a quick growing of the SPN regions, we use the scheme shown in Algorithm 2. This method maintains a set of merging candidates $\mathcal{M}$, which is initialized with the disjoint atomic regions. In each iteration of the overall Algorithm 1, the two most dependent regions are selected from $\mathcal{M}$ and merged to a parent region. The two selected regions are excluded from the merging

---

Algorithm 2. Select Regions

1: **if** Select Regions is called the first time or $|\mathcal{M}| = 1$ **then**
2:      $\mathcal{M} \leftarrow$ set of all atomic regions.
3: **end if**
4: Select $\boldsymbol{\mathcal{R}}_c = \{\mathcal{R}', \mathcal{R}''\} \in \mathcal{M}$ which minimize $\mathrm{BD}(W_{\mathcal{R}'}, W_{\mathcal{R}''})$, s.t. $\mathcal{P}(\mathcal{R}_p, \boldsymbol{\mathcal{R}}_c) \notin \mathcal{G}$.
5: $\mathcal{M} \leftarrow \mathcal{M} \setminus \boldsymbol{\mathcal{R}}_c$
6: $\mathcal{M} \leftarrow \mathcal{M} \cup \mathcal{R}(\mathrm{sc}(\mathcal{R}') \cup \mathrm{sc}(\mathcal{R}''))$
7: Return $\boldsymbol{\mathcal{R}}_c = \{\mathcal{R}', \mathcal{R}''\}$.

---

candidates $\mathcal{M}$ and the parent region is inserted. In this way, the generated region graph is grown to a binary tree. When $\mathcal{M} = 1$, i.e. when the root region has been reached, we start the process again, i.e. $\mathcal{M}$ is reset to the set of all atomic regions. Then a parallel, interleaved binary tree is grown, where the constraint $\mathcal{P}(\mathcal{R}_p, \mathbf{R}_c) \notin \mathcal{G}$ guarantees that this tree is different from the first tree. This process, growing interleaved binary trees, is repeated for several iterations (*tree-growing* iterations), where the maximal number of iterations is used as a stopping criterion for Algorithm 1. Note that due to the constraint $\mathcal{P}(\mathcal{R}_p, \mathbf{R}_c) \notin \mathcal{G}$ it can happen that in later iterations no more merging candidates can be found in step 4 of Algorithm 2. In this case, we also stop Algorithm 1.

3.3   Selecting Features and Learning Parameters

We now turn to the problem of selecting features and learning parameters. The most general approach for selecting features $\mathcal{F}$ is to take the Cartesian product of the node sets $\{\mathcal{R}_{c,a}\}_{a=1,\ldots,A}$, which for $A = 2$ grows quadratically in the number of nodes in the child regions. We reduce this number and use $\mathcal{F} = \{f = (N_k, N_l) | c_{k,l} > 0\}$, where $c_{k,l}$ is defined in (5). In words, we select those features, whose corresponding product node *wins* at least once against all other potential product nodes. The sum nodes in $\mathcal{R}_p$ can be regarded as a compression of the product nodes $\mathcal{P}(\mathcal{R}_p)$ corresponding to the features $\mathcal{F}$. A natural way to perform this compression is the *information bottleneck method* [15]. Recalling Definition 7, the aim is to maximize the *mutual information* between the winner variable of $\mathcal{R}_p$ and the winner variables of parts$(\mathcal{R}_p) := \{\mathcal{R}'_1, \ldots, \mathcal{R}'_{|\text{parts}(\mathcal{R}_p)|}\}$,

$$\underset{\{w_{kf}\}}{\text{maximize}} \; I(W_{\mathcal{R}'_1}, \ldots, W_{\mathcal{R}'_{|\text{parts}(\mathcal{R}_p)|}}; W_{\mathcal{R}_p}) \tag{6}$$

where $\{w_{kf}\}$ are the weights of all sum nodes $S_k \in \mathcal{R}_p$, i.e.

$$S_k(\mathbf{e}) = \sum_{f : P_f \in \mathcal{P}(\mathcal{R}_p)} w_{kf} P_f(\mathbf{e}), \tag{7}$$

where $\sum_f w_{kf} = 1$, $w_{kf} \geq 0$. Since this problem can be expected to be NP-hard[3], we restrict ourselves to a greedy solution, outlined in Algorithm 3 and illustrated in Fig. 4. Our method starts with a number of sum nodes identical to the number of product nodes $\mathcal{P}(\mathcal{R}_p)$, where each product node is the child of exactly one sum node. Then we iteratively combine a pair of sum nodes to a single sum node, such that the mutual information is reduced as little as possible in each iteration. The weights of the new sum node are updated according to the maximum likelihood estimate

$$w_{kf} = \frac{\sum_{m=1}^{M} P_f(\mathbf{x}^m)}{\sum_{f' : P_{f'} \in \text{ch}(S_k)} \sum_{m=1}^{M} P_{f'}(\mathbf{x}^m)}, \tag{8}$$

---

[3] In general, the information bottleneck method is NP-hard.

where $\mathbf{x}^m$ denotes the mth sample. Note that by this approach each product node becomes the child of exactly one sum node, i.e. the sum nodes have non-overlapping child sets.

Algorithm 3. Information Bottleneck Parameter Learning

1: Delete any sum nodes in $\mathcal{R}_p$.
2: For each product node P_i in $\mathcal{P}(\mathcal{R}_p)$ generate a sum node S_i.
3: Connect P_i as a child of S_i, set $w_{ii} = 1$.
4: **while** number of sum nodes $> K$ **do**
5:     $I_{\text{best}} \leftarrow -\infty$
6:     **for** all pairs S_i, S_j **do**
7:         Generate tentative sum node S_{tmp}.
8:         Connect $\text{ch}(S_i)$, $\text{ch}(S_j)$ as children of S_{tmp}.
9:         Set weights of S_{tmp} according to (8).
10:        $\mathcal{R}_{tmp} \leftarrow S_{\text{tmp}} \cup \mathcal{R}_p \setminus \{S_i, S_j\}$.
11:        **if** $I(W_{\mathcal{R}'_1}, \ldots, W_{\mathcal{R}'_{|\text{parts}(\mathcal{R}_p)|}}; W_{\mathcal{R}_{\text{tmp}}}) > I_{\text{best}}$ **then**
12:            $I_{\text{best}} = I(W_{\mathcal{R}'_1}, \ldots, W_{\mathcal{R}'_{|\text{parts}(\mathcal{R}_p)|}}; W_{\mathcal{R}_{\text{tmp}}})$
13:            $\mathcal{R}_{\text{best}} = \mathcal{R}_{tmp}$
14:        **end if**
15:    **end for**
16:    $\mathcal{R}_p \leftarrow \mathcal{R}_{\text{best}}$
17: **end while**

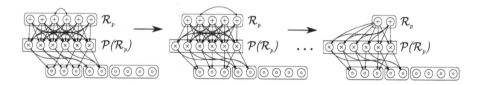

Fig. 4. Illustration of Information Bottleneck parameter learning ($K = 2$) for the first region merge in the trivial multi-root SPN (Fig. 3)

When we want to obtain a single-root SPN, we insert only a single sum node in the root region. We do not need to perform Algorithm 3 in this case, but merely apply (8). For non-root regions, we set K to a fixed value. However, [16] suggests a method to set K in data-driven way: Each merge will inevitably lead to a loss of information[4] w.r.t. the relevant variables $W_{\mathcal{R}'_1}, \ldots, W_{\mathcal{R}'_{|\text{parts}(\mathcal{R}_p)|}}$ – if for a specific number of sum nodes a merge causes a loss which is high compared to the previous merges, this suggests that a meaningful representation has been achieved.

---

[4] This rather informal formulation about information loss can be made rigorous: see [17] for a recent effort.

4   Experiments

Poon and Domingos [1] demonstrated that SPNs achieve convincing results on the ill-posed problem of image completion, i.e. reconstructing occluded parts of face images. To this end, they trained SPNs on the ORL face image data set [18] and used MPE-inference (*most probable explanation*) to recover missing pixel values, i.e. they inferred the most likely states of the occluded pixels and the states of the latent variables associated with the sum nodes. MPE-inference is efficient in SPNs and serves as an approximation for MAP-inference, which actually is appropriate for this task. We conjecture that, although marginalization and MPE-inference is easy in SPNs, the exact MAP-problem is still intractable, since MAP is inherently harder than MPE [19]. However, proving or disproving this conjecture is future work and out of the scope of this paper.

We trained SPNs with the method of Poon and Domingos (PD) [1], the method of Dennis and Ventura (DV) [7], and our method (Merge). As in PD and DV, we model single pixels with several Gaussian nodes, where the means are set by the averages of histogram quantiles, and the standard deviation is uniformly set to 1. Using the notions introduced in this paper, this means that single pixels are used as atomic regions, containing a set of Gaussian nodes.

The ORL faces contain 64×64-pixels, which yields more than 8 million evaluations of the BD score (5) in the first iteration of Merge. Although we cache evaluations of the BD score, the computational effort is still large. Therefore, although we emphasize in this paper that our algorithm does not need prior knowledge of the problem domain, we use a similar approach as the PD algorithm, and introduce a "coarser" resolution level. We show the advantage of our algorithm, when no prior knowledge can be assumed, in the experiment following below.

Fig. 5. Result of unsupervised segmentation of ORL faces using affinity propagation on the absolute value of correlation coefficients between pixels

To find the coarse resolution level, we apply affinity propagation [20] on the absolute value of the correlation coefficient matrix of the pixels, calculated on the training set. This process performs an unsupervised segmentation into image patches, which is shown in Fig. 5. For each image segment, we train a multi-root SPN with 20 roots using our Merge algorithm. These 20 roots serve in turn as atomic distributions for learning the SPN over the whole image. Within each image segment, pixels are treated as atomic regions. For each application of Merge learning, we used a single tree-growing iteration, i.e. the overall regions graph is constructed as a binary tree. In this experiment, we arbitrarily set the number of Gaussians $G = 10$ for all three algorithms. As in [1, 7] we use $K = 20$ sum nodes per composite region for all algorithms. Fig. 6 shows results on the face image completion task for PD, DV, and Merge. All three algorithms show convincing results but differ in the artifacts they produce. In Table 1 we summarize objective evaluation measures for this learning task; the signal-to-noise ratios show that Merge competes well with PD and DV, although no clear preference can be shown. However, we see that while Merge achieves the lowest training likelihood, it achieves the *highest* likelihood on the test set, stating that Merge generalized best in this task.

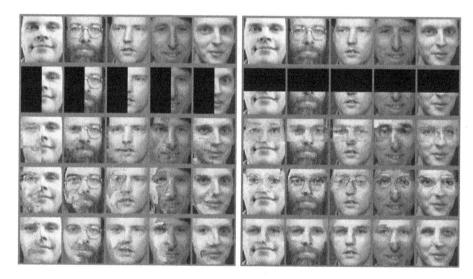

Fig. 6. Examples of face image reconstructions using MPE-inference. Rows from top to bottom: original image, covered image, PD [1], DV [7], Merge learning (this paper).

Table 1. Evaluation measures on ORL data. Left: reconstruction SNRs for the top, bottom, left, and right halves of face images covered. Right: Log-likelihoods on training and test set, normalized by number of samples.

	top	bottom	left	right		Train	Test
PD	12.34	10.18	11.58	11.72	PD	-4287.82	-5068.97
DV	11.69	9.29	10.43	10.83	DV	-4356.41	-4673.73
Merge	12.43	9.83	10.96	11.78	Merge	-4493.46	-4667.04

In this first experiment we used prior knowledge about the problem domain, by using AP for segmenting the image into atomic regions. To demonstrate that our method does not rely on the incorporation of prior knowledge, we generated an artificial modification of the ORL data set; First, we rescale the ORL images to size 16×16, yielding 256 variables. As in [7], we permute all pixels, i.e. we destroy the neighborhood information. Next, we discretized all pixels into 10 histogram bins and permute these randomly, i.e. we destroy the relation of the RVs by their magnitude. Then we again performed the same experiment as with the large ORL data, where for each method we again used 10 Gaussian nodes per pixel, setting their means on the discretized values of bins $1 - 10$, and setting the standard deviation uniformly to 1. Table 2 shows the result for the modified ORL data. We now see a clear trend: PD, relying on locality shows consistently the worst SNRs in the image reconstruction task. DV, not relying on locality, but on similar value trends, is consistently better than PD. Merge shows consistently the best SNRs. Similarly, looking at the log-likelihoods, we see that Merge shows the best test likelihood, i.e. it generalizes best in this task.

Table 2. Evaluation measures on down-scaled and permuted ORL data. Left: reconstruction SNRs for the top, bottom, left, and right halves of face images covered. Right: Log-likelihoods on training and test set, normalized by number of samples.

	top	bottom	left	right		Train	Test
PD	15.16	8.49	12.11	10.37	PD	-442.76	-893.18
DV	15.32	9.24	12.62	10.55	DV	-506.81	-623.05
Merge	17.95	10.53	13.22	12.48	Merge	-551.97	-595.66

5 Conclusion

In this paper, we introduced a method to learn SPNs in a greedy bottom-up manner, giving an alternative to the top-down approaches proposed so far. The main principle we follow is that SPNs simply build composite and complex models out of simple and small models in a *recursive* manner. The basis of this recursive principle is given by what we call *atomic* or input distributions. We adopted the notion of *regions* and interpret them as dictionaries of distributions over the same scope. Product nodes or partitions serve as *cross-overs* of dictionaries with non-overlapping scope, corresponding to the notion of decomposability. These cross-overs yield a quickly growing number of new features or product nodes. Sum nodes of the newly created region serve as *compression* of these newly created features. This process can be seen as *abstracting* information, when proceeding to higher levels, which motivates the use of the information bottleneck method for learning sum nodes.

We showed that our method competes well with existing generative approaches to train SPNs. Furthermore, we demonstrated that our method does not rely on assumptions of the image domain, and shows the best overall performance when these are not fulfilled. In future work, we want to explore potential engineering applications for our approach, such as signal, speech and audio processing. Furthermore, we consider the discriminative paradigm, e.g. applying maximum margin methods for classification. Finally, we want to investigate different structure and parameter learning techniques within our learning framework.

Acknowledgments. This work was supported by the Austrian Science Fund (project numbers P22488-N23 and P25244-N15). The authors like to thank the anonymous reviewers for their constructive commends.

References

1. Poon, H., Domingos, P.: Sum-product networks: A new deep architecture. In: Proceedings of the Twenty-Seventh Conference on Uncertainty in Artificial Intelligence, pp. 337–346 (2011)
2. Darwiche, A.: A differential approach to inference in bayesian networks. ACM 50(3), 280–305 (2003)
3. Lowd, D., Domingos, P.: Learning arithmetic circuits. In: Twenty Fourth Conference on Uncertainty in Artificial Intelligence, pp. 383–392 (2008)
4. Poon, H., Domingos, P.: (2011), http://alchemy.cs.washington.edu/spn/
5. Gens, R., Domingos, P.: Discriminative learning of sum-product networks. Advances in Neural Information Processing Systems 25, 3248–3256 (2012)
6. Coates, A., Lee, H., Ng, A.: An analysis of single-layer networks in unsupervised feature learning. In: Proceedings of the 14th International Conference on Artificial Intelligence and Statistics (2011)
7. Dennis, A., Ventura, D.: Learning the architecture of sum-product networks using clustering on variables. Advances in Neural Information Processing Systems 25, 2042–2050 (2012)
8. Gens, R., Domingos, P.: Learning the structure of sum-product networks. In: Proceedings of ICML, pp. 873–880 (2013)
9. Lowd, D., Rooshenas, A.: Learning markov networks with arithmetic circuits. In: Proceedings of AISTATS, pp. 406–414 (2013)
10. Hinton, G., Salakhutdinov, R.: Reducing the dimensionality of data with neural networks. Science 313(5786), 504–507 (2006)
11. Bengio, Y., Lamblin, P., Popovici, D., Larochelle, H.: Greedy layer-wise training of deep networks. Advances in Neural Information Processing Systems 19, 153–160 (2007)
12. Bengio, Y.: Learning Deep Architectures for AI. Foundations and Trends in Machine Learning, vol. 2 (2009)
13. Koller, D., Friedman, N.: Probabilistic Graphical Models: Principles and Techniques. MIT Press (2009)
14. Margaritis, D., Thrun, S.: A bayesian multiresolution independence test for continuous variables. In: 17th Conference on Uncertainty in Artificial Intelligence, pp. 346–353 (2001)
15. Tishby, N., Pereira, F.C., Bialek, W.: The information bottleneck method. In: Proc. Allerton Conf. on Communication, Control, and Computing, pp. 368–377 (1999)
16. Slonim, N., Tishby, N.: Agglomerative information bottleneck. In: Advances in Neural Information Processing Systems (NIPS), pp. 617–623. MIT Press (1999)

17. Geiger, B.C., Kubin, G.: Signal enhancement as minimization of relevant information loss. In: Proc. ITG Conf. on Systems, Communication and Coding, Munich, pp. 1–6 (2013); extended version available: arXiv:1205.6935 [cs.IT]
18. Samaria, F., Harter, A.: Parameterisation of a stochastic model for human face identification. In: Proceedings of the 2nd IEEE Workshop on Applications of Computer Vision, pp. 138–142 (1994)
19. Park, J.: Map complexity results and approximation methods. In: Proceedings of the Conference on Uncertainty in Artificial Intelligence, pp. 338–396 (2002)
20. Frey, B., Dueck, D.: Clustering by passing messages between data points. Science 315, 972–976 (2007)

From Topic Models to Semi-supervised Learning: Biasing Mixed-Membership Models to Exploit Topic-Indicative Features in Entity Clustering

Ramnath Balasubramanyan[1], Bhavana Dalvi[1], and William W. Cohen[1,2]

[1] Language Technologies Institute, Carnegie Mellon University
{rbalasub,bbd,wcohen}@cs.cmu.edu
[2] Machine Learning Department, Carnegie Mellon University

Abstract. We present methods to introduce different forms of supervision into mixed-membership latent variable models. Firstly, we introduce a technique to bias the models to exploit *topic-indicative* features, i.e. features which are *apriori* known to be good indicators of the latent topics that generated them. Next, we present methods to modify the Gibbs sampler used for approximate inference in such models to permit injection of stronger forms of supervision in the form of labels for features and documents, along with a description of the corresponding change in the underlying generative process. This ability allows us to span the range from unsupervised topic models to semi-supervised learning in the same mixed membership model. Experimental results from an entity-clustering task demonstrate that the biasing technique and the introduction of feature and document labels provide a significant increase in clustering performance over baseline mixed-membership methods.

1 Introduction

Topic modeling based on Latent Dirichlet Allocation (LDA) [6] has become a popular tool for data exploration, dimensionality reduction and for facilitating myriad other tasks [2,1,12]. As a fully unsupervised technique, however, topic models are unequipped to utilize limited supervisory information, e.g. feature labels and document cluster membership. In this paper, we introduce methods to incorporate progressively stronger forms of weak supervision to influence the formation of topics that respect information that we might have about the latent structure.

First, we present a method to bias mixed-membership models (such as topic models) to better exploit known *topic-indicative* features. Unsupervised topic models do not necessarily optimally utilize topic-indicative features, i.e. features that are known to be strongly indicative of the latent topics of the documents. The biasing towards topic-indicative features serves to control the latent role distribution of the features, i.e., the degree of *polysemy*, and its strength can be adjusted to control the degree of polysemy permitted.

The flexibility of the biased models is examined by using it to cluster entities found in HTML pages [9]. While our model can be used for a variety of tasks, we focus on the HTML entities clustering tasks since it requires the use of several kinds of features (obtained from semi-structured data from the tables) and permits us to demonstrate ways

H. Blockeel et al. (Eds.): ECML PKDD 2013, Part II, LNAI 8189, pp. 628–642, 2013.

in which intuition and limited supervision about different kinds of features can be incorporated. In this task, potentially useful features of an entity includes features like the headers of columns (e.g. the entity *apple* might be found under the headers *company* or *fruit*) and domains (e.g. food.com, finance.com, etc.). The biasing technique presented could be used to capture domain knowledge that features of a certain type are more topic-indicative than other features. When the bias term is set high, the features to which it is applied are deemed to be more strongly indicative of topic and are strongly discouraged from assuming multiple latent roles in the mixed membership model. The bias is accomplished using a regularization term in the model which represents a noisy copy of the entropy of the latent role distribution of the word. The polysemy is reduced by pushing the entropy towards a pre-specified desired value that is a hyperparameter to the model.

Next, we show that stronger forms of supervision to the model in the form of feature and document labels can be injected into the model to achieve modeling flexibility to obtain models that range from fully unsupervised topic models to semi-supervised models. This form of light supervision can be in the form of known latent roles for certain subsets of features or known latent roles for documents which exhibit very slight mixed-membership characteristics. The supervision is incorporated into the model by modifying the Gibbs sampling procedure used for approximate inference.

The rest of the paper is organized as follows. Section 2 describes the mixed-membership latent variable model based approach to the entity clustering task. Next, we describe the biasing technique to exploit topic-indicative features in Section 3 and the approach to incorporate feature and document labels in Section 4. Experimental results are presented in Section 5. Finally, we present a short survey of related work in Section 6, followed by the conclusion.

2 Entity Clustering

Latent-variable mixed-membership models based on LDA have been used for a variety of tasks in NLP. Here, we use it for the task of clustering entities that are extracted from tables in HTML documents crawled from the web. Dalvi et al. [9] describe the task in detail.

In this task, the dataset consists of tables of entities extracted from HTML pages. For instance, it could contain a table of companies, tables of American football teams, etc. The goal of the task is to cluster entities of the same semantic class together. Therefore, if the dataset includes a table of fruits with apples, grapes and oranges, and another table with oranges, peaches and bananas, the goal of the task is to recover a cluster of fruits which includes apples, grapes, oranges, peaches and bananas.

Surface terms in such HTML tables frequently have multiple senses. For example, consider the term *apple*, which is found in tables of companies and fruits among others. Therefore we require a model that is capable of distinguishing the sense of the term to prevent companies and fruits from being collapsed into one cluster based on the term *apple* co-occurring with both companies and fruits. Mixed-membership models can account for the multiple-sense problem by assigning partial membership in both clusters to the entity. Typically, entity clustering has been based on distributional similarity based approaches or by using Hearst patterns [13]. In this task however, since we

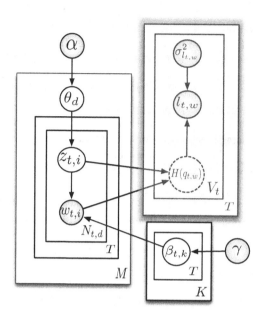

Fig. 1. Biased Link-LDA model to Exploit Topic Indicative Features

are dealing with entities in HTML tables as opposed to entity mentions in free text, we use a different set of features to assist in the clustering, namely:

a) co-occurring entities,
b) co-occuring entity pairs that the entity is observed with,
c) the tableid-columnid combinations under which the entity was observed,
d) web domains in which the entity was observed,
e) the hyponyms that are associated with an entity (extracted using Hearst patterns).

This task can therefore be seen as distributional clustering with a different set of contextual features than the free text features usually used. For every unique entry found in the collection of tables in a dataset, we construct a *"document"* in the LDA sense with the above five kinds of *"words"*. The document is represented by a set of bags of words, one for each kind of feature used. A document for the entity *apple*, for example might consists of the following bags -

a) co-occurring entities {*orange, apple, microsoft,* ... },
b) entity pairs {*orange:apple, google:apple* ... },
c) column ids {tab:326::colid::1 ...},
d) domains {business.com, produce.com ...},
e) hyponyms {*stocks, juice, tech companies* ...}.

These different classes of features are modeled using the Link-LDA model [10]. Figure 1 shows the plate diagram of the graphical model. The variables that are under the yellow shaded rectangle provide the bias that is introduced in later sections, and are not part of the regular Link-LDA model. In the generative story for the model, a document has T kinds of "words". For instance, in a corpus of academic papers, the kinds of words could be author names, words in the abstract, words in the body, references to other papers etc. For every document in the corpus of size M, a distribution over K topics θ is first drawn. Then the words of all T kinds are drawn by first sampling a topic indicator k for the word from θ and then drawing the word from the per-type topic word distributions $\beta_{t,k}$. Since exact inference is intractable for the model, we use a collapsed Gibbs sampler [19] for approximate inference. θ, the document topic distribution obtained after inference provides an estimate for the predicted cluster membership of an entity document.

The predicted clusters are evaluated using *Normalized Mutual Information (NMI)*. This information theory based score measures the amount of information about the true clusters that is encoded by the predicted topic/cluster distributions. NMI can be used in mixed-membership scenarios since the true cluster distribution and predicted topic distribution can have probability mass in more than one cluster. Additionally, the number of true clusters and topics do not have to be the same and therefore no mapping from topics to clusters is required. To compute NMI between the true cluster label distribution and predicted distributions for the test entity set, we first compute Ω the predicted distribution of topics which is equal to $\frac{\sum_{e \in \text{test set}} \theta_e}{|\text{test set}|}$. Let C be the distribution over true cluster labels, then NMI is defined as $\frac{I(\Omega;C)}{(H(\Omega)+H(C))/2}$, where I indicates mutual information. It should be noted that while the model returns mixed-membership assignments for entities, the human labeling scheme that was used provides only one true cluster assignment for an entity. We however present a qualitative analysis of the advantages of mixed-membership modeling in Section 5.

Entity clustering experiments were performed using the WebSets datasets [9], namely — the Asia NELL, Clueweb Sports, CSEAL Useful, Delicious Music, Delicious Sports and Toy Apple datasets. The Asia NELL dataset was collected using the ASIA system [24] using hypernyms of NELL [7] entities as queries. The Clueweb Sports dataset consists of tables extracted from Sports related pages in the Clueweb dataset. The Delicious music and sports datasets consist of tables from subsets of the DAI-Labor [25] Delicious corpus that were tagged as music and sports respectively. The Toy Apple dataset is a small toy dataset constructed using the SEAL [8] system to create set-expansion lists using the query "Apple", which is a typical example of a multi-sense entity (as a fruit and as a company). It is used primarily to illustrate the effects of clustering mixed membership entities. Statistics about the datasets are shown in Table 1.

In the WebSets approach by Dalvi et al., triplets of entities from HTML tables are extracted and then clustered. Their approach also proposes a method to propose labels for the clusters. It should be noted that their approach clusters triples of entities rather than individual entities which makes it hard to directly compare performance with the method proposed in this paper.

Table 1. Dataset Statistics

Dataset	entities	Size of vocabulary co-occurring entities	entity pairs	column ids	domains	hyponyms
Asia NELL	33455	18309	141352	9477	3207	31833
Clueweb Sports	29113	28891	354614	59117	8088	28618
CSEAL Useful	34565	24340	217328	7337	2118	28381
Delicious Music	18074	9748	106401	7564	1633	24934
Delicious Sports	6786	3183	24147	2050	509	16380
Toy Apple	2411	423	4737	109	53	2826

3   Biasing Topic Indicative Features Using Entropic Regularization

One of the attractive attributes of topic models is that they require no supervision in terms of data annotation. However, in many situations, limited amounts of labeled data may be available. We present an approach to bias topic models to utilize weak knowledge about features. Specifically, we aim to make the model exploit *topic indicative features*, which are a subset of features that are known beforehand to be strongly indicative of topic. For instance in the toy apple example, co-occurring entities of the ambiguous entity *apple* are topic indicative. Co-occurring entities such as *Google* and *Microsoft* are indicative of the company topic where as co-occurring entities like *grape* and *banana* indicate the fruit topic. The bias is introduced into the model via a regularization term that constrains the freedom of specific features to take on multiple latent roles.

The LDA model and its extensions allow the same word to belong to different topics when they are instantiated multiple times in the corpus. This freedom is essential in modeling *polysemy*. While this freedom is useful, we aim to control this freedom for features that are topic-indicative. Following the idea illustrated in Figure 1, we present a entropy based regularization technique based on pseudo-observed variables [4], which directly controls the freedom of words to take on different latent topics, by penalizing high entropies in their topic distributions. It should be noted that sparsity in a document's topic membership vector can be achieved using sparse priors, but sparsity in a words' latent role distribution cannot be similarly obtained since these distributions are not explicitly sampled in a topic model. The addition of the regularization term however allows us to impose such preferences by relaxing the conditional independence between topic multinomials in LDA-like models.

Let n_{tkw} be the number of times a word w of type t was observed with latent role k. The topic distribution of a word w of kind t in a topic model can be defined as $q_{t,w}^{(k)} = \frac{n_{tkw}}{\sum_{k'} n_{tk'w}}$, $k \in 1, \ldots, K$. $q_{t,w}$ therefore shows the degree of polysemy exhibited by a word in the model. The Shannon entropy of this distribution is denoted by $H(q_{t,w})$.

We now introduce word topic distribution entropy regularization by adding pseudo-observed variables, $l_{t,w}$ (Figure 1), one for each word of every kind t in the vocabulary V_t, which are noisy copies of $H(q_{t,w})$. These noisy copies are drawn from a one-sided truncated Gaussian, whose mass lies only on values between 0 and $\log_2 K$, with mean $H(q_{t,w})$ and variance $\sigma^2_{l_{t,w}}$, which is a hyperparameter to the model. The density function is given by

$$p(l_{t,w}|h, \sigma^2_{l_{t,w}}) = \begin{cases} \frac{1}{C} \exp\left(\frac{-(h-l_{t,w})^2}{2\sigma^2_{l_{t,w}}}\right) & \text{for } 0 \leq l_{t,w} \leq \log_2 K \\ 0, & \text{otherwise.} \end{cases}$$

$$C = \int_{h'=0}^{\log_2 K} \exp\left(\frac{-(h'-l_{t,w})^2}{2\sigma^2_{l_{t,w}}}\right) dh.$$

The joint distribution of the model with regularization is defined as:

$$\mathcal{L}(\beta, \theta, \mathbf{z}, \mathbf{w}|\alpha, \gamma, \mathbf{l_{t,w}}, \sigma^2_{\mathbf{l_{t,w}}}) = \prod_{d=1}^{M} \text{Dir}(\theta_d|\alpha)$$

$$\left(\prod_{t=1}^{T}\left(\prod_{i=1}^{N_{t,d}} \theta_d^{z_{t,i}} \beta_{t,z_{t,i}}^{(w_{t,i})}\right)\right) \prod_t \prod_k \text{Dir}(\beta_{t,k}|\gamma) \prod_t \prod_{w \in V_t} \exp\frac{-\left(l_{t,w} - H(q_{t,w})\right)^2}{2\sigma^2_{l_{t,w}}}/C$$

$$(1)$$

Approximate inference in the model is performed using a collapsed Gibbs sampler. Let n_{dk} be the number of words in document d that were assigned to topic k. The equation to sample a topic indicator for a word $w_{t,i}$ i.e. the i-th word of type t in d, is given by

$$p(z_{t,i} = k|\mathbf{l_{t,w}}, w_{t,i}, \mathbf{z}^{\neg t,i}, \mathbf{w}^{\neg t,i}, \alpha, \gamma, \sigma^2_{\mathbf{l_{t,w}}}) \propto$$

$$(n_{dk}^{\neg t,i} + \alpha)\frac{n_{tkw_{t,i}}^{\neg t,i} + \gamma}{\sum_{w'} n_{tkw'}^{\neg i} + |V_t|\gamma} \times \exp\left(\frac{-(H(q_{t,w_{t,i}}) - l_{t,w_{t,i}})^2}{2\sigma^2_{t,l_{t,w}}}\right) \quad (2)$$

During the Gibbs sampling process, the inference procedure tends to push the mean of the Gaussians i.e. $H(q_{t,w})$ close to the preset $l_{t,w}$ values. For topic-indicative features, we set $l_{t,w}$ to 0 which penalizes large entropies in the topic distributions of such features, therefore driving the inference procedure to return low entropy models. $\sigma^2_{l_w}$ dictates the strictness of the penalty.

It should be noted that an alternate method to achieve sparsity is to modify the priors. Replacing the Dirichlet priors to obtain preferences in word distribution characteristics however requires complicated priors that are capable of producing topic distributions that are not *iid*. The new prior will now need to generate a set of topics, which will no longer be independent of each other, instead of the Dirichlet prior from which multiple topics can be drawn in an iid manner.

When such priors are employed, they are no longer conjugate with the multinomial topic distributions necessitating sampling using computationally expensive methods like Metropolis-Hastings. The regularization technique described achieves a similar effect while requiring minimal additions to the existing Gibbs sampling inference procedure.

4 Injecting Labeled Features and Documents

In this section, we study how stronger prior knowledge in the form of labeled features and labeled documents can be incorporated into mixed-membership models by modifying the Gibbs sampling inference procedure. *Topic tables* are a commonly used method to display latent topics that are uncovered using models such as LDA. These tables depict topics using the top words of multinomials recovered after inference. Here, we use labeled features to indicate the topic a feature belongs to as a way to influence the formation of the topic tables. This is done by giving the inference procedure hints about the latent topic tables that we expect to see for the labeled features. Document labels, similarly bring the model closer to semi-supervised learning where a subset of the training data has known labels by providing apriori information about the latent topic assignment during inference.

Firstly, we look at a method to use labeled features by modifying the Gibbs sampler. As a concrete example, let us return to the task of clustering entities drawn from web tables. We might have domain knowledge that certain entities do not have multiple senses and should be assigned to a single pre-known latent cluster. An example is *Google* which in the context of our task is known to always be generated by the company topic. In general, we have pre-known latent cluster assignments for a small set of features which are strongly topic-indicative.

Let $\mathcal{L}$ be a set of pairs $\langle w, k_w \rangle$ where w is a feature i.e. $w \in V_t, t \in 1 \ldots T$ and $k_w \in 1, \ldots, K$. Each such pair indicates that the latent topic that generates an instance of w in the corpus is almost certainly k_w. Note that we do not have information about the nature of topic k_w at this stage before inference. We simply use the topic ids in $\mathcal{L}$ to separate and funnel features of different known clusters to different topics. During Gibbs sampling, when the topic indicator for a word is inferred, the procedure is modified to include a check to see if the word in question is present in $\mathcal{L}$. If yes, then instead of sampling a topic indicator for the word, the latent topic indicator is set to k_w with a probability of γ_f, where γ_f is a constant close to 1.

In terms of the generative story underlying LDA derived models, using labeled features implies that the topic multinomials $\beta_{t,k}$ are no longer drawn from the same symmetric Dirichlet priors parameterized by γ. Instead, the method implies that we use different asymmetric Dirichlet priors for each topic. For instance if $w \in V_t$ has a label k_w, then the prior for topic k_w is an asymmetric Dirichlet with parameters γ for all words other than w and a larger value γ^* for the word w. For all the other topics, the asymmetric Dirichlet has a lower value γ' for w to enforce our prior belief that w is more likely to be generated by topic k_w than any other topic.

Next, we examine how labeled data in the form of a-priori information about entity cluster membership that can be integrated into the inference procedure. While the motivation in using a LDA-derived approach for the entity clustering task lies in its ability to model mixed-membership, in the task of clustering entities, there are many entities that belong to only one cluster. In such a context, it would be useful to allow the inference procedure to use known cluster assignments for a small number of documents to influence the latent cluster formation. For instance, in the entity clustering task using the Toy Apple dataset, we might wish to use domain knowledge to say that "persimmon" belongs exclusively to the "fruit" cluster.

Let d be a document that is known to belong to cluster c_d. During inference using Gibbs sampling, for all words in the document, the cluster c_d is assigned with probability γ_d (≈ 1.0), and the usual Gibbs sampling procedure is used to determine the latent topic assignment with probability $1 - \gamma_d$.

Similar to the generative story underlying labeled features, the use of labeled documents implies a generative process where labeled documents' topic distributions i.e θ are drawn from asymmetric Dirichlet priors with higher parameter values for their topic labels instead of the symmetric Dirichlet priors that are usually used.

5   Experimental Results

First, we study the effect of biasing the model to better exploit topic-indicative features. Figure 2 shows the co-occurring entities perplexity of the biased Link-LDA model for the different datasets for different values of the variance parameter in the bias term. The reported values are averaged over 10 trials. For each trial, the Gibbs sampler ran for 100 iterations. The number of topics is set to 40 based on visual inspection of the clusters that were formed. The effect of regularization described below is however similar, when the number of topics is changed. It can be seen that the best perplexity is seen across all datasets when the variance is set to 0.2. We use this variance when using feature regularization (i.e. biasing) for the rest of the paper. When biasing is used, it is applied to the column id and entity-pair features: a column in a table is unlikely to contain entities from multiple clusters and is therefore strongly indicative of the topic; similarly, while an entity can belong to multiple topics, an entity-pair such as "apple:peach" is strongly indicative of a single topic.

Table 2 shows the difference in performance between the biased and baseline unbiased models as measured by NMI between predicted cluster distributions and known

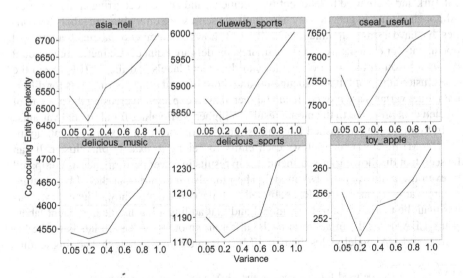

Fig. 2. Studying perplexity with feature regularization

Table 2. Feature regularization: Effect on NMI

Dataset	Regularization No Yes	Change
Asia-NELL	0.586 0.637	+8.70%
Clueweb-Sports	0.567 0.624	+10.05%
CSEAL-Useful	0.533 0.588	+10.31%
Delicious-Music	0.548 0.621	+13.32%
Delicious-Sports	0.609 0.615	+0.98%
Toy Apple	0.771 0.781	+1.29%

true cluster labels of labeled documents. For all the datasets, the biased models show a significant improvement over the unbiased variant. We note here that we cannot directly compare the entity clustering results from these experiments to the results from prior work in HTML table based entity clustering by Dalvi et al. [9] because the approach in that work clusters triplets of entities extracted from tables rather than individual entities. The biasing technique presented here is a general one and can be applied to any task that mixed-membership models are used for, whereas the WebSets approach specifically addresses the entity clustering task. For rough comparison however, the NMI value of clustering entities from the Delicious-Sports dataset is reported at 0.64 using the WebSets[9] approach whereas Table 2 indicates that the regularized model returns a NMI of 0.615 for the same dataset. It is worth re-emphasizing again that the results are not directly comparable.

Next, we study the effects of feature and document labeling in Figures 3 and 4. Feature and document labels are provided to the model for a subset of co-occurring entity features and entities. Labels for entities were obtained using Amazon's Mechanical Turk and were used to label entity documents and also co-occurring entity features. Although entities in general may have multiple senses, we only obtained labels for entities that have a single dominant sense. Table 3 shows the number of labeled features and documents for each dataset. In these figures, models are trained with increasing amount of supervision in the form of feature and document labels and the NMI between the true cluster labels of labeled documents and their inferred topic distributions for different model variants are plotted. It can be seen that as expected, increasing the amount of labeled data provided to the model results in higher NMI values for all model variants.

In figure 3, the red dashed line shows the performance of a mixture of multinomials (MoM) model[1] which allows each entity to belong to exactly one cluster. It can be seen that disallowing mixed-membership results in lower performance as compared to even the plain vanilla LDA model. The plot also indicates that the adding feature regularization (Link-LDA+FR) i.e. biased features consistently shows higher NMI values than the unbiased Link-LDA model and that adding all available document labels (Link-LDA+FR+DL) in addition to the different amounts of feature labels to the biased Link-LDA model yields the best NMI. It is interesting to note that adding feature

---

[1] While EM can be used for inference in the MoM model, we use Gibbs sampling for these experiments.

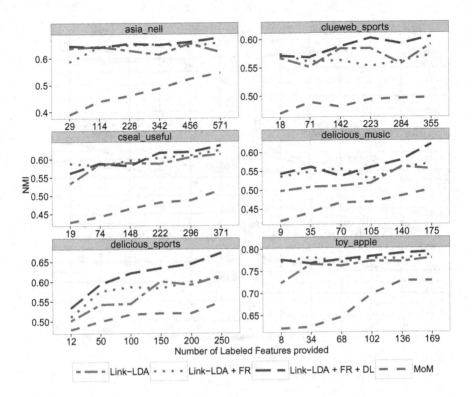

Fig. 3. Effect of injecting Feature Labels

labeling to the mixture of multinomials model, i.e., the points on the MoM line towards the right of the plot, describes a setting that is similar to DUALIST [21].

The entropy of θ can be subject to the same kind of regularization as the word topic distribution used in feature regularization, enabling us to restrict the degree to which entities are allowed to exhibit mixed-membership. In figure 4, it can be seen that adding such document regularization (+ DR), shows better performance than the regular Link-LDA model. Adding feature biasing (+ FR) and all available feature labels (+ FL), along

Table 3. Feature and Document Label statistics

Dataset	Co-occurring entities vocabulary size	#Labeled features	#Labeled documents
Asia-NELL	18309	571	411
Clueweb-Sports	28891	355	302
CSEAL-Useful	24240	371	600
Delicious-Music	9748	175	254
Delicious-Sports	3183	249	206
Toy Apple	423	169	177

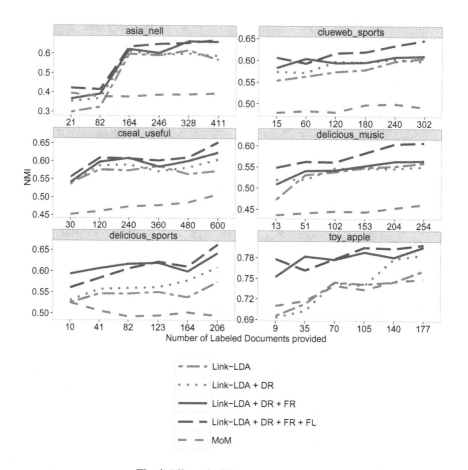

Fig. 4. Effect of adding Document Labels

with different degrees of document labels, shows progressively higher NMI across all datasets especially as the number of labeled documents provided is higher. The red dashed line in the plot representing the performance of the MoM model shows the performance of a single cluster membership model as we move from a fully unsupervised model to a semi-supervised model.

The above experiments show that introducing labeled documents and features consistently improves performance. While document labels have more impact, the labeling scheme used restricts us to only provide labels for entities with a single sense. We also see that for a fixed number of feature or document labels, adding feature regularization (i.e. biasing) and document regularization consistently improves the NMI scores.

In Table 4, we see illustrative examples of the advantage of the mixed-membership approach. For the ambiguous entities shown, the top two topics to which they are deemed to belong are shown using the top entries from the entity-pair multinomials. The results are from a biased Link-LDA model with no labeled features or documents. The topic titles in bold were added after inference by looking at the top entries for the

Table 4. Mixed-membership clustering results of ambiguous entities

Dataset: asia_nell, Entity: **franklin**
Names: (0.24) armstrong:brown, jennifer:jessica, chloe:gucci, brandon:joseph, benjamin:matthew, donald:edward, russell:stanley, benjamin:ethan, greg:gregg, angel:jose
Places: (0.21) montana:nebraska, dakotas:north_carolina, rock_island:san_francisco, atlanta:long_island, delaware:montana, montana:new_york, central_california:san_clemente_island, clearwater:cocoa_beach, sutter:tehama, oklahoma_city:salt_lake_city

Dataset:toy apple, Entity: **apple**
Food: (0.61) peaches:pears, cocoa:coconut_oil, apricots:avocados, sodium_carbonate:sodium_chloride, lactic_acid:lauric_acid, sugar_alcohols:sugars, coconut_oil:coffee, caffeine:calcium_carbonate, xanthan_gum:yeast, sodium_citrate:sodium_hydroxide, pears:pineapple
Companies: (0.16) nec:palmone, blackberry:google, sony:tomtom, asus:palm, philips:samsung, dell:ericsson, sagem:sharp, orange:philips, asus:google, sagem:samsung, asus:bosch

Dataset:delicious sports, Entity: **giants**
NFL teams: (0.26) chiefs:redskins, browns:raiders, cardinals:redskins, rams:saints, cowboys:redskins, cowboys:jaguars, bengals:eagles, bengals:patriots, falcons:patriots, saints:falcons, eagles:panthers
MLB teams: (0.21) arizona_diamondbacks:cincinnati_reds, pittsburgh_pirates:texas_rangers, cleveland_indians:minnesota_twins, milwaukee_brewers:san_diego_padres, boston_red_sox:los_angeles_dodgers, cincinnati_reds:new_york_yankees, minnesota_twins:pittsburgh_pirates, florida_marlins:houston_astros, chicago_cubs:los_angeles_dodgers, baltimore_orioles:montreal_expos, houston_astros:philadelphia_phillies

topic. The value in parentheses show the degree of membership that the entity has for the topic. It can be seen that the mixed-membership latent variable model approach is able to detect the multiple senses of ambiguous entities. The first entity in the table **Franklin** is ambiguous because it has multiple senses — as a common first or last name and as a name of a city in the state of Nebraska in the US, among others. The second example **apple** as discussed earlier could either refer to the fruit or the company. The top two topics returned for this entity denotes exactly these two concepts. The third example **giants** is from the sports domain and could refer to either the New York Giants who play in the National Football League (American Football) or the San Francisco Giants who play in Major League Baseball (MLB). The top two topics indicate these two concepts.

6 Related Work

Ganchev et al. [11] proposed Posterior Regularization (PR), a method to incorporate indirect supervision via constraints on posterior distributions of probabilistic models with latent variables. They demonstrate the use of the technique in models for several tasks such as POS induction, word alignment, etc. While the approach proposed in this paper is similar in spirit to PR in that both approaches provide a method for preferences for the posteriors of latent variables to be specified, there are significant differences. The PR framework is used in applications where exact inference is possible and the authors present a modified EM procedure to learn parameters for the model and incorporate constraints in an interleaved manner. In the approach introduced in this paper to bias the model, we focus on incorporating constraints on latent role distributions in models where exact inference is intractable by incorporating the constraints into the model instead of imposing them in a separate distinct step during inference.

Mann and McCallum [14] also proposed a general framework to introduce preferences in model expectations by adding terms called *generalized expectation (GE) criteria* to the objective function. Examples of such criteria were explored in the domain of log linear models. The approach in this paper is similar to the GE framework in that the regularization operates on entropies of distributions of inferred latent variables. The manner in which deviations from expectations are penalized, however differs from the criteria used by Mann and McCallum; the method introduced in this paper proposes that a desired value is drawn from a distribution parameterized by the inferred latent variables' values. The GE framework has not been applied to latent variable mixed-membership models as far as we know.

Newman et al. [16] presented a method to regularize topic models to produce coherent topics. In this approach, a pre-computed matrix of word-similarities from external data (Wikipedia) is used to construct a prior for the topic distributions. This regularization approach differs from the framework used in this paper in that it is aimed at producing topics that respect external word similarities. This is in contrast to our approach that is designed to control the latent structure properties without using external data.

Incorporating document labels into classifiers to obtain semi-supervised models is a well established technique in machine learning [17]. In the context of topic models, Labeled-LDA [20] uses tags attached to documents to limit the membership of the documents to specified topics. The labeled document injection technique discussed in this paper is closely related to Labeled-LDA. Supervised LDA [5] is a related model where supervision in the form of categorical or real-valued attributes of documents is provided. These attributes are derived from the topic distributions using regression models, which differs from the approach in this paper where the document labels directly indicate topic membership. Mimno et al. [15] propose a model where the Dirichlet prior for document topic proportion distribution is replaced with a log-linear prior that permits the distribution to be directly influenced by metadata. This work can be interpreted as a method to use metadata to tailor the latent structure formation. Settles [21] used labeled features for multinomial Naive Bayes classifiers. A similar approach was used by Attenberg et al. [3] in the context of active learning.

Steyvers et al. [22] present a related approach where they "pre-construct" some topics based on concepts obtained from Cambridge Advance Learner's Dictionary (CALD). This approach is similar to the labeled features idea presented in this paper. A concept topic as defined by this approach can be seen as a set of labeled words with the same topic indicator.

Entity clustering from semi-structured data has been addressed previously [18,23,9]. These approaches however do not address the issue of mixed-membership.

7 Conclusion

We presented a novel technique to bias latent variable mixed-membership models to exploit topic-indicative features and used the biased model for the task of clustering semi-structured data in the form of entities extracted from HTML tables. Our experiments show that the biased models outperform the baseline models in the cluster recovery task as measured by NMI. We then presented a method to allow for stronger supervision in the form of feature and document labels to move further along the spectrum toward semi-supervised learning from totally unsupervised learning. Results indicate that the stronger forms of supervision result in better cluster recovery. To summarize, we presented a framework in which mixed-membership models can be successfully used in a semi-supervised fashion to incorporate inexpensive weak prior domain knowledge.

References

1. Andrzejewski, D., Buttler, D.: Latent topic feedback for information retrieval. In: Proceedings of the 17th ACM SIGKDD International Conference on Knowledge Discovery and Data Mining, KDD 2011, pp. 600–608. ACM, New York (2011), http://doi.acm.org/10.1145/2020408.2020503
2. Arora, R., Ravindran, B.: Latent dirichlet allocation and singular value decomposition based multi-document summarization. In: ICDM, pp. 713–718. IEEE Computer Society (2008)
3. Attenberg, J., Melville, P., Provost, F.: A unified approach to active dual supervision for labeling features and examples. In: Balcázar, J.L., Bonchi, F., Gionis, A., Sebag, M. (eds.) ECML PKDD 2010, Part I. LNCS (LNAI), vol. 6321, pp. 40–55. Springer, Heidelberg (2010)
4. Balasubramanyan, R., Cohen, W.W.: Regularization of latent variable models to obtain sparsity. In: SDM (2013)
5. Blei, D., McAuliffe, J.: Supervised topic models. In: Platt, J., Koller, D., Singer, Y., Roweis, S. (eds.) Advances in Neural Information Processing Systems, vol. 20, pp. 121–128. MIT Press, Cambridge (2008)
6. Blei, D., Ng, A., Jordan, M.: Latent dirichlet allocation. The Journal of Machine Learning Research 3, 993–1022 (2003), http://www.cs.princeton.edu/~blei/papers/BleiNgJordan2003.pdf
7. Carlson, A., Betteridge, J., Kisiel, B., Settles, B., Hruschka Jr., E.R., Mitchell, T.M.: Toward an architecture for never-ending language learning. In: Proceedings of the Twenty-Fourth Conference on Artificial Intelligence, AAAI 2010 (2010)
8. Carlson, A., Betteridge, J., Wang, R.C., Jr. Hruschka, E.R., Mitchell, T.M.: Coupled semi-supervised learning for information extraction. In: Proceedings of the third ACM International Conference on Web Search and Data Mining, WSDM 2010, pp. 101–110. ACM, New York (2010), http://doi.acm.org/10.1145/1718487.1718501
9. Dalvi, B.B., Cohen, W.W., Callan, J.: Websets: extracting sets of entities from the web using unsupervised information extraction. In: Proceedings of the Fifth ACM International Conference on Web Search and Data Mining, WSDM 2012, pp. 243–252. ACM, New York (2012), http://doi.acm.org/10.1145/2124295.2124327
10. Erosheva, E.A., Fienberg, S., Lafferty, J.: Mixed-membership models of scientific publications. Proceedings of the National Academy of Sciences of the United States of America 101(suppl. 1), 5220 (2004)
11. Ganchev, K., Graça, J.A., Gillenwater, J., Taskar, B.: Posterior regularization for structured latent variable models. J. Mach. Learn. Res. 11, 2001–2049 (2010), http://dl.acm.org/citation.cfm?id=1756006.1859918

12. Griffiths, T.L., Steyvers, M., Blei, D.M., Tenenbaum, J.B.: Integrating topics and syntax. In: Advances in Neural Information Processing Systems, vol. 17, pp. 537–544. MIT Press (2005)
13. Hearst, M.A.: Automatic acquisition of hyponyms from large text corpora. In: Proceedings of the 14th Conference on Computational Linguistics, COLING 1992, vol. 2, pp. 539–545. Association for Computational Linguistics, Stroudsburg (1992), http://dx.doi.org/10.3115/992133.992154
14. Mann, G.S., Mccallum, A.: Generalized Expectation Criteria for Semi-Supervised Learning with Weakly Labeled Data. Journal of Machine Learning Research 11, 955–984 (2010), http://dl.acm.org/citation.cfm?id=1756038
15. Mimno, D.M., McCallum, A.: Topic models conditioned on arbitrary features with dirichlet-multinomial regression. In: McAllester, D.A., Myllymäki, P. (eds.) UAI, pp. 411–418. AUAI Press (2008)
16. Newman, D., Bonilla, E.V., Buntine, W.L.: Improving topic coherence with regularized topic models. In: Shawe-Taylor, J., Zemel, R.S., Bartlett, P.L., Pereira, F.C.N., Weinberger, K.Q. (eds.) NIPS, pp. 496–504 (2011)
17. Nigam, K., McCallum, A.K., Thrun, S., Mitchell, T.: Text classification from labeled and unlabeled documents using em. Mach. Learn. 39(2-3), 103–134 (2000), http://dx.doi.org/10.1023/A:1007692713085
18. Paca, M., Van Durme, B.: Weakly-Supervised Acquisition of Open-Domain Classes and Class Attributes from Web Documents and Query Logs, pp. 19–27. Association for Computational Linguistics, Columbus (2008), http://www.aclweb.org/anthology/P/P08/P08-1003
19. Porteous, I., Newman, D., Ihler, A., Asuncion, A., Smyth, P., Welling, M.: Fast collapsed gibbs sampling for latent dirichlet allocation. In: Proceeding of the 14th ACM SIGKDD International Conference on Knowledge Discovery and Data Mining, pp. 569–577 (2008)
20. Ramage, D., Hall, D., Nallapati, R., Manning, C.D.: Labeled LDA: a supervised topic model for credit attribution in multi-labeled corpora. In: Proceedings of the 2009 Conference on Empirical Methods in Natural Language Processing, pp. 248–256. Association for Computational Linguistics, Singapore (2009)
21. Settles, B.: Closing the loop: Fast, interactive semi-supervised annotation with queries on features and instances. In: Proceedings of the 2011 Conference on Empirical Methods in Natural Language Processing, pp. 1467–1478. Association for Computational Linguistics, Edinburgh (2011), http://www.aclweb.org/anthology/D11-1136
22. Steyvers, M., Smyth, P., Chemuduganta, C.: Combining Background Knowledge and Learned Topics. Topics in Cognitive Science 3(1), 18–47 (2011), http://doi.wiley.com/10.1111/j.1756-8765.2010.01097.x
23. Talukdar, P.P., Reisinger, J., Pasca, M., Ravichandran, D., Bhagat, R., Pereira, F.: Weakly-Supervised Acquisition of Labeled Class Instances using Graph Random Walks, pp. 582–590. Association for Computational Linguistics, Honolulu (2008), http://www.aclweb.org/anthology/D08-1061
24. Wang, R.C., Cohen, W.W.: Automatic set instance extraction using the web. In: Proceedings of the Joint Conference of the 47th Annual Meeting of the ACL and The 4th International Joint Conference on Natural Language Processing of the AFNLP, ACL 2009, vol. 1, pp. 441–449. Association for Computational Linguistics, Stroudsburg (2009), http://dl.acm.org/citation.cfm?id=1687878.1687941
25. Wetzker, R., Zimmermann, C., Bauckhage, C.: Analyzing social bookmarking systems: A del.icio.us cookbook. In: Mining Social Data (MSoDa) Workshop Proceedings, ECAI 2008, pp. 26–30 (July 2008), http://robertwetzker.com/wp-content/uploads/2008/06/wetzker_delicious_ecai2008_final.pdf

Hub Co-occurrence Modeling for
Robust High-Dimensional kNN Classification

Nenad Tomašev and Dunja Mladenić

Institute Jožef Stefan
Artificial Intelligence Laboratory
Jamova 39, 1000 Ljubljana, Slovenia
{nenad.tomasev,dunja.mladenic}@ijs.si

Abstract. The emergence of hubs in k-nearest neighbor (kNN) topologies of intrinsically high dimensional data has recently been shown to be quite detrimental to many standard machine learning tasks, including classification. Robust hubness-aware learning methods are required in order to overcome the impact of the highly uneven distribution of influence. In this paper, we have adapted the Hidden Naive Bayes (HNB) model to the problem of modeling neighbor occurrences and co-occurrences in high-dimensional data. Hidden nodes are used to aggregate all pairwise occurrence dependencies. The result is a novel kNN classification method tailored specifically for intrinsically high-dimensional data, the Augmented Naive Hubness Bayesian k nearest Neighbor (ANHBNN). Neighbor co-occurrence information forms an important part of the model and our analysis reveals some surprising results regarding the influence of hubness on the shape of the co-occurrence distribution in high-dimensional data. The proposed approach was tested in the context of object recognition from images in class imbalanced data and the results show that it offers clear benefits when compared to the other hubness-aware kNN baselines.

Keywords: hubs, k-nearest neighbor, classification, curse of dimensionality, Bayesian, co-occurrences.

1 Introduction

The basic k-nearest neighbor classification rule [1] is fairly simple, though often surprisingly effective, as it exhibits some favorable asymptotic properties [2]. Many extensions of the basic method have been proposed over the years. It is possible to use kNN in conjunction with kernels [3], perform large margin learning [4], multi-label classification [5], adaptively determine the neighborhood size [6], etc.

Even though kNN has mostly been replaced in general-purpose classification systems by support vector machines and some other modern classifiers [7], it is still very useful and quite effective in several important domains. Unlike many other methods, kNN has a relatively low generality bias and a rather high specificity bias. This makes it ideal for classification under class imbalance [8][9]. Many real-world class distributions are known to be very imbalanced and many examples can be found in medical diagnostic systems, spam filters, intrusion detection, etc. Nearest neighbor methods are

H. Blockeel et al. (Eds.): ECML PKDD 2013, Part II, LNAI 8189, pp. 643–659, 2013.

also currently considered as the state-of-the-art in time series classification when used in conjunction with the dynamic time-warping distance (DTW) [10]. Some recent experiments suggest that the kNN might also be quite appropriate for object recognition in images [11].

The curse of dimensionality [12] is an umbrella-term referring to many difficulties that are known to arise when dealing with high-dimensional feature representations. Many k-nearest neighbor methods are negatively affected by various aspects of the dimensionality curse. Most standard distance measures concentrate [13] and the overall contrast is reduced, which makes distinguishing between close/relevant and distant/irrelevant points difficult for any given query. The very concept of what constitutes nearest neighbors in high-dimensional data has rightfully been questioned in the past [14].

Hubness is a recently described consequence of high intrinsic dimensionality that is related specifically to k-nearest neighbor methods [15]. It was first noticed in music retrieval and recommendation systems [16], where some songs were appearing in the result sets of a surprisingly large proportion of queries [17]. Their occurrence frequency could not be explained by the semantics of the data alone and their apparent similarity to other songs was shown to be quite counter-intuitive. The initial thought was that this might be an artefact of the metric or the feature representation, though it was later shown [15][18] that hubness emerges naturally in most types of intrinsically high-dimensional data. *Hubs* become the centers of influence and the occurrence distribution asymptotically approaches a power law as the dimensionality increases. An illustrative example of the change in the distribution shape is shown in Figure 1. The almost scale-free topology of the k-nearest neighbor graph [18] and the skewed distribution of influence have profound implications for kNN learning under the assumption of hubness in high-dimensional data.

The hubness among neighbor occurrences was previously unknown and is not even implicitly taken into account in most standard kNN classifiers. This can lead to some

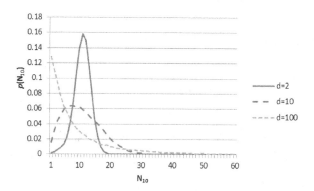

Fig. 1. The shape of the neighbor occurrence frequency distribution changes as the intrinsic dimensionality of the data increases. The example shows the distribution of 10-occurrences (N_{10}) of i.i.d. Gaussian data in case of 2, 10 and 100 dimensions.

problems in applying the standard methods for high-dimensional data analysis. Pathological cases have even been shown to exist [19][20] where the influence of hubs reduces the overall *k*NN performance below that of zero-rule. Such cases are rare, but warn of the danger that lurks in ignoring the underlying occurrence distribution.

The presence of hubs could in principle be beneficial in that it reduces the overall impact of noise, but the network of influence also becomes more vulnerable to any inaccuracies that are contained in hubs themselves, when errors can propagate much more easily. Therefore, the overall stability of the learning process is compromised. As hubness is a geometric property that results by an interplay of representational features and metrics, it does not necessarily reflect the underlying semantics well. Many hub points are in fact known to induce severe misclassification [18]. Consequently, there is a rising awareness of a need for novel approaches to algorithm design for properly handling high-dimensional data in *k*-nearest neighbor methods.

Recent research has shown that learning from past occurrences and hub profiling can be successfully employed for improving the overall *k*NN classifier performance [21][22][23][24][25]. Hubness-aware metric learning also seems to be helpful [26][27][20]. The consequences of data hubness have recently been examined in the unsupervised context as well [28][29].

The Naive Bayesian interpretation of the observed *k*-neighbor occurrences (NHBNN) [23] was shown to be quite promising in high-dimensional data classification, especially in the context of learning from class imbalanced data [20]. Yet, NHBNN naively assumes independence between neighbor occurrences in the same *k*-neighbor set, an assumption that is clearly severely violated in most cases, as close points tend to co-occur as neighbors.

1.1 Goal and Contributions

Our goal was to extend and augment the existing naive NHBNN approach by including the co-occurrence dependencies between the observed neighbors in the underlying Bayesian model. This was done by introducing hidden nodes in the augmented topology, as in the recently proposed Hidden Naive Bayes method [30]. This work represents the first attempt to exploit the neighbor co-occurrence dependencies in high-dimensional neighbor occurrence models and we propose a novel classification algorithm named the Augmented Naive Hubness-Bayesian *k*-nearest Neighbor (ANHBNN).

Additionally, we justify our approach by examining how the increase in the intrinsic dimensionality of the data affects the distribution of neighbor co-occurrences. Our tests on synthetic Gaussian data reveal some surprising results. We have shown that the distribution of the number of distinct co-occurring neighbor points becomes multi-modal with modes located approximately around the multiples of $(k-1)$. We have also shown that the tail of the distribution of neighbor pair occurrence frequencies becomes thicker with increasing dimensionality, which indicates *hub linkage*, as some hub points tend to co-occur frequently. Also, the number of distinct pairs of co-occurring neighbors increases. These phenomena seem beneficial for co-occurrence modeling and they explain why the proposed ANHBNN classifier works well in intrinsically high-dimensional data.

2 Related Work

As the emergence of hubs was shown to be potentially highly detrimental, hubness-aware classification of high-dimensional data has recently drawn some attention and several novel kNN classification methods have been proposed. The simplest approach (hw-kNN) was to include instance-specific weights that would reflect the nature of hubness of individual neighbor points [21]. This was later improved upon by including class-conditional occurrence profiles, as in [22][23][25]. In h-FNN [22], the occurrence profiles were used for forming fuzzy votes within the FNN [31] framework. HIKNN [25] was based on the information-theoretic re-interpretation of h-FNN, as less frequently occurring neighbors were judged to be more locally relevant and assigned higher weights, based on their occurrence self-information. On the other hand, the Naive Hubness Bayesian k-nearest Neighbor (NHBNN) [23] was based on a slightly different idea - interpreting the individual occurrences as random events and applying the Naive Bayes rule in order to perform classification. Prior tests have shown this to be a promising idea, so extending this basic approach will be the focus of this paper.

2.1 Naive Hubness-Bayesian kNN

In order to explain in detail the idea behind the Naive Hubness-Bayesian kNN (NHBNN) [23], it is necessary to introduce some formal notation.

Neighbor k-occurrence Models: Let $D = \{(x_1, y_1), (x_2, y_2) \ldots (x_n, y_n)\}$ be the data set, where x_i-s are the feature vectors and $y_i \in \{1 \ldots C\}$ the class labels. Also, let $D_k(x)$ be the set of k-nearest neighbors of x. The neighbor k-occurrence frequency of x will be denoted by $N_k(x) = |x_i : x \in D_k(x_i)|$ and will also sometimes be referred to as *point hubness*. The total *hubness of a dataset D* is defined as the third standard moment (skewness) of the neighbor occurrence degree distribution and will be denoted by $SN_k = \frac{\frac{1}{n}\sum_{i=1}^{n}(N_k(x_i)-k)^3}{(\frac{1}{n}\sum_{i=1}^{n}(N_k(x_i)-k)^2)^{3/2}}$. High skewness indicates the long-tailed distribution where most k-neighbor sets are dominated by occurrences of a limited number of highly frequent neighbors, while most other points occur very rarely or not at all. The very frequently occurring points are called *hubs*, the infrequently occurring points *anti-hubs* and the points that never occur as neighbors *orphans*.

The total occurrence frequency is often decomposed into either *good and bad hubness* or alternatively *class-conditional hubness* in the following way: $N_k(x) = GN_k(x) + BN_k(x) = \sum_{c \in C} N_{k,c}(x)$. Good hubness is defined as the number of neighbor occurrences where neighbors share the same class label and bad hubness the number of occurrences where there is label mismatch, i.e. $GN_k(x) = |x_i : x \in D_k(x_i) \wedge y = y_i|$ and $BN_k(x) = |x_i : x \in D_k(x_i) \wedge y \neq y_i|$. Similarly, class-conditional hubness measures the occurrence frequency within the neighbor sets of a specific class: $N_{k,c}(x) = |x_i : x \in D_k(x_i) \wedge y_i = c|$. These quantities are used to form an occurrence model from the training set that includes a neighbor occurrence profile for each neighbor point.

Naive Bayesian Interpretation of kNN: In the NHBNN method [23], the neighbor occurrences are interpreted as random events that can be used to deduce the class label in the point of interest. Equation 1 shows how the Naive Bayes rule [7] is used to deduce the class assignment probabilities for X based on its k-nearest neighbors from the training set. The label $y = \text{argmax}_{c \in \{1...C\}} p(c|D_k(x))$ is assigned to x by this rule.

$$p(Y|D_k(X)) \propto p(Y) \prod_{t=1}^{k} p(X_t \in D_k(X)|Y) \tag{1}$$

The order of neighbors is ignored in order to get better probability estimates in the model, so $p(X_t \in D_k(X)|Y)$ can be easily estimated from the class-specific hubness $N_{k,c}$ scores and the total class occurrences. Each point is trivially considered to be its own 0th nearest neighbor, for practical reasons.

$$p(X_t \in D_k(X)|Y) \approx \frac{N_{k,Y}(X_t) + \lambda}{n_Y \cdot (k+1) + \lambda|D|} \tag{2}$$

The actual algorithm is a bit more complex than this, mostly because there are points for which the $p(X_t \in D_k(X)|Y)$ can not be reliably estimated from the previous occurrences, orphans and anti-hubs. They need to be treated separately, as a special case. In analogy with the Naive Bayes classifier, it would be as if a completely new feature value was first encountered on the test data.

The obvious problem with this approach is that the Naive Bayes rule assumes independence between the random variables and this does not hold true among the k-neighbor occurrences, where close neighbors tend to co-occur together and there are clear dependencies between individual neighbor occurrences.

Naive Bayes sometimes works well even when the independence assumption does not hold [32] and the initial evaluation of the Naive Hubness-Bayesian k-nearest neighbor has shown it to be quite a promising approach to high-dimensional kNN classification. However, it was later observed that its performance quickly drops when the neighborhood size is increased and that it performs rather poorly for larger k values. It was hypothesized that this was a consequence of the independence assumption violation.

In order to test this hypothesis, we decided to proceed by including some sort of co-occurrence dependencies in the model, with the intent of increasing its robustness and overall performance. The extended algorithm was supposed be able to properly handle larger neighborhood sizes.

3   The Proposed Approach: Including the Co-occurrence Dependencies

Naive Bayes is the simplest among the Bayesian network models. The conditional independence assumption is often violated in practice, though its use can still be justified in some cases [32]. Learning the optimal Bayesian network from the data can sometimes be intractable, as it was shown to be an NP-complete problem [33]. As the structure

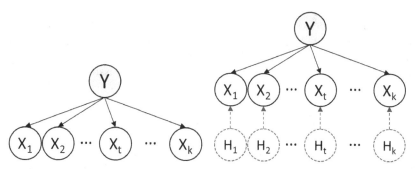

(a) The basic Naive Bayes model. All occurrences are conditioned only on the class
label Y.

(b) The Hidden Naive Bayes model. An additional hidden node is introduced for each
neighbor variable, modeling the dependencies on all other nodes.

Fig. 2. A comparison between the basic Naive Bayes and the Hidden Naive Bayes [30] models

learning is the most time consuming step, assuming a certain type of underlying structure is common. We base our extension of the hubness-aware NHBNN classifier on the
Hidden Naive Bayes model [30], shown in Figure 3. A hidden node is introduced for
each variable that accounts for the influence from all other variables. In our case, the
variables are the occurrences of points as neighbors in k-neighbor sets.

Hidden nodes help model the dependencies between neighbor co-occurrences.

Let $N_{k,c}(x_i, x_j)$ be the number of co-occurrences of x_i and x_j in neighborhoods of
elements from c, i.e. $N_{k,c}(x_i, x_j) = |x : y = c \wedge x_i \in D_k(x) \wedge x_j \in D_k(x)|$. Calcu-
lating all the $N_{k,c}(x_i, x_j)$ paired class-conditional co-occurrence frequencies is possible in $O(nk^2)$, as this is the time required to consider and count all co-occurrences
within the k-neighbor sets. In order to avoid the $O(Cn^2)$ memory complexity for
storing all the co-occurrence counts, C hash tables can be used to store only the non-
negative co-occurrence counts. Many $N_{k,c}(x_i, x_j)$ do equal zero, so this saves considerable memory space.

Classification in the extended Bayesian neighbor occurrence model is performed
based on the class probability estimate shown in Equation 3 and it forms a similar
expression as in NHBNN (Equation 1). The difference is that the probability of $X_t \in
D_k(X)$ is now also conditioned on the hidden variable $H_t(X, Y)$.

$$p(Y|D_k(X)) \propto p(Y) \prod_{t=1}^{k} p(X_t \in D_k(X)|Y, H_t(X, Y)) \qquad (3)$$

We will call the proposed algorithm that performs the k-nearest neighbor classification based on Equation 3 the Augmented Naive Hubness-Bayesian k-nearest Neighbor
(ANHBNN).

3.1 Modeling the Influence of Hubs and Regular Points

In order to infer reliable probability estimates, a certain number of observed occurrences is required. We will first derive the estimates for frequent neighbor points and then focus on approximations for anti-hubs and orphans.

Assuming $N_k(X_t) > 0$, the conditional probabilities are expressed as a weighted sum of separate one-dependence estimators, as shown is Equation 4. This is a standard approach to modeling the influence of the hidden nodes within the HNB framework [30].

$$p(X_t \in D_k(X)|Y, H_t(X, Y)) = \sum_{i=1, i\neq t}^{k} w_{it}^Y \cdot p(X_t \in D_k(X)|X_i \in D_k(X), Y)$$

$$ (4) $$

$$p(X_t \in D_k(X)|X_i \in D_k(X), Y) \approx \begin{cases} \frac{N_{k,Y}(X_t, X_i)}{N_{k,Y}(X_i)}, & \text{if } N_{k,Y}(X_i) > 0, \\ 0, & \text{if } N_{k,Y}(X_i) = 0. \end{cases}$$

The weights in Equation 4 sum up to one and correspond to the strengths of individual influences. It is possible to try optimizing the weights via cross-validation, but it is overly time-consuming and is usually avoided. We propose to extend the original idea [30] of expressing the weights by normalized mutual information by including the class-conditional occurrence self-information $I_{k,Y}(X_i)$ (Equation 5) and the occurrence profile non-homogeneity (Equation 6) which is expressed as the reverse neighbor set entropy. These quantities are supposed to account for differences in hubness between different points.

The class-conditional occurrence self-information measures how unexpected it is to observe X_i in neighborhoods of class Y. Including the self-information in the denominator in Equation 8 allows us to increase the influence of very frequent neighbors. This is beneficial, as there is more past occurrence data for these points and the probability estimates are thus somewhat more reliable. On the other hand, neighbor points with less homogenous occurrence profiles often act as bad hubs and exhibit a detrimental influence, so favoring neighbors with homogenous profiles tends to improve the overall performance.

$$I_{k,Y}(X_i) = \log \frac{n_Y}{N_{k,Y}(X_i)} \quad (5)$$

$$H_k(X_i) = \sum_{c \in C} \frac{N_{k,c}(X_i)}{n_c} \log \frac{n_c}{N_{k,c}(X_i)} \quad (6)$$

The class-conditional mutual information $I_P(X_j, X_t|Y)$ between two neighbor occurrences X_j and X_t is estimated based on the previously observed occurrence profiles on the training data as outlined in Equation 7. The four factors in the outer sum correspond to the two neighbor points occurring together or separately or not at all.

$$I_P(X_j, X_t|Y) = \sum_{c=1}^{C} \left(\frac{N_{k,c}(X_j, X_t)}{n} \cdot \log \frac{\frac{N_{k,c}(X_j, X_t)}{n_c}}{\frac{N_{k,c}(X_j)}{n_c} \cdot \frac{N_{k,c}(X_t)}{n_c}} \right) +$$

$$+ \sum_{c=1}^{C} \left(\frac{N_{k,c}(X_j) - N_{k,c}(X_j, X_t)}{n} \cdot \log \frac{\frac{N_{k,c}(X_j) - N_{k,c}(X_j, X_t)}{n_c}}{\frac{N_{k,c}(X_j)}{n_c} \cdot (1 - \frac{N_{k,c}(X_t)}{n_c})} \right) +$$

$$+ \sum_{c=1}^{C} \left(\frac{N_{k,c}(X_t) - N_{k,c}(X_j, X_t)}{n} \cdot \log \frac{\frac{N_{k,c}(X_t) - N_{k,c}(X_j, X_t)}{n_c}}{(1 - \frac{N_{k,c}(X_j)}{n_c}) \cdot \frac{N_{k,c}(X_t)}{n_c}} \right) +$$

$$+ \sum_{c=1}^{C} \left(\frac{n_c - N_{k,c}(X_j) - N_{k,c}(X_t) + N_{k,c}(X_j, X_t)}{n} \cdot \log \frac{\frac{n_c - N_{k,c}(X_j) - N_{k,c}(X_t) + N_{k,c}(X_j, X_t)}{n_c}}{(1 - \frac{N_{k,c}(X_j)}{n_c}) \cdot (1 - \frac{N_{k,c}(X_t)}{n_c})} \right)$$

$$(7)$$

Finally, the co-dependency weights from Equation 4 are obtained from the class-conditional occurrence self-information, homogeneity and class-conditional neighbor mutual information as shown in Equation 8. Unlike in the original Hidden Naive Bayes [30] model, the weights here are also conditioned on the class, because of the class-conditional self-information. Some smoothing is needed in order to avoid zero divisions in cases when the denominator goes to zero.

$$w_{it}^Y = \frac{\frac{I_P(X_i, X_t|Y)}{I_{k,Y}(X_i) \cdot H_k(X_i)}}{\sum_{j=1, j \neq t}^{k} \frac{I_P(X_j, X_t|Y)}{I_{k,Y}(X_j) \cdot H_k(X_i)}} \qquad (8)$$

The proposed extension of NHBNN embodied in the Augmented Naive Hubness-Bayesian k-nearest Neighbor (ANHBNN) does not have a significant impact on the overall computational complexity, as both algorithms are of the $O(n^2)$ complexity with respect to data size. Approximate k-neighbor set computations are possible and usually allow for considerable practical speed-ups in hubness-aware classifiers without sacrificing too much accuracy [25].

3.2   Dealing with Anti-hubs and Orphans

For infrequently occurring points X_t, the $p(X_t \in D_k(X)|Y, H_t(X, Y))$ can not be estimated from their past occurrences properly. In principle, it would be possible to model their conditioned influence by the average conditioned influence exhibited by other points from their class, as in Equation 9.

$$p(X_t \in D_k(X)|Y, H_t(X, Y)) \approx \frac{\sum_{X_i : Y_i = Y_t \wedge N_k(X_i) > 0} p(X_i \in D_k(X)|Y, H_i(X, Y))}{|X_i : Y_i = Y_t \wedge N_k(X_i) > 0|}$$

$$(9)$$

However, the exact $p(X_i \in D_k(X)|Y, H_i(X, Y))$ are not by default calculated during training, as they depend on the particular k-neighbor set and are inferred later from the pre-calculated one dependence estimators, mutual information and self-information. Therefore, approximating the influence of anti-hubs this way would require an additional time-consuming pass through the training data, as well as some initialization of $p(X_t \in D_k(X)|Y, H_i(X, Y))$ for anti-hubs anyway.

Luckily, points that never occur as neighbors on the training data very rarely occur as neighbors on the test data as well, so it is possible to employ very simple replacements in place of the actual conditional estimates, as it is not possible to arrive at a reliable proper estimate anyway [22][23][25]. As hubs account for most occurrences, this does not have a significant influence on the algorithm performance. Therefore, we propose to use the hidden nodes only for regular points and hubs and approximate the influence of anti-hubs and orphans by the average class-to-class occurrence probabilities as in Equation 10. Here $N_{k,Y}(Y_t)$ denotes the total number of occurrences of elements from class Y_t in neighborhoods of elements from Y. A similar global anti-hub modeling approach was previously shown to be acceptable in NHBNN [23].

$$N_k(X_t) = 0 : p(X_t \in D_k(X)|Y, H_t(X,Y)) \approx$$

$$p(X_t \in D_k(X)|Y) \approx \text{AVG}_{Y_i=Y_t} p(X_i \in D_k(X)|Y) = \frac{N_{k,Y}(Y_t)}{k \cdot n_Y \cdot n_{Y_t}} \qquad (10)$$

4 Neighbor Co-occurrences in High-Dimensional Data

We hypothesized that the emergence of hubs in the kNN topologies of intrinsically high-dimensional data might have some influence on the distribution of neighbor co-occurrences. As our proposed hubness-aware classifier learns from the observed co-occurrences, we have run extensive tests in order to establish whether the hypothesis holds.

To our knowledge, no previous research has been done on the impact of high intrinsic dimensionality on the neighbor co-occurrence distribution and its connection to the hubness phenomenon. Therefore, we hope that the results presented here might shed some light on the more subtle consequences of the curse of dimensionality.

We have run the tests for three different dimensionalities: 2, 10 and 100. For each number of dimensions, a series of 200 randomly generated hyper-spherical zero-centered Gaussian distributions was generated and 1000 points were randomly drawn from each distribution as sample data. We have run tests for several different neighborhood sizes and we give the results for $k = 5$ and $k = 10$ here for comparison.

Figure 3 shows how the number of distinct neighbors that points co-occur with changes with increasing dimensionality. For $d = 2$, the distribution of the number of distinct co-occurring neighbors has a single mode. However, surprisingly, when the number of dimensions is increased, multiple modes appear and are centered approximately around the multiples of $(k - 1)$. We believe that this is a direct consequence of hubness, as there are many points in intrinsically high-dimensional data that occur in k-neighbor sets very rarely. When these points do occur as neighbors, it is possible that most of their $(k - 1)$ co-neighbors co-occur with the anti-hub point for the first time, hence the observed distribution modes.

The emergence of hubs (and anti-hubs) also influences the distribution of the co-occurrence frequency of pairs of neighbor points, as shown in Figure 4. The number of very rarely co-occurring pairs increases significantly with increasing dimensionality,

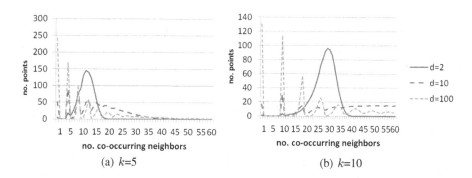

(a) k=5   (b) k=10

Fig. 3. The influence of increasing dimensionality on the distribution of number of different neighbors that points co-occur with. The distribution shape changes from a single modal to a multi-modal shape that has modes around multiples of $(k - 1)$.

due to a large number of rarely occurring neighbor points. On the other hand, the distribution tail also becomes thicker, as the number of pairs of points that co-occur very frequently increases with increasing dimensionality. These very frequently co-occurring pairs emerge as a consequence of what we will denote as *hub linkage*, pairs of hub points that co-occur together in many k-neighbor sets. The linked hub pairs enable the proposed ANHBNN classifier to infer more reliable class-conditional co-occurrence estimates, which is an essential part of the model.

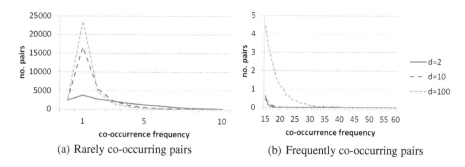

(a) Rarely co-occurring pairs   (b) Frequently co-occurring pairs

Fig. 4. The influence of increasing dimensionality on the distribution of co-occurrence frequency of pairs of neighbor points. The high-dimensional case shows two extremes: more very rarely co-occurring pairs and also more very frequently co-occurring pairs in the distribution tail. The results are given for $k = 10$.

The overall number of distinct co-occurring pairs of neighbor points increases with increasing dimensionality, as shown in Figure 5. From the perspective of co-occurrence modeling, this is a good thing. It is therefore expected that there would be more pairs of neighbor points for which we would be able to derive some estimates of co-occurrence dependencies in intrinsically high-dimensional data.

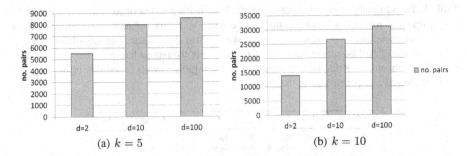

(a) $k = 5$ (b) $k = 10$

Fig. 5. Increasing the intrinsic dimensionality of the data increases the number of distinct co-occurring neighbor pairs

5   Experimental Evaluation

In order to evaluate whether the proposed approach offers any benefits, we have compared it with the other hubness-aware classifiers, namely NHBNN [23], hw-kNN [21], h-FNN [22] and HIKNN [25], as well as with the baseline kNN. Comparisons were performed on a series of intrinsically high-dimensional datasets that have been shown to exhibit very high hubness.

5.1   Data

In experimental evaluation, we have focused on the task of object recognition from images. Image data is high-dimensional and known to exhibit significant hubness [19]. The basic properties of the datasets are outlined in Table 1. Some of the data is imbalanced and the class imbalance is measured by the relative imbalance factor $\text{RImb} = \sqrt{(\sum_{c \in C} (p(c) - 1/C)^2)/((C - 1)/C)}$, which is merely the normalized standard deviation of the class probabilities from the absolutely homogenous mean value of $1/c$.

Datasets iNet3-iNet7 and iNet3Imb-iNet7Imb represent different subsets of the public ImageNet repository (http://www.image-net.org/). These particular subsets have previously been used in several hubness-aware classification benchmarks [22][19][24][20], so they have been selected here for easier comparisons. Images were processed as quantized SIFT [34] bag-of-visual-words representations, extended by binned color histogram information, normalized to the $[0, 1]$ range. This sort of feature representation is known to be quite prone to hubness [19].

Datasets WiM1-WiM5 represent five non-trivial imbalanced binary classification problems defined on top of the WIKImage data [35], a set of publicly available images crawled from Wikipedia (http://www.wikipedia.org/). These images are available along with the associated text and their labels. We present the results on the textual data obtained from the labels, represented in a standard bag-of-words format, weighted by TF-IDF. The five selected datasets correspond to the presence/absence of following types of objects in the images: buildings and constructions, documents and maps, logos and flags, nature and scenic, sports.

Table 1. The summary of high-hubness datasets. Each dataset is described both by a set of basic properties (size, number of features, number of classes) and some hubness-related quantities for two different neighborhood sizes, namely: the skewness of the k-occurrence distribution (S_{N_k}), the percentage of *bad* k-occurrences (BN_k), the degree of the largest hub-point (max N_k). Also, the relative imbalance of the label distribution is given [20], as well as the size of the majority class (expressed as a percentage of the total)

Data set	size	d	C	$S_{N_{15}}$	BN_{15}	max N_{15}	RImb	$p(c_M)$
iNet3	2731	416	3	9.27	29.7%	901	0.40	50.2%
iNet4	6054	416	4	8.99	48.9%	968	0.14	35.1%
iNet5	6555	416	5	12.10	57.2%	1888	0.20	32.4%
iNet6	6010	416	6	14.26	44.4%	1901	0.26	30.9%
iNet7	10544	416	7	12.29	59.2%	1741	0.09	19.2%
iNet3Imb	1681	416	3	2.22	18.8%	136	0.72	81.5%
iNet4Imb	3927	416	4	5.44	40.5%	374	0.39	54.1%
iNet5Imb	3619	416	5	7.35	44.4%	513	0.48	58.7%
iNet6Imb	3442	416	6	3.93	44.2%	268	0.46	54.0%
iNet7Imb	2671	416	7	4.35	45.6%	301	0.46	52.1%
WiM1	1007	3182	2	12.31	36.9%	997	0.26	62.8%
WiM2	1007	3182	2	12.31	7.0%	997	0.84	92.1%
WiM3	1007	3182	2	12.31	37.3%	997	0.91	95.7%
WiM4	1007	3182	2	12.31	22.4%	997	0.60	79.9%
WiM5	1007	3182	2	12.31	4%	997	0.93	96.9%
AVG	3484.6	1338	4	9.45	36.03%	931.73	0.48	59.70%

The quantities shown in Table 1 illustrate the consequences of high dimensionality and the hubness phenomenon. Neighbor k-occurrence distribution skewness is considerable, as anything above $S_{N_k} = 1$ is usually considered high-hubness data [18]. The most frequently occurring hub points dominate and appear in unexpectedly many k-neighbor sets. For instance, the major hub on iNet3 data appears in about 30% of all neighbor sets for $k = 15$, while the major hub in WiM1 appears in nearly all neighbor sets, 997 out of 1007 for $k = 15$. The situation is somewhat more bearable for smaller neighborhood sizes in a sense that the major hubs cover fewer neighbor sets, but the overall occurrence skewness is usually higher.

Removal of such frequently occurring hub-points is possible, but their positions in the k-neighbor sets are taken by other points and this often leads to emergence of new hubs and they exhibit their own detrimental influence on data analysis. Reducing the hubness of the data is, in general, a difficult task, though certain feature types, metrics and normalization methods are known to be somewhat less prone to the dimensionality curse [19]. As there is no guarantee that the preprocessing would significantly reduce the overall hubness of the data, robust hubness-aware learning methods are to be preferred.

5.2 Classification Experiments

All experiments and classifier comparisons were run as 10-times 10-fold cross-validation. Corrected re-sampled t-test was used to determine statistical significance. The L_1 Manhattan distance was used to measure the dissimilarity between quantized image pairs and cosine similarity to determine the distance between textual feature vectors.

All algorithms were run with standard parameter configurations, as given in the respective papers. As some datasets exhibit class imbalance, the macro-averaged F_1 score, denoted by F_1^M, was used to measure classifier performance [7]. The summary of results for neighborhood size $k = 15$ is given in Table 2. In principle, ANHBNN requires slightly larger neighborhood sizes, as it provides it with more co-occurrence information. Trivially, for $k = 1$, there would be no co-occurrences at all. The algorithm also performs rather poorly for $k = 2$ or $k = 3$, which is understandable. However, as the results show, it achieves very good results for larger k values.

Table 2. An overview of algorithm performance for $k = 15$. The macro-averaged F-score F_1^M percentage is given for Augmented Naive hubness-Bayesian kNN (ANHBNN), Naive hubness-Bayesian kNN (NHBNN), kNN, hubness-weighted kNN (hw-kNN), hubness-based fuzzy nearest neighbor (h-FNN) and hubness information k-nearest neighbor (HIKNN). The symbols •/∘ denote statistically significant worse/better performance ($p < 0.05$) compared to ANHBNN. The best result in each line is in bold.

Data set	ANHBNN	NHBNN	kNN	hw-kNN	h-FNN	HIKNN
iNet3	**81.1 ± 1.1**	77.3 ± 1.6 •	74.7 ± 1.7 •	78.3 ± 2.4 •	78.4 ± 1.7 •	80.3 ± 1.3 •
iNet4	65.9 ± 1.3	63.3 ± 1.4 •	62.4 ± 1.5 •	65.5 ± 1.7	63.4 ± 1.5 •	**66.9 ± 1.4** ∘
iNet5	**62.8 ± 1.2**	59.8 ± 1.3 •	47.5 ± 1.3 •	56.1 ± 2.5 •	53.7 ± 1.6 •	59.3 ± 1.3 •
iNet6	56.1 ± 1.3	**57.0 ± 1.4** ∘	56.2 ± 1.2	56.4 ± 1.3	51.3 ± 1.5 •	56.2 ± 1.2
iNet7	**59.9 ± 1.3**	56.3 ± 0.9 •	45.3 ± 1.0 •	55.5 ± 2.8 •	56.9 ± 1.0 •	59.1 ± 0.8 •
iNet3Imb	**71.9 ± 1.4**	67.6 ± 2.1 •	65.9 ± 2.0 •	65.3 ± 1.4 •	55.0 ± 1.6 •	64.7 ± 1.3 •
iNet4Imb	**67.1 ± 1.6**	60.1 ± 1.5 •	56.7 ± 1.4 •	57.9 ± 1.6 •	45.2 ± 1.5 •	54.6 ± 1.5 •
iNet5Imb	**56.8 ± 1.6**	52.7 ± 1.8 •	35.3 ± 1.9 •	43.2 ± 1.9 •	31.1 ± 1.6 •	38.1 ± 1.9 •
iNet6Imb	52.8 ± 1.3	52.4 ± 1.5	49.2 ± 1.6 •	52.7 ± 1.6	**50.5 ± 1.7** •	54.1 ± 1.4 ∘
iNet7Imb	**47.8 ± 1.3**	46.1 ± 1.2 •	33.3 ± 1.9 •	44.0 ± 2.1 •	35.7 ± 2.1 •	42.4 ± 2.2 •
WiM1	**69.1 ± 2.8**	64.4 ± 2.7 •	66.4 ± 2.2 •	53.9 ± 3.5 •	46.0 ± 3.1 •	54.3 ± 2.8 •
WiM2	75.2 ± 1.2	**75.7 ± 1.1**	58.1 ± 1.3 •	72.7 ± 1.2 •	69.1 ± 1.1 •	68.5 ± 1.2 •
WiM3	**72.1 ± 1.4**	72.0 ± 1.5	59.5 ± 1.3 •	67.6 ± 1.7 •	69.9 ± 1.3 •	**72.1 ± 1.4**
WiM4	**71.8 ± 3.0**	70.0 ± 2.8 •	69.8 ± 2.7 •	62.7 ± 2.9 •	54.1 ± 3.1 •	56.8 ± 2.6 •
WiM5	**54.2 ± 2.9**	49.9 ± 2.7 •	49.2 ± 2.7 •	49.2 ± 2.7 •	49.2 ± 2.7 •	49.2 ± 2.7 •
AVG	**64.30**	61.64	55.30	58.73	53.96	58.26

This is not the case with NHBNN, as it was already noticed that its performance drops significantly with increasing neighborhood size, as the independence assumption between different neighbor occurrences becomes more severely violated. As this is what ANHBNN aims at improving, the neighborhood size of $k = 15$ was used in most experiment runs. A more detailed comparison of algorithm performance under varying neighborhood size is shown in Figure 6, demonstrating that the performance of the proposed approach is not very sensitive to the choice of k, once it exceeds some lower threshold value. Its performance remains stable when k is increased, suggesting that it succeeds in modeling the hub co-occurrence dependencies.

The results in Table 2 suggest that the proposed ANHBNN does indeed outperform NHBNN in the evaluated context. Furthermore, it achieves the best overall F_1^M score on the examined data. Table 3 provides a summary of pairwise classifier comparisons by showing the number of wins and statistically significant wins in each individual comparison. The proposed approach achieves the highest number of wins against any given baseline, as well as the highest total number of wins (67) and statistically significant wins (63).

Table 3. Pairwise comparison of classifiers on the examined data: number of wins (with the statistically significant ones in parenthesis)

	ANHBNN	NHBNN	kNN	hw-kNN	h-FNN	HIKNN	Total Wins
ANHBNN	–	**13 (11)**	**14 (14)**	**14 (12)**	**15 (15)**	**11 (11)**	**67 (63)**
NHBNN	2 (1)	–	14 (10)	12 (9)	12 (11)	10 (7)	50 (38)
kNN	1 (0)	1 (1)	–	3 (2)	6 (6)	4 (4)	15 (13)
hw-kNN	1 (0)	3 (1)	11 (9)	–	11 (11)	8 (5)	34 (26)
h-FNN	0 (0)	3 (1)	8 (7)	3 (2)	–	1 (0)	15 (10)
HIKNN	**3 (2)**	5 (4)	9 (9)	6 (5)	13 (11)	–	36 (31)

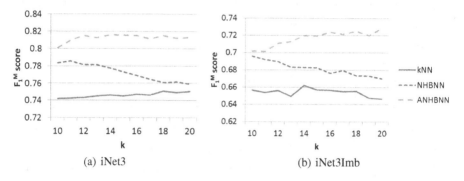

(a) iNet3   (b) iNet3Imb

Fig. 6. The influence of increasing the neighborhood size k. Neighbor occurrence dependencies induce a drop in NHBNN performance, while the ANHBNN performance slowly increases with additional neighbor occurrence and co-occurrence information.

Even though these results seem quite encouraging, some caution is still required when comparing different approaches. Namely, both NHBNN and ANHBNN assume high underlying hubness of the data and are not well suited for applications on datasets that exhibit low hubness or no hubness at all. In that sense, they are not general-purpose classification algorithms. Instead, they are tailored specifically for classifying intrinsically high-dimensional data. This is not the case with h-FNN, hw-kNN or HIKNN. Even though these remaining three methods are hubness-aware, they perform rather well even when the data exhibits only low to moderate k-occurrence distribution skewness [25]. In our initial experiments, we have determined that HIKNN is to be preferred in such cases, as for example on UCI datasets (http://archive.ics.uci.edu/ml/datasets.html).

In order to examine the nature of the observed differences in performance on the test data, we have analyzed the precision that the algorithms achieve on certain types of points. Not all points are equally hard to classify by k-nearest neighbor methods and a point characterization scheme based on the proportion of label mismatches in k-neighbor sets was recently proposed [36]. Four different point types were observed: *safe* points, *borderline* points, *rare* points and *outliers*, the latter being much more difficult to handle. A comparison between kNN, NHBNN and ANHBNN on two different datasets is shown in Figure 7. The proposed approach clearly outperforms NHBNN in

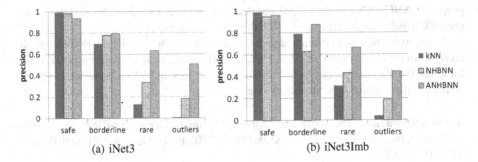

Fig. 7. Precision achieved by the classification algorithms on specific types of points: safe points, borderline points, rare points and outliers

terms of rare point and outlier classification precision and also achieves a slightly higher precision when classifying borderline points. In other words, ANHBNN achieves its improvements by being able to better handle very difficult points that lie far away from class interiors. This is a highly desired property.

6   Conclusions and Future Work

Hubness is an important aspect of the dimensionality curse that affects most k-nearest neighbor methods in severely negative ways, as hub points tend to dominate the k-neighbor sets and induce many label mismatches. Hubness-aware classification methods are required in order to properly deal with the emerging hubs.

We have proposed an extension of one such hubness-aware kNN classifier and have named it the Augmented Naive Hubness-Bayesian k-nearest neighbor (ANHBNN). The previous approach (NHBNN) failed to take the neighbor co-occurrences into account, which led to poor performance for larger neighborhood sizes. Our proposed approach (ANHBNN) overcomes this issue by adapting the Hidden Naive Bayes model to the problem of modeling neighbor k-occurrences. We have also proposed a novel set of hubness-aware weights for combining the one-dimensional estimators in the model.

We have performed an analysis of the high-dimensional neighbor co-occurrence distributions for Gaussian mixture data. The analysis has revealed several surprising facts. The distribution of the number of distinct co-occurring neighbor points becomes multimodal with modes located approximately around the multiples of $(k - 1)$. Additionally, there seems to be a phenomenon of *hub linkage*, as the tail of the co-occurrence frequency distribution becomes thicker with increasing dimensionality, indicating that some pairs of hub points co-occur frequently. The overall number of distinct co-occurring pairs also increases, which allows us to estimate more pairwise dependencies in high-dimensional data.

Our evaluation in the context of object recognition from images shows that the proposed approach clearly outperforms the compared baselines and offers additional benefits in achieving higher precision when classifying points that lie far from class interiors and are otherwise difficult to handle. Unlike NHBNN, the performance of the proposed ANHBNN classifier does not decrease when the neighborhood size k is increased, which was the main issue with the previous approach.

As many of the co-occurrence dependencies are somewhat difficult to estimate directly from the occurrence data, in our future work we intend to explore the possibilities for using the Poisson processes for neighbor occurrence modeling, in order to try and achieve a more robust k-occurrence model.

Acknowledgments. This work was supported by the Slovenian Research Agency, the ICT Programme of the EC under XLike (ICT-STREP-288342), and RENDER (ICT-257790-STREP).

References

1. Fix, E., Hodges, J.: Discriminatory analysis, nonparametric discrimination: consistency properties. Technical report, USAF School of Aviation Medicine, Randolph Field (1951)
2. Cover, T.M., Hart, P.E.: Nearest neighbor pattern classification. IEEE Transactions on Information Theory IT-13(1), 21–27 (1967)
3. Peng, J., Heisterkamp, D.R., Dai, H.K.: Adaptive quasiconformal kernel nearest neighbor classification. IEEE Trans. Pattern Anal. Mach. Intell. 26(5), 656–661 (2004)
4. Weinberger, K.Q., Saul, L.K.: Distance metric learning for large margin nearest neighbor classification. J. Mach. Learn. Res. 10, 207–244 (2009)
5. Ling Zhang, M., Hua Zhou, Z.: Ml-knn: A lazy learning approach to multi-label learning. Pattern Recognition 40, 2007 (2007)
6. Ougiaroglou, S., Nanopoulos, A., Papadopoulos, A.N., Manolopoulos, Y., Welzer-Druzovec, T.: Adaptive k-nearest neighbor classification based on a dynamic number of nearest neighbors. In: Ioannidis, Y., Novikov, B., Rachev, B. (eds.) ADBIS 2007. LNCS, vol. 4690, pp. 66–82. Springer, Heidelberg (2007)
7. Han, J.: Data Mining: Concepts and Techniques. Morgan Kaufmann Publishers Inc., San Francisco (2005)
8. Holte, R.C., Acker, L.E., Porter, B.W.: Concept learning and the problem of small disjuncts. In: Proc. 11th Int. Conf. AI, vol. 1, pp. 813–818. Morgan Kaufmann Publishers Inc. (1989)
9. van den Bosch, A., Weijters, T., Herik, H.J.V.D., Daelemans, W.: When small disjuncts abound, try lazy learning: A case study (1997)
10. Xing, Z., Pei, J., Yu, P.S.: Early prediction on time series: a nearest neighbor approach. In: Proceedings of the 21st International Jont Conference on Artifical Intelligence, IJCAI 2009, pp. 1297–1302. Morgan Kaufmann Publishers Inc., San Francisco (2009)
11. Boiman, O., Shechtman, E., Irani, M.: In Defense of Nearest-Neighbor Based Image Classification. In: CVPR (2008)
12. Bellman, R.E.: Adaptive control processes - A guided tour. Princeton University Press, Princeton (1961)
13. François, D., Wertz, V., Verleysen, M.: The concentration of fractional distances. IEEE Transactions on Knowledge and Data Engineering 19(7), 873–886 (2007)
14. Hinneburg, A., Aggarwal, C., Keim, D.A.: What is the nearest neighbor in high dimensional spaces?, pp. 506–515. Morgan Kaufmann (2000)
15. Radovanović, M., Nanopoulos, A., Ivanović, M.: Hubs in space: Popular nearest neighbors in high-dimensional data. Journal of Machine Learning Research 11, 2487–2531 (2010)
16. Aucouturier, J., Pachet, F.: Improving timbre similarity: How high is the sky? Journal of Negative Results in Speech and Audio Sciences 1 (2004)

17. Gasser, M., Flexer, A., Schnitzer, D.: Hubs and orphans - an explorative approach. In: Proceedings of the 7th Sound and Music Computing Conference, SMC 2010 (2010)
18. Miloš, R.: Representations and Metrics in High-Dimensional Data Mining. Izdavačka knjižarnica Zorana Stojanovića, Novi Sad, Serbia (2011)
19. Tomašev, N., Brehar, R., Mladenić, D., Nedevschi, S.: The influence of hubness on nearest-neighbor methods in object recognition. In: Proceedings of the 7th IEEE International Conference on Intelligent Computer Communication and Processing (ICCP), pp. 367–374 (2011)
20. Tomašev, N., Mladenić, D.: Hubness-aware shared neighbor distances for high-dimensional k-nearest neighbor classification. Knowledge and Information Systems, 1–34 (2013)
21. Radovanović, M., Nanopoulos, A., Ivanović, M.: Nearest neighbors in high-dimensional data: The emergence and influence of hubs. In: Proc. 26th Int. Conf. on Machine Learning (ICML), pp. 865–872 (2009)
22. Tomašev, N., Radovanović, M., Mladenić, D., Ivanović, M.: Hubness-based fuzzy measures for high-dimensional k-nearest neighbor classification. In: Perner, P. (ed.) MLDM 2011. LNCS (LNAI), vol. 6871, pp. 16–30. Springer, Heidelberg (2011)
23. Tomašev, N., Radovanović, M., Mladenić, D., Ivanovcć, M.: A probabilistic approach to nearest neighbor classification: Naive hubness bayesian k-nearest neighbor. In: Proceeding of the CIKM Conference (2011)
24. Tomašev, N., Mladenić, D.: Nearest neighbor voting in high-dimensional data: Learning from past occurrences. In: ICDM PhD Forum (2011)
25. Tomašev, N., Mladenić, D.: Nearest neighbor voting in high dimensional data: Learning from past occurrences. Computer Science and Information Systems 9, 691–712 (2012)
26. Schnitzer, D., Flexer, A., Schedl, M., Widmer, G.: Using mutual proximity to improve content-based audio similarity. In: ISMIR 2011, pp. 79–84 (2011)
27. Tomašev, N., Mladenić, D.: Hubness-aware shared neighbor distances for high-dimensional k-nearest neighbor classification. In: Corchado, E., Snášel, V., Abraham, A., Woźniak, M., Graña, M., Cho, S.-B. (eds.) HAIS 2012, Part II. LNCS, vol. 7209, pp. 116–127. Springer, Heidelberg (2012)
28. Tomašev, N., Radovanović, M., Mladenić, D., Ivanović, M.: The role of hubness in clustering high-dimensional data. In: Huang, J.Z., Cao, L., Srivastava, J. (eds.) PAKDD 2011, Part I. LNCS (LNAI), vol. 6634, pp. 183–195. Springer, Heidelberg (2011)
29. Tomašev, N., Radovanović, M., Mladenić, D., Ivanović, M.: The role of hubness in clustering high-dimensional data. IEEE Transactions on Knowledge and Data Engineering 99(PrePrints), 1 (2013)
30. Jiang, L., Zhang, H., Cai, Z.: A novel bayes model: Hidden naive bayes. IEEE Transactions on Knowledge and Data Engineering 21(10), 1361–1371 (October)
31. Keller, J.E., Gray, M.R., Givens, J.A.: A fuzzy k-nearest-neighbor algorithm. IEEE Transactions on Systems, Man and Cybernetics, 580–585 (1985)
32. Rish, I.: An empirical study of the naive Bayes classifier. In: Proc. IJCAI Workshop on Empirical Methods in Artificial Intelligence (2001)
33. Chickering, D.M.: Learning bayesian networks is np-complete. In: Learning from Data: Artificial Intelligence and Statistics V, pp. 121–130. Springer (1996)
34. Lowe, D.G.: Distinctive image features from scale-invariant keypoints. International Journal of Computer Vision 60(2), 91 (2004)
35. Pracner, D., Tomašev, N., Radovanović, M., Mladenić, D., Ivanović, M.: WIKImage: Correlated Image and Text Datasets. In: SiKDD: Information Society (2011)
36. Napierala, K., Stefanowski, J.: Identification of different types of minority class examples in imbalanced data. In: Corchado, E., Snášel, V., Abraham, A., Woźniak, M., Graña, M., Cho, S.-B. (eds.) HAIS 2012, Part II. LNCS, vol. 7209, pp. 139–150. Springer, Heidelberg (2012)

Fast kNN Graph Construction with Locality Sensitive Hashing

Yan-Ming Zhang[1], Kaizhu Huang[2], Guanggang Geng[3], and Cheng-Lin Liu[1]

[1] National Laboratory of Pattern Recognition, Institute of Automation,
Chinese Academy of Sciences, Beijing, China
{ymzhang,liucl}@nlpr.ia.ac.cn
[2] Department of EEE, Xi'an Jiaotong-Liverpool University, SuZhou, China
kaizhu.huang@xjtlu.edu.cn
[3] China Internet Network Information Center,
Chinese Academy of Sciences, Beijing, China
gengguanggang@cnnic.cn

Abstract. The k nearest neighbors (kNN) graph, perhaps the most popular graph in machine learning, plays an essential role for graph-based learning methods. Despite its many elegant properties, the brute force kNN graph construction method has computational complexity of $O(n^2)$, which is prohibitive for large scale data sets. In this paper, based on the divide-and-conquer strategy, we propose an efficient algorithm for approximating kNN graphs, which has the time complexity of $O(l(d + \log n)n)$ only (d is the dimensionality and l is usually a small number). This is much faster than most existing fast methods. Specifically, we engage the locality sensitive hashing technique to divide items into small subsets with equal size, and then build one kNN graph on each subset using the brute force method. To enhance the approximation quality, we repeat this procedure for several times to generate multiple basic approximate graphs, and combine them to yield a high quality graph. Compared with existing methods, the proposed approach has features that are: (1) much more efficient in speed (2) applicable to generic similarity measures; (3) easy to parallelize. Finally, on three benchmark large-scale data sets, our method beats existing fast methods with obvious advantages.

Keywords: graph construction, locality sensitive hashing, graph-based machine learning.

1 Introduction

Graph-based learning methods present an important category of machine learning methods, and have been widely used in areas like image processing, computer vision, and data mining. These methods first represent the data set by a similarity graph, and then perform on this graph the traditional learning tasks, such as clustering [25], dimensionality reduction [1], and classification [34]. As observed by many researchers [28,33], the graph construction step plays an extremely important role to this kind of methods, and has attracted much attention recently.

H. Blockeel et al. (Eds.): ECML PKDD 2013, Part II, LNAI 8189, pp. 660–674, 2013.
© Springer-Verlag Berlin Heidelberg 2013

Although various graph construction methods have been proposed to better describing the data set [19,9,7], the kNN graph still presents the most popular one in practice due to its robust performance. Given a data set, the kNN graph is constructed by connecting each item to k items which are the most similar to it under a given similarity measure. Despite its simplicity in concept, a direct implementation suffers from high computational cost. Since finding the kNN for each item needs $n - 1$ comparisons, it takes $O(n^2)$ time to construct the kNN graph. Obviously, it is too slow for large scale problems. In fact, after the recent development of fast graph-based learning methods [17,32], the speed of the graph construction method is becoming the bottleneck of graph-based methods.

To alleviate this problem, substantial efforts have been made to reduce the complexity of kNN graph construction. Early works [3,27] focused on constructing exact kNN graph. However, their complexity scales exponentially with data dimensionality. Recently, researchers switched their attention to construct approximate kNN graph and obtained encouraging results. These methods adopted techniques such as space partition tree [6,30], and local search [12]. A brief introduction to these methods can be seen in the next section.

Following the research direction, in this work, we propose a novel approximate kNN graph construction method which leads to significantly fast speed with high accuracy. Its basic idea is to divide the whole data set into small groups, and finds each item's kNN within the group it belongs to. Since the size of a group is much smaller than the size of the whole data set, the cost for finding the approximate kNN is much lower. To make this method work well, the key is to divide the data set in such a way that: (1) The pairwise similarity between items should be preserved such that similar items remain in the same group; (2) The group size should be kept as small as possible. We propose to group similar items by adopting locality sensitive hashing (LSH) [13], which enjoys a rigorous theoretical performance guarantee even in the worst case [18]. Further, we design a simple method to take control of the group size. As groups have no overlapping, the constructed graph is a union of multiple isolated small graphs. To improve the approximation quality, we repeat the division for several times to generate multiple basic approximate graphs. Finally, we combine them to yield a graph with high accuracy.

Compared with existing methods, we emphasize that the proposed method enjoys the following appealing advantages:

1. Fast and accurate. Our method has the time complexity of $O(l(d+\log n)n)$ (d is the dimensionality and l is usually a small number). This is much faster than most existing fast methods [12,6] (see the next section for details). Moreover, as shown by experiments, our method can generate good approximate kNN graphs by scanning a small proportion of pairwise distances. On several benchmarks, our method beats existing fast methods with obvious advantages.

2. Applicable to generic similarity measure. Thanks to the development of LSH, hash functions have been designed for different similarity measures, such as l_p [10], Mahalanobis distance [22], kernel similarity [21], and χ^2 distance [14].

Thus, we can conveniently use them to group data items according to the problem at hand; this presents one of the biggest advantages over the other methods such as [6] and [30], which can only be applied in Euclidean space.

3. Easy to parallelize. Since the construction of multiple basic approximate graphs are independent of each other, we can further speedup our method by constructing these basic graphs simultaneously.

The rest of this paper is organized as follows: the next section briefly reviews the existing approximation graph construction methods. Section 3 gives an introduction to the LSH technique, which provides the basic tools for our method. After that, section 4 presents a detailed description and complexity analysis of our method, while Section 5 gives comparison experimental results to validate the advantages of our method. Finally, we set out the conclusion in Section 7.

2   Related Work

The fast construction for kNN graph have been studied for a long time, and a comprehensive introduction to the graph construction methods can be found in [6]. In this section, we will briefly review several existing methods for approximate kNN graph construction, and a closely-related but different problem: the kNN search.

Chen, Fang, and Saad [6] proposed a divide-and-conquer style algorithm for constructing approximate kNN graph. The method recursively divides the data points into subsets with overlapping, then constructs one kNN graph on each small subset. The final graph is constructed by merging all the small graphs together using overlapping parts. Empirically, the authors reported their method had complexity of $O(dn^{1.22})$. Recently, Wang et al. [30] proposed another efficient algorithm using a similar idea. But data sets are recursively divided without overlapping. To increase the kNN recall, it constructs multiple basic graphs by repeating the division procedure for several times. To make good division, both methods use principle direction to partition data set. Thus, they need to compute $O(n)$ principle directions, one for each internal node. Although sub-sampling and Lanczos algorithm are adopted, it is still costly. Also, these methods can only be applied in Euclidean space.

Dong, Moses, and Li [12] proposed a fast kNN graph construction method based on local search. Its motivation lies in that a neighbor of a neighbor is also likely to be a neighbor. Initializing each node with a random set of neighbors, the method iteratively improves each node's neighborhood by exploring its neighbors' neighborhoods. Although the paper reported that its empirical cost was $O(n^{1.14})$, there is no formal guarantee on the algorithm complexity.

kNN search is a close-related but different problem which is extensively used in areas like information retrieval, pattern recognition. Given a query q, kNN search aims to find out the k most similar objects in the database. Thus, the construction of a kNN graph can be viewed as a kNN search problem where each data point itself is a query. Although many excellent works, such as space partition tree [2,26] and locality sensitive hashing (LSH) [13,10], have been done for performing efficient kNN search, the direct use of kNN search approach for the graph construction results in unfavorable results [12,30]. The differences between the two problems lie in two aspects:

1. Since the prime concern of kNN search is to reduce the query time, these methods typically build elaborate indexing structure in the training step. However, there is no separated training phase for the graph construction problem, and we can not afford the high cost for building complex indexing structure.
2. kNN search is an inductive problem, which means we can not acquire query points in the training phase. However, kNN graph construction is a transductive problem in which all query points are at hand. Thus, kNN graph construction is easier in general, and we could take advantages of its characteristic to design more efficient algorithm.

Recently, Goyal, Daumé and Guerra [15] proposed an approximate graph construction method for natural language processing problem by adopting LSH technique. Basically, it applies the method of [5] which returns k approximate neighbors for each query in constant time by using two hash tables.

3 Locality Sensitive Hashing

Locality sensitive hashing (LSH) is an efficient technique for approximate kNN search problem. Because it serves as a foundation to the proposed method, we briefly introduce it as follows.

The core idea of LSH is to hash items in a similarity preserving way, i.e., it tries to store similar items in the same buckets, while keeping dissimilar items in different buckets. In general, LSH method for kNN search has two steps: training and querying. In the training step, LSH first learns a hash function $h(x) = \{h_1(x), h_2(x), \ldots, h_m(x) : h_i(x) \in Z\}$ where m is the code length. For example, in the binary coding using linear projection, LSH adopts the hash function of form $h_i(x) = \text{sgn}(w_i^T x + b_i) \in \{-1, 1\}$, where $\{w_i, b_i\}_{i=1,\ldots,m}$ are parameters to be learned. Then, LSH represents each item in the database as a hash code by the hash mapping $h(x)$, and constructs a hash table by hashing each item into the bucket indexed by its code. In the querying step, LSH first converts the query into a hash code, and then finds its approximate kNN in the bucket indexed by the code. One attractive feature of the LSH method is that it enjoys a rigorous theoretical performance guarantee even in the worst case [18]. In practice, it can provide constant or sub-linear search time.

Although the classic LSH method builds hashing codes by random projection, recent works focus on learning data-dependent hash functions so as to generate more accurate and compact hash codes to accelerate the query. As many machine learning techniques, such LSH methods can be divided into three main categories: supervised methods [24,4], semi-supervised methods [29] and unsupervised methods [31,23,20].

Most of LSH methods described before only work for l_2 similarity measure. To apply LSH for different similarity measure, researchers have designed different hash functions, such as l_p [10], Mahalanobis distance [22], kernel similarity [21], and χ^2 distance [14]. This provides another motivation of our method: according to the problem at hand, we can conveniently choose from these methods to group data items.

4 kNN Graph Construction with LSH

4.1   Problem Definition

Given a set of n items $S = \{x_1, x_2, \ldots, x_n\}$ and a similarity measurement $\rho(x_i, x_j)$, the kNN graph for S is a directed graph that there is an edge from node i to j if and only if x_j is among x_i's k most similar items in S under ρ. Here, ρ could be any similarity measurement defined on domain S. For example, it can be cosine similarity, kernel similarity, Mahalanobis distance etc.

4.2   Algorithm

The key idea of our method is to divide the whole data set into small groups, then find each item's kNN within the group it belongs to. Since it is very hard to ensure that each item and its real kNN are located in the same group, the method only provides an approximate result. From the perspective of graph construction, it is equivalent to say the method constructs one kNN graph on each group, and then takes the union of all these small kNN graphs as the approximation kNN graph.

Since the kNN search is performed within a small subset, finding one item's kNN needs only block-sz comparisons, where block-sz is the size of the group it locates in. Assuming all groups are of equal size, the method's complexity is $O(\text{block-sz} \times n)$ instead of $O(n^2)$ for the brute-force manner.

To make the strategy work well, two conditions should be satisfied:

1. Similar items should be grouped together, which implies most of the real kNN of an item can be found in its group. Therefore, one can expect the resulting graph is a good approximation to the true kNN graph.
2. Since the complexity of the method is $O(\text{block-sz} \times n)$, to make the method efficient, each group should be as small as possible (block-sz $\ll n$).

Obviously, there is a contradiction between the two conditions: to make Condition 1 valid, we tend to use groups of big size which will violate Condition 2. Thus, a balance should be made by choosing a feasible block-sz.

In this work, we explore LSH to divide the data set with the hope that similar items will be grouped together. A straightforward way is to use LSH to hash all items into a hash table, then construct a kNN graph for each bucket. However, typical LSH methods yield highly un-even hash table which means some buckets contain a large number of items and some contain few items as shown in Fig. 1.

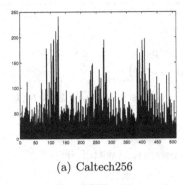 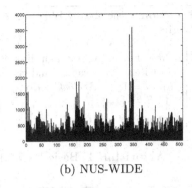

(a) Caltech256   (b) NUS-WIDE

Fig. 1. Distribution of the number of items in buckets. Experiments are performed on Caltech256 and NUS-WIDE data sets. We use the function $h(x) = \{\text{sgn}(w_i^T x + b_i)\}_{i=1,...,8}$ to hash all items into a hash table, where $\{w_i\}$ are sampled from a Gaussian distribution $N(0, I)$ and $\{b_i\}$ are the median values of the projections. Since $m = 8$, the hash table has 512 buckets.

The problems brought by this phenomenon are: (a) the *k*NN graph constructed on a small bucket will fail the Condition 1, and suffer a poor precision; (b) Constructing a *k*NN graph on a large bucket will fail the Condition 2, and suffer a high cost because of the complexity $O(\text{block-sz}^2)$. Actually, the direct use of LSH has been observed to result in unsatisfying performance[12,30], which is also verified by our experiments.

We propose an efficient way to overcome the problem and obtain equal size groups. Given the data set's hash code matrix $Y \in \{0,1\}^{n \times m}$ where the *i*th row $y_i \in \{0,1\}^m$ is the hash code of x_i, we first project items' hash codes onto a random direction $w \in \Re^m$, and get $p = Yw$. Then we sort items by their projection values to get the sequence $\{x_{\pi_1}, x_{\pi_2}, \ldots, x_{\pi_n}\}$ with $\{p_{\pi_1} \le p_{\pi_2} \cdots \le p_{\pi_n}\}$. Finally, we obtain $n/\text{block-sz}$ groups $\{S_i\}$ with equal size of block-sz by defining $S_i = \{x_{\pi_{(i-1) \times block-sz+1}}, \ldots, x_{\pi_{i \times block-sz}}\}$. Because items in the same bucket will have same projection values, they will remain in the same group with high probability. We summarize the procedure described above in Algorithm 1.

Algorithm 1 generates a basic approximation to the *k*NN graph. However, it is just an union of $n/\text{block-sz}$ isolated small graphs, and may suffer from a low accuracy. To improve it, we repeat Algorithm 1 using different hash functions for multiple times and then combine the resulting graphs. Denoting the approximate *k*NN of x found in the *i*-th iteration by $N_i(x)$, we can obtain at most $k \times l$ NN different candidates as $\{N_i(x)\}_{i=1,...,l}$ after l iterations. Obviously, by increasing l, it will cover more and more true *k*NN of x. The final graph is achieved by connecting each x to k items in $\{N_i(x)\}_{i=1,...,l}$ which are the nearest to it.

To further boost our method, we use a one-step neighbor propagation procedure which is widely used in efficient *k*NN graph construction methods [6,12,30]. It is based on the following observation: if x is similar to y and y is similar to z, then it is likely that x is similar to z. It implies that we could improve an

---

Function: basic_ann_by_lsh(X, k, m, block-sz)
begin
   $Y = LSH(X, m)$
   Project Y onto a random direction w, $p = Yw$
   Sort items by p values, and get $\{x_{\pi_1}, x_{\pi_2}, \ldots, x_{\pi_n}\}$
   for $i = 1, \ldots, n/block\text{-}sz$ **do**
      $S_i = \{x_{\pi_{(i-1)\times block-sz+1}}, \ldots, x_{\pi_{i\times block-sz}}\}$ $g_i = $ brute_force_kNN(S_i, k)
   return: $G = \bigcup\{g_i\}$

Algorithm 1. Basic kNN Graph Construction with LSH

Algorithm: Approximate kNN Graph Construction with LSH
Input: X, k, l, m, block-sz
begin
   for $i = 1, \ldots, l$ **do**
      $G_i = $ basic_ann_by_lsh(X, k, m, block-sz)
   Combine $\{G_1, G_2, \ldots, G_l\}$ to get G
   Refine G by one step neighbor propagation
   return: G

Algorithm 2. Main Algorithm

item's approximate kNN by selecting from its neighbors and the neighbors of its neighbors. Formally, denoting the approximate kNN of x found by now as $N(x)$, we update it by reselecting x's kNN from the set $N(x) \cup \{\cup_{v \in N(x)} N(v)\}$. As we will see in the next section, this simple local search procedure is a good complementary strategy with LSH, and gives significant improvement to the final results. The whole method is summarized in Algorithm 2.

In the proceeding of multiple basic graph constructions and neighbor propagation, some pairwise distances could be computed for many times. Thus, another technique to speedup the method is to store all the pairwise distances that have been computed so far in a hash table. When requiring the distance between two items, we first check the hash table if it has been evaluated before. However, we do not actually adopt this technique as the hash table may be too large to store. For example, assuming $n = 1,000,000$ and 10% of all the n^2 pairwise distances have been computed, the hash table needs about 8 GB memory to store these values.

4.3 Complexity Analysis

We briefly analyze the complexity of the proposed method as follows.

1. Complexity for computing LSH(S,k), projecting and sorting are $O(nmd)$, $O(nm)$ and $O(n \log n)$ respectively. Complexity for the construction of all g_i ($1 \leq i \leq n$/block-sz) is $O(nd$block-sz$)$. Thus, the complexity of Algorithm 1 is $O(n(md + m + d$block-sz$) + n \log n)$.

2. Combining l graphs needs $O(lnk)$ operations.
3. Each item has at most $k^2 + k$ candidate neighbors (k neighbors and k^2 neighbors of neighbor), and needs $O(dk^2)$ operations to find out the k nearest neighbors. Thus, the complexity of neighborhood propagation step is $O(ndk^2)$

Summarizing the above analysis, the complexity of our algorithm is $O(ln(md + d\text{block-sz} + \log n + k) + ndk^2)$. Since k, m and block-sz are small in practice, the actual complexity is $O(ln(d + \log n))$.

5 Experiments

To empirically validate the effectiveness of the proposed approximate kNN graph construction method, we provide results on several data sets. The primary goal is to verify: (1) The proposed method is much faster than existing ones; (2) Our method can generate good approximate graph by scanning only a small part of the total pairwise distances; (3) For many classification tasks, an approximate graph is enough to achieve good classification accuracy.

5.1 Experimental Setting

Data Set. We use 3 popular image data sets to evaluate the proposed method: Caltech256, Imagenet and NUS-WIDE. Caltech256 [16] is a benchmark for object classification, while Imagenet [11] and NUS-WIDE [8] are two real-world web image databases. Features are extracted by the bag-of-word (BOW) model based on SIFT descriptions. For Caltech256, we perform k-means clustering of SIFT descriptors to form a visual vocabulary of 1024 visual words. Then, SIFT descriptors are quantized into visual words using the nearest cluster center. BOW features for Imagenet and NUS-WIDE are directly downloaded from websites[1,2]. For the Imagenet database, we only adopt the first 100 categories of the training set. We summarize the size and dimensionality of the data sets in Table 1.

Table 1. Data sets description

	Caltech256	Imagenet	NUS-WIDE
Size	30607	120955	269643
Dimensionality	1024	1000	500

Comparison Methods. We compare our method, denoted by LSH, with 3 popular approximate kNN graph construction methods: OverTree [6], MultiTree [30] and NN-Descent [12]. We employ the direct use of LSH to construct approximate

---

[1] http://www.image-net.org/challenges/LSVRC/2010/download-public
[2] http://lms.comp.nus.edu.sg/research/NUS-WIDE.htm

kNN graph as the baseline method [3], denoted by DirectLSH. To examine the effect of the one-step neighbor propagation, we will also report the results of our algorithm without the neighbor propagation procedure, denoted by LSH–. The setting for different methods is given as follows.

1. For LSH, LSH– and DirectLSH, we adopt a LSH coding method proposed in [23] to divide data points. It is a nonlinear method based on spectral decomposition of a low-rank graph matrix. To efficiently generate different hash tables, we randomly sample 20 items as the anchor points for each iteration. The length of the hash code is set to $m = \lceil \log_2(n/\text{block-sz}) \rceil + 1$. For LSH and LSH–, block-sz (the maximum set size for performing brutal force kNN) is set to 100.
2. For OverTree, the specific algorithm we use is kNN-glue, and the codes are provided by authors [4]. block-sz is set to 100.
3. For MultiTree, we randomly sample 20 points on each internal node, compute the principal direction on the sampled set by Lanczos algorithm, then use it to perform the projection. block-sz is set to 100.
4. For NN-Descent, there is no hyperparameter to set, and the codes of NN-Descent are provided by authors [5].

Evaluation Criterion. To evaluate the quality of an approximate kNN graph G', we define the following accuracy measurement $acc(G') = \frac{|E(G') \cap E(G)|}{|E(G)|}$, where G is the exact kNN graph, $E(\cdot)$ denotes the set of the directed edges in the graph and $|\cdot|$ denotes the cardinality of the set. The exact kNN graph is computed by the brute-force method, and the time for constructing 10NN graph is shown in Table 2.

Table 2. 10NN graph construction time for the brute-force method

	Caltech256	Imagenet	NUS-WIDE
Time (sec.)	1108	17991	43696

Experiment Environment. All experiments are run on a Linux server with two 8-core 2.66GHz CPUs and 128G RAM. Parallelization is disabled for all method, and each method is restricted on one thread.

5.2   Experimental Results

Time versus Approximate Accuracy. We test various methods by comparing their time for constructing graphs with different accuracies. To generate

---

[3] In each iteration, we use LSH to group data, then construct one kNN graph for each bucket.

[4] http://www.mcs.anl.gov/~jiechen/software.html#knn

[5] http://code.google.com/p/nndes/

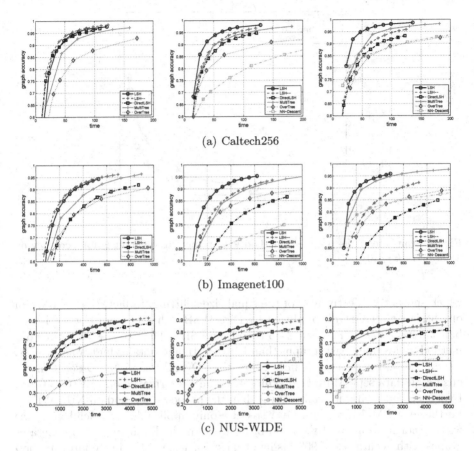

(a) Caltech256

(b) Imagenet100

(c) NUS-WIDE

Fig. 2. Graph accuracy versus construction time for different methods on 3 data sets. The 3 columns correspond to the results of building approximate 1NN, 10NN and 20NN graphs respectively.

different accuracy graphs, (1) For LSH, LSH–, DirectLSH and MultiTree, we use different times of divisions (parameter l). (2) For OverTree, we vary the overlapping factor from $[0.1\ 0.5]$. (3) For NN-Descent, we use different sample rates. The performance comparison is shown in Fig 2. The horizontal axis represents the time (in seconds) consumed for graph construction, and the vertical axis represents the graph accuracy defined above. Each row of the figure corresponds to one data set, and each column corresponds to a choice of k value. As the NN-Descend method fails for small k, its performance curve is not shown for $k = 1$.

From Fig. 2, we can clearly see the superiority of our method. For $k = 10$ and $k = 20$, LSH is consistently better than LSH–, and has similar performance for $k = 1$, which shows the effectiveness of the neighbor propagation procedure. On the other hand, the performance of LSH– is always better than DirectLSH,

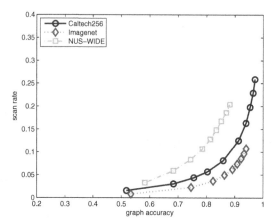

Fig. 3. Scan rate of the proposed method with respect to different graph accuracy on 3 data sets($k = 10$)

which shows the effectiveness of the equal size splitting. In fact, the accuracy of one basic isolated graph produced by DirectLSH is higher than that of LSH−, but it takes DirectLSH about 2 - 3 time than LSH−. Thus, in the same time our method builds more basic graphs than DirectLSH, and obtains higher accuracies.

Our method is at least 4 times faster than OverTree and NN-Descent methods on all data sets. MultiTree exhibits similar performance on Caltech256 and Imagenet, but is dramatically slower than the proposed method on NUS-WIDE. Furthermore, MultiTree and OverTree are only applicable for Euclidean space with l_2 norm. Compared with the brute force method, our approach generates graphs with accuracies of 90% using at most 5% time of the brute force method on Caltech256 and Imagenet, and using less than 10% time of the brute force method on NUS-WIDE.

Approximate Accuracy vs. Scan Rate. Because the dominant cost of kNN graph construction is the pairwise distance calculation, we evaluate our method by reporting the ratio of the number of actual distances calculated to the total number of pairwise distances $n(n − 1)/2$. The values for graphs of different accuracies under different data sets are shown in Fig. 3. We can observe that: (1) Our method generates good approximate kNN graphs by scanning only a small proportion of the distances, but to generate graphs with high accuracy, the method still has to scan a large number of distances. This is because: neighbors of some items are far away, and locality-sensitive hashing in nature is not good at searching for such points. (2) Different data sets have very different properties. For example, constructing approximate kNN graphs for Imagenet is much easier than for NUS-WIDE.

In experiments, we found the scan rate of MultiTree is even lower than ours. However, because of computing principle directions, in constructing a basic approximate graph, it takes about 3 times of the running time of Algorithm 1.

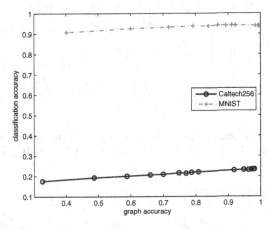

Fig. 4. Variance of classification accuracy with respect to graph accuracy

Thus, its overall speed is slow. Actually, by increasing the number of anchor points and using kmeans to choose anchor points (see experimental setting for LSH), our method can also achieve better scan rate. But this will make the cost of learning the hash coding much higher, and decrease the overall performance. Therefore, instead of pursuing the accurate but complex indexing structure as in *k*NN search, a trade-off must be made between the speed and accuracy in building the indexing structure for the graph construction problem.

Approximate Accuracy vs. Classification Accuracy. As we shown in the previous experiments, it is relatively easy to yield a good approximate graph, but hard to obtain graph with very high accuracy. The progress towards the real *k*NN graph becomes slow when *l* is large. Actually, this is a common problem shared by all approximate *k*NN graph construction methods. But fortunately for machine learning tasks, like classification, what we really care is not graph accuracy, but the classification accuracy. As long as edges found by our method are locally enough, we can expect the approximate graph leads to good description to the data set, and therefore yields similar results as the true *k*NN graph.

In this experiment, we examine the variance of classification accuracy with respect to the graph accuracy. First, approximate 20NN graphs with different accuracies are generated by our method on Caltech256 and MNIST(a popular handwritten digit data set with $n = 70000$, $d = 784$). We randomly sample 15420 nodes for Caltech256 and 700 nodes for MNIST as labeled data, and use Gaussian random fields [33] to classify the rest nodes. The classification accuracies are reported in Fig. 4.

As observed from the results, the classification accuracy is very robust with the graph accuracy. Actually, on Caltech256, the classification accuracy is 22.0% on graph of accuracy 80%, while is 23.4% on exact 20NN graph; on MNIST, the classification accuracy is 93.8% on graph of accuracy 80%, while is 93.9% on exact 20NN graph. This implies that for many applications, it is sufficient to construct a graph with a reasonable accuracy.

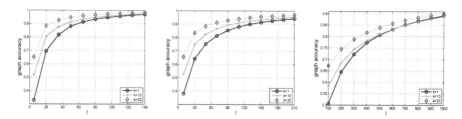

Fig. 5. Variance of classification accuracy with respect to the parameter l for building approximate 10NN graphs on Caltech256, Imagenet and NUS-WIDE respectively

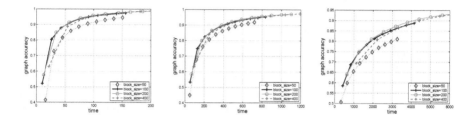

Fig. 6. Effect of the size of block for building approximate 10NN graphs on Caltech256, Imagenet and NUS-WIDE respectively

Approximate Accuracy vs. The Number of Divisions. In the proposed method, we use the number of divisions l to control the speed and accuracy of graphs. In this experiment, we examine how the graph accuracy varies with the number of divisions. For each data set, we construct $k = \{1, 10, 20\}$ NN graphs with different l, and report the accuracy of resulting graphs. The results are shown in Fig. 5.

Effect of the Size of Block. From the analysis in Section 4.2, we should choose a feasible block size to make a balance between the accuracy and the cost of basic graph. In this experiment, we examine how the block size affects the performance of our method. We construct approximate 10NN graphs on Caltech256, Imagenet and NUS-WIDE data sets with block-sz $\in \{50, 100, 200, 400\}$. For each data set, we build graphs with different accuracies by using different number of divisions, and report the building time and accuracies in Fig 6. As we can see, block-sz $\in \{100, 200\}$ are consistently better than $\{50, 400\}$ for all data sets. This is because block-sz$= 50$ is too small to generate a good basic graph while the cost for building basic graph with block-sz$= 400$ is too high. Actually, for all the previous experiments, we have fixed block-sz to 100.

6   Conclusion

In this paper, we have proposed a fast approximate kNN graph construction method for generic similarity measures. It engages the LSH technique to efficiently partition items into small groups, and then constructs kNN graph on each group. As shown by the experiments, our method can efficiently generate graphs with good accuracies by just computing a small proportion of distances, which is sufficient for many learning tasks. In addition to its high efficiency, the proposed algorithm is ready to be applied under many similarity measures, and is also natural to be parallelized.

Acknowledgements. This works have been supported in part by the National Basic Research Program of China (973 Program) Grant 2012CB316301, the National Natural Science Foundation of China (NSFC) Grants 61203296, 61075052 and 61005029, Tsinghua National Laboratory for Information Science and Technology (TNList) Cross-discipline Foundation.

References

1. Belkin, M., Niyogi, P.: Laplacian eigenmaps for dimensionality reduction and data representation. Neural Computation 15(6), 1373–1396 (2003)
2. Bentley, J.L.: Multidimensional binary search trees used for associative searching. Communications of the ACM 18(9), 509–517 (1975)
3. Bentley, J.L.: Multidimensional divide-and-conquer. Communications of the ACM 23(4), 214–229 (1980)
4. Bronstein, M.M., Fua, P.: LDAHash: Improved matching with smaller descriptors. IEEE Transactions on Pattern Analysis and Machine Intelligence 34(1), 66–78 (2012)
5. Charikar, M.S.: Similarity estimation techniques from rounding algorithms. In: Proceedings of the Thiry-Fourth Annual ACM Symposium on Theory of Computing (2002)
6. Chen, J., Fang, H., Saad, Y.: Fast approximate kNN graph construction for high dimensional data via recursive lanczos bisection. The Journal of Machine Learning Research 10, 1989–2012 (2009)
7. Cheng, B., Yang, J.C., Yan, S.C., Fu, Y., Huang, T.: Learning with l1-graph for image analysis. IEEE Transaction on Image Processing 19, 858–866 (2010)
8. Chua, T.-S., Tang, J., Hong, R., Li, H., Luo, Z., Zheng, Y.-T.: NUS-WIDE: A real-world web image database from national university of singapore. In: Proceedings of ACM Conference on Image and Video Retrieval (2009)
9. Daitch, S.I., Kelner, J.A., Spielman, D.A.: Fitting a graph to vector data. In: Proceedings of the International Conference on Machine Learning (2009)
10. Datar, M., Immorlica, N., Indyk, P., Mirrokni, V.S.: Locality-sensitive hashing scheme based on p-stable distributions. In: Proceedings of the Annual Symposium on Computational Geometry (2004)
11. Deng, J., Dong, W., Socher, R., Li, L.-J., Li, K., Fei-Fei, L.: ImageNet: A Large-Scale Hierarchical Image Database. In: IEEE Conference on Computer Vision and Pattern Recognition (2009)
12. Dong, W., Charikar, M., Li, K.: Efficient k-nearest neighbor graph construction for generic similarity measures. In: Proceedings of the International Conference on World Wide Web (2011)
13. Gionis, A., Indyk, P., Motwani, R.: Similarity search in high dimensions via hashing. In: Proceedings of the International Conference on Very Large Data Bases (1999)

14. Gorisse, D., Cord, M., Precioso, F.: Locality-sensitive hashing for chi2 distance. IEEE Transactions on Pattern Analysis and Machine Intelligence 34(2), 402–409 (2012)
15. Goyal, A., Daumé III, H., Guerra, R.: Fast large-scale approximate graph construction for nlp. In: Proceedings of the 2012 Joint Conference on Empirical Methods in Natural Language Processing and Computational Natural Language Learning (2012)
16. Griffin, G., Holub, A., Perona, P.: Caltech-256 object category dataset. Technical report, California Institute of Technology (2007)
17. Herbster, M., Pontil, M., Galeano, S.R.: Fast predciton on a tree. In: Advances in Neural Information Processing Systems (2008)
18. Indyk, P., Motwani, R.: Approximate nearest neighbors: towards removing the curse of dimensionality. In: Proceedings of the Annual ACM Symposium on Theory of Computing (1998)
19. Jebara, T., Wang, J., Chang, S.F.: Graph construction and b-matching for semi-supervised learning. In: Proceedings of the International Conference on Machine Learning (2009)
20. Kong, W., Li, W.J.: Isotropic hashing. In: Advances in Neural Information Processing Systems (2012)
21. Kulis, B., Grauman, K.: Kernelized locality-sensitive hashing for scalable image search. In: IEEE International Conference on Computer Vision (2009)
22. Kulis, B., Jain, P., Grauman, K.: Fast similarity search for learned metrics. IEEE Transactions on Pattern Analysis and Machine Intelligence 31(12), 2143–2157 (2009)
23. Liu, W., Wang, J., Kumar, S., Chang, S.F.: Hashing with graphs. In: Proceedings of the International Conference on Machine Learning (2011)
24. Salakhutdinov, R., Hinton, G.: Semantic hashing. International Journal of Approximate Reasoning 50(7), 969–978 (2009)
25. Shi, J., Malik, J.: Normalized cuts and image segmentation. IEEE Transactions on Pattern Analysis and Machine Intelligence 22(8), 888–905 (2002)
26. Uhlmann, J.K.: Satisfying general proximity/similarity queries with metric trees. Information Processing Letters 40(4), 175–179 (1991)
27. Vaidya, P.M.: An O(nlogn) algorithm for the all-nearest-neighbors problem. Discrete & Computational Geometry 4(1), 101–115 (1989)
28. Von Luxburg, U.: A tutorial on spectral clustering. Statistics and Computing 17(4), 395–416 (2007)
29. Wang, J., Kumar, S., Chang, S.F.: Semi-supervised hashing for scalable image retrieval. In: IEEE Conference on Computer Vision and Pattern Recognition (2010)
30. Wang, J., Wang, J., Zeng, G., Tu, Z., Gan, R., Li, S.: Scalable k-NN graph construction for visual descriptors. In: IEEE Conference on Computer Vision and Pattern Recognition (2012)
31. Weiss, Y., Torralba, A., Fergus, R.: Spectral hashing. In: Advances in Neural Information Processing Systems (2008)
32. Zhang, Y.M., Huang, K., Liu, C.L.: Fast and robust graph-based transductive learning via minimum tree cut. In: IEEE International Conference on Data Mining (2011)
33. Zhu, X.: Semi-supervised learning literature survey. Technical report, Computer Science, University of Wisconsin-Madison (2008)
34. Zhu, X., Ghahramani, Z., Lafferty, J.: Semi-supervised learning using Gaussian fields and harmonic functions. In: Proceedings of the International Conference on Machine Learning (2003)

Mixtures of Large Margin Nearest Neighbor Classifiers

Murat Semerci and Ethem Alpaydın

Department of Computer Engineering
Boğaziçi University
TR-34342 Istanbul, Turkey
{murat.semerci,alpaydin}@boun.edu.tr

Abstract. The accuracy of the k-nearest neighbor algorithm depends on the distance function used to measure similarity between instances. Methods have been proposed in the literature to learn a good distance function from a labelled training set. One such method is the large margin nearest neighbor classifier that learns a global Mahalanobis distance. We propose a mixture of such classifiers where a gating function divides the input space into regions and a separate distance function is learned in each region in a lower dimensional manifold. We show that such an extension improves accuracy and allows visualization.

Keywords: Nearest Neighbor Classifier, Margin Loss, Distance Learning.

1 Introduction

Nonparametric, memory-based methods, such as the k-nearest neighbor classifier, interpolates from past similar cases. This requires a good distance (or inversely, similarity) measure to determine the relevant subset of the training set. Given two d-dimensional instances $x_i, x_j \in \Re^d$, the Euclidean distance, or its square, is the best known:

$$D_E(x_i, x_j) = \|x_i - x_j\|_2^2 = (x_i - x_j)^\top (x_i - x_j)$$

The Euclidean distance assumes that all features have the same variance and that they are uncorrelated. If this is not the case and there is a covariance structure as given by a covariance matrix $\mathbf{S}$, one should use the Mahalanobis distance:

$$D_{\mathbf{M}} = (x_i - x_j)^\top \mathbf{M}(x_i - x_j)$$

where $\mathbf{M} \equiv \mathbf{S}^{-1}$. The Euclidean distance is a special case where $\mathbf{M} = \mathbf{S} = \mathbf{I}$, the identity matrix.

$\mathbf{M}$ is a $d \times d$ symmetric, positive semi-definite matrix and when d is large, not all features may be informative and/or there may be strong correlations

H. Blockeel et al. (Eds.): ECML PKDD 2013, Part II, LNAI 8189, pp. 675–688, 2013.

between features, and one may want to do dimensionality reduction by a low-rank approximation. Any symmetric Mahalanobis matrix can be factorized as $M = L^\top L$, where L is an $e \times d$ projection matrix and $e \leq d$:

$$
\begin{aligned}
D_M(x_i, x_j) &= (x_i - x_j)^\top M(x_i - x_j) = (x_i - x_j)^\top L^\top L(x_i - x_j) \\
&= (L(x_i - x_j))^\top L(x_i - x_j) = (Lx_i - Lx_j)^\top (Lx_i - Lx_j) \\
&= \|z_i - z_j\|_2^2 = D_E(z_i, z_j) \\
&= D_L(x_i, x_j)
\end{aligned}
\tag{1}
$$

That is, using such a low-rank ($e < d$) approximation is equivalent to projecting the data to this new e-dimensional space as, $z_i = Lx_i, z_i \in \Re^e$ and using Euclidean distance there.

In a high dimensional problem, different regions of the input space may exhibit different dependencies and variances and hence, instead of a single, global metric, it may be more appropriate to use different metrics in different regions. Besides, because regions have local structures, dimensionality can be further decreased. In this study, we propose a framework where the input space is partitioned into regions and different projection matrices are learnt in different regions.

The rest of this paper is organized as follows: We give a brief literature survey of related work in Section 2, and among these, the closest to our work are the Large Margin Nearest Neighbor (LMNN) algorithm—that learns M—and Large Margin Component Analysis (LMCA) algorithm—that learns L—which are discussed in more detail Section 3. Our proposed extension of mixtures of LMNN—that learns multiple M_m or L_m in different parts of the input space—is given in Section 4 which are discussed in more detail in Section 3. We discuss our experimental results in Section 5 and conclude in Section 6.

2   Related Work

In the literature, many methods have been proposed to train a Mahalanobis matrix M or a projection matrix L. Some methods train multiple Mahalanobis or projection matrices, which can be per-class or per-exemplar. Below, chronologically we give a brief summary of some methods.

One of the first distance metric learning algorithm is given by Xing et al. in [1] who define a convex optimization problem to find a Mahalanobis matrix. The instances in the data set form two disjoint subsets of similar and dissimilar pairs and a Mahalanobis matrix is trained such that the distance between similar points is minimized while the dissimilar points are at least 1 unit way from each other.

Neighborhood Components Analysis (NCA) is a stochastic gradient-based algorithm to find a linear projection matrix that minimizes the leave-one-out classification error of the nearest neighbor classifier in the new space (see [2]). A differentiable objective function is defined on the soft neighbor assignments in the linearly projected space. The projection matrix can be also used for dimensionality reduction. Slakhutdinov and Hinton extend NCA to embed a nonlinear projection in [3], where a multilayer neural network is trained for this purpose.

Frome et al. in [4] propose a method to train a weight vector for each image to calculate the global distance between images using the feature vector which is a concatenation of local patch distances between the images. A large margin classifier is trained over the local distance feature vectors in a convex programming problem. Since the trained distances are not guaranteed to be compatible, a logistic model is trained over them to estimate posterior probabilities. They are ranked and the query image is assigned to the class of the image with the highest rank. In [5], they improve the algorithm by training a globally consistent local distance functions such that no second-level classifier is required to be trained. They redefine the problem using a convex optimization formulation. Since all the weight vectors are trained together, the final distance estimates are compatible with each other.

Chang and Yeung in [6] train an affine function per instance that provides smooth transitions between instances. A variant of regularized moving least squares is applied in a semi-supervised setting. Although the objective function has a closed-form solution, it becomes intractable for data sets with many instances and an approximation algorithm is given.

Davis et al. in [7] study metric learning from an information-theoretic point of view. They define the optimal Gaussian distribution whose covariance matrix satisfies the distance constraints defined on the instance pairs; the distance between instance pairs belonging to the same class must be smaller than a predetermined threshold and the instance pairs from different classes must be away from each other by at least a specified distance. Then, the problem is converted into a LogDet (logarithm of determinant) optimization problem that is convex.

The Large Margin Nearest Neighbor algorithm (LMNN) (see [8] and [9]) defines a semi-definite programming problem over the squared Mahalanobis distances of target and impostor sets—the impostors are the closest instances with different class labels and targets are the closest instances with the same label. Distances to the target neighbors are minimized while the distances to impostors are penalized if they are within a margin, which is a safe distance away from the furthest target neighbor. This is a convex programming instance and hence has a unique solution. A multiple metrics version of LMNN where a metric is trained for each class is also studied in [9].

Large Margin Component Analysis (LMCA) in [10] is a variant of LMNN and finds a lower dimensional rectangular projection matrix $\mathbf{L}$ instead of a square Mahalanobis matrix $\mathbf{M}$ (see Equation 1). Both methods share the same objective function but since LMCA defines the squared distance in terms of the projection matrix, this is no longer a convex optimization problem and LMCA converges to a local optimum. Our proposed method is an extension of LMNN and LMCA, and these methods will be discussed in more detail in Section 3, before we discuss our method in Section 4.

Malisiewicz and Efros in [11] focus on training per-exemplar metric for image retrieval. They also work on the concatenated vector of segment distances. Their algorithm consists of two parts. Sequentially, they train metrics per-exemplar given the nearest neighbors and then they re-assign the nearest neighbors given the trained metrics. They specify a margin on the trained distance function values as used in Support Vector Machines (SVM). The neighbors that are away less than one unit distance are called the support set and the precision of classification result is determined by this support set.

Zhan et al. in [12] propose to learn instance-specific distances by using metric propagation. A smooth function (such as a Gaussian kernel) is propagated between the labelled and unlabelled instances. A regularized loss function is defined such that the distances between instances of the same label are minimized with respect to the given neighborhood relationships; the distance function trained using labeled instances can then be used for unlabeled instances. The proposed framework is formulated as a convex problem.

Chen and Sun in [13] propose a hierarchical LMNN algorithm. Overlapping ratio is defined to measure the confusion between classes and if this ratio is above a threshold, overlapping classes are grouped in the same cluster. The hierarchy describes how to map a test instance to a cluster. A Mahalanobis matrix is trained for each cluster and a given test instance is classified by using its cluster's metric matrix.

Noh et al. in [14], aim reducing the expectation error that the nearest neighbor has a different label. They show that if the distributions of the two classes are known, the difference between the empirical nearest neighbor error and the optimal nearest neighbor error based on asymptotic Bayes error is caused by finite sampling. They propose Generative Local Metric Learning which defines a convex problem if the divergence function used is also convex.

Chang in [15] proposes an iterative metric boosting method. An upper bound function on the leave-one-out error for the nearest neighbor classification is defined and is minimized. The misclassified instances are weighted and the Mahalanobis matrix is optimized with respect to these weights. An eigenvalue problem is solved to find the Mahalanobis matrix.

Bunte et al. in [16] propose Limited Rank Matrix Learning which is a recent algorithm that extends Learning Vector Quantization. It learns class prototypes and a low-rank projection matrix at the same time, iteratively. The projection matrix is trained to be discriminative by optimizing a cost function that maps instances close to their class prototypes and away from the other class prototypes, using a criterion similar to that of Linear Discriminant Analysis.

Wang et al., in [17], propose to combine multiple metric matrices to form per-exemplar local metrics. The algorithm consists of two steps. First, a weight matrix is trained such that each data point can be expressed as a linear combination of pre-defined anchor points. The cluster means are defined as the anchors and any clustering algorithm can be used for defining the anchors, i.e., k-means. Then, a metric learning algorithm, a modified version of Multiple-metric LMNN, is used to train a metric for each anchor point. The per-exemplar local metrics are combinations of these anchor metrics whose weights are determined by the weight matrix.

Table 1. The overall summary of distance metric learning methods

Method	Convexity	Type of Metric	Distance
Xing et al. [1]	Yes	Single	Mahalanobis
Golberger et al. [2]	No	Single	Projection
Salakhutdinov and Hinton [3]	No	Single	Nonlinear Projection
Frome et al. [4]	Partial	Per-Exemplar	Weight Vector
Frome et al. [5]	Yes	Per-Exemplar	Weight Vector
Chang and Yeung [6]	Yes	Per-Exemplar	Mahalanobis
Davis et al. [7]	Yes	Single	Mahalanobis
Weinberger and Saul [8]	Yes	Single	Mahalanobis
Weinberger and Saul [9]	Yes	Per-Class	Mahalanobis
Torresani and Lee [10]	No	Single	Projection
Malisiewicz and Efros [11]	Partial	Per-Exemplar	Weight Vector
Zhan et al. [12]	Yes	Per-Exemplar	Weight Vector
Chen and Sun [13]	Partial	Per-Cluster	Mahalanobis
Noh et al. [14]	Yes	Single	Mahalanobis
Chang [15]	Partial	Single	Mahalanobis
Bunte et al. [16]	No	Per-Class / Single	Vector / Projection
Wang et al. [17]	Partial	Multiple	Mahalanobis
Wu et al. [18]	Yes	Per-Exemplar	Mahalanobis

The Bregman distance functions are trained in a SVM-like manner in [18]. Since the general Bregman distances are not metrics, the authors work with a particular set of convex Bregman functions to ensure that they train a metric. Kernelizing the Bregman distances, they solve a quadratic problem.

The methods are summarized in Table 1. Note that partial convexity means that the algorithm consists of some sub-problems or steps and that not all of them are convex.

3   Large Margin Nearest Neighbor (LMNN) and Large Margin Component Analysis (LMCA) Algorithms

The Large Margin Nearest Neighbor (LMNN) trains a global Mahalanobis matrix that evaluates distances discriminatively (see [9] and [8]). Let us define our data set as pairs (x_i, y_i), where x_i is the input instance vector and y_i is the corresponding label. The notation $j \rightsquigarrow i$ (j leads to i) means x_j is a target neighbor of x_i. A *target* is a neighbor with the same (correct) class label whereas an *impostor* is a neighbor with different (wrong) class label. For accurate nearest neighbor classification, targets must be closer than the impostors.

Using the label information, a Mahalanobis matrix $\mathbf{M}$ can be trained to

$$\text{minimize} \quad (1 - \mu) \sum_{i,j \rightsquigarrow i} (\boldsymbol{x}_i - \boldsymbol{x}_j)^{\top} \mathbf{M}(\boldsymbol{x}_i - \boldsymbol{x}_j) + \mu \sum_{i,j \rightsquigarrow i,l} (1 - y_{il})\xi_{ijl}$$

$$\text{subject to} \quad (\boldsymbol{x}_i - \boldsymbol{x}_l)^{\top} \mathbf{M}(\boldsymbol{x}_i - \boldsymbol{x}_l) - (\boldsymbol{x}_i - \boldsymbol{x}_j)^{\top} \mathbf{M}(\boldsymbol{x}_i - \boldsymbol{x}_j) \geq 1 - \xi_{ijl}$$

$$\xi_{ijl} \geq 0$$

$$\mathbf{M} \succeq 0 \tag{2}$$

where $y_{il} = 1$ if $y_i = y_l$, which are the labels of $\boldsymbol{x}_i$ and $\boldsymbol{x}_l$, and $y_{il} = 0$ otherwise. The first term is the sum of distances of each instance to its target neighbors which we want to be minimum and the second term penalizes close impostors: For any instance i where l is an impostor and j is a target, we would like the distance to the impostor be at least one unit more than the distance to a target. If this is not satisfied, there is a slack and we minimize the sum of such slacks.

Equation 2 defines a positive semi-definite programming problem and there is a unique minimum. After some manipulations, the loss function can be rewritten as:

$$E = (1 - \mu) \sum_{i,j \rightsquigarrow i} \text{trace}(\mathbf{M}\mathbf{C}_{ij})$$

$$+\mu \sum_{i,j \rightsquigarrow i,l} (1 - y_{il}) \left[1 + \text{trace}(\mathbf{M}\mathbf{C}_{ij}) - \text{trace}(\mathbf{M}\mathbf{C}_{il})\right]_{+} \tag{3}$$

where $[a]_{+}$ is the hinge loss which is a when $a > 0$ and is 0 otherwise. The difference matrix, $\mathbf{C}_{ij}$, is defined as $\mathbf{C}_{ij} = (\boldsymbol{x}_i - \boldsymbol{x}_j)(\boldsymbol{x}_i - \boldsymbol{x}_j)^{\top}$. Though other solving methods such as alternating projection algorithms can also be used here, using iterative gradient descent is simple and the global solution can still be reached [8]. The gradient is:

$$\frac{\partial E}{\partial \mathbf{M}} = (1 - \mu) \sum_{i,j \rightsquigarrow i} \mathbf{C}_{ij} + \mu \sum_{(i,j,l)} (\mathbf{C}_{ij} - \mathbf{C}_{il}) \tag{4}$$

where (i, j, l) means active triples (that activate the hinge loss) in the current gradient update (the impostors can vary in each update).

As we discussed in Equation 1, the metric matrix learned can be factorized as $\mathbf{M} = \mathbf{L}^{\top}\mathbf{L}$, where $\mathbf{L}$ is the projection matrix. Large Margin Component Analysis (LMCA) in [10] is a variant of LMNN which uses this idea. It focuses on finding a lower dimensional rectangular projection matrix instead of a full square Mahalanobis matrix. LMCA also minimizes Equation 3, but when defined in terms of $\mathbf{L}$, this is no longer a convex optimization problem and gradient-descent is used. At each iteration, the projection matrix is updated in the negative direction of the gradient:

$$\frac{\partial E}{\partial \mathbf{L}} = 2(1 - \mu)\mathbf{L} \sum_{i,j \rightsquigarrow i} \mathbf{C}_{ij} + 2\mu\mathbf{L} \sum_{(i,j,l)} (\mathbf{C}_{ij} - \mathbf{C}_{il}) \tag{5}$$

4   Mixtures of Large Margin Nearest Neighbor Classifiers

LMNN uses the a single, global $\mathbf{M}$ and LMCA uses a single, global $\mathbf{L}$ over the whole input space. It may be the case that a data set has multiple locally varying distributions—features may have different variances and different correlations in different parts of the input space, defining multiple local manifolds. Our idea is to divide up the input space into local regions using a gating function and learn different metrics in different regions; we hence define a mixture of LMNNs. In doing this, we are inspired by the Mixture of Experts neural network model of Jacobs et al. in [19]. Previously, Gönen and Alpaydın in [20] used the same idea in multiple kernel learning where they write a kernel as a weighted sum of localized kernels.

The gating function that defines the region of expertise of a local metric can be *cooperative* or *competitive*, which is implemented respectively by the sigmoid or softmax function (P is the number of regions):

$$\text{Cooperative: } \eta_m(\boldsymbol{x}_i|\boldsymbol{w}_m) = \frac{1}{1 + \exp(-\boldsymbol{w}_m^{\top}\boldsymbol{x}_i - w_{m0})} \tag{6}$$

$$\text{Competitive: } \eta_m(\boldsymbol{x}_i|\boldsymbol{w}_m) = \frac{\exp(\boldsymbol{w}_m^{\top}\boldsymbol{x}_i + w_{m0})}{\sum_{h=1}^{P}\exp(\boldsymbol{w}_h^{\top}\boldsymbol{x}_i + w_{h0})} \tag{7}$$

Local model m becomes active if $\eta_m(\boldsymbol{x}_i) > 0$ and we say that $\boldsymbol{x}_i$ belongs to region m. The softmax function is competitive because it enforces a soft winner-take-all mechanism and for any input, we expect a single active local metric and the gating model works as a selector ($\sum_m \eta_m(\boldsymbol{x}_i) = 1$). The sigmoid function is cooperative because there can be more than one active local metric and the model takes a weighted sum ($\sum_m \eta_m(\boldsymbol{x}_i)$ need not be 1).

In each local region, using a full $\mathbf{M}$ may lead to overfitting and to regularize, we learn a local lower rank $\mathbf{L}$ in each: When $\boldsymbol{x}$ chooses local model m, $\mathbf{L}_m$ is the local projection used. The localized projection of $\boldsymbol{x}_i$ into region m is

$$z_{im} = \eta_m(\boldsymbol{x}_i|\boldsymbol{w}_m)\mathbf{L}_m\boldsymbol{x}_i$$

The *total distance* between a pair $(\boldsymbol{x}_i, \boldsymbol{x}_j)$ is calculated as the sum of the local distances:

$$D_{total}(\boldsymbol{x}_i, \boldsymbol{x}_j) = \sum_{m=1}^{P} D_{\mathbf{L}_m}(\boldsymbol{x}_i, \boldsymbol{x}_j)$$

where $D_{\mathbf{L}_m}(\boldsymbol{x}_i, \boldsymbol{x}_j)$ is the local distance in region m:

$$
\begin{aligned}
D_{\mathbf{L}_m}(\boldsymbol{x}_i, \boldsymbol{x}_j) &= \|z_{im} - z_{j,m}\|_2^2 \\
&= \|\eta_m(\boldsymbol{x}_i|\boldsymbol{w}_m)\mathbf{L}_m\boldsymbol{x}_i - \eta_m(\boldsymbol{x}_j|\boldsymbol{w}_m)\mathbf{L}_m\boldsymbol{x}_j\|_2^2 \\
&= \|\mathbf{L}_m(\eta_m(\boldsymbol{x}_i|\boldsymbol{w}_m)\boldsymbol{x}_i - \eta_m(\boldsymbol{x}_j|\boldsymbol{w}_m)\boldsymbol{x}_j)\|_2^2 \\
&= [\eta_m(\boldsymbol{x}_i|\boldsymbol{w}_m)\boldsymbol{x}_i - \eta_m(\boldsymbol{x}_j|\boldsymbol{w}_m)\boldsymbol{x}_j]^{\top}\mathbf{L}_m^{\top} \\
&\quad \mathbf{L}_m[\eta_m(\boldsymbol{x}_i|\boldsymbol{w}_m)\boldsymbol{x}_i - \eta_m(\boldsymbol{x}_j|\boldsymbol{w}_m)\boldsymbol{x}_j]
\end{aligned}
\tag{8}
$$

Hence, the total distance is a weighted combination of local distances and the contribution of local projections are determined by $\eta_m(\boldsymbol{x}_i)$. Thus, it is possible that multiple metrics are active, particularly in the cooperative setting.

The model parameters are the localized projection matrices $\mathbf{L}_m$ and the gating parameters $\boldsymbol{w}_m$. We use the same formulation of LMNN in Equation 2 by using D_{total} instead of the Mahalanobis distance $(\boldsymbol{x}_i - \boldsymbol{x}_j)^\top \mathbf{M} (\boldsymbol{x}_i - \boldsymbol{x}_j)$:

$$\text{minimize} \quad (1 - \mu) \sum_{i,j \rightsquigarrow i} D_{total}(\boldsymbol{x}_i, \boldsymbol{x}_j) + \mu \sum_{i,j \rightsquigarrow i,l} (1 - y_{il}) \xi_{ijl}$$

$$\text{subject to} \quad D_{total}(\boldsymbol{x}_i, \boldsymbol{x}_l) - D_{total}(\boldsymbol{x}_i, \boldsymbol{x}_j) \geq 1 - \xi_{ijl}$$

$$\xi_{ijl} \geq 0 \tag{9}$$

Let us rewrite the loss function:

$$E(\eta) = (1 - \mu) \sum_{i,j \rightsquigarrow i} \sum_{m=1}^{P} \text{trace}(\mathbf{L}_m^\top \mathbf{L}_m \mathbf{C}_{ij}^{(m)}(\eta)) + \mu \sum_{i,j \rightsquigarrow i,l} (1 - y_{il}) \left[1 + \zeta_{ijl} \right]_+ \tag{10}$$

where

$$\zeta_{ijl} = \sum_{m=1}^{P} \left(\text{trace}(\mathbf{L}_m^\top \mathbf{L}_m \mathbf{C}_{ij}^{(m)}(\eta)) - \text{trace}(\mathbf{L}_m^\top \mathbf{L}_m \mathbf{C}_{il}^{(m)}(\eta)) \right)$$

We can use the same trick and rewrite the gated loss function in terms of difference matrices. $\mathbf{C}_{ij}^{(m)}(\eta)$ is defined over the gated projections of $\boldsymbol{x}_i$ and $\boldsymbol{x}_j$ in region m:

$$\mathbf{C}_{ij}^{(m)}(\eta) = [\eta_m(\boldsymbol{x}_i | \boldsymbol{w}_m) \boldsymbol{x}_i - \eta_m(\boldsymbol{x}_j | \boldsymbol{w}_m) \boldsymbol{x}_j] \, [\eta_m(\boldsymbol{x}_i | \boldsymbol{w}_m) \boldsymbol{x}_i - \eta_m(\boldsymbol{x}_j | \boldsymbol{w}_m) \boldsymbol{x}_j]^\top$$

When we use a gating function, the problem is not convex anymore and we use gradient descent. The derivative of the loss function with respect to the local projection matrix $\mathbf{L}_m$ can then be derived:

$$\frac{\partial E(\eta)}{\partial \mathbf{L}_m} = 2(1 - \mu) \mathbf{L}_m \sum_{i,j \rightsquigarrow i} \mathbf{C}_{ij}^{(m)}(\eta) + 2\mu \mathbf{L}_m \sum_{(i,j,l)} (\mathbf{C}_{ij}^{(m)}(\eta) - \mathbf{C}_{il}^{(m)}(\eta)) \tag{11}$$

The derivative of objective function with respect to the gating parameters depends on the function used:

$$\frac{\partial E(\eta)}{\partial w_{hk}} = 2(1 - \mu) \sum_{i,j \rightsquigarrow i} \frac{\partial D_{total}(\boldsymbol{x}_i, \boldsymbol{x}_j)}{\partial w_{hk}}$$

$$+ \mu \sum_{(i,j,l)} (1 - y_{il}) \left(\frac{\partial D_{total}(\boldsymbol{x}_i, \boldsymbol{x}_j)}{\partial w_{hk}} - \frac{\partial D_{total}(\boldsymbol{x}_i, \boldsymbol{x}_l)}{\partial w_{hk}} \right) \tag{12}$$

Algorithm 1. Training a Mixture of Large Margin Nearest Neighbor Classifiers

1: Initialize w_{mk} and w_{m0} to small random numbers.
2: Initialize $\mathbf{L}_m$ matrices to the PCA projection matrix of the whole data.
3: **repeat**
4:     **repeat**
5:         Calculate $D(\boldsymbol{x}_i, \boldsymbol{x}_j)$ and find target neighbors and impostors.
6:         $w_{mk}^{(t+1)} \leftarrow w_{mk}^{(t)} - \gamma^{(t)} \frac{\partial E(\eta)}{\partial w_{mk}}$
7:     **until** convergence of gating parameters
8:     **repeat**
9:         Calculate $D(\boldsymbol{x}_i, \boldsymbol{x}_j)$ and find target neighbors and impostors.
10:        $\mathbf{L}_m^{(t+1)} \leftarrow \mathbf{L}_m^{(t)} - \gamma^{(t)} \frac{\partial E(\eta)}{\partial \mathbf{L}_m}$
11:    **until** convergence of local projections
12: **until** convergence

We can apply the chain rule to get the derivative of the total distance:

$$\frac{\partial D_{total}(\boldsymbol{x}_i, \boldsymbol{x}_j)}{\partial w_{hk}} = \sum_{m=1}^{P} 2 \left[\boldsymbol{x}_i \frac{\partial \eta_m(\boldsymbol{x}_i | \boldsymbol{w}_m)}{\partial w_{hk}} - \boldsymbol{x}_j \frac{\partial \eta_m(\boldsymbol{x}_j | \boldsymbol{w}_m)}{\partial w_{hk}} \right]^{\top}$$
$$\mathbf{L}_m^{\top} \mathbf{L}_m \left[\eta_m(\boldsymbol{x}_i | \boldsymbol{w}_m) \boldsymbol{x}_i - \eta_m(\boldsymbol{x}_j | \boldsymbol{w}_m) \boldsymbol{x}_j \right] \tag{13}$$

If the sigmoid gating is used, the derivatives are $(x_{i0} \equiv 1)$:

$$\frac{\partial \eta_m(\boldsymbol{x}_i | \boldsymbol{w}_m)}{\partial w_{hk}} = \delta_{mh} \left(1 - \eta_m(\boldsymbol{x}_i | \boldsymbol{w}_m) \right) \eta_m(\boldsymbol{x}_i | \boldsymbol{w}_m) x_{ik}, \ k = 0, 1, \ldots, d \tag{14}$$

For the softmax gating, we have $(x_{i,0} \equiv 1)$:

$$\frac{\partial \eta_m(\boldsymbol{x}_i | \boldsymbol{w}_m)}{\partial w_{hk}} = \left(\delta_{mh} - \eta_h(\boldsymbol{x}_i | \boldsymbol{w}_h) \right) \eta_m(\boldsymbol{x}_i | \boldsymbol{w}_m) x_{ik}, \ k = 0, 1, \ldots, d \tag{15}$$

where δ_{mh}, is 1 if $m = h$ and it is 0 otherwise.

The pseudo-code for the Mixture of LMNN (MoLMNN) is given in Algorithm 1. To have a meaningful starting projection direction we initialize the local projections $\mathbf{L}_m$ by using Principal Components Analysis (PCA) on the training data. At each iteration, we first apply gradient-descent to update the gating parameters, and then, using the trained gating model, the local projection matrices are updated. We apply these steps, until both the gating model and the projection matrices converge or the classification result does not improve any further. The learning rate, γ, is determined using linear search at each iteration.

5   Experiments and Results

We compare sigmoid and softmax-gated MoLMNN with LMNN and LMCA on 21 data sets, that are publicly available in [20, 21, 22, 23]. In Yeast, Faults and Segment data sets, two classes are used (nuc vs cyt, k_stratch vs bumps, and sky vs windows, respectively). In Musk data set, only the real valued features are used. The input data is z-normalized.

Our experimental methodology is as follows: Each data set is split into two subsets as one-third test data and two-thirds training and validation data. The two-thirds part is used to create ten training and validation folds using 5×2 cross-validation. The number of reduced dimensions, namely e, is chosen among the number of features that explain 90, 95 and 98 per cent of the variance; LMNN, LMCA and MoLMNNs models with $P = 1$ up to 10 regions are trained for $k = 3, 5, 7$ and 9 neighbors . We also try and choose the best of sigmoid and softmax gating. We do such a four-dimensional, $(k, e, P, \text{sigmoid/softmax})$ grid search and choose the combination that has the highest average validation accuracy—the other models are similarly trained and the best configuration is chosen for their parameter set. For the best setting, the corresponding model is trained on the ten different training folds and tested on the same left-out one-third test set. These ten test results are reported and compared with the parametric 5×2 cross-validation paired F test [24]. Table 2 shows the mean and standard deviation of the test results for each data set and the results of significance tests.

We see that on most data sets, a few regions $(P \leq 4)$ is enough. The number of regions correspond to the modalities of data with different input distributions. Increasing the number of regions does not improve accuracy beyond a certain value. Note that even with a single region, MoLMNN may be more accurate because it reassigns impostors and targets at each iteration while the other algorithms fix them at initialization.

We also find that MoLMNN uses more neighbors when compared with other algorithms. We believe this to be an indicator that MoLMNN trains a more suitable distance function which places more of the target neighbors nearby. Other algorithms use fewer nearest neighbors because due to inaccurate distance approximation, their performance degrade if more neighbors are used. In terms of sigmoid vs softmax gating, we do not notice one being always superior to the other—each has its use.

MoLMNN significantly outperforms both LMCA and LMNN on Arabidopsis, Musk, Yeast and Sonar. It outperforms LMCA on Splice and LMNN on Yale. LMCA gives higher accuracy results on Yale and Ionosphere data sets. Except Yeast, these data sets have more than 60 dimensions, which shows that MoLMNN can capture local information to improve performance.

On Arabidopsis data set, we can visualize the data by reducing dimensionality to two; this is a bioinformatics data set with $1,000$ dimensions. In Fig. 1(a), we see the plot using PCA; there we see that the data has significant structure but that PCA cannot capture the difference between the two classes. In Fig. 1(b), we see the plot using LMCA and the learned discriminant using $k = 3$. In Fig. 2, we see results with MoLMNN with two regions. Each data point is plotted in the region where its gating value is higher and the discriminants are plotted separately in each region with $k = 3$. We see that we get better discrimination

Table 2. The mean and standard deviation of test set accuracies of MoLMNN, LMCA and LMNN. The parameters of the best configuration are given in parantheses, where So is softmax and Si is sigmoid. Boldface indicates that the method is significantly better than the other two. In terms of pairwise comparisons (shown by '*'), on Splice, MoLMNN is more accurate than LMCA and on Ionosophere, LMCA is more accurate than MoLMNN.

Data set	MoLMNN (k,e,P,g)	LMCA (k,e)	LMNN (k,d)
Abalone	77.86 ± 1.01 (9,3,1,So)	78.21 ± 1.97 (9,3)	78.03 ± 1.33 (9,7)
Arabidopsis	**81.89 ± 0.97** (9,473,1,Si)	77.21 ± 1.92 (5,390)	69.24 ± 2.17 (3,1000)
Australian	86.17 ± 1.42 (9,14,7,So)	85.83 ± 2.15 (5,14)	86.43 ± 1.08 (9,14)
Bupa	61.65 ± 4.27 (5,6,4,Si)	58.61 ± 2.79 (9,5)	59.57 ± 3.13 (9,6)
Ctg	89.73 ± 0.80 (9,16,2,So)	89.96 ± 0.62 (5,11)	89.83 ± 0.71 (5,21)
Faults	98.52 ± 0.58 (9,14,4,Si)	98.48 ± 0.25 (9,14)	98.56 ± 0.16 (9,27)
German Numeric	70.21 ± 2.27 (9,20,1,Si)	72.10 ± 2.84 (9,18)	72.01 ± 1.95 (9,24)
Heart	83.89 ± 4.10 (5,3,5,Si)	82.44 ± 4.38 (5,4)	85.22 ± 3.10 (7,13)
Ionosphere	81.54 ± 2.36 (9,17,3,Si)	83.25 ± 2.65* (3,27)	82.91 ± 2.88 (3,34)
Mg	83.58 ± 0.79 (9,5,1,So)	82.34 ± 0.61 (3,6)	82.34 ± 0.61 (3,6)
Musk	**86.01 ± 2.66** (7,17,4,So)	80.82 ± 2.88 (5,28)	79.62 ± 4.00 (5,166)
Optdigits	96.98 ± 0.31 (9,51,3,Si)	97.26 ± 0.35 (3,41)	97.33 ± 0.35 (3,64)
Pendigits	97.33 ± 0.31 (5,11,2,So)	97.49 ± 0.15 (3,11)	97.31 ± 0.23 (5,16)
Pima	72.93 ± 0.96 (9,8,2,So)	73.32 ± 1.90 (9,8)	73.32 ± 1.90 (9,8)
Segment	99.95 ± 0.14 (7,8,9,Si)	100.00 ± 0.00 (7,8)	100.00 ± 0.00 (7,19)
Sonar	**76.52 ± 3.60** (7,36,10,Si)	71.16 ± 3.01 (3,28)	68.55 ± 5.11 (3,60)
Splice	89.03 ± 0.71* (9,50,1,Si)	84.97 ± 0.83 (5,58)	85.81 ± 0.76 (9,60)
Transfusion	78.67 ± 1.13 (9,3,2,Si)	79.40 ± 1.34 (9,3)	79.36 ± 1.37 (9,5)
Wdbc	94.97 ± 1.35 (9,14,4,So)	94.02 ± 2.23 (7,10)	94.29 ± 1.94 (3,30)
Yeast	**60.57 ± 2.58** (9,6,6,Si)	59.39 ± 2.38 (5,6)	59.33 ± 2.52 (5,8)
Yale	93.51 ± 0.77 (3,196,3,Si)	**94.90 ± 0.50** (3,88)	92.77 ± 0.56 (7,896)

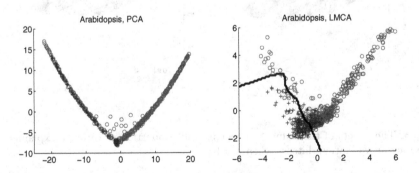

Fig. 1. The 2d mappings of Arabidopsis data set with (a) PCA and (b) LMCA

between the classes this way. Though we have not checked for this application, where the gating boundary lies and the dimensions in each region could also carry information.

We also check the relationship between the number of regions and the number of reduced dimensions. Figure 3 shows how test accuracy changes as we vary the number of dimensions and the number of regions. This is for $k = 9$, but we see similar behavior for other k. We see that it is more the number of regions that affect accuracy rather than the local dimensionality; we also see that sigmoid gating leads to more fluctuating performance—regions may overlap and hence may interfere more.

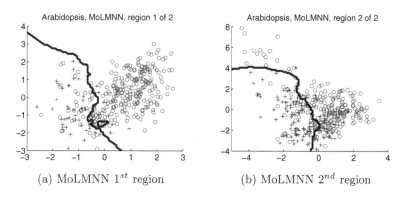

(a) MoLMNN 1st region   (b) MoLMNN 2nd region

Fig. 2. The 2d mappings of Arabidopsis data set with MoLMNN (softmax) with two regions

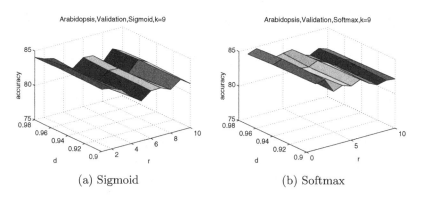

(a) Sigmoid   (b) Softmax

Fig. 3. The effects of the number of regions and proportion of variance explained on accuracy on validation dataset when $k = 9$

6 Conclusions

In this study, we propose the Mixture of LMNN (MoLMNN) method which softly partitions the input space and trains a separate projection matrix in each region to best discriminate the data. The partitioning of the space and the training of the projection matrices are coupled. Our experiments on real data sets show that compared with LMNN and LMCA proper, the mixture approach frequently performs better. Localization of the data and reducing dimensionality to two allows visualization. The boundary of the gating model and the projected dimensions could carry information which may help understand the data.

Acknowledgements. This work is supported by TÜBITAK Project EEEAG 109E186 and Boğaziçi University Research Funds BAP5701.

References

[1] Xing, P.E., Ng, A.Y., Jordan, M.I., Russell, S.: Distance Metric Learning, with Application to Clustering with Side-Information. In: Advances in Neural Information Processing Systems 15, pp. 505–512. MIT Press, Cambridge (2002)

[2] Goldberger, J., Roweis, S., Hinton, G., Salakhutdinov, R.: Neighbourhood Components Analysis. In: Advances in Neural Information Processing Systems 17, pp. 513–520. MIT Press (2004)

[3] Salakhutdinov, R., Hinton, G.: Learning a Nonlinear Embedding by Preserving Class Neighbourhood Structure. In: 11th International Conference on Artificial Intelligence and Statistics, vol. 2, pp. 412–419 (2007)

[4] Frome, A., Singer, Y., Malik, J.: Image Retrieval and Classification Using Local Distance Functions. In: Schölkopf, B., Platt, J., Hoffman, T. (eds.) Advances in Neural Information Processing Systems, vol. 19, pp. 417–424. MIT Press, Cambridge (2007)

[5] Frome, A., Singer, Y., Sha, F., Malik, J.: Learning Globally-Consistent Local Distance Functions for Shape-Based Image Retrieval and Classification. In: 11th IEEE International Conference on Computer Vision, pp. 1–8 (2007)

[6] Chang, H., Yeung, D.Y.: Locally Smooth Metric Learning with Application to Image Retrieval. In: 11th IEEE International Conference on Computer Vision, pp. 1–7 (2007)

[7] Davis, J.V., Kulis, B., Jain, P., Sra, S., Dhillon, I.S.: Information-Theoretic Metric Learning. In: 24th International Conference on Machine Learning, pp. 209–216 (2007)

[8] Weinberger, K.Q., Saul, L.K.: Distance Metric Learning for Large Margin Nearest Neighbor Classification. Journal of Machine Learning Research 10, 207–244 (2009)

[9] Weinberger, K.Q., Saul, L.K.: Fast Solvers and Efficient Implementations for Distance Metric Learning. In: 25th International Conference on Machine Learning, pp. 1160–1167 (2008)

[10] Torresani, L., Lee, K.C.: Large Margin Component Analysis. In: Advances in Neural Information Processing Systems, vol. 19, pp. 1385–1392 (2007)

[11] Malisiewicz, T., Efros, A.A.: Recognition by Association via Learning Per-Exemplar Distances. In: 21st IEEE Conference on Computer Vision and Pattern Recognition, pp. 1–8 (2008)

[12] Zhan, D.C., Li, M., Li, Y.F., Zhou, Z.H.: Learning Instance Specific Distances Using Metric Propagation. In: 26th International Conference on Machine Learning, pp. 1225–1232 (2009)

[13] Chen, Q., Sun, S.: Hierarchical Large Margin Nearest Neighbor Classification. In: 20th International Conference on Pattern Recognition. 906–909 (2010)

[14] Noh, Y.K., Zhang, B.T., Lee, D.D.: Generative Local Metric Learning for Nearest Neighbor Classification. In: Advances in Neural Information Processing Systems 23, pp. 1822–1830. MIT Press (2010)

[15] Chang, C.C.: A Boosting Approach for Supervised Mahalanobis Distance Metric Learning. Pattern Recognition 45(2), 844–862 (2012)

[16] Bunte, K., Schneider, P., Hammer, B., Schleif, F.M., Villmann, T., Biehl, M.: Limited Rank Matrix Learning, Discriminative Dimension Reduction and Visualization. Neural Networks 26, 159–173 (2012)

[17] Wang, Y., Woznica, A., Kalousis, A.: Parametric Local Metric Learning for Nearest Neighbor Classification. In: Advances in Neural Information Processing Systems, vol. 25, pp. 1610–1618. MIT Press (2012)

[18] Wu, L., Hoi, S.C.H., Jin, R., Zhu, J., Yu, N.: Learning Bregman Distance Functions for Semi-Supervised Clustering. IEEE Transactions on Knowledge and Data Engineering 24, 478–491 (2012)

[19] Jacobs, R.A., Jordan, M.I., Nowlan, S.J., Hinton, G.E.: Adaptive Mixtures of Experts. Neural Computation 3, 79–87 (1991)

[20] Gönen, M., Alpaydın, E.: Localized Multiple Kernel Learning. In: 25th International Conference on Machine Learning, pp. 352–359 (2008)

[21] Frank, A., Asuncion, A.: UCI Machine Learning Repository (2010), http://archive.ics.uci.edu/ml/

[22] Chang, C.C., Lin, C.J.: LIBSVM Data: Classification, Regression, and Multi-label (2011), http://www.csie.ntu.edu.tw/cjlin/libsvmtools/datasets/

[23] Sonnenburg, S., Ong, C.S., Henschel, S., Braun, M.: Machine Learning Data Set Repository (2011), http://mldata.org/

[24] Alpaydın, E.: Combined 5×2 cv F Test for Comparing Supervised Classification Learning Algorithms. Neural Computation 11, 1885–1892 (1999)

Author Index